SOME SI BASE UNITS

Physical Quantity	Name of Unit	Symbol
length	meter	m
mass	kilogram	kg
time	second	s
electric current	ampere	A
thermodynamic temperature	kelvin	K
amount of substance	mole	mol

SOME SI DERIVED UNITS

Physical Quantity	Name of Unit	Symbol	SI Unit
frequency	hertz	Hz	s^{-1}
energy	joule	J	$kg \cdot m^2/s^2$
force	newton	N	$kg \cdot m/s^2$
pressure	pascal	Pa	$kg/m \cdot s^2$
power	watt	W	$kg \cdot m^2/s^3$
electric charge	coulomb	C	$A \cdot s$
electric potential	volt	V	$kg \cdot m^2/A \cdot s^3$
electric resistance	ohm	Ω	$kg \cdot m^2/A^2 \cdot s^3$
capacitance	farad	F	$A^2 \cdot s^4/kg \cdot m^2$
inductance	henry	H	$kg \cdot m^2/A^2 \cdot s^2$
magnetic flux	weber	Wb	$kg \cdot m^2/A \cdot s^2$
magnetic flux density	tesla	T	$kg/A \cdot s^2$

SI UNITS OF SOME OTHER PHYSICAL QUANTITIES

Physical Quantity	SI Unit
speed	m/s
acceleration	m/s^2
angular speed	rad/s
angular acceleration	rad/s^2
torque	$kg \cdot m^2/s^2$, or $N \cdot m$
heat flow	J, or $kg \cdot m^2/s^2$, or $N \cdot m$
entropy	J/K, or $kg \cdot m^2/K \cdot s^2$, or $N \cdot m/K$
thermal conductivity	$W/m \cdot K$

SOME CONVERSIONS OF NON-SI UNITS TO SI UNITS

Energy:

1 electron-volt (eV) = 1.6022×10^{-19} J
1 erg = 10^{-7} J
1 British thermal unit (BTU) = 1055 J
1 calorie (cal) = 4.186 J
1 kilowatt-hour (kWh) = 3.6×10^6 J

Mass:

1 gram (g) = 10^{-3} kg
1 atomic mass unit (u) = 931.5 MeV/c^2
$\qquad\qquad\qquad\qquad = 1.661 \times 10^{-27}$ kg
1 MeV/c^2 = 1.783×10^{-30} kg

Force:

1 dyne = 10^{-5} N
1 pound (lb or #) = 4.448 N

Length:

1 centimeter (cm) = 10^{-2} m
1 kilometer (km) = 10^3 m
1 fermi = 10^{-15} m
1 Angstrom (Å) = 10^{-10} m
1 inch (in or ″) = 0.0254 m
1 foot (ft) = 0.3048 m
1 mile (mi) = 1609.3 m
1 astronomical unit (AU) = 1.496×10^{11} m
1 light-year (ly) = 9.46×10^{15} m
1 parsec (ps) = 3.09×10^{16} m

Angle:

1 degree (°) = 1.745×10^{-2} rad
1 min (′) = 2.909×10^{-4} rad
1 second (″) = 4.848×10^{-6} rad

Volume:

1 liter (L) = 10^{-3} m^3

Power:

1 kilowatt (kW) = 10^3 W
1 horsepower (hp) = 745.7 W

Pressure:

1 bar = 10^5 Pa
1 atmosphere (atm) = 1.013×10^5 Pa
1 pound per square inch (lb/in^2) = 6.895×10^3 Pa

Time:

1 year (yr) = 3.156×10^7 s
1 day (d) = 8.640×10^4 s
1 hour (h) = 3600 s
1 minute (min) = 60 s

Speed:

1 mile per hour (mi/h) = 0.447 m/s

Magnetic field:

1 gauss = 10^{-4} T

PHYSICS

FOR SCIENTISTS AND ENGINEERS

SECOND EDITION

PHYSICS

FOR SCIENTISTS AND ENGINEERS

SECOND EDITION

VOLUME I

PAUL M. FISHBANE

University of Virginia

STEPHEN GASIOROWICZ

University of Minnesota

STEPHEN T. THORNTON

University of Virginia

 PRENTICE HALL Upper Saddle River, New Jersey 07458

Library of Congress Cataloging-in-Publication Data

Fishbane, Paul M.
    Physics for scientists and engineers / Paul M. Fishbane, Stephen
Gasiorowicz, Stephen T. Thornton. — Extended version.
        p.   cm.
    Includes index.
    ISBN 0-13-231150-X
    1. Physics.   I. Gasiorowicz, Stephen.   II. Thornton, Stephen T.
III. Title.
QC23.F52   1996
530—dc20   95-25229
                                                                CIP

Acquisitions Editors: Ray Henderson/Alison Reeves
Senior Development Editor: Carol Trueheart
Managing Editor: Kathleen Schiaparelli
Production Editors: Barbara Mack and Jennifer Wenzel
Director of Marketing: Gary June
Marketing Manager: Leslie Cavaliere
Copy Editor: David George
Creative Director: Paula Maylahn
Art Director: Heather Scott
Interior and Cover Design: Lee Goldstein
Photo Editor: Lori Morris-Nantz
Photo Research: Mira Schachne
Artwork: Network Graphics and Rolando Corujo
Manufacturing Buyer: Trudy Pisciotti
Editorial Assistants: Pamela Holland-Moritz/Barbara Kraemer
Assistant Vice President of Production and Manufacturing: David W. Riccardi
Editor in Chief: Paul Corey
Editorial Director: Tim Bozik
Editor in Chief of Development: Ray Mullaney
Assistant Editor: Wendy Rivers
Cover Photo: Centre Audiovisuel SNCF

 © 1996, 1993 by Prentice-Hall, Inc.
Simon & Schuster/A Viacom Company
Upper Saddle River, New Jersey 07458

Printed in the United States of America
10 9 8 7 6 5 4 3 2

ISBN 0-13-231150-X

Prentice-Hall International (UK) Limited, *London*
Prentice-Hall of Australia Pty. Limited, *Sydney*
Prentice-Hall Canada Inc., *Toronto*
Prentice-Hall Hispanoamericana, S.A., *Mexico*
Prentice-Hall of India Private Limited, *New Delhi*
Prentice-Hall of Japan, Inc., *Tokyo*
Simon & Schuster Asia Pte. Ltd., *Singapore*
Editora Prentice-Hall do Brasil, Ltda., *Rio de Janeiro*

Dedication

To our students, the most important element in the making of this book

Brief Contents

Contents

12 GRAVITATION 319

13 SIMPLE HARMONIC MOTION 345

15 SUPERPOSITION AND INTERFERENCE OF WAVES 407

Preface

Our goals for the second edition of this book are primarily to increase its accessibility, relevance, and friendliness for the student, who needs a strong background in physics now more than ever. At the same time, we feel that it is important to maintain the standards that, if anything, have been raised higher in an increasingly technological world. Students must have a conceptual understanding as well as an understanding at the level of problem solving; we have worked hard to maintain and improve the balance between these two crucial aspects of learning physics.

This text is designed for a calculus-based physics course; such a course gives students a solid foundation in understanding the meaning of the fundamental physical laws and teaches them how these laws can be applied to solve problems. The students taking this course are typically in their first or second year and may be prospective engineers, scientists, physicians, or indeed any students who will use technology in their careers. They either have taken or are taking a course in calculus. We believe that it makes little difference which of the above profiles describe the student; it is the nature of physics that the few principles that unify the subject hold whatever the discipline in which they are applied.

Here, we take a look at some of the special features that make this text an effective teaching tool. Whatever these features are, we continue to hope that the most important "feature" of the text is the clarity and correctness with which the principles of physics, their relations, and their applications are set forth. Such an exposition must balance rigor with a more intuitive understanding of the subject, and it must balance explanation and a brevity that too often teaches physics as a set of rules to be memorized. We have tried to maintain these balances throughout.

WHAT'S NEW IN THIS EDITION?

- We have rewritten throughout for focus, clarity, and simplicity in language without loss of depth. In addition, we have made cuts, both large and small, for length control. Sometimes this means more compact writing; sometimes it means removing entire paragraphs of material deemed so far from the essential that the student will not miss it.

- We have rearranged some material that was not, on our own reflection or at the suggestion of users, properly placed or ordered. For example, conservative forces are now introduced in Chapter 6 rather than in Chapter 7. In addition, we have provided new and simpler derivations for certain subjects; for example, the derivation of the wave equation is a new one.

- Conceptual matters play a more prominent role. From the beginning, we have tried to take the tack that physics is more a matter of understanding certain ideas and methods than of being able to plug into a series of formulas. As part of an increased effort to help the student understand rather than simply parrot the solution of stereotypical problems, the questions at the end of each chapter have been strengthened.

- We have added more engineering-oriented application material. This is most noticeably realized in the new application boxes that now appear in each chapter and in new realistic problems. The application boxes are listed in the Table of Contents.

- An increased number of problem-solving boxes are presented throughout the text.

- Margin notes and short set-off remarks and signals have now been divided into four different categories: (i) ordinary side notes, which contain the important results, those that will "be on the test"; (ii) problem-solving notes, containing references to useful techniques for problems; (iii) application notes, signaling device or technology relevance; and (iv) cross references, indicated by the icon ∞, which remind the student that he or she has already seen the referenced material and just *where* he or she has seen it.

- In the context of the problem-solving techniques, a set of new figures take the form of "student sketches," which are drawn in a style that the student can realistically be expected to emulate. To carry through on this idea, all the figures in the end-of-chapter problems are also drawn in this style. In the sense that a sketch represents a first step in problem solving, the sketches in the end-of-chapter problems sometimes provide a crucial hint to the student; more generally, they help the student develop the habit of using sketches to help in problem solving.

- We have eliminated derivation boxes. A few of the derivations themselves have been eliminated entirely, and all the original derivations from these boxes have been simplified; where we feel the derivation can be skipped without great loss, the derivations simply form an optional portion of the running text.

- We have simplified our notation for multi-dimensional integrals with the consistent use of a single integral sign. To avoid possible misunderstanding, we have consistently added labeling on the integral sign that guards against the possibility that the student might confuse a multiple integral with a single integral.

- We have made our examples more useful to students by employing student sketches and changing the solutions to more clearly show students *how* and *why* we reach an answer. The text contains an average of more than 8 examples per chapter, at least 10% of the examples are new, and we have changed up to 30% of the remaining examples.

- Visualization is an important part of learning physics. In addition to redrafting all end-of-chapter illustrations as student sketches, we have added approximately 75 new in-text illustrations. Further, almost every figure has been redrafted for clarity, including the addition of grid lines to many graphs and, often, the presentation of a three-dimensional view instead of a two-dimensional view.

- We have made a major revision to the end-of-chapter problem sets. First, we have added more than ten new level I problems in each chapter to increase the possibility of student drill. Working toward our goal of providing a greater applications orientation, we have also added close to 20% more problems that are entirely new and changed almost 30% of existing problems. Finally, problems that have not proven useful have been eliminated.

Emphasis on Problem Solving. A student's ability to solve qualitative and quantitative problems is the best measure of whether the physics is truly understood. With this in mind, we have included certain features that will strengthen the student's problem-solving ability. Central to these is the end-of-chapter material. This material includes about 700 qualitative, conceptual **questions** under the heading "Understanding Key Concepts." These test the student's understanding of the ideas that underlie a situation. Such questions may not have simple or direct answers. They challenge students to consider concepts and to recognize the meaning of the principles taught in the text.

In addition to the questions, there are some 3000 **problems.** Some problems are keyed to text sections, and are presented in increasing order of difficulty within each section. Others are **general problems,** which bring together material from the entire chapter as well as from previous chapters. The general problems, with a lesser degree of guidance as to the subject being studied, resemble more closely the situations that are met in real-life science and engineering, and they also help to develop the student's appreciation of the links that exist throughout physics. All problems are labeled I, II, or III. Level I problems are "easy," not going much beyond plug-ins. These problems develop student recognition of particular physics concepts and build confidence. Level II problems are typically multistep and require an increased understanding of the interconnectedness of physics; the general problems carry this requirement a step further. Level III problems are especially challenging, in some cases demanding significant synthesis of concepts in the text. We know that you cannot assign every problem, even over many years. But the gradations in problem range and difficulty allow you to tailor the problems you assign to the capabilities of your class and to the subjects that interest you the most.

The text contains several significant aids to the development of the student's problem-solving capacities. First, there are numerous **examples,** an average of 8 or 9 per chapter (some chapters contain more than a dozen examples). Examples represent typical problems, generally at level II. Their solutions illustrate sound approaches to problem solving. We "think out loud" in the example solutions to reveal *how* and *why* we reach our answers rather than providing just the answers themselves. Many of the examples employ a technique that we refer to as **"student sketches,"** which reflect our advice that the student sketch the situation. These drawings are meant to be something that a student could reasonably emulate, not the sort of elaborate drawing that the student—not to mention the instructor—could not reproduce. Second, we have included **problem-solving techniques,** separately boxed throughout the text. This material appears in the key chapters that introduce new ways of looking at problems in physics. Third, we include brief **problem-solving margin notes** adjacent to text discussion of problem-solving methods. Finally, we have included answers to odd-numbered problems at the back of the book, and the new student solution manual contains solutions to a significant portion of all problems.

Mathematics.   Three remarks are worth making: First, we introduce the mathematics that students need to know the first time they need to know it, in the context of the physics being presented. We try to make that material self-contained, so that the student can understand the material without having to go elsewhere for mathematical help. In this way, the mathematics appears in progressive degrees of difficulty, and we believe that this approach fosters better understanding and less reliance on formula memorization. Second, we teach the correct usage of significant figures and vary the number of significant figures in examples and in problems—in keeping with the way problems arise in the real world. In this way, students must maintain an awareness of significant figures and are not lulled into thinking that all problems involve the same number of significant figures. Third, the ability to make quantitative estimates is one of the most important skills that a scientist or engineer can have. We have made the development of that ability an important part of our approach, both in the text and in problems.

Applications.   Students are interested in knowing how the principles that govern the physical world are associated with today's technology, or with the workings of devices that they see every day around them. With this in mind, we have included in every chapter at least one separate box that describes an application relevant to that chapter's subject. In addition, we have added margin notes that signal applications as they appear in the text.

Modern Physics.   Classical topics have lost none of their importance and must form the basis of any first course. However, what is traditionally called "modern" physics— the topics centered on relativity and quantum physics—began about a century ago. It hardly seems possible to ignore these topics in view of their importance for understanding today's world. Many of the ideas of modern physics are not mathematically difficult. However, they can be nonintuitive, and we think that it is important that students begin to develop intuition about this material as early as possible. Without sacrificing the essential aspects of classical physics, we have introduced modern notions from the beginning. Although much of this material appears in optional sections (marked with an asterisk), in many cases it is intertwined with the classical material. We conclude the text traditionally, with chapters on modern physics. We think that the preparation we have laid down for this material will make it more easily assimilated.

Other Study Aids. Effective keys and reviewing aids appear in several forms. A **summary** at the end of each chapter reminds students of the most important results of the chapter. For chapters that introduce numerous or especially difficult concepts, we offer **interim summaries** to give students extra assistance. In addition, two types of **margin notes** that appear adjacent to the corresponding text form study aids. These include margin notes that key important concepts or results, as well as cross references that recall where previously developed concepts are first used when they appear in a later chapter. **Important equations** are boxed, although we have resisted the temptation to box too many of them, a temptation which encourages a plug-in mentality.

Figures and Photographs. A number of features distinguish the figures and photographs in this text from those in other texts. Labeled photos clearly illustrate the physical principles present at each point in an object's movement. Sequence photos help "bring the physics to life" for the student. Student sketches are introduced in this edition; these sketches are used in the end-of-chapter problems to provide hints and are presented within the text to help students learn to draw sketches during problem solving. For clarity, a significant number of figures have been changed from a two-dimensional to a three-dimensional rendition. Finally, color is used in this text as a pedagogical aid, primarily where the consistent use of color to represent physical quantities and analytic elements in illustrations helps to avoid confusion. For example, color provides a visual reminder to students not to confuse acceleration vectors with force vectors.

SUPPLEMENTS

Quality supplements make a difference. Each of the supplements that accompany this text has been carefully developed to create a unique set of high-quality, tested tools that provide the means by which instructors and students can enliven and reinforce the teaching and learning of physics.

For the Instructor

Instructor's Solution Manual. Written by Professor Irv Miller of Drexel University, this manual contains complete solutions to all problems in the text. Considerable effort has been put into checking the accuracy of the solutions.

Transparencies. Approximately 200 four-color acetates are available in the transparency set.

Test-Item File and PH Test. The test questions that accompany this text have been developed by Professor Charles Scherr of the University of Texas and have been used at Austin by thousands of engineering and science students over the course of a decade. Approximately 500 questions have been constructed in algorithmic form with a broad range of input data, which allows each question to generate a significant and reliable number of versions. PH Test, the testing program, provides the means to create a variety of student testing devices, homework and drill sheets, and answer sheets in minutes. PH Test is available in Macintosh and DOS versions.

"Physics You Can See" Demonstration Experiments Video. This video is a collection of brief demonstrations for classroom use.

The Princeton Learning Guide. This study guide for our text has been adapted from guides developed at Princeton University during the past 20 years. The guide features complete problem assignments with aids in the form of leading questions and keyed answers. Help is given in the form of increasingly detailed hints and references to the text. The emphasis in this guide is on conceptual understanding, especially understanding of the links in the physical world.

Student Solution Manual. This manual provides complete solutions to many of the end-of-chapter problems. It also provides a discussion of answers to some of the end-of-chapter questions.

The Portable TA: A Problem-Solving Guide (Volumes I and II). A collection of problems with carefully detailed solutions to provide students with extra practice in problem-solving techniques. This collection is the result of two years of class testing.

The New York Times/Themes of the Times. A semiannual newspaper containing articles from *The New York Times*, this supplement demonstrates the connection between what goes on in the classroom and in the world around us. It is available in quantity, *free* for each student.

TECHNOLOGY-BASED SUPPLEMENTS

LOGAL Physics Explorer. Available only from Prentice Hall, this interactive "run time" version of LOGAL's award-winning simulation software contains over 100 simulations based on examples and problems from the text. Physics Explorer allows students to compare predictions with results by altering parameters of a problem. The simulation models cover mechanics, waves, diffraction and interference, and optics. Extension Exercises expand the boundaries of the initial problem, and Concept Checks offer immediate on-screen feedback. Available in Windows and Macintosh versions.

The Interactive Physics Player™. Available only from Prentice Hall, this "player" version of Knowledge Revolution's highly acclaimed Interactive Physics II computer program brings the text to life with real-time simulations and animations of over 100 of the examples and problems. The program allows the user to change values and view the outcomes, creating and observing "what if" scenarios. Tools showing numerical results, graphs, and vectors make it easy to analyze the simulation. The Interactive Physics Player provides students with a way to visualize the concepts they are trying to learn. Instructors can use it as a demonstration tool in class or in lab. The program is available in both Macintosh and Windows versions.

Interactive Physics II Workbook. Written by Cindy Schwarz of Vassar College, this highly interactive workbook/software package contains 40 simulation projects of varying degrees of difficulty. Each contains a physics review, simulation details, hints, explanation of results, math help, and a self-test. Available in Windows and Macintosh versions.

Mathematica Projects for Scientists and Engineers. This supplement, written by Rodney Varley of Hunter College, includes approximately 40 detailed modules that contain an explanation of the Mathematica model, reviews of the physical and mathematical concepts, "Learn by Doing" exercises that ask you to build or modify Mathematica simulations, and additional exercises with suggested hints.

Matlab Projects for Scientists and Engineers. This supplement, written by Alejandro Garcia of San Jose State University, contains approximately 20 detailed modules based on the numerical processing software Matlab. Each module contains reviews of the primary physical and mathematical concepts, an explanation of the Matlab model, "Learn by Doing" exercises that allow you to build or modify Matlab simulations, and additional exercises with suggested hints.

TEXT VERSIONS AND TEACHING ALTERNATIVES

This text provides a complete introduction to physics. It is available in a 41-chapter version, which ends with a chapter on special relativity and a chapter on quantum physics. For those who want more detailed coverage, a 46-chapter extended version containing additional coverage of quantum physics, nuclear physics, particles, and cosmology is also available. The text is published in split volumes as well. Volume I includes Chapters 1 to 21, covering mechanics, waves, and thermodynamics. Volume II includes Chapters 22 to 46 of the extended version, covering electricity and magnetism, light and optics, and modern physics.

The question of length is a knotty one; there is widespread agreement that a text such as this one is too long, but very little agreement on just what should be eliminated. For that reason, we have written this book in such a way that any instructor can make his or her own decisions on cuts. We nevertheless give two sorts of guidance:

- ***Optional Material.*** An asterisk (*) indicates material that only the most thorough courses will typically have time to cover. For the most part, such material is either self-contained modern physics or an in-depth discussion of material already introduced.

- ***A Lean Alternative.*** At many institutions, the constraints of time coupled with the degree of preparation of the students necessitate that the introductory physics course be compressed or that more time be spent on fewer topics. We propose here one possible way—out of many—in which this text could be used in a highly streamlined course. The criteria we use for suggesting what to retain and what to delete are associated with the degree to which the material is self-contained or will be studied later in a more applications-oriented course, not with how "hard" that material is. We would eliminate or de-emphasize the following topics or sets of topics: relative motion; the more complicated aspects of rotational motion; statics; the more advanced aspects of harmonic motion; the Doppler effect; the more advanced aspects of wave superposition, including diffraction; fluids and solids; much of statistical physics, including probability distributions and entropy; the effects of materials in electromagnetism; applied aspects of time-dependent circuits, including alternating currents; aspects of electromagnetic radiation; mirrors and lenses; the Lorentz transformations; and the more detailed applications of quantum mechanics.

Thus our compressed course would consist of the following:

Chapter	Sections	Chapter	Sections
1	1,2,3,5,6	20	1,2,3,4
2	1,2,3,4,5	22	1,2,3,4
3	1,2,3,4,5	23	1,2,3,4
4	1,2,3,4,5	24	1,2,3,4
5	1,2,3,4	25	1,2,3,4,6,7
6	1,2,3,4,5	26	1,2,3
7	1,2	27	1,2,3,7
8	1,3,4,6	28	1,2,3
9	1,2,3,5,6	29	1,2,3,4
10	1,2,3,4	30	1,2,3,5
12	2,3,4	31	2,3,4,5,6
13	1,3,4,5	33	1,2,3
14	1,2,3,5,6	35	1,2,3,4
15	1,3,4	36	1,2,3
17	1,2,3	38	1,2,3
18	1,2,3,4,5,6	40	1,2,3,4,7
19	1,2,3	41	1,2,3,4

ACKNOWLEDGMENTS

It is clear that a project of this magnitude cannot be accomplished by three authors alone. We are grateful to the many people who have contributed to making this a better text.

A special thanks goes to Professor Irv Miller of Drexel University. In addition to writing the Instructor's Solution Manual, Irv was an invaluable resource in creating, checking, and refining the problem sets and answers. Working with the input of Professor T. S. Venkataraman, also of Drexel, Irv coordinated the independent critique and feedback of the problems and answers from the following: Edward Adelson (Ohio State University); Howard Bale (University of North Dakota); John Barach (Vanderbilt University); William R. Cochran (Youngstown State University); George Dixon (Oklahoma State University); Brent Foy (Wright State University); Rex Gandy (Auburn University); Mark Heald (Swarthmore College); Paul Heckert (Western Carolina University): Larry Hmurcik (University of Bridgeport); Rex Joyner (Indiana Institute of Technology); Brij M. Khorana (Rose-Hulman Institute of Technology); Douglas A. Kurtze (North Dakota State University); David Markowitz (University of Connecticut); George Miner (University of Dayton); R. J. Peterson (University of Colorado, Boulder); and James H. Smith (University of Illinois at Urbana-Champaign). Equal thanks go to Laszlo Takacs of the University of Maryland at Baltimore County. Laszlo supplied many new problems and modified many first-edition problems for this edition. We are also exceedingly grateful to Tony Buffa, who was tremendously helpful in providing suggestions for updating and redrafting the art program.

We want to thank John Malone of the teaching laboratory at the University of Virginia for his help with demonstrations and photographs. John is the kind of person that no lecturer in physics should be without.

We would also like to acknowledge and thank the 70 physics instructors who provided valuable feedback. We took all of their comments very seriously and only regret that we could not satisfy them all.

V. K. Agarwal
Moorhead State University

Thomas Armstrong
University of Kansas

Philip S. Baringer
University of Kansas

John E. Bartelt
Vanderbilt University

William Bassichis
Texas A & M University

Benjamin F. Bayman
University of Minnesota

Michael Browne
University of Idaho

Timothy Burns
Leeward Community College

Alice Chance
Western Connecticut State University

Robert Clark
Texas A & M University

Lucien Cremaldi
University of Mississippi

W. Lawrence Croft
Mississippi State University

Chris L. Davis
University of Louisville

Robin Davis
University of Kansas

Jack Denson
Mississippi State University

John DiNardo
Drexel University

Robert J. Endorf
University of Cincinnati

Arnold Feldman
University of Hawaii at Manoa

Rex Gandy
Auburn University

Alexander B. Gardner
Howard University

Robert E. Gibbs
Eastern Washington University

Wallace L. Glab
Texas Tech University

James R. Goff
Pima Community College

Alan I. Goldman
Iowa State University

Phillip Gutierrez
University of Oklahoma

Frank Hagelberg
SUNY-Albany

Robert F. Harder
George Fox College

Warren W. Hein
South Dakota State University

Jerome Hosken
City College of San Francisco

Joey Houston
Michigan State University

Francis L. Howell
University of North Dakota

Evan W. Jones
Sierra College

Alain E. Kaloyeros
SUNY-Albany

Robert J. Kearney
University of Idaho

Arthur Z. Kovacs
Rochester Institute of Technology

Claude Laird
University of Kansas

Vance Gordon Lind
Utah State University

A. Eugene Livingston
University of Notre Dame

B. A. Logan
University of Ottawa

David Markowitz
University of Connecticut

Erwin Marquit
University of Minnesota

Marvin L. Marshak
University of Minnesota

Charles R. McKenzie
Salisbury State University

Norman McNeal
Sauk Valley Community College

Forrest Meiere
Indiana University-Purdue University

Irvin A. Miller
Drexel University

Thomas Muller
University of California-Los Angeles

Richard Murphy
University of Missouri-Kansas City

Peter Nemethy
New York University

David Ober
Ball State University

Gottlieb S. Oehrlein
SUNY-Albany

Micheal J. O'Shea
Kansas State University

Dan Overcash
Auburn University

Patrick Papin
San Diego State University

Robert A. Pelcovits
Brown University

Don D. Reeder
University of Wisconsin-Madison

John Lewis Robinson
University of British Columbia

Ernest Rost
University of Colorado

Mendel Sachs
SUNY-Buffalo

Francesca Sammarruca
University of Idaho

William G. Sturrus
Youngstown State University

Richard E. Swanson
Sandhills Community College

Laszlo Takacs
University of Maryland, Baltimore County

Leo H. Takahashi
Pennsylvania State University

Robert Tribble
Texas A & M University

Rod Varley
Hunter College

Gianfranco Vidali
Syracuse University

John Wahr
University of Colorado

Fa-chung Wang
Prairie View A & M University

Gail S. Welsh
Salisbury State University

We are also grateful to the many reviewers of our first edition:

Maris Abolins *(Michigan State University)*, Ricardo Alarcon *(Arizona State University)*, Bradley Antanaitis *(Lafayette College)*, Carl Bender *(Washington University)*, Hans-

Uno Bengtsson *(University of California, Los Angeles)*, Robert Bowden *(Virginia Polytechnic Institute and State University)*, Bennet Brabson *(Indiana University)*, Edward Chang *(University of Massachusetts)*, Albert Claus *(Loyola University)*, Robert Coakley *(University of Southern Maine)*, James Dicello *(Clarkson University)*, N. John DiNardo *(Drexel University)*, P. E. Eastman *(University of Waterloo)*, Gabor Forgacs *(Clarkson University)*, A. L. Ford *(Texas A & M University)*, William Fickinger *(Case Western Reserve University)*, Rex Gandy *(Auburn University)*, Simon George *(California State University, Long Beach)*, James Gerhart *(University of Washington)*, Bruce Harmon *(Iowa State University)*, Joseph Hemsky *(Wright State University)*, Alvin Jenkins *(North Carolina State University)*, Karen Johnston *(North Carolina State University)*, Garth Jones *(University of British Columbia)*, Leonard Kahn *(University of Rhode Island)*, Charles Kaufman *(University of Rhode Island)*, Thomas Keil *(Worcester Polytechnic University)*, Carl Kocher *(Oregon State University)*, Karl Ludwig *(Boston University)*, Robert Marande *(Pennsylvania State University)*, David Markowitz *(University of Connecticut)*, Roy Middleton *(University of Pennsylvania)*, George Miner *(University of Dayton)*, Lorenzo Narducci *(Drexel University)*, Jay Orear *(Cornell University)*, Patrick Papin *(San Diego State University)*, Kwangjai Park *(University of Oregon)*, R. J. Peterson *(University of Colorado)*, Frank Pinski *(University of Cincinnati)*, Lawrence Pinsky *(University of Houston)*, Stephen Pinsky *(Ohio State University)*, Richard Plano *(Rutgers University)*, Hans Plendl *(Florida State University)*, Shafigur Rahman *(Allegheny College)*, Peter Riley *(University of Texas, Austin)*, L. David Roper *(Virginia Polytechnic Institute and State University)*, Richard Roth *(Eastern Michigan University)*, Carl Rotter *(West Virginia University)*, Charles Scherr *(University of Texas)*, Eric Sheldon *(University of Lowell)*, Charles Shirkey *(Bowling Green State University)*, Marlin Simon *(Auburn University)*, Robert Simpson *(University of New Hampshire)*, James Smith *(University of Illinois)*, J. C. Sprott *(University of Wisconsin)*, Malcolm Steuer *(University of Georgia)*, Thor Stromberg *(New Mexico State University)*, Smio Tani *(Marquette University)*, William Walker *(University of California, Santa Barbara)*, George Williams *(University of Utah)*.

Finally, we would like to thank the publishing team at Prentice Hall who have carried this project through. Our editors, Ray Henderson and Tim Bozik, who directed the project, have been a constant source of ideas, encouragement, and material help. Carol Trueheart, our development editor, insisted on reminding us that this book is meant for students first, and that a failure to communicate at the appropriate level is the worst kind of failure. She has been encouraging from beginning to end, and we cannot imagine completing this work without her help. Director of Marketing Gary June has provided constant support and ideas, and we thank him. Last but not least, the production of a book such as this one is an enormous task demanding the most careful attention to detail. We are grateful to Tony Buffa, who was tremendously helpful in updating and redrafting the art program to increase clarity and improve perspective. David George and Karen Malley did an impossibly careful job of proofreading. We want to thank Rolando Corujo, who created many of the complex pieces of art for this edition. We especially want to thank Barbara Mack, Joanne Jimenez, and Jennifer Wenzel for the elaborate juggling job they did to bring this book to press. Finally, we would like to thank, in advance, Alison Reeves for the work she is doing to insure the continued success of this book. To all the individuals listed above, and to the many others at Prentice Hall who have worked to make this book a success, we extend our heartfelt thanks.

The cumulative and accelerating nature of science and technology make it more imperative than ever that our emerging scientists and engineers understand how few and how solid are the pillars of the enterprise. From this view, the distinctions between "science" and "engineering," and between "classical physics" and "modern physics," melt. We want in this book to make evident the pillars of physics as well as the highly interconnected structure that has been erected on those pillars.

About the Authors

Paul M. Fishbane

Paul Fishbane has been teaching undergraduate courses at the University of Virginia, where he is Professor of Physics, for some 25 years. He received his doctoral degree from Princeton University in 1967 and has published some 100 papers in his field, theoretical high energy physics. He has held visiting appointments at the State University of New York at Stony Brook, Los Alamos Scientific Lab, CERN laboratory in Switzerland, Amsterdam's NIKHEF laboratory, France's Institut de Physique Nucleaire, the University of Paris-Sud, and the Ecole Polytechnique. He has been active for many years at the Aspen Center for Physics, where current issues in physics are discussed with an international group of participants. His other interests include biking, music, and the physics of the kitchen. All of the rest of his time is spent trying to keep up with his family, especially his youngest son Nicholas.

Stephen Gasiorowicz

Stephen Gasiorowicz was born in Poland and received his Ph.D. in physics at the University of California, Los Angeles in 1952. After spending 8 years at the Lawrence Radiation Laboratory in Berkeley, California, he joined the faculty of the University of Minnesota, where his field of research is theoretical high energy physics. As a visiting professor, he has traveled to the Niels Bohr Institute, NORDITA in Copenhagen, the Max Planck Institute for Physics and Astrophysics in Munich, DESY in Hamburg, Fermilab in Batavia, and the Universities of Marseille and Tokyo. He has been a frequent visitor and an officer of the Aspen Center for Physics. He has written books on elementary particle physics and quantum physics. A relatively new occupation is that of grandfather, which still leaves some time for reading (history), biking, canoeing, and skiing.

Stephen T. Thornton

Stephen Thornton performed his doctoral research at Oak Ridge National Laboratory while completing his Ph.D. at the University of Tennessee. He joined the faculty at the University of Virginia in 1968, and became the first Director of the Institute of Nuclear and Particle Physics. He has held two Fulbright fellowships and has performed nuclear experiments at accelerators throughout the United States and Europe. His recent interests include teaching physical science to school teachers and performing science outreach to children. He helped establish the Center for Science Education at Virginia. He has published over 100 research papers as well as three textbooks. He has two grown sons and is married to Dr. Kathryn Thornton, NASA astronaut. They have three daughters. He owns and operates a Christmas tree farm, and his interests include snow skiing and scuba diving.

A Guided Tour

We hope more than anything else that we can convey to you the excitement of a vast body of knowledge. Like anything worthwhile, scientific knowledge takes some work to acquire, but the effort will be repaid manyfold. The most important characteristic of physics is that it describes and explains a vast set of phenomena in terms of just a few basic principles. You will have succeeded if you understand these principles well enough to use them to solve problems. The conceptual basis for solving those problems must also be present, however, and if it is, you can go very far with your knowledge.

This book is a tool, which, with the aid of your instructor and your fellow students, will allow you to acquire an understanding of physics. Tools half-used give only half value, and we encourage you to make use of every aspect of this book and of the ancillary materials that are available with it. For now, let's take a brief walk through the book itself.

Chapter Opener
Each chapter opens with an illustration that indicates how thinking about physics can explain an immense range of activities in the world around us.

CHAPTER

10

The large torque acting on the sailboat by wind forces on the sail tends to rotate the sailboat out of the water; the torque due to gravity tends to rotate the sailboat back to a normal position. The forces and torques on a sailboat such as this one are rather complex.

Introductory Text
Each chapter introduces the topic being presented and its connection to the physical concepts covered in previous chapters and in chapters to come.

Problem Solving Margin Note
This type of note contains references to useful techniques for problems.

More on Angular Momentum and Torque

We studied the rotations of rigid bodies in Chapter 9. Many real physical systems, such as the collections of stars that make up galaxies or a figure skater executing a complex maneuver, are not rigid, however. We will see that angular momentum and its conservation are far more useful concepts than might have been evident in Chapter 9—even for systems of particles that do not form rigid bodies and are not in pure rotational motion. We will extend to these systems the notions of angular momentum and torque. We will also explore situations in which the vector nature of these quantities helps in understanding more complicated motions. In doing so, we will be able to extend the correspondence between the dynamics of linear motion and rotational motion.

10–1 GENERALIZATION OF ANGULAR MOMENTUM

We have attributed an angular momentum to rigid systems in pure rotation. *But even a point mass moving in space has an angular momentum with respect to a fixed origin.* To assign that angular momentum, we will consider a free particle, which moves along a straight line with constant momentum **p**, together with a point that we label as the origin O. A point and a line form a plane and we will label this plane the xy-plane (Fig. 10–1a). We will insist that the angular momentum, **L**, that we assign to the par-

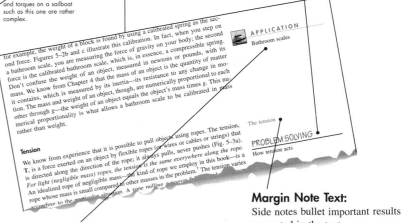

for example, the weight of a block is found by using a calibrated spring as the second force. Figures 5–2b and c illustrate this calibration. In fact, when you step on a bathroom scale, you are measuring the force of gravity on your body; the second force is the calibrated bathroom scale, which is, in essence, a compressible spring.

Don't confuse the weight of an object, measured in newtons or pounds, with its mass. We know from Chapter 4 that the mass of an object is the quantity of matter it contains, which is measured by its inertia—its resistance to any change in motion. The mass and weight of an object, though, are numerically proportional to each other through g—the weight of an object equals the object's mass times g. This numerical proportionality is what allows a bathroom scale to be calibrated in mass rather than weight.

Tension
We know from experience that it is possible to pull objects using ropes. The tension, **T**, is a force exerted on an object by flexible ropes (or wires or cables or strings) that is directed along the direction of the rope; it always pulls, never pushes (Fig. 5–3a). *For light (negligible mass) ropes, the tension is the same everywhere along the rope.* An idealized rope of negligible mass—the kind of rope we employ in this book—is a rope whose mass is small compared to other masses in the problem. The tension varies according to the particular situation. A rope pulling a wagon

APPLICATION

Bathroom scales

The tension

PROBLEM SOLVING

How tension acts

Application Margin Note
The train icon signals the presentation of technology or a relevant application of the text.

Margin Note Text:
Side notes bullet important results presented in the text.

Examples

Worked examples illustrate the use of concepts presented in the text. Most are accompanied, as this one is, by a figure or a student sketch. Solutions clearly show *how* and *why* an answer is reached.

amusement park ride
$= c_1 t^2 + c_2 t + c_3$
0.5 m, $d_1 = -1.0$
he locations of the
time in Figs. 3–3b

⊙⊙ The unit vectors **i**, **j**, and **k** are discussed in Chapter 1.

Cross Reference Marginal Note

Margin cross references are signaled by this icon and indicate where the referenced material has been presented earlier.

Chapter Review Text

At the end of each chapter, students have a chance to review the most important results.

EXAMPLE 10–7 A spool of mass M, rotational inertia I about its axis, and radius R unwinds under the force of gravity (Fig. 10–23). By using energy considerations, find the speed of the spool's center of mass after it has unwound a length h of thread.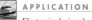

Solution: The total kinetic energy of the spool is the sum of the kinetic energy of motion of the center of mass and the kinetic energy of rotation about the center of mass. Thus

$$K = \frac{1}{2} Mv^2 + \frac{1}{2} I\omega^2.$$

The relation between the angular velocity and the velocity of the center of mass is identical to that of a cylinder rolling without sliding, so that

$$v = R\omega.$$

This implies that

$$K = \frac{1}{2} MR^2\omega^2 + \frac{1}{2} I\omega^2 = \frac{1}{2}(MR^2 + I)\omega^2.$$

Note that a simple application of the parallel-axis theorem shows that the quantity in parentheses is just the rotational inertia about the point at which the thread separates from the surface.

Another contribution to the energy is the potential energy of gravity. We can measure the potential energy from the initial height of the spool so that if the spool falls a distance h, the potential energy is $-Mgh$ and the spool has gained a kinetic energy Mgh. With a zero initial angular velocity, the kinetic energy after the spool has fallen a distance h is

$$K = Mgh.$$

Thus

$$\frac{1}{2}(MR^2 + I)\omega^2 = Mgh,$$

so

$$v = R\omega = R\sqrt{\frac{2Mgh}{MR^2 + I}}.$$

FIGURE 10–23 Example 10–7.

More Parallels Between Rotational and Linear Motion

Despite all the subtleties that appear when the full vector nature of angular momentum is involved, it is important to stress that there are no new laws of physics that describe rotational motion on the classical (the non-quantum-mechanical) level. *Every result on classical physics obtained in these last two chapters is a*

APPLICATION

Electronic devices based on tunneling

FIGURE 7–15 Individual atoms in a solar cell, shown here in false color, are made visible by using a scanning tunneling microscope. This image is part of a project to improve solar cells.

smaller. When the ^{2}H nuclei are raised to high temperatures (billions of degrees), the nuclei become agitated, move rapidly, have an increased energy E, and thus they have a smaller value of $U_{av} - E$. Effectively, by raising the energy of the nuclei, they are brought closer to the top of the barrier and both $U_{av} - E$ *and* the thickness of the barrier are reduced. Much research and development has been done since the early 1950s to bring the ^{2}H nuclei to these high temperatures. The technical difficulties are enormous but the reward for success would be great. The fusion reaction could yield a virtually unlimited amount of energy because seawater contains ^{2}H nuclei in abundance.

Other Applications

Several other manifestations of tunneling have technological application. The scanning tunneling microscope is a device that detects electrons that tunnel through a potential barrier at the surface of a piece of metal. It is used to study variations of atomic size in the surface of the metal in great detail (Fig. 7–15). Tunneling has also been used in studies of superconductivity. One of the most important applications is in the tunnel diode, a device with wide application in circuits. Quantum tunneling is not just some esoteric effect of interest to a small class of physicists. Many modern electronic devices rely in one way or another on electron tunneling. In later chapters, we will return to these applications.

SUMMARY

Conservative forces can be specified in terms of a single scalar function called the potential energy. In one-dimensional motion, the potential energy $U(x)$ depends on one coordinate only and is defined by

$$\text{for conservative forces:} \quad U(x) \equiv U(x_0) - \int_{x_0}^{x} F \, dx', \qquad (7\text{–}6)$$

where $U(x_0)$ $(= U_0)$ is an arbitrary constant and x_0 is an arbitrary point. In turn, the force is given in terms of U by

$$F(x) = -\frac{dU(x)}{dx}. \qquad (7\text{–}4)$$

Two important examples of potential energy are gravity and the spring. For gravity,

$$U(y) = mgy + U_0, \qquad (7\text{–}11)$$

PROBLEM-SOLVING TECHNIQUES

Integration

Integration of a function is the inverse of differentiation of that function. Put another way, if we integrate some function $f(x)$ and then differentiate the result, we get the function $f(x)$ back again. Expressed mathematically, if the functions $f(x)$ and $g(x)$ are related by

$$g = \int_{x_i}^{x_f} f(x)\,dx, \qquad (B2\text{-}1)$$

then

$$\frac{dg(x_f)}{dx_f} = f(x_f). \qquad (B2\text{-}2)$$

Moreover, we can interpret work as the area under the curve of F versus x (see Fig. 6–10). If the function F is negative and the displacements are positive, as in Fig. B2–1, the area and hence the work are negative. Conversely, if the displacement is negative and the force

positive, then the work will be negative. In the language of integrals, this comes about because the integral changes sign when we reverse the limits on it:

$$\int_{x_f}^{x_i} f(x)\,dx = -\int_{x_i}^{x_f} f(x)\,dx. \qquad (B2\text{-}3)$$

If the integral on the right-hand side is positive, then the left-hand side— which represents the work done in a displacement from x_f to x_i—is negative.

FIGURE B2–1 In the region between x_i and x_f, the shaded area under the curve is negative [part (a)], corresponding to negative values for the force $F(x)$ [part (b)].

formula for powers in Appendix IV–8, with $p = 1$, to find that

$$W = -k \int_0^L x\,dx = -\frac{1}{2}kx^2 \Big|_0^L = -\frac{1}{2}kL^2.$$

To find the extra work that must be done by the spring force as the mass moves from position L to position $2L$, we calculate

$$\int_L^{2L} (-kx)\,dx = -\frac{1}{2}kx^2 \Big|_L^{2L} = -\frac{1}{2}k[(2L)^2 - L^2] = -\frac{3}{2}kL^2.$$

The total work done by the spring as the mass moves from $x = 0$ to $x = L$ is the work done as the mass moves from $x = 0$ to $x = L$ plus this extra work

Spring

Equilibrium position

FIGURE 6–12 A spring exerts a force on an object that tends to bring the equilibrium

APPLICATION: CONSTRUCTION CRANES

The tower crane used at construction sites (Fig. 11AB–1) consists of a very flimsy tower constructed from three planes arranged in a triangle. Each plane consists of a lattice of triangular bars (this is called a geodetic construction). At the top of this tower is a normal crane boom, which can lift large masses, such as furnaces, air conditioners, and steel girders. The tower can rotate to place the mass at the correct angle, while the crane can move the mass radially inward or outward to position the object at the correct location; the object is finally set down. The strength of the tower is completely insufficient to withstand the bending moment of a large mass at the end of the boom, so there is a balance weight at the far side of the crane that moves in an opposite direction from the lifting equipment and the object being moved. The motion of the counterweight is controlled by strain sensors at the top of the tower and ensures that the crane boom produces no net torque at the top of the tower.

A similar idea is applied to the portable cranes that are carried by trucks. These cranes are provided with sen-

FIGURE 11AB–1 This type of

How to Prepare a Force Diagram

1. Make a simple sketch of the system.

2. Choose the body to be isolated (wagon).

3. Add convenient coordinate system.

4. Identify forces that act on wagon. Label them on diagram. Identify labels if necessary.

$\vec{F}_g$ = Gravitational force on cart from Earth
$\vec{F}$ = Force on cart handle from boy
$\vec{f}$ = Frictional force on cart wheels
$\vec{F}_N$ = Normal force on cart from ground

Problems

End-of-chapter problems answer the question: "How much have I learned?"

Student Sketch

All figures in the end-of-chapter material are drawn as student sketches and represent the first step in problem solving.

Level of Difficulty

A problem's level of difficulty is indicated: Level I problems are the easiest, Level II problems are usually multistep, and Level III problems are especially challenging.

PROBLEMS

3–1 Position and Displacement

1. (I) A car travels 21 km to the northeast, then 15 km to the east, before it travels 28 km to the north. Express the position vector from where the car starts to the point at which each turn occurs. What is the car's total displacement?

2. (I) A particle is located by the position described by the vector $\mathbf{r} = (c_1 - c_2t)\mathbf{i} + (d_1 + d_2t + d_3t^2)\mathbf{j}$, where $c_1 = 12$ m, $c_2 = 3$ m/s, $d_1 = -20$ m, $d_2 = -3$ m/s, and $d_3 = 0.5$ m/s^2. At what time(s) does the particle pass through the position $x = 0$ m? At what time(s), and where, does the particle cross the line $x = y$? Sketch the particle's trajectory from $t = -8$ s to $t = 10$ s.

3. (I) A gym teacher organizes a series of indoor races, which follow along the walls of the gym; the race starts from corner A, continues to corner B, and so forth. The gym is a rectangle with the distance $AB = 25$ m and the distance $BC = 35$ m. Suppose that the origin of a coordinate system is at point A, leg AB is in the $+x$-direction, and leg BC is in the $+y$-direction. Express the position vector of a running student at each of the four corners.

4. (I) A runner races with a uniform speed of 27.0 km/h around a circular track of radius 172 m. Draw the track and the runner's displacement vector after 20 s, 40 s, 60 s, and 2 min, assuming that at $t = 0$ s the runner is at the three o'clock position. Assume counterclockwise motion.

5. (I) A treasure map locates the site of a treasure by reference to two starting points, A and B. A is chosen to be the origin, and B is at the point (2.5 km)$\mathbf{i}$. The instructions state that the treasure lies at the intersection of two lines. One line starts at A and passes through the point (2.0 km)$\mathbf{i}$ + (4.0 km)$\mathbf{j}$; the other line, starting at B, passes through the point (6.0 km)$\mathbf{i}$ − (8.0 km)$\mathbf{j}$. Sketch the instructions and find the location of the intersection graphically as well as algebraically. Express the vector that gives the displacement of the intersection point from point C, whose location relative to the coordinate system specified by A and B is given by (1.2 km)$\mathbf{i}$ − (2.2 km)$\mathbf{j}$.

6. (II) The position of a particle in a given coordinate system is $\mathbf{r}(t) = (5 - 2t^2)\mathbf{i} + (-4 + 3t)\mathbf{j}$ where the distances are in meters when t is in seconds. At what time will the particle cross the y-axis? At what time will it cross the x-axis? Can you find an equation that relates the y-coordinate to the x-coordinate and therefore gives the trajectory in the xy-plane? Where would the x- and y-axes have to be moved so that at $t = 0$ s the tra...

FIGURE 3–29 Problem 7.

FIGURE 3–30 Problem 8.

10. (III) In a "shoot the coconut" lecture demonstration, the position of the coconut is given by the vector $(h_0 - \frac{1}{2}gt^2)\mathbf{j}$, whereas that of the projectile, aimed at the coconut at time $t = 0$, is given by $(-L + ut)\mathbf{i} + [(h_0ut/L) - (\frac{1}{2}gt^2)]\mathbf{j}$. Show that the two will always collide, and find the time at which this takes place (Fig. 3–32). Express the displacement vector of the coconut relative to the projectile.

Conceptual Questions

These questions check a student's understanding of the ideas that underlie a situation.

A law of physics has a wide range of applicability. Laws such as Newton's laws provide a framework for detailed predictions about what will happen under certain conditions. In their breadth and predictive power, Newton's laws are an ideal illustration of what constitutes a law of physics.

UNDERSTANDING KEY CONCEPTS

1. A small but dense mass is swinging freely at the end of a light string. A very sharp knife cuts the string when the mass is at the bottom of its swing; the knife does not disturb anything else. What is the subsequent flight of the mass?

2. Someone pushes on a wall. What experiment can you propose to determine the force with which the person pushes?

3. If you were in a freely falling elevator, the contact force on you due to the floor would drop to zero. If the elevator were to accelerate rapidly upward, the contact force between you and the floor would increase. Why?

4. An astronaut is working while in orbit. When the astronaut assembles a piece of equipment, will he or she notice a difference between working with components of large mass as opposed to components of small mass?

5. When a satellite travels around Earth in a circular orbit, it moves at a constant speed. Does Newton's first law apply in this situation? Is the velocity constant? Is there a force present?

6. There is a well-known parlor trick in which a tablecloth is pulled sharply from beneath a dinner setting, leaving the setting in place. Why does this work?

7. Is a spaceship heading from Earth to the Moon traveling in a force-free environment? Explain. What about a spaceship traveling from Earth to Mars that is currently in a region far away from either planet?

8. An adult sits on a child's table. The table is about to break. Is it correct to say that it is the weight of the adult that is causing the table to break?

General Problems

62. (I) Neutron stars are incredibly dense objects, compared to the Sun. One such (hypothetical) star has a radius of 40 km and rotates with a period of 26 s. What are the velocity and acceleration of a particle on the surface due to the rotation? Compare this acceleration with the acceleration of gravity on Earth.

63. (II) A golfer wants to land a golf ball on the green located 105 m away horizontally but 8.0 m down. The golfer chooses an eight iron that he knows will result in the ball leaving the tee at an elevation angle of 65°. (a) With what velocity should the ball leave the tee? (b) What is the maximum height of the ball above the green?

64. (II) A softball player hits the ball when it is 0.9 m above home plate. The ball leaves the bat at an elevation angle of 30°. What initial speed must the ball have to clear a fence 1.8 m high lo...

barrel makes with the ground? Unlike the projectile treated in Section 3–4, the cannonball is fired from the edge of a cliff of height h_0 above the level plain at which it is aimed. Show that the angle θ that gives the largest horizontal range is given by $\sin^2 \theta = v_0^2/2(v_0^2 + 2gh_0)$. [*Hint:* In calculus, you learn that a function of a variable such as θ has a maximum (or minimum) at an angle θ_0, the angle for which the derivative of the function with respect to θ is zero.]

69. (III) A juggler is able to handle four balls simultaneously. He takes 0.5 s to cycle each ball through his hands, throw the ball, and be ready to catch the next ball. (a) With what velocity must he throw each ball vertically? (b) What is the position of the other three balls when he has just caught one of the balls? (c) How high must he throw the balls if he is to juggle five balls?

70. (III) A wheel 72 cm in diameter rolls along a road, with the ... of 18 km/h.

General Problems

A comprehensive check of understanding involves problems that are not tied to a particular section of text.

P H Y S I C S

FOR SCIENTISTS AND ENGINEERS

SECOND EDITION

Tooling Up

A wide range of physical phenomena is presented to us by this rock-climbing scene. We can say that the force of gravity acts on the climber, who is saved by using the friction between himself and the rock. If we were to look at where friction comes from on an atomic scale, we would also discover a range of electromagnetic phenomena. Why does the rock have the properties it does? What makes the sky blue? Physics can help us to understand a huge variety of things in the world around us.

W hy physics? What is there about today's world that makes an understanding of the fundamentals of science so essential? To see forward, let's look backward. Curiosity and a need to make sense of what we see are distinctly human traits. The first steps toward the sciences arose out of the recognition of patterns of regularity. Many of these patterns—the sequence of day and night, the seasons, and the regular motion of the heavenly bodies—raised questions about their causes. It was not the proposed answers to these questions that moved us toward science, *it was the attention given to observation.* The accuracy of the observations of both position and time by ancient astronomers was surprisingly high. The Babylonians could predict the motion of planets. The description of the appearance of a brilliant "new star" in A.D. 1054— now known to be the Crab supernova—was recorded by Chinese astronomers in day-by-day detail. This description is accurate enough to confirm today's nuclear physics calculations of the star's brightness (Fig. 1–1).

Blind reverence for authority impedes scientific progress, and for a long period the work of the ancient Greeks was regarded with crippling reverence. For example, the idea that gravity makes heavier objects fall more quickly than lighter ones persisted from Greek times all the way to the Renaissance. An important component of the Renaissance was a departure from too much reverence for authority, and a reintroduction of the role of experiment in natural science. Among those who must be given credit for solid reestablishment of experiment are William Gilbert, for magnetism, and Galileo Galilei, for motion. Galileo had a more immediate impact on the rise of physics because of his influence on Isaac Newton, whose laws of motion form the underpinning of almost all of physics. The give and take between experimental observation, the mathematical formulation of descriptive and predictive theories, and further experi-

The scientific method

1

mental tests of predictions of these theories form the *scientific method*. In the scientific method, human imagination, through intuition and curiosity, is subject to the checks and balances of experiment, thought, and further testing. The scientific method has led to accelerated progress in our understanding of the physical world.

This progress has moved along two fronts. On one side, new concepts are built—sometimes in small steps, sometimes in large ones—on earlier concepts. These concepts summarize an ever-increasing body of experimental information, as well as permit an extrapolation of physical ideas into areas where patterns of regularity had not previously been seen or even suspected. We speak of opening up new fields. On the other side, progress is driven by an improvement of experimental techniques, which have allowed scientists to probe nature in domains hitherto inaccessible to experimentation. For example, much of modern technology can be traced to the development of pumps efficient enough to allow the creation of near-vacuum conditions in sealed vessels.

Scales of the Observable World

Physics underwent a series of revolutionary developments in the period 1900–1925, and the field is sometimes divided into classical (pre-1900) and modern (post-1900) physics. This division seems artificial to us, and we prefer to think of physics in terms of *scales*. The laws of physics and the structure of matter appear to have a hierarchical structure. Newton's laws of motion were thought for a long time to be universally applicable. Subsequent observation has shown, however, that there are extreme circumstances in which these laws must be replaced by a different picture of motion. We need quantum mechanics to describe matter at very small distances, and we need special relativity to describe motion at speeds close to that of light. The behavior of atoms and their constituents cannot be understood without these new laws of physics. These laws have the characteristic that for objects that are large on an atomic scale, or move slowly in comparison with the speed of light, they reduce to the simpler laws of motion discovered by Newton. Thus for the description of most of the world, and for many—but not all—engineering applications, Newton's laws apply.

This middle ground in the description of nature includes the great subjects of classical physics—mechanics, waves, thermodynamics, and electricity and magnetism—and forms the content of most of this book. In the middle ground, the effects of the detailed underlying structure of matter appear in the form of some numerical constants that can be taken from experiment. Numbers such as the conductivity of electricity or the index of refraction of light summarize a great deal of complexity in the structure of matter; as far as classical physics is concerned, their numerical value is the only connection with that structure, and these numbers can be taken from experiment.

To understand the realm of the very small, or the very fast, we need quantum mechanics, and we need relativity. This realm is what we might call the frontier, where we are continuing to build our understanding. A deeper understanding of the frontiers is one of the permanent goals of science. Even when we treat middle-ground physics, we shall indicate the underlying frontier physics. The reason that this is useful is obvious: technology is pushing engineering practice into these frontiers, and quantum physics—the physics of small scales—is becoming more and more essential to modern engineering. In the coming decades, few engineers will be able to function without some understanding of quantum mechanics and relativity. Devices such as the laser have moved from the physics research laboratory into everyday use in engineering practice.

Students sometimes feel that physics is somehow separate from engineering. In reality, it is only the goals of the two disciplines that differ. The aim of much of engineering is to use knowledge of the basic structure and functioning of materials to advance technology. That knowledge is the domain of physics. Both the advancement of technology and the attainment of the needed knowledge are extremely important as we move toward the twenty-first century. That is why a physics education is so essential to future engineers or to anyone who wishes to function successfully in the emerging world.

FIGURE 1–1 We know from historical record that the crab nebula is the result of a great stellar explosion. At its heart is a very compact remnant—a neutron star—of the original star. Objects such as these allow us to test our understanding of many aspects of the natural world.

One language of engineering and physics is that of numbers. The range of numbers that appear in the physical world is truly enormous. For example, the mass of Earth is about 5,980,000,000,000,000,000,000,000 kilograms (kg), and the diameter of a proton is about 0.000000000000001 meter (m). So many zeros are inconvenient, and we employ a shorthand method of writing very large and very small numbers. By using powers of 10, Earth's mass is more easily written as 5.98×10^{24} kg, and the diameter of a proton is written as 10^{-15} m. In this notation, 10^3 represents 1000 and 10^{-4} means 0.0001. We shall use this standard *scientific notation* throughout this book.

A considerable advantage of scientific notation is that multiplication and division are easily performed by adding and subtracting exponents of 10. Thus the product $100 \times 100 = 10,000$ can be written as $10^2 \times 10^2 = 10^{2+2} = 10^4$. The awkward multiplication $0.00000055 \times 24,000$ can be done more easily as $(5.5 \times 10^{-7}) \times (2.4 \times 10^4) = (5.5 \times 2.4) \times 10^{-7+4} = 13.2 \times 10^{-3} = 1.32 \times 10^1 \times 10^{-3} = 1.32 \times 10^{-2}$. Where division is involved, we simply change the sign of an exponent and use the multiplication rules. For example,

$$\frac{7.5 \times 10^{-3}}{2.5 \times 10^{-4}} = \frac{7.5}{2.5} \times 10^{-3} \times 10^{+4} = 3 \times 10 = 30.$$

1–2   LENGTH, TIME, AND MASS

In our exploration of the physical world, we have found that length, time, and mass play a fundamental role in measurement (Fig. 1–2). As Tables 1–1, 1–2, and 1–3 show, these quantities cover an enormous range of values in our universe. These three quantities are already intuitively familiar, and the idea that there is a system of units for specifying them is not a strange one. The tables contain a particular, standardized system of units. The importance of standardization is evident. You may wear a size 7 shoe in the United States, but this unit would not be of much use if you were to travel to Europe, where a different system is used; there, your shoe size would be 38.

Hundreds of years ago, people used what was readily available as standards for measurement. Length measurements such as the foot came into use in this manner. Over time, measurement systems have become both more precise and more universal. For an early example, French scientists established the forerunner of the International System of measurements in 1791. They defined the meter, the second, and the kilo-

FIGURE 1–2 (a) Measuring distance. The measurements of length needed to study a traffic accident have very different kinds of precision requirements than those involved in, say, the fabrication of integrated circuits. (b) It has become commonplace to see clocks and watches that display time in a digital fashion. (c) We must have an objective, repeatable way to measure mass.

(a)

(b)

(c)

gram: The meter—roughly one yard—was defined as one ten-millionth (10^{-7}) of the distance along Earth's surface between the equator and the North Pole; the second as 1/86,400 of a mean solar day; and the kilogram as the mass of a certain quantity of water. In 1889, an international organization called the General Conference on Weights and Measures was formed to meet periodically to refine these units of measure. In 1960, this organization decided to name a system of units based on the meter, kilogram, and second the International System, with the abbreviation **SI** (for the French words **Système International**). We use the term SI in this book. This system is also known as the *metric system* or *mks system* (after *m*eter, *k*ilogram, and *s*econd).

Length

The definition of the meter has been changed several times. In 1889, one meter was defined as the distance between two finely engraved marks on a bar of platinum–iridium that was kept in a vault outside Paris. Even though several copies of this bar were distributed throughout the world, such a standard of length had many shortcomings. For instance, with progress in optical techniques, the scratches on the bar were seen to be fuzzy and imprecise. In 1960, the standard of length was changed to depend upon an atomic constant—the wavelength of a particular orange-red light emitted by an isotope of krypton (^{86}Kr) gas. Because our ability (and need) to measure length has led us to require even greater accuracy, this standard also became insufficiently precise. Therefore, in 1983, the 17th General Conference on Weights and Measures established a standard of length based on the speed of light in vacuum (denoted by the letter c). A **meter** (m) is now defined as the distance light travels in vacuum during 1/299,792,458 second. Some orders of magnitude for lengths are given in Table 1–1.

Time

The second was originally defined as 1/86,400 of the mean solar day, which is the time interval, averaged over a year, from noon of one day to noon of the next. This definition is insufficient because Earth's rotation is both slightly irregular and gradually slowing down from year to year. Therefore, in 1967, a definition of the second was adopted that depends on an atomic standard. The **second** (s) is now defined as the duration of 9,192,631,770 periods of a particular vibration of a cesium atom isotope (^{133}Cs). Clocks based on this standard are, in effect, identical because all atoms of ^{133}Cs are indistinguishable and because frequency can be measured in the laboratory to an accuracy of about four parts in 10^{13}. Some orders of magnitude for time are given in Table 1–2.

Mass

The kilogram was originally defined as the mass of 1 liter of water under certain conditions of temperature and pressure. In 1901, the standard **kilogram** (kg) was defined as the mass of a particular cylinder of platinum–iridium alloy kept at the International

TABLE 1–1 Orders of Magnitude for Length

Parameter	*Length (m)*	*Parameter*	*Length (m)*
Proton	10^{-15}	Diameter of solar system	10^{13}
Hydrogen atom	10^{-10}	Distance to nearest star (Proxima Centauri)	10^{17}
Flu virus	10^{-7}		
Raindrop	10^{-3}	Diameter of our galaxy (Milky Way)	10^{21}
Height of person	10^{0}		
One mile	10^{3}	Distance to nearest galaxy	10^{22}
Diameter of Earth	10^{7}	Distance to edge of observable universe	10^{26}
Earth–Moon distance	10^{9}		
Earth–Sun distance	10^{11}		

TABLE 1–2 Orders of Magnitude for Time

5

1–2 Length, Time, and Mass

Parameter	Time (s)	Parameter	Time (s)
Time for light to cross proton	10^{-23}	Period of human heartbeat	10^0
Time for light to cross atom	10^{-19}	Class lecture	10^3
Period of visible light wave	10^{-15}	One Earth day	10^5
Period of vibration for standard cesium clock	10^{-10}	One Earth year	10^7
		Age of Greek antiquities	10^{11}
Half-life of muon	10^{-6}	Age of Earth	10^{17}
Period of highest audible sound	10^{-4}	Age of universe	10^{18}

Bureau of Weights and Measures in France. Duplicate copies of the cylinder made of this particularly stable alloy are kept in laboratories such as the National Institute of Standards and Technology in Maryland. Although the standards of time and length can be reproduced to precisions of one part in 10^{12}, the standard of mass can be reproduced only to perhaps one part in 10^8 or 10^9. This standard of mass leaves much to be desired. We would like to find an atomic or natural standard for mass but, even though we know that all atoms of the same type have the same mass, nobody knows how to count atoms with the required accuracy. Some orders of magnitude for mass are given in Table 1–3.

Other Systems of Units

The International System of units is by far the most important and widely accepted system used in the world today. Two other systems, however, are still in common use.

cgs.   The *cgs system* is based on the *c*entimeter, *g*ram, and *s*econd and is a metric system derived directly from SI. This system finds its major use in measurements of electricity and magnetism. Densities (mass per unit volume) normally are still quoted in grams per cubic centimeter (g/cm^3) because most densities are nearer unity in this system than in SI. For example, the density of water is 1 g/cm^3 in cgs but is 10^3 kg/m^3 in SI. The definition of the units of the cgs system is based on those of SI:

$$1 \text{ cm} \equiv 0.01 \text{ m and } 1 \text{ g} \equiv 0.001 \text{ kg},$$

where we have used the symbol $\equiv$ to indicate a definition.

British Engineering System.   The *British engineering system*, or British system, is based on units of the inch, pound, and second. This system is used only in the United States and in parts of the British Commonwealth, where it is in the process of being replaced by SI. Even in the United States, scientists seldom use the British system, but existing technology based on this system may require its use in engineering applications. Economic considerations—the desirability of international trade—suggest that even this limited use will eventually disappear.

The British system of units is now *defined* in terms of SI units. The unit of length, the **inch** (in), is defined as 1 in $\equiv$ 2.54 cm (Fig. 1–3). In the British system we must distinguish between "mass" and "weight," a topic we will discuss later. It is sufficient

(a)

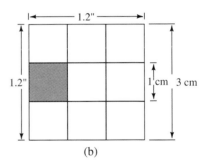

(b)

FIGURE 1–3 (a) One inch is exactly 2.54 cm. (b) A square 3 cm $\times$ 3 cm in size has an area 9 cm^2, or $(3/2.54)^2$ in^2 = 1.4 in^2.

TABLE 1–3 Orders of Magnitude for Mass

Parameter	Mass (kg)	Parameter	Mass (kg)
Electron	10^{-30}	Battleship	10^8
Hydrogen atom	10^{-27}	Moon	10^{23}
Uranium atom	10^{-24}	Earth	10^{25}
Dust particle	10^{-13}	Sun	10^{30}
Raindrop	10^{-6}	Our galaxy (Milky Way)	10^{41}
Piece of paper	10^{-2}	Observable universe	10^{52}
Man	10^2		

Why do we need high precision in our standards of length and time? As a practical application of high-precision time measurements, consider the *Global Positioning System* (Fig. 1AB–1). The system consists of a network of 24 satellites orbiting Earth at an altitude of 20,000 km. The satellites carry atomic clocks—accurate to about four parts in 10^{13}—that emit regular signals, which travel with the speed of light. Because the speed of light is finite, signals from different satellites arrive at a receiver at slightly different times, depending on the position of the receiver. The receiver's *position* can then be determined because the position of the satellites is very accurately tracked. The accuracy with which the receiver's position is determined has to do with the uncertainty in the time measurement and the uncertainty in the speed of light, known to three parts in 10^{10}. There are limitations on this accuracy, since the speed of light in air differs a little from the absolute speed of light; it is the meteorological variations in the speed in air that set limits on the accuracy of the position measurement. At present, the civilian users of hand-held receivers may determine their position to an accuracy of ±15 m. The Fed-

eral Aviation Authority has decided to use satellite positioning to track airplanes to airports, and the system has become indispensible for navigation. It is easy to imagine a host of applications of this technology in general transportation and geography.

FIGURE 1AB–1 With the GPS system, travelers—from airplane pilots to hikers—can confirm their positions with remarkable accuracy.

for now to state that a **weight** of 1 pound (lb) is a force (a push or pull) equivalent to the force exerted by a mass of exactly 0.45359237 kg on a bathroom scale—such scales read the weight of an object—at a particular point on Earth's surface. This means that a mass of 1 kg has the same weight as about 2.2 lb. (Notice that we do *not* write 1 kg = 2.2 lb, because mass and weight are *not* the same.) A ton is 2000 lb, but a metric ton is 1000 kg, which is equivalent to a weight of 2204.6 lb. The unit of mass in the British system is called a *slug*, but it is seldom used. A slug is equal to 14.5939 kg. Finally, the unit of time—the *second*—is the same in the British system as it is in SI.

Some Shorthand and History Concerning Units

A useful set of prefixes in SI replace certain powers of ten. These prefixes are listed in Table 1–4. For example, the distance between New York City and Los Angeles is 4,483,000 m, or 4.483×10^6 m. Neither of these quantities is convenient for everyday use. However, we can use the prefix *kilo-*, which stands for 10^3, to state this distance as 4483 kilometers (km). A finger's width, for example, would not normally be written as 0.015 m or as 1.5×10^{-2} m, but rather as 1.5 centimeters, or 1.5 cm. The standard of mass, the kilogram (kg), is often more convenient to use than the gram (g). In most countries of the world, meat and produce prices are quoted per kilogram because that is a convenient mass unit. You will soon appreciate the simplicity of the International System and obtain a feel for its units of length and mass.

As in every discipline, some special units are present because of historical precedent or simply because they are easy to use. For example, the quantity 10^{-6} m is equivalent to a micrometer, or μm, in SI, but is also known as a micron. This double name for the same quantity causes little confusion. Another example involves a unit often used in atomic physics for the wavelength of light, which is the angstrom, Å. One angstrom has the value 10^{-10} m or 0.1 nm (1 nanometer (nm) = 10^{-9} m). The range of visible light is about 3000–7000 Å, or 300–700 nm. Both the nanometer and the angstrom are commonly used.

At the other extreme of the length scale, astronomers and astrophysicists find the meter to be too small for practical use, and they employ three other distance measures:

TABLE 1–4 Unit Prefixes for Powers of 10

Prefix	Symbol	Multiple
exa[†]	E	10^{18}
peta[†]	P	10^{15}
tera	T	10^{12}
giga	G	10^9
mega	M	10^6
kilo	k	10^3
hecto[†]	h	10^2
deka[†]	da	10^1
deci[†]	d	10^{-1}
centi	c	10^{-2}
milli	m	10^{-3}
micro	μ	10^{-6}
nano	n	10^{-9}
pico[†]	f	10^{-15}
atto[†]	a	10^{-18}

[†]Except for centi, the prefixes near 1 (10^0) are rarely used in the United States. Similarly, the very small and very large multiples (powers of ±15, ±18) are not normally used or recognized. You should become familiar with all others.

(1) the *astronomical unit* (AU), which is the mean distance between Earth and the Sun; (2) the *light-year* (ly), the distance that light travels in one year; and (3) the *parsec* (pc), which is 3.0857×10^{16} m. The astronomical unit (1 AU = 1.496×10^{11} m) is useful for distances within our solar system. The light-year, 0.95×10^{16} m, is useful when dealing with interstellar distances, as is the parsec, equal to about 3.26 ly or 206,265 AU.

Unit Relations and Conversions

The angstrom and the astronomical unit are examples of *derived units,* units defined in terms of the basic set (the meter, the kilogram, and the second). These units are simple changes in scale of the basic units. A more complicated type of derived unit involves combinations of the fundamental units. Sometimes these combinations are given names, sometimes not. Speed is measured in meters per second (m/s), but it is not given a separate name. When we study force, however, we will find it to be measured in kilogram-meters per second squared $(kg \cdot m/s^2)$; this more complicated unit is given a name, the **newton** (N). The newton is a derived unit in SI.

In deriving relations between physical quantities within a *single* system of units, the units on both sides of such relations must match. This provides us with a very useful tool for checking results that we may derive. If units on both sides of an equation cannot be made to match, there is an error somewhere. Further, if a problem is stated using more than one unit for a particular quantity—such as meters and inches for length—then we must find a simple way to *convert* between different systems of units to solve the problem. As we will now see, this can be done conveniently and systematically with the primary conversion equations that relate one set of units to another. Examples of such primary equations are the expressions

$$1 \text{ in} = 2.54 \text{ cm and } 100 \text{ cm} = 1 \text{ m.}$$

> When a single unit system is employed, the units on one side of an equation must be the same as those on the other side.

PROBLEM SOLVING

Conversion of units

The first of these can be rewritten as

$$1 = \frac{2.54 \text{ cm}}{1 \text{ in}} \quad \text{or} \quad 1 = \frac{1 \text{ in}}{2.54 \text{ cm}}.$$

Any equation can be multiplied by the pure number 1 without change; by judicious choice of the factor 1, units can be canceled and replaced by others. For example,

$$15 \text{ in} = (15 \text{ in})(1) = (15 \text{ in})\left(\frac{2.54 \text{ cm}}{1 \text{ in}}\right).$$

The unit "in" cancels, and

$$15 \text{ in} = (15)(2.54 \text{ cm}) = 38.1 \text{ cm.}$$

Some other examples of conversions of this type are

$$1 \text{ yd} = 1 \text{ yd} \left(\frac{36 \text{ in}}{1 \text{ yd}}\right)\left(\frac{2.54 \text{ cm}}{1 \text{ in}}\right)\left(\frac{1 \text{ m}}{100 \text{ cm}}\right) = 0.9144 \text{ m}$$

and

$$1 \text{ mi} = 1 \text{ mi} \left(\frac{5280 \text{ ft}}{\text{mi}}\right)\left(\frac{12 \text{ in}}{\text{ft}}\right)\left(\frac{2.54 \text{ cm}}{\text{in}}\right)\left(\frac{1 \text{ m}}{100 \text{ cm}}\right)\left(\frac{1 \text{ km}}{1000 \text{ m}}\right) = 1.609 \text{ km.}$$

The technique described here is applicable to more complicated examples.

EXAMPLE 1–1 Given that the speed of light is 2.998×10^8 m/s, what is the distance that light travels in 1 yr? (This distance is the light-year.)

Solution: The primary relations involve time in this case; for example,

$$60 \text{ s} = 1 \text{ min.}$$

We can write this as

$$1 = \frac{60 \text{ s}}{1 \text{ min}}.$$

Similarly,

$$1 = \frac{365.25 \text{ d}}{1 \text{ yr}}.$$

For more precision, we use 365.25 d as an average year instead of just 365 d. The distance 1 ly is the speed of light times the time duration, 1 yr:

$$1 \text{ ly} = (2.998 \times 10^8 \text{ m/s})(1 \text{ yr}).$$

This equation has mixed units, a situation we can rectify by converting the units of time from years to seconds:

$$1 \text{ ly} = \left(2.988 \times 10^8 \frac{\text{m}}{\text{s}}\right)(1 \text{ yr})\left(\frac{365.25 \text{ d}}{1 \text{ yr}}\right)\left(\frac{24 \text{ h}}{1 \text{ d}}\right)\left(\frac{60 \text{ min}}{1 \text{ h}}\right)\left(\frac{60 \text{ s}}{1 \text{ min}}\right)$$

$$= 9.461 \times 10^{15} \text{ m}.$$

We have calculated that 1 ly = 9.461×10^{15} m. The units for second, minute, hour, day, and year all appear in both the numerator and denominator and, therefore, cancel. The only unit left is the meter.

1–3 ACCURACY AND SIGNIFICANT FIGURES

Uncertainty in Measurement

Physics rests on experiment, and experiment requires measurement. But measurements are, at best, only approximate. An **uncertainty** is an indication of the accuracy of a measurement. The uncertainty depends on the accuracy and calibration of the instrument that is making the measurement and on how well the instrument can be read. We can best illustrate the meaning of uncertainty with an example. If the width of a page of paper is measured with a ruler to be 21.6 cm with an uncertainty of 1 mm (or 0.1 cm), it would be correct to say that the width is 21.6 cm ± 0.1 cm, or 21.6 ± 0.1 cm. (The ± is read as "plus or minus.") Here, 21.6 cm is called the *central value* and 0.1 cm the *uncertainty* around that central value. In this case, the basis of the uncertainty lies in how well our eyes can read the ruler and on the precision with which the ruler was made. We often use the term *percentage uncertainty* as a measure of the ratio of the uncertainty of a quantity to its central value. The percentage uncertainty is found by multiplying this ratio by 100. The percentage uncertainty of our paper measurement is thus

$$100 \left(\frac{0.1 \text{ cm}}{21.6 \text{ cm}}\right) = 0.5\%.$$

We can find the area of the paper by measuring the length and multiplying by the width. Suppose that we measure the paper's length to be 27.9 ± 0.1 cm. The percentage uncertainty for the length is 0.4%. We find the area by multiplying (21.6 cm)(27.9 cm) = 603 cm². Because the measurements of width and length both contain uncertainties, the area is also uncertain. But what is this uncertainty? To get a rough idea, just add the percentage uncertainties when multiplying or dividing numbers.[†] In our

---

[†]A more precise measure is that if P_1 and P_2 are the percentage uncertainties of two quantities being multiplied, the net uncertainty is $\sqrt{P_1^2 + P_2^2}$.

case, the total area uncertainty is about 0.5% + 0.4%, or 0.9%. This means an uncertainty of (0.009)(603 cm²) = 5 cm², and the area of the paper is 603 ± 5 cm².

Significant Figures

Physical quantities are never known with exact precision (unless they are merely definitions). We imply a certain degree of uncertainty in a quantity when we assign a certain number of digits to its numerical value. Thus, when we say that an object is 2.00 m long, we mean that it is between 1.995 m and 2.005 m long. If we wanted to say that the length is somewhere between 1.9995 m and 2.0005 m, we would say that the length is 2.000 m. In the first case, 3 significant figures are used to describe the object's length; in the second case, the number of significant figures is 4. When we say that a sheet of paper has an area of 603 cm² we are using 3 significant figures, which means that the area lies between 602.5 cm² and 603.5 cm².

Zeros that are used only to set a decimal point are not part of our count of significant figures. Thus 0.00035 has 2 significant figures, not 6. To take a more extreme example, we mentioned in Section 1–1 that the mass of Earth is 5,980,000,000,000,000,000,000,000 kg. Surely we do not know Earth's mass to 25 significant figures! Scientific notation provides a way to avoid this ambiguity. When we write the mass of Earth as 5.98×10^{24} kg, we indicate unambiguously that we know the mass to 3 significant figures; if we knew the mass to only 2 significant figures, we would write 6.0×10^{24} kg.

There is one point you should keep in mind: *Carry a calculation only to as many significant figures as are contained in the input parameter with the fewest significant figures.* In other words, the quantity in a calculation that contains the most uncertainty largely dictates the accuracy of the final result. Watch out! Hand-held calculators make it all too easy to violate this rule. For example, if you are asked to calculate the ratio of 3.0 to 11.0 on your calculator, you might be tempted to write 0.27272727272.... But in this case, the number 3.0 has the fewest significant figures—2—and thus you should carry your calculation to only 2 significant figures, 0.27.

Resist the temptation to keep all the digits your calculator can supply.

PROBLEM SOLVING

Use good sense with significant figures

1–4   DIMENSIONAL ANALYSIS

Dimensions

Experience has shown that there are three basic ways to describe any physical quantity: the space it takes up, the matter it contains, and how long it persists. All descriptions of matter, relationships, and events are combinations of these three basic characteristics. All measurements can be reduced ultimately to the measurement of length, time, and mass. Any physical quantity, no matter how complex, can be expressed as an algebraic combination of these three basic quantities. Speed, for example, is a length per time.

Length, time, and mass therefore have a significance far beyond that of providing the basis of a system of units. They specify the three **primary dimensions**. We use the abbreviations $[L]$, $[T]$, and $[M]$ for these primary dimensions. The **dimension** of a physical quantity is the algebraic combination of $[L]$, $[T]$, and $[M]$ from which the quantity is formed. The speed v provides an example. The dimension of v is

$$[v] = [L/T] \quad \text{or} \quad [LT^{-1}].$$

Do not confuse the dimension of a quantity with the units in which it is measured. A speed may have units of meters per second, miles per hour, or, for that matter, lightyears per century. All of these different choices of units are consistent with the dimension $[LT^{-1}]$. In what follows, the square brackets, as used here, indicate that we are dealing with dimensions.

Any physical quantity has dimensions that are algebraic combinations $[L^q T^r M^s]$ of the primary dimensions, where the superscripts q, r, and s refer to the order (or power)

Any physical quantity has a dimension that can be formed from the three primary dimensions of length, mass, and time.

The dimension of a quantity should not be confused with the units in which it is measured.

of the dimension. Thus, for example, an area has dimension $[L^2]$. If all of the exponents q, r, and s are zero, the combination will be dimensionless. The exponents q, r, and s can be positive integers, negative integers, or even fractional powers.

Dimensional Analysis

The dimensions on one side of an equation must be the same as those on the other side.

Study of the dimensions of an equation—*dimensional analysis*—is an important exercise with several different uses in physics. Any equation that relates physical quantities must have consistent dimensions; that is, *the dimensions on one side of an equation must be the same as those on the other side.* This provides a valuable check for any calculation. Dimensional analysis can also reveal *scaling laws* (see Section 1–7), which describe how changing one quantity in a physical situation requires changes in others. Finally, when there is reason to believe that only certain physical quantities can enter into a physical situation, dimensional analysis can provide us with powerful insights.

Let's look at some examples of dimensional analysis. In Chapter 7, we derive a relation between the height h of a dropped object and the speed of that object. This relation involves the *acceleration of gravity*, g, a quantity whose dimension is $[g] = [LT^2]$. The relation reads

$$gh = \frac{1}{2} v^2.$$

Let's compare the dimensions on each side of this equation. The dimension of h is $[L]$, so the left-hand side has dimensions $[LT^{-2}][L] = [L^2 T^{-2}]$. The right-hand side has the dimensions of speed squared, $[LT^{-1}]^2 = [L^2 T^{-2}]$. Thus the dimensions match. If, through error, we had written a relation $gh^2 = \frac{1}{2}v^2$, then this check would have revealed the error. Note that dimensional analysis does not help us understand the numerical factor $\frac{1}{2}$.

EXAMPLE 1–2 Newton's law of universal gravitation gives the force between two objects of mass, m_1 and m_2, separated by a distance r, as

$$F = G\left(\frac{m_1 m_2}{r^2}\right).$$

Use dimensional analysis to find the units of the gravitational constant, G.

Solution: First, the dimensions of the two sides of the equation must match. In the previous section, we learned that the unit of force is the newton, equivalent to $kg \cdot m/s^2$. Using these units, the dimensions of force must be $[MLT^{-2}]$. We now know the dimensions of every quantity in the equation for gravitational force except G. Writing the dimensions for both sides gives

$$[MLT^{-2}] = [G][M][M][L^2] = [G][M^2 L^{-2}].$$

Note that the individual dimensions can be consolidated inside the square brackets or left within their own brackets—whichever is easiest. We solve for the dimension of G as

$$[G] = \frac{[MLT^{-2}]}{[M^2 L^{-2}]} = [MLT^{-2}][M^{-2}L^2] = [M^{-1}L^3 T^{-2}].$$

FIGURE 1–4 The square base of the great pyramid of Giza is 232 m on a side, and the peak is 138 m above the base. Napoleon's engineers estimated that the material that forms the pyramid could be used to build a wall 1 m high and 0.1 m thick all the way around France. Can you estimate the length of the perimeter of France?

1–5 HOW A LITTLE REASONING GOES A LONG WAY

Sometimes we must make a quick calculation. We may want to check a complicated numerical calculation to see if the answer is reasonable. Or we may not have access to all the data needed, and an estimate, or a very rough approximation, is in order (Fig. 1–4). In these cases, we perform an *order-of-magnitude* calculation, in which variables are rounded off to the nearest power of 10 or to some other easily handled number. The final result of a calculation with variables so dramatically rounded off is accurate only within an order of magnitude, but such an estimate can be extremely useful.

Some Examples

The sight of a Christmas tree lot may make us wonder how many natural trees are sold at Christmas time. There are roughly 100 million families in the United States, but perhaps only half of them—50 million—have a Christmas tree. It is reasonable to assume that about half of those families with a tree buy a natural tree. We thus arrive at an estimate of about 25 million natural trees sold.

Now, let's say that we have a friend who owns a 200-acre Christmas tree farm (Fig. 1–5), and we wonder how many trees he can plant. We reasonably suppose that trees are planted about 6 ft apart, so that each tree takes up about 36 ft². Considering space occupied by roads, buildings, or other uses, we might change this estimate to 1 tree/50 ft². But how big is 200 acres? Most of us do not remember the precise size of an acre, but we might hazard a guess that a typical suburban house lot is about a quarter-acre, and that it is perhaps 100 ft across by 100 ft deep, or 10⁴ ft². This indicates that an acre is about 40,000 ft². (In fact, 1 acre = 43,560 ft², so our guess is not far off.) If each tree requires about 50 ft², then our friend can plant 40,000/50 = 800 trees—let's take 1000 for simplicity—per acre, for a total of 200,000 trees. This is a reasonable order-of-magnitude calculation.

FIGURE 1–5 Images such as this one allow us to make useful estimates—in this case, a set of questions revolving around the number of Christmas trees that are planted.

EXAMPLE 1–3 Estimate the average area available to each person in the United States and then to each person on Earth.

Solution: There are some 250 million people in the United States. The United States is roughly a rectangle about 3000 mi (about 5000 km) from east to west and 2000 mi (about 3000 km) from north to south, giving an approximate total area of 15×10^6 km². With $(1 \text{ km})^2 = (10^3 \text{ m})^2 = 10^6$ m², this translates to 15×10^{12} m². Dividing 15×10^{12} m² by 3×10^8 people, we obtain 5×10^4 m² per person. This is a square roughly 200 m on a side.

For Earth, we might estimate a total of about 5 billion people. We remember from geography that the circumference of Earth is 25,000 mi, or about 40,000 km. We divide 40,000 km by 6 (circumference $= 2\pi r$, and we will approximate 2π as 6) to obtain a radius of 7000 km. The total surface area of Earth is $4\pi r^2$, $12 \times 7 \times 7 \times 10^6$ km², or roughly 5×10^8 km². However, only about one-third of Earth's surface area is land—roughly 2×10^8 km², or 2×10^{14} m². Dividing 2×10^{14} m² by 5×10^9 people, we estimate an area on Earth of about 5×10^4 m² per person. The United States has about the same population density as does Earth.

EXAMPLE 1–4 A human body consists mainly of water. One mole (mol) of water, which consists of about 6×10^{23} molecules, has a mass of 18 g. Assuming that the molecules of water in your body are closely packed together, make a rough estimate of the size of a molecule.

Solution: We want to find the number of molecules N in your body, then divide your total volume by N to determine the volume of a single molecule.

The number N is determined by finding out how many moles you contain.[†] For a mass, let's use 60 kg, which is equivalent to a weight of about 132 lb. Because 60 kg $= 60 \times 10^3$ g $= 6 \times 10^4$ g, the number of moles n is

$$n = \frac{\text{total mass}}{\text{mass of 1 mol}} = \frac{6 \times 10^4 \text{ g}}{18 \text{ g}} = 3.3 \times 10^3 \text{ mol.}$$

[†]A mole is about 6×10^{23} molecules. It is the quantity of a substance whose weight in grams equals the substance's molecular weight.

The total number of molecules is found by multiplying the number of moles by 6×10^{23}:

$$N = (6 \times 10^{23} \text{ molecules/mol})(3.3 \times 10^3 \text{ mol}) = 2 \times 10^{27} \text{ molecules}.$$

To find the volume of one molecule, we must divide your total volume by N. We might estimate that you form a solid with height 2 m, width 0.5 m, and depth 0.3 m, for a total volume

$$V = (2 \text{ m})(0.5 \text{ m})(0.3 \text{ m}) = 0.3 \text{ m}^3.$$

Thus the volume V_1 of a molecule is

$$V_1 = \frac{V}{N} = \frac{0.3 \text{ m}^3}{2 \times 10^{27} \text{ molecules}} = 0.15 \times 10^{-27} \text{ m}^3/\text{molecule}.$$

We can estimate the linear size of a molecule by assuming that it forms a little cube with sides of length ℓ; its volume is then $V_1 = \ell^3$. Thus

$$\ell = \sqrt[3]{V_1} = \sqrt[3]{0.15 \times 10^{-27} \text{ m}^3} \simeq 0.5 \times 10^{-9} \text{ m},$$

or about 5×10^{-10} m. Because 1 Å is 10^{-10} m, the size of a water molecule is estimated to be several angstroms. In fact, the separation between the hydrogen atoms in a water molecule is about 2 Å.

1–6 SCALARS AND VECTORS

Scalars have no direction associated with them.

The mathematical descriptions of physical systems in this book deal with two types of quantities. One type is an ordinary algebraic quantity called a **scalar**. A scalar does not have a direction associated with it. The statement that the mass of a ball is $\frac{1}{4}$ kg specifies all we need to know about its mass. In particular, no direction is specified when a mass is given. The same is true of the time it takes the ball to travel a certain distance. Some other examples of scalar quantities are temperature, the energy of a moving body, and electric charge. Some scalar quantities, such as mass, are always positive, whereas others, such as electric charge, can be positive or negative.

There are physical quantities, however, that cannot be described by scalars. Experiment shows that direction does matter at times: Objects fall down and not sideways. Specifying the velocity of a ball requires specifying not only the speed (how fast it is going), but also the direction in which it is traveling. Quantities that must be described by both a magnitude—always positive—and a direction are called **vectors**. Vectors describe velocity, force, electric fields, and numerous other quantities. They play an important role in physics, and this section summarizes their properties.

Vectors have both magnitude and direction.

An Example: The Displacement Vector

The definition of displacement

One important vector quantity is **displacement**, the difference between two positions of an object. We shall use displacement to describe many of the properties of vectors. To understand displacement, imagine a treasure hunt in which we must proceed from some point K to a second point L, which is 30 paces northeast of K. The displacement from K to L may be drawn as an arrow on a map (Fig. 1–6). That arrow is the pictorial representation of the displacement vector from K to L.

We can give the vector a name—**B** in Fig. 1–6. The vector **B** has two attributes: a length, or **magnitude** (30 paces), and a direction (to the northeast). We refer to the tail of the vector as the point where it starts; the tip of the vector represents the point where it ends. Although the length of the displacement vector and the direction in which it points are fixed, the position of the tail (or the tip) of the vector is not. We can shift a vector by moving it to another location in such a way that the vector retains its original direction. Thus vector **B** represents the displacement from *any* point to any other point that is 30 paces away from, and to the northeast of, the starting point (Fig. 1–7).

FIGURE 1–6 The displacement vector **B** from some point K to a second point L on the map for a treasure hunt.

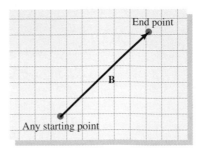

FIGURE 1–7 The same vector **B** represents the displacement 30 paces to the northeast from *any* point. Vectors can be shifted about without changing their characteristics, as long as their magnitude (length) and direction (orientation) are unchanged.

How To Draw

1. Draw first displacement

2. Draw second displacement

3. Draw resultant vector

FIGURE 1–8 The treasure hunt specifies a series of displacement vectors, which leads to a net displacement. Here, we use a series of steps to outline how to add vectors to arrive at a resultant vector.

In the text, we denote all vectors with a boldface letter, such as **B**. An alternative notation, useful for writing and lecturing, is to put an arrow over the quantity, $\vec{B}$. The magnitude of **B** is sometimes denoted by $|\mathbf{B}|$ but more usually by the unadorned symbol B.

Addition and Subtraction of Vectors

Addition. The result of two successive displacements is also a displacement, which we call a *net displacement*. For example, our treasure hunt may specify an initial displacement from the point J to the point K, which is 35 paces due east of J. We call this displacement **A** in Fig. 1–8. Our second displacement would be vector **B**, which starts at point K and proceeds 30 paces northeast to point L. The net displacement, then, takes us from point J to point L, and we denote this by the vector **R**. The vector **R** is the *sum of the two vectors* **A** and **B**:

$$\mathbf{R} = \mathbf{A} + \mathbf{B}. \tag{1–1}$$

The sum of the two vectors, which is known as the **resultant vector**, is formed as follows. Draw vector **A**, then place the tail of vector **B** on the tip of vector **A**. The line from the tail of **A** to the tip of **B** is vector **R**. The addition of vectors is *commutative*—that is, the order of the vectors does not matter, so

$$\mathbf{A} + \mathbf{B} = \mathbf{B} + \mathbf{A}. \tag{1–2}$$

This is easy to see in Fig. 1–9 and using our prescription of placing the tail of the second vector at the tip of the first. The figure contains the sum in both orders, and the result is the same.

FIGURE 1–9 (a) We can find the sum **R** of **A** and **B** by placing **A** and **B** tail to head *in either order*. (b) By combining the two graphs in a parallelogram, the equality of the sum in either order is apparent. We can pass either through point K or point I.

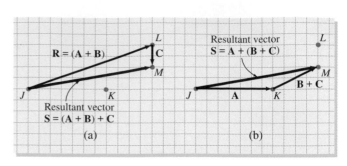

FIGURE 1–10 One more displacement vector, **C**, is added to the displacements in Fig. 1–9. The net displacement is **S**.

FIGURE 1–11 (a) The net displacement **S** is found by first adding **A** to **B**, then adding the result to **C**. (b) The net displacement **S** is found by first adding **B** to **C**, then adding the result to **A**. The net displacement is the same in both cases, which shows that vector addition is associative.

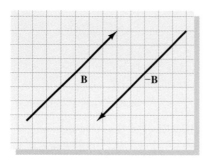

FIGURE 1–12 The vector −**B** is obtained from **B** by reversing the direction of **B** while leaving its magnitude unchanged.

The addition of vectors does not depend on how the vectors are grouped, or on their order.

Figure 1–10 shows the result of adding one more displacement vector to the series of displacements in Fig. 1–8: The vector **C** takes us from point *L* to point *M* on the figure. Vector **C** has magnitude 10 paces and points in a southerly direction. When **C** is added to **R** = **A** + **B**, we obtain a new vector, **S**, which represents the net displacement from point *J* to point *M*:

$$\mathbf{S} = \mathbf{A} + \mathbf{B} + \mathbf{C}. \tag{1–3}$$

With this displacement vector, we can see that vector addition is *associative*. This term means that we can group the addition in any way we find convenient (Fig. 1–11):

$$\mathbf{S} = (\mathbf{A} + \mathbf{B}) + \mathbf{C} = \mathbf{A} + (\mathbf{B} + \mathbf{C}). \tag{1–4}$$

Subtraction. The *null vector* **0** is a special vector with zero magnitude. With the help of this vector, we can define the negative, −**B**, of a vector **B**: When −**B** is added to the vector **B**, the sum is the null vector:

$$\mathbf{0} = \mathbf{B} + (-\mathbf{B}). \tag{1–5}$$

This means, as shown in Fig. 1–12, that −**B** is a vector that has the same magnitude as **B**, but −**B** points in the opposite direction. The subtraction **A** − **B** of two vectors is simply the addition of **A** and −**B**:

$$\mathbf{T} = \mathbf{A} - \mathbf{B} = \mathbf{A} + (-\mathbf{B}). \tag{1–6}$$

This vector is shown in Fig. 1–13.

Scalar Multiplication and Unit Vectors

When vectors are multiplied by scalars, the result is a vector. Twice the displacement **B** is just a displacement with twice the magnitude of **B**, but **B** continues to point in the same direction. More generally, $b\mathbf{B}$ has a length b times that of the vector **B** (Fig. 1–14). The vector 4**B** is formed by the *scalar multiplication* of 4 and **B**.

The product of a vector and a scalar is a vector.

Unit vectors

This allows us to define *unit vectors* as follows (Fig. 1–15): Any vector **U** is written as $\mathbf{U} = U\hat{\mathbf{u}}$, where U is the magnitude of **U**, and $\hat{\mathbf{u}}$ is a **unit vector** that points in the direction of **U** and has a magnitude of 1. We place all the units (m, m/s, and so on) into the magnitude U, so that the unit vector $\hat{\mathbf{u}}$ is *dimensionless*. In other words, the unit vector simply specifies a direction. Other notations for the unit vector will be introduced as needed.

The definition of the position vector

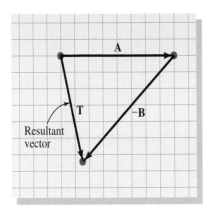

FIGURE 1–13 The vector difference $\mathbf{T} = \mathbf{A} - \mathbf{B}$.

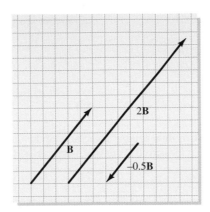

FIGURE 1–14 Scalar multiplication of a vector $\mathbf{B}$ by a scalar quantity b is a vector with the same (or opposite) direction as $\mathbf{B}$ but a magnitude or length scaled by the factor b.

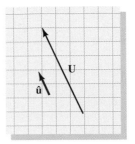

FIGURE 1–15 Any vector $\mathbf{U}$ can be broken into a scalar magnitude, U, times a unit vector (here labeled $\hat{\mathbf{u}}$), that points in the same direction as $\mathbf{U}$ ($\mathbf{U} = U\hat{\mathbf{u}}$). All units are placed into U so that $\hat{\mathbf{u}}$ is dimensionless.

Components

It is convenient to use coordinate axes to describe the location of points. In two dimensions, as on the surface of a table, we shall often specify two perpendicular axes, the x-axis and the y-axis (Fig. 1–16a). A point P can be located by giving its coordinates (x_1, y_1); that is, how far it is along the x-direction or the y-direction from the origin O. The point P may also be described with a *position vector* $\mathbf{D}$ that extends from the origin to P. We use the term **position vector** to denote the displacement *as measured from the origin of a particular coordinate frame*.

With our knowledge of vector addition and scalar multiplication, we arrive at this displacement in two steps: We first make a displacement of magnitude x_1 along the x-axis, then we follow it with a displacement of magnitude y_1 along the y-axis (Fig. 1–16b). To write this in vectorial form, we employ a unit vector $\mathbf{i}$ that points in the $+$ x-direction and another unit vector $\mathbf{j}$ that points in the $+$ y-direction. Then $x_1\mathbf{i}$ is a vector pointing in the x-direction whose magnitude is the absolute value of x_1, whereas $y_1\mathbf{j}$ is a vector pointing in the y-direction whose magnitude is the absolute value of y_1. Thus

$$\mathbf{D} = x_1\mathbf{i} + y_1\mathbf{j}. \tag{1–7}$$

Figure 1–17 applies this kind of reasoning to a more general vector $\mathbf{V}$. We draw the vector in the xy-plane. The vector has **components** V_x and V_y, the elements into which the vector can be resolved, defined such that

$$\mathbf{V} = V_x\mathbf{i} + V_y\mathbf{j}. \tag{1–8}$$

The **component vectors** of $\mathbf{V}$ are the vectors $V_x\mathbf{i}$ and $V_y\mathbf{j}$. The coordinate frame is specified in the figure, *as it always must be if we refer to particular components*. Figure 1–17 also shows another way to describe the vector $\mathbf{V}$. It follows from Pythagoras's theorem that the length of $\mathbf{V}$ is

$$V = \sqrt{V_x^2 + V_y^2}. \tag{1–9}$$

We see from the figure that $\mathbf{V}$ makes an angle θ with the $+$ x-direction. The angle θ is given by

$$\tan\theta = \frac{V_y}{V_x} \quad \text{or} \quad \theta = \tan^{-1}\left(\frac{V_y}{V_x}\right). \tag{1–10}$$

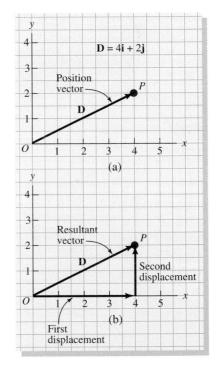

FIGURE 1–16 (a) A point P can be located in a two-dimensional coordinate system by specifying the two coordinates $x_1 = 4$ and $y_1 = 2$ along their x- and y-axes, respectively. The position of point P with respect to the origin, O, is the vector $\mathbf{D}$. (b) The displacement $\mathbf{D}$ from O to P can be produced with two successive displacements, taken in either order: a displacement in the x-direction and a displacement in the y-direction.

Unit vectors in the *x*- and *y*-directions

The definitions of components and component vectors

In vector addition, the components add algebraically to form the components of the resultant vector.

PROBLEM SOLVING

Use components with vector addition and subtraction.

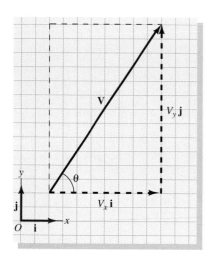

FIGURE 1–17 A vector **V** in two dimensions is described either by a magnitude *V* and an angle *θ* or by the components V_x and V_y.

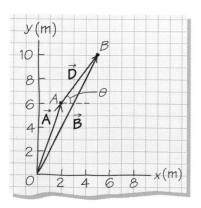

FIGURE 1–18 Example 1–5. The two points *A* and *B* have position vectors **A** and **B**, respectively. The vector **D** is the displacement from point *A* to point *B*.

16

Equivalently, the vector components are described in terms of *V* and *θ* by

$$V_x = V \cos \theta \qquad (1–11a)$$

and

$$V_y = V \sin \theta. \qquad (1–11b)$$

We have therefore shown that

> **In two dimensions, a vector may be described either with a magnitude *V* and an angle *θ* measured from the *x*-axis, or with components V_x and V_y.**

Addition (and subtraction) are especially simple to handle in terms of components. The component vectors add independently, so the components do too. Thus, for example, if **V** and **W** are two vectors with components (V_x, V_y) and (W_x, W_y), then

$$\mathbf{V} + \mathbf{W} = (V_x + W_x)\mathbf{i} + (V_y + W_y)\mathbf{j}. \qquad (1–12)$$

EXAMPLE 1–5 The displacement from a point *A* to point *B* describes a displacement vector **D**. The (*x*, *y*) coordinates of point *A* (measured in meters) are (2.0, 6.0); the (*x*, *y*) coordinates of *B* are (5.0, 10.0). (a) Describe **D** in terms of the position vectors **A** and **B** of the points *A* and *B*, respectively. (b) Calculate the length of **D** and the angle it makes with the *x*-axis.

Solution: First, we draw a figure of the situation, as shown in Fig. 1–18. The figure includes the specified coordinate system as well as the two points *A* and *B*, the position vectors **A** and **B**, and the displacement vector **D**.

(a) The figure tells us immediately that **A** + **D** = **B**. We can now solve this equation for **D**:

$$\mathbf{D} = \mathbf{B} - \mathbf{A}.$$

We have specified the coordinates of both **A** and **B**:

$$\mathbf{A} = (2.0 \text{ m})\mathbf{i} + (6.0 \text{ m})\mathbf{j} \quad \text{and} \quad \mathbf{B} = (5.0 \text{ m})\mathbf{i} + (10.0 \text{ m})\mathbf{j}.$$

Then, from Eq. (1–12),

$$\mathbf{D} = (5.0 \text{ m})\mathbf{i} - (2.0 \text{ m})\mathbf{i} + (10.0 \text{ m} - 6.0 \text{ m})\mathbf{j} = (3.0 \text{ m})\mathbf{i} + (4.0 \text{ m})\mathbf{j}.$$

The minus signs appear because, in **B** − **A**, the components of **A** are *subtracted* from those of **B**.

(b) Once we know the components of **D**, we can use Eqs. (1–9) and (1–10) to find the length *D* and the angle *θ* that **D** makes with the *x*-axis. We have, from Eq. (1–9),

$$D = \sqrt{D_x^2 + D_y^2} = \sqrt{(3.0 \text{ m})^2 + (4.0 \text{ m})^2} = 5.0 \text{ m},$$

and, from Eq. (1–10),

$$\tan \theta = \frac{D_y}{D_x} = \frac{4.0 \text{ m}}{3.0 \text{ m}} = 1.33;$$

that is, $θ = 53°$.

EXAMPLE 1–6 Figure 1–10 specifies a set of points *J*, *K*, *L*, and *M*, which refer to a treasure hunt. In the hunt, distances are measured in paces. Express the vectors **A**, **B**, and **C** by describing the displacements from *J* to *K*, from *K* to *L*, and from *L* to *M*, respectively. Describe these displacements in component form relative to a set of axes in which the *x*-axis points to the east and the *y*-axis points north. Use your description to calculate the sum of the three vectors. Give the length and inclination with respect to the *x*-axis of the resultant vector.

Solution: Figure 1–19 adds a set of *x*- and *y*-axes to Fig. 1–10 drawn in as specified in the problem. The vector **A** lies along the *x*-direction; thus, it has no *y*-component and is given by

$$\mathbf{A} = (35 \text{ paces})\mathbf{i}.$$

The vector **B** points in the northeast direction; that is, it makes an angle of $\phi = 45°$ with the *x*-axis. Thus $\sin \phi = \cos \phi = 1/\sqrt{2} \simeq 0.71$. From Eqs. (1–11a and b) we have $B_x = B \cos \phi = (30 \text{ paces})(0.71) = 21 \text{ paces}$. B_y has the same value. Thus

$$\mathbf{B} = B_x\mathbf{i} + B_y\mathbf{j} = (21 \text{ paces})\mathbf{i} + (21 \text{ paces})\mathbf{j}.$$

Finally, the vector **C** has a *y*-component only. Because it points to the $-y$-direction, it has a component that is -10 paces:

$$\mathbf{C} = (-10 \text{ paces})\mathbf{j}.$$

Now that we have the three vectors, we can find their sum **S** by adding the components:

$$\mathbf{S} = \mathbf{A} + \mathbf{B} + \mathbf{C} = (35 \text{ paces})\mathbf{i} + (21 \text{ paces})\mathbf{i} + (21 \text{ paces})\mathbf{j} - (10 \text{ paces})\mathbf{j}$$

$$= (35 \text{ paces} + 21 \text{ paces})\mathbf{i} + (21 \text{ paces} - 10 \text{ paces})\mathbf{j} = (56 \text{ paces})\mathbf{i} + (11 \text{ paces})\mathbf{j}.$$

To find the length and angle of **S**, given its components, we use Eqs. (1–9) and (1–10):

$$S = \sqrt{(56 \text{ paces})^2 + (11 \text{ paces})^2} = 57 \text{ paces};$$

$$\tan \theta = \frac{11 \text{ paces}}{56 \text{ paces}} = 0.20.$$

or $\theta = 11°$.

Vectors in Three-Dimensional Space

The vectors we have considered thus far are two-dimensional. Vectors can also represent quantities in three dimensions. A vector in three dimensions can be specified with a *Cartesian* or *Euclidean* set of axes *x*, *y*, and *z*. (The two-dimensional *x*- and *y*-axes are also called Cartesian.) Figure 1–20 illustrates such a three-dimensional system, showing the conventional orientation of the three axes. The orientation of the axes is best described using a *right-hand rule*. Start with the usual *x*- and *y*-axes (*x* to the east, say, and *y* to the north). To find the direction of the *z*-axis:

> **Point the fingers of your right hand along the *x*-axis. Now curl these fingers in the direction of the *y*-axis. Your thumb will point along the +*z*-axis.**

This right-hand rule has become a well-established convention and will appear in many places in this book. Figure 1–21 shows how a vector **V** is decomposed into three components along the three axes.

The three unit vectors for the three axes are denoted by **i**, **j**, and **k**; the unit vector **k** points in the +*z*-direction (Fig. 1–22). A point *P* in three-dimensional space is

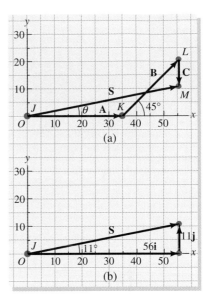

FIGURE 1–19 Example 1–6. (a) A set of coordinate axes has been added to the treasure hunt specified in Fig. 1–10. (b) **S** is constructed by adding the components of **A**, **B**, and **C**.

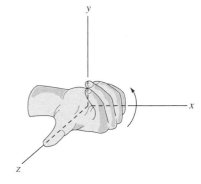

FIGURE 1–20 A Cartesian coordinate system for three dimensions. A right-hand rule specifies the orientation of the three axes. The *z*-axis points out of the plane of the paper. The thumb points along the *z*-axis; the fingers curl from the *x*-axis toward the *y*-axis.

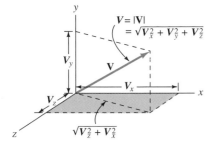

FIGURE 1–21 A vector **V** decomposed into three components in three-dimensional space.

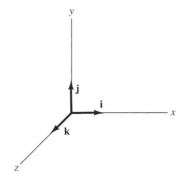

FIGURE 1–22 The unit vectors **i**, **j**, and **k** are aligned along the three Cartesian axes *x*, *y*, and *z*, respectively. Each has unit magnitude.

now assigned the coordinates (x, y, z) in a given frame, and its displacement vector from the origin—its position vector—is

$$\mathbf{D} = x\mathbf{i} + y\mathbf{j} + z\mathbf{k}. \tag{1–13}$$

Similarly, any vector $\mathbf{V}$ has the components (V_x, V_y, V_z), such that

$$\mathbf{V} = V_x\mathbf{i} + V_y\mathbf{j} + V_z\mathbf{k}. \tag{1–14}$$

Pythagoras's theorem again tells us the magnitude, or length, of $\mathbf{V}$:

$$V = \sqrt{V_x^2 + V_y^2 + V_z^2}. \tag{1–15}$$

Vector Equations

Vector equations are shorthand for a set of equations, one for each component.

Vector equations are equations in which vector quantities appear on both sides of the equal sign. For example, we may want to express the velocity of a projectile in terms of its position. Velocity is a vector, as is position. The equality in such an equation means either that the magnitudes and direction of the net vector on each side are the same, or, equivalently, that the respective components of the vector equation are equal. In two dimensions, a vector equation is really shorthand for two separate equations for the components; in three dimensions, a vector equation represents three separate equations.

*1–7   THE USES OF DIMENSIONAL ANALYSIS[†]

Scaling Laws

In Section 1–4, dimensional analysis was used as a check in problem solving. There are two other important applications of dimensional analysis. In the first of these, we can discover scaling laws. *Scaling laws* reveal how a change in one dimensional parameter of a problem leads to changes in another parameter with the same dimensions. Let us consider a problem that Galileo studied in his book *Dialogue Concerning Two New Sciences*; namely, the scaling properties of animals: What must giants look like? Consider a bone (Fig. 1–23) whose length is ℓ, characterizing the height, or linear size, of an animal. The width of the bone is b. It is a fact that the strength of such a bone is proportional to its cross-sectional area; that is,

$$\text{strength} = c_1 b^2.$$

FIGURE 1–23 The scaling of bones. (From Galileo, *Dialogue Concerning Two New Sciences*, 1638.)

The constant factor c_1 is characteristic of the material, and it does not change if the size of the animal changes. This strength must be such as to support the weight or, equivalently, the *mass m* of the animal. But the mass is proportional to the animal's volume, and the volume is, in turn, proportional to ℓ^3, just as the volume of a cube is equal to the length of the side cubed. Thus

$$m = c_2 \ell^3.$$

The precise value of c_2 depends on the precise shape of the animal and need not concern us. Now, if the strength of the bone must be proportional to the mass, we have

$$\text{strength} = c_3 m.$$

Combining these three equations, we have

$$c_1 b^2 = c_3 m = c_3 c_2 \ell^3;$$

$$c = \frac{\ell^3}{b^2}, \quad \text{where } c = \frac{c_1}{c_2 c_3}. \tag{1–16}$$

The constant c is characteristic of our animal's internal structure, but not of its size.

[†]An asterisk (*) indicates optional material that may be skipped without loss of continuity.

Now what happens if we try to change the size parameter ℓ by a factor f? The constant c cannot change, so b must change to compensate for the change in ℓ. Equation (1–16) can be satisfied when we change ℓ to $f \times \ell$ only if b changes by a factor f'. We find f' by insisting that the ratio ℓ^3/b^2 be constant:

$$\frac{f^3}{f'^2} = 1 \quad \text{or} \quad f' = f^{3/2}. \tag{1–17}$$

If the length of the bone changes by a factor f, then its width must change by a factor $f^{3/2}$. For example, if the length is changed by a factor 3, as in Fig. 1–23, then according to Eq. (1–17), the width of the bone must change by a factor $(3)^{3/2} \simeq 5$. The scaling argument demonstrates that the giant animal cannot have the same proportions as its smaller model and helps us to understand why hippopotamuses will never be mistaken for overgrown chihuahuas (Fig. 1–24).

Understanding the scaling behavior of physical systems is of primary importance for engineering. Such understanding allows filmmakers to film ripples in a bathtub in slow motion and convince us that they are mighty waves on the ocean, or tells us how the stiffness of a beam varies with its length and width. Scaling arguments tell us how to strengthen structures or how to test small models of large aircraft in wind tunnels.

FIGURE 1–24 The giant ants of the 1950s movie *Them* have the same proportions as their normal counterparts do. Galileo's scaling analysis shows that such creatures could not exist because their structure could not support their weight.

Relations Based on Dimensional Analysis

The second major application of dimensional analysis lies in the derivation of new relations between physical quantities. Consider the simple pendulum as an example. The simple pendulum consists of a small bob of mass m on the end of a light string of length ℓ. A pendulum swings because, when it is displaced away from the vertical direction, gravity pulls it back down. It overshoots the minimum, swings to the other side, then repeats its motion. One full cycle of this motion takes a time τ that we can call the *period*. The problem is this: How does the period depend on the mass of the pendulum bob?

To answer this question, we must gather a list of physical parameters that might be relevant to the period. This requires some knowledge. Could the period of the pendulum depend on the *internal* structure of either the bob or the string? We must answer no because if we change the string or change the bob from lead to wood (of the same mass) the period does not change. Nor do we expect air resistance to play a large role. The list we gather includes the bob mass m, with dimension $[M]$; the length ℓ of the string, with dimension $[L]$; and the acceleration of gravity, g, discussed in Section 1–4. The latter quantity has dimension $[g] = [LT^{-2}]$. There are no other dimensional quantities on which the period of the pendulum should depend.

The dimension of the period is time $[T]$. We now look for an algebraic combination of m, ℓ, and g that has the dimension of τ. We want to find q, r, and s so that

$$[\tau] = [m^q][\ell^r][g^s],$$

or, in terms of the dimensions,

$$[T] = [M^q][L^r][L^s T^{-2s}].$$

Now there are no powers of $[M]$ on the left-hand side, so $q = 0$. We see that *the mass does not enter at all into the period*. In Problem (1–53), we continue this treatment and show that, in fact, the only combination of the parameters with the same dimension as that of the period is

$$\tau \propto \sqrt{\frac{\ell}{g}}.$$

(The symbol $\propto$ indicates proportionality.) This is an illuminating result. It does not tell us the dimensionless numerical coefficient by which the square root is multiplied to give an equality for τ rather than a proportionality. But it does reveal the dependence of τ on ℓ and g, and it does say that the mass does not enter into the result. One mea-

surement of the period of a pendulum with known length would determine the unknown numerical coefficient. We might even hope that the coefficient is not too far
from unity. In fact, this is very often the case!

SUMMARY

The range of quantities relevant to our understanding of the physical world is so large
that it is useful to employ scientific notation. In this notation, any number can be represented by a decimal number from 1 to 10 multiplied by a power of 10.

The quantities that appear in physics and engineering have units as well as sizes.
The International System of units, or SI, provides reproducible and precise definitions
of mass, length, and time. The SI units are, respectively, kilograms (kg), meters (m),
and seconds (s). Some quantities are used so often that their units are given a special
name (for example, force is measured in newtons, N, in SI), but these units are derived
units: They can always be expressed in terms of the three primary units (for example,
$1 \text{ N} = 1 \text{ kg} \cdot \text{m/s}^2$). Units that appear in equations can be manipulated algebraically;
after conversion to a single unit system, the units on both sides of any correct equa-
tion will match.

Mass, length, and time are the quantities with the three primary dimensions, abbreviated *M, L, T.* Dimensions should not be confused with units, which refer to a particular choice of unit system. Any physical quantity has dimensions that are rational
combinations of the primary dimensions. Dimensions can be manipulated algebraically,
and both sides of any correct equation will have the same dimensions. When we analyze the dimensions of an equation, we are performing a dimensional analysis. Dimensional analysis is useful for checking the answers to problems, for learning scaling laws, and for discovering relations between physical quantities.

Numbers that represent physical quantities can be measured only to a certain accuracy. An explicit way to indicate this accuracy is to write a physical quantity x as a
central value ± an uncertainty. Calculations involving physical quantities are mean-
ingful only to within the known accuracy of those quantities. When several numbers
of different accuracies are involved in a calculation, the least accurate quantity primarily determines the accuracy of the result. A second way to indicate the known accuracy of a physical quantity is through the use of significant figures; that is, the number of digits between 1 and 10, which are then multiplied by the power of 10 in
scientific notation.

The ability to estimate is one that should be cultivated. An educated first guess is
a valuable start to the solution of any problem. An order-of-magnitude calculation is
such a guess. Similarly, when you arrive at an answer to a physical problem, it is always wise to ask yourself if it makes sense.

Some, but not all, physical quantities include directional information. Temperature
and time do not. Such quantities are scalars: They have magnitude (including the appropriate dimensions and units) only. Displacement and velocity do include directional
information; such quantities are represented by vectors. The vectors **A** and **B** are mathematical objects with both magnitude and direction. They obey the rule

$$\mathbf{A} + \mathbf{B} = \mathbf{B} + \mathbf{A}. \tag{1-2}$$

Vectors can be expressed in graphical form; we draw them as arrows of length equal
to their magnitude within a particular coordinate system. The simplest such system is
the Cartesian system, with mutually perpendicular x-, y-, and z-axes for three-dimensional space. We can express any vector **V** in terms of the unit vectors—vectors of unit
length—for a given coordinate system. Thus the unit vectors **i**, **j**, and **k** point along the
x-, y-, and z-axes, respectively. Then

$$\mathbf{V} = V_x\mathbf{i} + V_y\mathbf{j} + V_z\mathbf{k}. \tag{1-14}$$

The quantities V_x, V_y, and V_z are the components of **V**. The magnitude of **V** is

$$V = \sqrt{V_x^2 + V_y^2 + V_z^2}. \tag{1-15}$$

Vector equations are relations that equate different vectors. In such an equation, the components of the vectors are equal on both sides, so that, in three dimensions, a vector equation stands for three separate equations.

UNDERSTANDING KEY CONCEPTS

1. If space were somehow four-dimensional instead of three-dimensional, would the concept of displacement vectors still make sense? In what way could the idea of vector displacement be generalized?

2. How might the measurement of distance in centimeters, meters, and kilometers rather than in inches, feet, yards, and miles be more convenient? How might it be less convenient?

3. What is your height in centimeters and meters?

4. What is your mass (notice that we do not say *weight*) in kilograms?

5. The human pulse and the swing of a pendulum are possible time units. Are they ideal ones?

6. How can length be defined in terms of the speeds of light and time?

7. Three vectors all have the same magnitude. Is it possible to add them together to obtain a null vector?

8. Three vectors all have different magnitudes, given by 4 m, 5 m, and 7 m. Is it possible for the three vectors to add to zero? Prove your result by drawing a figure.

9. Is the vector sum of the two unit vectors **i** + **j** also a unit vector? Under what conditions is it possible for two unit vectors to add to a resultant vector that also has unit length?

10. Suppose we were in radio contact with some distant civilization. How would we communicate how large we are? What assumptions are you making in providing your suggestions?

11. A useful estimate is that there are $\pi \times 10^7$ s in a year. How accurate is this?

12. The next time you weigh yourself, consider the uncertainty of the numbers you read. What would you estimate the percentage uncertainty in this measurement of your weight to be?

13. A small worm absorbs the oxygen it needs through its surface. Assume that the oxygen needed by an animal is proportional to its mass, and estimate how much the absorption of oxygen per unit area would have to increase if the worm were to increase each of its dimensions by a factor of 10. You may find it interesting to know that the lungs of a human being have about 100 m^2 of surface available for the absorption of oxygen.

14. A mouse eats the equivalent of about one-quarter of its mass in food every day. You do not. Why?

PROBLEMS

1–1 Scientific Notation

1. (I) Ten thousand jelly beans are in a jar, and 25 percent of them are green. Express in scientific notation the number of green jelly beans.

2. (I) What is the product of 10^5 and 10^{-4}? The ratio $10^{-4}/10^5$?

3. (II) Calculate the cube root of the number 10^{21}, as well as the square of the resulting number.

1–2 Length, Time, and Mass

4. (I) The Empire State Building is 1472 ft high. Express this height in both meters and centimeters.

5. (I) What is your height in atomic diameters (see Table 1–1)?

6. (I) A 5-ft, 5-in-tall skier should use skis 5 cm longer than her height. How long should her skis be? Skis are made in 5-cm intervals (150 cm, 155 cm, and so on). What length skis should she buy if she rounds off to the nearest 5 cm?

7. (I) Grapes sell in Italy for 1000 Lire per kilogram. If the conversion rate between dollars and Lire is 1400 Lire = $1, what is the price of grapes in dollars per kilogram?

8. (II) Gasoline is heavily taxed in Europe, and at one time in West Germany cost 1.90 Deutschmark (DM) per liter. What price in dollars per gallon is this if the currency conversion is 1.7 DM = $1 US? One gallon (gal) = 3.8 liters (L).

9. (II) The acceleration due to gravity, g, is 9.80 m/s^2 in SI. Convert this to the British system, where length is measured in feet rather than meters.

10. (II) The gravitational constant G is 6.67×10^{-11} m$^3 \cdot$ s$^{-2} \cdot$ kg^{-1}. What is G in units of cm$^3 \cdot$ s$^2 \cdot$ g^{-1}?

11. (II) The radius of the Moon (assumed spherical) is 1.74×10^3 km, and its mass is 7.35×10^{22} kg. What is the density of the Moon in grams per cubic centimeter?

12. (II) A neutron star has a radius of 15 km and a mass of 1.4×10^{31} kg. What is the density of the neutron star in metric tons per cubic centimeter?

13. (II) Gasoline consumption in Europe is measured in liters per 100 km. For example, a small Opel uses 7.0 L per 100 km, while the gasoline consumption of a large Mercedes is 23 L per 100 km. Convert these to mi/gal.

1–3 Accuracy and Significant Figures

14. (I) A student wishes to make a measurement of the road distance from his dormitory to the physics building of his university. He uses his car's trip odometer, which measures distance only in units of a tenth of a mile. (a) He makes one trip, and the odometer reads 0.3 mi. What can he say is the distance, and, in particular, to how many significant figures?

(b) One day he has nothing better to do with his time and, adding trips both to and from, he makes 100 trips; his odometer measures 27.2 mi. What can he now say is the distance? How is this result consistent with the result of part (a)? (c) A friend challenges the student on his measurement and, upon reflection, the student is not sure whether he made 99, 100, or 101 trips. How should he modify his statement of the distance?

15. (II) A well-known approximation to π is $\pi \simeq \frac{22}{7}$. What percentage error does this result have? How much better is the approximation 355/113?

16. (II) The force F on a mass m moving at speed v in a circular path of radius r has magnitude $F = mv^2/r$. The mass is measured to be 0.00535 kg, the radius is 0.3 m, and the speed is 1.1 m/s. Give the magnitude of the force. Pay attention to the number of significant figures.

17. (II) A rectangular box is stated to have width $w = (1.25 \pm 0.03)$ m, length $\ell = (0.5 \pm 0.1)$ m, and height $h = (0.137 \pm 0.028)$ m. What is the volume, stated to the appropriate number of significant figures with an uncertainty?

18. (II) You wish to determine your density (mass/unit volume) by two measurements: by weighing yourself on a digital scale, and by submerging yourself in a tank of water with vertical sides and noting the rise in water level on a scale marked in centimeters (Fig. 1–25). The surface area of water in the tank is 1.5 m². The digital scale gives the weight in pounds at 0.5-lb intervals, and you cannot make a reading of the water level to better than 0.5-cm accuracy. Suppose that the weight reading is 213 lb, and the water level changes from 152 cm to 158.5 cm. What is your density? Express your result as a central value with a percentage error.

FIGURE 1–25 Problem 18.

19. (III) If you want to know the area of a circle to 5-percent accuracy, how accurately should you measure the diameter of the circle?

20. (III) Consider the infinite series $\sum_{n=0}^{\infty} (x^n/n!)$. The symbol $n!$ means the product $1 \times 2 \times 3 \times \cdots \times n$. By definition, $0! = 1$. If $x = 0.1$, how many terms in the series suffice to give a result correct to 6 significant figures?

1–4 Dimensional Analysis

21. (I) The kinetic energy of a baseball is denoted by $mv^2/2 = p^2/2m$, where m is the baseball's mass and v is its speed. This relation can be used to define p, the baseball's momentum. Use dimensional analysis to find the dimensions of momentum.

22. (I) One of Einstein's most famous results is contained in the formula $E = mc^2$, where E is the energy content of the mass m, and c is the speed of light. What are the dimensions of E?

23. (I) A length L that appears in atomic physics is given by the formula $L = h/m_e c$, where m_e is the mass of an electron, c is the speed of light, and h is a constant known as Planck's constant. What are the dimensions of h?

24. (II) What are the dimensions of h^2/m^3G, where h is a constant called Planck's constant, m is a mass, and G is the gravitational constant? The dimensions of the constants in this formula can be found in the list of physical constants given in Appendix II.

25. (III) A force F acting on a body of mass m a distance r from some origin has magnitude $F = Ame^{-\alpha r}/r^3$, where A and α are both constants. The constant $e = 2.718\ldots$ is Euler's constant. Given that the force has dimensions kg·m/s², what are the dimensions of (a) the constant α? (b) the constant A?

1–5 How a Little Reasoning Goes a Long Way

26. (I) How many times does an average person's heart beat in a lifetime? Estimate the number of times an automobile tire rotates in a trip across the United States.

27. (II) A criminal, posing as a tourist, wants to smuggle $100 million in gold across the U.S. border in his station wagon. Is he likely to make it? Use estimates, and explain your reasoning.

28. (II) Estimate the volume of concrete used for the construction of the tunnel under the English Channel. The system consists of two railway tunnels and a service tunnel, and is 30 km long.

29. (II) Estimate the area used for the storage of one bit on a 1.44 MB $3\frac{1}{2}''$ floppy disk (1 byte = 8 bits).

30. (II) What is the weight of water in a full 5-gal can? About what volume of water can a typical person carry?

31. (II) Make separate estimates of the number of automobile mechanics in the United States based on (a) your total bill for automobile repairs compared with a reasonable average salary for a mechanic; (b) how many hours it takes to repair a car; (c) the number of people you know who are automobile mechanics compared with the total number of people you know.

32. (II) The Sun is 93 million mi from Earth. What is the diameter of the Sun? (A dime held at arm's length will just about cover the surface of the Sun [Fig. 1–26].)

FIGURE 1–26 Problem 32.

33. (II) A typical cloud contains droplets of water with an average radius of 0.5×10^{-4} m. How many droplets are needed for a cloud that provides a rainfall of 1 cm in your city?

34. (II) A strong radioactive point source emits 10^8 gamma particles per second uniformly in all directions. What fraction of

these particles will hit a circular detector of diameter 4 cm that is 2 m away from the source?

35. (II) There are over 100 million automobiles in the United States, and each one is driven about 20,000 mi/yr. Estimate the number of automobiles on the road at any one moment.

36. (II) Suppose that all the eighteen-wheeler trucks in the United States line up bumper-to-bumper on I-80. Can they form a continuous line from New York to San Francisco?

37. (II) Suppose that the circumference of Earth is a perfect circle of exactly 25,000 mi. Somebody prepares a wire that is supposed to go around the equator completely, but makes it 1 m too long by mistake. If this 1-m-too-long wire were placed around the equator in a perfect circle with the ends of the wire just touching each other, by how much would the wire be off the ground?

38. (II) Make a rough estimate of the number of apples produced each year in a 10-acre orchard.

39. (II) The mass of one atom of hydrogen is 1.6×10^{-27} kg. Given that the mass of the Sun is 2×10^{30} kg, how many atoms of hydrogen would it contain if it consisted purely of hydrogen? Actually, it consists of 70 percent hydrogen and 30 percent helium by mass, and the mass of a helium atom is 6.4×10^{-27} kg. How many hydrogen atoms does the Sun have?

40. (II) Suppose that oil consists of molecules of CH_2. This means that 1 mol of oil, consisting of 6×10^{23} molecules, has a mass of 14 g. One milliliter (mL) of oil has a mass of 0.95 g. When this much oil is poured on water, it does not spread forever but spreads until it makes a circular film of area 1.5×10^7 cm^2. How does the fact that the oil does not spread forever support the idea of atoms and molecules? Assuming that the oil slick is one molecule thick (a *monomolecular layer*), with the molecules touching, and that the molecules are spheres, estimate the size of an oil molecule.

1–6 Scalars and Vectors

41. (I) A girl jogs around a circular lake. Devise a simple coordinate system to describe her position and direction of travel at any time, assuming that she begins at the south end of the lake and jogs clockwise at a speed of 2 m/s.

42. (I) Draw the vector $3\mathbf{i} + 4\mathbf{j}$ by first drawing the x-component vector, then the y-component vector, then adding them graphically. Multiply the vector by a factor of two and repeat the exercise.

43. (I) A drunken sailor stumbles 4 paces north, 6 paces northeast, 2 paces east, and 5 paces west. Describe the final location from the initial position by a single displacement vector.

44. (I) What is the resultant vector when the vectors $\mathbf{A} = 6\mathbf{i} - 5\mathbf{j}$ and $\mathbf{B} = 8\mathbf{i} + 3\mathbf{j}$ are added together? When $\mathbf{B}$ is subtracted from $\mathbf{A}$?

45. (II) A football player catches the kickoff on the 5-yd line and runs straight up the field for 20 yd, turns left for 15 yd, goes straight up the field for 10 yd, turns right for 25 yd, reverses his field (makes a 180° turn) for 10 yd, and then streaks straight up the field for a touchdown. Define a coordinate system and list the entire path in vector form.

46. (II) Draw a vector $\mathbf{V}$ that points in the northwesterly direction, making an angle α with the northerly direction, as in Fig. 1–27. If north is chosen as the $+y$-direction and east as the $+x$-direction, what is the x-component of $\mathbf{V}$?

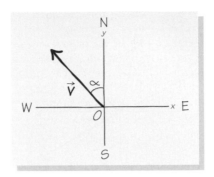

FIGURE 1–27 Problem 46.

47. (II) Suppose that in Problem 46 you choose north as the $+x$-direction and west as the $+y$-direction. What is the x-component of $\mathbf{V}$ in this case?

48. (II) Refer to the situation outlined in Problems 46 and 47. Choose the $+x$-axis as the line that makes an angle of 45° with the northerly direction and is inclined to the east, and the $+y$-axis as the line that makes a 45° angle with the westerly direction and is inclined to the north. What is the x-component of $\mathbf{V}$ in this case?

49. (II) In computer-aided drafting programs, lines can be specified in either rectangular or polar coordinates (Fig. 1–28). In such programs, if the coordinates (x, y) are given, the cursor draws a line from its current position a distance x to the right and a distance y up. If polar coordinates (r, θ) are given, the line is drawn from its current position through an angle θ in the counterclockwise direction from the positive x-axis, through a distance r. Give the instructions both in terms of (x, y) and (r, θ) for the drafting of the triangle shown in the figure, starting at the point A.

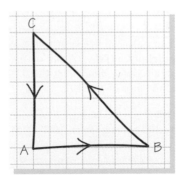

FIGURE 1–28 Problem 49.

50. (II) Consider the following vectors: $\mathbf{A} = -2\mathbf{i} - 3\mathbf{j}$; $\mathbf{B} = \mathbf{i} + 2\mathbf{j} + 3\mathbf{k}$; $\mathbf{C} = 3\mathbf{j} + 3\mathbf{k}$; and $\mathbf{D} = -2\mathbf{i} - \mathbf{k}$. Find (a) $\mathbf{A} + \mathbf{B} + \mathbf{C} + \mathbf{D}$; (b) $\mathbf{A} - \mathbf{D}$; (c) $\mathbf{A} + \mathbf{D} - \mathbf{B}$; and (d) $|\mathbf{A} - \mathbf{C}|$.

51. (II) Vectors $\mathbf{A}$, $\mathbf{B}$, $\mathbf{C}$, and $\mathbf{D}$ are shown in Fig. 1–29. (a) Give the vectors in component form. (b) Determine the following quantities both algebraically and graphically: $2\mathbf{A} + \mathbf{C} - \mathbf{D}$, $\mathbf{B} + \mathbf{C}/2$, $|\mathbf{D} - \mathbf{B}|$.

52. (II) Suppose that you have three vectors, $\mathbf{A} = 3\mathbf{i} + 4\mathbf{j}$, $\mathbf{B} = 2\mathbf{i} - 2\mathbf{j} + 4\mathbf{k}$, and $\mathbf{C} = -\mathbf{i} + 5\mathbf{j} - 3\mathbf{k}$. Show that the sum of these three vectors can alternatively be computed by first summing $\mathbf{A}$ and $\mathbf{B}$ and then summing the resultant with $\mathbf{C}$, or by first summing $\mathbf{B}$ and $\mathbf{C}$ and then summing the resultant with $\mathbf{A}$.

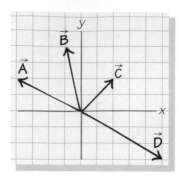

FIGURE 1–29 Problem 51.

*1–7 The Uses of Dimensional Analysis

53. (II) We have seen in the text that the period of a simple pendulum is independent of the mass of the pendulum bob. Further, we have seen that the dimensional relation between the period, τ, the length, ℓ, of the pendulum, and the acceleration of gravity, g, takes the form

$$[\tau] = [\ell^r][g^s].$$

Use the fact that the dimension of τ is $[T]$, that of ℓ is $[L]$, and that of g is $[L/T^2]$ to show that

$$\tau = \sqrt{\frac{\ell}{g}}.$$

54. (II) In quantum mechanics, the fundamental constant called Planck's constant, h, has dimensions of $[ML^2T^{-1}]$. Construct a quantity with the dimensions of length from h, a mass m, and c, the speed of light.

55. (II) It is known that the quantity Ke^2/hc is dimensionless (K is a numerical constant; h and c are as discussed in Problem 54). (a) What are the dimensions of e? (b) What are the dimensions of e^2/R, where R is a length?

56. (II) You are told that the speed of sound in a metal depends only on the density ρ ($[ML^{-3}]$) and on the bulk modulus of the metal, B, which has dimensions $[ML^{-1}T^{-2}]$. Express the sound speed in terms of ρ and B.

General Problems

57. (I) There are 100 cm in a meter, 1000 kg in a metric ton, and 1000 g in a kilogram. (a) How many metric tons are in 1 g? (b) How many cubic meters are in 1 cm³?

58. (I) A human hair has a diameter of about 10^{-4} m. Given that all of the hair on a head can be gathered into a lock of diameter 4 cm, estimate the fraction of the human scalp area from which hairs spring, and the number of hairs on the head.

59. (I) Given that the speed of light is 2.998×10^8 m/s and that the distance from the Moon to Earth is 2.4×10^5 mi, how long does it take light to travel from the Moon to Earth?

60. (I) Measure the height, width, and thickness of this book in centimeters. Estimate its mass. Calculate its density. An object will float in water if its density is less than that of water. Without doing the experiment, would you expect this book to float in water?

61. (I) Eighteen grams of water is known to contain 6.02×10^{23} molecules of H_2O. What is the mass in kilograms of one molecule of H_2O?

62. (II) The density of a human body is approximately 1 g/cm³, which is also the density of water. Use the result of the calculation in Problem 61 to estimate the number of molecules of water that a typical human body would contain if it were made up entirely of water.

63. (II) Imagine that molecules of H_2O are stacked up in a cubic array, like a large number of cubical boxes, with a water molecule at the center of each cube. Let the side of each cube be L. Given that the density of water is 1 g/cm³, estimate the distance L from the data given in Problem 61.

64. (II) A silver nucleus consists of 108 closely packed nucleons (protons and neutrons) and has a radius of 5×10^{-15} m. A neutron star is basically an overgrown nucleus, with neutrons only, closely packed in the same way as nucleons are in the silver nucleus. If the radius of a neutron star is 12 km, how many neutrons does it contain?

65. (II) A typical star has a mass of about 2×10^{30} kg, and there are about 10^{11} stars in a galaxy. What is the mass of this typical galaxy? Assume that stars are made primarily of hydrogen; the mass of a hydrogen atom is 1.67×10^{-27} kg. How many hydrogen atoms are there in a galaxy?

66. (II) The gasoline usage rate required to propel an automobile is very roughly proportional to the mass of the automobile. Assuming that the proportions and types of materials of an automobile do not change, calculate the percentage gasoline savings that would be realized if cars were reduced by 12 percent in all their dimensions.

67. In aquatic animals, the energy E available for motion is proportional to the mass of the animal, and the friction F with their skin is proportional to the surface area. All such animals have the same density, very close to that of water. If the maximum speed v such an animal can reach varies as $\sqrt{E/F}$, show that v is also proportional to $\sqrt{L}$, where L is some length characterizing the animal's size.

68. (II) Determine the thickness of a page of this book to an accuracy better than 10 percent. Explain your method and give your uncertainty.

69. (II) The water supply of Pittsburgh is contaminated by an oil spill. Make some reasonable assumptions in order to estimate how many trucks per day are needed to bring in a minimum supply of water. What if each person were allowed to take a bath every 3 days?

70. (II) Assume that houses are set on quarter-acre lots and each house receives four pieces of mail a day. Estimate how far a postal carrier walks in 1 day, and how much mail he or she carries. What is the mass of all this mail?

71. (II) According to Kepler's third law of planetary motion, the square of the period of a planet is proportional to the cube of its mean distance from the Sun. Given that Earth, whose period is 1 yr, is 1.5×10^8 km away from the Sun, calculate the distance from the Sun to Venus, whose period is 0.61 yr, and the period of Saturn, which is 14×10^8 km away from the Sun.

72. (II) A mouse is 10 cm in length, whereas an elephant is 4 m in length. The amount of food an animal must eat is proportional to its heat loss, and the heat loss is proportional to its surface area. Compare the percentage of body weight that a mouse and an elephant must eat each day. Ignore the detailed differences in shape between an elephant and a mouse.

73. (II) The mass of a cylinder cut from the entire atmosphere of area 1 cm² is approximately the same as the mass of a cylinder of the same area of water that is 30 ft high. Use this to estimate the number of "molecules of air" in the atmosphere, given the fact that, on average, one molecule of air is 1.6 times as massive as one molecule of water. (See Example 1–4.)

74. (II) Show that $(\mathbf{A} + \mathbf{B}) + \mathbf{C} = \mathbf{A} + (\mathbf{B} + \mathbf{C})$, where $\mathbf{A}$, $\mathbf{B}$, and $\mathbf{C}$ are vectors.

75. (II) A vector $\mathbf{u}$ in the xy-plane has x- and y-components $u \cos \theta$ and $u \sin \theta$, respectively, where θ is the angle that $\mathbf{u}$ makes with the $+x$-axis. A second vector, $\mathbf{v}$, also lies in the xy-plane, and it is perpendicular to $\mathbf{u}$. (a) Draw a figure to show that there are two possibilities for the vector $\mathbf{v}$: either its x- and y-components are $-v \sin \theta$ and $v \cos \theta$, or else they are $v \sin \theta$ and $-v \cos \theta$. (b) Show that $v_x u_x + v_y u_y = 0$, independent of θ.

76. (II) For objects that move in a circle about an origin O, it can be convenient to use the mutually perpendicular unit vectors $\mathbf{i}_r$ and $\mathbf{i}_t$, defined as in Fig. 1–30. If a Cartesian coordinate system has its origin at O, with an x-axis chosen so that the angle between it and the line OP is θ, then (a) show that $\mathbf{i}_r = \cos \theta \, \mathbf{i} + \sin \theta \, \mathbf{j}$; (b) calculate the y-component of $\mathbf{i}_r$; and (c) express $\mathbf{i}_t$ as a combination of $\mathbf{i}$ and $\mathbf{j}$.

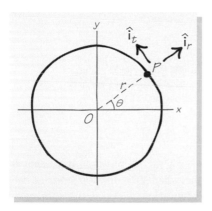

FIGURE 1–30 Problem 76.

77. (II) A vector $\mathbf{r}$ has length r and points in the direction shown in Fig. 1–31. The angles θ and ϕ are drawn in the figure. (a) Show that the x-component of $\mathbf{r}$ is $r \sin \theta \cos \phi$. (b) Show that the z-component of $\mathbf{r}$ is $r \cos \theta$. (c) Find the y-component of $\mathbf{r}$ in terms of r, θ, and ϕ.

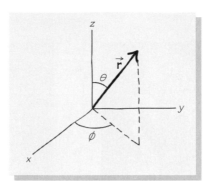

FIGURE 1–31 Problem 77.

78. (II) The number of molecules in 22.4 L of air at the surface of Earth is around 6×10^{23}. A liter is 10^{-3} m³. Let us say that the volume of air you take in and then expel with each breath is 2.5 L. Let us also say that the air around Earth is a uniform blanket 8 km thick. (Actually, it drops as height increases, but our assumption is not far off.) Finally, Michelangelo Buonarotti lived for 91 yr and died sufficiently long ago that all the "molecules of air" he ever breathed are thoroughly dispersed throughout the atmosphere. Under all these conditions, how many molecules of air breathed by Michelangelo are in your lungs right now?

79. (II) A long-legged bird, the stilt, has legs 20 cm long and a mass of 150 g. A flamingo has a mass of 2.5 kg. How long would you estimate the flamingo's legs to be, assuming that the density of the two birds is the same?

80. (II) An elephant has a mass of 4000 kg. What would you estimate the diameter of an elephant's leg to be, given that yours is around 12 cm? Does your result make sense?

81. (III) A stretched wire has three physical attributes: the density λ, or mass per unit length; the total length ℓ; and the tension τ. The latter is related to how hard the wire is being pulled to keep it stretched, and has dimensions of $[MLT^{-2}]$. Show by dimensional analysis that if the time t_0 of one back-and-forth vibration of the wire in a direction perpendicular to its length depends only on these three quantities, then t_0 has the form $t_0 = (\text{a constant}) \, \ell \, \sqrt{\lambda/\tau}$.

The Super High Altitude Research Project of the Lawrence Livermore National Laboratory is based on a two-stage gas gun meant to accelerate a projectile to very high speeds in very short periods of time. The white pipe in the center of the photo is the launch tube that directs the projectile; burning hydrogen gas that makes the projectile accelerate is seen here escaping from each end of the tube.

Straight-Line Motion

Understanding motion is one of the key goals of physical law. That is why we begin with a study of **mechanics**, the science of motion and its causes. The *description* of motion is the domain of **kinematics**, which is the subject of this chapter and the next. Once we know how to describe motion, we can explore the *causes* of motion, which come under the heading of **dynamics**. In this chapter, we study motion in a straight line; that is, one-dimensional motion. In the next chapter, we shall extend this study to two and three dimensions—motion on a plane and motion in space.

2-1   DISPLACEMENT

Consider the motion of athletes running a 100-m dash. Some runners are able to attain a tremendous advantage right at the beginning of the race—they leap ahead of the others. Other runners have a late kick that allows them to take the lead at the end of the race (Fig. 2–1). Suppose that we set up an electronic timing system along the path of the 100-m dash that records the times of the runners every 5 m. Data provided by this experiment would yield the distance each runner travels as a function of time.

We first want to quantify the notion of *distance traveled*. To do this, we construct a coordinate system with an x-axis, using the variable x to indicate the distance traveled. The choice of origin on the x-axis is up to us, but it is convenient to label the distance markers from the starting line, which we therefore set at $x = 0$ m. We also choose the direction that the runners take to be the $+x$-direction. We have invented a set of reasonable data points, shown in Fig. 2–2a, that represent the distance traveled by one runner—actually, the distance of a point on the runner's chest—plotted against the time it takes the runner to reach that x-value. We choose the starting time to be $t = 0$ s, again for convenience.

Figure 2–2b shows a curve that interpolates the data points of Fig. 2–2a to reasonable accuracy. It can be used to read the time taken to reach any distance x, even those values not given directly by the data. The curve, which we call $x(t)$, represents the distance from the starting point, x, as a function of time t. To understand this curve better, we use the notion of *displacement*, a concept we met in Chapter 1. The displacement is the *change* in the position of an object. If we denote the position at time t_1 by x_1, and the position at time t_2 by x_2, then the displacement is the difference between these two positions; this is defined mathematically by

$$\Delta x \equiv x_2 - x_1. \tag{2–1}$$

We use the capital Greek letter Δ (delta) to indicate a change (or difference) in a variable from one value to another. The time interval is, similarly,

$$\Delta t = t_2 - t_1. \tag{2–2}$$

There is an important distinction between Δx and Δt: the displacement Δx is really a *vector*, while the time interval Δt is a *scalar*. In our example of the runner, x increases steadily with time, and the vectorial property of the displacement does not play an important role. If there were motion in both the $+x$- and $-x$-directions, then we could have a negative displacement during some time interval and a positive one during some other time interval. For one-dimensional motion, just keeping track of the sign of x as well as its magnitude takes care of its specification. For two- or three-dimensional motion, however, the displacement has both a magnitude and a direction, and the full vectorial description is important. In anticipation, we define the vector displacement by

$$\Delta \mathbf{x} \equiv \mathbf{x}_2 - \mathbf{x}_1, \tag{2–3}$$

where $\mathbf{x}_2$ is the *position vector* at time t_2 (the vector from the starting line to the position of the runner at time t_2) and $\mathbf{x}_1$ is the position vector at t_1.

The vector aspect of our one-dimensional case is handled neatly with the unit vector $\mathbf{i}$ pointing along the $+x$-axis (Fig. 2–3a). The vector $\mathbf{x}_1$, for example, is written as $x_1\mathbf{i}$, where x_1 represents the distance from the origin, including the sign.

While the definition of the position vector depends on the choice of origin, the vector displacement $\Delta \mathbf{x}$ does not. Suppose that the origin were placed 10 m back of

FIGURE 2–1 The 100-m dash provides us with a test case in which we can refine our ideas of motion.

∞ Vectors and scalars were introduced in Chapter 1.

The displacement vector is independent of the origin.

FIGURE 2–2 (a) A set of data (invented for this illustration) of distance versus time at 5-m intervals for the 100-m dash. (b) The curve is an interpolation of the data shown in part (a). The runner is at position x_1 at time t_1 (event 1) and at position x_2 at time t_2 (event 2). In the time interval $\Delta t = t_2 - t_1$, the runner has had a displacement $\Delta x = x_2 - x_1$.

(a)

(b)

FIGURE 2-3 (a) The position vectors $\mathbf{x}_1$ and $\mathbf{x}_2$ run from the origin to the positions x_1 and x_2, respectively, of the runner. The displacement vector $\Delta\mathbf{x} = \mathbf{x}_2 - \mathbf{x}_1$ is also drawn. (b) The information contained in part (a) can be plotted as a curve of position versus time.

(a)

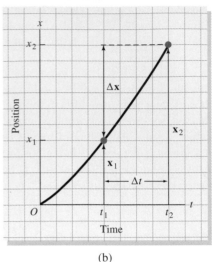

(b)

the starting blocks. Then the position at time t_1 would be given by $(x_1 + 10 \text{ m})\mathbf{i}$, and that at time t_2 would be given by $(x_2 + 10 \text{ m})\mathbf{i}$. The displacement would still be given by

$$\Delta\mathbf{x} = (x_2 - x_1)\mathbf{i}. \tag{2-4}$$

This relationship is shown in Fig. 2–3.

EXAMPLE 2–1 Use Fig. 2–2 to find the position vectors for the dash at times 2 s and 5 s. What is the displacement vector between these two times?

Solution: We can use the diagram in Fig. 2–2 to read the distance traveled for the two times 2 s and 5 s. To determine the displacement vector, defined in Eq. (2–3), we first find the position vectors for 2 s and 5 s. Because the origin is at the beginning of the race, the position vector at 2 s is $\mathbf{x}_1 = 10\mathbf{i}$ m, and at 5 s, $\mathbf{x}_2 = 40\mathbf{i}$ m. Substitution of the position vectors into Eq. (2–3) gives us the displacement vector $\Delta\mathbf{x}$:

$$\Delta\mathbf{x} = \mathbf{x}_2 - \mathbf{x}_1 = (40 \text{ m})\mathbf{i} - (10 \text{ m})\mathbf{i} = (30 \text{ m})\mathbf{i}.$$

We put the appropriate numerical values in at the last step of the problem to determine the displacement vector. The magnitude of the displacement vector between 2 s and 5 s is the magnitude of $\Delta\mathbf{x}$, or 30 m.

2-2 SPEED AND VELOCITY

Speed

The quantities *speed* and *velocity* describe how fast the position of an object changes. The speed of an automobile or airplane is a familiar concept; we want to refine it here. The **average speed** for a particular motion is defined as the total distance traveled divided by the time taken to travel that distance:

$$\text{average speed} \equiv \frac{\text{total distance traveled}}{\text{time interval}}. \tag{2-5}$$

The speed, a scalar quantity, is always positive. As the time interval in the definition of average speed is changed, the average speed may also change. For example, during the first 5.01 s of the 100-m dash, the runner has traveled 40 m. The average speed

PROBLEM-SOLVING TECHNIQUES

1. Problem solving is important. The techniques we present here represent the basis for a plan of action on *any* problem. As we proceed, other techniques will fit into the general scheme. Read the problem, then read it again. Failure to read the problem carefully is perhaps the source of more false starts and wrong answers than is any other cause.

2. Draw a sketch or diagram of the problem that will help you visualize the situation presented by the problem. We illustrate this aspect of the technique in Fig. B1–1.

3. Write down the given and known quantities.

4. Make sure you understand which quantities are to be found.

5. There are generally only a few principles applicable to the solution of a problem. Think about which principles link the quantities to be determined to those that are known.

6. Use the principles that apply to the situation to guide you to the equation or equations that contain the quantities in the problem. Pay attention—at times, certain equations apply to a given situation and others do not. The rest is mathematics! Several equations may need to be manipulated together at times. Count the number of equations available to see if there are enough equations to determine the unknowns.

7. When you solve for an unknown in terms of the known quantities, use symbols, not numbers. Wait until the end to replace symbols with numbers and units. It is important to include units; the answer may require them and the proper cancellation of units provides a valuable accuracy check.

8. When you arrive at a number, think about it. Does it make sense? If you find that it takes 3 min to drive from New York City to Los Angeles, you have probably made a mistake!

9. Use *any* checks you can find for your result.

FIGURE 2B1–1 Some suggestions for illustrations.

over this interval is

$$\text{for 0 s to 5.01 s:} \quad \text{average speed} = \frac{40 \text{ m}}{5.01 \text{ s}} = 8.0 \text{ m/s}.$$

For the last 5.49 s, however, the runner progresses 60 m, and the average speed is

$$\text{for 5.01 s to 10.5 s:} \quad \text{average speed} = \frac{60 \text{ m}}{5.49 \text{ s}} = 10.9 \text{ m/s}.$$

The runner completes the 100-m dash in 10.5 s. Over the entire 100-m dash, then, the runner's average speed is (100 m)/(10.5 s), or 9.5 m/s. Although the notion of speed may be useful for sports, it is of limited utility in mechanics, where velocity plays a much more important role.

Velocity

Whereas speed refers to the total distance traveled, **velocity** refers to how fast the *displacement* changes. Velocity, like speed, is measured over a certain time interval. If a car has a displacement $\Delta \mathbf{x}$ in a particular time interval Δt, then the car's **average velocity**, $\mathbf{v}_{av}$, over that time interval is defined by

$$\mathbf{v}_{av} \equiv \frac{\text{displacement}}{\text{time interval}} \tag{2-6}$$

$$= \frac{\mathbf{x}_2 - \mathbf{x}_1}{t_2 - t_1} \tag{2-7}$$

$$= \frac{\Delta \mathbf{x}}{\Delta t}. \tag{2-8}$$

Note that $\mathbf{v}_{av}$ is the ratio of a vector to an ordinary number (scalar), and *it is therefore a vector*. We write it in boldface for that reason. The dimensions of velocity are $[LT^{-1}]$, with SI units of meters per second.

Average Velocity. The average velocity, $\Delta \mathbf{x}/\Delta t$, provides us with only limited information. For example, consider a car being driven along a straight road for 1 h (so $\Delta t = 1$ h). If $\Delta \mathbf{x}$ is a displacement from the starting point to a point 30 mi down the road, then the above expression will read

$$\mathbf{v}_{av} = (30 \text{ mi/h})\mathbf{i} \tag{2-9}$$

no matter how the car traveled during that hour. It could have traveled at 60 mi/h to a point 45 mi down the road, turned around and traveled 15 mi back to the finishing point at 60 mi/h. It could have traveled a steady 30 mi/h, or it could have traveled the 30 mi at 120 mi/h, arriving in 15 min, and simply parked at the finish point until the hour was up. All that the definition in Eq. (2–8) has given us is the *average velocity* over the time interval in question. It is perfectly consistent to have a zero average velocity over a finite time interval even though a considerable distance may have been covered. For example, if an automobile starts from a given point and returns to that point at the end of a given time interval, the net displacement is zero, and the average velocity over the time interval is zero.

Instantaneous Velocity. The definition of the average velocity includes a time interval. We learn more about the motion when smaller time intervals are used. For example, if we had asked about the average velocity during 15-min intervals in the one-hour trip just discussed, we would already have a more detailed picture of the car's motion, even if the car is not moving steadily during those 15-min intervals. Dividing the hour into 60 1-min intervals and finding the average velocity during each minute would give us still more information: We would find the total displacement vector for a given minute and could then divide by 1 min to find the average velocity during that minute.

It is possible to continue to make the time interval Δt progressively smaller, finding the average velocity for each interval. Suppose that we look at the particular time interval from some time t to a time $t + \Delta t$. In this time interval, the displacement $\Delta \mathbf{x} = \mathbf{x}(t + \Delta t) - \mathbf{x}(t)$ occurs. The average velocity during this interval is the ratio of $\Delta \mathbf{x}$ to Δt. Now, if Δt shrinks to zero, so does $\Delta \mathbf{x}$, *but their ratio remains finite!* (In the example of the car trip, we could imagine that there is some finite average velocity in 1-min intervals, just as there is in 1-s intervals, or in intervals of 1/100 s.) We say that we are taking the *limit* as Δt approaches zero, symbolized by $\Delta t \to 0$. This limit refers to a particular time t and gives us the average velocity over a shorter and shorter time interval around that time. The **instantaneous velocity** at time t is the velocity of an object at that given instant of time, and it is defined as the limit of the average velocity as $\Delta t \to 0$:

The definition of instantaneous velocity, a vector

PROBLEM SOLVING

See Appendix IV–7 for rules for differentiation.

$$\mathbf{v}(t) \equiv \lim_{\Delta t \to 0} \frac{\mathbf{x}(t + \Delta t) - \mathbf{x}(t)}{\Delta t}. \tag{2-10}$$

APPLICATION: TIMED LIGHTS

You may have traveled roads with signs indicating that the traffic lights are timed for a certain average speed, v_{av}. To arrange such timing in any one direction, a traffic engineer must know only the distance between lights, and then arrange the timing of the lights accordingly. If the distance between the first light and a second light is Δx, then the lights should be timed so that the second light turns green at a time $\Delta x/v_{av}$ after the first has done so (Fig. 2AB–1). Note that it is the *average* vehicle speed that enters into such an arrangement. For the design of this system, it is irrelevant whether a driver achieves the average speed by driving rapidly for a portion of the distance and slowly for another portion, or whether he or she maintains v_{av} throughout. (Although it is more fuel efficient to do the latter!)

FIGURE 2AB–1 A close look at this photograph reveals several red lights in sequence, followed by green lights, followed by red lights. A motorist driving at the proper speed will meet only green lights.

Notice that we refer to the instantaneous velocity without subscript in Eq. (2–10). When we use the term "*velocity*," we mean instantaneous velocity, unless we state otherwise. The right-hand side of Eq. (2–10) is the *definition* of a derivative in calculus. Velocity is the derivative of **x** with respect to the variable t:

The instantaneous velocity is the derivative of the displacement vector with respect to time.

$$\mathbf{v} = \lim_{\Delta t \to 0} \frac{\Delta \mathbf{x}}{\Delta t} = \frac{d\mathbf{x}}{dt}. \qquad (2\text{--}11)$$

We shall soon see, in our discussion of an automobile's velocity, that the derivative of a function has a simple graphical interpretation in terms of the graph of the function.

We define the instantaneous speed v to be the magnitude of the instantaneous velocity:

$$v \equiv |\mathbf{v}|. \qquad (2\text{--}12)$$

This definition shows that there are some subtle points in the notion of speed. For example, the average speed is generally not equal to the magnitude of the average velocity. In the case of a car traveling at a uniform 60 mi/h for 1 h—30 mi straight out and 30 mi straight back—the magnitude of the average velocity for the period from 15 min to 45 min is zero, while the average speed is 60 mi/h. But in the first 15-min time interval, the quantities are identical. You may feel by now that the actual use of the term *speed* is confusing even if the definitions are precise. You are correct, and we will attempt to limit the use of the word *speed* in our discussion of mechanics to a synonym for the magnitude of the instantaneous velocity, as given in Eq. (2–12).

The derivative is the slope of a curve.

An Automobile's Velocity. Figure 2–4 shows an automobile's position as a function of time as it moves in a straight line from one traffic light to the next. The automobile starts when the first light turns green and comes to a halt at the second light, which is red, although this is perhaps not obvious from the curve. To learn such information, we must find the automobile's velocity as a function of time. Marked on Fig. 2–4 are two displacements, Δx, centered around the particular time $t = 3$ s. The larger interval corresponds to a time interval $\Delta t = 6$ s, while the smaller corresponds to the time interval $\Delta t = 4$ s. (The motion is one-dimensional, so we can dispense with the bold-faced notation here.) For each time interval, there is a corresponding displacement, Δx. In each case, the ratio $\Delta x/\Delta t$ gives us the value of v_{av} over the corresponding time interval Δt. This value is the slope of the line that connects the two points marking the

31

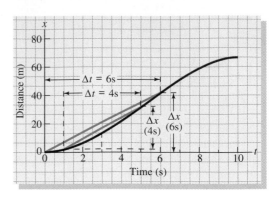

FIGURE 2–4 The displacements Δx corresponding to the two time intervals $\Delta t = 6$ s and $\Delta t = 4$ s, centered around the central time $t = 3$ s. The magnitude of the average velocity for each interval is given by the ratio $\Delta x/\Delta t$ and differs for the two intervals. The slopes of the straight, solid lines (in green) are the average velocities.

PROBLEM SOLVING

The derivative is a slope.

interval. You can see that, as the time interval Δt around the central value of $t = 3$ s becomes smaller, the slope of the line connecting the two points of the interval more nearly matches the slope of the tangent to the curve at that point (Fig. 2–5a). In the limit that Δt becomes infinitesimally small, we arrive at the instantaneous slope dx/dt at the particular time of interest. This is the instantaneous velocity at that point.

All this is completely consistent with what we know from calculus: *The derivative dx/dt at any given time t is the slope of the function $x(t)$ at that time.* Thus the slope of the tangent at any time is the instantaneous velocity of the automobile at that time. We can apply this idea to find the velocity of the car at any time t along its trajectory between the two lights. In Fig. 2–5b, we show the tangents to the curve for four separate time points. The slope of each of these four lines gives the instantaneous velocity of the automobile at the precise times when the lines are tangent to the curve. We can see, for example, that the car has come to rest at the second light, at roughly 65 m, because the slope is zero at that point. We had already seen how to find the derivative as an algebraic limit. Graphical methods like this give us an alternative way to find the derivative—in this case, the instantaneous velocity.

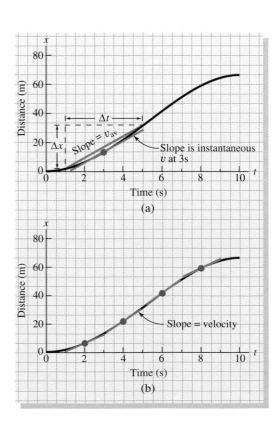

FIGURE 2–5 (a) In the limit that the time interval around a particular time becomes zero, the average velocity for that interval becomes the slope of the curve of displacement versus time at that time. (b) The slope of a line (in green) tangent to the curve of displacement versus time gives the instantaneous velocity at a time corresponding to the tangent point. For example, at the end of the curve, the line tangent to the curve is flat (has zero slope), so that the instantaneous velocity there is zero. The automobile has come to a stop at the light.

EXAMPLE 2–2 Using the data in Table 2–1, calculate the average velocity of a runner over the time intervals $\Delta t = 3.91$ s and $\Delta t = 8.20$ s, centered around $x = 30.0$ m. Also calculate the average velocity over the smallest time interval available near $x = 30.0$ m.

Solution: We must determine the runner's average velocity centered around the distance $x = 30.0$ m using time intervals of 3.91 s and 8.20 s. The average velocity can be calculated by using Eqs. (2–6) through (2–8). The direction of the average velocity is the direction of the runner's motion. According to Table 2–1, we find that the time points on either side of 30.0 m that give a time interval, Δt, of 3.91 s are $t_1 = 2.01$ s and $t_2 = 5.92$ s; for $\Delta t = 8.20$ s, the times are $t_1 = 0$ s and $t_2 = 8.20$ s. We obtain the distances that correspond to each of these time points from Table 2–1. The average velocities are then

$$\text{for } \Delta t = 3.91 \text{ s:} \quad \mathbf{v}_{av} = \frac{\mathbf{x}_2 - \mathbf{x}_1}{t_2 - t_1} = \left(\frac{50 \text{ m} - 10 \text{ m}}{5.92 \text{ s} - 2.01 \text{ s}} \right) \mathbf{i} = (10.2 \text{ m/s})\mathbf{i};$$

$$\text{for } \Delta t = 8.20 \text{ s:} \quad \mathbf{v}_{av} = \frac{\mathbf{x}_2 - \mathbf{x}_1}{t_2 - t_1} = \left(\frac{75 \text{ m} - 0 \text{ m}}{8.20 \text{ s} - 0 \text{ s}} \right) \mathbf{i} = (9.2 \text{ m/s})\mathbf{i}.$$

The smallest available time interval around 30.0 m in Table 2–1 is the interval from 25 m (3.60 s) to 35 m (4.55 s), and

$$\text{for } \Delta t = 0.95 \text{ s:} \quad \mathbf{v}_{av} = \frac{\mathbf{x}_2 - \mathbf{x}_1}{t_2 - t_1} = \left(\frac{35 \text{ m} - 25 \text{ m}}{4.55 \text{ s} - 3.60 \text{ s}} \right) \mathbf{i} = (10.5 \text{ m/s})\mathbf{i}.$$

EXAMPLE 2–3 Figure 2–6 displays the data shown in Table 2–1. Using graphical techniques, determine the velocity at times $t = 2$ s.

Solution: The instantaneous velocity is given by the slope of the curve shown in Fig. 2–6 at the given time $t = 2$ s. We carefully draw a line tangent to the curve at $t = 2$ s in Fig. 2–6. Using the scales of the axes, we are able to measure the slope by examining any two points on the straight line. We see that $x = 90$ m when $t = 12$ s and $x = 0$ m when $t = 0.8$ s. Therefore we have $\Delta x = 90$ m for $\Delta t = 11.2$ s. The slope is

$$\frac{\Delta x}{\Delta t} = \frac{90 \text{ m}}{11.2 \text{ s}} = 8.0 \text{ m/s}.$$

The instantaneous velocity is now given by

$$\mathbf{v} = (\text{slope})\mathbf{i} = (8.0 \text{ m/s})\mathbf{i}.$$

Note that the slope at $t = 2$ s is positive, which is why the coefficient of $\mathbf{i}$ in the expression for the velocity is positive.

TABLE 2–1 Times for a 100-m Dash

Distance (m)	Time(s)
0	0
5	1.36
10	2.01
15	2.57
20	3.09
25	3.60
30	4.09
35	4.55
40	5.01
45	5.47
50	5.92
55	6.37
60	6.83
65	7.28
70	7.74
75	8.20
80	8.65
85	9.11
90	9.57
95	10.04
100	10.50

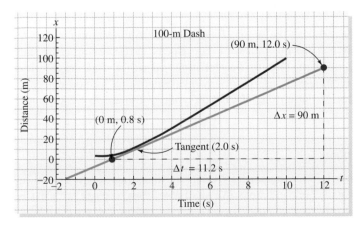

FIGURE 2–6 The straight green line is the tangent to the curve of position versus time at $t = 2$ s. The points at $x = 90$ m ($t = 12$ s) and $x = 0$ m ($t = 0.8$ s) determine the slope of this line and therefore the instantaneous velocity at $t = 2$ s.

Average Acceleration

The dimension and SI unit of acceleration

The term **acceleration** refers to the rate of change in velocity of an object with respect to time. The runner in the 100-m dash of Fig. 2–2 starts off with $v = 0$ m/s at $t = 0$ s. Two seconds later, he is moving with $v = 8$ m/s. His velocity has changed by $\Delta v = 8$ m/s over a time interval $\Delta t = 2$ s. We define the **average acceleration**, $\mathbf{a}_{av}$, in terms of velocity $\mathbf{v}_1$ at time t_1 and velocity $\mathbf{v}_2$ at time t_2:

$$\mathbf{a}_{av} \equiv \frac{\mathbf{v}_2 - \mathbf{v}_1}{t_2 - t_1} = \frac{\Delta \mathbf{v}}{\Delta t}. \tag{2–13}$$

Note that because velocity is a vector, so is acceleration. But because $\mathbf{v}$ and $\Delta \mathbf{v}$ need not point in the same direction, *the velocity vector and the acceleration vector are not necessarily parallel*. The dimensions of acceleration are $[LT^{-2}]$, with SI units of meters per second squared (m/s^2).

EXAMPLE 2–4 A runner in the 100-m dash accelerates uniformly to 10 m/s at 4 s and maintains this velocity for the next 4 s. She then realizes that she is going to win and slows uniformly over the next 4.7 s to reach a velocity of 8 m/s at the end of the race. She has run the 100-m dash in 12.7 s. What is the runner's average acceleration over the time periods 0 s to 4 s, 4 s to 8 s, and 8 s to 12.7 s?

Solution: The positive direction of the acceleration is the direction of the runner's motion, which is described by the unit vector **i.** We must be careful to keep track of signs. The magnitudes of all required velocities are given in the problem except for the additional magnitude $v = 0$ m/s at $t = 0$ s. We represent the speeds during the various time intervals in Fig. 2–7, where the changes in velocity between the times given are linear; that is, the acceleration is uniform. Equation (2–13) is used to find the average acceleration. We give the coefficient of **i** in each case:

$$\text{for 0 s to 4 s:} \quad a_{av} = \frac{v_2 - v_1}{t_2 - t_1} = \frac{10 \text{ m/s} - 0 \text{ m/s}}{4 \text{ s} - 0 \text{ s}} = 2.5 \text{ m/s}^2;$$

$$\text{for 4 s to 8 s:} \quad a_{av} = \frac{v_2 - v_1}{t_2 - t_1} = \frac{10 \text{ m/s} - 10 \text{ m/s}}{8 \text{ s} - 4 \text{ s}} = 0 \text{ m/s}^2;$$

$$\text{for 8 s to 12.7 s:} \quad a_{av} = \frac{v_2 - v_1}{t_2 - t_1} = \frac{8 \text{ m/s} - 10 \text{ m/s}}{12.7 \text{ s} - 8 \text{ s}} = -0.42 \text{ m/s}^2.$$

The greatest average acceleration occurs at the beginning of the race, when the runner is attempting to reach her greatest speed. Although she is running at her highest velocity during the middle part of the race, her average acceleration during this period is zero. During the time interval when she is slowing down at the end of the race, her average acceleration is negative.

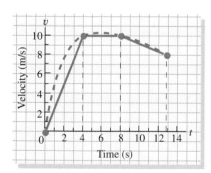

FIGURE 2–7 Example 2–4. The dashed line is the actual velocity; the segmented straight lines are the means of finding the average acceleration.

In Example 2–4, the fact that the runner's average acceleration at the end of the race was negative means that the magnitude of the velocity is decreasing rather than increasing. You will sometimes see the term **deceleration** to describe situations in which the *magnitude* of the velocity decreases. As for any one-dimensional vector, a negative value for the average acceleration means that the direction of the average acceleration is along $-\mathbf{i}$ rather than $+\mathbf{i}$.

Instantaneous Acceleration

The instantaneous acceleration is the derivative with respect to time of the velocity vector.

In physical phenomena, it is the instantaneous values of quantities that are most important. The velocity provides one example. We shall also want to work with the **instantaneous acceleration**. In fact, when we use the term *acceleration*, we shall be referring to the instantaneous acceleration, unless otherwise stated. We define the instantaneous acceleration as the limit of $\Delta\mathbf{v}/\Delta t$ as the time interval Δt goes to zero. This is again a derivative:

$$\mathbf{a} \equiv \lim_{\Delta t \to 0} \frac{\Delta\mathbf{v}}{\Delta t} = \frac{d\mathbf{v}}{dt}. \qquad (2\text{--}14)$$

What is the relation between acceleration and displacement? If we examine the derivatives presented in Eqs. (2–11) and (2–14), we find that

$$\mathbf{a} = \frac{d\mathbf{v}}{dt} = \frac{d}{dt}\left[\frac{d\mathbf{x}}{dt}\right] = \frac{d^2\mathbf{x}}{dt^2}. \qquad (2\text{--}15)$$

The acceleration is the time derivative of the velocity; equivalently, the acceleration is *the second time derivative of the displacement.*

We have discovered that we can find the velocity of a runner from a plot of position versus time; analogously, we can determine the acceleration from a plot of velocity versus time (Fig. 2–8a). We obtain the acceleration at any time t by finding the slope of the tangent to the curve of v as a function of t at the particular time t. In Fig. 2–8b, we show the tangent to the curve of v versus t at times 2 s and 4 s. We can determine the acceleration of our runner either by finding the slopes of tangents to the velocity curve or by taking the algebraic time derivative of the function $v(t)$. We show the acceleration curve in Fig. 2–8c. This curve was derived from the velocity curve in Fig. 2–8b. Notice that the acceleration is initially very high as the runner gains speed, but by 4 s, when the runner moves at a steady speed, the acceleration drops to zero.

EXAMPLE 2–5 The position x of an experimental rocket moving along a long rail is measured to be $x(t) = (5 \text{ m/s})t + (8 \text{ m/s}^2)t^2 + (4 \text{ m/s}^3)t^3 - (0.25 \text{ m/s}^4)t^4$ over the first 10 s of its motion, where t is in seconds and x is in meters. Find the velocity and acceleration of the rocket and display the results graphically.

Solution: The directions of vectors $\mathbf{x}$, $\mathbf{v}$, and $\mathbf{a}$ are taken as positive along the rail (which is one-dimensional, so we do not need boldfaced notation). The plot of the position x as a function of time is shown in Fig. 2–9a. The velocity and acceleration of the rocket can be determined by taking the time derivatives in Eqs. (2–11) and (2–15), respectively:

$$v = \frac{dx}{dt} = (5 \text{ m/s}) + (16 \text{ m/s}^2)t + (12 \text{ m/s}^3)t^2 - (1 \text{ m/s}^4)t^3;$$

$$a = \frac{dv}{dt} = (16 \text{ m/s}^2) + (24 \text{ m/s}^3)t - (3 \text{ m/s}^4)t^2.$$

Velocity and acceleration are measured in m/s and m/s², respectively, if time is measured in seconds. We plot these results over the time period 0 s to 10 s in Fig. 2–9b and c, respectively. Note that although the position x is zero at time $t = 0$, neither the velocity nor the acceleration is zero at this time.

If we wished, we could next discuss the rate of change of acceleration, $d\mathbf{a}/dt$. However, *we do not need to discuss further derivatives* for reasons that will become clear in Chapter 4. We shall see that it is acceleration, not its changes, that plays a primary role in Newton's laws of motion.

2–4 MOTION WITH CONSTANT ACCELERATION

The simplest example of acceleration is constant acceleration (Fig. 2–10). It is a physically important case for a variety of reasons, not the least of which is that we experience a constant acceleration near Earth's surface—that associated with gravity.

The instantaneous acceleration is the second time derivative of the displacement vector.

FIGURE 2–8 (a) The velocity of the runner whose position is shown in Fig. 2–2. This curve can be determined, for example, by finding the slope at each point along the position versus time curve and plotting it. Figure 2–6 illustrates the procedure for $t = 2$ s. (b) The instantaneous acceleration of the runner is found by measuring the slope of the tangents to the curve of velocity versus time. Two such tangents are drawn in blue; their slope gives the acceleration at $t = 2$ s and $t = 4$ s. The slope of the tangent at $t = 2$ s is calculated here. (c) The acceleration of the runner as a function of time. This curve can be found by plotting the slope of the tangents to the velocity versus time curve as a function of time.

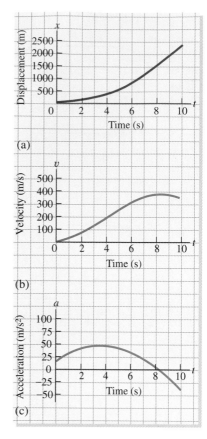

(a)

(b)

(c)

FIGURE 2–9 Example 2–5. We
have plotted the displacement, the
instantaneous velocity, and the
instantaneous acceleration as functions
of time.

FIGURE 2–10 The diver—or more
exactly, one special point in the
diver—accelerates uniformly under the
influence of gravity.

The position as a function of time for
constant acceleration

36

Although we might correctly argue that we do not accelerate when we are standing on
the ground, we certainly accelerate under the influence of gravity when, for example,
we dive into a pool.

In our discussion of displacement, velocity, and acceleration, we concentrated on
the *differences* between average and instantaneous quantities. But a *constant* acceleration means that $\mathbf{a} = \mathbf{a}_{av}$. Constant acceleration means also that the speed changes
steadily; that is, it increases linearly with time. Mathematically,

$$\text{for constant acceleration:} \quad \mathbf{a} = \mathbf{a}_{av} = \frac{\mathbf{v}(t) - \mathbf{v}(t_0)}{t - t_0}, \qquad (2\text{–}16)$$

where t_0 is *any* fixed time and $\mathbf{v}(t_0)$ is the velocity at that time. It is often convenient
to label the initial velocity $\mathbf{v}(t_0)$ as $\mathbf{v}_0$. Because we are dealing with only one dimension, we shall drop the boldfaced notation for vectors. We use a minus sign when the
vector is along the $-x$-direction. We also set $t_0 = 0$. Equation (2–16) can then be solved
for v:

$$\text{for constant acceleration:} \quad v = at + v_0. \qquad (2\text{–}17)$$

This result for the velocity has the linear dependence on time that we had foreseen.
(See also Section 2–6.)

When the velocity is constant, the displacement changes linearly with time.
Here, the velocity changes linearly with time, so we would expect the displacement to
change even more rapidly. To see how this is realized mathematically, we can write
Eq. (2–7) as

$$v_{av} = \frac{x(t) - x(t_0)}{t - t_0}.$$

If we let $x_0 = x(t_0)$, $x = x(t)$, and then let $t_0 = 0$ for convenience, we have

$$v_{av} = \frac{x - x_0}{t}.$$

This equation is solved for x to yield

$$x = v_{av}t + x_0. \qquad (2\text{–}18)$$

Figure 2–11 shows a plot of velocity as a function of time for the case of constant
acceleration. In this special case, the average velocity is simply the average of the initial and final velocities over the total time period t. The initial velocity is v_0 and the final velocity at time t is $v = v(t)$. The average velocity for the case of constant acceleration is then

$$\text{for constant acceleration:} \quad v_{av} = \frac{v_0 + v}{2}. \qquad (2\text{–}19)$$

For this case, v_{av} is just the velocity at time $t/2$. When we substitute this result for v_{av}
into Eq. (2–18), we find that

$$\text{for constant acceleration:} \quad x = \frac{(v_0 + v)t}{2} + x_0 = \frac{vt}{2} + \frac{v_0 t}{2} + x_0. \qquad (2\text{–}20)$$

We now substitute v from Eq. (2–17) into this equation and determine the position as
a function of time:

$$\text{for constant acceleration:} \quad x = \frac{at^2}{2} + v_0 t + x_0. \qquad (2\text{–}21)$$

As we had expected, the position varies more rapidly than linearly with time: it varies
quadratically.

We should discuss one more useful relation. The three boxed equations, which all
contain the time, allow us to find a relationship between displacement, velocity, and

acceleration that does not involve time. To do so, we first solve for the time t from Eq. (2–17):

$$\text{for constant acceleration:} \quad t = \frac{v - v_0}{a}. \qquad (2\text{–}22)$$

If we substitute this expression for time into the first part of Eq. (2–20), we have

$$x = \left(\frac{v_0 + v}{2}\right)\left(\frac{v - v_0}{a}\right) + x_0$$

for constant acceleration, or

$$x - x_0 = \frac{v^2 - v^2_0}{2a}. \qquad (2\text{–}23)$$

This equation can be written as

$$\text{for constant acceleration:} \quad v^2 = v_0^2 + 2a(x - x_0). \qquad (2\text{–}24)$$

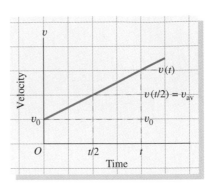

FIGURE 2–11 The velocity changes linearly with time for constant acceleration.

This result gives us the velocity at any position x from the constant acceleration a and the initial velocity and position. The time does not enter into this result. Of course, if we want to find the time t, we can find it by using Eq. (2–22).

The four boxed equations of this section are particularly useful, but you need not memorize them. We can get all the information in these equations from the simple statement that, in this case, the acceleration is constant; alternatively, simply remember that the position changes quadratically with time, Eq. (2–21).

EXAMPLE 2–6 An amateur bowler releases a ball with an initial velocity of 2 m/s; the ball slows down with a constant negative acceleration of -0.2 m/s^2. How far does the ball roll before stopping, and how long does it take to stop?

Solution: Choose a coordinate system with $x = 0$ m at the point where the ball leaves the bowler's hand, and start the clock at $t = 0$ s when the ball leaves his hand (Fig. 2–12). The x-axis is along the direction of the ball's motion. The initial conditions are then $x_0 = 0$ m, $t_0 = 0$ s, and $v_0 = 2$ m/s. The acceleration is negative along the $+x$-direction and has the constant value -0.2 m/s^2. In Fig. 2–12b, we have sketched the constant acceleration, as well as the linearly decreasing velocity and the quadratically varying position that correspond to the acceleration.

We know the initial and final velocities (zero) as well as the acceleration. We do not know the final x-position of the ball or the time elapsed. If we look at the available equations, we see that we know all the variables in Eq.(2–24) except x, which is the displacement of the ball. We can use Eq. (2–24) to find the x-value when the final velocity v has become zero. The solution for x is, from that equation,

$$x = x_0 + \frac{v^2 - v_0^2}{2a}.$$

All the quantities on the right-hand side are known, giving the value of x as:

$$x = (0 \text{ m}) + \frac{(0 \text{ m/s})^2 - (2 \text{ m/s})^2}{2(-0.2 \text{ m/s}^2)} = 10 \text{ m}.$$

Use Eq. (2–22), or Eq. (2–17), rearranged, to determine the time of motion:

$$t = \frac{(0 \text{ m/s}) - (2 \text{ m/s})}{-0.2 \text{ m/s}^2} = 10 \text{ s}.$$

FIGURE 2–12 (a) Example 2–6. The origin O marks the spot ($x = 0$ m) where the ball leaves the bowler's hand. (b) Sketches of acceleration, velocity, and displacement as functions of time.

Any checks you can find for your answer are helpful. In particular, use Eq. (2–21) with $t = 10$ s to determine the displacement again:

$$x = \frac{(-0.2 \text{ m/s}^2)(10 \text{ s})^2}{2} + (2 \text{ m/s})(10 \text{ s}) + (0 \text{ m}) = 10 \text{ m}.$$

So the displacement is again 10 m.

EXAMPLE 2–7 A runner bursts out of the starting blocks 0.1 s after the gun signals the start of a race. She runs at constant acceleration for the next 1.9 s of the race. If she has gone 8.0 m after 2.0 s, what are her acceleration and velocity at this time?

Solution: The runner is not moving during the first 0.1 s of the race, and thus her acceleration is zero during this period of time. During the next 1.9 s, she has an acceleration that is not zero. We can use the four boxed equations of this section *only if the acceleration is constant* during the entire time. Consider a time $t' = t - 0.1$ s, which is the interval of constant acceleration. For the times $t = 0.1$ s and 2.0 s, $t' = 0$ s and 1.9 s, respectively. Over the time period $t' = 0$ s to 1.9 s, we can use the four boxed equations because the acceleration is constant and the initial time t'_0 is 0. The initial conditions of the problem are therefore $t'_0 = 0$ s, $x_0 = 0$ m, $v_0 = 0$ m/s, and we want the acceleration and velocity at $t' = 1.9$ s and $x = 8.0$ m. An examination of the four boxed equations shows that there are two unknowns (v and a) in Eqs. (2–17) and (2–24). Equation (2–19) does not allow us to determine either v or a. However, Eq. (2–21) allows us to determine the acceleration. We insert the known values (with the primed values of time) into Eq. (2–21) and solve for the acceleration:

$$8.0 \text{ m} = \frac{a(1.9 \text{ s})^2}{2} + (0 \text{ m/s})(0 \text{ s}) + (0 \text{ m});$$

$$a = \frac{16 \text{ m}}{3.6 \text{ s}^2} = 4.4 \text{ m/s}^2.$$

Now we can use Eq. (2–17) to determine the velocity at $t' = 1.9$ s ($t = 2.0$ s) after the runner starts:

$$v = at' + v_0 = (4.4 \text{ m/s}^2)(1.9 \text{ s}) + 0 \text{ m/s} = 8.4 \text{ m/s}.$$

General Relations between Position, Velocity, and Acceleration for Constant Acceleration

There is nothing very special about the time $t = 0$ in motion under constant acceleration; in fact, Example 2–7 illustrated the difficulty with Eqs. (2–17) through (2–24) if the initial time must always be zero. That is why it is useful to allow t_0 to be an arbitrary time in the equations relating acceleration, velocity, and displacement. In this case, the four boxed equations of this section become

for constant acceleration:

$$v = a(t - t_0) + v_0, \tag{2–25a}$$

$$v_{av} = \frac{v_0 + v}{2}, \qquad \text{(unchanged)} \tag{2–25b}$$

$$x = \frac{a(t - t_0)^2}{2} + v_0(t - t_0) + x_0, \tag{2–25c}$$

$$v^2 = v_0^2 + 2a(x - x_0). \qquad \text{(unchanged)} \tag{2–25d}$$

The only difference between the four boxed equations (2–17, 2–19, 2–21, and 2–24) and Eqs. (2–25a–d) is that t is replaced by $t - t_0$ in the latter equations. Remember,

Eqs. (2–25a–d) are valid only when the acceleration is constant between the times t_0 and t. When $t_0 = 0$ in Eqs. (2–25a–d), we obtain the four boxed equations. Example 2–7 can now be worked more easily by letting $t_0 = 0.1$ s and $t = 2$ s because the acceleration is constant during this time interval.

EXAMPLE 2–8 On a given flight, a T-38 training jet has an acceleration of 3.6 m/s^2 that lasts 5.0 s during the initial phase of takeoff (Fig. 2–13). The afterburner engines are then turned up to full power for an acceleration of 5.1 m/s^2. The speed needed for takeoff is 164 knots (1 m/s = 1.94 knots). Calculate the length of runway needed and the total time of takeoff.

Solution: There are two different constant accelerations in this example; therefore, we need to divide the problem into two parts. Let us first look at the initial 5.0 s of takeoff. For this period, we have the values $x_0 = 0$ m and $v_0 = 0$ m/s at $t_0 = 0$ s. We use Eqs. (2–25a–d) to find the velocity and distance at 5.0 s. Equation (2–25a) gives the velocity,

for $t = 5.0$ s: $v(t) = (3.6 \text{ m/s}^2)(5.0 \text{ s} - 0 \text{ s}) + (0 \text{ m/s}) = 18$ m/s.

FIGURE 2–13 Example 2–8.

Next we use Eq. (2–25c) to determine the distance the jet has traveled:

for $t = 5.0$ s: $x(t) = \dfrac{(3.6 \text{ m/s}^2)(5.0 \text{ s} - 0 \text{ s})^2}{2} + (0 \text{ m/s})(5.0 \text{ s} - 0 \text{ s}) + 0 \text{ m} = 45$ m.

We now move to the second phase of takeoff, where full power is applied. We have a new set of initial conditions beginning with the time $t_0 = 5.0$ s; namely, $x_0 = 45$ m and $v_0 = 18$ m/s. We want to find the time and distance corresponding to a final velocity of 164 knots. Let us first convert this value into SI units. The final velocity of 164 knots is

$$(164 \text{ knots}) \frac{1 \text{ m/s}}{1.94 \text{ knots}} = 84.4 \text{ m/s}.$$

We use Eq. (2–25a) to find the time t at takeoff; the acceleration is now 5.1 m/s^2. From Eq. (2–25a) we have

$$a(t - t_0) = v - v_0,$$

$$t = \frac{v - v_0}{a} + t_0 = \frac{84.4 \text{ m/s} - 18 \text{ m/s}}{5.1 \text{ m/s}^2} + 5.0 \text{ s} = 18 \text{ s}.$$

Note that we waited until we had solved for the variable t before inserting the numerical values (with units) for v, v_0, a, and t_0. This technique also serves as a check when the cancellation of units gives the expected result, seconds in this case.

Equation (2–25c) can be used directly to determine the takeoff distance because all the variables for the second phase, except x, are now known:

$$x = \frac{(5.1 \text{ m/s}^2)(18 \text{ s} - 5.0 \text{ s})^2}{2} + (18 \text{ m/s})(18 \text{ s} - 5.0 \text{ s}) + 45 \text{ m} = 710 \text{ m}.$$

Because we have included $x_0 = 45$ m from the first phase, 710 m (or 2330 ft) is the total amount of runway used.

2–5   FREELY FALLING OBJECTS

In Section 2–4, we mentioned an important example of constant acceleration: gravity. The **acceleration due to gravity** is given the symbol **g**, and its magnitude is approximately 9.8 m/s^2.[†] Ignoring the effects of air resistance, any object dropped in the vicin-

Near Earth's surface, objects fall with constant acceleration.

[†]There are variations of the order of one percent in the magnitude and direction of **g** over Earth's surface.

ity of Earth's surface will move with constant acceleration **g**. The direction of **g** is down, toward Earth's center, a direction easily found using a plumb bob (a string with a mass at its end).

Experimental Tests of the Acceleration of Gravity

Galileo Galilei, who could be considered the first modern physicist, systematically investigated the motions of falling objects. Centuries earlier, Aristotle had suggested (incorrectly) that the speed of a falling object depends on the weight of the object, and that this speed is proportional to the distance fallen. For Galileo, such assertions were no longer sufficient. His approach was to test by experiment. Galileo performed precise measurements in planned experiments and mathematically described the results. He determined that the distance that objects fall after starting from rest is proportional to the square of the time; equivalently, the speed of a falling object is proportional to the square root of the distance fallen. As Eqs. (2–21) and (2–24) show, this type of motion is characteristic of constant acceleration.

If we drop a hard rubber ball and a sheet of paper simultaneously from the same height, we observe that the paper floats down and the ball reaches the floor first (Fig. 2–14a). The ball experiences a greater acceleration and a larger final velocity than does the sheet of paper. This is because air resistance affects the sheet of paper to a much greater degree than it affects the ball. If we wad the paper up and repeat the experiment (Fig. 2–14b), the effect of air resistance on the paper is decreased and the falling times for the paper and the ball in this experiment are more nearly equal. If the same experiment is done in a vacuum, the falling times are, as best as we can measure, equal. Although air resistance can be an important effect, we shall assume for now that we are dealing with small, heavy objects ("particles") with negligible air resistance. A falling particle near Earth's surface undergoes a constant acceleration that is the same for *all* particles.

Quantitative Description of the Motion of a Freely Falling Object

To study the effects of gravity, it is easiest to set up a coordinate system with a direction perpendicular to Earth's surface. Let us align the y-axis with the vertical direction. We have two choices for this direction. If we choose the positive direction toward the center of the Earth, then the acceleration of gravity is **g** = g**j**. (Recall that **j** is the unit vector in the y-direction.) If we choose the positive direction of the y-axis away from

(a)

(b)

FIGURE 2–14 (a) The rubber ball falls directly to the floor while the sheet of paper floats down more slowly. (b) After the sheet of paper is wadded up and the experiment is repeated, the ball and the paper fall almost together.

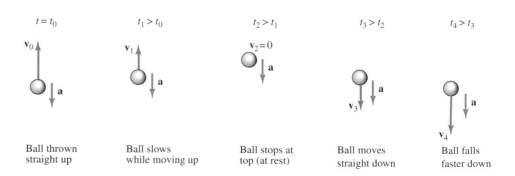

FIGURE 2–15 A ball tossed in the air slows, stops, and falls back. Although position and velocity change with time, the acceleration is constant.

Earth's center, then $\mathbf{g} = -g\mathbf{j}$. We could work problems by using either choice of axes, but let's choose the latter because it will facilitate our future discussion of motion in two or three dimensions. The vertical motion of any freely moving object for which air resistance can be ignored is then summarized by the four boxed equations of Section 2–4 for constant acceleration with $a = -g$,

PROBLEM SOLVING

A good choice of axes simplifies problem solving.

for a freely falling body:

$$v = -gt + v_0, \qquad (2\text{-}26a)$$

$$v_{av} = \frac{v_0 + v}{2}, \qquad \text{(unchanged)} \qquad (2\text{-}26b)$$

$$y = \frac{-gt^2}{2} + v_0t + y_0, \qquad (2\text{-}26c)$$

$$v^2 = v_0^2 - 2g(y - y_0). \qquad (2\text{-}26d)$$

Similar changes can be made to Eq. (2–25a–d). Remember that it is only the near constancy of the acceleration of gravity near the Earth's surface that makes these equations applicable. Figure 2–15 represents the up-and-down motion of an object, with the corresponding velocity and acceleration vectors.

EXAMPLE 2–9 How long does it take a ball to hit the ground if it is dropped from rest from a height of 100 m (Fig. 2–16)? What is the ball's velocity just before it hits the ground?

Solution: In Fig. 2–16, we indicate that the ground is at a level $y = 0$ m, while $y = 100$ m is the position from which the ball is dropped. Our initial conditions at $t_0 = 0$ s are $y_0 = 100$ m and $v_0 = 0$ m/s. We use Eq. (2–26c) to determine t, the time of flight:

$$0 \text{ m} = \frac{-(9.80 \text{ m/s}^2)t^2}{2} + (0 \text{ m/s})t + 100 \text{ m};$$

$$t^2 = \frac{200 \text{ m}}{9.80 \text{ m/s}^2} = 20.4 \text{ s}^2,$$

$$t = 4.52 \text{ s}.$$

We use Eq. (2–26a) to calculate the velocity:

$$\mathbf{v} = -(9.80 \text{ m/s}^2)(4.52 \text{ s})\mathbf{j} + 0 \text{ m/s} = -(44.3 \text{ m/s})\mathbf{j}.$$

Notice that the final velocity has a large magnitude and is *negative;* that is, it points downward. The negative sign appears because we chose the direction of the y-axis to be up, not down.

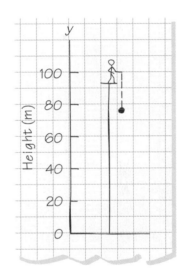

FIGURE 2–16 Example 2–9.

In Example 2–9, we calculated the velocity of an object falling freely from rest. Assume that an object starts at $y_0 = h$ with $v_0 = 0$ at $t_0 = 0$. What is the magnitude of the velocity when the object hits the ground ($y = 0$)? We use Eq. (2–26d) to determine

$$v^2 = 0 - 2g(0 - h) = 2gh;$$

$$\text{for constant acceleration:} \qquad v = \sqrt{2gh}. \qquad (2\text{-}27)$$

The positive sign of the square root is appropriate because it refers to the *magnitude* of the velocity. If the y-axis is up, the velocity is $\mathbf{v} = -(\sqrt{2gh})\mathbf{j}$.

EXAMPLE 2–10 Calculate the time elapsed for the ball of Example 2–9 to drop from 100 m to 75 m, from 75 m to 50 m, from 50 m to 25 m, and from 25 m to the ground (Fig. 2–17).

Solution: Let's calculate the times at 75 m, 50 m, and 25 m with value $y_0 = 100$ m at $t_0 = 0$ s. By the same procedure used in Example 2–9, we find, using Eq.

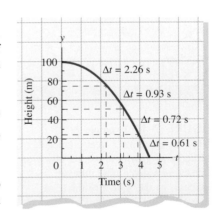

FIGURE 2–17 Example 2–10. The time for a ball to drop a given distance decreases as the ball gains velocity.

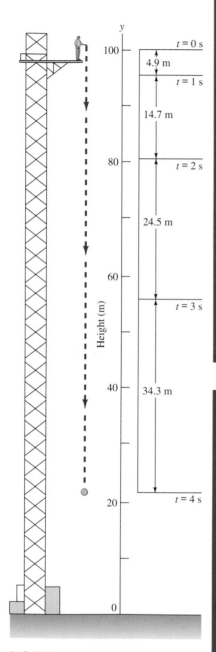

FIGURE 2–18 Example 2–11.
When an object falls freely under the influence of gravity alone, the distance traveled in a given time interval increases as the square of the total time elapsed.

(2–26c), that

$$y = \frac{-gt^2}{2} + y_0,$$

$$t^2 = \frac{2(y_0 - y)}{g};$$

fall from 100 m to 75 m: $t^2 = \dfrac{2(100 \text{ m} - 75 \text{ m})}{9.80 \text{ m/s}^2} = 5.10 \text{ s}^2;$

$$t = 2.26 \text{ s};$$

fall from 100 m to 50 m: $t^2 = \dfrac{2(100 \text{ m} - 50 \text{ m})}{9.80 \text{ m/s}^2} = 10.2 \text{ s}^2;$

$$t = 3.19 \text{ s};$$

fall from 100 m to 25 m: $t^2 = \dfrac{2(100 \text{ m} - 25 \text{ m})}{9.80 \text{ m/s}^2} = 15.3 \text{ s}^2;$

$$t = 3.91 \text{ s};$$

fall from 100 m to 0 m: $t = 4.52 \text{ s}$ (from Example 2–9).

Thus the times of flight are

$$100 \text{ m to } 75 \text{ m}, \quad t = 2.26 \text{ s} - 0 \text{ s} = 2.26 \text{ s};$$

$$75 \text{ m to } 50 \text{ m}, \quad t = 3.19 \text{ s} - 2.26 \text{ s} = 0.93 \text{ s};$$

$$50 \text{ m to } 25 \text{ m}, \quad t = 3.91 \text{ s} - 3.19 \text{ s} = 0.72 \text{ s};$$

$$25 \text{ m to } 0 \text{ m}, \quad t = 4.52 \text{ s} - 3.91 \text{ s} = 0.61 \text{ s}.$$

Due to the increasing velocity, the time needed for the ball to cover a distance of 25 m becomes shorter (Fig. 2–17).

EXAMPLE 2–11   Calculate the position of the ball at $t = 1$ s, 2 s, 3 s, and 4 s under the conditions stated in Example 2–9.

Solution: The initial conditions are the same as before; Fig. 2–18 illustrates the situation. We use Eq. (2–26c) to find y for each t:

$$y = \frac{-gt^2}{2} + 100 \text{ m};$$

$$t = 1 \text{ s}, \quad y = \frac{-(9.80 \text{ m/s}^2)(1 \text{ s})^2}{2} + 100 \text{ m} = 95.1 \text{ m};$$

$$t = 2 \text{ s}, \quad y = \frac{-(9.80 \text{ m/s}^2)(2 \text{ s})^2}{2} + 100 \text{ m} = 80.4 \text{ m};$$

$$t = 3 \text{ s}, \quad y = \frac{-(9.80 \text{ m/s}^2)(3 \text{ s})^2}{2} + 100 \text{ m} = 55.9 \text{ m};$$

$$t = 4 \text{ s}, \quad y = \frac{-(9.80 \text{ m/s}^2)(4 \text{ s})^2}{2} + 100 \text{ m} = 21.6 \text{ m}.$$

We show these distances in Fig. 2–18 for the 1-s equal time intervals. During the first second, the ball travels 100 m − 95.1 m = 4.9 m, but during the fourth second, the ball travels 55.9 m − 21.6 m = 34.3 m. Figure 2–19 clearly illustrates the increasing distance traveled in each time interval.

EXAMPLE 2–12   A ball is thrown straight up with a speed of 10.0 m/s from a third-floor window that is located 15.0 m above the ground. Calculate the maximum height of the ball, the ball's velocity when it hits the ground, and the total time it takes to reach the ground (Fig. 2–20).

Solution: A ball thrown upward reaches its maximum height when its velocity is zero. At this point, the ball's direction of motion changes. For the initial conditions at $t_0 = 0$ s, we know that $y_0 = 15.0$ m and $v_0 = +10.0$ m/s. We use Eq. (2–26a) to determine the time when the ball reaches its maximum height:

$$v = 0 \text{ m/s} = -(9.80 \text{ m/s}^2)t + 10.0 \text{ m/s};$$

$$t = \frac{10.0 \text{ m/s}}{9.80 \text{ m/s}^2} = 1.02 \text{ s}.$$

The maximum height above the ground, y_{max}, is found by using this value of t in Eq. (2–26c):

$$y_{max} = \frac{-(9.80 \text{ m/s}^2)(1.02 \text{ s})^2}{2} + (10.0 \text{ m/s})(1.02 \text{ s}) + 15.0 \text{ m} = 20.1 \text{ m}.$$

We have two choices to work the remainder of this example. We can either take a new initial condition to be at the maximum height of the ball, or we can keep the original initial conditions. Let's work the example using the second choice so the initial conditions are as given in the problem.

We use Eq. (2–26d) to calculate the final velocity v. When the ball hits the ground, the value of y is zero:

$$v^2 = (10.0 \text{ m/s})^2 - 2(9.80 \text{ m/s}^2)(0 \text{ m} - 15.0 \text{ m})$$

$$= 100 \text{ m}^2/\text{s}^2 + 294 \text{ m}^2/\text{s}^2 = 394 \text{ m}^2/\text{s}^2;$$

$$v = -19.9 \text{ m/s}.$$

Notice that the final velocity does not depend on whether the initial direction of the ball is up or down. The final value of the velocity, however, is negative, because the direction of the ball is down. We can choose either the + or − solution of the square root of v^2, whichever is physically sensible.

Equation (2–26a) is used to find the flight time:

$$-19.9 \text{ m/s} = -(9.80 \text{ m/s}^2)t + 10.0 \text{ m/s};$$

$$t = \frac{19.9 \text{ m/s} + 10.0 \text{ m/s}}{9.80 \text{ m/s}^2} = 3.05 \text{ s}.$$

FIGURE 2–19 This multiple image of a ball falling under the influence of gravity is made by flashing a stroboscope 22 times at equal time intervals over 0.5 s.

*2–6   MOTION WITH NONCONSTANT ACCELERATION

We have learned how to find the velocity if the displacement function $x(t)$ is known ($v = dx/dt$); we can also determine the acceleration if the velocity $v(t)$ is known ($a = dv/dt$). (Again we suppress the boldfaced notation for one-dimensional motion.) What about the inverse? Can we determine the displacement if the velocity is known or the velocity if the acceleration is known? What happens if the acceleration is not constant? You will learn in your calculus course (if you have not already) that the techniques of integration allow us to find the displacement if we are given the velocity and to find the velocity if we are given the acceleration.

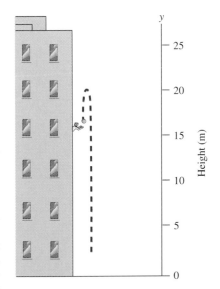

FIGURE 2–20 Example 2–12.

Displacement as an Integral of Velocity over Time

Assume that we know how the velocity depends on time, as shown in Fig. 2–21a. In this example, the velocity is not changing linearly with time, so we know that the acceleration is not constant. We divide the total time interval from an initial time t_i to the final time t_f into many small equal time intervals Δt. These new time intervals are so short that the velocity cannot change very much over their duration. Let's concentrate

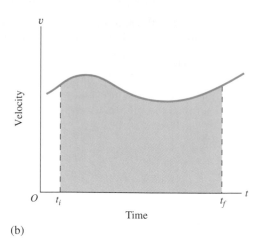

(a)　　　　　　　　　　　　　　　　　　　　(b)

FIGURE 2–21 (a) The displacement over the crosshatched area is $\Delta x_k = v_k \Delta t_k$, where Δt_k is a small time interval and v_k is the average velocity in this interval. If the interval is small enough, v_k is the value of the instantaneous velocity at time t. If all the Δx_k are summed, we find the net displacement between initial time t_i and final time t_f, which is the shaded area under the curve of velocity versus time. (b) This area is also found by integration, $x_f - x_i = \int_{t_i}^{t_f} v \, dt$.

on the tiny time interval from t to $t + \Delta t$, where t is some intermediate time between t_i and t_f. The velocity $v(t)$ at the beginning of the time interval Δt must be similar to the velocity $v(t + \Delta t)$ at the end; further, it should be nearly the same as the average value in the time interval. We therefore write this average value as just $v(t)$ itself. By definition, it is the net displacement Δx in the time interval divided by Δt, $v(t) = \Delta x/\Delta t$. Thus, for small intervals,

$$\Delta x = v(t)\Delta t.$$

Let's now label the tiny time intervals with an index k, and make the time intervals equal. Then the displacement of an object during the time period Δt_k is $\Delta x_k = v_k \Delta t_k$ and is shown graphically by the crosshatched area of the shaded curve in Fig. 2–21a.

The total displacement $x_f - x_i$ over the total time interval t_i to t_f is then a sum over the small displacements $\Delta x_k = v_k \Delta t_k$. There must be N equal tiny time intervals Δt_k to span the time interval from t_i to t_f.

$$N \Delta t_k = t_f - t_i;$$

$$x_f - x_i = \sum_{k=1}^{N} \Delta x_k = \sum_{k=1}^{N} v_k \Delta t_k. \tag{2–28}$$

The shaded area under the curve in Fig. 2–21a represents the displacement $x_f - x_i$. If we are concerned with accuracy because v changes significantly over the time interval Δt, we can simply make Δt smaller and smaller. In the limit that each time interval Δt goes to zero, we have an exact result:

$$x_f - x_i = \lim_{\Delta t \to 0} \sum_{k=1}^{N} v_k \Delta t_k. \tag{2–29}$$

The right-hand side of this expression is the definition of the integral of $v(t)$ over time:

$$x_f - x_i = \int_{t_i}^{t_f} v(t) \, dt. \tag{2–30}$$

This expression is called a **definite integral** because the limits of integration—here t_f and t_i—are specified. The symbol $\int$ replaces the summation sign Σ when we sum over an infinite number of infinitesimal intervals of Δt. The area under the curve of velocity versus time between t_i and t_f in Fig. 2–21b is exactly the displacement $x_f - x_i$.

If the integration limits are not specified, we have

$$x = \int v(t) \, dt + C, \tag{2–31}$$

where C is a constant of integration determined from the initial conditions discussed earlier (that is, x_0 and v_0 at t_0). Equation (2–31) is called an **indefinite integral**. You will find a table of some indefinite integrals in Appendix IV–8.

In Eqs. (2–30) and (2–31), we speak of taking the integral of $v(t)$ over the time. The function $v(t)$ in this case is called the *integrand*. The integral of a function between two limits is the area under the curve that represents a graph of the function versus its variable. For example, Fig. 2–21 shows that the integral of $v(t)$ over t is the total area under the curve $v(t)$ versus t.

Velocity as an Integral of Acceleration over Time

We can perform an analysis for the velocity and acceleration just as we have done for displacement and velocity. We find that velocity is an integral of the acceleration over time:

$$v_f - v_i = \int_{t_i}^{t_f} a(t)\, dt; \tag{2–32}$$

$$v = \int a(t)\, dt + C, \tag{2–33}$$

We have again written both the definite and indefinite integral forms.

More on Integration

Differentiation and integration are inverse processes. Take the derivative of a function. The integral of that derivative is the original function (except for an additive constant). To illustrate this, consider the function $f(t) = C_n t^n$, where C_n is a constant and n is an integer. The derivative of $f(t)$ is

$$\frac{df(t)}{dt} = \frac{d}{dt}(C_n t^n) = nC_n t^{n-1}. \tag{2–34}$$

A polynomial such as $C_0 + C_1 t + C_2 t^2 + \cdots + C_n t^n$ can be differentiated by means of this equation. The integral of $C_n t^n$ is

$$\int f(t)\, dt = \int C_n t^n\, dt = \frac{C_n t^{n+1}}{n+1} + C, \tag{2–35}$$

where C is an indefinite constant to be determined by the initial conditions. As a check, let us verify that the derivative of the right-hand side of Eq. (2–35) gives us $C_n t^n$; that is, perform the inverse operation. Using Eq. (2–34), we have

$$\frac{d}{dt}\left(\frac{C_n t^{n+1}}{n+1} + C\right) = \frac{(n+1)C_n t^{(n+1)-1}}{n+1} + 0 = C_n t^n. \tag{2–36}$$

Therefore, our results are consistent. Notice that the derivative of the constant C in Eq. (2–36) is zero, an application of Eq. (2–34) for $n = 0$.

The Special Case of Constant Acceleration

The value of Eqs. (2–30) and (2–32) is that they can be used to find the position and the velocity of an object whether its acceleration is constant or not. We can demonstrate the consistency of these integration techniques by using them to rederive the earlier constant-acceleration results. Suppose that the magnitude of the constant acceleration is a. Beginning with Eq. (2–32), with the initial conditions x_0 and v_0 for $t = t_0$, we have

$$v - v_0 = \int_{t_0}^{t} a\, dt.$$

The integral of a constant C_1 is, according to Eq. (2–35), $C_1 t$, so

for constant acceleration: $v - v_0 = at\big|_{t_0}^{t} = a(t - t_0),$

where the vertical line indicates that the total function is to be evaluated at the two limits of integration (t_0 and t) and the results subtracted. Rearranging this result gives Eq. (2–25a),

$$\text{for constant acceleration:} \quad v = v_0 + a(t - t_0).$$

We integrate this result once more to determine x:

for constant acceleration:

$$x - x_0 = \int_{t_0}^{t} v \, dt = \int_{t_0}^{t} (v_0 + at - at_0) \, dt = \left(v_0 t + \frac{at^2}{2} - at_0 t \right)\Big|_{t_0}^{t}$$

$$= v_0(t - t_0) + \frac{a}{2}(t^2 - t_0^2) - at_0(t - t_0)$$

$$= v_0(t - t_0) + \frac{at^2}{2} - at_0 t + \frac{at_0^2}{2} = v_0(t - t_0) + \frac{a}{2}(t - t_0)^2;$$

$$x = \frac{a(t - t_0)^2}{2} + v_0(t - t_0) + x_0,$$

which is Eq. (2–25c). Integration techniques allow us to obtain these results much more quickly than previously, and they allow us to dispense with the use of the average velocity.

SUMMARY

Kinematics is the mathematical description of motion. Quantities of particular physical significance are displacement, instantaneous velocity, and instantaneous acceleration. These are all vector quantities. The displacement $\Delta \mathbf{x}$ measures the change in an object's position; where an object moves from position $\mathbf{x}_1$ to position $\mathbf{x}_2$, its displacement is

$$\Delta \mathbf{x} \equiv \mathbf{x}_2 - \mathbf{x}_1. \tag{2–3}$$

The instantaneous velocity at any given time is defined as the limit of the average velocity as $\Delta t \to 0$:

$$\mathbf{v}(t) \equiv \lim_{\Delta t \to 0} \frac{\mathbf{x}(t + \Delta t) - \mathbf{x}(t)}{\Delta t}. \tag{2–10}$$

It is thus the rate of change, or derivative, of the displacement at that time:

$$\mathbf{v} = \lim_{\Delta t \to 0} \frac{\Delta \mathbf{x}}{\Delta t} = \frac{d\mathbf{x}}{dt}. \tag{2–11}$$

The instantaneous acceleration is defined as the rate of change of velocity:

$$\mathbf{a} \equiv \lim_{\Delta t \to 0} \frac{\Delta \mathbf{v}}{\Delta t} = \frac{d\mathbf{v}}{dt}. \tag{2–14}$$

We can also define the average speed and average velocity, but these concepts are used less frequently.

Useful relations between displacement, velocity, and the acceleration that are valid for constant acceleration in one dimension are for constant acceleration:

$$v = a(t - t_0) + v_0, \qquad\qquad (2\text{–}25a)$$

$$v_{av} = \frac{v_0 + v}{2}, \qquad\qquad (2\text{–}19,\ 2\text{–}25b)$$

$$x = \frac{a(t - t_0)^2}{2} + v_0(t - t_0) + x_0, \qquad\qquad (2\text{–}25c)$$

$$v^2 = v_0^2 + 2a(x - x_0). \qquad\qquad (2\text{–}24,\ 2\text{–}25d)$$

Graphical techniques are useful in determining velocity. Graphically, velocity is the slope of a curve of displacement versus time. Similar analysis can determine acceleration from a curve of velocity versus time.

The acceleration due to gravity, $\mathbf{g} = -g\mathbf{j}$, is an important example of a constant acceleration. In this case, the acceleration points to Earth's center. A particle falling a distance h from rest has the final velocity

for constant acceleration: $\quad v = \sqrt{2gh}. \qquad\qquad (2\text{–}27)$

By using integration techniques, the displacement and velocity can be determined from the velocity and acceleration, respectively:

$$x_f - x_i = \int_{t_i}^{t_f} v(t)\ dt. \qquad\qquad (2\text{–}30)$$

$$v_f - v_i = \int_{t_i}^{t_f} a(t)\ dt. \qquad\qquad (2\text{–}32)$$

UNDERSTANDING KEY CONCEPTS

1. Why is it a good idea to increase the space between your car and the car in front of you when the speed of the cars increases?

2. A piece of chalk is thrown straight up; at some point it stops and begins to drop. Can there be a nonzero acceleration at this point, even though the velocity falls to zero?

3. In a series of thought experiments, an object is dropped from rest from a given height on a variety of planets. Each of these planets has a different acceleration due to gravity, g_x. Describe how the time of fall varies with g_x. How does the speed of the object at the end of the fall vary with g_x?

4. You are in the unfortunate position of being in an elevator with 20 bowling balls when the elevator cable breaks, causing both you and the elevator to fall under the acceleration of gravity. The emergency brake has not yet cut in. What is happening inside the elevator?

5. If an object that is restricted to moving along a straight line has a positive initial velocity, and if the acceleration is always negative, can the velocity remain positive?

6. What is the role of an air bag placed where a falling object is expected to land? How can an air bag prevent injury to someone who jumps from a height?

7. For the data shown in Table 2–1, will there be any difference between the average speed over some interval and the magnitude of the average velocity over the same interval? How would you answer the same question if the motion were not on a straight track?

8. What should the velocity of the runner in Fig. 2–2 be at $t = 0$ s? Do the data justify your conclusion? Explain.

9. "Zeno's paradox" comes to us from ancient Greece. It con-

cerns the difficulty that a runner might have in catching a tortoise near the finish line of a race if the tortoise is ahead of the runner at one point, as follows. At some time, the tortoise is a distance of L in front of the runner. After a time interval Δt, the runner is $L/2$ behind the tortoise. After a later time $\Delta t/2$, the runner is $L/4$ behind the tortoise. After a time $\Delta t/4$, the runner is $L/8$ behind the tortoise. The runner always appears to be behind! Where did the Greeks go wrong? By the way, the correct answer to this question was given only in Newton's time and lies behind the crucial concepts of calculus.

10. If the velocity of an object is positive, is its acceleration necessarily positive? Is there any connection between the sign of the velocity and the sign of the acceleration?

11. A juggler tosses a beanbag straight up with initial speed v_0 under the influence of gravity, lets a second beanbag drop from rest, and tosses a third straight down with initial speed v_0. Compare the subsequent accelerations of the three beanbags.

12. A beanbag is tossed straight up. It rises, reaches a maximum height, then falls back down. What is the acceleration of the beanbag at its maximum height?

13. Given a stopwatch and a measuring rod, how would you determine the average acceleration experienced by someone jumping on a trampoline?

14. You are given a measuring rod and a movie camera, with a rather precisely known speed of the motion of the film. How would you use this to determine (to some degree of accuracy) the instantaneous velocity of a person jumping up and down on a trampoline? How would you use your apparatus to measure the instantaneous acceleration?

PROBLEMS

2–1 Displacement

1. (I) A grasshopper jumps along a groove aligned with the x-axis. Starting at the origin, the grasshopper's first jump has a displacement $+32$ cm, the second jump has a displacement -27 cm, the third a displacement -23 cm, and the fourth a displacement $+39$ cm. What is the net displacement? At what position is the grasshopper after all four jumps?

2. (I) Using the data in Table 2–1, draw position vectors to the runner for 40 m and 80 m. Write and draw the displacement from 40 m to 80 m.

3. (I) A gym teacher organizes a series of indoor races in the gym, which is 42 m in length. The students run from one end to the other and back again. After three round trips, what is the distance traveled by each student, and what is the displacement vector? Draw a graph of the magnitude of the displacement vector as a function of time if it takes 7 s to run each 42-m leg. Assume that the speed is constant.

2–2 Speed and Velocity

4. (I) In 1991, Carl Lewis edged out Leroy Burrell in the World Championships in Tokyo to set a new world record in the 100-m dash. Their times at 10-m intervals are given here. Calculate the average velocity for Lewis's world record for 0 m to 50 m, 50 m to 100 m, and 0 m to 100 m.

Distance(m)	Time(s)	
	Lewis	Burrell
10	1.88	1.83
20	2.96	2.89
30	3.88	3.79
40	4.77	4.68
50	5.61	5.55
60	6.46	6.41
70	7.30	7.28
80	8.13	8.12
90	9.00	9.01
100	9.86	9.88

5. (I) (a) Plot the path of an automobile that travels from a starting point to a point 15 km along a straight road at 75 km/h. It stops for 25 min, then continues on the same straight road for 40 km at 100 km/h. After a 5-min stop, it returns to its starting point at 60 km/h. Draw your position axis as horizontal and your time axis as vertical. (b) On the same plot, draw the path of an automobile that starts from the same spot 25 min after the first one and travels at 74 km/h in the original direction of the first automobile. Where, and how often, will the two cars meet?

6. (I) A car moving at 60 mi/h passes a pickup truck moving at 50 mi/h. The car goes on for 20 mi, then stops at a rest stop for 40 min. The car resumes its journey, again at 60 mi/h. Assuming the truck did not stop and maintained its speed, how long after the initial passing does it take for the car to catch the truck again? Solve this problem by graphical means.

7. (I) Redraw the same paths for the two vehicles of Problem 5 on a plot in which the horizontal axis is the time axis and the

position, x, is along the vertical axis. Suppose that you took this new plot and simply relabeled the axes, so that the vertical direction represents time and the horizontal direction represents position. Interpret the paths of the two vehicles.

8. (I) An automobile travels north, covering a distance of 30 mi in 40 min, stops for 20 min, and then continues north for 20 mi, taking 1/2 h. Assume that the car moves uniformly during each segment of the trip. Calculate the average velocity of the total trip. Calculate the average velocity for the first half (by time) and the last half of the trip.

9. (II) Using Fig. 2–5, find the average velocity at 60 m for $\Delta t = 1$ s and 2 s. Find the instantaneous velocity at 60 m.

10. (II) The position of a falling particle is given by $x = x_0 + v_0 t - \frac{1}{2} g t^2$. What is the velocity of the particle as a function of time? Calculate the average velocity during the time intervals $t = 0$ s to 1 s, $t = 1$ s to 2 s, and $t = \tau$ s to $(\tau + 1)$ s.

11. (II) Use the velocity of a particle as a function of time, tabulated below, to calculate the position of the particle at each of the times; assume that at $t = 0$ s, the particle was at the origin and at rest. [*Hint*: A graph is simplest.]

Time(s)	Velocity(m/s)
0.5	0.75
1.5	1.75
2.5	8.75
3.5	21.75
4.5	39.25
5.5	62.75
6.5	90.75
7.5	122.75

12. (II) The height of a bungee jumper above ground level is given as a function of time t by $y = (30 \text{ m})\cos[\pi t/(6 \text{ s})] + (42 \text{ m})$. (a) Sketch the function $y(t)$ from $t = 0$ s to $t = 9$ s. (b) Calculate the average velocity of the jumper between $t = 2$ s and $t = 4$ s and between $t = 4$ s and $t = 8$ s. (c) What is the instantaneous velocity of the jumper when he is closest to the ground?

13. (II) An old brain teaser reads as follows: Two trains leave different stations 80 km apart and travel toward each other on a straight track. One train has a speed of 80 km/h, and the other has a speed of 160 km/h. A very fast insect leaves the slower train and heads toward the faster train at a speed of 240 km/h. Upon encountering the second train, it turns around and just as rapidly returns to the first train. It continues these maneuvers until it is squashed between the two trains when they collide. Graph what has happened, and use your graph to determine how far the insect will have traveled. Can you think of a way to estimate your result or to calculate it rapidly?

14. (II) Traffic signals are placed along a straight road at positions $x = 0$ m, 600 m, and 1200 m (Fig. 2–22). The time intervals during which the signals are green are shown by the thick lines in the figure. (a) Draw the displacement versus time curves (fastest and slowest) for a car that passes through all the lights, when the car moves with constant speed. (b) Draw a similar set of lines for a car traveling in the opposite direction. (c) Assuming that the lights are timed such that a car passes

FIGURE 2-22 Problem 14.

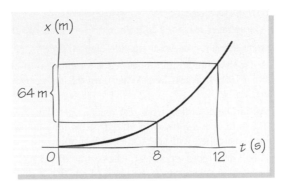

FIGURE 2-23 Problem 21.

through all lights in the middle of the time interval, what is the speed that the lights are timed for? (d) What is the fastest constant speed of a car that makes it through all the signals, assuming it arrives at the first light at the optimal moment?

15. (III) The distance an ant moves is given by $x = 0.02t^3 - 0.1t^2 + 2t$ cm, where t is in seconds. Calculate the velocities for $t = 1$ s, 5 s, and 10 s. What is the average velocity for the first 10 s? Why is the formula unrealistic for long times?

16. (III) The displacement of a particle as a function of t is described by the equation $x = \sqrt{(2 \text{ m}^2/\text{s})(t + 1 \text{ s})}$, where t is measured in seconds. (a) Plot $x(t)$ between $t = 0$ s and $t = 5$ s. (b) Calculate the average velocity between $t = 1$ s and $t = 5$ s; between $t = 2$ s and $t = 4$ s, and between $t = 2.8$ s and 3.2 s. (c) Compare these results with the instantaneous velocity at $t = 3$ s.

2-3 Acceleration

17. (I) An automobile badly in need of repairs is able to accelerate only in the forward direction at a constant value of 0.6 m/s². How long does it take the automobile to get to 40 mi/h?

18. (I) A bicyclist is pedaling at a constant speed of 10 m/s when she decides to slow down. She stops pedaling and sits up, and the combined effects of wind resistance and road friction cause a negative acceleration of -0.3 m/s². If this acceleration does not change, how long would it take her to slow to 5 m/s?

19. (I) A car is said to go from rest to 60 mi/h in 9.0 s. Assuming that the acceleration is uniform, what is its value in units of g?

20. (II) Car A leaves a city and travels along a straight road for 1.5 min at 60 km/h. It then accelerates uniformly for 0.25 min until it reaches a speed of 80 km/h. It proceeds at that speed for 2.0 min, then decelerates uniformly for 0.5 min until it comes to rest. Car B leaves the same city along the same road and accelerates uniformly for 1.6 min until it reaches a speed of 120 km/h. It then decelerates uniformly until it comes to rest again after 1.6 min. (a) Plot the curve of the cars' motions on a graph in which the vertical axis is the speed v and the horizontal axis is the time t. (b) Plot the trajectories of the cars on a graph in which the vertical axis is the distance x from the city and the horizontal axis is time t. (c) How far will the two cars have traveled during the different stages?

21. (II) An automobile starting from rest at $t = 0$ s undergoes constant acceleration on a straight line (Fig. 2-23). It is observed to pass two marks separated by 64 m, the first at $t = 8$ s and the second at $t = 12$ s. What is the value of the acceleration?

22. (II) Inclined planes are convenient tools to study motion under a constant acceleration. The time of passage of a ball rolling on an inclined plane is measured by three light gates positioned 60 cm apart. The ball passes the light gates at 0.30 s, 1.15 s, and 1.70 s. Find the acceleration of the ball.

23. (II) Suppose the position of a particle is described by $x = A \sin(\pi t/12)$. Calculate the velocity and acceleration of the particle as a function of time.

24. (II) Consider the motion of the particle whose velocity is tabulated in Problem 11. Use these data to make a table of approximate values of the acceleration for $t = 1$ s, 2 s, ..., 7 s.

25. (II) The position x of a block attached to a spring as a function of time is given by the formula $x = A \sin(\pi t/12)$, as shown in Fig. 2-24. Describe in words the motion of the block.

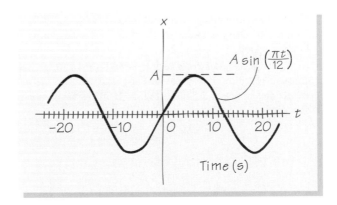

FIGURE 2-24 Problem 25.

26. (II) The position of a particle is given by $x = At^2 + Be^{\alpha t}$. The particle is initially at $x = -2.0$ m with $v = 4.0$ m/s at $t = 0$ s. After 0.2 s, the velocity is observed to be 5 m/s. What is the acceleration after 1.0 s?

27. (II) Consider an object whose acceleration is determined by its velocity, as in the equation $a = A - (v/t_0)$. Here t_0 and A are constants with the dimensions of time and acceleration, respectively. Assume that the object starts out at $t = 0$ s with an initial velocity v_0. Sketch the behavior of the acceleration and of the velocity. Describe the motion after a long time.

28. (I) A drag racer reaches 128 mi/h in a $\frac{1}{4}$-mi race. Assuming a constant acceleration, what was the elapsed time?

29. (I) A car traveling 25 mi/h must reach a minimum of 50 mi/h within a 1000-ft access lane. What must the car's constant acceleration be?

30. (I) An airplane starting from rest reaches its takeoff velocity of 212 mi/h over a runway of 6000 ft. How long did this take if the plane rolled with a constant acceleration?

31. (I) A ball undergoes a constant acceleration of 1.5 m/s². What is the average velocity over the period 1 s to 2 s, assuming the ball started from rest at $t = 0$ s?

32. (I) A rocket accelerates uniformly from rest to a speed of 4.2 × 10³ mi/h in 125 s. Over what distance does the rocket accelerate?

33. (I) A soccer ball rolls with an initial velocity of 8.0 m/s in an easterly direction across a flat field. Friction slows the ball down at the rate of 0.5 m/s². (a) Express the ball's velocity as a function of time. (b) Where is the ball 5.0 s after it starts to roll?

34. (II) The speed of a landing airplane is 60 m/s. After touching ground it rolls a distance of 300 m on the runway at a constant velocity. It then decelerates at 2.5 m/s² until it stops. (a) Sketch the displacement–time and velocity–time curves. (b) Calculate the distance traveled on the ground, and the time interval between touch-down and full stop.

35. (II) A car travels at a constant velocity of 20 m/s toward an intersection. When the car is 80 m from the intersection, the traffic light turns yellow. The driver continues with constant velocity for 1.2 s and then applies the brakes, with a constant acceleration such that the car stops just at the intersection. (a) Sketch the displacement–time and velocity–time curves for the motion. (b) Determine the acceleration of the car during the braking period.

36. (II) A car accelerates from rest at 3 m/s² for 4 s, travels at a constant speed for 7 s, accelerates at 1 m/s² for 15 s, and then decelerates to rest at 2.5 m/s² (Fig. 2–25). How far has the car traveled?

FIGURE 2–25 Problem 36.

37. (II) A child is in an open-cage elevator facing out on a hotel lobby. The elevator is descending at a constant speed of 1 m/s. The child lets a penny drop from his hand when the elevator is 20 m above the floor of the lobby. How much time does the penny spend in the air? Ignore air resistance.

38. (II) Suppose that a runner were capable of a constant acceleration of 3 m/s² for the entire length of a 100-m dash. (a) How long would it take the runner to run the first 10 m? (b) How long for the first 50 m? (c) For the second 50 m? (d) For the entire 100 m? (e) Compare to the times of Problem 4.

39. (II) In 1979, the Japanese tested a magnetically levitated train. The train is both suspended and propelled by magnetic forces. The train traveled on a straight 7000-m-long track starting from rest; it reached a peak speed of 144 m/s before it came to rest again. Both the acceleration and deceleration were constant and of the same magnitude. The entire length of the track was used. (a) What was the magnitude of the acceleration (and deceleration)? (b) How much time was spent on the trip from one end of the track to the other?

40. (II) Your bus is leaving the stop, accelerating at a constant rate of 0.6 m/s². You turn the corner to see the bus pulling out of the stop 30 m ahead of you. What is the minimum steady speed with which you must run to catch the bus? Olympic sprinters can run at 10 m/s.

41. (II) A speeder is traveling at 75 mi/h. He passes a standing police car, which starts to chase him. The police car accelerates from 0 mi/h to 85 mi/h in 13 s and travels at 85 mi/h thereafter. (a) Sketch the positions of both cars on the same x-versus-t graph. (b) How far from its starting point does the police car overtake the speeder? (c) What is the elapsed time?

42. (II) A car is moving at 40 mi/h when the driver sees a light turn red. She hits the brake pedal when she is 100 ft from the light, and the deceleration of the automobile is 0.5g, where g is the acceleration due to gravity. Does she stop before she arrives at the light? How far does she travel before stopping?

43. (II) In Problem 42, calculate the time it takes to stop from the moment the brakes are applied.

44. (II) Two automobiles are geared quite differently and are to be used for a drag race over a distance of 400 m. Car A accelerates at a constant value of 5.0 m/s² for the first 200 m, then at a constant value of 2.5 m/s² for the remaining 200 m. Car B accelerates at a constant rate of 4.5 m/s² for the first 200 m but at 3.0 m/s² for the remaining distance. (a) Give the value of the speed of each automobile at the 200-m mark and the time it took each to get there. (b) What are the finishing times for the race and the values of the respective speeds at the end of the race?

45. (II) An electron in the picture tube of a TV set, traveling in a straight line, accelerates uniformly from speed 3 × 10⁴ m/s to 5 × 10⁶ m/s along a length of 2 cm. (a) How much time does the electron spend in this 2-cm region? (b) What is the magnitude of the electron's acceleration?

46. (II) You have an old, heavy automobile that does not accelerate very rapidly, but can maintain acceleration for a long period. Suppose that your car has a maximum acceleration that takes it from 0 mi/h to 50 mi/h in 20 s. What would the speed be if this average acceleration were maintained for 1 min? How far does your car go during the first 20 s, and how far would it travel in 1 min under the above conditions?

47. (II) An elevator accelerates from the ground with a uniform acceleration a. After 3 s, an object is dropped out of an opening in the floor of the elevator and that object hits the ground 3.5 s later. How large is the acceleration? How high was the elevator when the object was dropped?

48. (II) Two small objects A and B are suspended from the ends of a rope thrown over a pulley (Fig. 2–26). A is 1.2 m above B when the system is released from rest. A descends with a downward acceleration of 0.3 m/s² and, because of the rope, B accelerates upward at the same rate. How much time elapses before the objects bump into each other?

FIGURE 2–26 Problem 48.

49. (II) A car moving at 60 mi/h can be brought to rest in 4 s. Assuming that the deceleration is uniform, how far will the car travel between the time the brakes are applied and the time the car stops?

50. (II) A mountain climber is attached to a rope. She slips, and after she has fallen 5 m, the rope starts to decelerate her. If the constant deceleration is 5g (five times the acceleration due to gravity), how much will the rope have to stretch? (In reality, the deceleration depends on the stretching of the rope.)

51. (II) A bullet traveling at 600 m/s penetrates a block of wood and comes to rest with a constant deceleration after traveling 20 cm (Fig. 2–27). What is the magnitude of the deceleration? How long does it take the bullet to stop?

FIGURE 2–27 Problem 51.

2–5 Freely Falling Objects

52. (I) The tower of Pisa is 54.5 m tall. Assuming that Galileo dropped his object from rest from the top of the tower and that the effects of air resistance were negligible, how long would it have taken the object to fall?

53. (I) A story claims that someone who fell off New York City's Empire State Building (which has approximately 100 floors) was overheard to say "so far, so good" as he passed a third-floor window. Make some estimates to see if this is possible.

54. (I) A string is to have a series of lead sinkers tried to it. The first is tied at the bottom, and the second is tied 10 cm up from the bottom. The string can be held at its top and dropped from a height onto the top of a drum on which the first sinker already rests; each time a sinker hits the drum, a tap is heard. How far above the bottom sinker must the third, fourth, and fifth sinkers be tied so that the series of four taps is spaced by equal time intervals when the string is dropped?

55. (I) The acceleration due to gravity on the surface of the Moon is only about one-sixth the acceleration due to gravity on Earth's surface. In the celebrated experiment of the dropped feather performed by an astronaut on the Moon, how long did it take for the feather to drop 1 m to the surface if it started from rest?

56. (I) The acceleration due to gravity on the surface of Jupiter is 25.9 m/s². If it were possible to perform the experiment, how long would it take for an object that is initially at rest to fall a distance of 10 m on Jupiter's surface? Ignore any effects due to "air" resistance.

57. (II) A rock is thrown upward from the top of a building at an initial speed of 10 m/s. How much later must a second rock be dropped from rest at the same initial height of 20 m so that the two rocks hit the ground at the same time?

58. (II) An astronaut shipwrecked on a distant planet with unknown characteristics is on top of a cliff, which he wishes to descend. He does not know the acceleration due to gravity on the planet, and he has only a good watch with which to make measurements. He wants to learn the height of the cliff, and to do this he makes two measurements (Fig. 2–28). First, he lets a rock fall from rest off the cliff edge; he finds that the rock takes 4.15 s to reach the distant ground. Second, he releases the rock from the same spot but tosses it upward so that it rises a height of what he estimates to be 2 m before it falls to the ground below. This time the rock takes 6.30 s to reach the ground. What is the height of the cliff?

FIGURE 2–28 Problem 58.

59. (II) A ball is dropped from the roof of a 25-m-tall building. What is the velocity of the object when it touches the ground? Suppose that the ball is a perfect golf ball, and it bounces such that the velocity as it leaves the ground has the same magnitude but the opposite direction as the velocity with which it reached the ground. How high will the ball bounce? Now suppose instead that the ball bounces back to a height of 20 m. What was the velocity with which it left the ground?

60. (II) A ball is thrown upward from the ground. It passes a window 10 m above the ground and is seen to descend past the window 2.2 s after it went by on its way up (Fig. 2–29). It reaches the ground 3.6 s after it was thrown. Use this information to calculate the acceleration due to gravity, g.

FIGURE 2–29 Problem 60.

*2–6 Motion with Nonconstant Acceleration

61. (II) An object moves with an acceleration that depends on time and has the form $a = -4\sqrt{t}$ m/s². Its velocity at $t = 0$ s is 15 m/s. How far will the object travel before it comes to a stop?

62. (I) A machine causes an object's speed to increase exponentially, $v(t) = v_0 \, e^{at}$, where $a = 0.5$ s⁻¹ and $v_0 = 1$ m/s. If the object starts from the origin, how far has it traveled after 2 s?

63. (II) The height of a mass suspended from a ceiling by a spring at different times t is given by the formula $y = (0.2 \text{ m}) \sin (4\pi t)$, where t is measured in seconds. (a) Plot the height as a function of time for times $t = 0$ s to $t = 1.0$ s (b) Use your graph to determine the instantaneous velocity at time $t = 0.15$ s. (c) A measurement of the slopes at different times shows that the instantaneous velocity can be represented by the formula $\mathbf{v} = A \cos (4\pi t)\mathbf{j}$, where again t is measured in seconds. Use your measurement from part (b) to determine A. (Do not forget the units of A.) (d) Plot v as a function of t, and use that graph to determine the instantaneous acceleration at time $t = 0.15$ s.

64. (III) When the effect of air resistance is taken into account, the acceleration of a falling object is described by the equation $a = ge^{-bt}$, where g is the acceleration due to gravity, $b = 0.5$ s⁻¹, and t is the time measured from the moment of release. (a) Calculate the velocity and displacement (from the place of release) of the object as a function of time. (b) Sketch these curves. (c) How long does it take for the object to fall 50 m? (d) Compare this time with the time it would take to fall 50 m without air resistance, i.e., with $b = 0$.

65. (III) A powerful rocket moves for a short time with an acceleration that grows with time according to the formula $a = \alpha t^2$. If the rocket is to accelerate from rest in this way until it reaches a speed v_f, how long must the acceleration be maintained?

General Problems

66. (I) A test of your reaction time is to catch a 12-in ruler held vertically by another person. Put your thumb and one finger near the bottom of the ruler, and as soon as the other person releases the ruler, squeeze your thumb and finger together to prevent the ruler from falling. Suppose that in such a test the ruler is grabbed after 5 in of it have passed your hand. What is the time interval between the visual detection of movement and the squeezing together of the fingers?

FIGURE 2–30 Problem 67.

67. (I) The simplest juggling act involves two objects, one of which is transferred from one hand to the other when the second object is tossed upward (Fig. 2–30). Perform an experiment that will tell you how fast you can transfer an object from one hand to the other (for example, by transferring something back and forth 20 times while a friend times you). Estimate from this how high you would have to toss the second object to perform the juggling act. How high would you have to toss the objects if you wanted to juggle three of them?

68. (I) You could probably jump off an 8-ft wall without hurting yourself (but do not try it!). Estimate what your deceleration would be when you hit the ground.

69. (I) A high jumper can jump 2 m on Earth. All other things being equal, how high could the same jumper jump on the surface of the Moon, where the acceleration of gravity is 1/6 that of Earth?

70. (II) Two long-separated friends, June and Bill, spot each other in an airport terminal from a distance of 30 m. They start to run toward each other (Fig. 2–31). Bill accelerates at a constant rate of 0.9 m/s², and June at a constant rate of 1.2 m/s². How far from June's initial position do they meet?

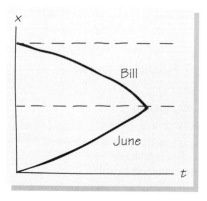

FIGURE 2–31 Problem 70.

71. (II) A Moon rock is thrown upward with velocity 7 m/s. After 7 s, it has a downward velocity of 4 m/s. What is the acceleration due to gravity on the Moon? How high above the starting point did the rock go before it began to fall?

72. (II) A water balloon is dropped from the top of a tower, 200 m off the ground. An alert archer at the base of the tower sees the balloon and shoots an arrow straight up toward the balloon 5 s after the balloon is dropped. The arrow's initial velocity is 40 m/s. Where does the arrow intercept the balloon?

73. (II) There are several known cases of paratroopers whose chutes did not open as they fell, but who survived by falling into brush or snow or onto a steep hillside. It is possible to survive a fall when the deceleration on impact is some 500 m/s^2, equivalent to about 50g. What is the distance traveled within a snowbank while a paratrooper comes to a stop if the deceleration is constant with a magnitude of 50g? The speed with which a paratrooper would enter the snowbank is about 40 m/s. Over how much time does the paratrooper decelerate?

74. (III) One test for the effects of the acceleration of gravity is to tie a set of weights to a string, with the second lowest separated from the lowest by L_0, the third lowest from the second lowest by L_1, and so forth, and to then drop the string (Fig. 2–32). Supposing that free fall corresponds to motion with constant acceleration, how should the separations $L_1, L_2, \ldots, L_n$ (where n is the number of weights) be related to L_0 if the sounds made by the weights as they land form a steady beat? The lowest weight starts at the surface onto which the other weights fall.

FIGURE 2–32 Problem 74.

The water droplets that fly from this fountain follow parabolic paths. Such paths are a consequence of the constant vertical acceleration associated with local gravity.

Motion in a Plane

We explored one-dimensional motion in Chapter 2. There, the vector nature of displacement, velocity, and acceleration plays a limited role. There are only two directions in straight-line motion—right or left—and these are adequately represented by assigning a plus or minus sign to the quantities used to describe such motion. In this chapter, we extend our study of motion into two and three dimensions—motion in a plane and motion in space, respectively—where vectors play a more elaborate role. We will concentrate on motion in two dimensions. This type of motion is simpler and, in most cases, is the motion of interest to us. For example, a rock thrown from a cliff moves in a vertical plane, as does a simple pendulum, and planetary orbits form planes.

3 – 1 POSITION AND DISPLACEMENT

The motion of a planet circling the Sun traces out a path, or an orbit, through space. Similarly, a rock thrown off a cliff follows a path, which is known as its **trajectory**. In fact, any pointlike object traces its own trajectory as it moves through space. Figure 3–1 depicts a particle moving in a two-dimensional plane. We label the plane as the xy-plane and introduce a Cartesian coordinate system that contains an origin and x- and y-axes. The particle is at the position P at time t; this position is described by the position vector $\mathbf{r}_P$, which points from the origin to the point P. At a later time $t + \Delta t$ (where Δt is not necessarily small), the particle is located at position Q and is described by the position vector $\mathbf{r}_Q$. This change in the particle's position between times t and

$t + \Delta t$ can be described by the displacement vector $\Delta \mathbf{r}$; this vector is defined by

$$\Delta \mathbf{r} \equiv \mathbf{r}_Q - \mathbf{r}_P. \qquad (3\text{-}1)$$

The vector $\Delta \mathbf{r}$ points from the tip of vector $\mathbf{r}_P$ to the tip of vector $\mathbf{r}_Q$. Whereas the position vectors $\mathbf{r}_P$ and $\mathbf{r}_Q$ depend on the choice of origin, the displacement vector $\Delta \mathbf{r}$ is independent of the choice of origin. To see this clearly, let's imagine that there is a new origin O' such that the vector from O' to O is the fixed vector $\mathbf{b}$ (Fig. 3-2). The position vector of point P in the new coordinate system is $\mathbf{r}_P + \mathbf{b}$, and that of point Q in the new system is $\mathbf{r}_Q + \mathbf{b}$. If we calculate the displacement $\Delta \mathbf{r}$, which is the difference $[\mathbf{r}_Q + \mathbf{b}] - [\mathbf{r}_P + \mathbf{b}]$, the vector $\mathbf{b}$ cancels. In other words, we have again arrived at the displacement vector defined in Eq. (3-1). This can also be seen in the graphical representation in Fig. 3-2. The choice of origin for the coordinate system did not affect our calculation of the change in the particle's position.

As the particle moves, its components with respect to the coordinate axes change. In other words, the components of the position vector $\mathbf{r}$ along Cartesian coordinate axes change with time:

$$\mathbf{r} = x(t)\mathbf{i} + y(t)\mathbf{j}. \qquad (3\text{-}2)$$

For three-dimensional motion, we would set up three axes, define three mutually perpendicular unit vectors $\mathbf{i}$, $\mathbf{j}$, and $\mathbf{k}$, and write a position vector in the form

$$\mathbf{r} = x(t)\mathbf{i} + y(t)\mathbf{j} + z(t)\mathbf{k}. \qquad (3\text{-}3)$$

The fact that there is more than one component is the only real difference between one-dimensional motion and two- or three-dimensional motion.

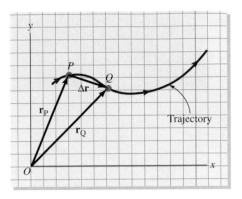

FIGURE 3-1 Position vectors $\mathbf{r}_P$ and $\mathbf{r}_Q$ point from the origin to the positions P and Q at the two times t and $t + \Delta t$, respectively, along the particle's trajectory. The displacement vector between these times is $\Delta \mathbf{r} = \mathbf{r}_Q - \mathbf{r}_P$.

PROBLEM SOLVING

The displacement is independent of the choice of origin.

⊙⊙ The unit vectors $\mathbf{i}$, $\mathbf{j}$, and $\mathbf{k}$ are discussed in Chapter 1.

EXAMPLE 3-1 The position of a small bumper car in an amusement park ride (Fig. 3-3a) is described as a function of time by the coordinates $x = c_1 t^2 + c_2 t + c_3$ and $y = d_1 t^2 + d_2 t + d_3$, where $c_1 = 0.2$ m/s², $c_2 = 5.0$ m/s, $c_3 = 0.5$ m, $d_1 = -1.0$ m/s², $d_2 = 10.0$ m/s, and $d_3 = 2.0$ m. These functions—that is, the locations of the car along the x-axis and the y-axis—are plotted as a function of time in Figs. 3-3b and 3-3c, respectively. Find the position vectors of the car at $t = 3.0$ s and $t = 6.0$ s, and the displacement vector between these times.

Solution: First, we find the x- and y-coordinates of the car at the two times—3.0 s and 6.0 s—by simply inserting the two values of time into the equations given for x and y:

for $t = 3.0$ s: $\quad x(t) = (0.2 \text{ m/s}^2)(3.0 \text{ s})^2 + (5.0 \text{ m/s})(3.0 \text{ s}) + 0.5 \text{ m} = 17 \text{ m}$,

$\qquad\qquad\qquad y(t) = (-1.0 \text{ m/s}^2)(3.0 \text{ s})^2 + (10.0 \text{ m/s})(3.0 \text{ s}) + 2.0 \text{ m} = 23 \text{ m}$;

for $t = 6.0$ s: $\quad x(t) = (0.2 \text{ m/s}^2)(6.0 \text{ s})^2 + (5.0 \text{ m/s})(6.0 \text{ s}) + 0.5 \text{ m} = 38 \text{ m}$,

$\qquad\qquad\qquad y(t) = (-1.0 \text{ m/s}^2)(6.0 \text{ s})^2 + (10.0 \text{ m/s})(6.0 \text{ s}) + 2.0 \text{ m} = 26 \text{ m}$.

Now that we have the components, Eq. (3-2) gives us the position vectors of the car at the two times:

for $t = 3.0$ s: $\quad \mathbf{r}(t) = (17\mathbf{i} + 23\mathbf{j})$ m;

for $t = 6.0$ s: $\quad \mathbf{r}(t) = (38\mathbf{i} + 26\mathbf{j})$ m.

Finally, the displacement vector of the car between 3 s and 6 s is [Eq. (3-1)]

$$\Delta \mathbf{r} = \mathbf{r}(t_{6s}) - \mathbf{r}(t_{3s})$$

$$= (38\mathbf{i} + 26\mathbf{j}) \text{ m} - (17\mathbf{i} + 23\mathbf{j}) \text{ m}$$

$$= (21\mathbf{i} + 3\mathbf{j}) \text{ m}.$$

One other graphical representation of this motion is useful here; we can plot x against y, moment by moment (Fig. 3-3d). This curve is the trajectory of the car.

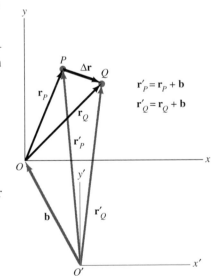

FIGURE 3-2 A displacement vector $\Delta \mathbf{r}$ is independent of the origin. O and O' are two origins and, although the initial and final position vectors to points P and Q do depend on the origins, the difference between these position vectors does not.

(a)

(b)

(c)

(d)

FIGURE 3-3 (a) Example 3–1. An *xy*-coordinate system can be laid out on the floor. (b) The location along the *x*-axis of the car's motion from 0 s to 5 s. (c) The car's location along the *y*-axis. (d) By combining the results of part (b) and part (c), we can plot the trajectory, a graph of the *y*-position versus the *x*-position.

The velocity vector

The component vectors of velocity

Velocity

As in Chapter 2, we want the velocity of a particle to describe the rate of change in the position of the particle as it moves on its trajectory. This description can be obtained using Eq. (3–1) for the particle's displacement as our starting point. The average velocity, $\mathbf{v}_{av}$, over the finite time interval from t to $t + \Delta t$ is then defined by

$$\mathbf{v}_{av} \equiv \frac{\mathbf{r}(t + \Delta t) - \mathbf{r}(t)}{\Delta t} = \frac{\Delta \mathbf{r}}{\Delta t}. \tag{3-4}$$

Equation (3–4) reveals that the direction of $\mathbf{v}_{av}$ points in the same direction as the displacement vector $\Delta \mathbf{r}$.

As the time interval Δt around a given time t decreases, the displacement vector $\Delta \mathbf{r}$ becomes tangent to the particle's trajectory at the location of the moving particle. Then, as in Eq. (2–11), the *instantaneous velocity* $\mathbf{v}(t)$ is obtained by letting Δt become infinitesimally small:

$$\mathbf{v}(t) \equiv \lim_{\Delta t \to 0} \frac{\mathbf{r}(t + \Delta t) - \mathbf{r}(t)}{\Delta t} = \frac{d\mathbf{r}}{dt}. \tag{3-5}$$

The instantaneous velocity $\mathbf{v}$ at time t is defined as the limit of the average velocity as $\Delta t \to 0$; in other words, the instantaneous velocity $\mathbf{v}$ is the time derivative of the position vector. The direction of $\mathbf{v}$ at time t is tangent to the trajectory curve at that time (Fig. 3–4).

We can write the velocity vector in terms of components by using Eqs. (3–2) and (3–5):

$$\mathbf{v} = \frac{d}{dt} \mathbf{r}(t) = \frac{d}{dt} (x\mathbf{i} + y\mathbf{j}) \tag{3-6}$$

$$= \frac{dx}{dt} \mathbf{i} + \frac{dy}{dt} \mathbf{j}. \tag{3-7}$$

This result follows because the unit vectors $\mathbf{i}$ and $\mathbf{j}$ are constant. We write Eq. (3–7) in the form

$$\mathbf{v} = v_x \mathbf{i} + v_y \mathbf{j} \tag{3-8}$$

$$= \mathbf{v}_x + \mathbf{v}_y, \tag{3-9}$$

where

$$v_x = \frac{dx}{dt}, \tag{3-10a}$$

$$v_y = \frac{dy}{dt}, \tag{3-10b}$$

and the component vectors $\mathbf{v}_x$ and $\mathbf{v}_y$ are

$$\mathbf{v}_x = \frac{dx}{dt} \mathbf{i}, \tag{3-11a}$$

$$\mathbf{v}_y = \frac{dy}{dt} \mathbf{j}. \tag{3-11b}$$

The component vectors $\mathbf{v}_x$ and $\mathbf{v}_y$ of the velocity vector $\mathbf{v}$ are drawn on Fig. 3–4. The magnitude of the velocity $\mathbf{v}$ can be written in terms of the components of $\mathbf{v}$:

$$v = |\mathbf{v}| = \sqrt{v_x^2 + v_y^2}. \tag{3-12}$$

The direction θ of the velocity vector $\mathbf{v}$ with respect to the x-axis is determined in terms of the components by

$$\tan \theta = \frac{v_y}{v_x}. \qquad (3\text{–}13)$$

EXAMPLE 3–2 Use the data presented in Example 3–1 to find the bumper car's average velocity over the period from 3.0 to 6.0 s, and the car's instantaneous velocity at $t = 3.0$ s.

Solution: We found in Example 3–1 that the displacement vector of the bumper car between $t = 3.0$ s and $t = 6.0$ s was

$$\Delta \mathbf{r} = (21\mathbf{i} + 3\mathbf{j}) \text{ m}.$$

The average velocity is given by Eq. (3–4),

$$\mathbf{v}_{av} = \frac{\Delta \mathbf{r}}{\Delta t} = \frac{(21\mathbf{i} + 3\mathbf{j}) \text{ m}}{3.0 \text{ s}}$$

$$= (7\mathbf{i} + \mathbf{j}) \text{ m/s}.$$

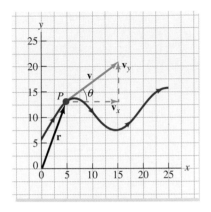

FIGURE 3–4 The velocity vector $\mathbf{v}$ at point P is tangent to the particle's trajectory at that point. The component vectors $\mathbf{v}_x$ and $\mathbf{v}_y$ of the velocity vector $\mathbf{v}$ at that point are also included.

Now, we can use Eq. (3–7) to evaluate the car's instantaneous velocity at $t = 3.0$ s by differentiating x and y given in Example 3–1:

$$\frac{dx}{dt} = \frac{d}{dt}(c_1 t^2 + c_2 t + c_3) = 2c_1 t + c_2;$$

$$\frac{dy}{dt} = \frac{d}{dt}(d_1 t^2 + d_2 t + d_3) = 2d_1 t + d_2.$$

Substituting the numerical values of c_1, c_2, d_1, and d_2 at $t = 3.0$ s from Example 3–1, we have

$$\frac{dx}{dt} = 2(0.2 \text{ m/s}^2)(3.0 \text{ s}) + 5.0 \text{ m/s} = 6.2 \text{ m/s};$$

$$\frac{dy}{dt} = 2(-1.0 \text{ m/s}^2)(3.0 \text{ s}) + 10.0 \text{ m/s} = 4.0 \text{ m/s}.$$

From these velocity components, we calculate the velocity at $t = 3.0$ s to be

$$\mathbf{v} = \frac{dx}{dt}\mathbf{i} + \frac{dy}{dt}\mathbf{j}$$

$$= (6.2\mathbf{i} + 4.0\mathbf{j}) \text{ m/s}.$$

This velocity vector is shown in Fig. 3–5.

Acceleration

Acceleration describes how rapidly velocity changes. Our discussion of how acceleration is found from velocity is analogous to the discussion of how velocity is found from displacement. For a finite time interval Δt, the average acceleration is defined as

$$\mathbf{a}_{av} \equiv \frac{\mathbf{v}(t + \Delta t) - \mathbf{v}(t)}{\Delta t} = \frac{\Delta \mathbf{v}}{\Delta t}. \qquad (3\text{–}14)$$

The instantaneous acceleration at time t is the limit of the average acceleration as Δt approaches zero, which is a derivative:

$$\mathbf{a} \equiv \lim_{\Delta t \to 0} \frac{\mathbf{v}(t + \Delta t) - \mathbf{v}(t)}{\Delta t} = \frac{d\mathbf{v}}{dt}. \qquad (3\text{–}15)$$

The acceleration vector

The components of acceleration

As we did for velocity, we can express acceleration in terms of its components:

$$\mathbf{a} = \frac{dv_x}{dt}\mathbf{i} + \frac{dv_y}{dt}\mathbf{j} \qquad (3\text{--}16)$$

$$= a_x\mathbf{i} + a_y\mathbf{j}. \qquad (3\text{--}17)$$

Here, the components of the acceleration vector are

$$a_x = \frac{dv_x}{dt} = \frac{d^2x}{dt^2}, \qquad (3\text{--}18a)$$

$$a_y = \frac{dv_y}{dt} = \frac{d^2y}{dt^2}. \qquad (3\text{--}18b)$$

EXAMPLE 3–3 Calculate the instantaneous acceleration of the bumper car in Example 3–1 at $t = 1.0$ s and $t = 3.0$ s. What are the magnitude and direction of the acceleration at these times?

Solution: Equations (3–18a and b) express the car's acceleration for any time t. The values for x and y as functions of time were given in Example 3–1, and we found the velocity components in Example 3–2 (see Fig. 3–5). Equations (3–18) then give

$$a_x = \frac{dv_x}{dt} = \frac{d}{dt}(2c_1t + c_2) = 2c_1$$

and

$$a_y = \frac{dv_y}{dt} = \frac{d}{dt}(2d_1t + d_2) = 2d_1.$$

Once we find the acceleration components, we can use Eq. (3–17) to determine the acceleration:

$$\mathbf{a} = 2c_1\mathbf{i} + 2d_1\mathbf{j}.$$

In this case, the car's acceleration is a constant—it is independent of time—and so it is exactly the same for $t = 1.0$ s and for $t = 3.0$ s. Given the values of c_1 and d_1 (0.2 m/s^2 and -1.0 m/s^2, respectively), the numerical value of the acceleration is

$$\mathbf{a} = (0.4\mathbf{i} - 2.0\mathbf{j}) \text{ m/s}^2.$$

The magnitude of the acceleration is

$$a = |\mathbf{a}| = \sqrt{a_x^2 + a_y^2}$$

$$= \sqrt{(0.4)^2 + (-2.0)^2} \text{ m/s}^2 = \sqrt{4.2} \text{ m/s}^2 = 2.0 \text{ m/s}^2.$$

The acceleration vector makes an angle θ with the x-axis, which is shown in Fig. 3–5; the angle θ is

$$\tan \theta = \frac{a_y}{a_x} = \frac{-2.0 \text{ m/s}^2}{0.4 \text{ m/s}^2} = -5.0,$$

$$\theta = -79°.$$

Representing Trajectories

Figure 3–5 shows the trajectory of the bumper car discussed in Examples 3–1, 3–2, and 3–3; this trajectory is a curve representing the car's x-position versus its y-position. In Fig. 3–5, we show a position vector $\mathbf{r}$, a velocity vector $\mathbf{v}$, and an acceleration vector $\mathbf{a}$, each at $t = 3$ s. Although the figure shows $\mathbf{a}$ at the particular point corresponding to $t = 3.0$ s, it is, in fact, the same vector everywhere along the curve. This is because we calculated that the acceleration vector of the car is independent of time.

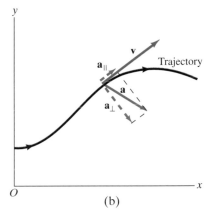

FIGURE 3–5 The trajectory of the car's path in Example 3–1 is plotted for times up to 10 s: the position vector **r**, velocity **v**, and acceleration **a** at $t = 3.0$ s.

FIGURE 3–6 The velocity **v** and acceleration **a** of a particle following some trajectory: (a) The acceleration of the particle is separated into x- and y-components; (b) the acceleration of the particle is separated into components parallel and perpendicular to the path.

We can apply a graphical representation like that of Fig. 3–5 to any motion. For example, consider a trajectory, formed by the locus of the position vector **r** as a function of time. The velocity vector **v** at any time t is a vector of magnitude $|\mathbf{v}|$ that is *tangential* to the trajectory at time t. A plot of the "locus" of **v** can similarly then be drawn; it is the curve of the points whose *horizontal coordinate* at time t is v_x and whose *vertical coordinate* at that time is v_y. The acceleration at time t is given by a vector whose magnitude is $|\mathbf{a}|$ and whose direction is tangential to the velocity curve at time t. Note: *The acceleration vector **a** is tangent to the velocity curve but not to the trajectory itself.*

Figure 3–6 shows the path of an object with **v** and **a** indicated at a particular time. In Fig. 3–6a, the acceleration is separated into its a_x and a_y components. Alternatively, we can separate the acceleration **a** into components that are parallel (tangential) and perpendicular (normal) to the velocity vector (Fig. 3–6b). We label these components $a_\parallel$ and $a_\perp$, respectively. The component $a_\parallel$ of **a** that is parallel to **v** affects the magnitude but not the direction of **v**. Similarly, the $a_\perp$ component changes the direction but not the magnitude of **v**. It is useful to refer separately to the parallel and perpendicular components of an object's acceleration because they affect the velocity differently.

3 – 3   MOTION WITH CONSTANT ACCELERATION

When an object moves under the effects of *constant acceleration*, it can move only in a plane. The plane is formed by the initial velocity vector and the acceleration vector; the object cannot acquire a component of velocity perpendicular to this plane. As we have noted, the classic case of constant acceleration is gravity. Because there are any number of examples of objects moving under the influence of gravity, a discussion of constant acceleration is one of the simplest and most important for us to undertake. When we say that the acceleration **a** remains constant, we mean it is constant in both magnitude and direction. That means that both components a_x and a_y must be constant. When both $\mathbf{a}_x$ and $\mathbf{a}_y$ are constant, our one-dimensional results from Chapter 2 *independently* determine the x- and y-components of position **r** and velocity **v** in terms of the constant acceleration. In other words, we can think of the x- and y-motions as separate ones, governed by separate constant accelerations. We use Eqs. (2–17) and (2–21)

PROBLEM SOLVING

We can use the results of Chapter 2 when the acceleration vector is constant.

The components of position and
velocity for a constant acceleration

to find:

x-component of **r**:
$$x = x_0 + v_{0x}t + \frac{1}{2}a_x t^2, \tag{3-19}$$

x-component of **v**:
$$v_x = v_{0x} + a_x t; \tag{3-20}$$

y-component of **r**:
$$y = y_0 + v_{0y}t + \frac{1}{2}a_y t^2, \tag{3-21}$$

y-component of **v**:
$$v_y = v_{0y} + a_y t. \tag{3-22}$$

Here, x_0 and y_0 are the specified components of $\mathbf{r} = \mathbf{r}_0$ at an initial time $t = 0$, and v_{0x} and v_{0y} are the components of $\mathbf{v} = \mathbf{v}_0$ at time $t = 0$. These specified values at the initial time $t = 0$ must be given; they are not determined simply because the acceleration is constant. We can now say that we have specified the **initial conditions**. In component form, the initial conditions are

$$\mathbf{r}_0 = x_0\mathbf{i} + y_0\mathbf{j} \tag{3-23}$$

and

$$\mathbf{v} = v_{0x}\mathbf{i} + v_{0y}\mathbf{j} \tag{3-24}$$

at $t = 0$.

Equations (3–19) through (3–22), which address position and velocity for motion with constant acceleration **a**, can be written more compactly in vector form:

$$\mathbf{r} = \mathbf{r}_0 + \mathbf{v}_0 t + \frac{1}{2}\mathbf{a}t^2 \tag{3-25}$$

$$\mathbf{v} = \mathbf{v}_0 + \mathbf{a}t. \tag{3-26}$$

These results are valid *only* when **a** is constant.

EXAMPLE 3–4 A wayward golf ball rolls off the edge of a vertical cliff overlooking the Pacific Ocean. The golf ball has a horizontal velocity component of 10 m/s and no vertical component when it leaves the cliff. Describe the subsequent motion.

Solution: Here, the displacement, the velocity, and the acceleration all lie in the same plane, which we assign to be the *xy*-plane. The ball's constant acceleration **a** is equal to that of gravity and thus $\mathbf{a} = \mathbf{g}$. The vector **g** points toward the center of Earth and has a magnitude of 9.8 m/s². Now we can draw a coordinate system as in Fig. 3–7a, placing the origin at the point where the ball leaves the cliff. For constant acceleration, we apply Eqs. (3–19) through (3–22).

We must specify the golf ball's initial position and velocity by using the given information. We set $t = 0$ s at the time when the motion starts; at that moment, then, we have $x_0 = 0$ m, $v_{0x} = 10$ m/s, $y_0 = 0$ m, and $v_{0y} = 0$ m/s. The acceleration due to gravity has components $a_x = 0$ m/s² and $a_y = -9.8$ m/s². Because there is no component of acceleration in the *x*-direction, the horizontal velocity is constant and will remain at 10 m/s. We determine the velocity components from Eqs. (3–20) and (3–22):

$$v_x = 10 \text{ m/s};$$

$$v_y = 0 \text{ m/s} + (-9.8 \text{ m/s}^2)t = (-9.8 \text{ m/s}^2)t. \tag{3-27}$$

The position of the golf ball is determined from Eqs. (3–19) and (3–21):

$$x = 0 \text{ m} + (10 \text{ m/s})t + \frac{1}{2}(0 \text{ m/s}^2)t^2 = (10 \text{ m/s})t;$$

$$y = 0 \text{ m} + (0 \text{ m/s})t + \left[-\frac{1}{2}(9.8 \text{ m/s}^2)t^2\right] = (-4.9 \text{ m/s}^2)t^2. \tag{3-28}$$

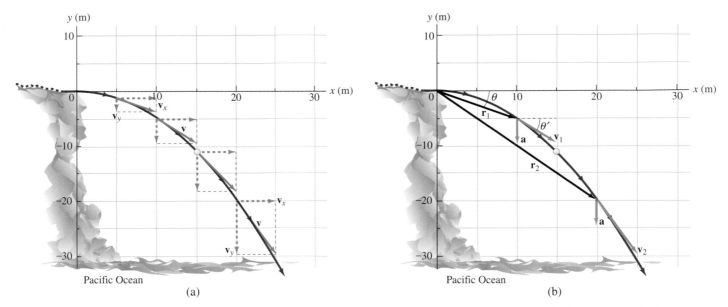

Pacific Ocean

(a)

Pacific Ocean

(b)

Figure 3–7a shows the trajectory of the golf ball. It also shows the velocity vector and its components every 0.5 s up to 2 s. While the horizontal component of the velocity stays constant, the vertical component changes linearly with time. Further, the total velocity vector is tangent to the ball's path of motion at each point along its trajectory.

Figure 3–7b shows the position vector $\mathbf{r}$, velocity $\mathbf{v}$, and acceleration $\mathbf{a}$ at $t = 1$ s and $t = 2$ s. While $\mathbf{a}$ remains constant, $\mathbf{r}$ and $\mathbf{v}$ change with time. The three vectors $\mathbf{r}$, $\mathbf{v}$, and $\mathbf{a}$ do not generally point in the same direction as one another at a given time during the golf ball's motion. The directions of $\mathbf{r}$ and $\mathbf{v}$ are determined by

$$\tan \theta = \frac{y}{x} = \frac{(-4.9 \text{ m/s}^2)t^2}{(10 \text{ m/s})t} = (-0.49 \text{ s}^{-1})t$$

and

$$\tan \theta' = \frac{v_y}{v_x} = \frac{(-9.8 \text{ m/s}^2)t}{(10 \text{ m/s})} = (-0.98 \text{ s}^{-1})t.$$

Both angles vary with time.

FIGURE 3–7 (a) Example 3–4. The velocity vector $\mathbf{v}$ and components $\mathbf{v}_x$ and $\mathbf{v}_y$ of the golf ball are shown at 0.5-s intervals up to 2.0 s. (b) The position $\mathbf{r}$, velocity $\mathbf{v}$, and acceleration $\mathbf{a}$ of the golf ball for $t_1 = 1$ s and $t_2 = 2$ s.

3–4 PROJECTILE MOTION

Let's consider the motion of a golf ball. The golf ball is an example of a *projectile* that moves under the effect of gravity. In the absence of air resistance, what is the trajectory of such a projectile? The motion is that of constant acceleration, which is due to gravity in this case. This constant acceleration, $\mathbf{g}$, has only a vertical component. We calculate the ball's trajectory by separating its motion into horizontal and vertical components and then applying the kinematic equations for constant acceleration (Sec. 3–3).

Our coordinate system is best designed by placing the origin at the starting point, assigning the y-direction vertically and the x-direction along the horizontal (Fig. 3–8). The initial position of the ball is $x_0 = y_0 = 0$; the initial velocity at $t = 0$ is $\mathbf{v}_0$. The trajectory of the golf ball makes an initial angle with the horizontal that is referred to as the elevation angle θ_0. The components of $\mathbf{v}_0$ are thus

$$v_{0x} = v_0 \cos \theta_0 \quad \text{and} \quad v_{0y} = v_0 \sin \theta_0. \tag{3–29}$$

The components of the acceleration are the constants

$$a_x = 0 \quad \text{and} \quad a_y = -g. \tag{3–30}$$

FIGURE 3–8 A golf ball leaves a tee with an initial velocity of magnitude v_0 at an elevation angle θ_0.

Using Eqs. (3–19) through (3–22), we find the components of **r** and **v** (the position and velocity of the ball, respectively) to be

$$x = 0 + (v_0 \cos \theta_0)t + \frac{1}{2}(0)t^2 = (v_0 \cos \theta_0)t, \tag{3–31}$$

$$y = 0 + (v_0 \sin \theta_0)t - \frac{1}{2}gt^2 = (v_0 \sin \theta_0)t - \frac{1}{2}gt^2, \tag{3–32}$$

and

$$v_x = v_0 \cos \theta_0 + (0)t = v_0 \cos \theta_0, \tag{3–33}$$

$$v_y = v_0 \sin \theta_0 - gt. \tag{3–34}$$

The Trajectory

We can find the trajectory of the golf ball by plotting its height y versus its x-position. We know both x and y as functions of time, and we can eliminate the time dependence by using Eq. (3–31) to find the time t as a function of x. We then insert the result for t into Eq. (3–32):

$$t = \frac{x}{v_0 \cos \theta_0} \,; \tag{3–35}$$

$$y = (v_0 \sin \theta_0)\frac{x}{v_0 \cos \theta_0} - \frac{1}{2}g\left(\frac{x}{v_0 \cos \theta_0}\right)^2$$

$$= (\tan \theta_0)x - \left(\frac{g}{2v_0^2 \cos^2 \theta_0}\right)x^2. \tag{3–36}$$

The coefficients of x and x^2 in Eq. (3–36) are both constants, so this equation has the form

$$y = C_1 x - C_2 x^2, \tag{3–37}$$

Objects moving with constant acceleration follow a parabolic trajectory.

This is the general equation of a parabola with its axis parallel to the y-axis. *The trajectory of all objects moving with constant acceleration is parabolic.* Take a look at the parabolic motion illustrated in the chapter-opening photograph or in Fig. 3–9; this motion is examined at various points of the trajectory in Fig. 3–10.

Range. We define the **range** R of a projectile launched from the ground ($y = 0$) to be the horizontal distance that the projectile travels over level ground. The quantity R is the value of x when the projectile has returned to the ground; that is, when y again equals zero. From Eq. (3–37), then, we have

$$0 = R(C_1 - C_2 R). \tag{3–38}$$

FIGURE 3–9 The path followed by a projectile is parabolic.

The value $R = 0$ clearly satisfies the condition $y = 0$ in this equation, but this is the starting point of the projectile's motion—it is launched from the ground and its x-position is zero at launch time. y can alternatively be zero if the factor $C_1 - C_2 R = 0$ in Eq. (3–38), or $R = C_1/C_2$; this case corresponds to the projectile having landed back on the ground after its flight. Inserting the values of C_1 and C_2 from Eq. (3–36) yields

$$R = \frac{C_1}{C_2} = \frac{\tan \theta_0 \,(2v_0^2 \cos^2 \theta_0)}{g} = \frac{2v_0^2}{g}\left(\frac{\sin \theta_0}{\cos \theta_0}\right)\cos^2 \theta_0 = \frac{v_0^2}{g}2 \sin \theta_0 \cos \theta_0.$$

From trigonometry, $\sin(2\theta_0) = 2 \sin \theta_0 \cos \theta_0$, and we find

$$\text{for motion over level ground:} \quad R = \frac{v_0^2 \sin(2\theta_0)}{g}. \tag{3–39}$$

FIGURE 3–10 The motion of a ball thrown in the air and moving under the influence of gravity. Air resistance is ignored. The ball moves with constant acceleration, which in this case is directed downward due to gravity. The velocity vector changes throughout the motion, but the vector component in the horizontal direction does not. The vector component in the vertical direction changes linearly with time. The resulting trajectory forms a parabola.

The range varies with the initial angle of the projectile. If, for example, $\theta_0 = 0°$, then $R = 0$; when a projectile moves off horizontally from ground level, the slightest drop makes it plow into the ground. If $\theta_0 = 90°$, Eq. (3–39) again gives $R = 0$; when a projectile is launched straight up, it comes back straight down. In fact, as θ_0 increases from 0° to 45° and then to 90°, $\sin(2\theta_0)$ first increases from 0 to 1, then decreases back down to 0, respectively. In other words, there are *two* initial angles at which to launch a projectile that give the same range for a given initial speed (Fig. 3–11). For example, a pop fly and a line drive in softball can both be caught by the shortstop. The range reaches a maximum value when the factor $\sin(2\theta_0)$ reaches its maximum value of 1. This occurs for $2\theta_0 = 90°$, or $\theta_0 = 45°$, in which case

$$R_{\max} = \frac{v_0^2}{g}.$$
(3–40)

If the projectile is shot at an angle higher or lower than 45°, the range is shorter (Fig. 3–11).

Flight Time. Let T be the total flight time of a golf ball. Figure 3–12 shows that the golf ball reaches its maximum height exactly halfway through its motion. At this point, its motion is horizontal; that is, the vertical component of velocity is zero. This occurs at time $t = T/2$. We can find $T/2$ by setting $v_y = 0$ in Eq. (3–34):

$$0 = v_0 \sin \theta_0 - \frac{gT}{2}.$$

We solve for T to find that

$$T = \frac{2v_0}{g} \sin \theta_0.$$
(3–41)

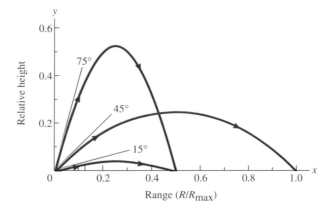

FIGURE 3–11 If the initial speed is fixed and if air resistance is ignored, a projectile's trajectory will have a maximum range for an elevation angle of 45°. The range is the horizontal distance the projectile travels over level ground.

FIGURE 3–12 A projectile (a golf ball) moving under the force of gravity is at its maximum height when $v_y = 0$. At that moment, the ball is traveling horizontally.

Maximum Height. The maximum height $y_{max} = h$ is reached at time $T/2$. From Eq. (3–32), we find the height at this time is

$$h = (v_0 \sin \theta_0) \frac{2v_0}{2g} \sin \theta_0 - \frac{1}{2} g \left(\frac{2v_0}{2g} \sin \theta_0 \right)^2 = v_0^2 \frac{\sin^2 \theta_0}{g} - g v_0^2 \frac{\sin^2 \theta_0}{2g^2}$$

$$= \frac{v_0^2 \sin^2 \theta_0}{2g}. \tag{3–42}$$

We use Eqs. (3–36), (3–39), (3–41), and (3–42) to determine a projectile's trajectory, range, flight time, and maximum height, respectively. The range and flight time refer to the special case where the golf ball returns to its original height. These equations need not be memorized; however, it is important to understand how these equations were obtained. We apply these methods again in Examples 3–5 through 3–8.

EXAMPLE 3–5 In order to win a bet that he can drive a golf ball a horizontal distance of 250 m, an amateur golfer goes to a cliff overlooking the ocean (Fig. 3–13). The cliff is 52 m above the ocean. The golfer strikes the golf ball so that the ball's initial speed is 48 m/s and the elevation angle (from the horizontal) is 36°. Does he win his bet? What is the horizontal distance actually covered by the ball?

Solution: We use the trajectory equation to find the value of x at which the golf ball reaches the ocean surface. We again place the origin of our coordinate system at the tee, letting $+y$ extend upward. The initial conditions are $x_0 = 0$ m, $y_0 = 0$ m, $v_0 = 48$ m/s, and $\theta_0 = 36°$. We cannot use Eq. (3–39) to calculate the range because that result applies to level ground. We want to determine the distance from the tee to the point at which the golf ball reaches the ocean ($y = -52$ m), not the distance when the ball returns to $y = 0$ m. However, we can still use Eq. (3–36) to find the value of x when $y = -52$ m. This value of x is the desired range R'. Equation (3–36) becomes

$$y = -52 \text{ m} = R' \tan \theta_0 - R'^2 \frac{g}{2v_0^2 \cos^2 \theta_0}.$$

Rearrange this equation to find that

$$R'^2 - \frac{2v_0^2 \cos^2 \theta_0 \tan \theta_0}{g} R' + \frac{2yv_0^2 \cos^2 \theta_0}{g} = 0;$$

$$R'^2 + bR' + c = 0,$$

where $\quad b = -\dfrac{2v_0^2 \cos^2 \theta_0 \tan \theta_0}{g} \quad$ and $\quad c = \dfrac{2yv_0^2 \cos^2 \theta_0}{g}.$

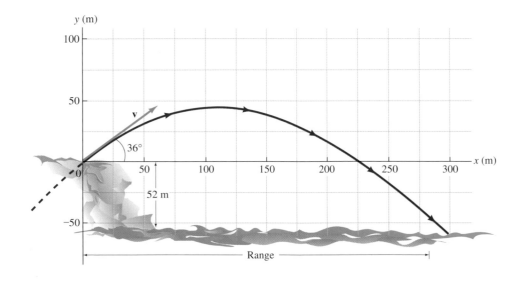

FIGURE 3–13 Examples 3–5 and 3–6.

Solve this quadratic equation to find R':

$$R' = \frac{-b \pm \sqrt{b^2 - 4c}}{2}.$$

Insert the values of b and c to obtain

$$R' = \frac{v_0^2 \cos^2 \theta_0 \tan \theta_0}{g} \pm \frac{1}{2} \sqrt{\frac{4v_0^4 \cos^4 \theta_0 \tan^2 \theta_0}{g^2} - \frac{8yv_0^2 \cos^2 \theta_0}{g}}.$$

Now insert the value of $y = -52$ m and the values of the constants to determine the range:

$$R' = \frac{(48 \text{ m/s})^2 (\cos 36°)^2 (\tan 36°)}{9.8 \text{ m/s}^2}$$

$$\pm \frac{1}{2} \sqrt{\frac{4(48 \text{ m/s})^4 (\cos 36°)^4 (\tan 36°)^2}{(9.8 \text{ m/s}^2)^2} - \frac{8(-52 \text{ m})(48 \text{ m/s})^2 (\cos 36°)^2}{9.8 \text{ m/s}^2}}$$

$$= 281 \text{ m or } -57 \text{ m}.$$

Now, did the golfer drive the ball a distance of 281 m or −57 m? The positive value must be correct. The negative result arises because it is a suitable solution to the quadratic equation that produces a parabola before $t = 0$ s (see the dashed line on the left in Fig. 3–13). The golfer wins his bet.

EXAMPLE 3–6 What was the maximum height above the ocean of the golf ball in Example 3–5, and how long was the golf ball in flight?

Solution: Let's refer again to Fig. 3–13. The maximum height of the golf ball occurs when the vertical component of the velocity is zero, and Eq. (3–42) gives the correct result, h, for the height above the original location. Notice, however, that the answer needed is $h + 52$ m, because we want to determine the maximum height *above the ocean*:

$$h = \frac{v_0^2 \sin^2 \theta_0}{2g} = \frac{(48 \text{ m/s})^2 (\sin 36°)^2}{2(9.8 \text{ m/s}^2)} = 41 \text{ m}.$$

The answer is 41 m + 52 m = 93 m.

We can find the total time T of the trip by using Eq. (3–31). We have already determined that the total horizontal distance traveled is 281 m. From Eq. (3–31):

$$281 \text{ m} = (48 \text{ m/s})(\cos 36°)T;$$

$$T = \frac{281 \text{ m}}{(48 \text{ m/s})(\cos 36°)} = 7.2 \text{ s}.$$

EXAMPLE 3–7 A group of engineering students constructs a slingshot device that lobs water balloons at a target. The device is constructed so that the launching speed is 12 m/s; the target is 14 m away at the same elevation on the other side of a fence. How can they accomplish their mission?

Solution: In this case, sketched in Fig. 3–14, the range equation for horizontal ground is relevant. The range varies with the initial angle, so the students need to find a value of θ_0 that will give a range of 14 m. Equation (3–39) applies here, and the known parameters are $R = 14$ m and $v_0 = 12$ m/s:

$$R = 14 \text{ m} = \frac{(12 \text{ m/s})^2 \sin(2\theta_0)}{9.8 \text{ m/s}^2};$$

$$\sin(2\theta_0) = 0.95.$$

FIGURE 3–14 Example 3–7. The students can orient their slingshot in two ways to get the same range for the same initial speed. One is shown here.

FIGURE 3–15 Two balls released simultaneously have two different trajectories. The difference lies in the (constant) *x*-component of velocity. But in a given time, each moves the same vertical distance.

FIGURE 3–16 Example 3–8.

FIGURE 3–17 A phonograph turntable provides an example of uniform circular motion.

An object following a circular path is accelerating, even when its speed is constant.

This equation has *two* solutions, $2\theta_0 = 72°$ and $2\theta_0 = 108°$, or $\theta_0 = 36°$ and $54°$. These are the two possible initial angles that result in a given range, as in Fig. 3–11. We have drawn one of these trajectories (Fig. 3–14). A reminder: There will always be two initial angles that generate the same range, except the 45° initial angle, which produces the longest range. Figure 3–15 shows a typical projectile trajectory.

EXAMPLE 3–8 A boy would rather shoot coconuts down from a tree than climb the tree or wait for the coconuts to drop. The boy aims his slingshot at a coconut, but just when his rock leaves the slingshot, the coconut falls from the tree. Show that the rock will hit the coconut.

Solution: Let's suppose that the rock starts out at the point $(x, y) = (0, 0)$ and that the coconut is at the point (x_0, y_0). The rock will have initial velocity (v_{x0}, v_{y0}). First let's see what would happen if there were no gravity acting. Without gravity, the properly aimed rock would follow a straight-line path that would place it at the point $(x_0 = v_{x0}t, y_0 = v_{y0}t)$ after a time t. This time is just the time necessary for the rock to reach the coconut, which is still at the tree since gravity has been ignored so far. Now let's factor in the effect of gravity. During the time t that the rock travels toward the coconut, the coconut falls the distance $gt^2/2$ (Fig. 3–16). In other words, the height of the coconut after time t is [Eq. (3–21)]

$$y = y_0 - \frac{1}{2}gt^2.$$

The rock's horizontal velocity component remains constant at v_{x0}. However, the vertical velocity component of the rock is changing under the effect of gravity and, after time t, Eq. (3–21) shows us that the rock's height is not $v_{y0}t$ but rather:

$$y = v_{y0}t - \frac{1}{2}gt^2.$$

The rock is a height $gt^2/2$ below the height it would have if it followed a straight-line motion, which is precisely the distance the coconut falls, as shown in Fig. 3–16. Thus the rock will hit the coconut at this point. This effect can be demonstrated in the lecture room or the laboratory. In effect, the parabolic path of the rock "tracks" the falling coconut.

3–5 CIRCULAR MOTION

The steady motion of any object on a spinning phonograph turntable is called **uniform circular motion** (Fig. 3–17). An object is experiencing uniform circular motion if it travels at a constant speed along a circular path. Perhaps surprisingly, this object is actually accelerating. Velocity is a vector quantity; although the object is not changing speed—the magnitude of the velocity remains the same—the *direction of its velocity is changing*. This means that an object undergoing uniform circular motion accelerates. The *x*- and *y*-motions are connected in uniform circular motion, so we will not want to separate these motions as we did for projectile motion.

An exciting application of simple parabolic motion in Earth's gravity is a NASA service that produces "zero gravity" for up to 20 seconds (Fig. 3AB–1). The NASA service is needed to test satellite and shuttle components for proper operation in a zero-gravity environment. Note that "zero gravity" does not mean that there is no force of gravity; rather, it means that there are none of the associated "contact forces" that otherwise act on an object (you normally feel such forces on your seat when you are in a chair, for example). Thus an object that falls in a vacuum (to eliminate air resistance) is said to be in an environment of zero gravity, as is an orbiting satellite. NASA achieves the zero-gravity effect with an airplane (suitably furnished for whatever experiment is desired) that is flown to a reasonable height and then allowed to dive to reach its maximum safe speed. At that point, control of the airplane is taken over by a computer system that provides an accurate parabolic flight path; the top of the flight path's parabola is limited by the lowest safe air speed. The motion of the airplane tracks the motion of a projectile under the influence of gravity, and so is that of an object in free fall. Anything

FIGURE 3AB-1 The interior of the airplane used by NASA for a zero-gravity environment, during that part of the flight where the plane follows the same parabolic path taken by a projectile in free fall. In this photograph, astronauts in training are experiencing some of the same effects they will feel during a shuttle flight.

inside the airplane also moves on a parabolic projectile trajectory, so it is as if the airplane were not present for the object: The object is in free fall.

Plane Polar Coordinates

The description of circular motion is very much simplified if we use **plane polar coordinates**. Figure 3–18 shows the two polar coordinates that locate a point in a plane, the radial coordinate ρ, and the angular coordinate ϕ. Conventionally, the angle ϕ is chosen to be measured from the $+x$-axis and to increase in the counterclockwise direction. We can measure it in *radians* (rad), defined by the relation 2π radians equals the circumference of any circle divided by the radius of that circle. If the angle ϕ is measured in radians, then the arc length intercepted by any circle is simply the product of the radius and the angle:

$$\text{arc length} = R\phi. \tag{3–43}$$

Notice that radians are dimensionless because they represent the ratio (arc length)/(radius); that is, the ratio of two lengths. Nevertheless, we will use radians as a unit. You may want to note that the relation $360° = 2\pi$ rad gives us the relation between degrees and radians: $1° \cong 0.0175$ rad and 1 rad $\cong 57.3°$.

For an object moving along a circle, the radial coordinate is fixed, $\rho = R$; thus, the motion is described by a single variable, the angle ϕ, which may have a time dependence $\phi(t)$. Suppose that during a time interval dt, the change in the angle is dϕ (Fig. 3–19). The arc length ds traversed during that time interval is given by

$$ds = R \, d\phi. \tag{3–44}$$

Division by the time interval dt gives us an expression for the speed of motion:

$$\frac{ds}{dt} = v = R \frac{d\phi}{dt}. \tag{3–45}$$

We are thus led to define the **angular speed**, ω, of the object moving along the circle as *the rate of change of the angle ϕ*, or

$$\omega \equiv \frac{d\phi}{dt}. \tag{3–46}$$

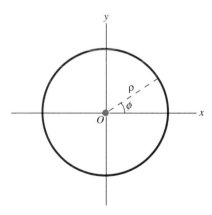

FIGURE 3–18 The radial variable ρ and angular variable ϕ of plane polar coordinates.

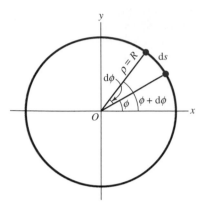

FIGURE 3–19 During the time interval d*t*, a particle in circular motion traverses an arc length d*s* and an angle dϕ.

The relations between period, frequency, and angular frequency

SI unit of frequency

FIGURE 3–20 (a) During time Δt, a particle in motion has changed its position vector from $\mathbf{r}(t)$ to $\mathbf{r}(t + \Delta t)$ and its velocity vector from $\mathbf{v}(t)$ to $\mathbf{v}(t + \Delta t)$. For uniform circular motion, the magnitude of $\mathbf{v}$ is constant; however, the direction is always perpendicular to $\mathbf{r}$ and is therefore changing continuously. (b) In order to form the vector difference $\Delta\mathbf{v} = \mathbf{v}(t + \Delta t) - \mathbf{v}(t)$, we translate the vector $\mathbf{v}(t + \Delta t)$ so that its tail meets the tail of the vector $\mathbf{v}(t)$. We do this at the midway point in the particle's path from time t to time $t + \Delta t$. In part (a), the angle between $\mathbf{r}(t)$ and $\mathbf{r}(t + \Delta t)$ is $\Delta\phi$, the same angle as that between $\mathbf{v}(t)$ and $\mathbf{v}(t + \Delta t)$. We see from the figure that the average change in velocity, which is proportional to $\mathbf{a}_{av}$, points toward the center of the circle.

In terms of ω, we have, for the object's speed, v:

$$v = R\omega. \tag{3–47}$$

The quantity ω has dimensions $[T^{-1}]$ and is measured in radians per second (rad/s) in SI.

Uniform Circular Motion

We say that circular motion is *uniform* when the speed v and the angular speed ω of an object moving along the circle are constant. However, if these two quantities are not constant—as in the case of an old phonograph turntable that varies in angular speed—then we refer to nonuniform circular motion. An important quantity that characterizes uniform circular motion is the time the moving object takes to make one complete revolution. This time is called the **period**, T. Because the distance traveled in one revolution is $2\pi R$, the period is given by

$$2\pi R = vT; \tag{3–48}$$

$$T = \frac{2\pi R}{v} = \frac{2\pi \cancel{R}}{\omega \cancel{R}} = \frac{2\pi}{\omega}. \tag{3–49}$$

The **frequency**, f, counts the number of revolutions that the particle makes per unit time. The frequency is the inverse of the period T:

$$f \equiv \frac{1}{T}; \tag{3–50}$$

and the relation between angular speed and frequency is

$$\omega = 2\pi f. \tag{3–51}$$

The SI unit of frequency is the hertz (Hz), defined as one cycle per second (cps). Another and perhaps more familiar unit, revolutions per minute (rev/min or rpm), is also useful; 60 rev/min = 1 Hz.

Acceleration in Uniform Circular Motion. Although the speed of an object in uniform circular motion is constant, the *direction* of the velocity changes, and thus the acceleration is not zero. Let's first consider the directional aspects of uniform circular motion. Specifically, we want to study the direction of the acceleration. We already know that acceleration is the change in velocity per unit time, so we first examine how the velocity changes. In Fig. 3–20a, we imagine a particle located at the point determined by angle ϕ at time t. At a later time $t + \Delta t$, the particle's angular position is given by $\phi + \Delta\phi$. The *direction* of the velocity vector is always tangential to the circle and therefore changes continuously with time. Figure 3–20a shows the velocity $\mathbf{v}(t)$ at time t. At time $t + \Delta t$, the velocity vector is $\mathbf{v}(t + \Delta t)$. These two vectors have the

(a)   (b)

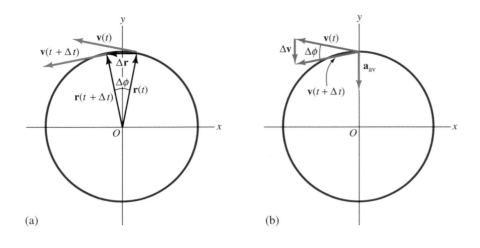

(a)   (b)

FIGURE 3–21 (a) The time
difference Δt is smaller than in Figs.
3–20a and 3–20b; so too are $\Delta\phi$, $\Delta\mathbf{r}$,
and $\Delta\mathbf{v}$. (b) The resulting $\mathbf{a}_{av}$ continues
to point to the center of the circle. In
the limit that $\Delta t \rightarrow 0$, $\mathbf{a}_{av}$ becomes the
instantaneous acceleration $\mathbf{a}(t)$. It
points to the center of the circle.

same magnitude; the difference between them lies in the fact that they point in different directions. The *change* in the velocity vector over the time period Δt is given by $\Delta\mathbf{v} = \mathbf{v}(t + \Delta t) - \mathbf{v}(t)$. This vector difference, which we have drawn to occur at the midway point (time $t + \Delta t/2$) is shown in Fig. 3–20b. We see that $\Delta\mathbf{v}$, and hence $\mathbf{a}_{av} = \Delta\mathbf{v}/\Delta t$, points toward the center of the circle. Let's now make the time interval Δt smaller, as in Fig. 3–21a. The points corresponding to ϕ and $\phi + \Delta\phi$ move closer together, and $\Delta\mathbf{v}$—or equivalently, $\mathbf{a}_{av}$—continues to point to the center of the circle (Fig. 3–21b). In the limit in which Δt goes to zero, the ratio $\Delta\mathbf{v}/\Delta t$ gives us the instantaneous acceleration. *The instantaneous acceleration points precisely to the center of the circle.*

To obtain the magnitude of the acceleration, notice that the angle between $\mathbf{v}$ and $\mathbf{v} + \Delta\mathbf{v}$ is the same as the angle between $\mathbf{r}$ and $\mathbf{r} + \Delta\mathbf{r}$, so they form similar triangles (see Fig. 3–20b). We then have

$$\frac{\Delta v}{v} = \frac{\Delta r}{r}. \tag{3–52}$$

We write this in the form $\Delta v = (v/r)\,\Delta r$, and, after dividing by Δt, we obtain

$$\frac{\Delta v}{\Delta t} = \frac{v}{r}\frac{\Delta r}{\Delta t}.$$

If we continue this process for smaller and smaller time intervals Δt, then $\Delta\phi$, $\Delta\mathbf{r}$, and $\Delta\mathbf{v}$ all become smaller and smaller. When we use $\Delta r = v\,\Delta t$, we are led automatically to the instantaneous acceleration:

$$a = \lim_{\Delta t \to 0}\frac{\Delta v}{\Delta t} = \frac{v}{r}\lim_{\Delta t \to 0}\frac{\Delta r}{\Delta t}$$

$$= \frac{v}{r}v = \frac{v^2}{r}. \tag{3–53}$$

An alternative form for the acceleration is expressed in terms of ω rather than v:

$$a = \frac{v^2}{r} = \frac{(\omega r)^2}{r}$$

$$= r\omega^2. \tag{3–54}$$

The acceleration of an object in uniform circular motion

As previously noted, the instantaneous acceleration of an object undergoing uniform circular motion is directed exactly inward toward the center of the circle. We can summarize this information with a unit vector in the radial direction, defined by

$$\hat{\mathbf{r}} \equiv \frac{\mathbf{r}}{r}, \tag{3–55}$$

The acceleration of an object in uniform circular motion is always directed toward the center of the circle.

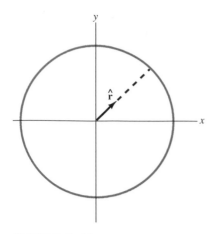

FIGURE 3–22 The unit vector $\hat{\mathbf{r}}$ has length one and points away from the origin.

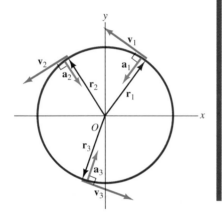

FIGURE 3–23 The instantaneous position $\mathbf{r}$, instantaneous velocity $\mathbf{v}$, and instantaneous acceleration $\mathbf{a}$ are shown for several positions around the path of a particle in uniform circular motion.

where the caret over the unit vector $\hat{\mathbf{r}}$ distinguishes this vector from the position vector $\mathbf{r}$. In contrast to the Cartesian unit vectors $\mathbf{i}$ and $\mathbf{j}$, which are fixed in space, the unit vector in the radial direction varies with the angle ϕ. Notice that the unit vector $\hat{\mathbf{r}}$ points outward from the origin (Fig. 3–22). In terms of this vector, the centripetal acceleration is

$$\mathbf{a} = -\frac{v^2}{r}\hat{\mathbf{r}}. \tag{3–56}$$

We should reemphasize the sign here: The unit vector $\hat{\mathbf{r}}$ points *away* from the origin, so the acceleration points *toward* the origin.

The relationship of $\mathbf{r}$, $\mathbf{v}$, and $\mathbf{a}$ is shown in Fig. 3–23 for various times during the particle's circular orbit. (Note that $\mathbf{a}$ is always perpendicular to $\mathbf{v}$.) All the vectors are constant in magnitude but *vary continuously in direction*. Quantities that point to the center of a circle, such as the acceleration in circular motion, are described as **centripetal**, meaning center-seeking.

EXAMPLE 3–9 A child's top spins uniformly at 16 Hz. What is the centripetal acceleration on the outside surface if the radius of the top is 3 cm?

Solution: The angular speed is determined from $\omega = 2\pi f$:

$$\omega = (2\pi \text{ rad})(16 \text{ Hz}) = 101 \text{ rad/s}.$$

The acceleration a has magnitude

$$a = \omega^2 r = (101 \text{ rad/s})^2(3 \text{ cm})\left(\frac{1 \text{ m}}{100 \text{ cm}}\right) = 306 \text{ m/s}^2.$$

Remember that the unit *radian* is a dimensionless quantity. It is a measure of angle and we use it in the measurement of angular speed. However, the unit is dropped in this example during the calculation of the acceleration; it would not be appropriate to write the answer for acceleration as $306 \text{ rad}^2 \cdot \text{m/s}^2$.

3–6 RELATIVE MOTION

When we drive with velocity $\mathbf{v}$ past a tree, we normally think of the tree as stationary and ourselves as moving. Yet this way of thinking can be reversed: If we measure the position of the tree with respect to ourselves, it is also perfectly correct to think of the tree moving past us with a velocity $-\mathbf{v}$. At the same time, someone picnicking in the shade of the tree would say that the tree is stationary; that is, the picnicker measures the tree to have zero velocity. It is all a question of the **frame of reference**, and it is necessary for us to be able to see how descriptions can vary according to the frame of reference of the observers. Although it is not quite so obvious in the example of the tree, we shall discuss situations in which the laws of physics are more easily understood in one frame of reference than in another. The laws of planetary orbits and those of electricity and magnetism are good examples of such situations that we'll be exploring in later chapters. Our aim here is to see how the observations of some simple situations made from different reference frames are related to one another.

Kinematic Quantities as Seen from Two Frames of Reference

An object may appear to have one motion to one observer and a different motion to a second observer, depending on how the two observers are moving with respect to one another. The two observers are said to have a **relative motion**. Figure 3–24 illustrates how the view of a scene depends on the observer. The position of an object—for example, the person marked as A—is measured differently by the two observers B and

(a) According to Observer *A* at rest

(b) According to Observer *B* at rest

(c) According to Observer *C* at rest

FIGURE 3–24 The scene in part (a) is observed from the point of view of (b) observer B and (c) observer C, who are in relative motion.

C, who are each in a different frame of reference. If we specify the relative motion of the observers, then we also specify the velocity at which one observer sees the other moving. For example, if observer A sees observer B moving with velocity **u**, then B sees A moving with velocity $-\mathbf{u}$ (in the opposite direction at the same speed). Now suppose that observer A sees object *P* moving with velocity $\mathbf{v}_A$ and observer B sees object *P* moving with velocity $\mathbf{v}_B$. Velocities $\mathbf{v}_A$ and $\mathbf{v}_B$ must be related through the relative velocity **u**. From Fig. 3–24, which refers to one-dimensional motion,

$$\mathbf{v}_B = \mathbf{v}_A - \mathbf{u}. \tag{3–57}$$

As the vector notation indicates, Eq. (3–57) holds even when we refer to more than one dimension. We can see this more formally as follows. Suppose the position of an object were measured from reference frame A as $\mathbf{r}_A$ and from frame B as $\mathbf{r}_B$. As Figure 3–25 shows, the relation is

$$\mathbf{r}_B = \mathbf{r}_A - \mathbf{R}, \tag{3–58}$$

where **R** is the position of the origin of frame B as measured in frame A. The relative movement of the two frames is specified according to

$$d\mathbf{R}/dt = \mathbf{u}. \tag{3–59}$$

Then, by measuring the time rate of change of Eq. (3–60), we see how the velocities in the two frames compare:

$$d\mathbf{r}_B/dt = d\mathbf{r}_A/dt - d\mathbf{R}/dt = d\mathbf{r}_A/dt - \mathbf{u}. \tag{3–60}$$

This is exactly Eq. (3–57).

Will our two observers agree on the *acceleration* of an object *P*? That depends on

How two observers each measure the velocity of an object

FIGURE 3–25 The position of an object can be measured by an observer in the A frame or by an observer in the B frame. The positions in the two frames can be related once the relative position of the two frames is known.

Will our two observers agree on the *acceleration* of an object *P*? That depends on whether the relative motion of the two observers has a constant velocity or not. If we differentiate Eq. (3–57) with respect to time, we obtain the relation between the two accelerations of the object as seen by the two observers. We have

$$\frac{d\mathbf{v}_B}{dt} = \frac{d\mathbf{v}_A}{dt} - \frac{d\mathbf{u}}{dt},$$

or

How two observers measure the acceleration of an object

$$\mathbf{a}_B = \mathbf{a}_A - \frac{d\mathbf{u}}{dt}, \tag{3–61}$$

where $\mathbf{a}_A$ and $\mathbf{a}_B$ are the accelerations of the object as seen by observers A and B, respectively. If $\mathbf{u}$ is a constant—and thus its time derivative is zero—then the two observers agree on the object's acceleration if not on its velocity. However, it is not always the case that $\mathbf{u}$ is a constant. Consider a tricycle standing motionless relative to a woman we will call observer A, while her child (observer B) spins on a nearby merry-go-round (Fig. 3–26). The child observes the tricycle to be moving—but not in a straight line. Motion in which the direction and/or the speed of an object changes is accelerated motion. Because $\mathbf{v}_A$ (the velocity of the tricycle as measured by the mother) is zero at all times, so too is $\mathbf{a}_A$. Equation (3–61) shows that the child would measure the acceleration of the tricycle to be

$$\mathbf{a}_B = - \frac{d\mathbf{u}}{dt}, \tag{3–62}$$

where $\mathbf{u}$ is the velocity that the mother measures her child to have on the merry-go-round; this velocity is not constant, so its time derivative is not zero.

EXAMPLE 3–10 A boat must cross a river that is 150 m wide. The river has a current of 3 km/h, and the boat can be rowed through the water with a uniform speed of 4 km/h (Fig. 3–27a). Set up two coordinate systems in which to describe the displacement of the boat: one fixed to the bank and the other fixed to a spot moving with the current of the river. Using these coordinate systems, express the position vector of the boat at time *t*; assume that the boat leaves the dock at the angle θ with respect to a point moving with the water, as shown in Fig. 3–27b. Calculate θ such that the boat lands at a point exactly opposite the starting point. How long will the trip take?

Solution: The two frames of reference in this case refer to a frame whose origin, *O*, is fixed on the dock on the bank of the river (Fig. 3–27a), and a frame whose

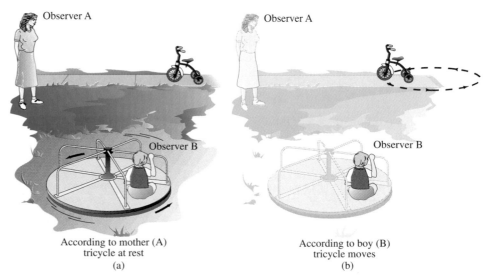

FIGURE 3–26 A tricycle and a mother are at rest relative to each other. A child is sitting on a rotating merry-go-round. The mother and child, observers A and B, respectively, are not moving in steady straight-line motion with respect to one another; they measure the tricycle as having different accelerations. (a) The mother sees the tricycle at rest. (b) From the point of view of a frame at rest relative to the child, the tricycle moves in circles—hence it accelerates.

Observer A

Observer A

Observer B

Observer B

According to mother (A)
tricycle at rest
(a)

According to boy (B)
tricycle moves
(b)

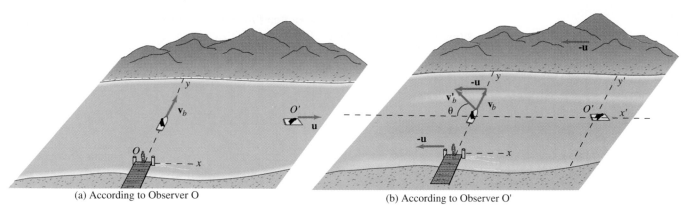

(a) According to Observer O (b) According to Observer O'

FIGURE 3–27 Example 3–10. (a) As seen by an observer O standing on the dock, the boat moves straight across the river. (b) As seen by an observer O′ moving with the water, the boat moves at an angle θ.

origin, O', is fixed to a hypothetical spot in the water—a raft that moves with the current (Fig. 3–27b). In each frame of reference, we set the x-direction in the direction of the current, with the y-direction pointing directly across the river. The origin O' is seen by the observer on the fixed dock to move with the velocity

$$\mathbf{u} = u\mathbf{i},$$

where u is the speed of the current (3 km/h).

If the boat can move at speed $v = 4$ km/h with respect to the water, then, as observed in the frame moving with the water, the boat moves with velocity

$$\mathbf{v'}_B = -v\cos\theta\,\mathbf{i} + v\sin\theta\,\mathbf{j}.$$

As long as the rower does not look at the bank and fixes his vision on a spot in the water, he sees himself moving with speed v at the angle θ through the water. An observer on the dock, however, sees the boat move with velocity

$$\mathbf{v}_B = \mathbf{v'}_B + \mathbf{u} = (-v\cos\theta + u)\mathbf{i} + v\sin\theta\,\mathbf{j},$$

which is consistent with Eq. (3–57). All velocities are constant, so the position of the boat as seen by the observer on the dock is

$$\mathbf{r}_A = (-v\cos\theta + u)t\,\mathbf{i} + vt\sin\theta\,\mathbf{j}.$$

If angle θ is chosen such that the dockside observer sees no horizontal velocity (x-component of velocity), he will see the boat moving straight across the water to the point on the opposite shore. This occurs when the angle is given by

$$\cos\theta = \frac{u}{v} = \frac{3\ \cancel{\text{km/h}}}{4\ \cancel{\text{km/h}}} = 0.75;$$

$$\theta = 41°.$$

The boat is then seen by the dockside observer to move uniformly across the river with speed $v\sin\theta$, and the trip takes a time

$$t = \frac{y}{v\sin\theta} = \frac{0.15\ \cancel{\text{km}}}{(4\ \cancel{\text{km/h}})(\sin 41°)} = 0.06\ \text{h} \simeq 3\ \text{min}.$$

With no river current, the boat could be rowed directly across the water; in this case, the time to cross would be (0.15 km)/(4 km/h) = 0.04 h $\simeq$ 2 min.

SUMMARY

A particle moving in space follows a trajectory, or path. In three dimensions, the position vector of such a particle is

$$\mathbf{r} = x(t)\mathbf{i} + y(t)\mathbf{j} + z(t)\mathbf{k}. \qquad (3\text{–}3)$$

The displacement vector between times t and $t + \Delta t$ is the difference between the position vectors at these times. It is given by

$$\Delta \mathbf{r} = \mathbf{r}(t + \Delta t) - \mathbf{r}(t). \tag{3-1}$$

The instantaneous velocity $\mathbf{v}$ is found from the displacement vector $\Delta \mathbf{r}$ over small time intervals:

$$\mathbf{v} \equiv \lim_{\Delta t \to 0} \frac{\mathbf{r}(t + \Delta t) - \mathbf{r}(t)}{\Delta t} = \frac{d\mathbf{r}}{dt}. \tag{3-5}$$

In two dimensions, the velocity is expressed in terms of its component vectors as

$$\mathbf{v} = \mathbf{v}_x + \mathbf{v}_y, \tag{3-9}$$

where, in terms of unit vectors,

$$\mathbf{v}_x = \frac{dx}{dt}\,\mathbf{i}, \tag{3-11a}$$

$$\mathbf{v}_y = \frac{dy}{dt}\,\mathbf{j}. \tag{3-11b}$$

In terms of its components, the magnitude of the velocity $\mathbf{v}$ is

$$v = \left| \mathbf{v} \right| = \sqrt{v_x^2 + v_y^2}, \tag{3-12}$$

Further, the angle θ that $\mathbf{v}$ makes with the x-axis is given by

$$\tan \theta = \frac{v_y}{v_x}. \tag{3-13}$$

The instantaneous acceleration is a derivative of the velocity:

$$\mathbf{a} \equiv \lim_{\Delta t \to 0} \frac{\mathbf{v}(t + \Delta t) - \mathbf{v}(t)}{\Delta t} = \frac{d\mathbf{v}}{dt}. \tag{3-15}$$

Like velocity, acceleration can also be expressed in vector components. The acceleration has components a_x and a_y, which are derivatives of the x- and y-components of the velocity.

For constant acceleration in the xy-plane, we have

$$x = x_0 + v_{0x}t + \frac{1}{2} a_x t^2, \tag{3-19}$$

$$v_x = v_{0x} + a_x t; \tag{3-20}$$

$$y = y_0 + v_{0y}t + \frac{1}{2} a_y t^2, \tag{3-21}$$

$$v_y = v_{0y} + a_y t, \tag{3-22}$$

which can be written more concisely as

$$\mathbf{r} = \mathbf{r}_0 + \mathbf{v}_0 t + \frac{1}{2} \mathbf{a} t^2, \tag{3-25}$$

$$\mathbf{v} = \mathbf{v}_0 + \mathbf{a} t. \tag{3-26}$$

In the absence of air resistance, a projectile moves under the influence of gravity with a constant acceleration vector $\mathbf{a} = \mathbf{g}$. The trajectory of such a projectile is a parabola. The range (the horizontal distance a projectile launched from the ground travels over level ground) and maximum height of the trajectory can be calculated, as can the projectile's flight time.

Circular motion is most simply described by means of plane polar coordinates, with ρ as the radial coordinate and ϕ as the angular coordinate. Angles are measured in radians (which are dimensionless), so that the arc length formed by the angle ϕ in a circle of radius R is

$$\text{arc length} = R\phi. \tag{3-43}$$

The angular speed of an object in circular motion is

$$\omega \equiv \frac{d\phi}{dt}. \tag{3-46}$$

The speed of a particle moving in circular motion with angular speed ω is

$$v = R\omega. \tag{3-47}$$

The period T is the time to make one complete revolution, and the frequency f is the inverse of T. The relation between angular speed and frequency is

$$\omega = 2\pi f. \tag{3-51}$$

The velocity of a particle in uniform (or constant) circular motion is tangent to the circle. The acceleration of this particle has magnitude

$$a = \frac{dv}{dt} = \frac{v^2}{r} = r\omega^2. \tag{3-53, 3-54}$$

The direction of the particle's acceleration is centripetal; that is, it points inward along a radius toward the center of the circle.

Suppose that two observers, A and B, are in relative motion. Observer A measures the velocity of observer B to be $\mathbf{u}$, the relative velocity. Then, if the velocities of an object as measured by observers A and B are $\mathbf{v}_A$ and $\mathbf{v}_B$, respectively, these velocities are related by

$$\mathbf{v}_B = \mathbf{v}_A - \mathbf{u}. \tag{3-57}$$

The acceleration of the object according to the two observers obeys the relation

$$\mathbf{a}_B = \mathbf{a}_A - \frac{d\mathbf{u}}{dt}. \tag{3-61}$$

UNDERSTANDING KEY CONCEPTS

1. What factors determine how far an athlete travels in the long jump? Which factor do you think must be the most important?

2. Why is it important to have limits on wind speed for world-record track and field events?

3. Under what conditions is it possible to have a constant speed yet a nonzero acceleration?

4. You wish to row a boat to a point directly opposite your starting point on the bank of a swiftly moving river. The best speed you can make through the water—meaning with respect to the water—is 2 mi/h. The downstream current of the river is also 2 mi/h. Can you reach your desired goal in a finite time without doing some walking? Explain your answer.

5. There are two initial angles for which the range of a projectile is the same (except in the instance where the initial angle is 45°). What are the differences in the two trajectories that might lead a football quarterback to choose one of them over the other in throwing a pass? If you are familiar with football, you will know that the quarterback usually chooses one of these two trajectories.

6. A ball is thrown straight up and comes straight down. What is the value of its velocity at the top of its flight? the value of its acceleration?

7. Your lecture demonstrations may include a car that, even while moving, shoots a ball vertically off a platform on the car. If the car is moving with constant speed on a horizontal surface, the ball ends up at the same position on the car that it left. What would happen if the car were decelerating as it released the ball?

8. Suppose that you measure the period for successive rotations of a phonograph turntable as being unchanging. Is this enough to assure you that the turntable has uniform circular motion?

9. With a given initial speed, there are two initial firing angles for which a projectile has the same range. How many initial velocities give the same maximum height? The same time of flight?

10. You are running in the rain. At what angle should you hold your umbrella for the most protection? Assume that no wind is blowing, so that the raindrops fall straight down. You may need to assume some value for the raindrops' speed as well as your own.

11. You are on the rotating platform of a merry-go-round and throw a ball straight up. Describe the motion of the ball as you see it.

12. Earth is a rotating platform because it turns on an axis. If a projectile initially aimed at a particular point on the equator is shot from the North Pole, will it arrive at that point, assuming that it could travel such a great distance? Describe the motion of the projectile as seen by someone on Earth.

13. Suppose a projectile is fired from Washington, D.C. toward the equator. How does this situation differ from the one discussed in Question 12?

14. A motorcyclist rides on a large, rotating platform in an amusement park. The cyclist starts at the center of the rotating platform and travels in a radial direction with a constant speed. Is it correct to say that, when he or she has traveled some distance from the platform center, the acceleration due to the circular motion (which is directed in the inward radial direction) will stop the motorcycle? Analyze the motion.

3–1 Position and Displacement

1. (I) A car travels 21 km to the northeast, then 15 km to the east, before it travels 28 km to the north. Express the position vector from where the car starts to the point at which each turn occurs. What is the car's total displacement?

2. (I) A particle is located by the position described by the vector $\mathbf{r} = (c_1 - c_2 t)\mathbf{i} + (d_1 + d_2 t + d_3 t^2)\mathbf{j}$, where $c_1 = 12$ m, $c_2 = 3$ m/s, $d_1 = -20$ m, $d_2 = -3$ m/s, and $d_3 = 0.5$ m/s². At what time(s) does the particle pass through the position $x = 0$ m? At what time(s), and where, does the particle cross the line $x = y$? Sketch the particle's trajectory from $t = -8$ s to $t = 10$ s.

3. (I) A gym teacher organizes a series of indoor races, which follow along the walls of the gym; the race starts from corner A, continues to corner B, and so forth. The gym is a rectangle with the distance $AB = 25$ m and the distance $BC = 35$ m. Suppose that the origin of a coordinate system is at point A, leg AB is in the $+x$-direction, and leg BC is in the $+y$-direction. Express the position vector of a running student at each of the four corners.

4. (I) A runner races with a uniform speed of 27.0 km/h around a circular track of radius 172 m. Draw the track and the runner's displacement vector after 20 s, 40 s, 60 s, and 2 min, assuming that at $t = 0$ s the runner is at the three o'clock position. Assume counterclockwise motion.

5. (I) A treasure map locates the site of a treasure by reference to two starting points, A and B. A is chosen to be the origin, and B is at the point $(2.5\ \text{km})\mathbf{i}$. The instructions state that the treasure lies at the intersection of two lines. One line starts at A and passes through the point $(2.0\ \text{km})\mathbf{i} + (4.0\ \text{km})\mathbf{j}$; the other line, starting at B, passes through the point $(6.0\ \text{km})\mathbf{i} - (8.0\ \text{km})\mathbf{j}$. Sketch the instructions and find the location of the intersection graphically as well as algebraically. Express the vector that gives the displacement of the intersection point from point C, whose location relative to the coordinate system specified by A and B is given by $(1.2\ \text{km})\mathbf{i} - (2.2\ \text{km})\mathbf{j}$.

6. (II) The position of a particle in a given coordinate system is $\mathbf{r}(t) = (5 - 2t^2)\mathbf{i} + (-4 + 3t)\mathbf{j}$ where the distances are in meters when t is in seconds. At what time will the particle cross the y-axis? At what time will it cross the x-axis? Can you find an equation that relates the y-coordinate to the x-coordinate and therefore gives the trajectory in the xy-plane? Where would the x- and y-axes have to be moved so that at $t = 0$ s the trajectory passes through the origin?

7. (II) Figure 3–28 shows the trajectory of a bicyclist traveling with a constant speed of 25.2 km/h. Express the position vector of the bicyclist as a function of time.

8. (II) The x- and y-coordinates of the position of a land surveyor are shown on Fig. 3–29 as a function of time. (a) Plot the trajectory of the surveyor. (b) Calculate the area of the plot that she surveyed.

9. (II) The position of a particle is given in a certain coordinate system by the vector $\mathbf{r}(t) = (4\ \text{m})\cos(\pi t/T)\mathbf{i} - (4\ \text{m})\sin(\pi t/T)\mathbf{j}$. Find the displacement vector at times $t = (T/3)$, $t = (T/2)$, and $t = 2T$. In each case, compute the distance to the origin. What is the angle that the position vector makes with the $+x$-axis for arbitrary t?

FIGURE 3–28 Problem 7.

FIGURE 3–29 Problem 8.

10. (III) In a "shoot the coconut" lecture demonstration, the position of the coconut is given by the vector $(h_0 - \frac{1}{2}gt^2)\mathbf{j}$, whereas that of the projectile, aimed at the coconut at time $t = 0$, is given by $(-L + ut)\mathbf{i} + [(h_0 ut/L) - (\frac{1}{2}gt^2)]\mathbf{j}$. Show that the two will always collide, and find the time at which this takes place (Fig. 3–30). Express the displacement vector of the coconut relative to the projectile.

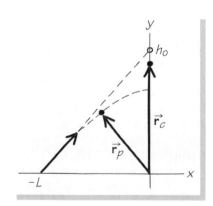

FIGURE 3–30 Problem 10.

3–2 Velocity and Acceleration

11. (I) A tennis ball leaves a racquet with an initial speed of 15 m/s. After 2, it has slowed down to 13 m/s. What is the mag-

nitude of the average acceleration? What would you need to know to determine the direction of the acceleration?

12. (I) A particle is observed to move with the coordinates $x(t) = 4\text{ m} + (3\text{ m/s})t + (1\text{ m/s}^2)t^2$ and $y(t) = 6\text{ m} - (4\text{ m/s})t + (0.5\text{ m/s}^2)t^2$. What are the particle's position, velocity, and acceleration? At what time are the velocity's horizontal and vertical components equal?

13. (I) At a given moment, a fly moving through the air has a velocity vector that changes with time according to $v_x = 2.2$ m/s, $v_y = (3.7\text{ m/s}^2)t$ and $v_z = (-1.2\text{ m/s}^3)t^2 + 3.3$ m/s, where t is measured in seconds. What is the fly's acceleration?

14. (I) A particle moves in such a way that its coordinates are

$$x(t) = A\cos\omega t;\ y(t) = A\sin\omega t$$

Calculate the x- and y-components of the velocity and the acceleration of the particle.

15. (II) A whale traveling southwest at 7.0 km/h is spotted 5.0 km to the northwest off the coast of Malibu. Photographers jump into a boat that can move at 30 km/h. With what velocity will the photographers intercept the whale, assuming that their boat travels a straight-line path? What is the position vector of the whale from the original point on the coast when the photographers reach the whale?

16. (II) A lifeguard standing on a tower throws a buoy to a swimmer 20 m from the tower (Fig. 3–31). The lifeguard, positioned 3 m above the water, pulls in the rope at a speed of 1 m/s. How fast is the swimmer coming to the shore when he is (a) 15 m and (b) 5 m from the water's edge?

FIGURE 3–31 Problem 16.

17. (II) An engineering student holds her open compass perpendicular to the drafting board, touching the board with both tips of the compass (Fig. 3–32). She slowly closes the compass so

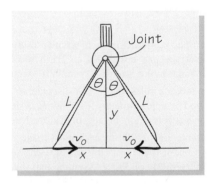

FIGURE 3–32 Problem 17.

that the tips move toward each other with a speed of $2v_0 = 0.06$ m/s. Initially, the angle between the arms of the compass is $2\theta = 60°$. The arms of the compass are $L = 15$ cm long. (a) Calculate the velocity of the joint of the compass as a function of time. (b) Give the velocity of the joint at the moment that she starts to close the compass, and at the moment that the two tips reach each other.

18. (II) Calculate the velocity vectors for the coconut and projectile in Problem 10. Express the difference between the two velocity vectors (the relative velocity). What is the magnitude of the relative velocity? What is the angle that its direction makes with the $+x$-axis?

19. (II) Calculate the velocity vector for the particle described by the position vector given in Problem 9. Calculate the angle that the velocity vector makes with the $+x$-axis, and show that the velocity vector is always perpendicular to the position vector.

20. (II) A particle moving with an initial velocity $\mathbf{v} = (50\text{ m/s})\mathbf{j}$ undergoes an acceleration $\mathbf{a} = [35\text{ m/s}^2 + (2\text{ m/s}^5)t^3]\mathbf{i} + [4\text{ m/s}^2 - (1\text{ m/s}^4)t^2]\mathbf{j}$. What are the particle's position and velocity after 3 s, assuming that it starts at the origin?

21. (II) The motion of a planet about a star may be described by the vector $\mathbf{r}_1 = R\cos(2\pi ft)\mathbf{i} + R\sin(2\pi ft)\mathbf{j}$. The position vector of another planet about the same star is given by $\mathbf{r}_2 = 4R\cos(\pi ft/4)\mathbf{i} + 4R\sin(\pi ft/4)\mathbf{j}$. Show that both of these describe circular motion with constant speed. Calculate the position vector of the second planet relative to the first planet. Sketch the path by noting that one circular motion is eight times faster than the other. This is like the motion of a planet whose year is 8 Earth years.

22. (II) Calculate the acceleration vectors for the two planets in Problem 21. Express the vector that describes the acceleration of the second planet relative to the first one.

23. (II) A bag is dropped from a hot-air balloon. Its height is given by the formula $h = H - ut - (u/B)e^{-Bt}$. What are the dimensions of B? What is the initial velocity? What is the velocity as $t \to \infty$? Calculate the accelerations at $t = 0$ and at $t = \infty$.

3–3 Motion with Constant Acceleration

24. (I) An airplane is flying due south on a level course at a speed of 400 km/h. At an altitude of 2.50×10^3 m directly above a hospital, the airplane meets severe turbulence and descends with a vertical acceleration of 5.00 m/s² for 8 s. (a) What is the total displacement of the plane in the 8 s? (b) What is the plane's velocity at the end of the 8 s? (c) What is the plane's position vector with respect to the hospital at the end of the 8 s?

25. (I) A launching mechanism accelerates a baseball horizontally at 24 m/s² for 0.5 s. The baseball's initial velocity is zero. With what velocity will it leave the launcher? The baseball leaves the launcher at the same speed when the launcher is turned in the vertical direction. How high will the baseball go?

26. (I) A gymnast works out on a trampoline. At the instant that she leaves the trampoline, a point on her waist is 1.1 m above the floor and at the center of the trampoline. At that instant, the point has an upward velocity of 7.8 m/s and a horizontal velocity of 3.0 m/s. Write equations that describe the subsequent motion of that point, and find its maximum height.

27. (II) A boy shoots a rock with an initial velocity of 21 m/s straight up from his slingshot. He quickly reloads and shoots

another rock in the same way 3.0 s later. (a) At what time and (b) at what height do the rocks meet? (c) What is the velocity of each rock when they meet?

28. (II) A man in the crow's nest of a sailing ship moving through smooth seas at a steady 12 km/h accidentally lets a cannonball drop from his station, which is 8.5 m above the deck at the top of the mainmast. (a) Assuming that he dropped the ball from a position immediately adjacent to the vertical mast, where does the ball land with respect to the mast? (b) How long does it take for the ball to fall to the deck? (c) In the time it takes the ball to fall, how far has the ball moved with respect to an observer fixed on the shore?

3–4 Projectile Motion

29. (I) A runner attempting a broad jump leaves the ground with a horizontal velocity of magnitude 9.0 m/s. Assuming the horizontal component of velocity is unaffected, what vertical component of velocity must the runner acquire to jump 9.5 m?

30. (I) An engineering student wants to throw a ball out a third-story dorm window (10 m off the ground) onto a target on the ground placed 8.0 m away from the building. (a) If the student throws the ball horizontally, with what velocity must it be thrown? (b) What must the velocity of the ball be if it is thrown up at an elevation angle of 29°? (c) What is the ball's time of flight in case (b)?

31. (I) A projectile is shot at an angle of 34° to the horizontal with an initial speed of 225 m/s. What is the speed at the maximum height of the trajectory?

32. (I) At what points in a projectile's trajectory above level ground is the magnitude of the velocity a maximum and a minimum? What are these velocities in terms of the initial speed v_0 and elevation angle θ_0?

33. (I) A projectile is shot at an angle of 40° to the horizontal over level ground. Assuming air resistance plays no role, what angle does the projectile make with the horizontal when it lands?

34. (I) Find the initial angle if the range of a projectile is twice its maximum height.

35. (II) (a) Show that the range R can be expressed in terms of the maximum height h, and in particular that $R = 4h \cot \theta_0$. (b) Show that, when the range is a maximum, $h = R/4$.

36. (II) Galileo throws a rock from the top of the Leaning Tower of Pisa at an upward angle of 45° with speed v_0. The rock is in flight for 4.0 s and hits the ground 20 m from the base of the building. Ignore air resistance and ignore the fact that the tower tilts a bit. (a) What is the speed v_0? (b) How high off the ground is the top of the tower? (c) What is the speed of the rock just before it hits the ground?

37. (II) In the Battle of Hastings in A.D. 1066, during which the Normans of France defeated the Saxons in England, an important role was played by Norman archers who shot arrows over a wall of shields erected by the Saxons. If the Norman bows had a maximum range of 350 m and the arrows were shot at an elevation of 55°, how close were the Normans to the Saxons? Assume that the arrows reached their target.

38. (II) A punter kicks a football during a critical football game. The ball leaves his foot from ground level with a speed of 28 m/s at an angle of 50° to the horizontal. At the very top of its flight, the ball hits a wandering seagull. The ball and the seagull each stop dead and fall vertically from the point of collision (Fig. 3–33). In the following, ignore air resistance.

FIGURE 3–33 Problem 38.

(a) With what speed is the ball moving when it strikes the seagull? (b) How high was the unfortunate seagull when it met the ball? (c) What is the speed of the seagull when it hits the ground?

39. (II) A place kicker attempts an extra point, giving the ball an initial velocity of 15 m/s at an angle of 37° with the field. The uprights are 15 m from the point at which the ball is kicked, and the horizontal bar is 4.0 m from the ground. (a) At what time after the kick will the ball pass the goal posts? (b) Is the kick successful, and by how many meters does the ball clear or pass beneath the bar?

40. (II) Astronaut John Q. Hero had the good fortune to play golf on the Moon. The acceleration of gravity on the surface of the Moon is only about one-sixth of that on the surface of Earth. Assuming that Hero was not noticeably hampered by his space suit, and assuming that his best drive on Earth (unhampered by air resistance) is some 210 yd, how far can he drive a ball on the Moon? Derive a general answer to this question for planet X if g_X is a known fraction of g.

41. (II) A projectile is launched over flat ground and the effects of air resistance are minimal. At what angles with respect to the ground should the launcher be oriented so that the projectile's range is half its maximum range? Why are there two possible angles? What are the angles so that the range is zero; that is, the projectile lands at the foot of the launcher?

42. (II) You must throw a baseball to hit a target on the ground 50 m from the base of a building that is 20 m in height. You are standing at a point on the edge of the roof nearest the target. (a) With what velocity must you throw the baseball if it is to leave the hand horizontally? (b) With what velocity must you throw the baseball if it is to leave the hand at an angle of 45° up from the horizontal? (c) What is the horizontal component of the initial value of the velocity in case (b)?

3–5 Circular Motion

43. (I) The Space Shuttle is in a circular orbit 220 km above Earth's surface and completes an Earth revolution every 89 min. (a) What is the Shuttle's speed? (b) Acceleration?

44. (I) The Moon circles Earth at a distance of 3.84×10^5 km. The period is approximately 28 d. What is the magnitude of the moon's acceleration, in units of g, as the Moon orbits Earth?

45. (I) A runner in the 200-m dash must make part of the dash around a curve that forms the arc of a circle. This arc has a radius of curvature of 30 m. Assuming that she runs at a steady speed and completes the 200 m in 24.7 s, what is her centripetal acceleration while she is running the curve?

46. (I) A rock placed in a plastic bag is tied to a rope 1.2 m long. The rock is whirled in a horizontal circle. (a) What is the rock's centripetal acceleration if the period of motion is 1.8 s? (b) The plastic bag will break if the radial acceleration exceeds 56 m/s^2. With what speed must the rock be whirled if the plastic bag is to be broken?

47. (I) A passenger on the outer edge of a merry-go-round, 7.5 m from the central pivot, learns that when the merry-go-round is in steady motion, his centripetal acceleration is 3.3 m/s^2. How long does it take to make one revolution?

48. (II) The shaft of the engine of a car rotates at 4000 rev/min. A flywheel, 20 cm in diameter, rotates with the shaft. Calculate the centripetal acceleration of a point on the rim of the flywheel, and express it in units of g, the acceleration of gravity.

49. (II) Safety requires that the centripetal acceleration of cars traveling along highway curves may not exceed one-tenth of the acceleration of gravity, even when traveling at the posted speed limit of 55 mi/h. How small can the radius of curvature of a curve be?

50. (II) Suppose that a point object is in uniform circular motion, moving steadily at a distance R from some central point. The time for one revolution is T. Use dimensional analysis to find the dependence of the centripetal acceleration on T and R. Compare this result to the acceleration derived from a detailed analysis of uniform circular motion performed in Section 3–6.

51. (II) A mass is tethered to a post and moves in a circular path of radius $r = 0.5$ m on an air table—friction free—at a constant speed $v = 6.3$ m/s. We employ the coordinate system shown in Fig. 3–34. (a) If at $t = 0$ s, the mass is at $\theta = 0°$, what are the coordinates (x, y) of the mass at $t = \frac{1}{24}$ s? (b) What is the acceleration vector of the mass at $t = 0$ s? (c) What is the acceleration vector of the mass when $\theta = 90°$?

FIGURE 3–34 Problem 51.

52. (II) The Space Shuttle is moving in a circular orbit with a speed of 7.8 km/s and a period of 87 min. In order to return to Earth, the Shuttle fires its retro engines opposite to its direction of motion. The engines provide a deceleration of 6 m/s^2 that is constant in magnitude and direction. What is the total acceleration of the Shuttle?

53. (II) An electron in a research apparatus follows a circular path. On the electron's first circuit of the apparatus, its speed is v_0, and the radius of its circular path is R. Each time it makes one circuit, it passes a short region where it receives a "kick" and gains an additional speed of $v_0/100$. The electron follows a circular path such that the magnitude of its acceleration is always the same. What is the radius of the circular path after the electron has received 10 kicks?

54. (II) An automobile moves on a circular track of radius 1.00 km. It starts from rest from the point $(x, y) = (1.00$ km, 0 km) and moves counterclockwise with a steady *tangential* acceleration such that it returns to the starting point with a speed of 30.0 m/s after one lap. (The origin of the Cartesian coordinate system is at the center of the circular track.) What is the car's velocity (magnitude and direction) when it is one-eighth of the way around the track? Express the position and velocity at this point in terms of the unit vectors along the x- and y-axes.

3–6 Relative Motion

55. (I) A sailor wants to travel due east from Miami at a velocity of 15 km/h with respect to a coordinate system fixed on land. The sailor must contend with the Gulf Stream, which moves north at 5 km/h. With what velocity with respect to the water should the sailboat proceed under sail?

56. (I) During an uphill portion of a bicycle race, a cyclist reads a message on a board informing him that the leader is 15 s ahead and that the leader is traveling at 4 m/s. The cyclist's speedometer informs him that he is traveling at 3 m/s. (a) What is the speed of the leader with respect to the cyclist? (b) How far in front of the cyclist is the leader, assuming that the speeds have not changed in the last half minute or so?

57. (I) Rain is falling steadily, but there is no wind. You are in an automobile that moves at 80 km/h, and you see from the drops on a side window that the rain makes streaks at a 58° angle with respect to the vertical. What is the vertical velocity of the raindrops?

58. (II) A cyclist's top speed on a flat road is v. This is an "air speed" because the limiting speed for a cyclist is determined by the wind resistance. In other words, this is her top speed with respect to the air. She cycles a flat course straight north for a distance L, turns around, and cycles straight south for the same distance. In the following, ignore the time it takes her to turn around, and assume that the cyclist can maintain her top speed with respect to the air. (a) Write a formula for the total course time t_0 in terms of L and v. (b) There is a north wind blowing at speed v_w. Write a new formula for the course time t_1 of the cyclist, including the effect of v_w. (c) Show that for $v_w \ll v$, the course time can be approximated by $t_1 = t_0 [1 + (v_w^2/v^2)]$. To show this result you may want to use the approximation $(1 - x)^{-1} \simeq 1 + x$ for $x \ll 1$. (d) Plot the time t_1 as a function of v_w and show that it is always greater than t_0. What happens at $v_w = v$, and why?

59. (II) An airplane flies due south with respect to the ground at an air speed of 900 km/h for 2 h before turning and moving southwest with respect to the ground for 3 h. During the entire trip, a wind blows in the easterly direction at 120 km/h. (a) What is the plane's average speed with respect to the ground? (b) What is the plane's average velocity with respect to the ground? (c) What is the final position vector?

60. (II) An airplane is to fly due north from New Orleans to St. Louis, a distance of 673 mi. On that day and at the altitude of the flight, a wind blows from the west at a steady speed of 65 mi/h. The airplane can maintain an air speed of 180 mi/h. Ignore the periods of takeoff and landing. (a) In what direction must the airplane fly in order to arrive at St. Louis without changing direction? Draw a diagram and label this direction with an angle. Would this calculation change if the distance between the cities were twice as great? (b) What is the flying time for this flight? (c) Recalculate the flying time if the airplane heads due north until it reaches the latitude of St. Louis, and then flies due west, into the wind to reach the city.

61. (III) Earth has a radius of 6.4×10^6 m, and its orbit around the Sun has a radius of some 1.5×10^{11} m. Earth simultaneously rotates about its own axis and moves around the Sun (Fig. 3–35). Assume a circular orbit and that Earth's axis of rotation is perpendicular to its orbital plane. (a) What is the speed with respect to the Sun of the point on Earth's equator nearest the Sun? (b) of the point on Earth's equator farthest from the Sun? (c) of the two points on Earth's equator, midway between the points in parts (a) and (b)?

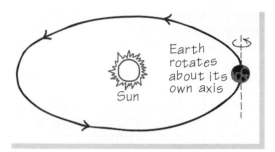

FIGURE 3–35 Problem 61.

General Problems

62. (I) Neutron stars are incredibly dense objects, compared to the Sun. One such (hypothetical) star has a radius of 40 km and rotates with a period of 26 s. What are the velocity and acceleration of a particle on the surface due to the rotation? Compare this acceleration with the acceleration of gravity on Earth.

63. (II) A golfer wants to land a golf ball on the green located 105 m away horizontally but 8.0 m down. The golfer chooses an eight iron that he knows will result in the ball leaving the tee at an elevation angle of 65°. (a) With what velocity should the ball leave the tee? (b) What is the maximum height of the ball above the green?

64. (II) A softball player hits the ball when it is 0.9 m above home plate. The ball leaves the bat at an elevation angle of 30°. What initial speed must the ball have to clear a fence 1.8 m high located 65 m away in left field?

65. (II) A boat is required to traverse a river that is 150 m wide. The current in the river moves with a speed of 6 km/h. The boat can be rowed on still water with a speed of 10 km/h. Set up a convenient coordinate system in which to describe the various displacements. Using this coordinate system, write down the position vector of the boat at time t, assuming that the boat moves with uniform speed and that it leaves one side with the velocity vector making an angle θ with the direction of the river. Calculate θ such that the boat lands at a point exactly opposite the starting point. How long will the trip take?

66. (II) A possible way to measure g is to send a mass traveling horizontally with a known speed off the edge of a vertical drop, such as a table. The table height can be measured, as can the distance from the base of the table where the mass strikes the floor. (a) If the speed of the mass just as it leaves the table is 2.50 m/s, and the distance from the table base to the point at which the mass strikes the floor is 108 cm, how long was the mass in flight? (b) The table height is 86 cm. What is the value of g obtained from this data? (c) Use $g = 9.80$ m/s^2 and neglect the air resistance to calculate the speed of the mass when it arrives at the floor.

67. (II) A sailor on top of a mast 26 m high drops a hammer. The ship is rolling with a maximum angle away from the vertical of no more than 15°. At the moment that the hammer is dropped, the mast is exactly vertical while the top of the mast is moving laterally at a speed of 3.6 m/s. Will the hammer fall into the sea or onto the deck, given that the ship is 19 m wide?

68. (III) A cannon can project a cannonball from its barrel with a certain muzzle speed v_0. Ignoring the effects of air resistance, what formula expresses the distance the cannonball travels before it reaches the ground as a function of the angle θ that the barrel makes with the ground? Unlike the projectile treated in Section 3–4, the cannonball is fired from the edge of a cliff of height h_0 above the level plain at which it is aimed. Show that the angle θ that gives the largest horizontal range is given by $\sin^2 \theta = v_0^2/2(v_0^2 + 2gh_0)$. [*Hint*: In calculus, you learn that a function of a variable such as θ has a maximum (or minimum) at an angle θ_0, the angle for which the derivative of the function with respect to θ is zero.]

69. (III) A juggler is able to handle four balls simultaneously. He takes 0.5 s to cycle each ball through his hands, throw the ball, and be ready to catch the next ball. (a) With what velocity must he throw each ball vertically? (b) What is the position of the other three balls when he has just caught one of the balls? (c) How high must he throw the balls if he is to juggle five balls?

70. (III) A wheel 72 cm in diameter rolls along a road, with the center moving in a straight line at a uniform speed of 18 km/h. What are the position vector, the velocity vector, and the acceleration vector of a fixed point on the rim of the wheel, relative to a fixed point on the straight line followed by the wheel on the road?

The idea that the apples on these trees might have something in common with the Moon originated with Isaac Newton, who showed that the same force governs the falling of an object and the Moon's movement around Earth. This observation, popularly linked with Newton watching an apple fall, is one of the truly significant discoveries in science.

Newton's Laws

In the previous chapters, we explored quantities that *describe* motion, such as velocity and acceleration. Such a description is called kinematics. We cannot, however, *predict* the motion of an object using kinematics alone. As we shall see in this chapter, forces cause changes in motion and enable us to predict the subsequent motion of an object. The study of the *causes* of motion is called **dynamics**.

The laws that govern the motion of an object were described by Isaac Newton in 1687. Now known as Newton's laws, they are based on careful and extensive observation of motion and its changes. These laws provide an extraordinarily accurate description of the motions of all objects—small or large, simple or complicated—on the everyday scale. There are, however, some motions that cannot be characterized using these laws. We must go beyond Newton's laws to describe motion within systems as small as the atom and for motion at speeds near the speed of light (300,000 km/s). Nevertheless, Newton's laws represent a tremendous achievement in their simplicity and breadth. Much of this book treats Newton's laws as applied to the motion of the heavenly bodies, fluids, springs, projectiles, electric charges, and many other systems. We use Newton's laws to calculate the motion of an object given the forces acting on it.

4-1   FORCES AND NEWTON'S FIRST LAW

Newton's first law of motion describes what happens to atoms, apples, asteroids, and any other objects—*moving or at rest*—when they are left alone. We might at first think that when a moving object is left alone, it will eventually come to rest. Indeed, the an-

Newton's first law

FIGURE 4–1 The force **F** acts on the wagon. Other forces act on the wagon as well.

FIGURE 4–2 (a) A mass attached to the end of a relaxed spring is at rest. (b) The spring has been compressed to the left and now pushes the mass to the right.

Newton's concept of an object being left alone is that there is no net force acting on it.

cient Greeks believed this; however, they were wrong. Galileo challenged this prejudice in the first half of the seventeenth century with experiments on the motion of objects on smooth planes. Isaac Newton built upon Galileo's work and, with great insight and power of abstraction, correctly and simply stated what happens:

> **When an object is left alone, it maintains a constant velocity.**

This law is *Newton's first law*, or the *law of inertia*. Notice that an object at rest is a special case of an object with constant velocity!

Constant velocity means that an object has both a constant speed and an unchanging direction of motion,

$$\mathbf{v} = \mathbf{v}_0 = \text{constant vector.} \tag{4–1}$$

The subscript on $\mathbf{v}_0$ indicates the value of the velocity $\mathbf{v}$ at time $t = 0$ or any other starting time. It is important to realize that Eq. (4–1) is a vector equation, which means that each vectorial *component* of the velocity is a constant. We know from Chapter 2 that this means that the acceleration **a** of the object is zero. Mathematically, we can write this as

$$\mathbf{a} = \frac{d\mathbf{v}}{dt} = 0. \tag{4–2}$$

Further, we know from Chapter 3 that constant velocity means that an object's position vector **r** changes linearly with time; that is, $d\mathbf{r}/dt = $ a constant. In other words, all three position coordinates change linearly with time:

$$\mathbf{r} = \mathbf{r}_0 + \mathbf{v}_0 t. \tag{4–3}$$

Again, an object at rest is just a special case, with $\mathbf{v}_0 = 0$, of this general result.

According to Newton's approach, an object that is left alone—whether it is moving or at rest—has a very specific relation to the concept of **force**. Although we will investigate a precise definition of force in Section 4–2, for now we will use the intuitive notion that a force is something that acts to push or pull an object. Let's take a look at Fig. 4–1, in which an adult is pulling a child's wagon. Forces acting on the object will have an identifiable source—in this case, the adult. A good example of a force source is a compressed spring. In Fig. 4–2, we show a mass attached to the end of a relaxed spring. The end of the spring is then pushed to the left and released. It exerts a force on the mass and pushes it to the right.

A force has a magnitude and a direction and is thus described by a *vector*, denoted by **F**. The **net force** exerted on an object is the overall push or pull that comes from separate pushes or pulls; the net force is the vector sum of all the different forces that act on that object. We will use the notation $\mathbf{F}_{\text{net}}$ when it is important to distinguish the net force from a particular force. For example, a cart is being pulled by two ropes in Fig. 4–3. If we ignore gravity, there are two forces acting on the cart: $\mathbf{F}_1$ and $\mathbf{F}_2$. The net force exerted on the cart is the vector sum of $\mathbf{F}_1$ and $\mathbf{F}_2$; that is, $\mathbf{F}_{\text{net}} = \mathbf{F}_1 + \mathbf{F}_2$. Of course, more than two forces can act on an object; each push and pull on an object can be labeled as an individual force $\mathbf{F}_i$. These forces add vectorially to yield a net force, $\mathbf{F}_{\text{net}}$, acting on the object:

$$\mathbf{F}_{\text{net}} = \sum_i \mathbf{F}_i. \tag{4–4}$$

It is an experimental fact that forces add as vectors, and experiment is the ultimate judge of statements in science. Forces that are equal in magnitude but oppositely directed add vectorially to zero. In Fig. 4–3, if the magnitude of the forces exerted by the ropes were equal and if the forces were oppositely directed, there would be no net force on the cart. In fact, any number of forces can add vectorially to zero net force. In Newton's approach, an object is left alone when there is zero *net* force acting on it; thus, an object that is left alone means more than it may seem at first. Zero net force could mean either that there are *no* forces acting on an object at all or that there are

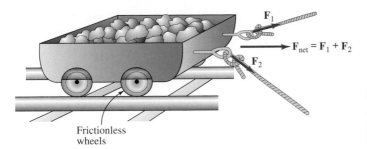
Frictionless wheels

$$F_{net} = F_1 + F_2$$

FIGURE 4–3 Two forces, F_1 and F_2, act on a mass, resulting in the net force, F_{net}. We are ignoring gravity here.

several forces acting on the object that add vectorially to zero. If two ropes exerting equal and opposite forces on a cart were simultaneously cut, there would be no discernable effect on the cart.

A question may spring to mind: Is an object ever really left alone? Is there a "force-free" environment, in which there is *no net force* and where we can really perform experiments to test Newton's first law? In the special case where the constant velocity of an object is zero and the object is at rest, the answer is yes. The object remains at rest because there is zero net force acting on it. The answer, however, becomes less clear when the velocity of an object is not zero. In fact, our common experience appears to contradict Newton's first law; we continually observe moving objects gradually slow down and come to a halt. This is because a moving object is seldom left alone. Try as we might to eliminate the overall forces acting on an object, there are always small residual forces—called *frictional forces*—that act on an object to slow it down and ultimately bring it to rest (Fig. 4–4). For example, a brick sliding on a table comes to a stop quite rapidly. We can try to approximate a force-free environment and reduce the frictional forces for the brick by pouring a thin layer of oil on the table. The brick might travel farther but would nevertheless still come to a stop fairly rapidly; obviously, this is not a very good approximation to the ideal "force-free" environment. Sliding a car with smooth tires on an icy road brings us much closer to friction-free motion. As we approach closer and closer to the idealization of a force-free environment, experiment confirms Newton's first law, which we can rephrase as

FIGURE 4–4 In the game of curling, the broom melts a layer of ice and reduces the friction for the sliding rock, but friction eventually brings the rock to rest.

When the net force, F_{net}, acting on an object is zero, that object moves with constant velocity.

A more mathematical form of Newton's first law

There's Nothing Special about Being at Rest

The formulation of Newton's first law includes its converse:

An object moving with constant velocity has no net force acting on it.

A consequence of Newton's first law

This version of the first law has some interesting consequences. We have already indicated that "being at rest" is just a special case of moving with a constant velocity. No net force is acting in either situation; thus, the object retains its initial velocity. To approach this situation from another perspective, let's consider how observers in different reference frames see the motion of an object.

Suppose that a bus moves with uniform velocity. An observer A watches the bus moving with what he measures to be velocity $v_{bus} = v_1$. Let's also imagine an observer B who is walking parallel to the street at the same velocity as the bus. Observer A measures observer B to have velocity v_1 as well. Figure 4–5a shows the situation from the point of view of observer A, whereas Fig. 4–5b depicts observer B's point of view. We can see from Fig. 4–5b that observer B measures the velocity of the bus relative to her to be $v_{bus} = 0$. Both observers see uniform velocities, albeit different ones. At a given time, both observers report uniform velocities and therefore agree that there is no net force acting on the bus according to Newton's first law.

(a) According to Person *A* at rest

FIGURE 4–5 Observer A stands watching a bus moving down a street, while observer B moves at the same speed as the moving bus, but on the sidewalk. In (a) we see things from the point of view of observer A's frame; in (b) we see things from the point of view of observer B's (and C's) frame.

(b) According to Person *B* (at rest)

(a)

(b)

FIGURE 4–6 A parachutist falling to Earth is acted upon by gravity and by a drag force. In (a) these forces are not equal and opposite, whereas in (b) they are.

The reference frame of an observer for whom Newton's first law holds is an **inertial reference frame**. What we have realized is that there is no fundamental way to distinguish between different inertial reference frames. Observer A and observer B are both in inertial reference frames and they both agree that there is no net force on the bus—even if they disagree on the bus's velocity. So, what is the bus's true velocity: v_1 (A's measurement) or 0 (B's measurement)? Which observer is truly at rest: observer A, who is at rest relative to the street, or observer B, who is at rest relative to the bus? The first law provides no clue as to who is at rest. *When observers A and B try to determine which one of them is at rest or exactly what his absolute velocity is by looking at the bus, they cannot do so.* Observer B might go so far as to say that the street is also moving; in fact, there is no real way to make the distinction. If you have ever had the experience of sitting in a very slowly moving train leaving a station on a track adjacent to another train, you will recall a disorientation as to whether your own train or the other train was moving.

In summary, when we say that there is no way to distinguish between inertial frames in the absence of forces, we are also saying that there is nothing sacred about the particular inertial frame in which an object is at rest. In other words, an object can be said to be at rest only in a particular inertial frame, not at rest in an absolute sense. Further, there is no way to say if any particular inertial frame is at rest in an absolute sense. We shall see in Section 4–2 that, *even in the presence of forces*, there is no fundamental way to distinguish between different inertial frames.

Net Forces

As we remarked earlier, the net force acting on an object may be zero either because there are no forces acting at all or because several forces are acting that add vectorially to zero. In fact, these two situations amount to the same thing in determining the object's motion. We observe all around us that constant motion is normally the result of the cancellation of many forces. When a parachutist first leaps from a plane, as in Fig. 4–6a, there is a net force accelerating him downward because the upward force of air drag, $\mathbf{F}_d$, does not cancel the downward force of gravity, $\mathbf{F}_g$. When the parachute is fully deployed, as in Fig. 4–6b, the parachutist falls at a constant velocity because $\mathbf{F}_d$ and $\mathbf{F}_g$ cancel. Similarly, an apple resting motionless in the palm of your hand remains there because the downward-directed gravitational force acting on it is exactly canceled by an upward contact force from your hand.

EXAMPLE 4–1   Three children each tug at the same sled (Fig. 4–7a). (All the forces are in the horizontal plane.) The three forces on the sled have the vectorial decomposition $\mathbf{F}_1 = -5\mathbf{k}$ units, $\mathbf{F}_2 = 5\mathbf{i}$ units, and $\mathbf{F}_3 = (-5\mathbf{i} + 5\mathbf{k})$ units. What is the net force on the sled? What can you say about its consequent motion? Ignore the force of gravity on the sled.

Solution: The net force on the sled is the vectorial sum of the individual forces (Fig. 4–7b). By adding the components, we see that the three forces add vectorially to zero:

$$\mathbf{F}_{net} = \mathbf{F}_1 + \mathbf{F}_2 + \mathbf{F}_3 = -5\mathbf{k} + 5\mathbf{i} + (-5\mathbf{i} + 5\mathbf{k})$$

$$= (-5 + 5)\mathbf{k} + (5 - 5)\mathbf{i} = 0.$$

The net force is zero; according to the first law, then, the sled's velocity is unchanging. If it is at rest when the three children apply their forces, it remains at rest.

Note: we have not specified the units of force in this example. When the net force is zero, the units are irrelevant. We shall return to the question of dimensions and units in the next section.

FIGURE 4–7 (a) Example 4–1. Three children pull on the sled in the *xz*-plane (horizontal).(b) Force diagram for forces on the sled. We have ignored the force of gravity.

Contact Forces and Forces at a Distance

In Example 4–1, the children must grab the sled in order to tug on it. Our qualitative notions about pushes and pulls arise to a large extent from *contact forces* that are associated with *physical* contact. Let's consider the wagon shown in Fig. 4–1. The child sitting in the wagon exerts a force on the wagon due to contact with it. The child's mother pulls the wagon by a handle that is in contact with both the mother's hand and the wagon. We may think of the wagon as being an isolated object with external forces acting on it due to the mother, the sitting child, its contact with the ground, and the pull of Earth. The first three of these forces are contact forces.

Physical contact, however, is not necessary for forces to act. Forces can also act from a distance. Gravity, the fourth force acting on the wagon in our example, is not a contact force; a softball thrown into the air is acted upon by gravity throughout its path of motion. One bar magnet can repel or attract another without actually touching it. Interestingly, a closer examination of forces reveals that *all* forces act at a distance. The distance involved in the contact forces we have discussed just happens to be of atomic size. The atoms of the mother's hand and those of the wagon handle interact—forces act between them—over imperceptibly small distances. Contact forces refer to forces that act over distances too short to be visible to the eye.

Now, we know that there are forces acting at a distance upon objects, so how can we be *sure* that there is no net force acting on an object moving with constant velocity? If we answer that there is no net external force acting on the object because the object's motion is constant, we present a circular argument. In fact, the first law makes sense only because we have performed many experiments with forces in different situations. These results depend heavily on Newton's second law, which describes *changes* in motion as a response to forces. Physicists know a good deal about the sources of force: We know that magnets and electric charges, for example, lead to forces that act at a distance, and we know how to detect contact or friction forces of various kinds. We know that there is a force acting on a mass when that mass is attached to a compressed spring. For the first law to be useful to us, we must have *independent* means of determining the forces.

PROBLEM SOLVING

We must identify the forces to apply Newton's laws.

4-2 NEWTON'S SECOND LAW OF MOTION

Newton's first law expresses the fact that when the net force acting on an object is zero, the object will maintain a constant velocity. Simply put, an object maintains a constant velocity when it is left alone. What happens to this object when it is not left alone; that is, when a net force acts on it? Newton's second law answers this question:

An object acted upon by a net force accelerates.

Further, the object will accelerate, or change its velocity, in the same direction as the net force.

We mentioned in Section 4–1 that a car on a horizontal surface of slick ice approaches a friction-free environment. If a car were at rest under these conditions and you could push it with a given force—say by bracing your back against a tree and pushing with your feet—the car would move with a given acceleration. In fact, the acceleration of the car is proportional to the magnitude of the push you give it; if somehow you were to push the car with twice as much force, its acceleration would double.

As our own experience tells us, the specific response of an object to a given net force also depends on the mass of the object. Mass is related to the property known as *inertia*, which describes an object's resistance to a change in its motion; an object's resistance to change can be either its resistance to slowing down if the object is already moving or its resistance to starting to move if it is at rest. For this reason, the mass that appears in Newton's second law is often called the **inertial mass**. If the mass of the car in our example were doubled and you were to push it with the same force you used the first time, the car would move with half of the initial acceleration. The greater the object's mass, the less the acceleration (Fig. 4–8); the less the mass, the greater the acceleration (Fig. 4–9). In fact, for a given force, experiment describes an *inverse* relation between mass and acceleration.

The basic relation we have just described between the net force on an object, $\mathbf{F}_{net}$, the mass of the object, m, and the object's acceleration, $\mathbf{a}$, was discovered by Newton and has come to be known as **Newton's second law**. It is stated mathematically as[†]

$$\mathbf{F}_{net} = m\mathbf{a}. \tag{4-5}$$

Earlier, we discussed how acceleration is the rate of change of velocity, $\mathbf{a} = d\mathbf{v}/dt$. In terms of the velocity, then, Newton's second law becomes

$$\mathbf{F}_{net} = m\frac{d\mathbf{v}}{dt}. \tag{4-6}$$

We must emphasize that the force appearing here is the net force; that is, the vector sum of all the forces acting on the object. As with any vectorial expression, Newton's second law is equivalent to three scalar equations for the components:

$$F_{x,net} = ma_x = m\frac{dv_x}{dt}; \tag{4-7a}$$

$$F_{y,net} = ma_y = m\frac{dv_y}{dt}; \tag{4-7b}$$

$$F_{z,net} = ma_z = m\frac{dv_z}{dt}. \tag{4-7c}$$

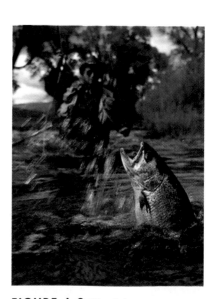

FIGURE 4-8 An enormous force would be required to provide substantial acceleration and change this oil tanker's direction or bring it to a stop quickly. Because its engines (or those of tugs) can exert relatively limited forces, the tanker's acceleration will be small. It is necessary to think far ahead when operating such tankers.

FIGURE 4-9 The fisherman causes the fish to accelerate out of the water with the help of the tension in the fishing line.

[†]When the object has a mass that changes with time (as in rocket motion), we shall modify the second law.

We shall refer interchangeably to Newton's second law in its vector form, Eq. (4–5) or (4–6), or in its equivalent component form, Eqs. (4–7).

More About Mass

Under the influence of a given force, it seems intuitively reasonable that an object with a small mass accelerates more briskly than does an object with a large mass. Mass is an *intrinsic* property of an object; a given object has a given mass—regardless of the forces acting on it. Let's look at an example in which mass enters into Newton's second law: Imagine that you are again braced against a tree while you push a sleigh that is sitting on a horizontal icy surface (Fig. 4–10). This surface is so smooth that we can consider it to be friction-free. You push the sleigh three times, each time with exactly the same effort; once when the sleigh is empty (Fig. 4–10a), again when it contains your little sister (Fig. 4–10b), and a final time when it contains your entire family (Fig. 4–10c). With each push, the sleigh will undergo successively smaller accelerations. The empty sleigh will show the largest acceleration while the fully loaded sleigh will have the smallest. In each case, the sleigh responds quite differently to your efforts even though the force you exert on it is the same. Now let's consider the mathematical expression of this property. Let the mass of the empty sleigh be m_1 and the mass of the fully loaded sleigh be m_2. When they are subject to the same net force, the respective accelerations a_1 and a_2 of the empty and fully loaded sleigh are related to their masses by

$$\text{for a given force:} \quad \frac{m_1}{m_2} = \frac{a_2}{a_1}. \tag{4–8}$$

Mass is a scalar quantity; it has no direction associated with it. It is always positive. As a consequence, the acceleration of an object is in the same direction as the force exerted on it. Furthermore, masses are additive; that is, if an object contains two parts with masses m_1 and m_2, the total mass of that object is[†]

$$m = m_1 + m_2. \tag{4–9}$$

Is the Second Law Merely a Definition of Force?

In Chapter 5 and beyond, we will explore many applications of Newton's second law. At this point, however, we must consider what the second law really means. Does Eq. (4–5) do more than simply define the force? We can measure an object's acceleration with a meter stick and a watch. Do we use that measurement merely to define the quantity $\mathbf{F}/m$? The power of the second law lies in the fact that it goes beyond this mere definition. In this subsection, we shall explain how forces and masses are defined and how Newton's second law makes predictions.

Suppose that we have a large supply of lumps of gold. We will label one of them A and call it our standard mass, $m_A = 1$ kg. We also have available a spring and a very smooth (frictionless) table. We compress the spring, place lump A against it, release the spring, and measure the acceleration of the lump as it is pushed away by the spring (Fig. 4–11). Suppose that when the spring is compressed by 1 cm, lump A has an acceleration of $a_A = 2$ m/s², as in Fig. 4–11b. At this point, *Newton's second law provides a precise definition of the force exerted by this spring* when it is compressed by 1 cm. Because $F = m_A a_A$, the force is $F = (1 \text{ kg})(2 \text{ m/s}^2) = 2 \text{ kg} \cdot \text{m/s}^2$. Every time the spring is compressed by 1 cm, it will exert this force.

We can now calculate the masses of our other lumps of gold. Place a second lump, B, in front of the spring and compress the spring by the same 1 cm (Fig. 4–11c). When the spring is released, the acceleration of lump B is measured to be $a_B = 3$ m/s². By

(a)

(b)

(c)

FIGURE 4–10 By bracing yourself against a tree, you can exert a fixed force against a sleigh. You do so three times, as follows: (a) the sleigh is empty; (b) the sleigh contains your little sister; (c) the sleigh contains your whole family. Even though you push the same way each time, the results are quite different. The empty sleigh leaves with a much greater speed than does the full sleigh.

[†]When the forces holding the system together are strong, special relativity implies a modification. In most of the situations we meet, the effect is very small.

FIGURE 4–11 (a)–(c) A series of experiments with a spring and a set of masses allows us to construct a mass scale and to learn about the forces exerted by the spring. (d)–(e) Once we have established masses and learned the force law, we can predict the motion of a mass.

A force law describes the force due to one source and how that force can vary.

Newton's second law, the mass m_B is

$$m_B = \frac{F}{a_B} = m_A \frac{a_A}{a_B} = m_A \frac{2 \text{ m/s}^2}{3 \text{ m/s}^2} = \frac{2}{3} m_A = \frac{2}{3} \text{ kg}.$$

In this way, we can determine the masses of each of our lumps of gold—or of any object for that matter.

The force of the spring is always the same whenever it is compressed by 1 cm. Let's now compress the spring by a different amount, place lump A in front of it, and measure the acceleration of lump A when the spring is released (Fig. 4–11d). This time, we measure $a'_A = 5$ m/s². We can again use Newton's second law to define the force that the spring exerts when it is compressed by the new amount; it is $F' = m_A a'_A = (1 \text{ kg})(5 \text{ m/s}^2) = 5 \text{ kg} \cdot \text{m/s}^2$. We can continue this process to define the force exerted by the spring for each different amount of compression. We have found a *force law*, which in this case is just a catalog of how much force the particular spring exerts at any given compression. A force law describes the force due to a particular source and how that force depends on variables such as position or time. Throughout this book, we shall study the force laws describing friction, gravity, springs, and electric charges, among others.

At this point, Newton's second law enables us to make predictions for the first time. If we compress the spring to the position shown in Fig. 4–11d and place lump B against it, we can predict lump B's acceleration. We have already determined that the force exerted by the spring in this second position has magnitude $F' = 5 \text{ kg} \cdot \text{m/s}^2$. When the spring is released, as in Fig. 4–11e, the second law *predicts* the acceleration to be

$$a'_B = \frac{F'}{m_B} = \frac{5 \text{ kg} \cdot \text{m/s}^2}{2/3 \text{ kg}} = 7.5 \text{ m/s}^2.$$

FIGURE 4–12 By using Newton's first law and a known force, such as that exerted by the stretched spring, we can measure other forces—in this case, the force of gravity acting on the (stationary) block.

Let's suppose that we know just how much force is exerted by the spring for a given compression or stretch. We can then use the spring to learn about *other* forces. Instead of studying the motions these other forces cause, we can just balance them against the spring and use the first law. For example, if we turn our spring to the vertical and suspend a block of known mass from it so that the block is motionless (Fig. 4–12), the force of gravity acting on the block must exactly balance the upward force of the stretched spring. By observing how much the spring is stretched, we have measured the force of gravity on the block. We could now predict how the block would accelerate if there were no spring and only gravity acted on it. In this book, we are most interested in using the second law for its ability to predict motion.

EXAMPLE 4–2 You must deliver a box of bowling balls to a bowling alley. The balls will be placed in a box that is initially at rest (its mass is very small compared to even one bowling ball). You start with one ball in the box, exert a given force upon the box, and, at the end of a time period Δt, the box moves at a speed of 3.2 m/s. You then add more bowling balls to the box; you exert the same amount of force on the box for the same period of time (Δt) and find the box to have a final speed of 0.40 m/s. How many balls are in the box now?

Solution: In the two cases described (which we will call 1 and 2), the box containing the bowling ball(s) is subject to the same force. Moreover, we can ignore the mass of the box. The mass of the box with one ball in it is then approximately $m_1 = m$, and its mass with the unknown number of balls is $m_2 = nm$, where m is the mass of one bowling ball and n is the number of balls, which we are trying to find. Using Newton's second law, we can find the accelerations a_1 and a_2 during the period Δt when the force (of constant magnitude F) operates. These accelerations are

$$a_1 = \frac{F}{m_1} = \frac{F}{m} \quad \text{and} \quad a_2 = \frac{F}{m_2} = \frac{F}{nm}.$$

Although we do not know the numerical values of the two accelerations, we do know the speeds v_1 and v_2 after a fixed period of acceleration. Further, we learned in Chapter 2 that an object that starts at rest and undergoes a fixed acceleration **a** for a given period of time Δt has the velocity $\Delta \mathbf{v} = \mathbf{v} = \mathbf{a}\,\Delta t$. In our one-dimensional case, then, we have

$$v_1 = a_1 \Delta t = \frac{F\,\Delta t}{m} \quad \text{and} \quad v_2 = a_2 \Delta t = \frac{F\,\Delta t}{nm}.$$

We now have enough information to solve for the unknown, n. We can solve for the ratio F/m in terms of v_1 and Δt and substitute it into the equation for v_2, which we could solve for n. Alternatively, we can simply take the ratio of the two speeds:

$$\frac{v_1}{v_2} = \frac{\cancel{F}\,\cancel{\Delta t}/\cancel{m}}{\cancel{F}\,\cancel{\Delta t}/n\cancel{m}} = n.$$

Numerical substitution gives $n = (3.2 \text{ m/s})/(0.40 \text{ m/s}) = 8$ bowling balls.

Units. We first discussed the dimensions and units of mass, length, and time in Chapter 1. Because acceleration has dimensions of $[LT^{-2}]$ and units of m/s^2 in SI, force has dimensions of $[M \cdot LT^{-2}]$ and, in SI, units of kg $\cdot$ m/s^2, or **newtons** (N):

Units of force

$$1 \text{ N} \equiv 1 \text{ kg} \cdot \text{m/s}^2. \tag{4–10}$$

In other words, a 1-N force exerted upon an object with a mass of 1 kg will produce an acceleration of 1 m/s^2.

In the cgs system, the force 1 g $\cdot$ cm/s^2 is called the *dyne*:

$$1 \text{ dyne} \equiv 1 \text{ g} \cdot \text{cm/s}^2 = 10^{-5} \text{ N}. \tag{4–11}$$

Here again, a force of 1 dyne acting on a mass of 1 g causes an acceleration of 1 cm/s^2.

Another force unit in everyday use is the pound (lb); 1 lb = 4.448 N. The pound is used in the British engineering system, in which mass is measured in slugs.

How Do Different Observers See Newton's Second Law?

Let's revisit the two streetside observers A and B (Fig. 4–5). They are in different inertial frames, moving with fixed velocity with respect to one another. To be more general, let's label this fixed velocity **u**. These observers measure a moving object to have different velocities, $\mathbf{v}_A$ and $\mathbf{v}_B$. But what if there is a net force on the object? Do each of the observers see Newton's second law in the same way? We have already seen in Eq. (3–57) how the two observers see the velocity of a bus:

$$\mathbf{v}_B = \mathbf{v}_A - \mathbf{u}.$$

Let's now see how observers A and B measure the *rate of change* in the bus's velocity. The rate of change is a derivative with respect to time, so we take the derivative with respect to time of both sides of the equation:

$$\frac{d\mathbf{v}_B}{dt} = \frac{d\mathbf{v}_A}{dt} - \frac{d\mathbf{u}}{dt}. \tag{4–12}$$

The ability to construct devices that provide constant acceleration has an important application in the design of the aircraft emergency ejection seat. Early ejection seats consisted of an explosive charge placed under the seat assembly. In an emergency, the crew member would send an electric current to the detonator, which fired the charge in milliseconds. Because of the short duration of the acceleration, it had to be very large, in order to provide the crew member with enough height to clear the tail of the airplane. Collapsed spinal disks, crushed vertebrae, and heart tears were the not uncommon by-products of this method of ejection.

In modern ejection assemblies, the explosive charge is replaced by a small rocket engine, which provides a smaller acceleration over longer periods (Fig. 4AB–1). In this way, the crew member can gain the needed height without the danger of the high accelerations of the explosive force. The crew member not only clears the tail assembly, but

FIGURE 4AB–1 Test ejection of a pilot from F-14 Tomcat cockpit. Note the jet plumes below the seat and the release of the parachute.

can gain enough height to deploy a parachute before the rocket has burned out. This means that the seat can even be used at ground level in the event something goes wrong on take-off or landing.

The observers are in inertial frames, and thus their relative velocity **u** is constant and its derivative is zero. Therefore

$$\frac{d\mathbf{v}_B}{dt} = \frac{d\mathbf{v}_A}{dt}. \tag{4–13}$$

Because the mass m of the object under observation—the bus—does not depend on the velocity of the observer,[†] we multiply both sides of Eq. (4–13) by m to find that

$$m\frac{d\mathbf{v}_B}{dt} = m\frac{d\mathbf{v}_A}{dt},$$

or, according to the second law,

$$\mathbf{F}_B = \mathbf{F}_A. \tag{4–14}$$

$\mathbf{F}_B$ is the force on an accelerating object, such as bus, as measured by observer B, and $\mathbf{F}_A$ is the force measured by observer A.

Observers in different inertial frames agree on the net force acting on an object.

A different way of stating this result is that *two observers in different inertial frames cannot by experiment tell which of them is moving and which of them is at rest.* Remarkably, this remains true in situations where speeds are comparable to the speed of light and Newton's second law in the form we have stated is no longer useful (see Chapter 40).

INTERIM SUMMARY

We have learned how to measure masses and forces in terms of a set of standard masses and forces. Force has units of newtons in SI. Newton's second law states that $\mathbf{F}_{net} = m\mathbf{a}$. This law predicts the acceleration of an object of known mass if the net force acting on it is known. We can view Newton's first law as a special case of the second law:

[†]Actually, experiment shows that at speeds close to the speed of light these remarks must be reexamined.

When the net force acting on an object is zero, the object's acceleration is zero, and it maintains a constant velocity.

The second law is used initially to calibrate mass and to define a single force law, such as the force exerted by a particular spring at different compressions or stretches. After that, the first law can be used to define all other forces. Once these forces are known, the second law allows us to predict the motion of any object of known mass. Neither the first nor the second law provides the means to tell observers in different inertial frames which one is moving and which is standing still. (See Fig. 4–13 for an example of these two laws.)

4 – 3   NEWTON'S THIRD LAW OF MOTION

The first and second laws describe forces acting *on* objects, and such forces have a source. For example, the force acting on the lumps of gold in Section 4–2 was due to the compression of a spring, and the Sun is the source of the gravitational force that causes Earth to circle the Sun. *When a force due to object B acts on object A, then a force due to object A also acts on object B* (Fig. 4–14). When you push on a wall, the wall pushes back on you. It is less obvious that when Earth tugs on an apple, causing it to fall (the force is gravity), the apple also tugs on Earth, causing Earth to accelerate toward the apple. Earth exerts a force on the Moon, making it orbit us overhead, while the Moon exerts a gravitational pull on Earth. In fact, ocean tides are associated closely with the force the Moon exerts on Earth. Earth and the Moon exert gravitational forces *on each other*.

Newton quantified this phenomenon in his third law. According to Newton's third law, the force on Earth due to the apple is equal in magnitude but opposite in direction to the force on the apple due to Earth. Forces do not simply act *on* objects; rather, forces act *between* two objects or between an object and its surroundings. Objects are said to *interact* when forces act between them. This effect does not depend on the observer's inertial frame of reference.

A mathematical statement of the third law is the following: If the force on object A due to object B is $\mathbf{F}_{AB}$, then there exists an equal and opposite force $\mathbf{F}_{BA}$ that acts on object B due to object A. The mathematical relation is[†]

$$\mathbf{F}_{BA} = -\mathbf{F}_{AB}. \qquad (4\text{–}15)$$

The third law is sometimes called *the law of equal action and reaction*. Do not let this particular phrasing mislead you into believing that the *accelerations* of the two objects are the same. Each object accelerates according to the second law, which means that the response depends on the mass of the object.

The third law is illustrated in Examples 4–3 and 4–4. In these examples, we imagine an outer space environment where we can isolate astronauts and satellites from all forces except the forces they exert on one another. Such a situation is more difficult to arrange on Earth.

EXAMPLE 4–3 An astronaut and a satellite are in an environment where they can be considered to form an isolated system with no external forces acting on that system. The astronaut tugs on the satellite with a force of 10.0 N to the right (and down) (Fig. 4–15). What is the force on the astronaut?

Solution: We label the two objects involved as the astronaut A and the satellite S. The given force on the satellite due to the astronaut is $\mathbf{F}_{SA} = 10.0$ N to the right (and down). According to the third law, the force on the astronaut due to the satellite, $\mathbf{F}_{AS}$, is

$$\mathbf{F}_{AS} = -\mathbf{F}_{SA} = 10.0 \text{ N to the left.}$$

[†]We shall see that Newton's third law must be reformulated when we study electromagnetic forces.

(a)

(b)

FIGURE 4–13 (a) Two equal and opposite forces act on this object, the tension due to the string and gravity. The net force is zero, and the object maintains its velocity, which has the value zero in this case. (b) When the string is cut, only gravity acts, and the object accelerates toward Earth according to Newton's second law.

Newton's third law

PROBLEM SOLVING

Equal action and reaction does not mean equal motion.

FIGURE 4–14 A demonstration of Newton's third law. Action causes reaction; in this case, the release of CO_2 from a fire extinguisher causes the initially stationary cart to be propelled in a direction opposite that of the released gas.

(a)

(b)   (c)

FIGURE 4–15 (a) Example 4–3. (b) According to Newton's third law, the force the astronaut exerts on the satellite is equal and opposite to the force the satellite exerts on the astronaut. (c) Even though the force on the astronaut has the same magnitude as the force on the satellite, the accelerations of astronaut and satellite are quite different; they have unequal masses.

The force on the astronaut due to the satellite has a magnitude of 10.0 N, which is the same magnitude as the force exerted on the satellite due to the astronaut; however, the force on the astronaut is directed to the left (and up), whereas the force on the satellite is directed to the right (and down) (Fig. 4–15b).

EXAMPLE 4–4 Let's now assume that the mass of the astronaut in Example 4–3 is 75.5 kg and that of the satellite is 755 kg. What is the acceleration of each?

Solution: The force, and thus the acceleration, is along the line between the astronaut and satellite and therefore we can drop the vector notation. We use the second law to determine the acceleration of each object. According to this law, the satellite has an acceleration of magnitude

$$a_S = \frac{F_{SA}}{m_S} = \frac{10.0 \text{ N}}{755 \text{ kg}} = 0.0132 \text{ m/s}^2.$$

This acceleration of the satellite is directed to the right (and down). For the astronaut, the acceleration has magnitude

$$a_A = \frac{F_{AS}}{m_A} = \frac{F_{SA}}{m_A} = \frac{10.0 \text{ N}}{75.5 \text{ kg}} = 0.132 \text{ m/s}^2.$$

This acceleration is directed to the left (and up). Thus, as Fig. 4–15c shows, the astronaut experiences an acceleration whose magnitude is 10 times larger than that of the satellite. This difference is the result of the difference in the masses of the satellite and the astronaut.

These examples demonstrate why the force that the apple exerts on Earth is not directly observable. With a small apple of mass $m_a = 0.1$ kg, the force on the apple due to Earth ($\mathbf{F}_{aE}$) is approximately 1 N, directed downward. This leads to an acceleration of magnitude $F_{aE}/m_a = 10$ m/s^2 when the apple falls from the tree. According to Newton's third law, the upward force $\mathbf{F}_{Ea}$ that the apple exerts on Earth also has magnitude 1 N. Because Earth's mass, m_E, is approximately 6×10^{24} kg, its upward acceleration has magnitude $F_{Ea}/m_E \simeq 1.7 \times 10^{-25}$ m/s^2, much too small to be observable. The evidence that led Newton to the third law involved the Earth–Moon system, where the law's effects are much more significant.

---

* **4 – 4   THE EFFECTS OF NONINERTIAL FRAMES**

Let's return to our streetside observers one more time and reconsider observers A and B moving with respect to one another. We found that if they are in inertial frames—so that their relative velocity **u** is constant—they agree on the forces acting on *any object* they observe.

FIGURE 4–16 The woman is at rest with respect to the stop sign. Because she does not see the sign accelerating, she would say that there is no net force acting on it. The boy on the bicycle, however, is in an accelerating frame. As seen from his frame of reference, the stop sign is accelerating, and he may think that forces of unknown origin act on the stop sign.

Suppose, however, that **u** is not a constant velocity. One or both of the observers are in a **noninertial frame of reference**. According to Eq. (4–12), observer B will measure an object to have a different acceleration from the acceleration observer A measures the object to have, and the two observers will disagree on the forces acting. To take a specific example, consider a stop sign at rest relative to a woman standing nearby; however, the sign is not at rest relative to a boy on a bicycle, as shown in Fig. 4–16. Assume the boy could somehow block out the surroundings and concentrate solely on the stop sign. As he goes around the corner, the boy would observe the sign to be moving—and not in a straight line. A change in either the observed direction of motion of the stop sign or in its speed (relative to the boy) implies acceleration. The boy could say, "I am accelerating, so the stop sign I see only *seems* to be accelerating, and there is actually no net force on it." Or he could say, "I am at rest, but some un-usual force acts on the stop sign to make it move and accelerate in the odd way that I observe. I cannot identify the source of this odd force on the stop sign."

The observant child's second interpretation is an interesting one. His failure to find and identify the source of the "force"[†] is an indication that the acceleration is associ-ated with an observation that was made from a *noninertial frame of reference*. In ef-fect, Newton's second law does not apply to motions observed from noninertial frames. A certain amount of experience is necessary to determine whether all possible force sources acting on an object have been identified, and whether an observer is in an in-ertial or noninertial frame. We shall normally deal with real forces in inertial frames and give ample warning when accelerating (noninertial) frames are involved.

EXAMPLE 4–5 Professor A is at rest on a train platform; her friend, Professor B, is leaving the station in a train with acceleration α in the $+x$-direction (Fig. 4–17). Professor A considers herself to be at rest and states that there is no net force act-ing on her. What does Professor B observe Professor A's motion to be, and how might he interpret that motion?

Solution: The motion in this example is in one dimension and so we can drop our boldfaced notation for vectors. We then say that Professor B is accelerating in the positive direction, which is to the right. Professor B wants to measure the mo-

---

[†]"Forces" associated with noninertial frames—forces with no identifiable sources—are referred to as *ficti-tious forces, pseudo-forces,* or *noninertial forces.*

(a) According to Professor A

(b) According to Professor B

FIGURE 4–17 (a) Example 4–5. (b) Because Professor B is in a noninertial frame, Newton's second law does not apply to his measurements of Professor A's motion, and he sees Professor A accelerating.

tion of Professor A. The variables are

$$v_A = \text{velocity of Professor A as measured by A;}$$

$$v_B = \text{velocity of Professor A as measured by B;}$$

$$u = \text{velocity of Professor B as measured by A.}$$

We first applied Eq. (4–12) in the case of the bus's velocity, and it continues to relate the rates of change in the various velocities in this example:

$$\frac{dv_B}{dt} = \frac{dv_A}{dt} - \frac{du}{dt}.$$

Because Professor A measures herself to be standing still, $dv_A/dt = 0$. In addition, the acceleration of the train as seen by Professor A is given by $du/dt = \alpha$. Thus

$$\frac{dv_B}{dt} = 0 - \alpha = -\alpha.$$

Professor B sees Professor A accelerating *backward* (minus sign) with an acceleration of magnitude α. He would then say that, according to the second law, there must be a force on Professor A of $-m\alpha$ that is responsible for giving her this motion. He would not, however, be able to find an identifiable agent for this force and might in this way decide that he is in a noninertial frame. Indeed, Professor B sees the entire train platform accelerating backward!

4–5 USING NEWTON'S LAWS I: FORCE DIAGRAMS

PROBLEM SOLVING

How to prepare a force diagram

Ordinarily, we take two different steps to apply Newton's second law. In this section, we will discuss the first step, in which we identify all the forces that act on an object and we express the second law explicitly. In the second step, the topic of Section 4–6, we find the subsequent motion of the object.

Newton's second law, $\mathbf{F}_{net} = m\mathbf{a}$, relates the mass of an object, the forces acting on it, and its acceleration. We must be able to distinguish these quantities in order to use the second law. Therefore, *we must know exactly what object we are talking about.* This step may be less obvious than it seems. For example, if we want to analyze the motion of a wagon being pulled by a child, we must consider *only* the forces acting *on the wagon*—not those forces that act on the child. Further, if we want to find the motion of the wagon, it is only the forces on the wagon and the mass of the wagon itself that enter Newton's second law. We want to *isolate* the wagon.

The wagon is most easily isolated in a sketch that we shall refer to as a **force diagram** for the wagon. The force diagram indicates each individual force that acts on the wagon and indicates the direction of these forces. These forces will appear as arrows along the directions of the force vectors (Fig. 4–18). The force diagram should also indicate a set of coordinate axes so that we can use this diagram to help us separate the vectors into their vector components. In order to avoid the effects of noninertial frames, place the axes in an inertial frame—usually attached to some fixed, stationary point—rather than attached to an accelerating point. In Fig. 4–18, we have placed the origin of the coordinate system at a spot on the ground adjacent to the wagon. It is often convenient to use a set of Cartesian axes: (x, y) for planar figures or (x, y, z) in space. Newton's second law breaks down into separate equations for the vector components along these mutually perpendicular axes.

Any orientation of the axes is acceptable, but certain choices will be easier to use. For example, it is often convenient to orient the y-axis vertically in the study of falling objects so that the force of gravity has only a y-component.

Sometimes it is convenient to draw the resulting acceleration of the object in question on the force diagram, but *it is most important that we do not confuse the acceleration with a force*. The acceleration is the object's *response* to the net force acting on it; *it is not a force itself*. In figures, we single out the acceleration vector in bright blue when we include it in the force diagram, whereas force vectors are always drawn in bright pink.

EXAMPLE 4–6 Block 1 is glued to the top of block 2 (Fig. 4–19a). The masses of the blocks are m_1 and m_2, respectively. A rope is attached to block 2, pulling it horizontally to the right with a force of constant magnitude T along a perfectly smooth horizontal surface. What equation describes the motion of block 1? Solve this problem in two ways: (a) Consider the system of the two blocks glued together; and (b) consider block 1 isolated.

Solution: Before we find the equation describing the motion, let's note that, whatever forces act in the vertical direction (gravity, for example), *they must cancel out entirely* because there is no motion of either block 1 or the whole system in the vertical direction. Thus we ignore all vertical forces. (a) We can consider the system of two blocks glued together to find the motion of block 1 alone, because the motion of block 1 will also be the motion of the two blocks together. Figure 4–19b shows the system isolated, together with the net force that acts on it. That net force is the force **T** due to the rope because any vertical forces cancel. The mass of the system is $m_1 + m_2$. Thus Newton's second law is, for the system,

$$(m_1 + m_2)\mathbf{a}_{sys} = \mathbf{T}.$$

This equation allows us to find the acceleration of the system; hence, because $\mathbf{a}_1 = \mathbf{a}_{sys}$, it also gives the acceleration of block 1. (b) Figure 4–19c is a force diagram for block 1 alone. This time, the net (horizontal) force on block 1 is exclusively a contact force due to block 2. This contact force is written as $\mathbf{F}_{12}$, where the subscript specifies that we have a force *on* block 1 *due to* block 2. Thus we have

$$m_1\mathbf{a}_1 = \mathbf{F}_{12}.$$

This certainly does not appear to be equivalent to our result in part (a) yet. To go further, let's look at the force diagram for block 2 (Fig. 4–19d). From that diagram, we find

$$m_2\mathbf{a}_2 = \mathbf{T} + \mathbf{F}_{21}.$$

Here, $\mathbf{F}_{21}$ is the force on block 2 due to block 1. According to Newton's third law, however, $\mathbf{F}_{21} = -\mathbf{F}_{12}$; moreover, the expression for Newton's second law applied to

How to Prepare a Force Diagram

1. Make a simple sketch of the system.

2. Choose the body to be isolated (wagon).

3. Add convenient coordinate system.

4. Identify forces that act on wagon. Label them on diagram. Identify labels if necessary.

$\vec{F}_g$ = Gravitational force on cart from Earth
$\vec{F}$ = Force on cart handle from boy
$\vec{f}$ = Frictional force on cart wheels
$\vec{F}_N$ = Normal force on cart from ground

FIGURE 4–18 How to draw a force diagram for a wagon being pulled by a child.

(a)

(b) Isolate m_1 and m_2 together

(c) Isolate only m_1

(d) Isolate only m_2

FIGURE 4–19 (a) Example 4–6. Two blocks that are glued together are pulled by a rope. Force diagrams for (b) the two-block system, (c) block 1 alone, and (d) block 2 alone.

(a) Internal forces

(b) Net external force

FIGURE 4–20 (a) The internal forces within the system of two blocks studied in Example 4–6. (b) The net external force on that same system. Some vertical forces act (gravity and contact forces), but these do not contribute to the net external forces because they cancel.

block 2 is

$$m_2\mathbf{a}_2 = \mathbf{T} - \mathbf{F}_{12}.$$

We then solve for the unknown force $\mathbf{F}_{12}$, with result $\mathbf{F}_{12} = \mathbf{T} - m_2\mathbf{a}_2$. We substitute this back into our equation for block 1:

$$m_1\mathbf{a}_1 = \mathbf{F}_{12} = \mathbf{T} - m_2\mathbf{a}_2.$$

Finally, the blocks move together and thus $\mathbf{a}_2 = \mathbf{a}_1 = \mathbf{a}_{sys}$, so that

$$m_1\mathbf{a}_1 = \mathbf{T} - m_2\mathbf{a}_1,$$

or $(m_1 + m_2)\mathbf{a}_{sys} = \mathbf{T}$. We can now recognize the equation we found in part (a) and discover that the two methods give the same answer.

External and Internal Forces

Example 4–6 illustrates the important difference between *external* and *internal* forces. Internal forces are those that act within a system we are isolating. These forces may act to hold the system rigidly together or they may simply act between different parts of the system. If we consider the two blocks in Example 4–6 as a single, isolated system, the internal forces would be associated with the glue holding the two masses together. We can examine these internal forces, $\mathbf{F}_{12}$ and $\mathbf{F}_{21}$ in Fig. 4–20a.

External forces, in contrast, are forces that act on the isolated system from outside. In Example 4–6, the net force $\mathbf{T}$ is an external force acting on the two-block system (Fig. 4–20b). Once a system has been isolated, only the external forces acting on it influence its overall motion. As we shall see in more detail in Chapter 8, internal forces do not enter into the second law because by Newton's third law they cancel in pairs. For example, we saw that the internal forces $\mathbf{F}_{12}$ and $\mathbf{F}_{21}$ of the two-block system canceled each other and did not enter into the solution of the problem. *In the statement of the second law, the net force is actually the net external force.* All of the objects we deal with are, in fact, complicated systems composed of many atoms held together by internal forces. But even though these internal forces act within the object, they cannot influence its overall acceleration.

If a given system is broken up into separate pieces, the internal forces for the original system may become external forces for the pieces. The motion of the smaller pieces is then governed by these external forces. In Example 4–6, the isolation of block 1 compelled us to treat $\mathbf{F}_{12}$ as an external force on block 1. In that same example, we made two different choices of the objects to be isolated; hence, we made different choices as to which forces to treat as external and which ones as internal. Each time, we found that only the external forces affected the motion of the blocks. We saw that a judicious choice of which object to isolate can simplify the calculation considerably. Take a careful look at the force diagrams in Fig. 4–19 to learn how to isolate a given object and recognize the forces on it.

Some Common Forces

We will use a familiar example to illustrate the preparation of force diagrams and introduce some forces that enter into many physical situations. Let's consider a sled of mass m moving down a snow-covered hill. What is the force diagram for the sled?

Three forces act on the sled, two of which involve contact between the sled and the hill (Fig. 4–21). One of the forces, however, is not a contact force. It is the force due to Earth—the **force of gravity**, $\mathbf{F}_g$. It acts vertically downward in Fig. 4–21.

Figure 4–21 shows a second force, one exerted by the hill on the sled. We know that this force must be present because, in its absence, the force of gravity would cause the sled to accelerate down into the surface of the hill. We call this force the **normal force** and label it $\mathbf{F}_N$. It is a contact force and acts in a direction *normal*, or perpen-

PROBLEM-SOLVING TECHNIQUES

Let's refine some aspects of the general problem-solving strategy laid out in Chapter 2. To prepare a force diagram for a given object (see Fig. 4–18):

1. Identify and isolate the object in question. Make a sketch with the object clearly labeled.

2. Identify all the forces acting on the isolated object. Draw each force on the force diagram as a labeled arrow; include rough approximations of the direction and magnitude of each force.

3. If more than one direction is involved, draw a set of coordinate axes with the origin at a fixed point of the diagram; that is, the origin should not be attached to the object itself. Choose these axes so that you can easily pick out the components of the various forces along them.

4. Separate all forces into their components with respect to the coordinate axes you have chosen.

5. If you include an arrow that represents your guess as to the acceleration of the object, distinguish it clearly from the arrows representing forces. Remember, the acceleration is not a force, it is the response to a force.

Figure 4B1–1 summarizes these steps and gives some further hints for finding the net force.

How to Draw
More on Drawing Force Diagrams

(a) Draw force diagram (see Fig. 4–18)

Add
$$\vec{F}_N + \vec{F} + \vec{F}_g + \vec{f}$$
Sum = $\vec{F}_{net}$

(b) Add forces, tail to head

$\vec{F}_x$ (net): →
$$\vec{F}_{net} = \vec{F}_x (net) + \vec{F}_y (net)$$
$\vec{F}_{net}$: →

(c) Add components

FIGURE 4B1–1 More hints on drawing force diagrams. Parts (b) and (c) show two ways to find the net force, the first a graphical method and the second a method in which components are used.

---

dicular, to the surface of the hill. This force prevents an object from penetrating into another surface; in this case, the hill. The magnitude of the normal force is not a prefixed constant, but adjusts itself to cancel any forces that might make the sled penetrate the hill. The normal force is simple to use in practice—just as simple as gravity. In contrast to gravity, though, which is a direct and fundamental force between two massive objects (here, Earth and the sled), a normal force actually results from the addition of many complicated intermolecular forces within and between the materials making up the object and the surface on which it rests. Fortunately, we do not have to worry about these complications.

The net force in Newton's second law comes from external forces only.

PROBLEM SOLVING

Your choice of which object to isolate can simplify problem solving.

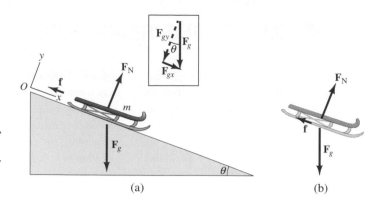

FIGURE 4–21 Force diagram for a mass (sled) sliding on an inclined plane. In the inset, the force of gravity, $\mathbf{F}_g$, is decomposed into components perpendicular and parallel to the plane. Part (b) is a simplified sketch that isolates the forces.

The force of gravity, $\mathbf{F}_g$

The normal force, $\mathbf{F}_N$

The friction force, $\mathbf{f}$

The third force is the **friction force**, which we label $\mathbf{f}$. Friction, like the normal force, is an approximation to a complicated interaction between the sled rails and the snow. Friction is a type of contact force as well. Its direction here is opposite to the motion at the surface, parallel to the hill and uphill (Fig. 4–21). If the sled were moving uphill, the force of friction would act downhill.

Friction is always parallel to the surface—perpendicular to the normal force. The second law as applied to the sled, $\mathbf{F}_{net} = m\mathbf{a}$, becomes

$$\mathbf{F}_g + \mathbf{F}_N + \mathbf{f} = m\mathbf{a}. \tag{4–16}$$

PROBLEM SOLVING

Choose your axes carefully.

⚬⚬ The unit vectors $\mathbf{i}$, $\mathbf{j}$, and $\mathbf{k}$ in the x-, y-, and z-directions were introduced in Chapter 1.

We now break this equation into component form. To do so, we must choose a set of axes. These axes should be placed in an inertial (nonaccelerating) frame to avoid the complications associated with noninertial frames. A convenient choice is one in which the decomposition of the force vectors is simplest. Another convenient choice, often equivalent, is one in which the acceleration is parallel to one axis. When there are several forces and they cannot *all* point along an axis, as in this example, it is best to align the axes with as many forces as possible. A good choice for this example is shown in Fig. 4–21, where the y-axis is perpendicular to the hill and the x-axis points downhill. Then two of the three forces—$\mathbf{F}_N$ and $\mathbf{f}$—are along these axes.

In terms of our coordinate axes, the forces $\mathbf{F}_N$ and $\mathbf{f}$ have the component form

$$\mathbf{F}_N = F_N\mathbf{j},$$

$$\mathbf{f} = -f\mathbf{i}.$$

Finally, we must find the components of $\mathbf{F}_g$ in the x- and y-directions. To do this, we have drawn a useful diagram for $\mathbf{F}_g$ inset into Fig. 4–21. From Fig. 4–21, then, we find that

$$\mathbf{F}_g = F_g \sin\theta\,\mathbf{i} - F_g \cos\theta\,\mathbf{j},$$

where F_g stands for the (positive) magnitude of the force of gravity. Because all three forces have only x- or y-components, the interesting part of the motion occurs entirely in the xy-plane. Any motion in the z-direction (perpendicular to the page) is unchanging.

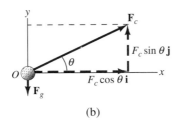

(a)   (b)

FIGURE 4–22 (a) Example 4–7. (b) Force diagram for the golf ball being hit by a golf club. The ball has already left the tee, so there is no contact force from the tee.

Equation (4–16) can now be written as

$$F_g \sin \theta\, \mathbf{i} - F_g \cos \theta\, \mathbf{j} + F_N \mathbf{j} - f\mathbf{i} = ma_x\mathbf{i} + ma_y\mathbf{j}.$$

The equations for the separate components of this equation are

$$\text{in the } x\text{-direction:} \qquad F_g \sin \theta - f = ma_x; \qquad (4\text{–}17)$$

$$\text{in the } y\text{-direction:} \qquad -F_g \cos \theta + F_N = ma_y. \qquad (4\text{–}18a)$$

We can immediately simplify Eq. (4–18a) by setting $a_y = 0$: We know that the sled does not leave the hill surface. It is this constraint that determines the magnitude of $\mathbf{F}_N$. With $a_y = 0$, Eq. (4–18a) becomes

$$\text{in the } y\text{-direction:} \qquad -F_g \cos \theta + F_N = 0. \qquad (4\text{–}18b)$$

Equation (4–17) is the dynamical equation for the unknown sled acceleration a_x. That equation depends on the magnitude of the friction force, f, and as we shall learn in Chapter 5, f depends on the value of F_N, which, in turn, we can find from Eq. (4–18a). We shall return to this in Chapter 5.

EXAMPLE 4–7 A swinging golf club strikes a golf ball. During contact, the force of the club (magnitude F_c) on the ball makes an angle θ with respect to the horizontal. You may assume that the ball is no longer in contact with the tee. Draw a force diagram for the golf ball and specify the forces in the coordinate system that you choose.

Solution: We must sketch a force diagram for the golf ball. The ball in Fig. 4–22a has left the tee, so there is no contact force from it. The only forces acting on the ball are therefore the force due to the club, $\mathbf{F}_c$—this is a contact force—and the force of gravity, $\mathbf{F}_g$. Figure 4–22b isolates the ball and shows the forces that act on it. It also contains a coordinate system for which the forces are decomposed as

$$\mathbf{F}_c = F_c \cos \theta\, \mathbf{i} + F_c \sin \theta\, \mathbf{j}$$

and

$$\mathbf{F}_g = -mg\mathbf{j},$$

where m is the mass of the golf ball.

4–6   USING NEWTON'S LAWS II: FINDING THE MOTION

We have learned how to use force diagrams to write Newton's second law for an object that is acted on by forces. This is a very important step along the way to finding the motion of that object. That motion is described by *solving* the equations expressing Newton's second law. It is for this reason that we call those equations the *equations of motion*. We can solve them only if we know the particular *force law*; that is, the force expressed as a function of variables such as position, velocity, time, or any other parameters of the problem. Force laws are determined in experiments that measure these forces in controlled conditions. Many of the most common forces depend only on the position of the object—and the simplest case of all occurs when the force is constant. Let's take a look at how we solve the equations of motion for a constant force in Example 4–8.

EXAMPLE 4–8 The constant net force **F** shown in Fig. 4–23 acts on a nugget of gold whose mass is m. What is the motion of the nugget?

Solution: Because **F** is a constant vector, it points in a fixed direction. For convenience, we choose the x-axis to point in the same direction as the force, so that

$$\mathbf{F} = F\mathbf{i}. \tag{4–19}$$

The force has no components in the y- and z-directions so the nugget of gold will not have a component of acceleration in these directions. The nugget can, of course, have motion with constant velocity in these directions. Motion of constant velocity in the y- and z-directions means that these components of the nugget's position vector change linearly with time:

$$y(t) = y_0 + v_{0y}t; \tag{4–20a}$$

$$z(t) = z_0 + v_{0z}t. \tag{4–20b}$$

The quantities v_{0y} and v_{0z} are the constant components of the velocity vector in the y- and z-directions, and y_0 and z_0 are the values of the y- and z-components of the position at time $t = 0$.

We must still describe the motion in the x-direction, though. The x-component of Newton's second law is

$$F_x = F = ma_x, \tag{4–21}$$

or, equivalently,

$$F = m\frac{d^2x}{dt^2}. \tag{4–22}$$

We want to find the function $x(t)$ that satisfies this equation. We have already done so in Chapter 2, and we can turn to that chapter to find the answer. Once the *initial conditions* are stated, the answer is given by Eq. (2–16). We restate the result below in Eq. (4–23). For the initial conditions, we place the location of the object at $t = 0$ to be $x = x_0$, and the x-component of the velocity at that time is $dx/dt = v_{0x}$. The result is

$$x(t) = x_0 + v_{0x}t + \frac{F}{2m}t^2. \tag{4–23}$$

This result obeys both the equation of motion—Eq. (4–22)—and the initial conditions at $t = 0$. We can verify this by direct differentiation and direct substitution.

The motion described in Example 4–8 is merely a review of our study of kinematics under constant acceleration from Chapters 2 and 3. What is new here is the idea

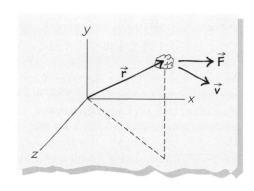

FIGURE 4–23 Example 4–8. A gold nugget is located at a displacement **r** from the origin of a coordinate system and moves with velocity **v**.

that constant acceleration is associated with a constant force. We will be returning to this example in Chapter 5. It is especially important because it is applicable to the force law of *gravity* near Earth's surface.

When the force on an object is not constant but depends explicitly on its position, the equations of motion become more complex. In fact, we can find simple expressions for such motion only for a few important cases of the force law. For example, we shall see later that the force law for a spring that has been compressed or stretched by an amount *x* from a relaxed position is proportional to *x*. The larger the compression or stretch, the stronger the spring force (Fig. 4–24). Mathematically, the proportionality is expressed as[†]

$$\text{force law:} \qquad F \propto x.$$

We also have Newton's second law,

$$F = m \frac{d^2x}{dt^2}.$$

The equation of motion therefore has the form

$$x \propto \frac{d^2x}{dt^2},$$

where we have left off all the constants. When the other components of the force are known, and when all the constants and signs are properly included, we are left with a differential equation for *x*. The meaning of the equation in this case is that *x* must be a function of time whose second derivative is proportional to the function itself. Only certain functions will satisfy such an equation[‡].

When the equations of motion have simple mathematical solutions, as for the spring, we say we have an *analytic solution*. In fact, there are not many problems that we can solve in this way. Problems with analytic solutions populate textbooks such as this one. You might get the mistaken impression that all problems have such solutions. Although there are indeed important, real-life problems that can be solved analytically, the equations of motion frequently have to be solved numerically with the aid of computers. While the idealized motion of a rock thrown up in the air is simple, the problem of the exact motion of a rocket launched from Cape Canaveral is quite a different matter. The force laws cannot be written so simply because air friction, the acceleration of gravity, and wind forces all vary with altitude, and the propulsion forces need to be adjusted to compensate for them. Sophisticated computer calculations are needed to learn just how the Space Shuttle will behave as it rises into orbit.

Chaos

Some problems are quite different from those we can solve in terms of simple functions. One consequence of the computer revolution has been a much deeper understanding of systems with motions that are quite different from the very regular ones that appear in analytic solutions. These systems are called **chaotic**, and examples occur all around us: from planetary atmospheres—the great red spot of Jupiter as well as Earthly hurricanes exhibit chaotic phenomena—to aspects of the flying capability of advanced aircraft to the turbulent flow of fluids. Systems with chaotic behavior still obey Newton's second law, but their motions are highly dependent on initial conditions; that is, their behavior may change radically if the initial conditions change just a little. Such systems have a rich and complex behavior (Fig. 4–25).

Prediction of the behavior of our chaotic atmosphere from Newton's laws must be done numerically. Here we face the problem that we may not have sufficient knowledge of the initial conditions. Even if we do, the computer may be unable to solve the

FIGURE 4–24 The stretched bungee cord acts like a spring; the more it is stretched, the stronger the force it exerts on the jumper. Here, the force on the jumper is upward.

 APPLICATION

Real projectile motion

FIGURE 4–25 (a) The great red spot of Jupiter, a gigantic storm within Jupiter's atmosphere, exhibits chaotic behavior. The spot is about two Earth diameters across. This picture, with computer-enhanced colors, was taken from one of the Voyager satellites.

[†]Here we are assuming one-dimensional motion.

[‡]We shall see later that these solutions are sinusoidal: the object moves back and forth.

problem rapidly. Although we are sure that Newton's laws apply to the problem of predicting the weather, no one has succeeded in finding either an analytical or a numerical solution that will accurately predict tomorrow's weather today!

4 – 7   WHAT IS A LAW OF PHYSICS?

All natural phenomena are governed by the laws of physics. But what do we mean by a *law of physics*? The word "law" is loosely applied to many statements; for example, we used it when we talked of a "force law" describing how a particular force varies as a function of position or velocity. In this context, we could equally well have used the word "prescription." When we explore electricity in Chapter 27, we will study "Ohm's law," which relates voltage, V, to current, I, in a resistor. For many materials, these two quantities are a constant multiple of one another ($V = IR$), and this isolated fact is given the name of a "law." It is not valid, however, for many materials; it is generally not true when temperatures vary; further, it is certainly not true for currents that vary rapidly in time. Thus Ohm's law or the spring force law are not "laws" in the same sense as Newton's laws. What characterizes a true physical law such as Newton's second law is the breadth of its applicability. But even Newton's laws cease to be accurate for systems that move with extremely high speeds (near the speed of light, when special relativity applies) or for systems that are very tiny (where the laws of quantum mechanics apply). In those cases, however, there is still a clear connection to the laws postulated by Newton.

Newton's descriptions have proven to be so universally applicable that they have repeatedly earned their label as *laws of physics*. They played a crucial role in the explanation of most of physical science up to the latter half of the nineteenth century.

The events associated with the discovery of the planet Neptune provide us with a spectacular example of the power of Newton's laws. The motion of astronomical objects is sufficiently frictionless that the fundamental forces can be studied under ideal conditions. Newton himself deduced the force law for the interaction between the Sun and the planets. This law is called the *law of universal gravitation*. The motion of a planet is accurately predicted by taking into consideration the presence of the Sun alone. Even more accurate predictions use the same fundamental law but also take into account the presence of other planets. In the nineteenth century, elaborate mathematical methods were developed to solve the full equations of motion for astronomical objects—remember, this was before the era of computers. Calculations of the motion of the planet Uranus did not agree with observations. By this time, the laws of motion and the law of gravitation had been tested in so many cases that the most conservative view of the discrepancy in Uranus's position was that there must be an additional, unobserved astronomical object somewhere in the solar system that interacts with Uranus by gravitation and thereby perturbs Uranus's orbit. This view was preferable to the alternative, in which the well-established laws of physics were violated.

In 1845, the astronomers Urbain Le Verrier and John Adams independently and simultaneously showed that the tiny discrepancy in the orbit of Uranus could indeed be explained by the presence of an undiscovered planet that was about twice as far from the Sun as was Uranus. Their calculations were based only on Newton's laws and on the law of universal gravitation. Both men predicted the details of the new planet's orbit. When, in 1846, Le Verrier convinced the astronomer Johann Galle to look for the new planet at a location he had predicted, Neptune was discovered.

SUMMARY

Newton's three laws express the dynamics of motion: how forces acting between objects determine the subsequent motion of those objects. The first law states what happens to an object—moving or at rest—when it is left alone:

F_{net} is the vector sum of any individual forces that act on an object. When **F**_{net} is zero, the object moves with constant velocity.

103

Understanding Key Concepts

Forces, which behave as vectors, act on objects and cause them to accelerate. For a given force, this acceleration is inversely proportional to the mass m of the object in question. This is expressed in Newton's second law,

$$\mathbf{F}_{net} = m\mathbf{a}. \qquad (4\text{--}5)$$

In SI, the force is measured in newtons, abbreviated N, where $1\text{ N} \equiv 1\text{ kg}\cdot\text{m/s}^2$.

Forces act between objects. If objects A and B interact—if there are forces acting between them—then Newton's third law states that the force on object A due to object B, $\mathbf{F}_{AB}$, is equal and opposite to the force on object B due to object A, $\mathbf{F}_{BA}$:

$$\mathbf{F}_{BA} = -\mathbf{F}_{AB}. \qquad (4\text{--}15)$$

Observers in reference frames moving with respect to one another observe the motion of a given object differently. An observer who verifies that Newton's second law holds, with known or identifiable sources of forces, is said to be in an inertial frame. If a second observer moves with constant velocity relative to the first, the second observer is also in an inertial frame; if there is nonconstant relative motion, the second observer is in a noninertial frame. Observers in inertial frames agree on the forces they see acting on an object. There is no experiment they can perform to decide who is moving in an absolute sense. Observers in noninertial frames disagree on the forces that act on an object and, in effect, Newton's second law fails for an accelerating observer.

Newton's laws help us to determine the motion of an object if we know the nature of the forces that act on it. Conversely, the laws allow us to measure forces acting on an object by measuring the object's motion. If we want to determine the motion of an object, we must independently know the forces that act on it. A force law describes the force due to a particular source and how that force varies with position, time, or other variables. We must also distinguish between external forces that act on an object and internal forces that act within the object. Only the external forces determine an object's motion. To use Newton's laws, we draw force diagrams that conceptually isolate the object. We include all the forces acting on it, keeping in mind that a force is a vector. Once a force diagram has been prepared, we choose a convenient set of axes to write the three components of the second law. These equations can be solved, either analytically or numerically, to find the object's motion.

A law of physics has a wide range of applicability. Laws such as Newton's laws provide a framework for detailed predictions about what will happen under certain conditions. In their breadth and predictive power, Newton's laws are an ideal illustration of what constitutes a law of physics.

UNDERSTANDING KEY CONCEPTS

1. A small but dense mass is swinging freely at the end of a light string. A very sharp knife cuts the string when the mass is at the bottom of its swing; the knife does not disturb anything else. What is the subsequent flight of the mass?

2. Someone pushes on a wall. What experiment can you propose to determine the force with which the person pushes?

3. If you were in a freely falling elevator, the contact force on you due to the floor would drop to zero. If the elevator were to accelerate rapidly upward, the contact force between you and the floor would increase. Why?

4. An astronaut is working while in orbit. When the astronaut assembles a piece of equipment, will he or she notice a difference between working with components of large mass as opposed to components of small mass?

5. When a satellite travels around Earth in a circular orbit, it moves at a constant speed. Does Newton's first law apply in this situation? Is the velocity constant? Is there a force present?

6. There is a well-known parlor trick in which a tablecloth is pulled sharply from beneath a dinner setting, leaving the setting in place. Why does this work?

7. Is a spaceship heading from Earth to the Moon traveling in a force-free environment? Explain. What about a spaceship traveling from Earth to Mars that is currently in a region far away from either planet?

8. An adult sits on a child's table. The table is about to break. Is it correct to say that it is the weight of the adult that is causing the table to break?

9. From your experience with forces, which of these are contact forces: friction, the force of gravity, the normal force on an object on the floor, the force due to a magnet, the pull on the rope used in a pulley?

10. A box is placed on a table. The box's weight and the normal force on it are equal and opposite. Is this a good example of Newton's third law?

11. Shortly after jumping from an airplane, a parachutist will descend with constant velocity. Why is this?

12. Using Newton's second law, devise a system (other than the compressed spring described in Section 4–3) that could be used to measure masses.

13. Newton stated that a reference frame at rest with respect to the distant stars would be a good inertial system. Comment on this. Suggest systems that would be good inertial systems for experiments conducted in, or on (a) a physics lab; (b) the Space Shuttle; (c) a ship at sea; and (d) Mars.

14. A rubber ball and a golf ball have the same mass but the rubber ball has a larger radius than the golf ball. If they are identically accelerated with the same initial force, why might the golf ball go farther in the atmosphere?

15. Describe the forces that are responsible for the acceleration of an automobile. In particular, in what way is friction between the tires and the road responsible for the acceleration? What, then, is the role of the engine?

16. If you tie a rock to one end of a piece of string and swing it in a circle, you experience a force on your hand. Why?

17. When a baseball hits a bat, a force must be exerted on the ball to change its direction of motion. Describe the forces between the ball and the bat. What effect does the ball have on the bat? Does the batter feel the effects of any of these forces?

18. Give three examples of noninertial forces.

19. A marble is placed on the side of a bowl. It rolls down into the bowl, then continues to roll up and down the bowl's side until it finally comes to rest at the bottom. Explain this in terms of the first and second laws.

20. Describe qualitatively how you might go about constructing an accelerometer, which measures the acceleration of a ship. [Hint: Think about how a spring placed between your back and the back of a seat would react when an automobile in which you are sitting accelerates forward. Such devices are of great importance for "blind" navigation, which might be required

for a submarine. They allow us to follow the path when we know the acceleration as a function of time and then reconstruct the velocity and the position as functions of time. (This procedure is described in Chapter 2.) It is, in fact, difficult to construct an accurate version of such a device without knowing about rotational motion.]

21. You are standing on a scale in an elevator. The elevator suddenly starts to move upward. What happens to the reading of the scale, and why?

22. A diver jumps from a high platform and experiences a feeling of weightlessness. Is the force of gravity no longer acting on the diver?

23. A 5-kg mass is placed on a table. A professor states that the normal force on the mass due to the table is $(5 \text{ kg}) (9.8 \text{ m/s}^2)$ and that this is a consequence of Newton's third law. Is this a sound analysis?

24. In Example 4–5, we discussed Professor A, who is at rest on a train platform, and Professor B, in the accelerating train, to illustrate the consequences of the noninertial nature of Professor B's reference frame. How do we know that Professor A's frame is inertial?

25. Consider a horse that pulls on a cart. By Newton's third law, the cart pulls on the horse with a force of equal magnitude but in the opposite direction. How can there be any motion?

26. What makes a car go forward when the engine is turned on and the transmission is engaged?

27. You are standing in a stationary bus and suddenly find yourself thrown backward. What does Newton's second law say about that?

28. A car is stationary on a flat parking lot. The force of gravity acts downward and an equal and opposite normal force acts upward. Is it correct to say that these forces are equal and opposite because of Newton's third law?

29. In a tug of war, one side is stronger than the other. Assuming that both sides exert themselves to the maximum, discuss the motion of a handkerchief tied to the rope being pulled.

30. A fellow student states that forces cause an object to move. Criticize this statement.

31. Baron Munchausen claimed that it is possible for a very strong man to pull himself off the ground and rise into the air by pulling on his bootstraps. Discuss this mode of liftoff in the cold light of Newton's laws.

PROBLEMS

4–1 Forces and Newton's First Law

1. (I) In applying Newton's laws, we must identify the forces acting on an object. Are there any forces acting on the following objects? If so, list them: (a) the Space Shuttle in Earth orbit; (b) an ice skater coasting on ice; (c) the *Voyager I* spacecraft far past the orbit of the planet Pluto.

2. (I) The forces acting on an airplane are the following:

    gravity: $F_g = 4 \times 10^5 \text{ N}$ down;

    engine thrust: $F_E = 10^5 \text{ N}$ forward;

    lift: $F_L = 5 \times 10^5 \text{ N}$ up;

    air drag: $F_D = 5 \times 10^4 \text{ N}$ backward.

    What is the net force on the airplane?

3. (I) In a tug of war, a red ribbon tied around a point on the rope between the two teams moves with a uniform velocity of 0.1 m/s in the *y*-direction. One team exerts a force on the rope of 600 N in the +*y*-direction. What force does the other team exert on the rope?

4. (I) A boat sailing in the northeasterly direction with constant speed experiences a wind force of magnitude 10^4 N from the east. What is the force on the sailboat due to the resistance of the water to motion through it?

5. (I) In a classic demonstration, Otto von Guericke used 16 horses—8 on each side—to try to pull apart two hemispheres forming a sphere from which air had been evacuated. Could he as well have used only 8 horses on one side, with the other side tied to a sturdy tree?

6. (II) Three nonzero forces act on a particle at the origin of a coordinate system. $\mathbf{F}_1$ is in the z-direction, whereas $\mathbf{F}_2$ and $\mathbf{F}_3$ lie in the xy-plane. Can you arrange the magnitudes and directions of $\mathbf{F}_2$ and $\mathbf{F}_3$ (keeping them always in the xy-plane) so that the particle does not accelerate?

7. (II) (a) A spider is suspended from a single vertical thread; the spider has a mass of $m = 30$ mg. The spider is acted upon by the force of gravity, which is directed downward and has magnitude 3×10^{-4} N, and by the *tension*, T, in the thread, a common type of contact force that always acts in the direction of the thread, and is directed away from the point at which it is attached. In this case, the tension acts in the upward direction. What is the magnitude of the tension? (b) The spider is now attached to two threads of equal length, which make a 120° angle with each other, as in Fig. 4–26. The spider is motionless, waiting for a victim. What is the tension in each thread? [*Hint*: The tension is a vector directed away from the attachment point. Newton's first law must be satisfied in its vector form.]

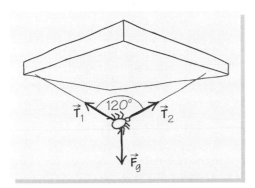

FIGURE 4–26 Problem 7.

8. (II) A very large—consider it to be infinitely large—mesh of stiff wires makes a horizontal plane and large electric charges are placed at the intersections of the wires. A mass with another charge is placed below one of the charges on the mesh. Electric charges exert forces on one another that are proportional to the strength of the charges. Each charge on the mesh repels the lone charge with a force that varies with the distance between the charges. The forces are directed along the line between the charges. Show that the net force on the lone charge is a repulsion directed straight downward.

4–2 Newton's Second Law of Motion

9. (I) A car coasts along a road with initial velocity $\mathbf{v}_0$. It inevitably slows down and finally comes to rest. (a) Describe why this is so. (b) An observer traveling with uniform velocity $\mathbf{v}_0$ starts out at rest relative to the car. What does she see with the passage of time? How does she explain what happens?

10. (I) The force of gravity on an apple (mass 0.1 kg) has a magnitude of about 1 N. What is the acceleration of the apple as it falls toward Earth? How large is the force of gravity on a falling automobile, mass 1000 kg, if it has the same acceleration as the apple? In each case, assume that only the force of gravity acts.

11. (I) The force of gravitation attracts two masses m_1 and m_2 to each other. If the masses are separated by a distance d, the magnitude of the force on each mass is

$$F = Gm_1m_2/d^2,$$

where G is a constant. Suppose $m_2 = 3m_1$. Make a sketch of the two masses with vectors that indicate the direction of the forces on the two bodies. Draw the lengths of the vectors to correspond to the magnitude of the two forces. Repeat the sketch but replace the force vectors with acceleration vectors.

12. (II) A spring exerts a force when it is compressed or extended. The force is proportional to the distance x by which it is compressed or stretched away from its equilibrium position. The direction of the spring's force is toward its equilibrium position. Draw diagrams with the spring compressed and extended. Choose a direction for x. Write mathematical expressions for the force in both situations.

13. (II) Five forces, all of the same magnitude F, act on an object of mass m at the origin of the coordinate system shown in Fig. 4–27. Two of the forces are aligned along the x-axis, one is oriented in the $+x$-direction and one in the $-x$-direction. Two other forces are similarly aligned with the z-axis. The fifth force points in the $+y$-direction. What are the direction and magnitude of the acceleration of the object? If we had specified the forces in a different order, would the answer have been different?

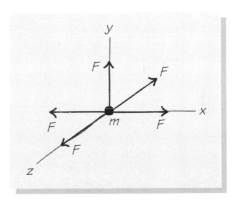

FIGURE 4–27 Problem 13.

14. (II) A sports car of mass 820 kg accelerates from 0 mi/h to 60 mi/h in 6.00 s. What is the average force, in newtons, that the road exerts on the wheels of the car? A huge station wagon of mass 1900 kg can accelerate from 0 mi/h to 60 mi/h in 9.5 s. What force does the road exert on the station wagon? What would the station wagon's acceleration be if it were acted on by the same external force as the sports car experiences?

15. (II) A car of mass 1150 kg accelerates from rest to 100 km/h in 11 s. With additional streamlining, the same car undergoes acceleration to the same speed in 9 s. What is the difference in the force exerted by the air (the *drag force*) on the car in the two cases? For this problem, assume the drag force is constant. (This assumption is a poor one in practice.)

16. (II) A common type of contact force is that provided by a taut rope. Suppose that a taut rope is attached to an object. In what direction can the rope apply a force to the object? This force is known as the *tension*. (a) A cart being pulled horizontally by a light rope (the word "light" in this context means that you can ignore the mass of the rope in all your considerations) has a mass of 25 kg and accelerates in the horizontal direction at 2.4 m/s^2. What is the tension in the rope? If the same cart is pulled so that it accelerates at 0.65 m/s^2, what is the tension? Assume that the only force acting on the cart is the tension. (b) Suppose that the rope passes over a fixed pulley (Fig. 4–28) and the cart accelerates horizontally at 1.4 m/s^2. What is the upward force on the pulley?

FIGURE 4–28 Problem 16.

17. (II) A father pulls his identical twins on identical sleds tied one after the other (Fig. 4–29). He exerts a force **F** that makes an angle of 30° with the horizontal and that leads to an acceleration of the two sleds. What is the tension in the rope that he is pulling? What is the tension in the rope that connects the front sled to the rear sled? The mass of each sled plus its twin is *m*. Assume that there is no friction between the sleds' runners and the snow surface.

FIGURE 4–29 Problem 17.

18. (II) Electrons, mass about 10^{-30} kg, are constituents of atoms and respond to electrical forces. Such forces, of varying strengths, can be generated in the laboratory. Suppose that a constant electrical force of 10^{-15} N acts on an electron. What is the speed of the electron after 10^{-10} s? after 10^{-9} s? after 10^{-7} s?

4–3 Newton's Third Law of Motion

19. (I) A hook is screwed into a ceiling; one end of a string is tied to the hook and the other end has an 8-kg mass attached to it. Assuming that the force of gravity acting on a mass *m* has magnitude *mg*, where *g* ≈ 9.8 m/s^2, what force does the hook exert on the string?

20. (I) A 3000-kg pickup truck pulls a 1200-kg boat on a trailer, and they are accelerating together at 1.2 m/s^2. What is the horizontal force that the truck and boat trailer exert on the road?

21. (I) A falling automobile, mass 950 kg, has an acceleration of magnitude 9.8 m/s^2 when only the force of gravity acts on it. What is the magnitude of the upward acceleration of Earth? Take the mass of Earth to be 6.0 × 10^{24} kg.

22. (I) The force exerted on a satellite by an astronaut is (0.5**i** + 0.7**j** + 5.1**k**) N. What is the force exerted by the satellite on the astronaut? What is the magnitude of this force?

23. (II) In Problem 1, you were asked to find the forces acting on certain objects. By Newton's third law, these original objects are the sources for forces that act on other objects. What are these other objects for each of the following original objects: (a) the Space Shuttle in an Earth orbit, (b) an ice skater coasting on ice, (c) the *Voyager I* spacecraft far past the orbit of the planet Pluto?

24. (II) A force of magnitude 8.0 N pushes on a horizontally stacked set of blocks on a frictionless surface (Fig. 4–30) with masses $m_1 = 2.0$ kg, $m_2 = 3.0$ kg, and $m_3 = 4.0$ kg. (a) What is the acceleration of the stack? (b) What are the forces on block 1, as well as the net force on this block? (c) Repeat part (b) for block 2. (d) Repeat part (b) for block 3.

FIGURE 4–30 Problem 24.

25. (II) The engine of a train pulls five cars, each of mass $m = 30,000$ kg. During a period of acceleration, the force between the engine and the first car is 45,000 N. (a) What is the acceleration of the train? (b) Sketch the forces acting on the first car. (c) How large is the tension in the hook between the first and second cars?

26. (II) Repeat Problem 24, but this time with the blocks stacked in the reverse order; that is, block 3 to the left and block 1 to the right.

27. (II) Three charges move through space with no forces acting on them except the electric forces that they exert on each other. In an appropriate coordinate system, some of the forces can be broken down as follows: The force that charge 1 exerts on charge 2 is $\mathbf{F}_{21} = (2\mathbf{i} - 3\mathbf{j} + \mathbf{k})$ N; and the force that charge 1 exerts on charge 3 is $\mathbf{F}_{31} = (-3\mathbf{i} + 2\mathbf{j} - 3\mathbf{k})$ N. What is the total, or net, force on charge 1?

*4–4 The Effects of Noninertial Frames

28. (I) An observer inside an elevator that is in free fall will see any object that was initially at rest in midair inside the elevator remain in that position. How does he explain this fact, assuming that he knows about the existence of gravity?

29. (I) Consider the situation described in Example 4–5, with $v_A = 0$ and $v_B = 10$ m/s. The train is accelerating in the *x*-direction at 2 m/s^2, according to Professor A. Can Professor B tell that he is accelerating? Does he think that Professor A is accelerating? Do our physical senses help us solve these questions? What is the acceleration of Professor A according to Professor B? Is this a real acceleration?

30. (II) An observer sits in a cylindrical bathtub that can rotate about a vertical axis. At first, the tub is at rest—relative to a neighboring tree for example. A little later, the tub is rotating at high speed around its vertical axis. The observer is still at rest relative to the tub, but the water level is no longer flat; it runs up on the sides of the tub. How does the observer explain this?

31. (II) A simple demonstration of the effect of a noninertial frame can be made by dangling a watch from a chain while you are taking a trip in a jet. What is the angle of the chain with respect to a vertical window edge when (a) the jet is parked at the gate; (b) the jet is accelerating with some acceleration a' along the runway just before takeoff; (c) the jet is at cruising speed at 30,000 ft? You can try this type of experiment in an accelerating automobile.

32. (II) A mass of 2 kg lies on a horizontal table that is placed in the back of a truck (Fig. 4–31a). The mass is held in position by a string and a force of 6 N acts on it in the $-y$-direction. The truck then accelerates in the $+x$-direction. As a result, an observer in the truck sees that the string holding the mass on the table makes an angle of 20° with respect to its initial orientation, as shown in Fig. 4–31b. What is the acceleration of the truck? Describe what the observer in the truck sees and how he might interpret his observations.

FIGURE 4–31 Problem 32.

4–5 Using Newton's Laws I: Force Diagrams

33. (I) A car is accelerating straight ahead on a flat road. Draw the force diagram for the car.

34. (I) A child slides down a water slide at an amusement park. What forces are present? Draw a force diagram for the child.

35. (I) Consider an astronaut sitting in a rocket just after blastoff from Cape Canaveral. Draw a force diagram showing the forces present on the astronaut sitting in her seat.

FIGURE 4–32 Problem 36.

36. (I) Consider the arrangement shown in Fig. 4–32. The whole system is at rest. Draw all the forces, draw the force diagram for each mass, and give the equations of motion for each mass.

37. (I) A person throws a medicine ball into the air. Draw separate force diagrams for the ball and for the person (a) for the time just before the ball leaves the hands, and (b) after the ball has been thrown.

38. (I) A piano sits within a large freight elevator. (a) Draw a force diagram for the piano alone. (b) Draw a force diagram for the elevator alone.

39. (II) Using a rope, a horse pulls a wagon that is initially at rest. (a) Draw a force diagram for the horse. Include friction, Earth's gravity, the influence of the wagon, and any other forces that may be operating. Draw the direction of the forces as accurately as possible. (b) Repeat part (a) for the wagon. (c) Repeat part (a) for Earth, including only the horse and wagon as influences. (d) Why doesn't the horse accelerate backward toward the wagon? In answering, think about the different forces acting on horse and wagon.

40. (II) A block of mass M sits on a rough horizontal surface. A rope inclined upward at an angle θ with the horizontal exerts a force of magnitude T on the block. The block remains stationary. Draw a force diagram and express all the forces on the block in terms of M, g, θ, and T.

41. (II) Consider a brick sliding down an inclined plane of 30°. Draw a force diagram for the brick that includes the gravity, friction, and normal forces. Choose a coordinate system and give the force components in equation form.

42. (II) Draw two force diagrams for a tug of war in which the strengths of the two sides are unbalanced. (The two objects in the diagram will be the two groups of contestants.) Be sure to include all forces!

43. (II) A heavy man stands on a ladder that is leaning against a rough-surfaced wall and rests on the ground (Fig. 4–33). Draw the force diagram for the system of man and ladder. Ignore the mass of the ladder. If the ground is very smooth, the ladder may start to slip. Why?

44. (II) Two astronauts, Joe and Moe, are taking a space walk. They are connected to the ship only by a slack rope. They have a light rope between them and decide to have an outer space tug of war. In their space suits, Joe has a mass of 100 kg, whereas Moe has a mass of 110 kg. The loser of the tug of war is the first astronaut to cross an imaginary line midway between them. (a) Draw a force diagram for each astronaut.

FIGURE 4-33 Problem 43.

What is the constraint of Newton's third law? (b) Show that Moe wins the tug of war. (c) Miffed at his loss, Joe decides to challenge Moe to a tug of war with their feet firmly planted on Earth. Draw the force diagram for each astronaut. (d) Joe wins the tug of war this time. How is this possible?

45. (II) An electrically charged mass, when placed between two charged parallel plates, experiences a constant force F_e perpendicular to the plates. Suppose that such a mass moving vertically with some speed v_y enters the region between two such plates, which are oriented vertically. Draw a force diagram for the mass while it is between the plates. Include the force of gravity. Note that there are two possibilities for the direction of F_e; choose one or the other.

46. (II) Objects made of iron are attracted by an electromagnet when a switch is thrown activating it. Suppose that a small piece of iron is suspended by a thread in the vicinity of an electromagnet, as shown in Fig. 4–34. (a) Draw a force diagram for the iron object immediately after the switch is closed and the electromagnet is activated, and (b) for the situation when the switch has been closed for a long time.

FIGURE 4-34 Problem 46.

47. (II) A person is standing in an elevator that is moving upward with a constant speed. (a) What are the forces acting on the person? (b) Suppose that the elevator is accelerating upward with an acceleration g that is equal in magnitude to the acceleration that the person would have if the only force acting on him were gravity. What are the forces acting on the person, and how large are they? (c) Suppose that the cable breaks and the elevator, together with its unfortunate passenger, is falling freely. What forces are acting on the person?

4-6 Using Newton's Laws II: Finding the Motion

48. (I) An object of mass 25.43 kg that can move only in the xy-plane is at rest at the origin of the xy-coordinate system at time

$t = 0$ s. Constant forces $\mathbf{F}_1 = (1.715, 0, 0)$ N and $\mathbf{F}_2 = (0, 1.128, 0)$ N act on the object. (a) Draw the force diagram for the object. (b) State the initial conditions; that is, the position and velocity vectors at $t = 0$ s. (c) What are the position and velocity of the object at time $t = 1.200$ s? (d) at $t = 3.600$ s?

49. (II) An object of mass M is subject to two constant forces: $\mathbf{F}_1$, pointing in the x-direction, and $\mathbf{F}_2$, pointing in the y-direction. Show that the motion is one of constant acceleration. Find the magnitude and direction of this uniform acceleration.

50. (II) A third constant force $\mathbf{F}_3 = (F_{3x}, F_{3y}, 0)$ acts on the object in Problem 48 at $t = 0$ s in addition to the forces $\mathbf{F}_1$ and $\mathbf{F}_2$. (a) What must F_{3x} and F_{3y} be so that the object does not accelerate? (b) What must F_{3x} and F_{3y} be so that at $t = 5.000$ s, the object is at position $\mathbf{r} = (1.000, 1.000, 0)$ m?

51. (III) After some time, the displacement of an object of $m = 2.0$ kg is measured to be $x = At^{3/2}$, where $A = 6.0$ m/s$^{3/2}$. Determine the net force acting on this object. Note that the force will depend on the time.

52. (III) When a small sphere, starting from rest, falls through a liquid, it experiences a so-called linear drag force $\mathbf{F}_D$, in addition to the force of gravity mg directed downward. $\mathbf{F}_D$ is directed upward and has a magnitude proportional to the velocity, $F_D = bv$, where b is a constant. The equation of motion—Newton's second law—for motion in the vertical direction z (measured with the +direction downward) reads $m \, dv/dt = mg - bv$. (a) Show that this equation is solved when the velocity takes the form $v(t) = mg(1 - e^{-bt/m})/b$. (b) Sketch the speed as a function of time. (c) Sketch the position as a function of time.

53. (III) The force on a mass m that can move along the x-axis is given by $F = -kx$. (This force is characteristic of the spring; k is a property of this particular spring and is called the spring constant.) (a) Show that, with an appropriate choice of the constant c, $x = A \cos(ct) + B \sin(ct)$ is a solution of the equation of motion. What is the value of c? (b) If the object is at $x = 0$ at $t = 0$ and if its velocity at that time is v_0, what are A and B?

General Problems

54. (I) The gravitational force that the Sun exerts on Earth has a magnitude of 3.5×10^{22} N and is directed toward the Sun. What force does Earth exert on the sun?

55. (II) A load of mass 53.2 kg is to be lifted by two ropes, each going over pulleys. The ropes are separated in such a way that both ropes initially make an angle of 65° with the horizontal (Fig. 4–35). The magnitude of the forces exerted by each of

FIGURE 4-35 Problem 55.

the ropes is 333 N. (a) Draw a force diagram for the load. The only forces acting on it are the forces of the ropes, directed along the ropes, and the force of gravity, which points down and has a magnitude given by the mass of the load times the constant acceleration $g = 9.80$ m/s^2. (b) What is the net force on the load? (c) What is the initial acceleration (magnitude and direction) of the load?

56. (II) An object of weight $F_g = 20$ N is suspended as shown in Fig. 4–36. The rope and pulleys are very light and friction is negligible. Determine the weight of the object that maintains the equilibrium, given that the angle between the rope and the horizontal is 35° on both sides of the movable pulley.

FIGURE 4–36 Problem 56.

57. (II) A window washer sits on a board supported by two ropes that go over pulleys attached to a scaffold. He holds the other ends of the ropes in his hands (Fig. 4–37). He pulls on them in such a way that he moves upward at constant speed. Draw a force diagram for the window washer only and give all the equations; assume that the tension in a given rope is the same everywhere in the rope, even after it wraps around a pulley. Suppose that the effect of the window washer's efforts leads to an upward acceleration of both the window washer and the board. How are the equations modified?

FIGURE 4–37 Problem 57.

58. (II) A tennis ball is dropped onto a granite slab and it bounces back to about 90 percent of the height from which it was dropped. What forces acted on the tennis ball for it to exhibit this behavior? In what direction do these forces point? Estimate their magnitude.

59. (II) A train consists of an engine and three cars tied closely together. The mass of the whole train is 10^5 kg; the engine alone has a mass of 4×10^4 kg and each car has a mass of 2×10^4 kg. Through its wheels, the engine can exert a horizontal force of 2×10^4 N on Earth. (a) When the engineer wishes to accelerate forward, in which direction should the engine exert its force on Earth? What is the magnitude of the force exerted by Earth on the engine then? (b) When the engineer acts to accelerate the train, what is the force on the set of three cars and what is the subsequent acceleration of these cars? (Treat the three cars as a unified whole.) (c) When the engineer acts to accelerate the train, what are all the forces on, and subsequent acceleration of, the second car of the train?

60. (II) A boat is being pulled up the middle of a canal at a steady speed by two horses, one on each side of the canal (Fig. 4–38). The ropes tying the boat to the horses each make an angle of 30° with the lengthwise direction of the canal. (a) Draw a force diagram for the boat. Include the friction force due to the boat sliding through the water. (b) If the force exerted on each horse by its rope is 2000 N, what is the force of friction that the boat experiences due to the water?

FIGURE 4–38 Problem 60.

61. (II) A passenger in a stationary elevator is acted on both by the force of gravity, which is directed downward, and by a contact force, directed upward by the floor. The contact force from the floor cancels the force of gravity, and in this way there is no net force on the passenger, which explains why she measures herself to be at rest with respect to the walls. (a) What is the force diagram on the passenger when the elevator descends at constant velocity? (b) The elevator cable breaks. (Do not be alarmed, because the emergency brake will act if you finish this problem in time.) As seen from the outside, the passenger and elevator fall together and accelerate under the force of gravity. A force diagram for the passenger would show the single force of gravity acting. What does the passenger observe, and how would she draw a force diagram for herself?

62. (II) A parachutist has jumped from a plane. (a) What are the forces acting on the parachutist when the parachute opens? Suppose that the force due to air resistance opposes the parachutist's motion and is proportional to the speed of the system; that is, $\mathbf{F}_{\text{air resistance}} = -A\mathbf{v}$. (b) What are the dimensions of the constant A? (c) Sketch the downward acceleration of the parachutist as a function of velocity. Show that the velocity of the system becomes constant at some point. This constant velocity is called the *terminal velocity*. (d) If the force of gravity on the system is 1000 N and the terminal velocity is 6.0 m/s, what is the value of the constant A?

63. (II) A parachutist experiences two forces: One is the force of gravity, which is of the form $\mathbf{F}_g = -mg\mathbf{j}$. Here, m is the mass of the parachutist and g is the acceleration due to gravity. The other force is a drag force, and, in contrast to Problem 62, it has the form $\mathbf{F}_d = Av^2\mathbf{j}$ in this problem, where the velocity of the parachutist is given by $-v\mathbf{j}$. (a) What are the dimensions (and units) of A? (b) Complete the equation $dv/dt = \ldots$ (c) At some point, the parachutist reaches a terminal velocity that is constant. What is it?

64. (II) An egg of mass 60 g is in a stiff box of mass 500 g, surrounded on all sides by padding that keeps the egg in place. A force of $\mathbf{F} = (3.2\ \mathrm{N})\mathbf{i} + (2.8\ \mathrm{N})\mathbf{j}$ acts on the outside of the box, in addition to the force of gravity (which is oriented in the $-y$-direction). (a) Draw force diagrams for the egg–box system and for the egg. (b) Express Newton's second law as it applies to the egg. (c) What are the acceleration of the egg–box system and net force on the egg?

65. (II) A bicyclist rides a flat course at constant speed v with respect to the ground into a stiff but steady headwind of speed v_w, also with respect to the ground. (a) Draw a force diagram for the cyclist–cycle system. List the forces acting on this system, including the wind resistance, which pushes in his face, and the frictional forces with the ground, which propel him forward. (b) If you are observing the cyclist from a reference frame at rest with respect to the ground, what is the vector sum of all the forces? (c) If you are observing the cyclist from a frame at rest with respect to the bicycle, what are the speed of the cyclist, the speed of the wind, and the vector sum of all the forces? (d) If you are observing the cyclist from a frame at rest with respect to the air, what are the speed of the cyclist, the speed of the wind, and the vector sum of all the forces?

66. (II) An electron is placed between two charged parallel plates, oriented vertically. The electron experiences a constant force perpendicular to the plates. Suppose that the electron, whose mass is 9×10^{-31} kg, is moving vertically with $v_y = 2 \times 10^6$ m/s, and enters the region between the two plates (Fig. 4–39). This region has a vertical height of 1 cm, and while in it, the electron experiences a force of 3×10^{-18} N. (a) How long does the electron stay in the region? (b) After the electron has passed through the region, what is its horizontal velocity component?

67. (III) As shown in Fig. 4–40, a plumb bob suspended from a frame settles to a steady position as the frame slides down a frictionless inclined plane. What is the angle the plumb bob makes with the vertical during the slide?

FIGURE 4–40 Problem 67.

68. (III) The acceleration of an object dropped from a height is smaller at the equator than at the poles, even when the force of gravity is assumed to be the same everywhere on the surface of Earth. (a) Assuming that the difference has something to do with the rotation of Earth, obtain an expression for the difference between the accelerations in terms of the radius of Earth and its rotation frequency. (b) Given that the radius of Earth is $R = 6.38 \times 10^6$ m and that the acceleration at the poles is $g_0 = 9.80$ m/s^2, calculate the acceleration of a falling object at the equator. (c) Calculate the acceleration of a falling object as a function of the angle of latitude, in terms of g_0. (d) If experiment shows a deviation from your prediction, what effects could be responsible for that deviation?

FIGURE 4–39 Problem 66.

The crawler that transports the Space Shuttle to its launch pad uses friction between its treads and the ground to move the mammoth assembly. The combined weight of the crawler and shuttle assembly is some 80 million Newtons.

Applications of Newton's Laws

In this chapter, we apply Newton's laws to a variety of situations in which several forces act, including gravity, tension, normal and contact forces, friction, and drag. These forces are ones that frequently act on us and the objects all around us. We shall also look at the role of forces in circular motion, as well as the features of motion in rotating frames. Finally, we shall take the time to see how the forces we look at in this chapter are ultimately described in terms of more fundamental forces—in particular, ones that act at a microscopic level.

5–1 SOME SIMPLE CONSTANT FORCES

Constant forces—forces whose magnitude and direction are independent of time and position—describe a wide variety of phenomena. We discussed some examples in Chapter 4: the force of *gravity*, $\mathbf{F}_g$, which acts on an apple falling from a tree, and the *normal force*, $\mathbf{F}_N$, which acts on the apple when it sits on a kitchen table. Although a normal force is not necessarily constant, it is indeed constant in the situations to be illustrated here.

Gravity

The cause of projectile motion near Earth is the force of gravity. As we saw in Chapter 3, projectile motion is motion with a constant acceleration **a**, which points downward and has magnitude g. The acceleration vector provides a complete description of projectile motion, which is generally parabolic (Fig. 5–1). Having now studied the *causes* of acceleration, we can attribute this constant acceleration to the force of gravity—a force that is constant and directed down toward the center of Earth. Newton's second law tells us that an object accelerates if a force acts on it. According to Newton's second law, then, the acceleration **a** of an object on which gravity alone acts can be written in terms of the force of gravity, $\mathbf{F}_g$, and the inertial mass m of an object:

$$\mathbf{a} = \frac{\mathbf{F}_g}{m}. \tag{5-1}$$

All objects at the same location accelerate under gravity in the same way, regardless of their mass.

A very special characteristic of the force of gravity is that, at any given location, *it causes all objects to accelerate in the same way—no matter what their mass.* In other words, the right-hand side of Eq. (5–1) is independent of the mass m; *the only way this can happen is for the force itself to be proportional to the mass*, so that the factor m in the denominator cancels with another such factor in the force itself.

Let's look more carefully at the meaning of this remarkable cancellation. That an object has constant acceleration under the influence of gravity means only that this force takes the form

$$\mathbf{F}_g = m_g\mathbf{g}, \tag{5-2}$$

where **g** is a constant vector with dimensions of acceleration, pointing to the center of Earth.[†] The "gravitational mass" m_g is the property of the object that determines the strength of the gravitational force acting on it. However, the acceleration of *all* objects under the influence of gravity is *precisely the same*; mathematically, this statement is

$$m_g = m. \tag{5-3}$$

FIGURE 5–1 A number of forces act on the child playing on a rope swing, including gravity and the tension of the rope.

This relation is remarkable because we have no reason to think that the force of gravity has anything special to do with the inertial mass. We recall from Chapter 4 that the inertial mass determines the *response* to a force, and tests to determine the inertial mass are possible without using gravity at all. Conversely, tests to determine the force of gravity on an object can be performed without using motion at all—say, by observing the equilibrium stretch of a spring while the object hangs from it under the influence of gravity. Nevertheless, the equality of Eq. (5–3) has been experimentally verified to a very high degree of accuracy.[‡] Thus the force of gravity has the simple form

The force of gravity

$$\mathbf{F}_g = m_g\mathbf{g} = m\mathbf{g}. \tag{5-4}$$

The acceleration **a** of any object at Earth's surface under the influence of gravity alone is given by the constant vector **g**:

$$\mathbf{a} = \frac{\mathbf{F}_g}{m} = \mathbf{g}. \tag{5-5}$$

As described in Chapter 3, the magnitude of **g** on the surface of Earth is roughly

$$g = 9.8 \text{ m/s}^2. \tag{5-6}$$

Do not confuse weight with mass. Weight is a force.

[†]This force is present near any large spherical object, but the magnitude of **g** is different.

[‡]The so-called Eötvös experiments verify that the inertial and gravitational masses are the same to one part in 10^{12}.

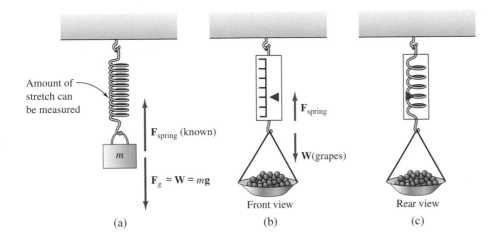

FIGURE 5–2 The weight of an object is found by balancing it against a spring, for which the force law is known. (b) and (c) illustrate the calibration.

Experiments show that this value varies by about 1 percent over Earth's surface, with the higher values occurring at the poles. This observed variance is due to irregularities in the shape and density of Earth and to Earth's rotation.

The force of gravity on an object is commonly called the object's **weight, W,** where

$$\mathbf{W} = \mathbf{F}_g = m\mathbf{g}. \tag{5–7}$$

We can experimentally determine the weight of an object by balancing the force of gravity against a second calibrated force that is acting on the object. In Fig. 5–2a, for example, the weight of a block is found by using a calibrated spring as the second force. Figures 5–2b and c illustrate this calibration. In fact, when you step on a bathroom scale, you are measuring the force of gravity on your body; the second force is the calibrated bathroom scale, which is, in essence, a compressible spring. Don't confuse the weight of an object, measured in newtons or pounds, with its mass. We know from Chapter 4 that the mass of an object is the quantity of matter it contains, which is measured by its inertia—its resistance to any change in motion. The mass and weight of an object, though, are numerically proportional to each other through *g*—the weight of an object equals the object's mass times *g*. This numerical proportionality is what allows a bathroom scale to be calibrated in mass rather than weight.

Tension

We know from experience that it is possible to pull objects using ropes. The tension, **T,** is a force exerted on an object by flexible ropes (or wires or cables or strings) that is directed along the direction of the rope; it always pulls, never pushes (Fig. 5–3a). *For light (negligible mass) ropes, the tension is the same everywhere along the rope.* An idealized rope of negligible mass—the kind of rope we employ in this book[†]—is a rope whose mass is small compared to other masses in the problem.[†] The tension varies according to the particular situation. A rope pulling a wagon full of rocks has more tension than does the same rope pulling an empty wagon. Tension originates in the molecular forces that hold a rope together and give it flexibility. Note that the tension is the same everywhere in the rope, even when the rope turns corners, provided that it turns the corner by means of (idealized—negligible mass) pulleys turning on frictionless bearings (Figs. 5–3b, c).

[†]When the mass of the rope is not negligible, then the tension will vary along the rope.

APPLICATION

Bathroom scales

FIGURE 5–3 (a) The tension force pulls but cannot push. (b, c) A rope of negligible mass with a given tension can maintain that tension even if the direction of the rope is changed: Because the rope is passed over an ideal pulley, the tension can be transmitted around corners.

The tension

PROBLEM SOLVING

How tension acts

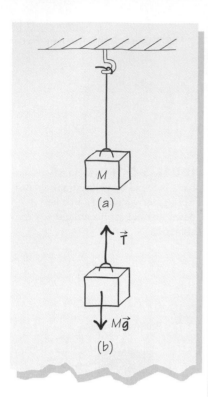

FIGURE 5–4 (a) Example 5–1. (b) Force diagram for a 2.0-kg package suspended from a line.

EXAMPLE 5–1 Let's consider Fig. 5–4a, which shows a fishing line hanging from a stationary support. This fishing line is 10-lb-test line, which means that it may break when the tension within it exceeds 10 lb. A package of mass $M = 2.0$ kg is suspended from the line. Find the tension in the line. Will the line hold? What will happen with a package of mass 5.0 kg?

Solution: We want to find the tension, T, in the line for the two different masses. We draw a force diagram for mass M (Fig. 5–4b) to determine the tension: As long as the line holds, **T** and M**g** balance, and there is no acceleration of the package. Without acceleration, Newton's second law reads

$$\mathbf{F}_{net} = \mathbf{T} + M\mathbf{g} = 0.$$

The tension is directed upward while the vector **g** is directed downward, so the second law is represented by a one-dimensional equation:

$$T - Mg = 0.$$

Thus, when the mass is suspended, the tension in the line has magnitude

$$T = Mg.$$

When $M = 2.0$ kg, we have

$$T = Mg = (2.0 \text{ kg})(9.8 \text{ m/s}^2) = 19.6 \text{ N} = (19.6\,\cancel{N})\left(\frac{1 \text{ lb}}{4.45\,\cancel{N}}\right) = 4.4 \text{ lb.}$$

The line will hold.
   When $M = 5.0$ kg,

$$T = (5.0 \text{ kg})(9.8 \text{ m/s}^2) = 49 \text{ N} = (49\,\cancel{N})\left(\frac{1 \text{ lb}}{4.45\,\cancel{N}}\right) = 11 \text{ lb.}$$

The line may not hold with the heavier package suspended from it because the tension within it exceeds its breaking tension.

How could we measure the tension within a rope at a given point? One way would be to cut the rope at the given point and insert a calibrated spring scale. The scale will stretch by an amount corresponding to the tension in the rope.

Normal Force

The normal force

Let's take a look at the forces acting on an apple that has been placed on a table (Fig. 5–5). In addition to the force of gravity, a *normal force*, $\mathbf{F}_N$, acts on the apple. In Chapter 4, we were introduced to this force and acknowledged that such a normal force must exist—or gravity would cause the apple to simply accelerate into the surface of the table. We also learned that the normal force acts perpendicular to, and away from, the surface at the object's point of contact. We know that the normal force acts in a direction perpendicular to the table top because the apple does not accelerate to one side of the table or the other. In fact, whatever the tilt of the table, the apple does not move into its surface because *the normal force always acts perpendicular to, and away from, the surface*. The normal force pushes but never pulls.

As for tension, the magnitude of $\mathbf{F}_N$ is determined by the situation, often by the use of Newton's first law. An apple of 0.25 kg sitting on a table must experience an upwardly directed normal force that just cancels the downward force of gravity; the magnitude of this force would be: $mg = (0.25 \text{ kg})(9.8 \text{ m/s}^2) = 2.45 \text{ N}$.

How can we measure the normal force? An ordinary bathroom scale suffices. The spring in the scale opposes the force of gravity in exactly the same way as does the normal force and therefore gives the value of the normal force. In Fig. 5–6, a scale measures the normal force of 2.5 N that acts on the apple (although the force is often

PROBLEM SOLVING

Normal forces are perpendicular to the surface.

(a)   (b)

(c)

FIGURE 5–5 Some normal forces. These forces are always perpendicular to the surface involved.

expressed in units other than newtons!). The scale could equally well be placed on other surfaces that exert normal forces. For example, the normal force exerted on a hand by a wall can be measured directly if a scale is inserted between the wall and the hand.

There are other types of forces that are similar to the tension and the normal force; namely, the support forces supplied by hooks, connection points, bearings, and so forth. The direction and magnitude of such forces are determined by the requirement that the attachment point in question does not move. For example, a hook from which a 20-lb weight is suspended must supply an upward force of 20 lb if the mass is not to fall.

Some Examples

In many everyday situations, the tension and the normal force are constant. Examples involving constant tensions, constant normal forces, and gravity provide us with solvable equations of motion. We shall investigate several such examples here.

It may be helpful for us to recall the motion that results when *any* constant force **F** acts on an object of mass m. The object's velocity and position as a function of time represent the solutions to Newton's second law, $\mathbf{F} = m\mathbf{a}$. In one dimension, where the acceleration, the velocity, and the position have only x-components, these solutions are

$$x = x_0 + v_0 t + \frac{1}{2}at^2 = x_0 + v_0 t + \frac{F}{2m}t^2, \quad (5\text{–}8)$$

for constant force:

$$v = v_0 + at = v_0 + \frac{F}{m}t. \quad (5\text{–}9)$$

The constants x_0 and v_0 refer here to the values of position and velocity at time $t = 0$ and are usually specified in the statement of the problem. Sometimes measurements of position and velocity at two different times allow us to determine x_0 and v_0. For a force with three constant components, we must apply the solution to each component separately. Remember that these formulas apply only when **F** is constant.

EXAMPLE 5–2 A coffee cup of mass 75 g is placed on a slippery (frictionless) ramp tilted at an angle $\theta = 20°$ (Fig. 5–7a). The coffee cup starts from rest at $t = 0$ s and slides down the ramp. How far down the ramp has the cup moved after 2.0 s?

Solution: As emphasized in Chapter 4, we first prepare a force diagram, choose a convenient set of axes, and decompose the forces into their components along these

(a)

FIGURE 5–6 (a) A scale can be used to measure the normal force on an apple. (b) We can draw the forces acting on the apple.

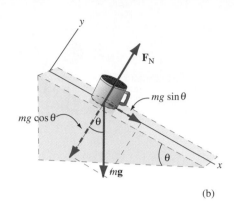

FIGURE 5–7 (a) Example 5–2. A coffee cup on a frictionless, tilted ramp. (b) Force diagram for the coffee cup, including the vector decomposition of the forces.

(a)

(b)

axes (Fig. 5–7b). We then write Newton's second law and solve the equations of motion. Because there is no friction, the only forces acting on the cup are gravity, $\mathbf{F}_g$, and the normal force, $\mathbf{F}_N$. *Note*: $\mathbf{F}_N$ is perpendicular to the surface of the inclined plane and thus it is *not* oriented in the vertical direction.

We choose the axes shown in Fig. 5–7b because we know from experience that the cup will not "jump" off the ramp; in other words, the acceleration $\mathbf{a}$ of the cup will be along the ramp and will not have a component perpendicular to the ramp. Accordingly, the *x*-direction points down along the ramp and the origin is placed at the starting point of the coffee cup. The acceleration $\mathbf{a}$ will have only an *x*-component a_x:

$$\mathbf{a} = a_x\,\mathbf{i}.$$

This choice of axes is also convenient for the normal force, which has only a *y*-component:

$$\mathbf{F}_N = F_N\,\mathbf{j}.$$

The magnitude F_N is as yet unknown. We will find it below from the requirement that the acceleration has no *y*-component. The force of gravity has both *x*- and *y*-components (Fig. 5–7b):

$$\mathbf{F}_g = mg \sin \theta\,\mathbf{i} - mg \cos \theta\,\mathbf{j}.$$

where *m* is the mass of the cup.

Newton's second law has the vector form

$$\mathbf{F}_g + \mathbf{F}_N = m\mathbf{a},$$

or

$$mg \sin \theta\,\mathbf{i} - mg \cos \theta\,\mathbf{j} + F_N\,\mathbf{j} = ma_x\,\mathbf{i}.$$

Each component of this equation is a separate equation of motion:

$$\text{for the } x\text{-component:} \qquad mg \sin \theta = ma_x; \tag{5–10}$$

$$\text{for the } y\text{-component:} \qquad -mg \cos \theta + F_N = 0. \tag{5–11}$$

Now we solve the equations of motion. Equation (5–10) gives

$$a_x = g \sin \theta. \tag{5–12}$$

Equation (5–11) shows that $\mathbf{F}_N$ has magnitude $mg \cos \theta$.

To find how the position of the coffee cup varies with time, we use the kinematical equation for constant acceleration in one dimension: Eq. (5–8). In this equation, the cup starts at the origin so $x_0 = 0$; moreover, $v_0 = 0$ because the cup starts at rest. We then insert Eq. (5–12) into Eq. (5–8), with the result

$$x = \frac{1}{2}a_x t^2 = \frac{1}{2}g \sin \theta\, t^2.$$

At $t = 2.0$ s,

$$x = (0.5)(9.8 \text{ m/s}^2)(\sin 20°)(2.0 \text{ s})^2 = 6.7 \text{ m}.$$

We find the cup has moved 6.7 m after 2.0 s. As is typical for problems involving gravity, the answer is independent of the mass of the cup. The angle $\theta = 0°$ presents a special limit; in this case, the plane is flat, and there should be no acceleration whatsoever. This is ensured by the $\sin \theta$ factor in the acceleration, which is zero when $\theta = 0°$. The normal force has magnitude mg, as expected. A second limit is the case $\theta = 90°$, when $\sin \theta = 1$. Here, $a_x = g$ and $F_N = 0$, as expected.

EXAMPLE 5–3 Because of a wager, a woman wishes to lift a professional football player off his feet. The player is a huge interior lineman with a mass of 149 kg. (He weighs 328 lb.) The woman has devised a system for the task, which is shown in Fig. 5–8a. We will assume that all pulleys, ropes, and miscellaneous gear in the apparatus have negligible mass and are frictionless. What is the magnitude of the downward force the woman must exert on the end of the rope in order to lift the lineman?

Solution: Even a smooth lift at constant velocity is sufficient; we do not ask that the lineman accelerate continuously. This condition is met when the net force on the lineman balances his weight. The tension in the rope is transmitted all the way through the rope, and the tension is the force that must be supplied by the woman pulling on the rope. Thus, we want to find the tension, T.

If we examine Fig. 5–8a, we see that all four rope segments, 1 through 4, pull upward on the lineman. Figure 5–8b is the force diagram for the lineman and the two lower pulleys. As the rope becomes taut when the woman begins to pull, the sum of the tensions $T_1 + T_2 + T_3 + T_4$ increases from zero. As long as this sum is less than the weight of the lineman, Mg, he will remain on the ground. But when this sum becomes equal to Mg, there is no net force on the lineman, and any additional tension—no matter how small—will start him accelerating upward. Thus the condition to lift the lineman is

$$T_1 + T_2 + T_3 + T_4 = Mg.$$

Because T is the same everywhere in the rope,

$$4T = Mg;$$

$$T = \frac{Mg}{4} = \frac{(149 \text{ kg})(9.80 \text{ m/s}^2)}{4} = 365 \text{ N}.$$

This is the magnitude of the smallest downward force that the woman must apply to her end of the rope. Provided that she weighs more than 365 N (82.1 lb), which corresponds to a mass of $m = W/g = 37.3$ kg, she can apply at least this force just by hanging on the rope.

The arrangement described here is called a *block and tackle*.

In Examples 5–4 and 5–5, we will investigate all three forces discussed so far in this section. These examples also introduce something new: Two masses are involved. Each mass requires an identification of the forces acting on it, a force diagram, and an equation of motion.

EXAMPLE 5–4 Masses $m_1 = 1.0$ kg and $m_2 = 1.1$ kg are attached to opposite ends of a massless rope draped over a pulley (Fig. 5–9a). (This device is called an *Atwood machine*.) Mass m_2 rests on a scale that measures the normal force exerted on m_2. There is no motion. What is the reading on the scale, and what is the tension in the rope?

(a)

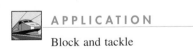

(b)

FIGURE 5–8 (a) Example 5–3. (b) Force diagram for the lineman.

APPLICATION

Block and tackle

(a)

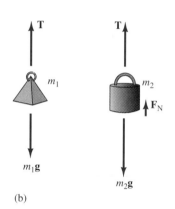

(b)

FIGURE 5–9 (a) Example 5–4. An Atwood machine. (b) Force diagrams for the masses.

PROBLEM SOLVING

It is often useful to choose different coordinates for each part of a system.

Solution: Let's first prepare a force diagram for each mass (Fig. 5–9b) and find the equations of motion for each mass. Because there is no acceleration, the net force on each mass is zero. All three forces—gravity, the tension, and the normal force—are vertical, so we have only one component (*y*-component) to consider; thus, we drop the vector (boldface) notation.

Take the upward direction to be positive. The forces acting on m_1 are gravity $(-m_1 g)$ and the rope tension (T). Newton's first law gives

$$F_{\text{net}} = -m_1 g + T = 0. \tag{5–13}$$

Three forces act on mass m_2: $-m_2 g$, T, and F_N. Notice that the same value of T acts on each mass, pulling (acting upward) in each case. Newton's first law for mass m_2 is then

$$F_{\text{net}} = -m_2 g + T + F_N = 0. \tag{5–14}$$

We can immediately solve Eq. (5–13) for the tension:

$$T = m_1 g = (1.0 \text{ kg})(9.8 \text{ m/s}^2) = 9.8 \text{ N}.$$

The tension is one of the quantities we were asked to find. If we insert this value of T into Eq. (5–14), we can solve for the normal force and hence the scale reading:

$$F_N = m_2 g - T = m_2 g - m_1 g = (m_2 - m_1)g$$

$$= (1.1 \text{ kg} - 1.0 \text{ kg})(9.8 \text{ m/s}^2) = (0.1 \text{ kg})(9.8 \text{ m/s}^2) \simeq 1 \text{ N}.$$

For comparison, the weight of m_2 alone is $(1.1 \text{ kg})(9.8 \text{ m/s}^2) = 10.8 \text{ N}$.

EXAMPLE 5–5 Consider two masses $m_1 = 1.00$ kg and $m_2 = 2.00$ kg connected by a rope that passes over an ideal pulley. Mass m_1 hangs straight down, while m_2 slides without friction on a ramp inclined at an angle θ. At $t = 0$, the system is started from rest in the position shown in Fig. 5–10a. Describe the motion of the two masses for $\theta = 25.0°$. Find the angle θ' for which the system remains motionless.

Solution: In this problem, we again have two masses for which we must write the equations of motion: $\mathbf{F}_{\text{net on 1}} = m_1 \mathbf{a}_1$ and $\mathbf{F}_{\text{net on 2}} = m_2 \mathbf{a}_2$. Connected as they are by a rope, the motion of one mass is correlated with the motion of the other. *The magnitudes of the accelerations are the same.* We begin by drawing a force diagram for each mass (Fig. 5–10b). A force with common magnitude acts on each mass; this force is the tension, **T**. **T** and gravity $m_1 \mathbf{g}$ act on m_1. Three forces act on m_2: the normal force, $\mathbf{F}_N$, tension, **T**, and gravity, $m_2 \mathbf{g}$.

Once the force diagrams are drawn, we choose coordinate systems and write Newton's second law. *We are not compelled to choose the same coordinate system for the two masses.* In fact, choosing different systems makes this problem easier to solve (Fig. 5–10b). We choose the origin of each coordinate system to be at the location of the respective mass at $t = 0$. Because m_1 moves only in the vertical direction, labeled y_1, its acceleration $\mathbf{a}_1$ is aligned with y_1. For m_1, then, we need look only at the component of Newton's second law along y_1,

$$T - m_1 g = m_1 a_1. \tag{5–15}$$

Although the mass m_2 has forces acting in two directions, it moves only in the direction labeled x_2 in Fig. 5–10b. Thus its acceleration vector $\mathbf{a}_2$ is aligned with x_2. We must now decompose the forces acting on m_2 into components. $\mathbf{F}_N$ is in the $+y_2$-direction and **T** is in the $-x_2$-direction. The third force, gravity, has two components; we can determine these components by recalling from geometry that the angle θ indicated in Fig. 5–10b is the same as the ramp angle θ in Fig. 5–10a. Then the force of gravity $m_2 \mathbf{g}$ has x_2-component $m_2 g \sin \theta$ and y_2-component $-m_2 g \cos \theta$. Thus Newton's second law for mass m_2, $\mathbf{F}_{\text{net}} = \mathbf{F}_N + \mathbf{T} + m_2 \mathbf{g} = m_2 \mathbf{a}_2$, breaks down into the two component equations for the x_2-component: $\quad -T + m_2g \sin\theta = m_2a_2;$ $\qquad$ (5–16)

for the y_2-component: $\quad F_N - m_2g \cos\theta = 0.$ $\qquad$ (5–17)

The three equations (5–15), (5–16), and (5–17) are not enough to solve for the four unknowns T, F_N, a_1, and a_2. The missing fourth equation expresses the connection between the motion of m_1 and that of m_2; that is, the motion of m_1 in the $+y$-direction is matched by the motion of m_2 in the $+x$-direction, so

$$a_1 = a_2 \equiv a. \qquad (5\text{–}18)$$

The single acceleration magnitude a is then substituted in Eqs. (5–15), (5–16), and (5–17), which become three equations for the three unknowns a, T, and F_N. Equation (5–17) gives the normal force,

$$F_N = m_2g \cos\theta.$$

The tension cancels in the sum of Eqs. (5–15) and (5–16):

$$T - m_1g - T + m_2g \sin\theta = m_1a + m_2a.$$

We are left with an equation for a with solution

$$a = \frac{m_2 \sin\theta - m_1}{m_1 + m_2} g. \qquad (5\text{–}19)$$

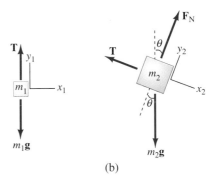

(a)

(b)

FIGURE 5–10 (a) Example 5–5. (b) Force diagrams for masses m_1 and m_2.

Now that we have found the accelerations, we can find positions as a function of time by using the constant acceleration equations: Eq. (5–8). Each mass starts from rest at the origin of its respective coordinate system, so $v_0 = 0$ for both masses, and both y_{10} ($\equiv y_1$ at $t = 0$) and x_{20} ($\equiv x_2$ at $t = 0$) are zero. Thus

for m_1: $\qquad y_1 = \dfrac{1}{2}at^2,$

for m_2: $\qquad x_2 = \dfrac{1}{2}at^2.$

The masses move only in these directions. Together with Eq. (5–19), these equations describe the motion fully.

For the special case $\theta = 25.0°$, we use $\sin 25.0 = 0.423$, so

$$a = \frac{(2.00 \,\cancel{kg})(0.423) - 1.00 \,\cancel{kg}}{1.00 \,\cancel{kg} + 2.00 \,\cancel{kg}} (9.80 \text{ m/s}^2) = -0.503 \text{ m/s}^2.$$

Thus $y_1 = (-0.503 \text{ m/s}^2)t^2/2 = x_2$. Note the minus sign in the acceleration: The sign indicates that m_1 drops and m_2 moves up the ramp. Equation (5–19) shows that the acceleration—including its sign—depends on the masses and the ramp angle.

We must also find the angle for which the acceleration is zero given m_1 and m_2. According to Eq. (5–19), the acceleration is zero at an angle θ' for which

$$m_2 \sin\theta' - m_1 = 0, \qquad (5\text{–}20)$$

or $\sin\theta' = m_1/m_2$. For this particular problem, $\sin\theta' = \dfrac{1.00}{2.00}$, or $\theta' = 30.0°$.

5–2 FRICTION

Friction is a familiar concept to all of us. It is a contact force that impedes sliding. It allows us to walk, it allows cars to move along the road, and it even holds nails and screws in place (Fig. 5–11). The need to reduce friction in machinery requires us to learn how to lubricate rubbing surfaces. Some 20 percent of the power of an automobile is spent overcoming internal friction—and where there is friction, there is surface wear.

FIGURE 5–11 Nails are held in place by the force of friction, which can be quite substantial.

PROBLEM-SOLVING TECHNIQUES

When more than one mass appears in a problem, there are two useful techniques to keep in mind:

1. There may be a constraint equation that connects the motion of the masses, as for the two masses connected by a taut rope in Example 5–5.

2. You can choose different coordinate systems for the different masses. As in Example 5–5, a problem can be made simpler if the coordinate systems are well chosen.

Let's suppose that you want to slide a crate from one place to another. You push on it with a small horizontal force, but nothing happens. Even when you push as hard as you can, the crate does not move. Why not? **Static friction** acts between the floor and the crate in the absence of motion in such a way as to *prevent* motion. This force must be variable because it balances each of your own different pushes. Suppose that you finally get the crate moving with the help of another person. The combined force overcame the static friction because *static friction has a maximum magnitude.*

Once the crate is moving, it is easier to keep it moving at a constant speed. There is still friction opposing your push, but it is now **kinetic** (or **sliding**) **friction**; that is, friction associated with motion. The magnitude of kinetic friction is smaller than the maximum value of static friction. The entire sequence of getting the crate started and keeping it moving is illustrated in Fig. 5–12. Both static and kinetic friction act along the surface (Fig. 5–13). Static friction acts in a direction opposite to the component of an applied force along the surface; sliding friction acts opposite to the direction of the velocity of a sliding object at its point or points of contact.

The Force of Friction: A Quantitative Treatment

Studies of the nature of friction were made by Leonardo da Vinci some 200 years before Newton's work on dynamics. Leonardo experimented with a set of blocks of varying sizes sliding on table tops (Fig. 5–14) and learned some surprising facts. He found that *both static and kinetic friction are independent of the apparent contact area of the block on the table top. Moreover, both static and kinetic friction are proportional to the magnitude F_N of the normal force exerted by the table top on the blocks.* To determine that the force is independent of the apparent area that is in contact with the surface, Leonardo simply took a given block and turned it so that faces of different areas were in contact with the table top. He found that the friction force on the block was the same no matter what face of the block was in contact with the table top.

FIGURE 5–12 A sequence in which an increasing pushing force is opposed by an equal and opposite static friction force. Static friction can increase only to a certain point, after which the crate accelerates. Kinetic friction, which applies when the crate actually moves, is smaller than the maximum size of static friction, so it takes less pushing force to move the crate at that point.

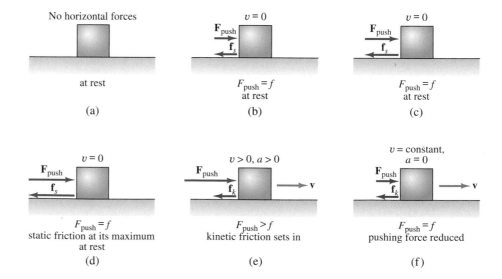

The proportionality constant that relates the friction force and the normal force is the **coefficient of friction**, $\boldsymbol{\mu}$. The (unitless) constant μ is determined experimentally. The maximum value of static friction is generally not equal to the force of kinetic friction, so we distinguish two coefficients: μ_s for static friction and μ_k for kinetic friction.

If we write the force of static friction as $\mathbf{f}_s$ and that of kinetic friction as $\mathbf{f}_k$, their magnitudes are given by

$$\text{static friction:} \qquad 0 \le f_s \le \mu_s F_N \qquad (5\text{--}21)$$

and

$$\text{kinetic friction:} \qquad f_s = \mu_k F_N. \qquad (5\text{--}22)$$

Equation (5–21) expresses a range because static friction is variable. The experimental fact that the maximum value of static friction exceeds kinetic friction implies the inequality

$$\mu_s > \mu_k. \qquad (5\text{--}23)$$

We also make the assumption, reasonably well satisfied, that μ_k *is independent of the relative speed of the two surfaces.*

The coefficients of friction depend on the two surfaces involved. We know from everyday experience that a basketball shoe on a basketball court involves a larger coefficient of friction than does the blade of an ice skate on a frozen lake. A lubricating material—such as sweat—between the basketball shoe and the court will drastically reduce the coefficient of friction. Table 5–1 shows some typical values of coefficients of static friction. The materials listed are generally unlubricated ("dry"). Coefficients of kinetic friction can be anywhere from roughly 25 percent to 100 percent of the corresponding coefficients of static friction.

The values in Table 5–1 are meant only to be indicative—the coefficients of friction are sharply dependent on such things as the cleanliness of the surfaces, their roughness, and so forth. Two very rough objects may have a large coefficient of friction that can be reduced once the objects are smoothed. But if they are smoothed too much and are free of dirt and oxidation as well, the coefficient of friction may rise virtually to infinity because the surfaces weld together!

In Example 5–6, we will learn one way to measure the coefficient of static friction.

EXAMPLE 5–6 A mass m is set at rest on a horizontal ramp; there is friction between the mass and the ramp. The ramp is slowly raised from its horizontal orientation. By analyzing the forces on the mass, find the critical ramp angle θ_c at which the mass will start to slide.

Solution: In Fig. 5–15, we can see the ramp and the mass. The forces on the mass are drawn and a suitable coordinate system is shown, with the origin at the initial position of the mass. The $+x$-direction is oriented down along the ramp, and the y-direction is perpendicular to the ramp. This figure thus serves as a force diagram. The three forces acting on the mass before it starts to move are the force of gravity (vertically down), the normal force, $\mathbf{F}_N$ (perpendicular to the surface), and the force of static friction, $\mathbf{f}_s$ (up the ramp). The friction force opposes the motion that would take place if there were no friction. The force magnitudes F_N and f_s are each determined by the fact that the mass remains motionless for a sufficiently shallow ramp angle. Note that we must find F_N because the maximum value of static friction depends on F_N [see Eq. (5–21)].

If the mass is stationary, Newton's second law with zero acceleration applies: $\mathbf{F}_g + \mathbf{F}_N + \mathbf{f}_s = 0$, or, in Cartesian form,

$$mg \sin \theta \, \mathbf{i} - mg \cos \theta \, \mathbf{j} + F_N \mathbf{j} - f_s \mathbf{i} = 0.$$

FIGURE 5–13 Friction opposes the motion of the crate.

The forces of static and kinetic friction

FIGURE 5–14 A page from Leonardo's notebooks. By simply turning the faces of blocks in contact with a table top, Leonardo showed that the friction between the blocks and the table top is independent of the apparent surface area in contact.

TABLE 5–1
Some Coefficients of Static Friction[†]

Materials	μ_s
Brake material on a brake drum	1.2
A dry tire on dry asphalt	1.0
Hard steel on hard steel	0.8
Oak on oak, parallel to the grain	0.6
A book on a table	0.3
A wet tire on wet asphalt	0.2
Ice on wood	0.05
Teflon on steel	0.04

[†]The coefficients of kinetic friction are from 25 percent to 100 percent of the corresponding coefficients of static friction.

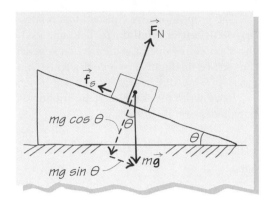

FIGURE 5–15 Example 5–6.
A mass m experiencing a friction force
on a ramp. The angle at which the
mass starts to slide tells us the
coefficient of static friction.

In component form, the equations of motion are

$$\text{for the } x\text{-component:} \quad mg \sin \theta - f_s = 0, \quad (5\text{--}24a)$$

$$\text{for the } y\text{-component:} \quad F_N - mg \cos \theta = 0. \quad (5\text{--}24b)$$

The x-component equation relates the force of static friction to the force of gravity,

$$f_s = mg \sin \theta. \quad (5\text{--}25)$$

The y-component equation determines F_N as a function of θ:

$$F_N = mg \cos \theta. \quad (5\text{--}26)$$

As θ (and $\sin \theta$) increases, the force of friction from Eq. (5–25) required to hold the mass in static equilibrium also increases. Eventually, static friction reaches its maximum value, $\mu_s N$. Beyond that point, the mass will start to slide. When we set static friction to its maximum value, $f_s = \mu_s F_N$, Eq. (5–25) determines the critical angle θ_c at which the mass starts to slide. Equation (5–25) then becomes

$$\mu_s F_N = mg \sin \theta_c,$$

or, from Eq. (5–26),

$$\mu_s mg \cos \theta_c = mg \sin \theta_c.$$

Cancel the factor mg from this equation to find θ_c:

$$\frac{\sin \theta_c}{\cos \theta_c} = \tan \theta_c = \mu_s. \quad (5\text{--}27)$$

Thus if we measure the angle at which the mass begins to slip, we measure μ_s.

With an extension of the reasoning used in Example 5–6, we can figure out how to measure the coefficient of kinetic friction. Once the mass begins to slip, kinetic friction acts, given by $f_k = \mu_k F_N = \mu_k mg \cos \theta$. If we use Newton's second law and apply it along the x-direction, then, instead of Eq. (5–24a), we find:

$$mg \sin \theta - \mu_k mg \cos \theta = ma_x. \quad (5\text{--}28)$$

Because μ_k is smaller than μ_s, the ramp can be lowered back down, decreasing θ while the mass is still sliding. There is a second critical value of θ—call it θ'_c—for which the forces in Eq. (5–28) cancel and the object no longer accelerates. Instead, the object slides at constant velocity. This critical angle is given by

$$\frac{\sin \theta'_c}{\cos \theta'_c} = \tan \theta'_c = \mu_k, \quad (5\text{--}29)$$

Thus θ'_c measures the coefficient of kinetic friction.

Let's now take a look at Example 5–7, which is a straightforward application of the concept of static friction.

EXAMPLE 5–7 A professor with a light eraser in her hand leans against a blackboard (Fig. 5–16a). Her arm makes an angle of 30° with the horizontal, and the force $\mathbf{F}_{\text{prof}}$ exerted by her arm on the eraser has magnitude $F_{\text{prof}} = 50$ N. The coefficient of static friction between the eraser and the blackboard is $\mu_s = 0.15$. Does the eraser slip?

Solution: Figure 5–16b is a force diagram for the eraser, which also indicates a useful coordinate system. If nothing moves, Newton's first law applies: $\mathbf{F}_N + \mathbf{f}_s + \mathbf{F}_{\text{prof}} = 0$. In the coordinate system on the figure,

$$-F_N \mathbf{i} - f_s \mathbf{j} + F_{\text{prof}} \cos\theta \, \mathbf{i} + F_{\text{prof}} \sin\theta \, \mathbf{j} = 0.$$

The two component equations are

for the x-component: $\qquad -F_N + F_{\text{prof}} \cos\theta = 0,$

for the y-component: $\qquad -f_s + F_{\text{prof}} \sin\theta = 0.$

The x-component equation determines F_N from the requirement that it balances the perpendicular component of the force the professor exerts. Once we have found that $F_N = F_{\text{prof}} \cos\theta$, we can determine the *maximum* value of static friction, $f_{s,\text{max}} = \mu_s F_N = \mu_s F_{\text{prof}} \cos\theta$. When this maximum value is exceeded, the eraser begins to slip. Thus, when we substitute the maximum value of static friction into the y-component equation, we find a condition for the critical angle θ_c for which the eraser begins to slip:

$$-\mu_s F_{\text{prof}} \cos\theta_c + F_{\text{prof}} \sin\theta_c = 0;$$

$$\frac{\sin\theta_c}{\cos\theta_c} = \tan\theta_c = \mu_s.$$

Note the striking feature that the critical angle is independent of the force the professor exerts! Numerical substitution yields $\tan\theta_c = 0.15$, or $\theta_c = 8.5°$. This angle is less than the 30° angle made by the arm, so the eraser indeed slips.

Static friction does not always act to impede motion. If we are not dealing with a pointlike object or with a rigid system, then static friction can act to accelerate a system. Walking is perhaps the most familiar example of this phenomenon (Fig. 5–17): Static friction between our feet and Earth is the force that propels us forward when we walk. By Newton's third law, there is a corresponding backward-directed force on the planet. Without friction, our feet would fruitlessly slip *backward* when we tried to walk, an experiment that is easy to perform on a frozen puddle. This point is illustrated by Example 5–8.

EXAMPLE 5–8 An automobile with four-wheel drive and a powerful engine has a mass of 1000 kg. Its weight is evenly distributed on its four wheels, whose coefficient of static friction with the dry road is $\mu_s = 0.8$. If the car starts from rest on a horizontal surface, what is the greatest forward acceleration that it can attain without spinning its wheels?

Solution: The automobile has its greatest forward acceleration when the static friction between the wheels and the road is at its maximum. The engine of the car serves to impart the rotational motion shown in Fig. 5–18a to the wheels. If there were no friction between the wheels and the road, the wheels would simply spin. Because the wheels do not slip, it is *static* friction that acts here, and μ_s, not μ_k, enters the problem. The motion of the tires at the point of contact with the road is to the rear in the absence of friction. The force of friction opposes this rearward motion, so the direction of $\mathbf{f}_s$ is toward the front of the car. The forward frictional force

(a)

(b)

FIGURE 5–16 (a) Example 5–7. A professor leaning against an eraser at a blackboard. (b) Force diagram for the eraser at the blackboard.

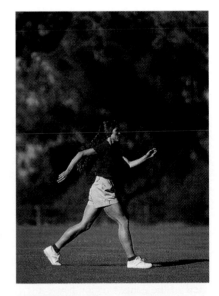

FIGURE 5–17 A walker uses the force of static friction between shoe and ground to move forward.

Force of static
friction on tires

(a)

(b)

FIGURE 5–18 (a) Example 5–8. A car accelerates forward under the influence of static friction. (b) Force diagram for the car.

is the *only* external horizontal force acting on the car and hence determines its forward acceleration. We draw the forces acting on the car in Fig. 5–18b.

Figure 5–18b is a force diagram for the car, and Newton's second law is

$$\mathbf{F}_{\text{net}} = \mathbf{F}_{\text{N}} + m\mathbf{g} + \mathbf{f}_s = m\mathbf{a},$$

where m is the mass of the car. (Actually, there is a separate normal force and friction force at each wheel, but because the weight is distributed evenly over each wheel, all the normal and friction forces at each wheel are equal.) With the coordinate system in the figure, $\mathbf{F}_N = F_N\mathbf{j}$, $m\mathbf{g} = -mg\,\mathbf{j}$, $\mathbf{f}_s = f_s\,\mathbf{i}$, and the acceleration is forward, so $\mathbf{a} = a\mathbf{i}$. Newton's second law is then

$$F_N\mathbf{j} - mg\,\mathbf{j} + f_s\,\mathbf{i} = ma\,\mathbf{i},$$

and the component equations are

for the x-component: $f_s = ma,$

for the y-component: $N - mg = 0.$

The acceleration is a maximum a_{max} when static friction reaches its maximum value $\mu_s F_N = \mu_s mg$, where we have used $F_N = mg$ from the y-component equation. Thus

$$\mu_s mg = ma_{\text{max}}.$$

The mass of the car cancels out of this expression, leaving

$$a_{\text{max}} = \mu_s g = (0.8)(9.8 \text{ m/s}^2) \approx 8 \text{ m/s}^2.$$

This is quite a significant forward acceleration. Note that as μ_s decreases, the maximum acceleration decreases; in other words, when μ_s is zero, the automobile can only spin its wheels.

Where Does Friction Come From?

To understand the origins of friction, let's look closely at two surfaces that rub against one another. Figure 5–19a shows a microscopic view of two such surfaces in contact. Because of the hills and valleys present on any rough surface—and all surfaces are rough when viewed closely enough (Fig. 5–19b)—the area on two surfaces that actually touches together is a small fraction of the area that appears to be in contact. Friction forces are due to three major effects: the interlocking of surface irregularities; the attraction between the contact points due to forces between the molecules of the two objects (the objects "adhere"); and the "plowing out" of softer materials by harder ones. The coefficient of static friction can be greater than that of kinetic friction because the materials have a longer time to "settle in" together.

This description helps us understand why the friction force is independent of the apparent surface area that is in contact, while it is dependent on the normal force. The normal force is a measure of how strongly the two surfaces are pressed together; when the normal force is large, the two surfaces are pressed strongly together. The rough surfaces shown in Fig. 5–19a mesh more closely when pressed strongly together, and the

FIGURE 5–19 (a) A microscopic view of the contact region between two rough surfaces. (b) At lower magnification, the same surfaces are apparently smooth.

(a)

(b)

actual surface area in contact increases. In fact, the normal force is a good measure of the actual surface contact area! Whether we place the broad side or the narrow end of a brick on a table, approximately the same surface areas are in actual contact, even if the apparent contact areas are vastly different. So the friction force is, finally, proportional to the real contact area.

The study of friction, wear, and lubrication is called *tribology*. Despite much effort, a truly fundamental understanding of friction remains elusive. The discovery of Teflon™ was a happy accident, not the result of a planned development program. The discoverer has stated that he was lucky he was not blown up in the process.

5–3   DRAG FORCES

A spoon dropping through molasses, an automobile moving at highway speeds, and a plane using a parachute to slow down are all subject to a substantial *drag force*, which is a resistive force somewhat like friction. However, in motion *through* media such as gases or liquids, there is no normal force, yet there is still a drag force acting to impede the motion. Drag forces act like sliding friction in that *they act in a direction opposite to that of the motion*, but they differ from sliding friction in that they depend on the *speed v* of the object that is moving through the medium.

In many everyday situations—an automobile moving on a highway is one—the drag force, $\mathbf{F}_D$, is found by experiment to have magnitude

$$F_D = \frac{1}{2}\rho A C_D v^2, \tag{5–30}$$

where ρ is the mass density (the mass per unit volume) of the medium through which the object moves, A is the maximum cross-sectional area presented by the moving object, and C_D is the *drag coefficient*. The area A is the area found by cutting across the object in a plane perpendicular to the direction of motion. The drag coefficient, C_D, is dimensionless and depends on the shape of the object. A very highly streamlined object might have a drag coefficient as small as 0.1, whereas a particularly awkward shape will have a drag coefficient greater than 1. The most streamlined automobiles have drag coefficients around 0.25.

Terminal Velocity

The fact that the drag force on an object increases with the speed has an important consequence. Consider a falling parachutist who is acted on by both gravity and a drag force like that of Eq. (5–30) due to the air (Fig. 5–20). When the parachutist first starts to fall, his or her speed is slow, so the drag force is small and there is an acceleration of $\mathbf{g}$. As the parachutist's speed increases, so does the drag; at some point, the drag will actually cancel the force of gravity. Then the parachutist no longer accelerates and his or her speed remains constant at a maximum value. The maximum speed v_t is called the *terminal velocity*.[†] Figure 5–20 illustrates a parachutist who has reached terminal velocity.

Let's calculate the terminal velocity in the case of falling objects, where the applied force $\mathbf{F}_1$ is the force of gravity. We assume that the drag force is given by Eq. (5–30), which is a good approximation for all but very small falling objects. The terminal velocity is found from the condition that the forces of gravity and drag cancel: $F_g = F_D$, or

$$mg = \frac{1}{2}\rho A C_D v_t^2.$$

[†]Some typical values are: for a parachutist, 5 m/s; for a falling raindrop, 8 m/s; for a baseball, 40 m/s; for a skydiver, 60 m/s.

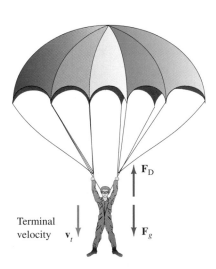

FIGURE 5–20 When the drag force and the force of gravity acting on a parachutist cancel, the parachutist has reached his terminal velocity.

FIGURE 5–21 (a) Example 5–9. (b) Force diagram for an automobile that moves under the influence of the force provided by the engine and the drag of the air through which it moves.

The drag force

We solve for the terminal velocity:

$$v_t = \sqrt{\frac{2mg}{\rho A C_D}}. \tag{5–31}$$

We can apply similar reasoning to the calculation of the terminal velocity for other external forces.

EXAMPLE 5–9 The maximum force with which a certain automobile engine can accelerate a car is 3200 N. At highway speeds, Eq. (5–30) is a good approximation to the true drag force. The density of air is about 1.2 kg/m³, the cross-sectional area of the automobile is 3.4 m², and the drag coefficient is 0.50. Assuming that the drag force is the only force resisting the motion, what is the maximum speed of the automobile?

Solution: Figure 5–21 shows the forces and a force diagram for the automobile. It includes the forces $\mathbf{F}_{engine}$ and $\mathbf{F}_{drag}$. (Friction from the road actually propels the car, but this occurs because of forces provided by the engine so we will refer to this force as $\mathbf{F}_{engine}$.) The maximum speed, or terminal velocity v_t, of the car occurs when the net force on the car is zero. This occurs when the vector sum of the engine and drag forces is zero:

$$\mathbf{F}_{engine} + \mathbf{F}_{drag} = 0.$$

Because the two forces are equal in magnitude but opposite in direction, we can write $F_{engine} = F_{drag}$. We then use Eq. (5–30) for the drag force to find that

$$F_{engine} = \frac{1}{2} \rho A C_p v_t^2.$$

We solve this equation for the terminal velocity squared:

$$v_t^2 = \frac{2 F_{engine}}{\rho A C_D} = \frac{(2)(3200 \text{ N})}{(1.2 \text{ kg/m}^3)(3.4 \text{ m}^2)(0.50)} = 3100 \text{ m}^2/\text{s}^2.$$

The terminal velocity of the automobile is then

$$v_t = 56 \text{ m/s, or } 125 \text{ mi/h.}$$

In a more realistic situation, road friction, sometimes referred to as rolling friction, also plays an important role in opposing the motion of an automobile. Road friction is fairly constant over a large range of speeds and, as a rule of thumb, road friction is as important as drag at about 40 mi/h. As the speed increases, the relative size of drag compared to road friction increases.

A Better Approximation to the Drag Force

The experimental force law for the drag force, $\mathbf{F}_D$, depends in a more complicated way on the medium through which an object moves and on the shape and size of the moving object than Eq. (5–30) suggests. At low speeds, the drag force is linearly dependent upon the speed. As the speed increases, however, new effects involving the onset of turbulence in the medium set in, and the drag force becomes proportional to the square of the object's speed. Thus, a better approximation to the drag force is

$$F_D = bv + cv^2. \tag{5–32}$$

The coefficients b and c contain information on the shape of the moving object as well as on the medium in which it moves. The first term always dominates for sufficiently low speeds, while the second term dominates for higher speeds.

EXAMPLE 5–10 A marble of mass 5 g falls into a jar of oil (Fig. 5–22). The drag force on the marble is given by Eq. (5–32) with $b = 0.2$ kg/s and $c = 0.1$ kg/m. (These values are typical for real fluids such as oil and for an object the size and shape of a marble.) Find the value of the speed for which the two terms in the drag

A simple application of the use of drag forces in low-level technology is the door closer used on storm doors. The device consists of a piston fitting into a closed cylinder with a very small hole at the closed end. A strong spring is placed behind the piston to move it to the closed end of the cylinder (Fig. 5AB–1). The device is attached to the storm door so that, when the door is opened, the piston is pulled to the bottom of the cylinder. There is a valve-like seal between the piston and the cylinder that allows air to fill the cylinder; this makes it easy to open the door. However, the seal does not allow air to leave the cylinder. When the open door is released, the spring force tries to push the piston into the cylinder, thus closing the door. This force is opposed by the drag force of the air leaving by the very small hole in the closed end of the cylinder. This causes the door to close quietly and smoothly.

FIGURE 5AB–1 Drag forces are effectively used in this common device.

force are equal. Which of the two terms is dominant when the drag force is comparable to the force of gravity on the marble?

Solution: Let the value of the speed for which the two terms in the drag force are equal be v'. Then v' is determined by

$$bv' = cv'^2.$$

This equation is solved by

$$v' = \frac{b}{c} = \frac{0.2 \text{ kg/s}}{0.1 \text{ kg/m}} = 2 \text{ m/s}.$$

At terminal speed, magnitudes of the force of gravity and the drag force are equal; that is, $F_D = mg$. For now, let's compare only the first term in the drag, bv:

$$bv = mg;$$

$$v = \frac{mg}{b} = \frac{(5 \times 10^{-3} \text{ kg})(9.8 \text{ m/s}^2)}{0.2 \text{ kg/s}} = 0.25 \text{ m/s}.$$

When the values of the two terms in the drag force are equal—when $bv = cv^2$—the speed of the marble is 2 m/s. However, when the gravity and drag are the same size, the b term in the drag force dominates; this yields a speed of 0.25 m/s.

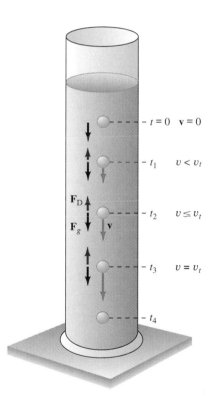

FIGURE 5–22 Example 5–10. When the marble has reached the terminal velocity, the magnitudes of the drag force and gravity are equal.

5 - 4   FORCES   AND   CIRCULAR   MOTION

If you were to tie this book to the end of a rope and swing it smoothly over your head in a nearly horizontal circle, keeping your hand as close to the center of the circle as possible (Fig. 5–23a), you would feel the rope become taut. This tautness is just the tension in the rope. As we discussed in Section 5–1, a rope under tension pulls on whatever it is connected to—in this case, your hand at one end and the book at the other. In fact, the pull of the rope would be the only horizontal force on the book. Thus, there would be a net horizontal force on the book, and the book would accelerate in the horizontal plane. This isn't surprising, though; in Chapter 3, we realized that an object that is in circular motion is accelerating—even an object in circular motion at constant speed (*uniform circular motion*). There is acceleration because the *direction* of the velocity vector is changing. If the constant speed of the uniform circular motion of the book is v and the radius of its circular motion is r, then Eq. (3–56) tells us that the acceleration vector $\mathbf{a}$ of the book is

$$\mathbf{a} = -\frac{v^2}{r}\hat{\mathbf{r}}. \tag{5–33}$$

Newton's second law as applied to circular motion

The unit vector $\hat{\mathbf{r}}$ points outward from the center of the circular motion. The minus sign in Eq. (5–33) indicates that the acceleration is *toward* the center of the circle.

Newton's second law, $\mathbf{F} = m\mathbf{a}$, tells us that the force $\mathbf{F}$ required to keep a mass m in uniform circular motion is

$$\mathbf{F} = -\frac{mv^2}{r}\hat{\mathbf{r}}. \tag{5–34}$$

Again, the minus sign reminds us that the force vector points to the center of the circle. Forces pointing in toward the center of a circle are called *centripetal forces*; the acceleration toward the center of a circle is a *centripetal acceleration*. In the case of the swinging book, tension is the force that keeps the book in its circular path and Eq. (5–34) tells us the *magnitude* of the tension in the rope. Because the rope leads to the center of the circle, the *direction* of the tension force on the book is indeed toward the center.

If the rope were cut, the tension that provided the centripetal force would no longer be present. According to the first law, then, the book would move without acceleration in the horizontal plane; that is, in a straight line (Fig. 5–23b).

Centripetal forces that lead to uniform circular motion *are not constant forces*. They are constantly changing in direction, just like the centripetal acceleration.

The tension of a rope is not the only centripetal force that can maintain an object in uniform circular motion. A planet in circular orbit about the Sun or a satellite in circular orbit about Earth are maintained in their circular motion by gravitational forces. Friction between the wheels and the road is the force that takes an automobile around a curve or, as Example 5–11 illustrates, that acts on a die resting on a rotating turntable.

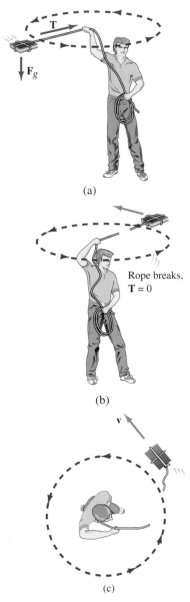

FIGURE 5–23 (a) The tension, $\mathbf{T}$, in the rope holding a whirling book is centripetal force, which maintains the circular motion of the book. (b) If the rope breaks, there is no force on the book, which then moves in a straight line tangent to the circle. (c) Top view of (b).

EXAMPLE 5–11 A small die sits 0.15 m from the center of a rotating horizontal turntable (Fig. 5–24a). If the coefficient of static friction between the die and the turntable is 0.55, what is the largest possible angular speed such that the die will not slide off?

Solution: Figure 5–24b is a force diagram for the die. The centripetal force maintaining the die in uniform circular motion is static friction, $\mathbf{f}_s$, between the die and the turntable. The largest rotation frequency that is possible without the die slipping occurs when the friction force reaches its maximum value, $f_s = \mu_s F_N$.

Newton's second law, $\mathbf{F}_{net} = \mathbf{F}_N + \mathbf{F}_g + \mathbf{f}_s = m\mathbf{a}$, must be decomposed into equations for the vertical direction and for the horizontal plane. For uniform circular motion, the horizontal motion equations are represented by a single radial component equation, Eq. (5–34).

There is no acceleration component in the vertical direction; that is, the direction in which $\mathbf{F}_N$ and $\mathbf{F}_g = m\mathbf{g}$ act. Thus the vertical component of Newton's second law is

$$F_N - mg = 0.$$

In turn, the maximum value of static friction is

$$f_s = \mu_s F_N = \mu_s mg.$$

The maximum radial acceleration $\mathbf{a}_{max}$ occurs for the maximum value of static friction. The radial component of Newton's second law, the coefficient of $\hat{\mathbf{r}}$ in Eq. (5–34), is then

$$\mu_s mg = \frac{m(v_{max})^2}{r}.$$

Note that the mass of the die cancels. We want to express this result in terms of a frequency of rotation. From our work in Chapter 3, we know that if the turntable is rotating with angular speed ω, then a point on the turntable that is a distance r from the center moves with speed $v = \omega r$. Thus our radial equation of motion becomes

$$\mu_s g = \frac{(v_{max})^2}{r} = \frac{(\omega_{max} r)^2}{r} = (\omega_{max})^2 r.$$

We solve for ω_{max} and find that

$$\omega_{max} = \sqrt{\frac{\mu_s g}{r}} = \sqrt{\frac{(0.55)(9.8 \text{ m/s}^2)}{0.15 \text{ m}}} = 6.0 \text{ rad/s}.$$

Normal forces can also act centripetally, as in Example 5–12.

FIGURE 5–24 (a) Example 5–11. (b) Force diagram for the die.

EXAMPLE 5–12 A space station designed to keep its inhabitants comfortable consists of a hollow circular tube that is rotating around its central axis (Fig. 5–25a). The astronauts move on the inner surface of the outermost wall. If the distance from the outer wall to the axis is 50 m, what must be the speed v of a point on the outer wall such that a bathroom scale will read what it would read on Earth?

SOLUTION: Rather than setting up Newton's second law in what might be a rather complicated coordinate system, let's try to think through this problem to gain more insight. An astronaut of mass m who stands on a scale in the space station moves in a circular path of radius R at constant speed v. He therefore accelerates toward the center of the circle (the axis) with magnitude v^2/R from Eq. (5–33). The force that causes this acceleration is the normal force, $\mathbf{F}_N$, of magnitude mv^2/R, which is supplied by the inner surface of the tube (Fig. 5–25b). The bathroom scale indicates this force, just as it would indicate the magnitude of the (upwardly directed) normal force on the astronaut if he were standing on the scale on Earth. Because the normal force on Earth balances gravity, it would have magnitude mg. Thus the scale on the space station reads as it would read on Earth provided that

$$\frac{mv^2}{R} = mg.$$

The mass m cancels out from this equation and we can solve for the speed v:

$$v = \sqrt{gR} = \sqrt{(9.8 \text{ m/s}^2)(50 \text{ m})} = 22 \text{ m/s}.$$

This is a fairly considerable speed—about 50 mi/h. Motion such as this involves internal forces that the station must be constructed to withstand. These internal forces are proportional to v^2. It would thus be a great luxury to have a large enough rotation to give the astronauts the illusion of being on Earth. Actual space stations will have to make do with much less of a normal force—if any at all.

We have discussed how tension, a normal force, and a friction force can each act as the centripetal force responsible for circular motion. In particular, static friction (the tires holding to the road rather than skidding) is the centripetal force that causes an automobile to make a turn. But static friction is limited to a maximum value. Above this value, sliding friction takes over and the automobile starts to skid. Thus the speed and/or radius with which an automobile can make a turn is limited. This fact explains why it is useful to "bank" a curve. In a banked curve, the road is tilted so that the normal force, perpendicular to the road surface, has a component pointing to the center of the circle of the curve. In this way, the normal force acts together with static friction to make up the centripetal force that accelerates the automobile through the curve. Example 5–13 illustrates this point in the extreme case where there is no friction.

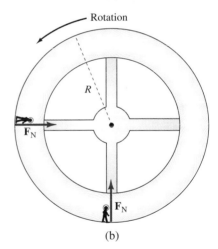

FIGURE 5–25 (a) Example 5–12. The space station from the movie *2001, A Space Odyssey*. (b) A normal force, $\mathbf{F}_N$, pointing to the circle's center acts on each astronaut. The existence of this force gives the astronauts the illusion of weight.

APPLICATION

Banking turns

EXAMPLE 5–13 It is a dark and stormy night, and a driver advancing along an icy road must negotiate a turn. The road traces out an arc of a circle with radius

(a)

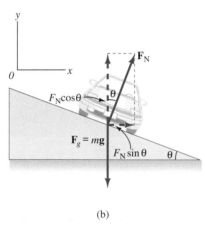

(b)

FIGURE 5–26 (a) Example 5–13. (b) The automobile is seen face on. The center of the circle described by the road curve is to the right.

320 m and is banked at an angle of 5.1°. We will assume that the friction between the tires and the road is zero. At what speed should the driver take the curve to avoid sliding off the road?

Solution: Figure 5–26 is a force diagram for the car as seen from the front, with the center of the circle toward the $+x$-direction. In the absence of friction, there are only two forces acting on the car. The force of gravity is purely in the y-direction, and the normal force, perpendicular to the banked road, has components in the y- and x-directions. Newton's second law is $\mathbf{F}_{net} = \mathbf{F}_N + \mathbf{F}_g = m\mathbf{a}$, or

$$F_N \cos\theta\,\mathbf{j} + F_N \sin\theta\,\mathbf{i} - mg\,\mathbf{j} = \frac{mv^2}{r}\,\mathbf{i}.$$

In component form, we have

for the x-component: $\qquad F_N \sin\theta = \dfrac{mv^2}{r},$ $\qquad$ (5–35)

for the y-component: $\quad F_N \cos\theta - mg = 0.$ $\qquad$ (5–36)

Only the x-component of F_N acts as the centripetal force to carry the car around the curve. When the automobile moves without sliding, the vertical component of F_N just balances the force of gravity, and the horizontal component provides the centripetal force, as in Eq. (5–34).

Equation (5–36) determines F_N:

$$F_N = \frac{mg}{\cos\theta},$$

and this is inserted into Eq. (5–35), which is solved for v:

$$\frac{mg}{\cos\theta}\sin\theta = \frac{mv^2}{r};$$

$$v = \sqrt{gr\tan\theta}.$$

With $r = 320$ m and $\theta = 5.1°$ ($\tan 5.1° = 0.089$),

$$v = \sqrt{(9.8 \text{ m/s}^2)(320 \text{ m})(0.089)} = 17 \text{ m/s} \approx 60 \text{ km/h}.$$

This system is balanced quite delicately. For a given banking, curvature, and speed, Eq. (5–35) uniquely determines F_N. If F_N is too small—if the speed is too slow—the product $F_N \cos\theta$ cannot balance gravity, and the car slips *down* the slope. If F_N is too large—if the speed is too fast—$F_N \cos\theta$ would exceed mg, and the car would skid *up* the slope and off the road. The driver must negotiate the curve just right! When there is some static friction, there is some leeway for error because static friction can take on an appropriate value within a range and hold the car on the road (see Problem 49).

Circular Motion with Changing Speed

What happens when the speed changes in circular motion? Figure 5–27a illustrates a *simple pendulum*: A point mass m suspended from a string of negligible mass that has a length ℓ. The pendulum swings in a plane under the influence of gravity. During this swing, the mass on the end follows a circular path determined by the length of the string, but it is *not* uniform circular motion because the speed is not constant. To understand this, we draw a force diagram (Fig. 5–27b). Two forces act on the mass: gravity and tension. The mass accelerates according to Newton's second law, $\mathbf{F}_{net} = \mathbf{F}_g + \mathbf{T} = m\mathbf{a}$, because these two forces are unbalanced. When the mass is at the particular angle θ shown, the forces can be broken into components perpendicular to the path (radial forces), and components along the path (tangential forces). The tension is purely

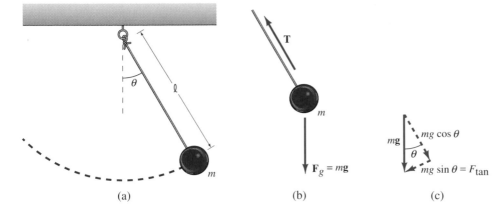

(a)   (b)   (c)

FIGURE 5–27 (a) A simple pendulum of mass m and length ℓ. The path followed traces a circle, but the speed along this path varies with angle θ. (b) Force diagram for the simple pendulum. (c) The components of the force of gravity on the simple pendulum.

radial and inwardly directed, whereas the force of gravity has an outward radial component $mg \cos \theta$ and a tangential component $mg \sin \theta$ (Fig. 5–27c). Thus the component breakdown of the second law is

for the radial (inward) component: $T - mg \cos \theta = \dfrac{mv^2}{r},$

for the tangential component: $mg \sin \theta = ma_{\text{tan}} = m\dfrac{dv}{dt},$

where the tangential component of the acceleration, a_{tan}, is the rate of change of speed.

The speed of the pendulum changes because a tangential force is present. The radial forces simply guide the mass on a circular path and *do not change its speed.* Figure 5–28 enables us to see that the pendulum changes its speed.

For *any* curving path, there will generally be forces acting with components perpendicular to the path and tangential to the path. Force components perpendicular to the path change the direction of the object's motion but not its speed. The tangential components of the net force change the magnitude of the velocity but not its direction.

Tangential forces are required to change the speed of an object.

EXAMPLE 5–14   A skater of mass $m = 52$ kg is swinging in a circle of radius $r = 2.0$ m at a speed $v = 3.0$ m/s. He maintains this motion by holding onto a rope attached through a frictionless bearing to a fixed pole at the center of the circle (Fig. 5–29). The tension of the rope provides the centripetal force for his circular motion and we can regard the ice as being frictionless. As he swings around the pole, however, he runs into a sandy patch of ice 48 cm long, where the coefficient of friction is $\mu_k = 0.10$. (a) What is the tension, T, in the rope before he hits the sandy patch? (b) What is his speed just after he passes the sandy patch? (c) How must the tension change to keep him on the same circular path after he has passed the sandy patch?

Solution: (a) Before he reaches the sandy patch, the skater is in uniform circular motion. The tension is the unique centripetal force,

$$T = \frac{mv^2}{r} = \frac{(52 \text{ kg})(3.0 \text{ m/s})^2}{2.0 \text{ m}} = 230 \text{ N}.$$

Compare this to his weight, which is $mg = (52 \text{ kg})(9.8 \text{ m/s}^2) = 510$ N.

(b) The friction of the sandy patch is a constant tangential force, which slows the skater down. The patch is small enough so that we can assume that the motion of the skater through the patch is a straight line and use our formulas for one-dimensional motion. The magnitude of the friction force is

$$f_k = \mu_k F_N = \mu_k mg = (0.10)(510 \text{ N}) = 51 \text{ N}.$$

FIGURE 5–28 As the pendulum bob moves under the influence of gravity and the string tension, it speeds up to a maximum at the bottom of the swing and moves most slowly at the top. This is revealed here by photographing the swing with light flashes at equal time intervals, here 1/10 s.

FIGURE 5–29 Example 5–14.

Then, we use the one-dimensional formula relating speed change to distance traveled for a constant force, Eq. (2–24):

$$v^2 = v_0^2 + 2a\,\Delta x = v_0^2 - \frac{2f_k}{m}\,\Delta x$$

$$= (3.0 \text{ m/s})^2 - \frac{2(51 \text{ N})}{52 \text{ kg}}(48 \text{ cm})\left(\frac{1 \text{ m}}{10^2 \text{ cm}}\right) = 8.1 \text{ m}^2/\text{s}^2.$$

Note that the acceleration is negative because friction slows the skater down. Thus, after passing the sandy patch, the skater has slowed to

$$v = \sqrt{8.1 \text{ m}^2/\text{s}^2} = 2.8 \text{ m/s}.$$

(c) We simply follow the first procedure and compute the tension with the new speed:

$$T = \frac{mv^2}{r} = \frac{(52 \text{ kg})(8.1 \text{ m}^2/\text{s}^2)}{2.0 \text{ m}} = 210 \text{ N}.$$

This is 20 N less than the tension in the rope before the sandy patch is encountered.

Circular Motion and Noninertial Frames

We have examined the centripetal forces illustrated in the doughnut-shaped space station of Example 5–12. This system also illustrates the effects of noninertial frames, first mentioned in Section 4–4. An astronaut in the space station is an observer in a noninertial frame. Seen from outside, the astronaut is in uniform circular motion and is therefore accelerating. The centripetal force responsible for this acceleration is the normal force, which can be read by a scale at the astronaut's feet. As the astronaut perceives it, however, everything around him within the station is at rest. The scale, of course, continues to read the same normal force. This force is directed toward the center of the circle. Thus Newton's first law no longer holds for the astronaut: He is at rest and there is no identifiable source for the force that keeps him on the scale. That "force" is directed *outward* ("centrifugal," meaning "directed away from the center") because it balances the inward-directed normal force that the scale reads.

While the notion of a centrifugal force is a commonplace one, we should keep in mind that it is not a force in the same sense as the forces that appear in Newton's laws. It is merely the consequence of an observer being in a rotating reference frame. We use the language of the centrifugal force to interpret the invisible hand that forces us against the door when an automobile makes a high-speed turn. Seen from outside the automobile, a passenger is accelerating centripetally. A centripetal normal force from the door of the automobile pushing on the passenger is the force responsible for moving the passenger in his curving path. The automobile passenger views things differently. The passenger perceives herself to be at rest with respect to the car, yet she feels the normal force from the door. The passenger then invents a "centrifugal force" to balance the normal force from the door.

5–5 FUNDAMENTAL FORCES

Certain forces are *fundamental forces*. They differ from what we might call secondary forces: weight, the normal force, friction, drag, tension, spring forces, and so forth. The fundamental forces play a privileged role among the various forces that we investigate. In fact, the secondary forces are ultimately derived from these forces. For example, the force of gravity (the term we normally use when an object moves near Earth's surface) is just a special case of the fundamental force of *universal gravitation* (which describes the force on objects located near or far from any mass) first understood by Newton. The force exerted by a stiff coil of wire forming a spring depends on the fundamental electric forces between the molecules that make up the material of the spring. It is easy to derive the connection between gravity and universal gravitation. Even with the largest computers at our disposal, however, we can only dream of calculating the spring force from fundamental forces. In any case, such a calculation would not clarify any aspect of physics.

What we refer to today as the fundamental forces are not necessarily the forces that were called fundamental 200 or 100 or even 20 years ago. Our progress on the understanding of what is truly fundamental has been steady. There have been a series of *unifications* in the historical development of fundamental forces: What were previously thought to be distinct fundamental forces have become just different aspects of the same fundamental force.

We now believe that there are only three fundamental forces in nature. They are the force of **universal gravitation**, the **electroweak** force, and the **strong** force (often called the **nuclear** force). The expression of the force of universal gravitation (see Chapter 12) was one of Newton's greatest discoveries. Albert Einstein replaced Newton's expression in his law of general relativity, making some new predictions of tiny experimental effects that are not predicted by Newtonian gravitation. Einstein also made persistent but unsuccessful efforts to unify gravitation with the other fundamental forces. The electroweak force is a recent discovery, dating from a theory proposed by Sheldon L. Glashow, Abdus Salam, and Steven Weinberg in the mid-1970s that was verified experimentally in the early 1980s. This work unified two forces formerly thought to be independent and fundamental: the *weak* force, responsible mainly for some types of radioactive processes in nuclei and important in the evolution of the universe, and the forces of *electromagnetism* (Chapters 22–35). On the scale appropriate to the secondary forces that we deal with every day, the electromagnetism aspect of the electroweak force is dominant, and it is often convenient to refer simply to the electromagnetic force. Electromagnetism is itself the result of a nineteenth-century unification of the forces of *electricity* and *magnetism*; these two forces had previously been thought to be independent. The strong force is responsible for holding together the nuclei of atoms. Both the strong force and the electroweak force have been the object of more recent attempts at unification, which have yet to bear fruit. Since the 1980s, there has also been a new effort to bring gravitation into the unification program. Unification continues to be a fascinating beacon that will allow us to describe more of the universe in a simpler way.

On the everyday scale of the secondary forces that we have discussed in this chapter, only the force of gravity is a direct aspect of a fundamental force: universal gravitation. All the other forces that we have investigated are ultimately due to the electromagnetic force, which binds atoms and molecules together into ordinary matter. We do not directly see the strong force on our scale; it holds nuclei together so tightly that, for most practical purposes, we can think of them as indivisible lumps of matter. Only when we discuss nuclear energy or the composition of stars do strong forces come into practical play.

It is customary to speak of the relative size of the fundamental forces. This cannot be done without talking about a particular situation; that is, placing the fundamental forces in context. We can say the following: Within the nucleus, the typical strong force is about 100 times larger than the electromagnetic forces. The gravitational force between two protons in the nucleus is many, many orders of magnitude smaller than either the strong or the electroweak force. Thus the forces are ordered in decreasing strength as strong, electroweak, and gravitational. It is only because the strong force acts over such a limited range that the electroweak and gravitational forces ever come into view. The gravitational force dominates on the astronomical scale because matter is arranged into an electrically neutral combination, in which the electromagnetic forces cancel out to zero.

Three fundamental forces

Except for gravity, all the forces discussed in this chapter are electromagnetic in origin.

This chapter is devoted to describing various forces that occur in nature and to applying the problem-solving techniques developed in Chapter 4. These techniques are very important and we will use them throughout the book.

The force of gravity (or weight), tension, and the normal force are all common forces for which it is often possible to solve the equations of motion given by Newton's second law. The force of gravity is expressed as

$$\mathbf{F}_g = m\mathbf{g}, \tag{5–4}$$

where $\mathbf{g}$ is a constant vector that points down toward Earth's center.

The tension, $\mathbf{T}$, is a variable force that is determined according to the circumstance and exerted by ropes (or wires or cables or strings). Tension always pulls on a mass in the direction taken by the rope; it is transmitted everywhere along the rope, taking a single constant value when the mass of the rope is negligible.

The normal force, $\mathbf{F}_N$, is also variable. It is directed perpendicularly to a surface and acts to cancel any other forces that might make a mass accelerate into the surface.

The friction force acts when two surfaces slide or attempt to slide across one another. Static friction, $\mathbf{f}_s$, is variable and acts in a direction that would oppose any sliding that would occur if there were no friction. It can increase up to a maximum value proportional to the magnitude of the normal force:

$$0 \le f_s \le \mu_s F_N. \tag{5–21}$$

Kinetic friction, $\mathbf{f}_k$, acts when sliding actually is occurring and is also proportional to F_N:

$$f_k = \mu_k F_N. \tag{5–22}$$

The constants μ_s and μ_k are the coefficients of friction, and generally

$$\mu_s > \mu_k. \tag{5–23}$$

Another type of friction, the drag force, occurs when objects move through media. For all but small objects, the drag most frequently encountered varies with the speed squared:

$$F_D = \frac{1}{2} \rho A C_D v^2. \tag{5–30}$$

It implies that objects accelerating within media can be accelerated only up to a terminal velocity.

Forces are responsible for accelerating objects moving in a curved path. When the motion is uniform circular motion, then Newton's second law takes the form

$$\mathbf{F} = -\frac{mv^2}{r}\hat{\mathbf{r}}. \tag{5–34}$$

The vector $-\hat{\mathbf{r}}$ is directed to the center of the circle and the force is said to be centripetal. When both the direction and the magnitude of the velocity change, then $\mathbf{F}$ has a component tangential to the motion. This component causes the magnitude of the velocity to change, while a force component perpendicular to the motion causes the direction of the motion to change.

All the forces of nature are ultimately described in terms of three fundamental forces: the force of universal gravitation, the electroweak force, and the strong force. The electromagnetic force, which is part of the electroweak force, is responsible for most of the secondary forces, including tension, friction, drag, normal forces, and spring forces.

1. A tightrope walker moves to the center of a thin wire that was initially stretched taut to a horizontal position. Why is it that the wire cannot remain horizontal?

2. We have said that it is actually the force of friction that is responsible for both the acceleration of automobiles and our ability to walk. What is the role of the engine or of the muscles in these processes?

3. An observer sees a mass hanging motionless from a vertical string. Under what circumstances is the tension in the string greater than or less than the weight of the mass?

4. What are some factors that could limit how fast a hot rod can go in a $\frac{1}{4}$-mi race?

5. How does the fact that a rope has mass complicate solving a problem about lifting a load with a pulley? What is the effect of friction in the pulley?

6. Why is it helpful for an automobile with an engine in the front to have front-wheel drive? Why is it useful to put sand in the trunk of your car in winter if your car has rear-wheel drive?

7. Why do bicyclists or motorcyclists "lean into" a curve? In explaining why, make use of the fact that we are able to balance ourselves best when we feel that the net force on us is coming from directly beneath our feet. Also think about the forces that friction must cancel.

8. How might the result of Example 5–8 vary if the weight was not evenly distributed over the wheels and/or the car was equipped with rear-wheel drive instead of four-wheel drive?

9. Suppose that a rope has tension because a mass is suspended from its end. Let's say that the rope is now cut, a spring is inserted at the cut, and we observe the stretch of the spring as a measure of the rope's tension. If the mass of the rope is negligible, does the observed tension of the rope depend on where the cut is made along the string? Does your answer change if the rope cannot be considered massless?

10. A bowl of water with floating ice cubes is placed on a scale. The ice cubes melt. Does the reading of the scale change? Why or why not?

11. What is the role of the keel, which runs along the center of the bottom of a sailboat? Some sailboats have centerboards rather than keels. These are simply large boards that can be lowered or raised in the position of a keel. What is the role of a centerboard? Why might you prefer a keel to a centerboard or vice versa?

12. We have referred to massless ropes and the tension in them. What physical considerations allow you to think of a rope as massless?

13. The speed of the boats (shells) used in scull racing is, to a good approximation, independent of the number of people rowing (provided that the number is larger than three or four). At first, this might appear strange: The more people there are, the larger the forward propelling force available to overcome the drag of the water. Can you explain this seeming contradiction?

14. Describe a series of experiments that would have allowed Leonardo da Vinci to decide that the force of friction depends only on the normal force on an object.

15. At the moment a car in a loop-the-loop roller coaster is at the top of the loop (directly below the track), can there be a normal force on it? Such a force would point straight down, in the same direction as that of gravity.

16. The riders on the loop-the-loop of Question 15 experience a feeling of near-weightlessness close to the top of the loop. Why? Would the coins in their pockets fall out?

17. Suppose that it were possible for a ship to sail all the way around the world along a great circle. Is a centripetal force necessary to keep the ship moving in this circle? What force or forces would act centripetally?

18. We have said that the tension in a string is the result of the intermolecular forces within it. Suppose that you had some idea of the forces between individual, separate molecules within the string. How could you estimate the tension at which the string breaks?

19. When you sit in the passenger seat of a car that makes a tight turn to the left, you could be thrown out of the car if the door should open (and you are not belted in). Why?

20. At the beginning of Section 5–4, we spoke of swinging a book on the end of a rope in a nearly horizontal circle. Could the plane of the circle be *perfectly* horizontal?

21. Imagine that you and your partner are on skates on a perfectly frictionless ice surface. You hold on to opposite ends of a rope and pull toward each other. What happens if you each have exactly the same mass? If you have half your partner's mass?

22. Why is it hard to run when the ground is icy?

23. Tarzan swings from tree to tree on a vine. At which point in his swing is the vine most likely to break, and why?

24. A person on a rapidly moving Ferris wheel feels that she is about to fly off the seat when her seat reaches the top of the circular path. Why is that?

25. A hemispherical bowl is placed open end up on a table and rotates around its own vertical axis. A die is allowed to slide down from the edge into the bowl. Describe the motion of the die as seen by a hypothetical observer at the bottom of the bowl, assuming that there is no friction between the bowl and the die.

PROBLEMS

5–1 Some Simple Constant Forces

1. (I) An 8-g bullet that travels at 500 m/s is fired into a rigidly fixed block of wood. The bullet is found 7 cm into the wood. What was the average force exerted by the wood opposing the bullet's motion? Assume that the deceleration was uniform.

2. (I) In Example 5–3, a woman lifts a football lineman (mass 149 kg) by pulling down on one end of a rope with 365 N of force. Suppose that the woman has a mass of 50 kg. If a bathroom scale were beneath her feet as she lifted the lineman, what would the scale read in pounds? Analyze the force dia-

gram for the woman and remember that a scale of this type reads the upward normal force that the floor exerts on her.

3. (I) A helium balloon just manages to lift 100 kg (including the mass of the balloon, the helium it contains, and its payload) off the ground and then hovers 1 m off the ground. The upward force that maintains the balloon is *buoyancy* (we shall treat this force in Chapter 16). What is the magnitude of the buoyancy in this case?

4. (I) A woman of mass 56 kg sits in a racing car. When she depresses the accelerator, the car accelerates in a straight line to 170 km/h in 9.3 s. What are the direction and magnitude of the force she experiences? Where is the force applied?

5. (I) A hockey puck of mass 0.10 kg slides without friction on ice. In an appropriate coordinate system, its velocity is $\mathbf{v}_1 = (1.4\mathbf{i} + 3.0\mathbf{j})$ m/s. A constant force $\mathbf{F} = 4.0\mathbf{i}$ N is then applied to the puck. After how many seconds will the puck have a speed of 6 m/s?

6. (II) A man of mass 85 kg is escaping a burning building using a rope that will break if the tension exceeds 700 N. (a) With what acceleration must he slide down the rope if it is not to break? (b) How far down the rope is he, and what is his velocity, after 4.0 s, assuming he drops with the minimum acceleration of part (a)?

7. (II) A metal rod of mass 5.6 kg and length 3.5 m is suspended from the ceiling. (a) What is the tension in the rod at a distance of 2.0 m from the top? (b) What is it at a distance of 3.0 m from the top?

8. (II) A brick hangs from a string attached to the ceiling. When a horizontal force of 12 N is applied to the brick, the string makes an angle of 25° with the vertical (Fig. 5–30). What is the mass of the brick?

FIGURE 5–30 Problem 8.

9. (II) Figure 5–31 shows a person applying a horizontal force in trying to push a 25-kg block up a frictionless plane inclined at an angle of 15°. (a) Calculate the force needed just to keep the block in equilibrium. (b) Suppose that she applies three times that force. What will be the acceleration of the block?

10. (II) Two blocks, of mass m_1 and m_2, are placed in contact on a smooth surface, with the more massive one (m_1) on the left. A force of magnitude F, pointing to the right, is applied to the block on the left. (a) What is the acceleration of the system? (b) What force acts on the block on the right?

FIGURE 5–31 Problem 9.

11. (II) An automobile of mass 1200 kg pulls another automobile of mass 1400 kg with a tow rope. (a) In order to pull out onto a highway, the automobile must accelerate to 55 mi/h in an access lane that is only 120 m long. What must its acceleration be? (b) What is the tension in the tow rope?

12. (II) An Atwood machine consists of a massless string connecting two masses over a massless, frictionless pulley (Fig. 5–32). In this case, the masses are 0.800 kg and 0.650 kg. The system is released from rest with the 0.8-kg mass 2.15 m above the floor and the 0.65-kg mass on the floor. (a) What is the acceleration of the 0.8-kg mass? of the 0.65-kg mass? (b) What is the speed of the 0.8-kg mass just before it hits the floor? (c) How long does it take the 0.8-kg mass to reach the floor?

FIGURE 5–32 Problem 12.

13. (II) Two blocks of masses M and m are connected by a light rope which passes over a frictionless pulley. Mass M sits on

FIGURE 5–33 Problem 13.

an inclined plane with an angle of inclination of 30° (Fig. 5–33). The coefficient of static friction between mass M and the inclined plane is 0.20, while $m = 3.0$ kg. Determine the largest and smallest possible values of M for which the system remains in equilibrium. Calculate the force of static friction on the block of mass M if $M = 6$ kg.

14. (II) Consider a variation on an Atwood machine in which the masses are each on an incline (Fig. 5–34). The mass sliding on incline 1, m_1, is 1.50 kg, and the angle of this incline is $\theta_1 = 62°$. If the mass on the second incline, m_2, is 2.50 kg, what is the angle θ_2 so that the system does not accelerate?

FIGURE 5–34 Problem 14.

15. (III) Consider the three-pulley arrangement shown in Fig. 5–35. The three masses m_1, m_2, and m_3 have the values 2.00 kg, 5.00 kg, and 4.00 kg, respectively. All the pulleys are frictionless, and the strings are massless. What are the tensions in all the strings, and what are the accelerations of the masses?

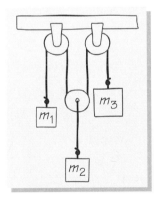

FIGURE 5–35 Problem 15.

16. (III) In the text, we stated that the tension is the same everywhere in a massless rope. Give the equation that describes the tension as a function of position for a stationary hanging rope, as in Fig. 5–36a, with constant mass density $\lambda =$ mass/unit length. Figure 5–36b shows the forces on an element of rope. For this problem assume that no separate mass is attached at the bottom. [*Hint*: Measure the height h up from the bottom point of the rope to any point along the rope. The bottom of an isolated slice is at height h, and its top is at height $h + \Delta h$. This means that its mass Δm is given by $\lambda \Delta h$. The tension T_{up} acting at the top of the slice is $T(h + \Delta h)$, and the tension T_{down} acting at the bottom of the slice is $T(h)$. The resulting equation can be written as a *differen-*

tial equation for T as a function of h by dividing by Δh and recognizing that in the limit of $\Delta h \to 0$, $d[T(h)]/dh = [T(h + \Delta h) - T(h)]/\Delta h$. The equation's solution is the *function $T(h)$* that satisfies the equation.]

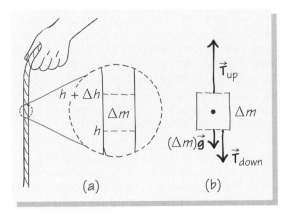

FIGURE 5–36 Problem 16.

17. (III) A compound Atwood machine is constructed by replacing one of the masses of a conventional Atwood machine (see Problem 12) with the pulley of another Atwood machine (Fig. 5–37). Altogether there are three masses, two ropes, and two pulleys; the ropes and pulleys are to be considered massless and friction free. Describe the method by which the motions of the masses of this machine can be analyzed. (*Hint*: There are two independent tensions and two independent accelerations, which must be found by analyzing the equations of motion for the three masses simultaneously.) Solve these equations for arbitrary masses. What happens if $m_2 = m_3 \neq m_1$? [*Hint*: If m_1 accelerates, so does the pulley, so m_2 and m_3 share an additional acceleration.]

FIGURE 5–37 Problem 17.

18. (III) A double-pulley system—like that described in Problem 17—has masses $m_1 = 2.00$ kg, $m_2 = 1.20$ kg, and $m_3 = 0.800$ kg. (a) What are the accelerations of all the masses? (b) What are the tensions in all the ropes?

19. (I) A truck of mass 5000 kg accelerates on a straight road at 1.2 m/s². Assuming that air resistance is negligible, what is the coefficient of static friction between the road and the tires? (Don't worry about differences in normal forces, on each wheel, and assume 4-wheel drive.)

20. (I) A boy of mass 35 kg runs into an ice-skating rink in street shoes and begins a smooth slide. He starts with a speed of 4.5 m/s and comes to a stop after sliding for 3.4 s. What is the coefficient of kinetic friction of his shoes on ice?

21. (I) A car of mass 1200 kg is moving at 25 m/s. The driver suddenly sees a dog crossing the road, slams on the brakes, and manages to stop the car in 4.2 s. What is the coefficient of static friction between the tires and the road? (Assume that the acceleration is constant and that there is no skidding).

22. (I) The coefficient of static friction between a worker's shoes and the floor is 0.81, while the coefficient of static friction between the floor and a crate is 0.43. The worker, mass 80 kg, pushes the crate, which has a mass of 140 kg. What is the friction force on the worker due to friction between the shoes and the floor when the crate starts to slide?

23. (I) A worker must push an 85-kg crate across a floor. The coefficient of kinetic friction between the crate and the floor is $\mu_k = 0.4$. What is the minimum force that the worker must exert to keep the crate moving once the crate starts moving?

24. (I) A rope is connected to a 25-kg cement block placed on a board leaning against a wall at an angle of 25° with respect to the horizontal (Fig. 5–38). The coefficient of kinetic friction between the cement block and board is $\mu_k = 0.4$. (a) What is the tension in the rope if it is pulled at constant speed straight up the board? (b) What is the tension if the rope is pulled up at constant speed at an angle 40° from the horizontal?

26. (II) A man wants to push a package of shingles of total mass 25 kg up a roof being built at an angle of 35°. The coefficient of kinetic friction between the package and the roofing paper already in place is $\mu_k = 0.32$. (a) How much force does the man have to exert on the package directly along the slope of the roof to cause the package to accelerate at 0.20 m/s²? (b) If the coefficient of static friction is 0.42, will the package remain on the roof?

27. (II) The coefficient of static friction between a car of mass 1500 kg and an asphalt road is $\mu_k = 0.7$. (a) What is the shortest distance over which the car can accelerate from rest to a speed of 96 km/h? (b) How long will this take? (Because we neglect drag and rolling friction, the distance and time will be unusually short.)

28. (II) A crate of mass 250 kg is loaded on the back of a truck. The coefficient of static friction between the crate and the truck bed is μ_s. The truck decelerates such that it comes to a stop from a speed of 60 mi/h (26.7 m/s) in a distance of 140 m. How large must μ_s be so that the crate does not slide forward on the truck bed?

29. (II) A 500-g box is placed on a board at a 30° incline and accelerates from rest down the board at 0.3 m/s². (a) How long does the box take to travel down the board of length 1.2 m? (b) What frictional force opposes the motion of the box? (c) What is the coefficient of friction between box and board?

30. (II) A block of mass $m = 2.0$ kg is placed on a horizontal surface. The coefficient of static friction between the block and the surface is $\mu_s = 0.4$. A light rope is tied to the block and thrown over a frictionless pulley (Fig. 5–39). The free end of the rope is pulled with a slowly increasing force T. At what value of T will the block start to move if the angle that the rope makes with the horizontal is 50°?

FIGURE 5–38 Problem 24.

FIGURE 5–39 Problem 30.

25. (II) A person learning to snow ski will use the snowplow position; it is a rudimentary way of keeping one's skiing speed under control. Let's imagine that a beginning skier finds herself on an icy slope of 22°. Only by setting her skis in a good snowplow position, with the tips of both skis pointed inward and the inner edges dug in, is she able to keep from accelerating. Effectively, what coefficient of sliding friction is created by the snowplow?

31. (II) A 50-kg box rests on a rough horizontal surface with which the box has a large coefficient of static friction, $\mu_s = 0.75$. The box is pulled by means of a light rope with a force of magnitude F, making an angle θ with the horizontal. (a) Find the magnitude of the force F that will just start the box moving horizontally as a function of θ. (b) Show that there is some angle θ for which F takes a minimum value. What is this value for our case, and what is the force F corresponding to this value? Explain why, physically, there is such a minimum value.

32. (II) A block of mass 1.2 kg rests on top of another block of mass 1.8 kg, which rests on a frictionless surface. The coefficient of static friction between the blocks is $\mu_s = 0.3$. What is the maximum horizontal force that can be applied to the upper block so that the blocks accelerate together, without the upper block sliding on the lower one? If the horizontal force is applied to the lower block instead, what is the maximum force that will give rise to the same motion?

33. (II) Consider again the professor of Example 5–7. Her mass is 55 kg, and the coefficient of static friction between her and the floor is μ_1. What is the minimum value of μ_1 for which she will not slip on the floor?

34. (II) A pile of snow at the crest of a roof with a slope of 40° from the horizontal starts to slide off. The distance from the crest to the edge of the roof is 8 m, and the coefficient of kinetic friction for the snow on the roof is 0.1. (a) What is the speed of the pile of snow when it reaches the edge of the roof? (b) Assuming that it is 6 m from the edge of the roof to the ground, how far out from the base of the building does the snow land?

35. (III) A mass m_1 rests on top of another mass, m_2, which in turn rests on a frictionless horizontal surface (Fig. 5–40). A light cord is attached to m_2, which is used to pull on it with a force F. (a) Find the acceleration of each object when the surface between the two objects is frictionless. (b) Find the acceleration of each object when the surface between the two objects is rough enough to ensure that m_1 does not slide on m_2. (c) What are the magnitude and direction of the contact forces exerted by the lower object on the upper one, assuming that the upper object is sliding on the lower object with a nonzero coefficient of kinetic friction μ_k? (d) Find the acceleration of each object if the surface between the two objects is such that the upper object is sliding on the lower one under the influence of kinetic friction, with a coefficient of kinetic friction μ_k.

FIGURE 5–40 Problem 35.

5–3   Drag Forces

36. (I) A parachute is rigged so that a parachutist of total mass 116 kg with full gear reaches the ground at a terminal velocity of 4.9 m/s. Assuming that the drag force on the parachutist, moving with speed v, has a magnitude equal to kv^2, what is the value of k?

37. (I) Estimate the drag force on an automobile cruising at 55 mi/h. Assume that the drag coefficient, C_D, is 0.3 and that the car's cross-sectional area is 5 m². Take air to have a density of 1.25 kg/m³.

38. (I) A race car of mass 800 kg has a maximum acceleration from rest of 4.8 m/s². Assume that the car's engine is such that the force on the tires is constant and that the car's effective cross-sectional area into the air is 1.8 m². If the car's top speed is observed to be 90 m/s, what is its drag coefficient, C_D? Take the density of air as 1.25 kg/m³.

39. (I) A ball of mass 500 g is observed to reach its terminal velocity of 18 m/s after being dropped from the top of a tall building. Assume that the density of air is 1.25 kg/m³ and the drag coefficient, C_D, is 0.4. What is the effective cross-sectional area of the ball? You can use Eq. (5–30) here.

40. (II) A sphere of radius r_1 and mass m_1 falls through the air and is found to have a terminal velocity v_1. Equation (5–30) applies. (a) What is the terminal velocity v_2 for a sphere with twice the radius and the same mass density as the first sphere? (b) Generalize the result of part (a) to find the terminal velocity for a sphere with a radius z times the first radius but the same mass density.

41. (II) A barge, moving at uniform speed, is pulled by two horses moving on opposite sides of a canal in which the barge floats. The ropes connecting the horses to the barge make an angle of 28° with the line of motion of the barge. If the resistance to the motion of the barge is characterized by a frictional force given by $F = -(160 \text{ N} \cdot \text{s/m})v$, where v is the speed of the barge in meters per second and the tension in each of the ropes is 42 N, what is the speed of the barge?

42. (II) Assume that the drag force on a parachute is given by Eq. (5–30) in the text. The effective area of the parachute is 30 m² and the density of air is 1.25 kg/m³. If a 90-kg parachutist finds that his terminal velocity is 6 m/s, what is the drag coefficient?

43. (II) The terminal velocity of a skydiver of mass 75 kg can be controlled by the orientation of her body and can range from 40 m/s to 60 m/s. Assume that she can change the area presented to the ground by a factor of 1.5 in going from the minimum to the maximum terminal speed, and that the larger area presented to the ground slows the skydiver down. Now, how does the drag coefficient change? (Express your answer with a proportion.)

44. (II) A marble of mass m falls through a fluid and is subject to the drag force $\mathbf{F}_D = -A\mathbf{v}$, where $\mathbf{v}$ is the velocity of the marble. The marble will reach a terminal velocity given by $\mathbf{v}_t = m\mathbf{g}/A$. Use dimensional analysis to estimate how long it will take to reach the terminal velocity. [*Hint*: A characteristic "time" can be constructed from A, g, and m.]

5–4   Forces and Circular Motion

45. (I) An airplane of mass 2×10^4 kg executes a banked turn of radius 30 km while flying at 200 m/s. What acceleration will the passengers feel as a result of the turn?

46. (I) A rock swings in a nearly horizontal circle at the end of a string whose breaking tension is 12 N. The circular path is 0.25 m in radius, and the rock's mass is 150 g. What is the maximum speed the rock can have before the string breaks?

47. (I) A man of mass 65 kg stands at the edge of a merry-go-round of radius 5.3 m. The merry-go-round turns at 6.0 rev/min. What are the magnitude and direction of the net force on the man?

48. (I) An accelerometer shows that an airplane flying at 850 km/h undergoes a vertical acceleration of 0.17 g's (1 $g = 9.8$ m/s²) at a certain moment. What is the radius of curvature of the airplane's (horizontal) path at that point?

49. (II) An automobile makes a turn whose radius is 150 m (Fig. 5–41). The road is banked at an angle of 18°, and the coefficient of friction between the wheels and road is 0.3. Find the maximum and minimum speeds for the car to stay on the road without skidding up or down the banked road.

FIGURE 5–41 Problem 49.

50. (II) A merry-go-round has a circular platform that is 1 m from the central axis at its inner edge and is 5 m from the central axis at its outer edge. The ride turns at a rate of one full rotation every 10 s. A passenger holds himself to the surface with a pair of very sticky shoes and is most comfortable when he orients his body length along the line of the net force on him. What angle does his body make to the vertical (a) 1 m from the axis? (b) 3 m from the axis? (c) 5 m from the axis?

51. (II) The coefficient of static friction between a small stone and a horizontal turntable is measured in the following way. The stone is placed on the turntable at a distance R from the axis, and the speed of rotation is slowly increased to 45 rev/min. When the experiment is repeated for several different values of R, it is found that the stone remains on the turntable if $R <$ 26 cm, and that it slides off with increasing speed of rotation if $R >$ 26 cm. Determine μ_s from these data.

52. (II) A Ferris wheel in an amusement park has a radius of 30 m and makes one complete turn every 75 s. Calculate the normal force that a passenger of mass 60 kg experiences through the seat of the pants (the seat bottom is parallel to the ground) when the passenger is (a) at the bottom of the path, nearest the ground; (b) at the maximum height of the path.

53. (II) Assume that the acceleration of the Moon due to Earth's gravity is 0.0027 m/s². What is the velocity of the Moon with respect to Earth if the period of the Moon's motion around Earth is 28 d? Do not look up the distance between Earth and the Moon; instead, calculate it. Compare this result with the distance given in Appendix III–1.

54. (II) A fighter pilot makes a dive almost vertically down and pulls up while traveling at 1500 km/h in a turn of radius 2.50 km. How many g's will the fighter pilot feel at the bottom of the dive? Because the pilot will black out if the number of g's is greater than 11, is this a safe maneuver? (The number of g's is the acceleration in units of $g =$ 9.80 m/s².)

55. (II) A mass of 1.00 kg hangs from a rope placed through a hole in a smooth, frictionless table. At the other end of the rope is attached a puck of mass 400 g, 80 cm from the hole in the table. The puck swings in a circular orbit around the hole (Fig.

FIGURE 5–42 Problem 55.

5–42). With what speed must the puck rotate to keep it 80 cm from the hole?

56. (II) A mass m at the end of a string of length r moves at constant speed v in a circle on a frictionless table. (a) Find the tension in the string in terms of m, r, and v. (b) A second mass identical to the first is attached at the midpoint of the string and the two are whirled; the speed of the outer mass is again v. Draw force diagrams for the two masses and calculate the tensions in terms of m, r, and v.

57. (II) Consider the *conical pendulum*, a mass on the end of a massless string, with the other end of the string fixed on a ceiling. Given the proper push, this pendulum can swing in a circle at a given angle θ, maintaining the same height h throughout its swing, as shown in Fig. 5–43. (a) What is the force diagram for such a pendulum? (b) If the mass of the pendulum is $m =$ 0.2 kg, the length of the pendulum is $\ell =$ 0.5 m, and the angle at which it swings is $\theta =$ 10°, what is the speed of the mass as it swings?

FIGURE 5–43 Problem 57.

58. (II) A small mass slides without friction in a horizontal circular path around the sides of a circular bowl. The bottom of the bowl may be described as a parabola, with the height h above the bottom varying quadratically with the distance r from the axis: $h = br^2$. The mass is observed to move in its circular path with a speed v. What is the height of the path?

59. (II) A small block slides in a horizontal circle on the inside of a conical surface, with the cone making an angle of 50° with the vertical (Fig. 5–44). Assuming that there is no friction between the block and the surface, and the block slides with an angular speed of 6 rad/s, at what vertical height above the apex of the cone does the block slide?

FIGURE 5–44 Problem 59.

60. (II) Two light strings 1.0 m in length are attached to a vertical support 1.0 m apart, and a mass of 5.0 kg at the end of the two strings is whirled about the vertical z-axis (Fig. 5–45). Both strings are taut, so that they and the vertical support form an equilateral triangle. The tension in the upper string is measured to be 150 N. (a) What is the tension in the lower string? (b) How much time does it take for the apparatus to make one complete circuit around the vertical support?

FIGURE 5–45 Problem 60.

61. (II) The road in Example 5–13 is now in better condition, and the coefficient of static friction between the tires and the road is 0.8. With what maximum constant speed can the motorist now negotiate the curve? Friction acts here both to provide a centripetal force and to help cancel the vertical forces. Is there one and only one speed with which the curve can be taken, as in Example 5–13? Why or why not?

62. (II) A train of mass 1.5×10^5 kg is traveling horizontally at 80 km/h and rounding a bend whose radius of curvature is 2 km. At the same time it is decelerating at a rate of 0.2 m/s². The length of the train is negligible compared with the size of the bend, and the train can be treated as a point. What net force does the track exert on the train? Give an approximate answer in which the net change in speed is small compared to the speed itself. Such questions help engineers decide how "robustly" a track must be constructed, or how much to bank it.

63. (II) A small puck of mass $m = 0.2$ kg moves in a circle of radius 0.5 m on a table top; the puck is tied with a massless string to a tether at the origin. The coefficient of kinetic friction between the puck and the table top is $\mu_k = 0.2$. At $t = 0$ s, the puck is at the point shown with a velocity in the $+y$-direction of magnitude 10 m/s. (a) What is the tension in the string at $t = 0$ s? (b) What is the tension in the string at the end of one revolution?

64. (II) A pendulum hangs at rest from a hook in a ceiling of a building. The building is located at a latitude such that the radius vector from the center of Earth to the building makes an angle θ with Earth's axis of rotation. Assume that the force of gravity is the same everywhere and points directly to the center of Earth. Where will the pendulum point? (By definition, Earth rotates about its axis once a day.)

65. (III) A satellite of mass 3000 kg travels in a circular orbit 180 km above Earth, where the acceleration due to gravity is 5 percent smaller than on Earth's surface. Assume that, in a year, the satellite loses 5 km in altitude because of the drag of the extremely thin atmosphere at that altitude. What would you estimate the density of air to be at that altitude, given that the effective area of the satellite is 6 m² and the drag coefficient, C_D, in Eq. (5–30) is 1.0?

66. (III) Consider a ball thrown outward from the center of a platform that rotates counterclockwise with uniform angular velocity ω (Fig. 5–46). An observer standing off the platform (in an inertial reference frame) will describe the ball as moving with uniform velocity in a straight line. (Ignore the effect of gravity; imagine looking down on the platform from above, so that you do not see the up-and-down motion of the ball.) The inertial observer will see that the ball reaches a horizontal distance r from the center in time $t = r/v$, where v is the speed of the ball. (a) Show that, in time t, a point at a radius r on the platform will have moved a distance $d = v\omega t^2$. (b) An observer moving with the platform will see the ball curve away to the right (as seen from the center of the platform). Show that the perceived acceleration is perpendicular to the velocity vector, and that its magnitude is $2\omega v$. (c) What is the direction of the

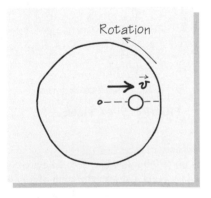

FIGURE 5–46 Problem 66.

perceived acceleration of a ball thrown by a person on the platform from the rim toward the center? It is not surprising that an observer in an accelerated frame sees force-free motion that nevertheless undergoes acceleration. Under certain circumstances—for example, in the study of global wind motion on a rotating Earth—it is convenient to study motion as seen in an accelerating frame. The frame-imposed acceleration is then attributed to a purely fictitious force, which, for rotating systems like the one treated in this problem, is called the *Coriolis force*.

General Problems

67. (I) A 5.0-kg sphere at the end of a 1.2-m cable swings in a horizontal circle on a frictionless surface at the rate of one revolution every 1.4 s. What is the tension in the cable?

68. (II) A small die that is placed 5 in from the center of a turntable begins to slide. The turntable is rotating at $33\frac{1}{3}$ rev/min. What is the coefficient of static friction between the die and the turntable?

69. (II) Three masses (from left to right: 0.3 kg, 0.4 kg, and 0.2 kg) are connected by light cords to make a "train" sliding on a frictionless surface. They are accelerated by a constant horizontal force $F = 1.5$ N that pulls the rightmost mass to the right. What is the tension, T, in the cord (a) between the 0.3-kg and the 0.4-kg masses? (b) between the 0.4-kg and 0.2-kg masses?

70. (II) A string 6.95 m long is strung between two pegs (4.96 m apart) on a ceiling. A mass of 3.88 kg is attached to a point 2.96 m along the string. What are the tensions in the two segments of the string?

71. (II) Consider a system of masses connected by light ropes that pass over massless and frictionless pulleys (Fig. 5–47). (a) When m_1 is displaced vertically by Δx_1, what is the displacement Δx_2 of m_2? (b) For $m_1 = 1.2$ kg and $m_2 = 1.8$ kg, calculate the respective accelerations of the two masses. (c) What is the tension in the string for the masses given in part (b)?

FIGURE 5–47 Problem 71.

72. (II) Masses $m_1 = 0.6$ kg and $m_2 = 1.2$ kg are connected by a taut rope. Mass m_2 is just over the edge of a ramp inclined at an angle of $\theta = 35°$, as in Fig. 5–48, and the masses have a coefficient of kinetic friction of $\mu_k = 0.25$ with the surface. At

$t = 0$ s, the system is given an initial speed of $v_0 = 1.5$ m/s, which starts mass m_2 down the ramp. (a) Draw the force diagram for each mass. (b) Solve the equations of motion to predict the motion of the system with time. Assume that the rope is long enough so that mass m_1 does not go over the edge.

FIGURE 5–48 Problem 72.

73. (II) One of the entertainments at the carnival is a rotating cylinder. The participants step in and place themselves against the interior wall. The cylinder starts to rotate more and more rapidly, and at some point the floor falls away, leaving the customers stuck like so many flies to the wall. If the cylinder were to slow down without the floor coming back up, the participants would begin to slip down. In terms of the relevant parameters, express the rotational speed, ω, at which this happens.

74. (II) A motorcycle moves in a horizontal circular path on the inside surface of a vertical cylinder of radius 8 m. Assuming that the coefficient of static friction between the wheels of the motorcycle and the wall is 0.9, how fast must the motorcycle move so that it stays in the horizontal path?

75. (II) Consider the conical pendulum described in Problem 57. Express the angular velocity in terms of the string angle θ and the string length ℓ.

76. (II) A tractor of mass 1200 kg is pulling a sled loaded with 450 kg of hay bales. The coefficient of kinetic friction between the sled and the ground is 0.50. (a) What horizontal force must the tractor exert to move at constant speed? (b) What is the tension in the rope between the tractor and sled? (c) If the tractor stops, how much horizontal force must it exert to get the sled moving again if $\mu_s = 0.60$?

77. (II) A bicyclist traveling at 10 m/s rides around an unbanked curve. If the coefficient of friction between the tires and the road is $\mu = 0.4$, what is the shortest turn the bicyclist can safely make? Is the coefficient of friction here static or kinetic?

78. (II) Two cars are traveling at 60 mi/h, one behind the other. The driver of the second car reacts by braking 0.8 s after she observes the sudden braking of the car ahead of her. The front car has a mass of 1200 kg, and the coefficient of friction between the tires and the road with the brakes applied is 0.8. The second car has mass 1600 kg, and the coefficient of friction with the brakes applied is 0.7. How far behind must the second car have been in order to avoid hitting the first car?

79. (II) A stunt motorcyclist rides with uniform speed on the inside rim of a vertical circular ramp of radius 12 m (Fig. 5–49). How fast must the motorcyclist travel to avoid leaving the surface at the top of the loop?

FIGURE 5–49 Problem 79.

bon is placed on the belt with zero initial velocity relative to the ground. The block will slip a bit before moving with the belt, leaving a black mark on the belt (Fig. 5–50). How long is that mark if the coefficient of kinetic friction between the carbon block and the belt is 0.20 and the coefficient of static friction is 0.30?

FIGURE 5–50 Problem 80.

80. (III) The new white belt of a long horizontal conveyor is moving with a constant speed of $v = 3.0$ m/s. A small block of car-

Work is done to lift these crushed stones against gravity as they move through this conveyor belt assembly. As the contact force from the conveyor belt does positive work on the stones, gravity is doing an equal magnitude of negative work on them.

Work and Kinetic Energy

A baseball flying through the air will change speed due to the forces of gravity and air drag. We have solved the equations of motion—Newton's second law—when gravity alone acts on the ball. If we included air drag in that same approach, however, it would present formidable technical obstacles. We can use another approach to this problem that is much simpler. It involves two new concepts that are the subjects of this chapter: work and kinetic energy. The notion of energy is one of the most basic concepts in physics and indeed in all the sciences. Energy takes many forms and, in this chapter, we will focus on the energy contained in a moving object, which we call its energy of motion, or *kinetic energy*. The *work* done on an object involves the force acting on it as it moves. We can relate the change in the kinetic energy of an object to the work done on it as it moved; this relation is called the work–energy theorem. The change in speed of a baseball, which is related to its change in kinetic energy, may thereby be related to the work done on it by the forces of gravity and air drag. Once we learn how to calculate the work done on an object, we have a powerful tool for the understanding of motion.

Nowadays, no matter how complex a force might be, a computer can generate solutions to Newton's second law with any given set of initial conditions. Powerful computers were not available in the nineteenth century, however, and it is fortunate that the scientists of that time made an effort to extract information from the general structure of the equations of motion. Their legacy, which includes the notion of energy, is a much deeper understanding of the properties of systems in which forces act than reams of computer output could possibly provide.

In this section, we want to introduce the idea that moving objects have kinetic energy and we want to relate that kinetic energy to the forces acting on the object. There are two reasons for exploring this idea. First, the energy approach can provide a significant shortcut for many aspects of problem solving. Second, the idea of energy turns out to be a truly fundamental one in physical science. Let's consider a one-dimensional problem in which a *constant* net force of magnitude F_{net} acts on an object of mass m in the $+x$-direction. In this case, the acceleration (a) of the object is also a constant, with magnitude F_{net}/m. Equation (2–24) gives us a relation between the object's acceleration and its displacement:

$$v^2 = v_0^2 = 2a(\Delta x) = 2(F_{\text{net}}/m)(\Delta x). \qquad (6\text{–}1)$$

∞ We introduced this relation in Eq. (2–24).

Here, $\Delta x = x - x_0$ represents the change in the object's position—the displacement from an initial position x_0 to the present position x. Similarly, $v^2 - v_0^2$ is the change in the speed squared, where v_0 is the initial speed.

If we multiply by m and divide by 2, we find

$$F_{\text{net}}\Delta x = \frac{1}{2}mv^2 - \frac{1}{2}mv_0^2. \qquad (6\text{–}2)$$

If an object is subjected to a constant net force, this type of relation is all we need to determine how the speed of the object has changed after it has traveled a certain distance. Notice that the time dependence of the motion does not appear here.

The definition of the work done by a constant force in one dimension

Work

The quantity on the left-hand side of Eq. (6–2) is the **net work**, W_{net}, done by the net force on the object:

$$\text{for constant net force in one dimension:} \quad W_{\text{net}} \equiv F_{\text{net}}\Delta x. \qquad (6\text{–}3)$$

The SI unit of work

The definition of kinetic energy

The unit of work is so important that a special name has been reserved for it: the **joule** (J). One joule is the work done by a force of 1 N in pushing an object a distance of 1 m. Thus $1\text{ J} = 1\text{ N} \cdot \text{m}$. In the cgs system, the unit of work is the **erg**, which is the work done by a force of 1 dyne in moving an object a distance of 1 cm. We have

$$1\text{ erg} = (1\text{ dyne})(1\text{ cm}) = (1\text{ g} \cdot \text{cm/s}^2)(1\text{ cm}) = (1\text{ g} \cdot \text{cm}^2/\text{s}^2)\left(\frac{1\text{ kg}}{10^3\text{ g}}\right)\left[\frac{(1\text{ m})^2}{(10^2\text{ cm})^2}\right]$$

$$= 10^{-7}\text{ kg} \cdot \text{m}^2/\text{s}^2 = 10^{-7}\text{ J}. \qquad (6\text{–}4)$$

The word *work* has several connotations in our language; the more colloquial uses of the word should not be confused with the precise definition presented in Eq. (6–3). We all agree that pushing a crate across a rough floor requires us to do work on the crate. This is because a force must be applied for work to be done on an object. Less in accord with our everyday use of the word is the idea that *there must be displacement if there is to be work* (Fig. 6–1). In holding a bag of groceries stationary, no work is ever done, even though colloquial usage of the word suggests that it is a lot of "work" to hold the bag stationary.

Kinetic Energy

The quantity $mv^2/2$, whose change is expressed on the right-hand side of Eq. (6–2), is the **kinetic energy**, K:

$$K \equiv \frac{1}{2}mv^2. \qquad (6\text{–}5)$$

(a)

(b)

FIGURE 6-1 In raising the barbell from a position (a) near the floor to a position (b) above his head, this weightlifter has done positive work. When holding the barbell above his head, the weightlifter is doing no work.

Notice that the kinetic energy cannot be a negative number. There is no directionality in the kinetic energy; it is a scalar quantity with dimensions of [mass][velocity]^2:

$$[K] = [M][L^2T^{-2}] = [ML^2T^{-2}].$$

The combination of SI units with this dimension is kg·m^2/s^2. According to Eq. (6–2), kinetic energy has the same dimensions as work. Thus we also measure kinetic energy (and, as we shall see, all forms of energy) in joules. Note that work has the units N·m, whereas kinetic energy has the units kg·m^2/s^2. We can verify the consistency of these units by recalling that 1 N = 1 kg·m/s^2, so 1 N·m = 1 kg·m^2/s^2. See Table 6–1 for typical values of kinetic energy.

The Work–Energy Theorem

Let's look again at Eq. (6–2) in terms of our definitions of kinetic energy and work. Suppose that an object moves under the influence of a constant net force from an initial point x_i to a final point x_f (and thus undergoes a displacement $\Delta x = x_f - x_i$). Its speed changes, and hence so does its kinetic energy. This change ΔK in kinetic energy is given by

$$\Delta K = K_f - K_i. \qquad (6-6)$$

In terms of ΔK and W_{net}, Eq. (6–2) becomes

$$W_{net} = \Delta K. \qquad (6-7)$$

This equation is known as the **work–energy theorem**.

The *sign* of work and therefore of the change in kinetic energy in Eq. (6–7) can be either positive or negative. We see from Eq. (6–3) that the sign of W_{net} is determined by the relative signs of F_{net} and Δx. If the force is directed in the *same* direction as the displacement (they have the same sign), the net work done during that displacement is positive. If the force is directed in a direction *opposite* to the displacement (they have opposite signs), the work done is negative. Suppose that a baseball moves upward. The displacement is upward, with a sign opposite to that of the force of gravity acting on it. The product of force and displacement is negative. Thus the net work done by gravity on the ball as it rises is negative. According to the work–energy theorem, the *change* in kinetic energy is then negative; that is, the baseball loses speed. Similarly, when the ball falls, the work done on it is positive, and it gains kinetic energy.

TABLE 6–1
Some Orders of Magnitude for Kinetic Energies

System	Kinetic Energy (J)
Electron in orbit around a nucleus	10^{-18}
Molecule of air at room temperature	10^{-17}
Electron in a TV tube	10^{-15}
Falling raindrop	10^{-3}
Baseball pitch	10^{2}
A running human	10^{3}
Automobile on a highway	10^{5}
Large earthquake	10^{17}
Earth in orbital motion around the Sun	10^{33}

EXAMPLE 6–1 A car of mass 1200 kg falls a vertical distance of 24 m starting from rest (Fig. 6–2). What is the work done by the force of gravity on the car? Use the work–energy theorem to find the final velocity of the car just before it hits the water. (Treat the car as a pointlike object.)

Solution: We measure the position of the car along a vertical y-axis, conventionally pointing upward. Then the downward displacement is $\Delta y = -24$ m. The force of gravity is the only force acting, so it makes the only contribution to the net work. Gravity has magnitude mg and is also oriented downward, so the y-component of the net force is $F_{net} = -mg$. Therefore the net work is

$$W_{net} = F_{net}\Delta y = (-mg)\,\Delta y = (-(1200 \text{ kg})(9.8 \text{ m/s}^2))(-24 \text{ m}) = 2.8 \times 10^5 \text{ J}.$$

We can now use the work–energy theorem to find the final speed v. The initial kinetic energy, K_i, is zero because the car starts from rest. Thus Eqs. (6–5), (6–7), and (6–8) give

$$\Delta K = K_f = W_{net} = mg\,\Delta y;$$

$$\tfrac{1}{2}mv^2 = mg\,\Delta y.$$

The mass cancels and we find that

$$v = \sqrt{2g\,\Delta y} = \sqrt{2(9.8 \text{ m/s}^2)(24 \text{ m})} = 22 \text{ m/s, nearly 50 mi/h.}$$

(This problem is one we could have solved in Chapter 3; we treat it here to bring in the concepts of work and kinetic energy.)

Calculating the Work

It is possible to calculate the work done by each force acting on an object. For example, if each force were constant and acted in one dimension, then the work done on an object by a particular force F as the object moves through a displacement Δx is given by

$$W = F\Delta x.$$

Remember, however, that the work–energy theorem specifies the *net* work. Often, several forces act on an object. When a piano is slowly lowered at constant speed by a rope, the force of gravity is canceled by the rope tension; that is why there is no acceleration. Here, the net force is zero, so the *net work* is also zero. This is consistent with the fact that there is no change in kinetic energy and no change in speed.

Because the net force is the sum of the individual forces acting, the net work can be decomposed into a sum of all the work done by individual forces. *This is a result that will hold for the most general definition of work.* It is often simpler to find the work W_i done by individual forces F_i and take the algebraic sum of the W_i to find the net work. In the case of the piano lowered by a rope at constant speed, positive work is done by gravity as the piano is lowered. The tension of the rope, however, which points upward, does the same magnitude of *negative* work on the piano. The forces cancel and the net work is zero in this case.

> **EXAMPLE 6–2** A box of books of mass 100 kg is pushed with constant speed in a straight line across a rough floor with a coefficient of kinetic friction $\mu_k = 0.2$. Find the work done by the force that pushes the box if the box is moved a distance $d = 3$ m.
>
> **Solution:** We approach this problem by drawing a force diagram (Fig. 6–3). With no vertical displacement, no work is done by gravity or by the normal force. The forces in the vertical direction must cancel and so $F_N = mg$. Because the box moves with a constant velocity, the *net* horizontal force must vanish. Thus the pushing force **F** must be equal in magnitude but opposite in direction to the force of friction **f**, whose magnitude is given by $f = \mu_k F_N = \mu_k mg$. Hence the magnitude of **F** is also $\mu_k mg$. The direction of **F** is the same direction as the displacement d. Thus the work done by the pushing force is positive. We have for the work done by the pushing force
>
> $$W = Fd = \mu_k mgd = (0.2)(100 \text{ kg})(9.8 \text{ m/s}^2)(3 \text{ m}) = 6 \times 10^2 \text{ J.}$$

FIGURE 6–2 Example 6–1.

PROBLEM SOLVING

The work done by the sum of forces is equal to the sum of the work done by individual forces.

FIGURE 6–3 Example 6–2. Force diagram for a crate being pushed across a floor.

A jet engine is a good example of a system that works by employing the work–energy theorem (Fig. 6AB–1). In the reference frame of the engine, air of mass m is brought into the engine at velocity v by the intake fans, which compress the air—indeed, the fans are also called compressors. The air, consisting mainly of oxygen and nitrogen molecules, is then mixed with fuel. Several things then happen to the air as it burns. One chemical reaction that occurs during combustion produces two water molecules ($2 \times$ H_2O) for each oxygen molecule (O_2). This doubles the volume of that part of the oxygen from the air that combines with the hydrogen. In another chemical reaction, part of the oxygen (O_2) combines with carbon to form CO_2; this portion of the oxygen maintains its volume. The nitrogen does not burn. Because of the water-producing reaction, a bigger volume of gas leaves the engine than entered it. To be able to keep up the continuous action of the engine, the outgoing gas must therefore leave with a velocity greater than the incoming velocity. Furthermore, the mass of the departing gas is larger since the mass of the fuel that has

been burned has been added to it. Thus the outgoing gas has substantially higher kinetic energy than does the incoming gas. By the work–energy theorem, the engine has done work to provide this kinetic energy increase. In effect, this work is associated with a force that is applied to the outgoing gas to give it increased kinetic energy; by the third law, then, a corresponding force acts on the airplane to move it forward against the frictional drag forces of the air. In this way, the work done by the engine propels the airplane.

FIGURE 6AB–1 The jet engines you see here are devices that use energy to do work and thereby propel the airplane.

EXAMPLE 6–3 A crate of mass 96 kg is pushed across a horizontal floor by a force $\mathbf{F}$. The coefficient of kinetic friction between the crate and the floor is $\mu_k = 0.27$ (Fig. 6–4a). The crate moves with uniform velocity. What is the magnitude of force $\mathbf{F}$? Suppose that at some point the crate passes onto a new section of floor, where $\mu'_k = 0.085$. The pushing force on the crate is unchanged. After 1.25 m on the new section of the floor, the crate moves with a speed of $v_f = 2.3$ m/s. What was the original speed of the crate, v_i?

Solution: We start with the force diagram drawn in Fig. 6–4b. This diagram permits us to find the force, which must balance the force of friction because the crate is moving with constant speed on the first section of floor. The force of fric-

(a)

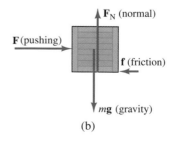

(b)

FIGURE 6-4 (a) Example 6–3. The floor changes composition from left to right to a more slippery surface at $x = 0$ m. (b) Force diagram for the crate.

tion that opposes **F** has magnitude

$$f = \mu_k mg = (0.27)(96 \text{ kg})(9.8 \text{ m/s}^2) = 2.5 \times 10^2 \text{ N}.$$

Force **F** must have just this magnitude: $F = 2.5 \times 10^2$ N.

The force of friction is decreased on the new section of floor, whereas the pushing force remains the same. Thus there is a (constant) net force in the direction of motion and the crate accelerates uniformly. The work done on the crate can be calculated. Then the work–energy theorem will allow us to find how much the crate's kinetic energy has changed and hence to find the original speed.

On the new section of floor, the force of friction is given by

$$f' = \mu_k' mg.$$

The net force on the crate then has magnitude

$$F_{net} = F - f' = (\mu_k - \mu_k')mg$$

and acts in the direction of motion of the crate. The net work done on the crate when it moves a distance Δx on the new section of floor is then

$$W_{net} = F_{net}\Delta x = (\mu_k - \mu_k')mg \, \Delta x.$$

According to the work–energy theorem, this is the increase in kinetic energy of the crate as it moves over the new section of floor:

$$W_{net} = K_f - K_i = \tfrac{1}{2}mv_f^2 - \tfrac{1}{2}mv_i^2.$$

We solve this equation for the initial speed:

$$v_i^2 = v_f^2 - \frac{2W_{net}}{m} = v_f^2 - \frac{2(\mu_k - \mu_k')mg \, \Delta x}{m}$$

$$= (2.3 \text{ m/s})^2 - 2(0.27 - 0.085)(9.8 \text{ m/s}^2)(1.25 \text{ m}) = 0.76 \text{ m}^2/\text{s}^2;$$

$$v_i = 0.87 \text{ m/s}.$$

Note that we did not need to find the kinetic energy of the crate over the entire distance where it was changing.

We can do a fixed amount of work by applying a large force over a small distance or a small force over a large distance. We are all familiar with the classic simple machines: the ramp, the lever, the block and tackle (see Example 5–3), the screw thread, and so forth (Fig. 6–5). The simple machines are devices that do a given amount of work by employing a small force over a compensatingly longer distance. For example, the lever lifts a very heavy object a small distance by applying a force (a small force compared to the object's weight) to the other end of the lever, which must move a good deal. The woman in Example 5–3 pulls quite a bit of rope with a relatively small force to lift a football player a short distance. In each case, we say we are using a "mechanical advantage," or that we are using "leverage." The simple machines are therefore force "amplifiers." They represent engineering at its very best: simple, elegant, and practical.

FIGURE 6–5 A simple machine is a kind of force amplifier. It does the same work that a large force would do over a small distance by applying a small force over a large distance. Thus a lever lifts a large mass a height h by applying a force much less than the mass's weight over a distance much greater than h. Here, we can see a lever, a ramp, a wedge, and various screw-driven devices.

APPLICATION

Simple machines

6–2   CONSTANT FORCES IN SPACE

Thus far, we have discussed the concepts of kinetic energy, work, and the work–energy theorem as they apply to one direction only. We now want to extend these concepts to two and three dimensions. We continue to assume that the net force **F** is a constant, but now it is a vector, as are the displacement, velocity, and acceleration.[†] (For simplicity, we have dropped the subscript "net" on the net force).

[†]A constant force is a vector that does not change with time and has the same value at every point in space.

The application of energy and work to two and three dimensions is a straightforward one; we know that Newton's laws include the information that motion and the effects of forces on motion are *independently applicable to the different Cartesian directions.* Let's apply these concepts to two dimensions for simplicity. Newton's second law, $\mathbf{F} = m\mathbf{a}$, now represents two separate equations for the *x*- and *y*-directions. If we label the components of all vector quantities with the appropriate axes, then we can derive an equation like the work–energy theorem—Eq. (6–2)—for each direction:

$$F_x \, \Delta x = \tfrac{1}{2}mv_x^2 - \tfrac{1}{2}mv_{x0}^2; \qquad (6\text{--}8a)$$

$$F_y \, \Delta y = \tfrac{1}{2}mv_y^2 - \tfrac{1}{2}mv_{y0}^2. \qquad (6\text{--}8b)$$

Here, the displacement vector $\Delta \mathbf{r} = \Delta x\mathbf{i} + \Delta y\mathbf{j}$. The velocity $\mathbf{v}$ and the initial velocity $\mathbf{v}_0$ have each been separated into their components, as has the net force $\mathbf{F}$.

Let's now take the sum of these two equations. We recognize that the square of the *magnitude* of the velocity—speed squared (v^2)—is the sum of the *x*- and *y*-components of the velocity squared:

$$v^2 = v_x^2 + v_y^2.$$

The summed equations then form the generalization of the work–energy theorem:

$$F_x \, \Delta x + F_y \, \Delta y = \tfrac{1}{2}mv^2 - \tfrac{1}{2}mv_0^2. \qquad (6\text{--}9)$$

If we continue to define kinetic energy as in Eq. (6–5)—$K = mv^2/2$—then the change in kinetic energy appears on the right-hand side of this equation.

The left-hand side of Eq. (6–9) generalizes the definition of the work done on an object in two dimensions. The work done easily generalizes to three dimensions:

$$\text{for constant } \mathbf{F}: \quad W \equiv F_x \, \Delta x + F_y \, \Delta y + F_z \, \Delta z. \qquad (6\text{--}10)$$

This definition for the work contains the components of the two vectors $\mathbf{F}$ and $\Delta \mathbf{r}$ multiplied, dimension by dimension, with the resulting products added together. This combination of two vectors occurs frequently and is called the **scalar product** (or **dot product**) of these two vectors. The work, then, is the scalar product of $\mathbf{F}$ and $\Delta \mathbf{r}$. A scalar product, although it is the product of two vectors, is itself a scalar quantity. If you are not already familiar with this concept, refer to the separate box explaining the scalar product.

From Eq. (6–10) and the expression Eq. (B1–6) for the scalar product, the work W done by a constant force $\mathbf{F}$ acting on an object that moves through a displacement $\Delta \mathbf{r}$ is given by

$$W = \mathbf{F} \cdot \Delta \mathbf{r}. \qquad (6\text{--}11)$$

With this definition of work, the work–energy theorem takes exactly the same form as before—$W_{\text{net}} = \Delta K$—as long as the net force is a constant vector. It does not matter that the force and the displacement point in different directions (Fig. 6–6).

Suppose the vectors $\mathbf{F}$ and $\Delta \mathbf{r}$ at any given position have an angle θ between them (Fig. 6–7). From Eq. (B1–1) in the Problem-Solving Techniques box on page 151 and Eq. (6–11), the work then takes the form

$$W = F \, \Delta r \cos \theta. \qquad (6\text{--}12)$$

Equivalently, the work done is the product of the magnitude of force $\mathbf{F}$ and the magnitude of the displacement $\Delta \mathbf{r}$ *in the direction of the force.* Put another way, *only the component of the displacement along the force counts in the work.* In particular, a force perpendicular to the motion does no work. A frequently occurring example is the normal force, $\mathbf{F}_{\text{N}}$, which is perpendicular to the surface on which an object moves; $\mathbf{F}_{\text{N}}$ does no work on that object. Another important example of a force that does no work in this way is the centripetal force responsible for uniform circular motion (see Section 6–4).

FIGURE 6–6 The tugboat is doing positive work on the freighter even though the direction of the tugboat force on the freighter is different than the direction of the freighter's velocity and displacement.

The scalar product of two vectors

FIGURE 6–7 The two vectors $\Delta \mathbf{r}$ and $\mathbf{F}$ always form a plane. We have labeled that plane as the *xy*-plane. The angle between these vectors determines the work done by the constant force on an object undergoing displacement $\Delta \mathbf{r}$, namely $W = F \, \Delta r \cos \theta$.

The work done by a constant force

PROBLEM SOLVING
_____

Only the component of the displacement along the force counts in the work.

PROBLEM–SOLVING TECHNIQUES

The Scalar Product

In Chapter 1, we discussed the definition of vectors and the multiplication of vectors by scalars. The product $b\mathbf{A}$ of a scalar b and a vector $\mathbf{A}$ is a vector. It points in the same direction as $\mathbf{A}$ and has magnitude bA, where A is the magnitude of $\mathbf{A}$. One way to multiply two vectors—$\mathbf{A}$ and $\mathbf{B}$, for example—is the **scalar product** $\mathbf{A} \cdot \mathbf{B}$. The scalar product is a scalar quantity whose value is

$$\mathbf{A} \cdot \mathbf{B} \equiv AB \cos \theta. \quad \text{(B1–1)}$$

Here, θ is the angle between the directions of the two vectors (Fig. 6B1–1a). The scalar product has the properties that

$$\mathbf{A} \cdot \mathbf{B} = \mathbf{B} \cdot \mathbf{A} \quad \text{(B1–2)}$$

and

$$\mathbf{A} \cdot (\mathbf{B} + \mathbf{C}) = (\mathbf{A} \cdot \mathbf{B}) + (\mathbf{A} \cdot \mathbf{C}). \quad \text{(B1–3)}$$

If two vectors are perpendicular (*orthogonal*) to each other, then $\theta = 90°$ and $\cos \theta = 0$, and their scalar product is zero. If the vectors are parallel to each other, then the scalar product takes on its maximum value; that is, the product of the magnitudes of the two vectors. The scalar product of a vector with itself is the square of its magnitude, $\mathbf{A} \cdot \mathbf{A} = A^2$. The unit vectors $\mathbf{i}$, $\mathbf{j}$, and $\mathbf{k}$ along some set of orthogonal axes x, y, and z have the property that

$$\mathbf{i} \cdot \mathbf{i} = \mathbf{j} \cdot \mathbf{j} = \mathbf{k} \cdot \mathbf{k} = 1. \quad \text{(B1–4)}$$

Because they are orthogonal to each other,

$$\mathbf{i} \cdot \mathbf{j} = \mathbf{j} \cdot \mathbf{k} = \mathbf{i} \cdot \mathbf{k} = 0. \quad \text{(B1–5)}$$

Two vectors $\mathbf{A}$ and $\mathbf{B}$ can be decomposed into their vector components: $\mathbf{A} = A_x\mathbf{i} + A_y\mathbf{j} + A_z\mathbf{k}$ and $\mathbf{B} = B_x\mathbf{i} + B_y\mathbf{j} + B_z\mathbf{k}$. The rules in Eqs. (B1–4) and (B1–5) allow us to write the scalar product of $\mathbf{A}$ and $\mathbf{B}$ as

$$\mathbf{A} \cdot \mathbf{B} = (A_x\mathbf{i} + A_y\mathbf{j} + A_z\mathbf{k}) \cdot (B_x\mathbf{i} + B_y\mathbf{j} + B_z\mathbf{k})$$

$$= A_xB_x + A_yB_y + A_zB_z. \quad \text{(B1–6)}$$

Thus the scalar product of two vectors is *the sum of the product of the components of the two vectors.*

The scalar product is a scalar quantity, so it remains the same even if the axes of our coordinate system are rotated. If we consider two vectors $\mathbf{A}$ and $\mathbf{B}$, we may choose our coordinate frame in such a way that $\mathbf{A}$ lies along the x-axis, $\mathbf{A} = A\mathbf{i}$. The other axes can be arranged so that the vector $\mathbf{B}$ has only x- and y-components, $\mathbf{B} = B_x\mathbf{i} + B_y\mathbf{j}$. These vectors are shown in Fig. 6B1–1b, which is a view looking down on the plane formed by $\mathbf{A}$ and $\mathbf{B}$. According to Eq. (B1–6), the scalar product is then given by

$$\mathbf{A} \cdot \mathbf{B} = AB_x.$$

Thus the scalar product of two vectors may be described as the *product of the length of one vector and the projection of the other vector along the direction of the first one.* Because $B_x = B \cos \theta$, we recover here our original definition: Eq. (B1–1).

(a)   (b)

FIGURE 6B1–1 (a) Two vectors $\mathbf{A}$ and $\mathbf{B}$ have been displaced so that their tails meet at the same point. They are oriented in space with the angle θ between them. (b) The x-axis of our coordinate system has been redefined so that $\mathbf{A}$ lies along the $+x$-direction. The scalar product is independent of the orientation of the axes.

EXAMPLE 6–4 A block of mass m is placed on an inclined plane that makes an angle θ with the horizontal (Fig. 6–8a). A horizontal force $\mathbf{F}$ is applied to the block. Find the magnitude of $\mathbf{F}$ such that the block moves up the plane with acceleration $\mathbf{a}$. What is the work done by $\mathbf{F}$? (Assume that the plane is rough, with coefficient of kinetic friction μ_k, and that the distance the block is moved along the plane is d).

Solution: Start with a force diagram for the block (Fig. 6–8b). We will first determine the magnitude F of the horizontal force. Once we have found this value, the work done by this force on the block will be

$$W = Fd \cos \theta.$$

In Fig. 6–8c, we choose axes and decompose all the forces into components. The direction up along the plane is the x-direction, and the direction pointing directly away

(a)

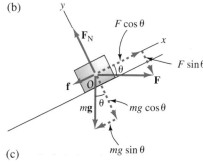

(b)

(c)

FIGURE 6–8 (a) Example 6–4. (b) Force diagram for the block on the inclined plane. (c) The forces are decomposed into components along the plane surface and perpendicular to it.

from the surface of the plane is the *y*-direction. The *y*-components of the forces must add up to zero because there is no acceleration in the direction perpendicular to the plane. We thus have

$$F_N - F \sin \theta - mg \cos \theta = 0.$$

From this equation, $F_N = F \sin \theta + mg \cos \theta$. There is also an acceleration up the plane, which is determined by Newton's second law applied to the *x*-direction:

$$ma = F \cos \theta - mg \sin \theta - \mu_k F_N$$

$$= F \cos \theta - mg \sin \theta - \mu_k(F \sin \theta + mg \cos \theta).$$

Hence

$$F = \frac{ma + mg \sin \theta + \mu_k mg \cos \theta}{\cos \theta - \mu_k \sin \theta}.$$

The work done by the force in accelerating the block is then

$$W = Fd \cos \theta = md \frac{a + g \sin \theta + \mu_k g \cos \theta}{1 - \mu_k \tan \theta}.$$

Application of the Work–Energy Theorem to Projectile Motion

A particularly simple application of the work–energy theorem for constant forces concerns projectile motion. Let's suppose that an object of mass m is thrown with initial velocity $\mathbf{v}_1 = v_{1x}\mathbf{i} + v_{1y}\mathbf{j}$ off a tall building. We choose the upward vertical direction to be the *y*-direction. After the object has fallen a vertical distance h (if we choose the top of the building to be $y = 0$, then the mass falls from $y = 0$ to $y = -h$), it has acquired a velocity $\mathbf{v}_2 = v_{2x}\mathbf{i} + v_{2y}\mathbf{j}$ (Fig. 6–9). The speed v_2 at this point is determined from the kinetic energy K_2, and the work–energy theorem gives K_2:

$$K_2 = K_1 + W. \tag{6–13}$$

Here, W is the work done by gravity on the falling object as it moves from its initial point to its final point. The force of gravity is given by $\mathbf{F} = F_y\mathbf{j} = -mg\mathbf{j}$. The work done is therefore

$$W = \mathbf{F} \cdot \Delta\mathbf{r} = (F_y\mathbf{j}) \cdot [(\Delta x)\mathbf{i} + (\Delta y)\mathbf{j} + (\Delta z)\mathbf{k}]$$

$$= F_y \, \Delta y = (-mg)(-h) = mgh. \tag{6–14}$$

The force has only a *y*-component, so only the *y*-component of the displacement enters into the work. Substituting into Eq. (6–13), we have

$$\tfrac{1}{2}mv_2^2 = \tfrac{1}{2}mv_1^2 + mgh. \tag{6–15}$$

We have arrived at this simple expression without solving Newton's second law. Note that the increase in speed depends only on the vertical distance fallen; it does *not* depend on the horizontal distance covered, even though the speed has contributions from both the vertical and horizontal components of the velocity.

6–3 FORCES THAT VARY WITH POSITION

Thus far in our study of the work–energy theorem, the force acting on an object has been constant. In contrast, we may have a force whose magnitude varies from point to point, one whose direction varies from point to point, or one for which both the magnitude and direction vary. We want to generalize the definition of work to include these cases so that the work–energy theorem continues to hold.

Variable Forces in One Dimension

Let's again consider motion in one dimension—we will drop the vector notation temporarily—but this time, let's suppose that the magnitude of the net force varies with position. We can use our earlier development of the work–energy theorem by subdividing the total interval over which the object moves (Fig. 6–10a). We make the subintervals of the object's displacement small enough so that the force may be considered to be constant in each of them. For example, if a force changes smoothly by some factor over a distance of 1 m, then it may change by only 1 percent of this factor over distances of 1 cm. But if, as in Fig. 6–10b, the force changes more abruptly during one portion of the object's movement, we can always make the intervals smaller in this region so that we can regard the force as constant in each interval.

Let's take a look at how we might partition a varying force into intervals (Fig. 6–11). In each interval shown in Fig. 6–11, the force (acting in one dimension) may be viewed as constant. We suppose that the force takes on the approximately constant value F_1 in the interval from initial position x_i to x_1, the value F_2 in the interval x_1 to x_2, and, generally, the value F_k in the interval x_{k-1} to x_k. There are a total of N intervals. Let's calculate the total work done by the force in moving an object of mass m from initial position x_i to final position x_f; we will use the expression for work appropriate for constant force in each interval. In each of the intervals, we apply Eqs. (6–2) and (6–3):

$$W_1 = F_1(x_1 - x_1) = \tfrac{1}{2}m(v_1^2 - v_i^2) = K_1 - K_i, \qquad (6\text{–}16a)$$

$$W_2 = F_2(x_2 - x_1) = \tfrac{1}{2}m(v_2^2 - v_1^2) = K_2 - K_1, \qquad (6\text{–}16b)$$

$$W_3 = F_3(x_3 - x_2) = \tfrac{1}{2}m(v_3^2 - v_2^2) = K_3 - K_2, \qquad (6\text{–}16c)$$

and so on, finishing with the last interval, which is expressed as

$$W_N = F_N(x_f - x_{N-1}) = \tfrac{1}{2}m(v_f^2 - v_{N-1}^2) = K_f - K_{N-1}. \qquad (6\text{–}16d)$$

Now let's sum all these equations. The work done in moving the object through each interval of displacement is placed on the left-hand side of the equation; this is the total work done. On the right-hand side, all of the terms that involve $v_1^2, v_2^2, \ldots, v_{N-1}^2$ cancel, leaving

$$W = \sum_{k=1}^{N} W_k = \tfrac{1}{2}m(v_f^2 - v_i^2)$$

$$= K_f - K_i = \Delta K. \qquad (6\text{–}17)$$

We find that Equation (6–17) is the work–energy theorem again.

The left-hand side of Eq. (6–17) is just the sum of the areas of individual rectangles—each one of the form $F_k \Delta x_k$. In the limit that the intervals Δx_k become infinitesimally small, the sum of their areas is the total area under the curve of $F(x)$ versus x. This area is the *integral* of $F(x)$ over x. We denote it mathematically as

$$W = \lim_{N \to 0} \sum_{j=1}^{N} F(x_j)\Delta x_j = \int_{x_i}^{x_f} F(x)\,dx. \qquad (6\text{–}18)$$

Equation (6–17) is useful to find ΔK only if we can evaluate the work, and this is done by using the techniques of **integration**. If you are not familiar with the properties of integrals and the techniques of integration, you may want to refer back to Section 2–6 or consult the Problem-Solving Techniques box included here on page 155.

Just as work is given by an integral of the force, force is given by a derivative of the work, because differentiation and integration are inverse processes. In particular, Eq. (B2–2) from the Problem-Solving Techniques box allows us to find the force if the work done by that force is known:

$$\frac{dW(x_f)}{dx_f} = F(x_f). \qquad (6\text{–}19)$$

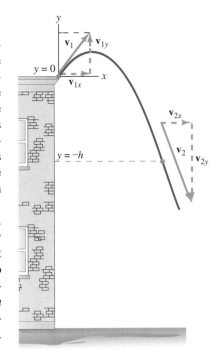

FIGURE 6–9 A projectile is launched off the top of a building. The work that gravity does on the object as it moves along its trajectory determines its kinetic energy and thus its speed at various points on the trajectory.

(a)

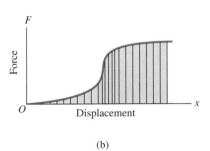

(b)

FIGURE 6–10 (a) A smoothly varying force is approximately constant in small, equally spaced intervals of x. (b) If there is a region where the force changes more rapidly, the intervals in that region must be made smaller, so that the force can be thought of as constant in each interval.

153

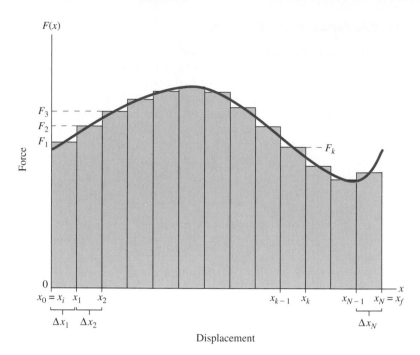

F(x)

Force

F_3
F_2
F_1

$-F_k$

0

$x_0 = x_i$ x_1 x_2 x_{k-1} x_k x_{N-1} $x_N = x_f$

Δx_1 Δx_2 Δx_N

Displacement

FIGURE 6–11 A variable force $F(x)$ has an approximately constant value ($F(x_j) = F_j$) within each of the small intervals Δx_j in the region between x_i and x_f. The work done by the force on an object as the object moves between these limits is the sum of the work W_k done in each interval, for which we can use the constant force formula for work, $W_k = F_k \, \Delta x_k$. This sum, W, is the integral of $F(x)$ over x, and it is equal to the area under the curve of F versus x.

Finally, we can recover our earlier result for constant force, Eq. (6–4). Multiplicative constants (in this case, F) can be removed from beneath an integral sign, so that

$$\text{for constant } F: \quad W = F \int_{x_i}^{x_f} dx = F(x_f - x_i) = F \, \Delta x.$$

The Work Done by a Spring

One of the most important examples of a one-dimensional variable force is the force exerted on a mass by a spring attached to it (Fig. 6–12). This force takes the form

$$F = -kx \qquad (6\text{–}20)$$

Here, x measures the displacement of the mass from an equilibrium position; k is a constant characteristic of the particular spring known as the **spring constant.** This force law is known as **Hooke's law,** after its seventeenth-century discoverer, Robert Hooke. Note the sign here: When x is positive, the mass is on the right side and the force points to the left; when x is negative, the mass is on the left side and the force points to the right. The spring force always acts to bring the mass back to $x = 0$.

Let's now calculate the work done on the mass by the spring force when the mass moves from the equilibrium position to the position $x = L$ (Fig. 6–13a), then calculate the extra work that is done as the mass moves still further to a position $2L$ beyond the equilibrium point. (Note that if the mass is at rest, it cannot move away from $x = 0$ under the influence of the spring force alone; the spring force is zero there. But the mass may have been set in motion by other forces. We assume that is the case.) To calculate the work, we must integrate the (nonconstant) force over the displacement, as in Eq. (6–18):

$$W = \int_0^L (-kx) \, dx.$$

Note the sign: The spring force is in the direction opposite to the displacement of the mass, so the work done by the spring will be negative. We use the general integration

Hooke's law

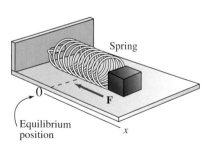

Spring

0

F

Equilibrium position

x

FIGURE 6–12 A spring exerts a force on an object that tends to bring the object back to the equilibrium position. Here the spring is stretched out past the equilibrium position.

PROBLEM-SOLVING TECHNIQUES

Integration

Integration of a function is the inverse of differentiation of that function. Put another way, if we integrate some function $f(x)$ and then differentiate the result, we get the function $f(x)$ back again. Expressed mathematically, if the functions $f(x)$ and $g(x)$ are related by

$$g = \int_{x_i}^{x_f} f(x)\mathrm{d}x, \qquad (\text{B2-1})$$

then

$$\frac{\mathrm{d}g(x_f)}{\mathrm{d}x_f} = f(x_f). \qquad (\text{B2-2})$$

Moreover, we can interpret work as the area under the curve of F versus x (see Fig. 6–11). If the function F is negative and the displacements are positive, as in Fig. 6B2–1, the area and hence the work are negative. Conversely, if the displacement is negative and the force positive, then the work will be negative. In the language of integrals, this comes about because the integral changes sign when we reverse the limits on it:

$$\int_{x_f}^{x_i} f(x)\mathrm{d}x = -\int_{x_i}^{x_f} f(x)\mathrm{d}x. \qquad (\text{B2-3})$$

If the integral on the right-hand side is positive, then the left-hand side—which represents the work done in a displacement from x_f to x_i—is negative.

FIGURE 6B2–1 In the region between x_i and x_f, the shaded area under the curve is negative [part (a)], corresponding to negative values for the force $F(x)$ [part (b)].

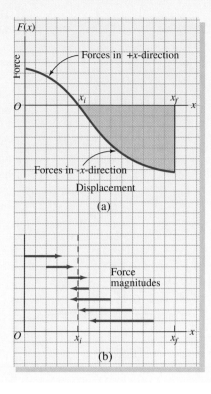

formula for powers in Appendix IV–8, with $p = 1$, to find that

$$W = -k\int_{0}^{L} x\,\mathrm{d}x = -\frac{1}{2}kx^2\,\bigg|_{0}^{L} = -\frac{1}{2}kL^2.$$

To find the extra work that must be done by the spring force as the mass moves from position L to position $2L$, we calculate

FIGURE 6–13 (a) The force exerted by a spring changes as the spring stretches (or compresses). Twice the stretch corresponds to twice the force. (b) The area under the curve of force versus x is the work the spring force performs on a mass attached to its end. The work done as the spring stretches from $x = 0$ to $x = 2L$ is four times the work done as the spring stretches from $x = 0$ to $x = L$.

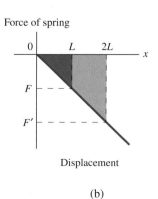

$$\int_{L}^{2L}(-kx)dx = -\frac{1}{2}kx^2 \bigg|_{0}^{2L} = -\frac{1}{2}k[(2L)^2) - L^2] = -\frac{3}{2}kL^2.$$

The total work done by the spring as the mass moves from $x = 0$ to $x = 2L$ is the work done as the mass moves from $x = 0$ to $x = L$ plus this extra work: $-\frac{1}{2}kL^2 - \frac{3}{2}kL^2 = -2kL^2$ (compared to $-\frac{1}{2}kL^2$ for the first part of the motion). The spring does four times as much work when the mass moves twice the distance.

It is easy to understand these results graphically. Figure 6–13b is a graph of the spring force F as a function of x. The work done by this force as the mass moves from one value of x to another is the area under the curve of F versus x, between the two values of x. In particular, we will consider the shaded triangle under the force curve (here, a straight line with slope k) from $x = 0$ to $x = L$ and a second shaded triangle under the force curve from $x = 0$ to $x = 2L$. These triangles are similar but the second has twice the base length and twice the height. The second triangle has four times the area of the first, so the spring force does four times as much work when the mass moves twice as far.

FIGURE 6–14 Example 6–5.

EXAMPLE 6–5 A mass $m = 50$ g is placed on the end of a spring with a spring constant k = 6.0 N/m. The mass is released from rest at $x = 10$ cm (Fig. 6–14). Opposing the force of the spring is an unknown drag force. Find the work done on the mass by the drag force if the mass's velocity at the equilibrium position $x = 0$ is 0.85 m/s.

Solution: We must put several elements into place to be able to use the work–energy theorem and find the work done by the drag force. The net work is the difference between the final kinetic energy K_f and the initial kinetic energy K_i; we know both K_i (0 because the mass starts from rest) and K_f. The net work, however, is also the algebraic sum of the work done by the spring, W_s, and the work done by the drag force, W_D. Thus if we can find W_s we can solve for W_D.

We know that the force exerted by the spring on the mass is $F = -kx$. The work done by the spring is then

$$W_s = \int_{-x_0}^{0}(-kx)dx = -\int_{0}^{x_0}(-kx)dx = \frac{1}{2}kx^2 \bigg|_{0}^{x_0} = \frac{1}{2}kx_0^2.$$

We now apply the work–energy theorem. We denote the speed of the mass at the equilibrium point as v_1, and thus

$$W_{\text{net}} = W_s + W_D = K_f - K_i$$

or

$$W_D = K_f - K_i - W_s = \frac{1}{2}mv_1^2 - 0 - \frac{1}{2}kx_0^2$$

$$= (0.5)(50 \times 10^{-3} \text{ kg})(0.85 \text{ m/s})^2 - (0.5)(6.0 \text{ N/m})(0.10 \text{ m})^2 = -0.012 \text{ J}.$$

Why is the sign negative? It is because the drag acts in a direction opposite to the displacement of the mass.

Forces that Vary in Both Magnitude and Direction

We must make one further generalization in order to define work that will satisfy the work–energy theorem. It is the case where the force acts not only with varying magnitude but also with varying direction. The net displacement $\Delta \mathbf{r}$ of an object under the influence of a net force may result from a rather complicated *path* in space (Fig. 6–15). (Here we have drawn the path in a plane, but the path could move through three-dimensional space.) The path is described with a changing position vector $\mathbf{r}$. At the distance r from the origin, the force on the object is approximately a constant vector

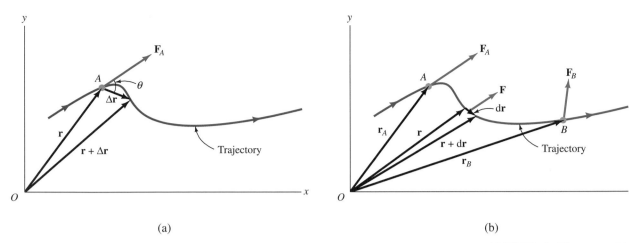

(a)

(b)

F$_A$: It does not change very much as the object moves through a small displacement Δ**r** from the point labeled by position vector **r** to an adjacent point labeled by position vector **r** + Δ**r** (Fig. 6–15a). As Δ**r** → 0, this displacement is *tangential* to the curve at the tip of **r**, while the force vector points in some other direction. The work done on the object in moving through the small interval is then **F**$_A$·Δ**r**. We find the total work done for a displacement from point A with position vector **r**$_A$ to point B with position vector **r**$_B$ (Fig. 6–15b) by summing these infinitesimal contributions:

$$W = \int_{\mathbf{r}_A}^{\mathbf{r}_B} \mathbf{F} \cdot d\mathbf{r}. \tag{6–21}$$

The work done by a force that varies over space

This formula is the general definition of work and is consistent with all our earlier definitions. The integral that appears here is called a **line integral** because it depends not only on the beginning and ending points A and B, but, in general, on the path, or line, taken to move between these points. We shall explore the properties of this integral in Section 6–4.

No Work Is Done in Uniform Circular Motion

We conclude this section with the important observation that *no net work is done on a particle that undergoes uniform circular motion* (Fig. 6–16). This remark holds for *any* part of a circular trajectory at constant speed. To see this result, it is enough to recall from Chapter 5 that an object undergoing uniform circular motion experiences an acceleration that is directed along the radius to the center of the circle. Thus the force is directed in the (negative) radial direction and is always *perpendicular* to the direction of motion, which is tangential. If the infinitesimal displacement along an arc is d**s**, as in Fig. 6–17, then the scalar product **F**·d**s** is zero, and thus no work is done.

If the motion is circular with varying speed, a tangential force **F**$_t$ must be present (one that is parallel to the direction of the displacement). Work is done for such motion. Because the force is directed along (or opposite to) the displacement d**s**, the scalar product **F**$_t$·d**s** for the infinitesimal work done has nonzero magnitude $F_t ds$. We integrate this quantity to find the total work.

EXAMPLE 6–6 A metallic sphere with a mass of 8.0 kg is attached to a vertical shaft by a rod 1.5 m long of negligible mass that is perpendicular to the shaft (Fig. 6–18). A tangential force **F** is applied to the sphere for one-quarter turn. As a result, the sphere rotates about the frictionless shaft with an angular speed of 1.2 rev/s. If the sphere was initially at rest, what was the magnitude of the tangential force?

FIGURE 6–16 The net work done on a Ferris wheel passenger in moving from the bottom to the top and back again is zero. In particular, no work is done by the forces, a combination of friction and normal forces, that uniformly move the passenger in a circle.

⚭ Angular variables were introduced in Chapter 3.

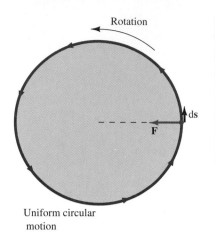

FIGURE 6–17 In the uniform circular motion of an object, the infinitesimal displacement is tangent to the circle. The force responsible for this motion is directed toward the center of the circle and is perpendicular to the displacement. The work done by this force on the object is therefore zero.

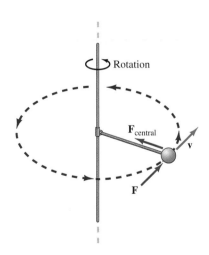

FIGURE 6–18 Example 6–6.

Solution: The angular speed after the force has acted is 1.2 rev/s = (1.2 rev/s) $(2\pi$ rad/rev) = 7.5 rad/s, so the speed of the sphere at the end of the push is $v = \omega r = (7.5 \text{ s}^{-1})(1.5 \text{ m}) = 11$ m/s. From the work–energy theorem, then,

$$K_f - K_i = \tfrac{1}{2}mv_f^2 - \tfrac{1}{2}mv_i^2 = \tfrac{1}{2}(8.0 \text{ kg})(11 \text{ m/s})^2 = 4.8 \times 10^2 \text{ J}.$$

The force **F**, applied for a quarter-turn, acts over a distance $\frac{1}{4}(2\pi r) = \frac{1}{4}\pi(3.0 \text{ m}) = 2.4$ m. Thus the force has magnitude

$$F = \frac{K_f - K_i}{\text{distance}} = \frac{4.8 \times 10^2 \text{ J}}{2.4 \text{ m}} = 2.0 \times 10^2 \text{ N}.$$

6–4 CONSERVATIVE AND NONCONSERVATIVE FORCES

One of the most important questions that arises when an object undergoes a certain displacement is: How does the work done depend on the path taken by the object during its displacement? We call this a question of *path dependence*. We illustrate what is involved by considering, in order, gravity, friction, and the spring force.

Gravity. Let us consider the work that gravity does in moving an object of mass m from the top of a ramp of length L to the bottom. Figure 6–19a shows the geometry of the ramp, which makes an angle θ with the horizontal. The force of gravity acts vertically downward with magnitude mg on the object. This force can be decomposed into a component $mg \cos \theta$ perpendicular to the ramp and a component $mg \sin \theta$ pointing along the ramp toward its bottom. When an object slides down the ramp from an initial point to a final point, the motion is perpendicular to the force $mg \cos \theta$, and the work done by that component of the force is therefore zero. The work done by the force of gravity to move the object the total length L along the direction of motion is thus

$$W = (mg \sin \theta)L. \tag{6–22}$$

Consider now a second path from the top of the ramp to the bottom. This second displacement is achieved by dropping the object vertically off the back of the ramp and then moving it horizontally to the previous end point (Fig. 6–19b). The force downward is mg, but the distance through which the force acts is $L \sin \theta$. Thus the work done for this part of the second path of the displacement is $(mg)L \sin \theta$. The object is then moved horizontally; no work is done by gravity during this portion of the second displacement because the force of gravity is perpendicular to this motion. Thus the total work done by gravity is the same as that calculated in Eq. (6–22). We would find the same result for other paths: The work done by gravity depends only on the difference between the final and initial heights of the object being moved; it does not depend on the path by which this height difference was reached.

Friction. Let us next consider the work done by friction on a coffee cup of mass m that is pushed across a rough horizontal surface. In this case, friction is not the only force acting on the cup, but it is not the work–energy theorem that is at issue here—that theorem involves the *net* work—rather, it is only the work done by friction that interests us here. The force of friction on the cup has magnitude $\mu_k mg$ and it is directed *opposite* to the motion. Suppose that the cup in Fig. 6–20 is moved (with the aid of some external force) along the +x-axis from $x = 0$ to $x = x_{max}$ and then back again. For the first half of the motion, the displacement of the cup is +$x_{max}\mathbf{i}$, while the force of friction is directed back toward $x = 0$: $\mathbf{f} = -\mu_k mg\mathbf{i}$. In the motion back to the origin, the displacement is $-x_{max}\mathbf{i}$, while the force of friction is directed to +x: $\mathbf{f} = \mu_k mg\mathbf{i}$. Then

$$W = W_{out} + W_{back} = (-\mu_k mg)(x_{max}) + (\mu_k mg)(-x_{max}) = -2\mu_k mgx_{max}.$$

Rather than canceling, the same amount of negative work is done on the cup by friction both out and back, and there is work done even though the net displacement is zero. Moreover, the amount of work depends on x_{max}; that is, it depends on the path taken by the cup.

Spring Force. Equation (6–18) is an expression for the work done that applies to any one-dimensional force that depends only on the position of the object. It is a general property of integration that integrals such as these depend only on x_f and x_i, not on any intermediate points. This is illustrated with the spring force in the following example.

EXAMPLE 6–7   A mass m is attached to the end of a spring with spring constant k (Fig. 6–21a). The mass moves from the point $x = L$ to the point $x = 0$ in two ways: (1) Path 1 is the direct motion (Fig. 6–21b), and (2) path 2 is the displacement from $x = L$ to $x = a$, with $a < 0$, followed by a displacement to $x = b$ with $b > L$, completed with a final displacement to $x = 0$ (Fig. 6–21c). Find the work done on the mass by the spring for each path. (You should find the same answer in each case.)

Solution: Unlike gravity, the spring exerts a force that varies with position. For the work, we must therefore use the form $W = \int F\,dx$ between suitable limits. In the case of path 1, we have

$$W = \int_L^0 (-kx)\,dx = -\frac{kx^2}{2}\Big|_0^L = -\left(0 - \frac{kL^2}{2}\right) = \frac{kL^2}{2}.$$

The work done by the spring on the mass is positive.

In the case of path 2, there are three contributions to W—corresponding to the three displacement steps described—and their sum is

$$W = \int_L^a (-kx)\,dx + \int_a^b (-kx)\,dx + \int_b^0 (-kx)\,dx = -\frac{kx^2}{2}\Big|_L^a - \frac{kx^2}{2}\Big|_a^b - \frac{kx^2}{2}\Big|_b^0$$

$$= -\frac{ka^2}{2} + \frac{kL^2}{2} - \frac{kb^2}{2} + \frac{ka^2}{2} - \frac{k0^2}{2} + \frac{kb^2}{2} = \frac{kL^2}{2},$$

the same work as for path 1.

Conservative and Nonconservative Forces

We have seen that the work done is independent of the path taken by the object for both gravity and the spring force; however, that is not the case for friction. In two and three dimensions, it is not enough to say that the work done on an object during displacement will be path independent if the force depends solely on the object's position. The integral in Eq. (6–21) may or may not depend on the path. In any case, we can generally divide forces into two classes: Forces for which the work is independent of the path are called **conservative forces**; those for which the work depends on the path are called **nonconservative forces**. We have seen that gravity and the spring force are conservative, whereas friction is nonconservative.

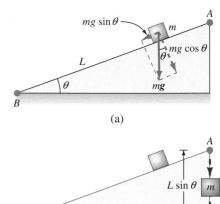

(a)

(b)

FIGURE 6–19 (a) An object of mass m moves from point A to point B on an inclined plane under the influence of gravity. Gravity does positive (or negative) work on the object as it moves down (or up) the plane. (b) The object now moves from point A to point B by a different path: a vertical motion from point A to point C, followed by a horizontal movement from C to B. The work done by gravity is exactly the same as in part (a).

The meaning and definition of conservative forces

FIGURE 6–20 A cup moves along a horizontal surface from $x = 0$ to $x = x_{max}$ and back again. One of the forces acting is friction. The work done by friction in the back-and-forth motion is not zero.

(a)

x = 0

Equilibrium position

(b)

x = L

Path 1

(c)

x = a

x = b

Path 2

0

x

FIGURE 6–21 (a) Example 6–7. The equilibrium position of the mass attached to the spring. (b) The spring is pulled to *x = L* and then moves from this point back to *x = 0*. (c) A more complicated path for motion from *x = L* to *x = 0*.

An important property of conservative forces

The work done by a conservative force in moving an object from one position to another is independent of the path taken by the object.

The statement that conservative forces do work that is independent of the path of the object can be reformulated in terms of closed paths, paths that end at the same point at which they start:

The work done by a conservative force in moving an object along any closed path is zero.

The relation between these two statements is easy to understand. One possible closed path is the path in which the object does not move at all, in which case *W* = 0. But the statement that the work done is path independent means that the same amount of work is done along *any* path for which the object starts and finishes at the same point. Thus no work is done in moving along any closed path. This statement may be used as a definition of conservative forces.

Any one-dimensional force that depends only on position is conservative.

We can now restate our conclusions about forces in one dimension, which were earlier stated in terms of path independence. *Any* force that can be written in the form *F = F(x)* is a conservative force. This includes constant forces as a special case.

Friction is Nonconservative

You might think from our discussion that forces with constant magnitude are always conservative, a conclusion that friction shows is patently false. This apparent difficulty is resolved by noting that a nonconservative force such as friction cannot, in fact, be represented by a function of the form *F(x)*. In fact, friction depends on the direction of the velocity as well as the position (Fig. 6–22):

$$\mathbf{F} = \mu_k \mathbf{F}_N \text{ in the direction opposite to } \mathbf{v},$$

which means that, strictly speaking, **F** = **F**(*x*, **v**). The work done by friction when an object moves out cannot possibly cancel the work done by friction as the object comes back along a reversed path, as was the case for the spring.

FIGURE 6–22 The snowplow position increases the effect of kinetic friction and thereby controls the speed of the skier. This force is doing negative work on the skier.

The importance of conservative forces is twofold. First, we shall see in Chapter 7 that we can dispense with the work itself for these forces and instead use the notion of a *potential energy*. Second, *most of the forces we deal with are conservative*. This is true of all the fundamental forces in nature and, in particular, for those forces known as *central forces*, in which the force is directed along a line from a fixed center. The gravitational force between the Sun and the planets is one example.

6 – 5   P O W E R

Up to this point, we have said nothing about how *fast* work is done. **Power**—symbol *P*— is *the rate at which work is done*:

The definition of power

$$P \equiv \frac{dW}{dt}. \qquad (6\text{–}23)$$

It is easy to calculate the power for a constant force. In one dimension, we have

$$W = F \, \Delta x, \qquad (6\text{–}24)$$

where F is the force and Δx is the displacement from some fixed starting point. In two or three dimensions, we have

$$W = \mathbf{F} \cdot \Delta \mathbf{r}. \qquad (6\text{–}25)$$

Let's now calculate the power. The amount of work done is a function of time because the displacement changes with time. The general calculus rule states that $d(fg)/dt = g(df/dt) + f(dg/dt)$; therefore, we have in one dimension

$$P = \frac{dW}{dt} = F \frac{d(\Delta x)}{dt} + \Delta x \frac{dF}{dt}.$$

As long as the force is constant in time, the second term on the right is zero. The rate of change of displacement $d(\Delta x)/dt$ is the velocity, so we find

$$P = Fv. \qquad (6\text{–}26)$$

In two or three dimensions, we can generalize this result to

$$P = \mathbf{F} \cdot \frac{d\Delta \mathbf{r}}{dt} = \mathbf{F} \cdot \mathbf{v}. \qquad (6\text{–}27)$$

We have derived this result for a constant force but for a velocity that can change with time. The power P in Eq. (6–27) is therefore more appropriately called the **instantaneous power**. It can be shown with a little more calculus that Eq. (6–27) is the correct expression for the instantaneous power even when the forces depend on time.

The instantaneous power

The units of power are joules per second (J/s) and the unit has been given its own name. A **watt** (W for short; do not confuse this unit with the algebraic symbol *W* that we use for work!) is the power generated when a force of 1 N displaces an object with a speed of 1 m/s. Another commonly used unit of power is the **horsepower** (hp):

The SI unit of power

$$1 \text{ hp} = 746 \text{ W}. \qquad (6\text{–}28)$$

A useful measure of *electrical energy* is based on the watt or, more conveniently, the kilowatt (kW). The **kilowatt-hour** (kWh) is the amount of work done when 1 kW of power is generated for 1 h.[†] Because there are 3600 s in 1 h,

$$1 \text{ kWh} = 3.6 \times 10^6 \text{ J}. \qquad (6\text{–}29)$$

[†]A typical large hydroelectric project generates a power of 10^{10} W; household energy consumption in the northern United States typically runs from 250 to 1000 kWh per month.

FIGURE 6–23 As this nineteenth-century representation shows, James Watt's steam engine could do the same work as many horses.

EXAMPLE 6–8 Early in the nineteenth century, James Watt wanted to market his newly discovered steam engine to a society that until then had relied heavily on horses. So Watt invented a unit that made it clear how useful a steam engine could be. He conducted a demonstration in which a horse lifted water from a well over a certain period of time and called the corresponding power expended "one horsepower" (Fig. 6–23). Assume that water has a mass density of 1.0×10^3 kg/m³, the well was 20.0 m deep, and the horse worked for 8.0 h. How many liters of water did the horse raise from the well?

Solution: Suppose that the mass density of water is denoted by ρ. This means that a volume V of water has mass $m = \rho V$. Then the work done in lifting a mass m of water from the bottom of the well is

$$W = F\,\Delta y = mg\,\Delta y,$$

where Δy is the depth of the well. Thus the work done in lifting a volume V from the well in a time t is $\rho V g\,\Delta y$, and the power is

$$P = \frac{\text{work}}{\text{time}} = \frac{\rho V g\,\Delta y}{t}.$$

The unknown here is the volume V, and we can solve for it:

$$V = \frac{Pt}{\rho g\,\Delta y}.$$

All the quantities on the right-hand side are known. The power is, by definition, 1 hp = 746 W, and

$$V = \frac{(746\ \text{W})(8.0\ \text{h} \times 3600\ \text{s/h})}{(1.0 \times 10^3\ \text{kg/m}^3)(9.8\ \text{m/s}^2)(20.0\ \text{m})} = 1.1 \times 10^2\ \text{m}^3.$$

Because there are 10^3 L in 1 m³, the number of liters lifted by the horse is 1.1×10^5 L.

---

*6–6   KINETIC ENERGY AT VERY HIGH SPEEDS

We mentioned in Chapter 5 that Newton's laws cease to be applicable in two domains. One is the domain in which speeds approach the speed of light and the other is the domain of quantum physics, which applies largely to atoms and smaller entities. The first domain is discussed in Chapter 40, whose subject is special relativity. In Chapter 40, we will show that the maximum speed attainable by any particle is the speed of light itself. We will also show that the precise expression for the kinetic energy of a particle of mass m is

$$K = mc^2\left(\frac{1}{\sqrt{1 - (v/c)^2}} - 1\right). \tag{6–30}$$

Here, c is the speed of light with a value $c = 3 \times 10^8$ m/s. In this section, we explore the expression for K.

Equation (6–30) reduces to the usual expression $K = mv^2/2$ when $(v/c)^2$ is very small (see Problem 71). We call a quantity x "very small" when x^2 can be neglected in comparison with x, and we denote this by $x \ll 1$. For example, if $v/c = 10^{-2}$, then $(v/c)^2 = 10^{-4}$; this is indeed much less than 10^{-2} if we maintain 1-percent accuracy. Under normal conditions, v/c is much smaller than 10^{-2}. The speed v_E of Earth around the Sun is a large number, but it is still true that $v_E/c \simeq 10^{-4}$. Molecular speeds in air are also of this order of magnitude.

For particles in cosmic-ray showers and in particle accelerators (atom smashers), v can be very close to c, and the factor $1/\sqrt{1 - (v/c)^2}$ can be very large. In the highest energy accelerators, this factor is as large as 10^4, and the relativistic kinetic energy is very much larger than the Newtonian kinetic energy formula would indicate. We note that the formula for K ceases to make sense when $(v/c) \geq 1$. It is indeed an essential tenet of the theory of relativity, which improves on Newton's laws for high-speed particles, that no information—that is, no particles—can be sent at a speed greater than the speed of light. The case of $v = c$ is delicate: With $m = 0$, the expression for K in Eq. (6–30) is ambiguous but not manifestly wrong. The theory of relativity states that massless particles not only can move with the speed of light, but always *must* move with the speed of light. An example of such a particle is the elementary particle called the *neutrino*, which plays an important role in the fundamental structure of matter, in astrophysics, and in cosmology. Experiment shows the neutrino has a mass at most 10^{-5} times the electron mass. A second example is the photon, which is the particle that represents light itself in the quantum physics description of nature.

SUMMARY

The central result of this chapter is the work–energy theorem as it applies to an object of mass m under the influence of one or more forces. It is given by

$$W_{net} = \Delta K, \tag{6–7}$$

where W_{net} is the net work done by the net force on the object. In this expression, the quantity K is the kinetic energy of the object:

$$K \equiv \tfrac{1}{2}mv^2. \tag{6–5}$$

This is a scalar quantity formed from the velocity vector, with v^2 (the speed squared) $= v_x^2 + v_y^2 + v_z^2$. The change ΔK is the change in the kinetic energy of the object as it moves from an initial position $\mathbf{r}_i$ to a final position $\mathbf{r}_f$ through a displacement $\Delta\mathbf{r} = \mathbf{r}_f - \mathbf{r}_i$:

$$\Delta K = K_f - K_i. \tag{6–6}$$

Here, K_i is the kinetic energy at the initial position and K_f is the kinetic energy at the final position. The net work, W_{net}, depends on the net force acting on an object as well as its displacement and, in general, on how the displacement is made. The net work is the work done by the net force; equivalently, it is the sum of the works done by the individual forces that make up the net force. The work done by a force can be expressed in different ways according to the form the force takes. We can enumerate these forms:

for a constant net force in one dimension:

$$W = F\,\Delta x.$$

for a constant force in three dimensions:

$$W = F_x\,\Delta x + F_y\,\Delta y + F_z\,\Delta z \tag{6–10}$$

$$= \mathbf{F}\cdot\Delta\mathbf{r} \tag{6–11}$$

$$= F\,\Delta r\cos\theta, \tag{6–12}$$

where θ is the angle between the vectors $\mathbf{F}$ and $\Delta\mathbf{r}$. For a nonconstant force in one dimension:

$$W = \int_{x_i}^{x_f} F(x)\,\mathrm{d}x. \tag{6–18}$$

For nonconstant, space-dependent force in three dimensions,

$$W = \int_{\mathbf{r}_A}^{\mathbf{r}_B} \mathbf{F} \cdot d\mathbf{r}. \tag{6–21}$$

In Eq. (6–21), the displacement is between points A and B. This form for work is the most general one and reduces to the other forms in the appropriate limit. By definition, we conclude that no work is done on an object that is in uniform circular motion. Both work and kinetic energy are measured in units of joules (J) in SI.

In some cases, the work done depends on the path the object takes; in others, the work done is independent of the path. When the work done is path dependent, we say that the force is nonconservative; friction provides an example. When the work done is path independent, we say that the force is conservative; gravity or the spring force provide examples. Conservative forces are important because all the fundamental forces of nature are conservative.

The work–energy theorem allows us to calculate the speeds of objects when the work done by the forces is known, and it allows us to calculate the net work that must be done if a certain speed is to be achieved. This theorem is often much simpler to use for these purposes than is Newton's second law.

Power is the rate at which work is done:

$$P \equiv \frac{dW}{dt} \tag{6–23}$$

From the definition of the instantaneous power, we find that

in one dimension: $\qquad P = Fv; \tag{6–26}$

in two or three dimensions: $\quad P = \mathbf{F} \cdot \mathbf{v}. \tag{6–27}$

The unit of power in SI is the watt (W).

UNDERSTANDING KEY CONCEPTS

1. We mentioned in the introduction that the work–energy theorem might help us to analyze the motion of a baseball under the influence of both gravity and drag force from the air. How would you do so?

2. Does it make sense to refer to a force doing negative work when an object moves under its influence? What does negative work mean? In answering, consider what happens when an object is stopped by a force.

3. It certainly seems like work to us when we hold a bag of groceries for a long period of time. Are we expending energy when we hold a bag of groceries for a long period of time? How is the answer to this question consistent with the work–energy theorem?

4. The centripetal forces that cause uniform circular motion do no work because they are perpendicular to the motion. How do such forces fit into the work–energy theorem?

5. A piano can be lifted to the third story of a building by having a crew carry it up the stairs or by using some type of pulley system. Is the same work done in both cases? Assume that friction can be neglected.

6. If the moving crew of Question 5 uses a rope and pulley, it pulls on the rope in the same direction as the force of gravity. Because the crew pulls the rope in the direction opposite to the displacement of the load, is the crew doing negative work on the load?

7. A man pushes against the smokestack on a cruise boat. When the ship is stationary, he does no work. When the ship starts to move in the direction in which he is pushing, he appears to be doing work, yet he experiences no change in the level of his exertion. Why is this? Keep in mind that the man does not fall down because the force of friction keeps his shoes from sliding backward. What is the work done on the man by the rough deck?

8. You do no net work when you walk at a constant speed. Why do you get tired?

9. A parachutist jumps from a plane and lands safely in a field. Does the net work done on the parachutist depend on the height from which he or she jumps? (The work done by gravity does depend on that height.)

10. No work is done in uniform circular motion. Suppose that you observe circular motion in which the moving object first speeds up and then slows down to its original speed. Is any net work done?

11. No work is done on a bag of groceries while you are holding it stationary. Is work done on the same bag if you are holding it while you move steadily upward in an elevator? What is the difference in the two situations? Are your hands still the origin of the force that does work on the bag?

12. A car with cruise control transports you at a constant speed. The engine does work. How much of that work is done on you?

13. One of the entertainments at the carnival is a rotating (vertical) cylinder. The participants step in and place themselves against the interior wall. The cylinder starts to rotate more and more rapidly, and at some point the floor falls away, leaving the customers stuck like so many flies to a wall. Is any work done on the participants? If so, what force does the work?

14. Tarzan swings from tree to tree on a jungle vine. Is any work done on him during the motion? If so, what forces do the work?

15. A stunt consists of one acrobat standing on the short end of a seesaw whose pivot point is not at its midpoint. A second acrobat leaps down on the long end of the board and flips the first acrobat several meters into the air. How would you determine the work done by the second acrobat in flipping the first one into the air?

16. Two identical twins work side-by-side as butchers. They use identical motions and identical hatchets. One brings a hatchet down on some very tender meat, and the other on a large bone. Which one does more work per swing?

17. You tow a small child on a sled at a constant speed by pulling the sled with a rope. The rope makes an angle θ with respect to the horizontal. What forces act on the sled and which ones do work?

18. A one-dimensional force acts on an object, changing its velocity from 0 to $\mathbf{v}$. By the work–energy theorem, the work done is $W = \frac{1}{2}mv_f^2 - \frac{1}{2}mv_i^2 = \frac{1}{2}mv^2$. An observer moving with velocity $\mathbf{v}$ with respect to the original system sees the initial velocity as $-\mathbf{v}$ and the final velocity as 0. This observer would conclude that $W = -\frac{1}{2}mv^2$. What accounts for the difference?

19. A diver plunges from a 10-m high diving board into water. How would you determine the average force of resistance of the water that slows down and stops the diver? (Neglect the force of gravity on the diver while she is in the water. We shall see in Chapter 16 how the water's buoyancy takes care of that.)

20. A parachutist jumps off a tower. What measurements would you have to make to determine the work done by the drag force of the air during the entire fall? The drag force is a rather complicated function of the velocity of the jumper.

21. Are the following forces conservative or nonconservative? (a) Air drag on a parachute. (b) The force opposing the fall of a steel ball bearing in a beaker of water. (c) The explosive force causing a bullet to leave a rifle barrel. (d) The force of an ideal trampoline that propels you into the air.

22. How do you know that the drag forces you experience when you swim are not conservative?

PROBLEMS

6–1 Kinetic Energy and Work

1. (I) An automobile of mass 10^3 kg moves at 1.0 km/h = 0.28 m/s. (a) What is its kinetic energy? (b) At what speeds must a person of mass 80 kg and a bullet of mass 10 g move to have the same kinetic energy as the automobile? (c) What would the speed of the automobile be if the kinetic energy doubled?

2. (I) A construction worker of mass 93 kg rides in an elevator up to the tenth floor, which is 33 m above the ground. The elevator travels with uniform speed. (a) What is the net work done on the worker? (b) What is the work done on the worker by the contact force of the elevator? (c) What is the work done on the worker by gravity?

3. (I) A person lifts a suitcase of mass 10 kg from the floor. Ignore the initial acceleration of the suitcase, and suppose that it moves upward at a constant speed between a height $h = 0$ m and $h = 1$ m. (a) What are the forces acting on the suitcase, as well as the net force? (b) What is the net work done on the suitcase? (c) Find the work done on the suitcase by the person.

4. (I) A bedroom bureau of mass 42 kg is moved from the first floor of an apartment building to the penthouse on the ninth floor, 32 m higher. (a) What is the work done on the bureau by three men in carrying it up the steps? (b) If the three men take it up on an elevator, how much work is done on the bureau by the normal force of the floor of the elevator?

5. (I) A truck carrying a 66-kg crate accelerates uniformly from rest to 63 km/h in 15 s. Calculate the work done on the crate by the truck.

6. (I) An old piano of mass 120 kg is removed from an apartment building being converted into condominiums. The previous owners found it too much trouble to remove and left it. The workmen decide the easiest thing to do is to drop it out of a double-width window to the ground 30 m below. (a) How much work do the workmen do if they just push it out the window? (b) If the men slowly lower the piano by rope, what is the work done on the piano by the rope's tension? (c) How much work does gravity do in each case?

7. (I) A man pushes a refrigerator of mass 40 kg at uniform speed for a distance of 1.5 m to the kitchen wall. The coefficient of friction between the refrigerator and the floor is $\mu_k = 0.4$. (a) How much work does the man do in moving the refrigerator? (b) What other sources of work done are there? (c) What is the net work done in this process?

8. (I) A person pulls a heavy load of mass 37 kg up the side of a building by using a frictionless pulley. The load travels up a distance of 7.5 m. Take the load to move with constant velocity and ignore any acceleration at the beginning or end of the move. (a) How much work is done on the load by gravity? (b) by the tension of the rope? (c) by the person?

9. (II) Consider the woman who lifts the huge interior lineman in Example 5–3. How much work does she do while pulling down a 2 m length of rope? What is the work done by gravity on the lineman while this is going on?

10. (II) The mass $M = 60$ kg is lifted to a height of $h = 2$ m using the system of pulleys shown in Fig. 6–24. The motion is slow and the initial acceleration is negligible. (a) Find the force that must be applied at the free end of the rope. (b) Find the work done on the mass by this force. (c) Calculate the work done on the mass by gravity during the process.

11. (I) A baseball of mass 145 g leaves a pitcher's hand at 96.6 mi/h, but, due to air resistance, it arrives at home plate 60.0 ft away traveling at 95.3 mi/h. Assume that the magnitude of the

FIGURE 6–24 Problem 10.

ball's acceleration is constant and that the ball travels in a straight line (ignore gravity). How much work is done by friction during the flight of the ball?

12. (II) A ball of mass 240 g is dropped from a height of 2 m. (a) What is the work done on the ball by gravity? (b) Suppose that the ball bounces to a height of only 1.5 m. How much work is done by gravity on the ball as it moves from ground level to 1.5 m?

13. (II) A construction worker of mass 75 kg hoists a load of bricks of mass 42 kg by throwing a rope attached to the load over a pulley and letting his weight lift the load. Assuming that there is no friction, what is the work done by gravity during a 2.0-s period?

14. (II) Two masses are connected by a light string over a light, frictionless pulley, as in Fig. 6–25. The table surface is also frictionless. (a) Apply the work–energy theorem for this system to calculate the speed of the masses after the masses have moved a distance Δx, starting from rest. Note that the work of the tensions drops out. (b) Use this result to obtain the acceleration of the system.

FIGURE 6–25 Problem 14.

15. (II) A waterfall, whose height is 40 m, has 200 m³ of water falling every second. How many joules of work are done by gravity every hour? (The mass of 1 m³ of water is 10³ kg.)

16. (II) A construction crew is required to pull up a load of mass 58 kg by means of a rope thrown over a pulley. They are to lift the load from rest on the ground to a height of 12 m, and the load should arrive at the end point with a speed of 2.0 m/s.

(a) Calculate the work done by the crew if it accelerates the load uniformly over the whole distance. (b) Repeat the calculation for the case in which the acceleration takes place in the first 5 m and the load is pulled with uniform speed the rest of the way.

17. (II) A child has three different sets of cubical blocks (Fig. 6–26). The first set consists of 3 blocks, each 12 cm on a side and of mass 36 g; the second set consists of 6 blocks, each 6 cm on a side and of mass 18 g; the third is a set of 12 blocks, each 3 cm on a side and of mass 9 g. For each set, what is the work the child must do to stack the blocks into a tower 36 cm high? The blocks can be treated as point objects at their *centers* in calculating the work.

FIGURE 6–26 Problem 17.

18. (III) Find the work that must be done by a force lifting against gravity to raise a coiled rope of length L and mass M entirely off a level surface. [*Hint*: Use the method of Problem 17 and divide the rope into more and more segments.]

6–2 Constant Forces in Space

19. (I) What is the scalar product of $\mathbf{A} = -2\mathbf{i} + 3\mathbf{j} - 5\mathbf{k}$ and $\mathbf{B} = 5\mathbf{i} + \mathbf{j} - 2\mathbf{k}$?

20. (I) A force $\mathbf{F} = (2.7\text{ N})\,\mathbf{i} - (6.8\text{ N})\,\mathbf{j}$ is used to displace an object of mass 2.6 kg by an amount $\mathbf{r} = (3.4\text{ m})\,\mathbf{i} + (1.2\text{ m})\,\mathbf{j}$. What is the work done by the force on the object?

21. (I) Assuming that the object of Problem 20 is initially at rest, what is its final speed?

22. (I) Show that the vector $\mathbf{v} = -y\mathbf{i} + x\mathbf{j}$ is always perpendicular to the vector $\mathbf{u} = x\mathbf{i} + y\mathbf{j}$.

23. (I) Consider two vectors, $\mathbf{u} = 3\mathbf{i} - 4\mathbf{j} + 7\mathbf{k}$ and $\mathbf{v} = -2\mathbf{i} + 3\mathbf{j} + z\mathbf{k}$. What must z be so that $\mathbf{u}$ and $\mathbf{v}$ are orthogonal?

24. (I) A person puts a suitcase of mass 15 kg into a van, moving the suitcase a total distance of 1.6 m: 0.6 m up and 1 m horizontally. How much work is done by the person?

25. (I) A man pulls a sled by a rope, moving his two daughters to the top of a 15° slope. He holds the rope parallel to the slope. If the daughters and the sled have a total mass of 43 kg and the length of the slope is 36 m, how much work does the man do on the sled, assuming that he pulls the sled with uniform velocity? Ignore all friction on the sled.

26. (I) A block of material with mass 560 kg is used in the construction of a building. During one part of the process of setting the block in place, a complex network of cables acts on

it and its motion is transformed from a horizontal motion with speed 10 cm/s to a vertical motion with speed 38 cm/s. What is the net work done on the block during this motion?

27. (II) A skier of mass 72 kg (including skis), starting from rest, slides down a slope at an angle of 18° with the horizontal. The coefficient of kinetic friction is $\mu_k = 0.12$. What is the net work done on the skier in the first 7.0 s of descent?

28. (II) Consider a vector **A** in the xy-plane. Its x- and y-components are A_1 and A_2, respectively. Show that any vector **B** in the same plane that points in a direction perpendicular to **A** must have components $-cA_2$ and cA_1, respectively, where the magnitude of c is the ratio of the lengths ($|c| = B/A$).

29. (II) Sketch the direction of the vector $\mathbf{e} = \cos\theta\mathbf{i} + \sin\theta\mathbf{j}$. Show that it has unit length, and use your sketch to give an expression for the unit vectors **f** that are perpendicular to **e** and that lie in the xy-plane. How many such vectors **f** are there?

30. (II) Consider the vector $\mathbf{A} = 3\mathbf{i} - 4\mathbf{j} + 5\mathbf{k}$. Find the most general vector in the yz-plane that is perpendicular to **A**.

31. (II) Consider the unit vector $\mathbf{e} = -0.6\mathbf{i} + 0.8\mathbf{j}$. What is the magnitude of the projection of vector $\mathbf{A} = 3\mathbf{i} - 2\mathbf{j}$ onto the line along which **e** points?

32. (II) A stone is thrown from a height h_0 above a level field, leaving the hand at a 40° angle. Ignore all effects of air resistance. (a) Compute the work done by gravity as the stone follows its trajectory back to the height h_0. Recall that the motion can be divided into motion in the vertical direction and motion in the horizontal direction. (b) Show, by applying the work–energy theorem, that the speed of the stone when it reaches h_0 again is identical to the speed it had when it left the hand.

33. (II) A force given by $\mathbf{F} = (2\mathbf{i} - 5\mathbf{j})$ N acts on an object that moves from $\mathbf{r}_1 = (7\mathbf{i} - 8\mathbf{j} + 2\mathbf{k})$ m to a new position $\mathbf{r}_2 = (5\mathbf{i} - 4\mathbf{j} + 5\mathbf{k})$ m. How much work does this force do on the object?

34. (II) A 25-kg crate slides down a plane that makes an angle of 30° with the horizontal, starting from rest at the top. The speed of the crate when it reaches the bottom of the 10-m-long slide is 8 m/s. What is the coefficient of friction? How much work is done by the force of friction?

6–3 Forces that Vary with Position

35. (I) A small gizmo is confined to a groove that is aligned with the x-direction. A rod pulls the gizmo in the +x-direction. The rod is attached to an apparatus such that the pulling force is 0.3 N when the gizmo is to the left of a point in the groove we label as the origin and 0.7 N when the gizmo is to the right of the origin. What is the work the pulling force does on the gizmo as the gizmo moves from $x = -6$ cm to $x = +7$ cm?

36. (I) A spring with spring constant $k = 12$ N/m is attached to a wall at ground level. The end of the relaxed spring is on the floor at a location that we take to be the origin. A mass of 3.0 kg is attached to the end of the spring, and the spring is stretched by 50 cm and released. How much work has the spring done on the mass by the time the mass passes through the origin?

37. (I) A one-dimensional force F depends on the position x of a particle on which it acts as $F = g_1x - g_2x^3$, where g_1 and g_2 are constants. What is the work done in moving the particle from the origin to $x = 2.0$ m?

38. (I) A one-dimensional force on a particle is given by $F = \alpha x$, where $\alpha = -5.00$ N/m for $x < 0$ and $\alpha = +3.00$ N/m for $x > 0$ (Fig. 6–27). Calculate the work done by the force on a block when the block is moved from $x = -3.00$ m to $x = +4.00$ m.

FIGURE 6–27 Problem 38.

39. (I) A spring gun is made by compressing a spring (assumed to be perfect) and latching it. A spring of constant $k = 60$ N/m is used and the latch is located at a distance of 7 cm from equilibrium. The pellets have mass 4 g. What is the muzzle velocity of the gun?

40. (II) A man pushing a 50-kg crate up a slope that makes an angle of 30° with the horizontal exerts a force parallel to the plane. The coefficient of kinetic friction varies along the slope and is given by $\mu_k = \mu_1 + [(\mu_2 - \mu_1)s/L]$, where s is the distance along the slope starting at the bottom. The largest value of s is L, where $L = 10$ m is the length of the slope; $\mu_1 = 0.2$, and $\mu_2 = 0.3$. What is the work done on the crate by the force as a function of s if the force varies such that the crate is pushed at a constant speed?

41. (II) A nonstandard spring exerts a force $F = -k_1x - k_2x^3$ to restore itself to equilibrium, where x is the distance from equilibrium. The values of k_1 and k_2 are 5.0 N/m and 15 N/m³, respectively. Calculate the work done to stretch the spring from 0.10 m to 0.20 m.

42. (II) A rocket is scheduled to blast off from Cape Canaveral to study a neighboring solar system. Earth's gravitational force is $F = K/r^2$, where r is the distance from Earth's center and K is a negative constant. What is the minimum work the rocket engine must do so that the rocket leaves the gravitational force of Earth? Assume that the mass of the rocket does not vary in the process. (This assumption is actually very poor, but a more exact treatment must wait until Chapter 8.)

6–4 Conservative and Nonconservative Forces

43. (I) A child swings a streamer toy over her head in a nearly horizontal plane. The toy, of mass 85 g, is at the end of a massless string of length 1.5 m. She starts twirling the toy from rest while she slowly lets out the string to full length, and gets the angular speed up to 2 rev/s. How much work has she done?

44. (I) A 100-g ball is tossed straight up in the air, rises to a maximum point, then falls back until it is 1 m below the position of the hand that tossed it up. A second ball is simply dropped from the same hand position, and also lands 1 m below that position. What is the net work done by gravity in the two cases?

45. (I) You are in the process of moving. A 54-kg bed can be brought from ground level to the second floor (4.0 m above ground level) either by pulling it straight up by means of a rope or by dragging it up a frictionless plane inclined at 30° to the horizontal. Calculate the work done in each case by those who move the bed.

46. (I) An object of mass 1.5 kg is suspended from a vertical spring that is attached to the ceiling. Without the mass, the spring is 30 cm long. When the mass is attached to it, the spring is extended to a length of 110 cm. What is the work done by the force of gravity during the extension of the spring?

47. (I) A constant force of 10 N pushes a particle along the x-axis. The position of the particle is represented by $x = 11$ m $- (2$ m/s$)t + (0.5$ m/s$^2)t^2$. Find the work done by the force between $t = 0$ s and $t = 1$ s, and between $t = 1$ s and $t = 2$ s. Is the force conservative?

48. (II) A force $\mathbf{F}$ has components $F_x = axy - by^2$, $F_y = -axy + bx^2$, where $a = 2$ N/m^2 and $b = 2$ N/m^2 (Fig. 6–28). Calculate the work done on an object of mass 4 kg if it is moved in a closed path from $(x, y) = (0, 1)$ to $(4, 1)$, to $(4, 3)$, to $(0, 3)$, and back to $(0, 1)$. The path between the points is always the shortest straight one, and all the distances are given in meters.

FIGURE 6–28 Problem 48.

49. (II) A force with an x-component acts on a mass with a strength that varies only with the position x of the mass on the x-axis, according to $|F(x)| = Ax^2$. The sign of the force is negative when x is positive and positive when x is negative, indicating that the force always attracts the mass toward the origin. Compute the work done by this force when $A = 1500$ N/m^2 and the mass moves along the x-axis (a) from $x = -5.0$ cm to the origin; (b) from $x = -5.0$ cm to $+5.0$ cm; (c) from $x = +5.0$ cm to $+2.0$ cm; (d) from $x = -2.0$ cm to -5.0 cm.

50. (II) A child's playground ride consists of four seats, of mass 5 kg each, connected to a vertical axle with spokes of small mass. The seats are placed equidistant in a circle of radius 1.4 m and rotate about the vertical axle. A child of mass 15 kg sits in one of the seats, and his friend pushes the ride to accelerate him from rest to 0.4 rev/s. How much work does the friend do?

51. (II) The net force acting on a particle depends on the position of the particle on the x-axis according to the relation $F = F_0 + Cx$, where $F_0 = 5$ N and $C = -2$ N/m. The particle is initially at rest at the point $x = 0$ m when the force begins to act. (a) Calculate the work done by the force when the particle

reaches $x = 1$ m, 2 m, 3 m, and 4 m. (b) Determine any positions (other than at $x = 0$ m) where the work done is zero. (c) Is the force conservative?

52. (II) An object of mass m is to be moved from the top of a building of height h to a point on the ground a horizontal distance h from its original location, so that the position vector at the beginning may be chosen to be $h\mathbf{j}$, and at the end, $h\mathbf{i}$. Two possible paths are: (a) the object is lowered at constant speed by rope and, after it reaches the ground, it is moved horizontally to the final location; and (b) the object is allowed to slide along a straight support that runs from the initial point to the final point. Show that the work done by the force of gravity is the same in both cases.

53. (III) Prove that a force acting in one dimension is conservative if it is a function of position only, and not a function of any other information about the motion of an object under its influence. Does this include forces with constant magnitude? In view of your answer, how does the friction force manage to be nonconservative?

54. (III) A small object of mass m is moved up along a track that forms one quarter of a circle of radius R in the vertical plane (Fig. 6–29). The object moves with a small uniform speed maintained by a tangential force (that varies with the angle) along the track. Calculate the work done by this force in moving the object through a 90° arc from the lowest point to the highest point (a) by direct use of the definition of work, and (b) by using the work–energy theorem.

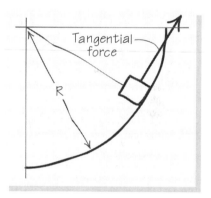

FIGURE 6–29 Problem 54.

55. (III) The force acting on an object of mass m is given by the expression $F(x) = C|x|$, where $|x|$ is measured in centimeters. (a) How much work is done by the force when the object is moved from $x = -4.0$ cm to $x = +4.0$ cm? Compare your result with the amount of work done if the form of the force law were $F(x) = Cx$. (b) Repeat the calculation if the object moves from $x = 0.0$ cm to $x = 8.0$ cm.

56. (III) Consider a force that acts on an object of mass m, which moves in the xy-plane. The force is given by $\mathbf{F}(x, y) = k_1 x\mathbf{i} + k_2 y\mathbf{j}$. Calculate the work done by the object if it moves in a circle of unit radius (given by $x^2 + y^2 = 1$ m^2) starting at $x = 1$ m, $y = 0$ m, and ending at a point that makes an angle of (a) 90°, (b) 180°, and (c) 360° with the original direction of the position radius vector. [*Hint:* The problem is simplified with polar coordinates r and θ, where $x = r \cos \theta$ and $y = r \sin \theta$.]

57. (I) Electricity costs about $0.08 per kWh. Your monthly electric bill is $26.00. Assuming that your only use of electricity is for light and that you keep your house lit 5 h/d, how many 100-W bulbs do you keep going?

58. (I) How much energy is used by running a fleet of 100 80-hp cars around the clock for 1 month?

59. (I) How long does it take you to climb four flights of steps? Assume that your body is 20 percent efficient, and estimate the power you have to generate.

60. (I) Two engines are used to move a mass of 18 kg, starting from rest, a distance of 20 m in a straight line along a frictionless flat surface. Engine 1 exerts a constant force of 2.0 N, and engine 2 exerts a constant force of 3.0 N. (a) What is the work done by each engine? (b) What is the average power expended by each engine during the process?

61. (I) A test car of mass 700 kg is moving at a speed of 15 mi/h when it crashes into a wall to test its bumper. If the car comes to rest in 0.3 s, how much average power is expended in the process? (To find the average power, simply imagine Eq. (6–23) for a finite time interval: $P_{av} = \Delta W/\Delta t$, where ΔW is the work done during the time interval Δt.)

62. (I) Assume that a car of mass 1200 kg has an engine with power output of 80 hp. How long would it take to accelerate such a car to a speed of 100 km/h? (Neglect air resistance, which would make this time much larger.)

63. (II) The maximum power of a particular horse is 1 hp. With what speed can this horse pull a sled on level ground if the weight of the sled with its load is 5000 N and if the coefficient of kinetic friction is $\mu_k = 0.03$? What is the maximum speed on a 5° upward incline?

64. (II) An accelerator accelerates a proton to 0.999c, where c is the speed of light. If 1.35×10^{10} protons are accelerated every minute, how much power is expended by the accelerator, assuming 9.76 percent efficiency? (For this problem, with $v \approx c$, use Eq. (6–30) for the kinetic energy. The *efficiency* is the fraction of the total power that goes into changing the kinetic energy of the protons.)

65. (II) Consider the waterfall in Problem 15. If the waterfall is used to produce electricity in a power station, and if the efficiency of conversion of kinetic energy of falling water to electrical energy is 60 percent, what is the power production of the station?

66. (II) An escalator moves people from one floor up to another. The height difference between the floors is 5.2 m, and the angle that the escalator makes with the horizontal is 33°. The speed of the escalator is 1.5 m/s, and it is supposed to carry a maximum of 60 passengers, with an average mass of 75 kg. How much power must be generated by the motor that runs the escalator?

67. (II) Trained athletes can exert power for their movements ranging from around 5 hp for a second to 0.4 hp or less for periods extending over several hours. (a) A bicyclist is limited by wind resistance, which is roughly of the form $F = Av^2$, where $A \approx 0.08$ kg/m. Estimate the speed a cyclist can maintain for 1 h. (b) Estimate the time it takes a weight lifter to lift 100 kg a distance of 2 m. (c) Assuming that not too much time is taken up turning the corners, estimate how fast it is possible to climb three flights of steps, a vertical distance of 12 m. You can easily try this one!

68. (I) A proton is accelerated from rest to a final speed of 0.85c, where c is the speed of light. How much work is done by the accelerator on the proton, given that a proton's mass is 1.6×10^{-27} kg?

69. (I) What is the kinetic energy of an electron (mass = 9.1×10^{-31} kg) moving at a speed of 0.9999c? of 0.9999999999c?

70. (I) (a) How much work does it take to accelerate an electron from a speed of 0.1c to 0.8c? (b) from 0.8c to 0.998c? (c) from 0.998c to 0.99998c?

71. (II) In order to show that Eq. (6–30) reduces to the usual form for small v/c, we need an approximation for $1/\sqrt{1-x}$ for small values of x, namely, $1/\sqrt{1-x} \approx 1 + \frac{1}{2}x + \frac{3}{8}x^2$. Check this approximation by calculating both sides of the expression on your calculator. For what value of x is the approximation correct to within 10 percent? 1 percent?

General Problems

72. (I) A grocery store pays a monthly power bill of $475. Electricity costs $0.09/kWh. How many joules of energy were used in the month?

73. (I) A ball with mass 100 g is set in motion inside a bowl, with initial speed $v = 2$ m/s. Due to friction, the ball ultimately comes to rest. How much net work was done by the external forces acting on the ball?

74. (I) A pile driver works by lifting a large mass and dropping it to the ground. A driver used to put pilings in the ground for a tall building has a mass of 1400 kg and is raised to a height of 6 m in 4 s. (a) How much work is done by the engine each time the weight is lifted? (b) What horsepower engine must be used to run the pile driver?

75. (I) The so-called Domesday Book recorded a general census carried out in England in the year 1086. It catalogued all 6000 waterwheels in the country, each of which had a power output of roughly 2 hp. Waterwheels were the main source of nonanimal energy at the time. By contrast, the power of a Boeing 747 jet airplane at maximum thrust is approximately 1 MW, and at cruise level the power output is roughly 0.3 MW. What fraction of the total nonanimal power of eleventh-century England does a cruising 747 represent?

76. (II) A mass $M = 3$ kg moving without friction in the xy-plane starts at the point labeled by the position vector $\mathbf{r}_i = 0\mathbf{i} + 0\mathbf{j}$ with velocity $\mathbf{v}_i = (2\mathbf{i} + 1\mathbf{j})$ m/s. Two forces, $\mathbf{F}_1 = (2\mathbf{i} + 7\mathbf{j})$ N and $\mathbf{F}_2 = (2\mathbf{i} - 5\mathbf{j})$ N, act on the mass as it moves in a straight line to the point labeled by the position vector $\mathbf{r}_f = (10\mathbf{i} + 5\mathbf{j})$ m (Fig. 6–30). (a) How much work is done by $\mathbf{F}_1$

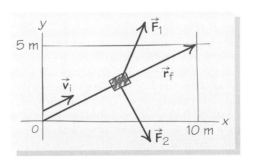

FIGURE 6–30 Problem 76.

as the mass moves from $\mathbf{r}_i$ to $\mathbf{r}_f$? (b) What power is provided by $\mathbf{F}_1$ at the instant the mass is at $\mathbf{r}_i$? (c) What is the kinetic energy of the mass when it reaches $\mathbf{r}_f$?

77. (II) A rocket in outer space, initially at rest, is accelerated uniformly at the rate of 2.0 m/s² for 33 s. (Neglect gravity throughout.) (a) If the mass of the rocket, assumed here to be constant, is 1.0×10^4 kg, how much work is done by the rocket engine? (b) Suppose that the rocket decelerates at such a rate that it comes to rest in 55 s. How much work is done on the rocket during the deceleration?

78. (II) A worker pushes a box of mass $m = 40.5$ kg in a straight line along a rough floor. The applied force $\mathbf{F}$ has magnitude 285 N and acts downward at an angle $\theta = 25.5°$ with respect to the horizontal. The box is initially at rest at the position $x_1 = 0$ m, and it has speed $v_2 = 1.72$ m/s at position $x_2 = 3.50$ m. (a) Use the data to calculate the coefficient of friction. (b) What is the net work done? (c) How much work is done to overcome friction? (d) What is the instantaneous power generated by the worker at $x = x_2$?

79. (II) A block of mass $m = 2.6$ kg is placed on an inclined plane that makes an angle $\theta = 32°$ with the horizontal. It is given an initial speed v_0 up the ramp and slides a distance 1.3 m up the ramp before it comes to a stop. The coefficient of kinetic friction between the block and the ramp is $\mu_k = 0.25$. (a) What are the three forces acting on the block? Give the magnitude and direction of each. (b) What is the work done by each of the three forces during the motion of the block? (c) What was the initial speed v_0?

80. (II) The engine of an automobile requires 45 hp to maintain a constant speed of 80 km/h. (a) What is the resistive force against the automobile? (b) If the resistive force is proportional to the velocity, what must the engine power be to drive at constant speeds of 60 km/h? (c) at 140 km/h?

81. (II) A mass of 4.0 kg is attached to a string tied to a hook in the ceiling. The length of the string is 1.0 m, and the mass is released from rest in an initial position in which the string makes an angle of 30° with the vertical. Calculate the work done by gravity by the time the string is in a vertical position for the first time. [*Hint*: Recall that you can use any path you like to get from the initial to the final point, because the work done by gravity is path independent.]

82. (II) A mass of 1 kg is accelerated by applying a force. In each case, calculate the work, in joules, that this force must do on the mass. (a) The mass is brought from rest to a speed of 1 m/s. (b) The mass is accelerated from 100 m/s to 101 m/s. (c) The speed of the mass is increased from v m/s to $(v + 1)$ m/s. (d) If $v \gg 1$ m/s, find an approximation that simplifies the calculation in part (c).

83. (II) A puck of mass $m = 0.2$ kg moves in a circle of radius 0.8 m on a table top and is tied with a massless rope to a tether at the origin. The coefficient of kinetic friction between the puck and the table top is $\mu_k = 0.02$. At $t = 0$ s, the puck is at the point shown in Fig. 6–31 with a velocity in the $+y$-direction of magnitude 10 m/s. (a) How much work is done by the rope on the first revolution? (b) How much work is done by friction on the first revolution? (c) What is the kinetic energy at the end of one revolution?

FIGURE 6–31 Problem 83.

84. (II) An automobile of mass 1100 kg is brought to a halt by applying the brakes, which lock the wheels. The coefficient of kinetic friction between the wheels and the road is 0.55. The car leaves skid marks 48 m long. (a) What is the force of friction between the car and the road? (b) What is the work done on the car by friction in bringing the car to a halt? Include the sign of the work. (c) What was the speed of the automobile when the brakes were first applied?

85. (II) A mass m is hauled from ground level up an inclined plane that makes an angle θ with the horizontal by means of a rope passing over a frictionless pulley. The mass is pulled along until it reaches a height H. The building of the Egyptian pyramids and many other ancient construction jobs used such ramps. Logs were sometimes used as rollers to reduce friction. (a) Show that if the contact between the mass and the ramp is frictionless, the work done by the tension in the rope (or, equivalently, by the person hauling the rope) is independent of the angle θ. (b) Calculate the work done by the tension in the rope as a function of the ramp angle θ if the coefficient of kinetic friction between the mass and the surface is μ_k.

86. (II) A large laser designed for nuclear fusion can produce 2×10^4 J of energy over a time period of 10^{-9} s. (a) How much power can such a laser produce during its discharge? (b) How much power is required to reenergize it over a period of 45 min?

87. (II) A pendulum of length L and mass m starts from an initial position in which it makes an angle θ_i with the vertical. Calculate the work done by the force of gravity as the mass moves from θ_i to θ_f and use your result to calculate the speed of the pendulum at the bottom of the swing. Why is it possible to ignore the tension in the string of the pendulum? [*Hint*: The work done by gravity is path independent.]

88. (II) A small dam produces electrical power. The water falls a distance of 18 m to turn a turbine. If the efficiency to produce electrical energy is only 68 percent, at what rate must water flow over the dam to produce 850 kW of electrical energy? (See Problem 64 for the meaning of efficiency.) The power system costs $3.5 million. How many years will it take for the power plant to pay for itself if electricity can be sold for $0.10/kWh? Ignore effects such as inflation and the cost of borrowing money.

89. (II) The force the Sun exerts on a planet in a circular orbit of radius x around the Sun is given by an expression of the form $F(x) = -mK/x^2$, where m is the mass of the planet and K is a constant. How much work must be done by a passing celestial body to move the planet to a radius that is 1 percent larger?

90. (II) Early twentieth-century investigations of the structure of atoms carried out by Ernest Rutherford involved the collisions of alpha-particles (α-particles) and gold atoms. The most interesting results concern the cases in which an α-particle is scattered at a large angle from a gold nucleus. Assume that an α-particle approaches a gold nucleus (at rest) head on and that the force between these two objects is repulsive, with magnitude $F = k/r^2$, where $k = 3.65 \times 10^{-26}$ N·m². Suppose that the α-particle moves in toward the nucleus from far away under the sole influence of the repulsive force of the nucleus. What is the minimum kinetic energy the α-particle must have initially in order for it to get as close as 1.00×10^{-14} m?

The motion of a roller coaster car illustrates the conversion of potential energy into kinetic energy. As the car rises, it slows; as it drops, it speeds up. The total energy of the car—the sum of the kinetic and potential energies—is conserved if there is no friction.

Potential Energy and Conservation of Energy

A roller coaster gains kinetic energy as it descends. Where did the energy come from? When the roller coaster rises from a low point, it loses kinetic energy. What happened to that energy? Conservative forces such as gravity have a special property: the work they do to make an object gain or lose energy can be recovered. The work is "stored" and we can think of it as a form of energy. That energy, *potential energy*, is a function of the position of the object, and the sum of the potential energy and the kinetic energy—the total energy—remains constant throughout the object's motion. As we will see, energy is conserved when conservative forces act. Thus when the roller coaster descends, potential energy is converted into kinetic energy; that kinetic energy is converted back into potential energy when the roller coaster climbs. Conservation of energy is a fundamental property that can also significantly simplify the solution of problems involving conservative forces. What about nonconservative forces such as friction? Nonconservative forces cannot be characterized by a potential energy. When such forces are present, energy is lost—converted into heat, abrasion, and noise.

A close look at the mechanisms of energy loss for nonconservative forces reveals that we can account for *all* the energy at the microscopic level. Thermal energy, for example, is actually related to the kinetic energy of molecular motion. Indeed, *all* the fundamental forces (see Chapter 5) are conservative. The concept of the conservation of energy is one of the central concepts of science.

Here, we develop the concepts of potential energy and total energy. The results of this section *apply only to conservative forces*. The work–energy theorem shows us how the kinetic energy of an object changes when there is net work done on it. Now we shall see that we can reexpress the work done by conservative forces in terms of a simple function of position called the potential energy.

By definition, the work done on a given object by a conservative force when the object moves from point A to point B depends only on those points; it does not depend on any points in between: $W = W(A, B)$. We can actually say more than this. Suppose we go from A to B, then from B to C. In this case, the total work does not depend on the intermediate point B and thus

$$W(A, C) = W(A, B) + W(B, C).$$

But this, in turn, can be true only if

$$W(A, B) = U(A) - U(B), \qquad (7\text{–}1)$$

where U is some scalar function that depends only on position. Not only does W depend on the end points only, but it does so as a difference of two values of a function that depends only on position. If we know this function for a given conservative force, then we know once and for all how much work that force does as an object moves from one spot to another. We have already worked out some examples of this result. One occurs when an object is moved from height y_1 to height y_2. As we saw in the last chapter, the work done by gravity is

$$W(y_1, y_2) = mg(y_1 - y_2).$$

Comparing the work done by gravity with Eq. (7–1) we find

$$\text{for gravity:} \quad U(y) = mgy + U_0, \qquad (7\text{–}2)$$

where U_0 is any constant. We call the function U the **potential energy**.

We can easily see how this idea works for a general one-dimensional conservative force $F(x)$. The work done in moving from point x_0 to point x_1 is

$$W(x_0, x_1) = \int_{x_0}^{x_1} F(x) \, dx. \qquad (7\text{–}3)$$

As we show next, we can guarantee that $W(x_0, x_1) = U(x_0) - U(x_1)$ as stated in Eq. (7–1) if and only if

$$F(x) = -\frac{dU(x)}{dx}. \qquad (7\text{–}4)$$

The force in one dimension in terms of the potential energy

The function U is the potential energy. To check that Eq. (7–4) leads to Eq. (7–1), use Eq. (7–4) in Eq. (7–3) along with the fact that integration and differentiation are inverse processes to find

$$W(x_0, x_1) = \int_{x_0}^{x_1} \left[-\frac{dU(x)}{dx} \right] dx = -U(x_1) + U(x_0). \qquad (7\text{–}5)$$

This is indeed Eq. (7–1).

Note that we now have a way to find the one-dimensional conservative force from the corresponding potential energy function, $U(x)$; we simply use Eq. (7–4). We can check that this works for the case of gravity, where we have already found $U(y) = mgy + U_0$, or Eq. (7–2).

$$F(y) = -\frac{dU}{dy} = -\frac{d}{dy}(mgy) - \frac{dU_0}{dy} = -mg.$$

The Definition of Potential Energy

If we compare Eq. (7–3) with Eq. (7–5), we find an expression for the potential energy $U(x)$:

for conservative forces: $$U(x) = U(x_0) - \int_{x_0}^{x} F(x') \, dx'. \qquad (7\text{–}6)$$

Note that the dependence on the initial point x_0 does not appear on the left-hand side of the equation; that dependence cancels on the right-hand side. Usually, we simply write $U_0 = U(x_0)$. The value of U_0, as well as the point x_0 itself, can be chosen for convenience.

Equation (7–6) gives us a way to interpret the potential energy. The change in the potential energy associated with a given force as an object moves from one point to another is the negative of the work done by that force over that displacement. The potential energy represents the possibility of getting that work back (Fig. 7–1).

The Conservation of Energy

Let's now apply our expression for work in terms of the potential energy to the work–energy theorem. We have

$$\frac{1}{2}mv_1^2 - \frac{1}{2}mv_0^2 = W(x_0, x_1) = -U(x_1) + U(x_0).$$

Simple rearrangement gives

$$\frac{1}{2}mv_1^2 + U(x_1) = \frac{1}{2}mv_0^2 + U(x_0). \qquad (7\text{–}7)$$

This is a remarkable relation: It expresses the conservation of energy. Because this is true for any two points along the path of an object that is moving under the influence of the conservative force $F(x)$, Eq. (7–7) shows that *the quantity $\frac{1}{2}mv^2 + U(x)$ must have the same value at every point in the trajectory of the object.* We call this quantity E:

$$E \equiv \frac{1}{2}mv^2 + U(x). \qquad (7\text{–}8)$$

FIGURE 7–1 The stretched bow and arrow system contains potential energy, much of which can be converted into the kinetic energy of the arrow.

The quantity E is called the **total energy** (or more simply, the **energy**) of the object, and we have shown that it is *conserved* during the motion, no matter how that energy is transformed from potential energy to kinetic energy and back again. A thrown ball starts with a certain kinetic energy. This kinetic energy is converted to potential energy as the ball rises to its maximum height; the potential energy is converted back to kinetic energy as the ball falls. Throughout the ball's flight, the total energy is always the same. The idea that the total energy is conserved is one of the most powerful ideas in physics and is one of the most useful (Fig. 7–2).

At this point, you should feel more comfortable with the idea that the potential energy is the negative of the work. As a force does positive work, which increases the kinetic energy, the potential energy decreases; as the force does negative work, which decreases the kinetic energy, the potential energy increases.

Applications of the Conservation of Energy

The conservation of E means that any change ΔK in the kinetic energy is compensated by an equal and opposite change ΔU in the potential energy:

$$\Delta K = -\Delta U. \qquad (7\text{–}9)$$

This equation explains the origin of the term potential energy. Potential energy is energy that has the potential to be converted into energy of motion; that is, kinetic energy.

Energy conservation is a very useful tool for determining the speed of an object. If the function $U(x)$ is known and the initial conditions are given, we can determine the speed at any point x on the trajectory of an object. We simply use the fact that the total energy is constant.

It is clear from Eq. (7–7) that we can add an arbitrary constant U_0 to the potential energy $U(x)$ because this constant would cancel on both sides of Eq. (7–7). This arbitrary constant is chosen for convenience in practical applications. This choice does not affect *changes* in the potential energy or the total energy. The process of choosing a particular point x_0 where $U(x_0)$ is defined to be zero is sometimes referred to as "choosing the zero of the potential energy."

Application to Gravity. We have already seen from Eq. (7–2) that for a mass m under the influence of gravity

$$U(y) - U(y_0) = mgy - mgy_0. \qquad (7\text{–}10)$$

(Here, we measure y with the positive direction up.) If we choose the reference height to be ground level in this case ($y_0 = 0$), and write $U(y_0) = U(0)$ as the arbitrary constant U_0, we recover Eq. (7–2):

$$U(y) = mgy + U_0. \qquad (7\text{–}11)$$

We may further define the potential energy at $y = 0$ to be $U_0 = 0$; we have chosen the zero of the potential energy to be at height zero. Then we find the simple result

$$U(y) = mgy, \qquad (7\text{–}12)$$

and the conserved energy is

$$E = \frac{1}{2}mv^2 + mgy. \qquad (7\text{–}13)$$

Let's see how energy conservations help us to find the ground-level speed of a ball dropped from rest. At the beginning of the ball's motion, the height is y and $v = 0$, so $E_{\text{beginning}} = mgy$. At the end, $y = 0$ and the speed is v, so $E_{\text{end}} = \frac{1}{2}mv^2$. Equating the two values of E—that is, $E_{\text{end}} = E_{\text{beginning}}$—we have

$$\frac{1}{2}mv^2 = mgy;$$

$$v = \sqrt{2gy}.$$

We derived this result using Newton's equation of motion in Chapter 2; here, we found the same result in a much simpler way.

We defined U to be zero at height $y = 0$. Alternatively, we can define U at, say, the top of a building ($y = H$) so that E is zero there. To do so, choose the arbitrary constant U_0 so that

$$E = 0 = mgH + U_0;$$

(Notice there is no kinetic energy at $y = H$.) This equation determines U_0:

$$U_0 = -mgH.$$

With this choice, the total energy is given at any height y by

$$E = \frac{1}{2}mv^2 + mgy - mgH = \frac{1}{2}mv^2 + mg(y - H). \qquad (7\text{–}14)$$

In this equation, $y - H$ is the displacement of the ball from a starting point on the roof of our building to its new position at height y. Whatever the choice of U_0, the total energy is unchanged during the fall (Fig. 7–3).

PROBLEM SOLVING

We use the conservation of energy to find speeds.

The point where the potential energy is zero does not affect predictions about observable quantities.

The potential energy of a mass acted on by gravity

FIGURE 7–2 The conservation of energy gives us an immediate way to calculate the speed of each ball as a function of height, given its initial speed.

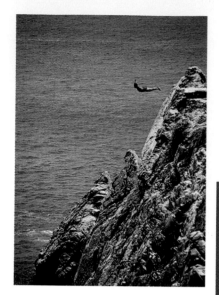

FIGURE 7–3 The total energy of this Mexican cliff diver is conserved when he jumps from rest off the top of the cliff, where his kinetic energy is zero, to the water below, where his potential energy is taken to be zero. As he drops, his kinetic energy grows as his potential energy drops.

FIGURE 7–4 Example 7–1.

The potential energy of a mass at the end of a spring

If we use the form of Eq. (7–14), an object starting at $y = H$ with $v = 0$ has $E = 0$. Thus $E = 0$ later; that is,

$$\frac{1}{2}mv^2 + mg(y - H) = 0;$$

$$v^2 = -2g(y - H).$$

If we take into account that $y - H$ is always negative for an object falling from rest from an initial height H, we see that this result is also in agreement with the formula obtained in Chapter 2.

EXAMPLE 7–1   A brick is thrown nearly straight up from 18 m above the ground by a person standing at the edge of a roof (Fig. 7–4). Use the fact that energy is conserved to determine the velocity with which the brick is thrown, given that it reaches the ground with a velocity of 24 m/s.

Solution: Let's choose the potential energy to be zero at ground level. Using Eq. (7–11), we have

$$U(y) = mgy + U_0 = mgy.$$

where y is the height measured from ground level. The total energy, E, is unchanging and it is

$$E = \frac{1}{2}mv^2 + mgy.$$

In particular, if the final speed is v_f, the final height is $y_f = 0$ m, and the initial velocity v_i is set when the brick is at an initial height $y_i = 18$ m, then we find

$$\frac{1}{2}mv_f^2 + mgy_f = \frac{1}{2}mv_f^2 + 0 = \frac{1}{2}mv_i^2 + mgy_i.$$

We divide out the unknown mass m and multiply by 2, solving for v_i:

$$v_i^2 = v_f^2 - 2gy_i = (24 \text{ m/s})^2 - 2(9.8 \text{ m/s}^2)(18 \text{ m}) = 220 \text{ m}^2/\text{s}^2;$$

$$v_i = \pm 15 \text{ m/s}.$$

Note that energy conservation does not determine the sign of v_i. It does not matter whether the brick is thrown upward or downward with that velocity. In fact, when the brick is thrown upward, it reaches the top of its trajectory and then comes down, passing the roof with the same speed v_i with which it was thrown, except that now the velocity is negative instead of positive.

Application to a Spring Force.   An ideal spring provides another example of a conservative force. The displacement from the equilibrium position of a mass m attached to the spring is labeled as x. The force on the mass is a restoring force, pulling the mass back to equilibrium:

$$F(x) = -kx, \tag{7–15}$$

where k is the spring constant. Equation (7–10) takes the form

$$U(x) = U(x_0) - \int_{x_0}^{x} (-kx) \, dx = U(x_0) + \frac{1}{2}kx^2 - \frac{1}{2}kx_0^2.$$

We can now proceed as we did for gravity. We set the arbitrary point x_0 to be zero, and denote $U(0)$ by U_0. We have

$$U(x) = \frac{1}{2}kx^2 + U_0. \tag{7–16}$$

It is customary to take U_0 to be zero, which is equivalent to choosing the zero of potential energy at $x = 0$. With this choice, Eq. (7–16) becomes

$$U(x) = \frac{1}{2}kx^2. \tag{7–17}$$

The total conserved energy is given by the sum of the kinetic and potential energies: $E = (mv^2/2) + (kx^2/2)$.

EXAMPLE 7–2 One end of a massless spring is placed on a flat surface; the other end points upward (Fig. 7–5a). A mass of 1.0 kg is gently set down on top of the spring, until the spring is compressed by 17 cm to a new equilibrium position (Fig. 7–5b). What is the spring constant?

Now, the 1.0-kg mass is removed and a 2.0-kg mass is set on top of the spring. The spring is then compressed by hand so that the end of the spring is 42 cm lower than the position of the spring with no mass on top (Fig. 7–5c). The spring is then suddenly released. What is the maximum kinetic energy of the 2.0-kg mass?

Solution: Use the coordinate y to measure the position of the masses. The top of the unloaded spring defines $y = 0$ (Fig. 7–5a), with $+y$ directed upward. We set both the potential energy of gravity and the potential energy of the spring to be zero at that point. Thus

$$U_g = mgy, \quad U_{\text{spring}} = \frac{1}{2}ky^2,$$

where k is the spring constant.

We find the spring constant by using Newton's first law. As usual, we drop vector notation for this one-dimensional problem. When the 1.0-kg mass, m_1, is resting motionless on the spring, the net force on the mass is zero. The force due to the spring acts upward (in the positive direction), $F_{\text{spring}} = -ky_1$, where the compression is $y_1 = -17$ cm (Fig. 7–5b). Thus

$$F_{\text{net}} = -ky_1 - m_1g = 0.$$

We then solve for the unknown spring constant k:

$$k = -\frac{m_1g}{y_1} = -\frac{(1.0\text{ kg})(9.8\text{ m/s}^2)}{-0.17\text{ m}} = 58\text{ kg/s}^2 = 58\text{ N/m}.$$

We can determine the potential energy of the mass because we know the compressed spring position $y_2 = -0.42$ m (Fig. 7–5c) with the 2.0-kg mass, m_2. To find the mass's speed at any position, we then use the conservation of total energy, E,

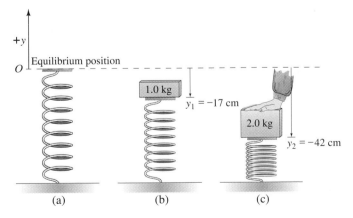

(a) (b) (c)

FIGURE 7–5 (a) Example 7–2. In our coordinate system, the top of the relaxed spring is at $y = 0$. (b) When a 1.0-kg mass is placed gently on top of the spring, the new equilibrium position is $y = -17$ cm. (c) A 2.0-kg mass is attached to the end of the spring, and the mass is then pushed down to $y = -42$ cm. The mass is then released.

which is the sum of the kinetic energy, the spring potential energy, and the gravitational potential energy:

$$E = \frac{1}{2}m_2v^2 + \frac{1}{2}ky^2 + m_2gy.$$

We evaluate E by noting that $v = 0$ m/s at $y = y_2$. Thus

$$E = \frac{1}{2}(2.0 \text{ kg})(0 \text{ m/s})^2 + \frac{1}{2}(58 \text{ N/m})y_2^2 + (2.0 \text{ kg})(9.8 \text{ m/s}^2)y_2$$

$$= (29 \text{ N/m})(-0.42 \text{ m})^2 + (19.6 \text{ kg} \cdot \text{m/s}^2)(-0.42 \text{ m}) = -3.1 \text{ J}.$$

The potential energy has a minimum—and therefore the kinetic energy has a maximum—when the force is zero. This occurs at y_3, where the spring force and the force of gravity cancel; namely, at $-ky_3 - m_2g = 0$, or

$$y_3 = -\frac{m_2g}{k} = -\frac{(2.0 \text{ kg})(9.8 \text{ m/s}^2)}{58 \text{ N/m}} = -0.34 \text{ m}.$$

At this point, where we have seen that the kinetic energy has a maximum, $E = K_{max} + \frac{1}{2}ky_3^2 + m_2gy_3$. Because E is constant, we just set this to the value of E we have already calculated, $E = -3.1$ J, or

$$K_{max} = -\frac{1}{2}ky_3^2 - m_2gy_3 + E$$

$$= -\frac{1}{2}(58 \text{ N/m})(-0.34 \text{ m})^2 - (2.0 \text{ kg})(9.8 \text{ m/s}^2)(-0.34 \text{ m}) - 3.1 \text{ J}$$

$$= 0.21 \text{ J}.$$

The Energy of Systems

Up to this point, we have concentrated on how energy applies to problems involving single objects. The scalar nature of energy makes it easy to apply it to entire systems—as long as those systems involve conservative forces. Let's look at a problem involving a system.

EXAMPLE 7–3 Consider the Atwood machine introduced in Example 5–4 (Fig. 7–6a). The masses have the values $m_1 = 1.37$ kg and $m_2 = 1.51$ kg. The system is released from rest with m_2 84 cm from the floor. Use energy conservation to find the speed of m_2 just before it hits the floor.

Solution: We consider the two masses tied together as a whole. Only gravity acts on this system, and gravity is conservative, so we can use the conservation of energy with the potential energy due to gravity. Figure 7–6a shows the system in its initial state, with no kinetic energy:

$$E_{init} = m_1gh_1 + m_2gh_2$$

The final state of the system is illustrated in Fig. 7–6b. There is movement at this time, so the system contains kinetic energy as well as potential energy. Each mass has the same speed, v, because they are tied together. Thus

$$E_{final} = m_1g(h_1 + h_2) + \frac{1}{2}m_1v_1^2 + m_2g(0) + \frac{1}{2}m_2v_2^2 = m_1g(h_1 + h_2) + \frac{1}{2}(m_1 + m_2)v^2.$$

When we set the final and initial energies equal we find

$$(m_2 - m_1)gh_2 = \frac{1}{2}(m_2 + m_1)v^2.$$

(a)

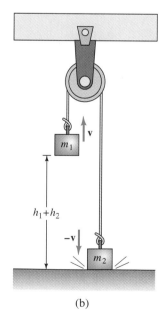

(b)

FIGURE 7–6 (a) Initial conditions for Example 7–3. (b) The situation just before mass m_2 hits the floor.

Solving for v, we find

$$v = \sqrt{\frac{2(m_2 - m_1)gh_2}{(m_2 + m_1)}} = \sqrt{\frac{2(1.51\ \text{kg} - 1.37\ \text{kg})(9.81\ \text{m/s}^2)(0.84\ \text{m})}{(1.51\ \text{kg} + 1.37\ \text{kg})}} = 0.895\ \text{m/s}.$$

APPLICATION

Balancing energy

Conservation of energy is a powerful tool in the study of complicated systems, when accompanied by the realization that energy comes in many forms. It may be possible to account for energy entering or leaving the system by certain routes and at certain rates. For example, if a certain amount of electric lighting is necessary in a building, how does that affect the heating or air conditioning that must be installed? In constructing a refinery, how do the heights at which various pipes must be placed affect the necessary pumping apparatus? The conservation of energy is a bookkeeping tool that allows engineers to design and analyze the behavior of a system without knowing every last detail.

INTERIM SUMMARY

The work done by a conservative force acting on an object as it moves from a point x_0 to a point x_1 is independent of the path taken by the object. For such forces, the work done can be expressed as a function of one end point of the path minus a function of the other. We can then express this work in terms of a scalar function of position: the potential energy—again, *only* for conservative forces. For one-dimensional conservative forces, the potential energy is defined by

$$\text{for conservative forces:} \qquad U(x) = U(x_0) - \int_{x_0}^{x} F(x')\,dx'. \qquad (7\text{–}6)$$

The potential energy difference between two points is thus the negative of the work done between those points. Because we only ever deal with differences of potential energy, it is only defined up to an arbitrary constant, which is expressed here by the fact that both x_0 and the value U_0 of U at the point x_0 are arbitrary. If the potential energy is known, then the (one-dimensional) force is found by taking a derivative:

$$F(x) = -\frac{dU(x)}{dx}. \qquad (7\text{–}4)$$

The work–energy theorem takes a simple useful form for conservative forces. The energy E, defined by

$$E = \frac{1}{2}mv^2 + U(x), \qquad (7\text{–}8)$$

is conserved, meaning it has the same value everywhere along the trajectory of an object that is acted on by that force only. The conservation of energy is a powerful tool for solving problems.

7–2 ENERGY CONSERVATION AND ALLOWED MOTION

Let's take a look at the energy conservation equation for a mass attached to a spring with spring constant k. To do so, we will use an **energy diagram** of the system. For such a diagram, we place the position of the mass (the object whose motion is of interest to us) on the horizontal axis and we place any of the energies—potential energy, $U = kx^2/2$; kinetic energy, $K = mv^2/2$; or total energy, $E = U + K$—on the vertical axis. Suppose that the spring is initially compressed by an amount x_0, then released

The energy diagram

(a)

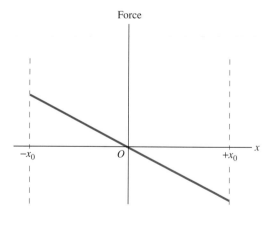

(b)

FIGURE 7–7 A mass on the end of a spring; equilibrium is $x = 0$. (a) At the top, the spring is compressed by an amount x_0. During one period (or cycle), the mass moves through $x = 0$ to an extension $x = x_0$ and back to $x = -x_0$. As the mass moves, the kinetic and potential energies associated with the mass-spring system change: One increases as the other decreases. (b) A plot of the force of a spring on a mass attached to its end, together with an energy diagram for the mass. The potential energy $U = \frac{1}{2}kx^2$ is drawn. The straight line represents the constant total energy, E; the kinetic energy, K, is the difference between E and U. The curve of K is determined purely by the conservation of energy. Points $+x_0$ and $-x_0$ are turning points for the motion, where the kinetic energy and therefore the speed are zero. For x outside the range $-x_0 < x < +x_0$, the kinetic energy would be negative, so the mass cannot go outside this range.

(Fig. 7–7a). Initially, $v = 0$ and $x = -x_0$; at this point, then, $E = kx_0^2/2$. Because E is unchanging, this is the value of E once and for all. Figure 7–7b is an energy diagram for the system, in which the spring at its initial, compressed position corresponds to $-x_0$. Equation (7–4) states that the force is the negative of the slope of the potential energy. The slope of $U(x)$ at $-x_0$ is negative, which means that the force is positive according to Eq. (7–4); the mass thus accelerates to the right, back to the equilibrium position. As it moves to the right past $x = 0$, the potential energy curve falls below the horizontal line that represents the constant total energy, E.

Energy conservation, as expressed in Eq. (7–8), implies that the difference between E and the potential energy curve is equal to the kinetic energy—$K = \frac{1}{2}mv^2$—as seen in Fig. 7–7b. The kinetic energy and the potential energy always add up to the total energy. In other words, if the kinetic energy increases, the potential energy decreases by an equal amount. The kinetic energy increases to a maximum at $x = 0$ as the mass moves to the right. The mass slows down as it continues to the right because the potential energy curve rises and the kinetic energy decreases. Finally, at $+x_0$, $U(x)$ is equal to E again. The mass cannot go any farther to the right; there, $U(x) > E$ and this would require a negative kinetic energy, an impossibility for the quantity $\frac{1}{2}mv^2$. Thus $+x_0$ and (for exactly the same reason) $-x_0$ are *turning points* for the motion. **Turning points** are those points where the speed drops to zero and the mass changes direction. The mass must therefore remain between $-x_0$ and $+x_0$. If the energy were larger, then the horizontal line at E in Fig. 7–7b would just lie higher and the turning points would correspond to points C and D. A turning point is always characterized by the fact that $E = U(x)$ at that point so that $K = \frac{1}{2}mv^2 = 0$.

Let's suppose that an object moves from $-x_0$ to $+x_0$ and then back to $-x_0$. It then restarts its motion from $-x_0$ and there is nothing to distinguish this second traversal from the first. The motion of the mass attached to the end of a spring is repetitive or, to use more technical language, it is **periodic**. Periodic motion appears everywhere in the physical world and is characteristic of the motion near *any* minimum of a potential energy function.

Energy Conservation and Equilibrium

We can use the energy diagram technique to explore the stability of systems. The mass on a spring is a stable system because the spring force always accelerates the mass back toward the equilibrium point. Let's now look at a rather different system—a marble balanced on top of a hill. Physically, we know that once the marble starts rolling, it rolls all the way down the hill. Suppose that, at any point on the hill, the height $y = y_0 - cx^2$, where x is the horizontal distance from the top of the hill, y_0 is the height of the top of the hill, and c is a constant. Because the potential energy of the marble is that of gravity and is therefore proportional to y, the marble has a potential energy superficially like that of the spring—$U = U(y) \propto -x^2$—*where the coefficient of x^2 is negative*. This reflects a force whose magnitude grows with the marble's displacement but whose sign, unlike the spring force, tends to send the marble *away* from $x = 0$. This type of force is relevant for *unstable* situations such as the marble on the top of the hill; its potential has the form

$$U(x) = -\frac{1}{2}\kappa x^2, \tag{7–18}$$

where the new constant κ is positive. The total energy is then $E = (mv^2/2) - (\kappa x^2/2)$.

How do we see the behavior of the marble on the hill in this language? We draw an energy diagram that includes the potential energy of the marble (Fig. 7–8). We observe that if the marble starts from rest at some positive point x_0, then the energy is negative: $v = 0$ there, so $E = -\frac{1}{2}\kappa x_0^2$. (A negative total energy is perfectly acceptable, because we have seen that we can add or subtract any constant amount to the energy

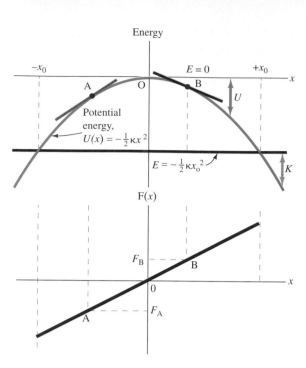

FIGURE 7–8 Energy diagram for the potential energy function $U(x) = -\frac{1}{2}\kappa x^2$, where κ is positive. We have also drawn the force that leads to this potential energy. For the total energy $E = -\frac{1}{2}\kappa x_0^2$, the particle can exist only where $x \geq x_0$ or $x \leq -x_0$; otherwise, the kinetic energy would be negative. For $E = 0$, the position $x = 0$ is an unstable equilibrium point. A slight displacement either way from $x = 0$ will send the particle hurtling away from the equilibrium point.

Stable and unstable equilibrium points

(a)

(b)

FIGURE 7–9 This skateboarding trough has a stable equilibrium point at the lowest height. (a) The forces on the skateboarder tend to move him to the equilibrium point. (b) When the skateboarder moves farther to the right, he will find a rise that will move him back to the equilibrium point.

without affecting the behavior of the marble.) We notice that the slope of the potential energy is negative at $x = x_0$, so the force is positive and the marble accelerates in the direction of increasing x. As it moves to the right, the potential energy becomes more and more negative and thus the kinetic energy (given by the difference between the total and potential energies) becomes larger and larger. We call this *runaway motion.* Once started, the marble accelerates down the hill.

If the marble were to start at the very top of the hill ($x = 0$), there would be no acceleration because the slope of the curve $U(x)$ vanishes there—and the force is the slope of the curve $U(x)$. With an initial velocity of zero, nothing happens. In principle, the marble could stay there forever. Points such as this occur frequently in physical systems and we call them **equilibrium points**. Because the slightest perturbation would send the marble rolling one way or another, $x = 0$ is an **unstable** equilibrium point. Unstable equilibrium points are characterized by peaks, or maxima, in the potential energy. We can contrast this with the potential energy curve $U(x) = kx^2/2$ for the spring (see Fig. 7–7b). There, too, $x = 0$ is an equilibrium point; in this case, however, a slight displacement from $x = 0$ brings a restoring force into play that accelerates the particle *back* toward $x = 0$. A perturbation from this equilibrium point leads to the periodic behavior discussed previously, and not to runaway motion. When the potential energy has a trough, or minimum—as for a mass on a spring, a skateboarder on a ramp, or a marble in a bowl—the equilibrium point is **stable** (Fig. 7–9).

More on Energy Diagrams

To illustrate some new features of motion under the influence of conservative forces, let us return to the potential energy $U(x)$ of Eq. (7–18) but *add* the term qx^4. This potential energy is shown in the energy diagram in Fig. 7–10. The qx^4 term may be negligible for small $|x|$, but as $|x|$ becomes larger, it begins to dominate the negative x^2 term of Eq. (7–18). We analyze the motion of a particle with such a potential energy by first choosing a total energy, this time E_0.

The motion of a particle with energy E_0 that starts from rest at x_0 (point B) is periodic. The motion is described by the kinetic energy $K = E_0 - U(x)$. Starting from zero at x_0, K grows until it reaches a maximum above the right-hand-side minimum ($x = x_1$) in $U(x)$. The kinetic energy then decreases, reaching a minimum at $x = 0$; it

PROBLEM-SOLVING TECHNIQUES

Assuming that we know the potential energy, the energy diagram is a useful tool for analyzing motion in one dimension under the influence of a conservative force. The choice of the zero of the potential energy should already have been made.

1. Motion in regions of the energy diagram where the kinetic energy would be negative are excluded.

2. A single point where the object is known to be at rest (zero kinetic energy) determines the total energy E, which is a constant (by conservation of energy) and can be drawn as such on the diagram. The object is then at rest at other places—turning points—where the flat line of E intersects the curve of potential energy. Two such points, A and B, are shown in Fig. 7B1–1.

3. With both U and E known, the speed at any point—such as x_0 on Fig. 7B1–1—is determined by finding the kinetic energy there, which is the difference $E - U$.

4. The shape of the curve of U versus position carries useful information. If the curve has a minimum, and if the total energy is such that the turning points are on either side of that minimum, then that minimum is a

stable equilibrium point. If the curve has a maximum, and if the total energy is such as to allow the object to start from rest at that point, then the maximum is an unstable equilibrium point.

5. For motion in more than one dimension, the energy diagram is more complicated but can still prove useful. Figure 7B1–2 shows a possible potential energy for an object that can move in two dimensions; this plot looks like a Mexican hat. If the total energy is less than the peak of the hat, then the motion of the object will be restricted to lie within the circle formed by the well of the potential energy curve.

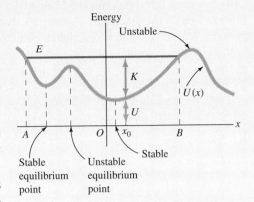

FIGURE 7B1–1 An energy diagram for one-dimensional motion: The solid curve is a plot of potential energy, U, versus position. The total energy, E, is a constant as the position changes, expressing conservation of energy. The kinetic energy, K, is then the difference between E and U. The points A and B represent turning points for the given total energy. Stable and unstable equilibrium points are also shown.

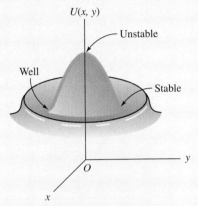

FIGURE 7B1–2 An energy diagram for two-dimensional motion, with a potential energy curve known as the Mexican hat potential. The circular well is a locus of stable equilibrium points for motion toward or away from the origin. The tip of the hat is an unstable equilibrium point.

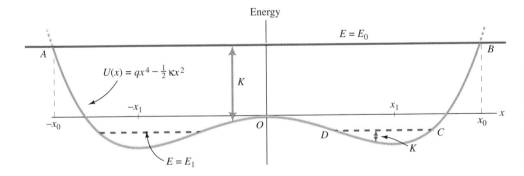

FIGURE 7–10 The term qx^4 is added to the potential energy $U(x)$ of Fig. 7–8. The graph of U shows that two minima of potential energy occur at $x = \pm x_1$. A particle with energy E_1 can move only within either of these pockets.

increases again until the left-hand-side minimum in $U(x)$ is reached ($x = -x_1$), and then decreases to zero at the point $x = -x_0$, where the line $E = E_0$ intersects the potential energy curve. The particle then moves to the right, retracing its motion, speeding up near $x = -x_1$, slowing down near $x = 0$, speeding up as x approaches x_1, and coming to a stop at $x = x_0$; starting from this time, the whole motion is repeated.

If the energy of the particle is E_1 and the particle starts between C and D, the particle moves between the two turning points C and D. The particle never reaches the left side of the vertical axis. We describe the peak between the two troughs as a **potential energy barrier**—so called because it may act to keep a particle in one trough from getting to the other. For an energy value such as E_1, $x = 0$ is inaccessible to the particle. We see that there is a stable equilibrium point located at the bottom of each trough. There, a particle displaced from the bottom of the trough will experience a restoring force that pulls it back toward the bottom.

Our discussion has shown that the potential energy curve $U(x)$ contains a great deal of information about the motion of the particle. Energy diagram techniques are useful even when the equation of motion is a very complicated differential equation.

PROBLEM SOLVING

Uses of the potential energy curve

7 – 3   MOTION IN TWO OR THREE DIMENSIONS

Just as for one dimension, there exists a potential energy that is a function of position for conservative forces acting in two or three dimensions. This potential energy is again defined so that the change in its value in going from one point to another is the negative of the work done by the force between those points. The potential energy is now a function of all three variables, x, y, and z; equivalently, it is a function of the position vector $\mathbf{r}$: $U(x, y, z) = U(\mathbf{r})$. The total energy,

$$E \equiv \frac{1}{2}m(v_x^2 + v_y^2 + v_z^2) + U(x, y, z)$$

$$= \frac{1}{2}mv^2 + U(\mathbf{r}), \tag{7–19}$$

is then independent of time; that is, it is a constant of the motion. The value of the constant E is fixed by the initial value of U (at the starting point) and by the initial velocity. The potential energy function $U(\mathbf{r})$ determines the force in a manner analogous to Eq. (7–4), which gives $F(x) = -dU/dx$ for a force acting in one dimension. In particular, the force may have two (or three) components in two- (or three-) dimensional space but, *for conservative forces, all components of the force—the complete force vector—can be found from a single potential energy function $U(x, y, z)$.* The potential energy is certainly a more economical way for us to describe the force than is a listing of three components.

The conservation of energy holds for conservative forces in any number of dimensions.

Potential Energy for Projectile Motion

The potential energy due to gravity for projectile motion

In Section 7–1, we derived and applied the expression $U = mgy + U_0$ [Eq. (7–11)] for up-and-down motion. *This expression is valid also for arbitrary projectile motion under the influence of gravity.* We studied the motion of a mass on a ramp in Section 6–4 to show that the force of gravity is conservative in that the work does not depend on the path taken—even when that path contains horizontal components. Thus there is a potential energy. To find that potential energy, we must evaluate the work done when gravity acts on an object. We have seen that this work, and hence the potential energy change, depends only on the change in height of the object; it does not depend on the object's horizontal motion. For example, $W = mgL \sin \theta$ [Eq. (6–22)] is the work done by gravity in moving a mass a total length L along a ramp that makes an angle θ with the horizontal. But the quantity $L \sin \theta$ is the height through which the mass drops (see Fig. 6–22), h. The work done is thus mgh, which is just the expression that holds when we consider vertical motion only. We conclude that *the potential energy function $U(y) = mgy + U_0$ applies to general projectile motion under the influence of gravity.*

EXAMPLE 7–4 An inexperienced golfer hits a bad shot and the ball leaves the tee with an initial speed of 28 m/s at an angle of 84° with respect to the horizontal. A bee is cruising innocently at a height of 37 m when it has the bad luck to meet the golf ball. What is the speed of the ball when it hits the bee? Ignore all effects of air resistance.

Solution: We use the conservation of energy to find the speed of the golf ball at any height given an initial speed v_i of 0 m/s. We let the potential energy function be zero at ground level height; that is $h = 0$ m, so $U_0 = 0$ and $U(h) = mgh$. The golf ball has an initial energy of $E_i = K_i + U(0) = mv_i^2/2 + mg(0) = mv_i^2/2$, where m is the mass of the ball and v_i is the initial speed. The energy of the ball when it meets the bee at height h is $E_f = K_f + U(h) = mv_f^2/2 + mgh$, where v_f is the desired information. Conservation of energy allows us to equate these two energies:

$$E_i = \frac{mv_i^2}{2} = E_f = \frac{mv_f^2}{2} + mgh.$$

Cancel the factor m and solve for v_f^2:

$$v_f^2 = v_i^2 - 2gh = (28 \text{ m/s})^2 - 2(9.8 \text{ m/s}^2)(37 \text{ m}) = 58.8 \text{ m}^2/\text{s}^2;$$

$$v_f = 7.67 \text{ m/s}.$$

The golf ball is fairly close to the top of its trajectory and has slowed down considerably. Note that we do not need to use the detailed information about the initial angle of the ball nor calculate the angle at which the ball is moving when it hits the bee. That is part of the utility of working with the conservation of energy.

Central Forces

There is a most interesting case that occurs frequently in nature: a force with a magnitude that depends only on the radial distance from a fixed point and has a direction aligned along the corresponding radius vector. For example, the force of gravitational attraction on Earth due to the Sun depends on the distance between Earth and the Sun and is aligned along the line between the two objects. Such forces are called **central forces** and *all central forces are conservative.*[†] Such a force takes the form

$$\mathbf{F}(\mathbf{r}) = F(r)\left(\frac{\mathbf{r}}{r}\right) = F(r)\hat{\mathbf{r}}, \tag{7–20}$$

where $\hat{\mathbf{r}}$ is a unit vector pointing away from the origin. This force is reminiscent of a one-dimensional force. The potential energy function of a central force depends only on the radial distance r; the expression for the force in terms of a potential energy function $U(r)$ is

$$\mathbf{F} = -\frac{dU(r)}{dr}\hat{\mathbf{r}}. \tag{7–21}$$

Example 7–5 illustrates how we can apply Eq. (7–21).

EXAMPLE 7–5 The potential energy describing the gravitational interaction (the *gravitational potential energy*) between two point masses m_1 and m_2 is given by the expression

$$U(r) = -\frac{Gm_1m_2}{r},$$

PROBLEM SOLVING

Central forces are conservative.

A central force is expressed in terms of a potential energy that depends only on the radial distance.

[†]More precisely, a conservative central force must depend *only* on r.

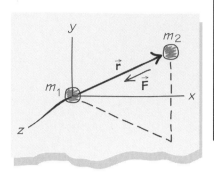

FIGURE 7–11 Example 7–5. The vector **r** pointing outward radially from the origin (the location of mass m_1) is the position vector of mass m_2. The force vector on m_2 acts inward radially.

where G is a constant and r is the distance between the two point masses. Calculate the force experienced by mass m_2 due to the presence of m_1. Is the force attractive or repulsive?

Solution: We define the vector **r** to be in the direction from m_1 (taken to be located at the origin) to m_2 (Fig. 7–11). With the help of Eq. (7–21) and $(d/dr)(1/r) = -1/r^2$, we find

$$\mathbf{F} = -\frac{Gm_1 m_2}{r^2}\,\hat{\mathbf{r}}.$$

The force is proportional to $1/r^2$ (an inverse-square form), and it is directed opposite to the position vector **r** from m_1 to m_2. It therefore acts to pull m_2 back toward m_1 and is an *attractive* force.

*A General Form

If the force is not a central force, it will still be a conservative force if the work it does is path independent. There is, then, a potential energy function that depends on x, y, and z separately. The rule that takes us from the potential energy function to the force itself is a simple generalization of Eq. (7–4) or (7–21). To express this rule, we use **partial derivatives**. If we have a function $U(x, y)$ of two variables, we define partial derivatives—denoted by $\partial U/\partial x$ and $\partial U/\partial y$—as follows: The partial derivative $\partial U/\partial x$ is obtained by differentiating $U(x, y)$ with respect to x *while treating y as if it were a constant*; the partial derivative $\partial U/\partial y$ is obtained by differentiating $U(x, y)$ with respect to y *while treating x as if it were a constant*. Given this simple definition, the force **F** has components given by

$$F_x = -\frac{\partial U}{\partial x}, \; F_y = -\frac{\partial U}{\partial y}, \text{ and } F_z = -\frac{\partial U}{\partial z}. \tag{7–22}$$

In particular, if the potential energy depends on, say, the variable y alone, then Eq. (7–22) shows that the force has a component only in the y-direction. The force of gravity provides a good example: In this case, U depends on y (the height) alone, and **F** has only a y-component.

7–4 IS ENERGY CONSERVATION A GENERAL PRINCIPLE?

At first glance, it appears that most systems involve nonconservative forces because forces that dissipate energy, such as friction or drag, are almost everywhere. In the presence of friction, energy does not appear to be conserved. Actually, the energy that is lost in friction can be identified through experiment as **thermal energy**; that is, the type of energy that raises the temperature of two objects when they are rubbed together. When thermal energy is measured and included as still another component of the total energy, *the total energy is then found to be conserved.* Thermal energy is not a form of potential energy because it is not completely retrievable. A system that contains thermal energy therefore differs from a system for which kinetic and potential energies can be freely converted back and forth. The ways in which mechanical and thermal energy can be converted back and forth are important, and we shall devote a great deal of effort to this subject when we study thermodynamics.

The conservation of energy is a general principle.

PROBLEM SOLVING

Conservation of energy and nonconservative forces

The work–energy theorem when both conservative and nonconservative forces are present

In fact, when we look closely enough, *all the fundamental forces in nature are conservative*, and conservation of energy always holds. On the atomic level, there is no such thing as friction. *The conservation of energy is one of the most fundamental principles of physics.* It has received experimental support from all fields of physics. We shall encounter energy conservation in a variety of applications, including thermodynamics, electromagnetism, relativity, and quantum physics (Fig. 7–12).

Energy Conservation and Nonconservative Forces

While it is important to realize that all fundamental forces are conservative, we do not often operate at the level of the fundamental forces. How then do we deal with the non-fundamental and often nonconservative forces, such as friction, that appear in every-day situations?

Suppose both conservative and nonconservative forces act on an object. A skier descending a mountain moves under the influence of both gravity and friction. In an example such as this, we can first construct the potential energy as if there were no friction. We then ask how much energy is lost due to friction; the loss of energy is attributable to any work done by friction. All this is possible because we can divide the work as it appears in the work–energy theorem into a part that can be associated with a potential energy (the work of the conservative forces) and another part that cannot. Thus, if we write W_{nc} as the work done by the net nonconservative force and U as the potential energy of the net conservative force, the work–energy theorem becomes

$$W_{nc} = \Delta E = \Delta(K + U). \tag{7–23}$$

Here, we have continued to think of E as $K + U$, but E is no longer conserved.

Let's consider our skier. If friction is ignored, we have $E = K + mgy =$ constant, where $K = mv^2/2$ is the kinetic energy at a height y above some zero level. When friction is introduced, the initial and final energies, E_i and E_f respectively, differ, with $E_f < E_i$. The deficit is due to the work done by friction. We have

$$W_{friction} = E_f - E_i = (K_f + mgy_f) - (K_i + mgy_i). \tag{7–24}$$

The work done by friction is negative, so that the final energy is less than the initial energy.

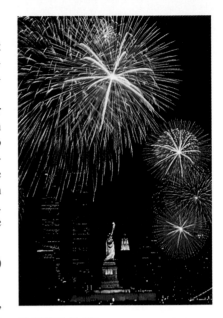

FIGURE 7–12 A fireworks display represents the conversion of chemical energy into the kinetic energy of the shooting components and the energy of the excited atoms, providing electromagnetic energy—light.

 APPLICATION: CONVERSION OF ENERGY IN ELECTRICAL POWER GENERATION

The conversion of one form of energy into another finds an application in the electrical power industry; it is necessary for economic reasons to smooth out the load on electrical generating plants. One way in which this is done is to build an upper and a lower water storage reservoir, separated by a reasonable vertical distance and connected with a hydro-electric generating plant (Fig. 7AB–1). The plant operates by converting the mechanical energy of falling water to electrical energy; this is done through fundamental laws of electricity and magnetism. During periods when there is a low requirement for electrical energy, the excess electrical energy generated by the plant can be used in reverse, driving motors that pump water from the lower to the upper reservoir. This work done on the water by the motors gives it potential energy. During periods of high demand for electrical energy, the water is allowed to flow down from the upper reservoir through the electrical generation apparatus to the lower reservoir. This transfers the water's potential energy back again to electrical energy. What is important about this process is that it is easily and entirely reversible on a very large scale, in contrast to the conversion of chemical energy to electrical energy in the burning of fossil fuel. Such a scheme has been implemented in several places, most notably in Luxembourg.

FIGURE 7AB–1 The Bath County pumped storage power generation system in Virginia illustrates how the energy associated with lifting water can later be re-used in the form of electric energy. Water is pumped into the upper reservoir during times of low demand for electricity and is released during times of high demand to generate hydroelectric power.

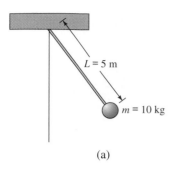

$L = 5$ m

$m = 10$ kg

(a)

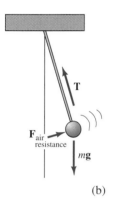

T

F_{air}
resistance

$m\mathbf{g}$

(b)

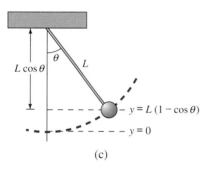

$L \cos \theta$ θ L

$y = L(1 - \cos \theta)$

$y = 0$

(c)

FIGURE 7-13 (a) Example 7-6.
(b) Force diagram for the mass.
(c) Geometry for the determination of
the potential energy.

EXAMPLE 7-6   A ball of mass 10 kg is attached to a 5-m-long wire that swings freely from a support. In other words, we have a *simple pendulum* (Fig 7-13a). The ball is pulled aside so that the wire makes an angle of 31° to the vertical and it is released from rest. After 10 swings, the maximum angle that the ball reaches is 25° from the vertical. What is the work done by air resistance and any other nonconservative forces acting on the ball during these 10 swings?

Solution: As Fig. 7-13b shows, the forces acting are gravity, the tension of the wire, and the air resistance, which is nonconservative. Tension acts in a direction perpendicular to the motion, so it does no work and plays no role in any energy considerations. Gravity is conservative and so it has a potential energy. We will first express conservation of energy as if air resistance were not present. If we take the zero of potential energy at the minimum point of the swing, and measure the height y from that level, then $U(y) = mgy$. In terms of the angle θ of the wire, we see from Fig. 7-13c that

$$y = L(1 - \cos \theta),$$

where L is the wire's length. There is no kinetic energy at the top of the swing, so the total energy is

$$E = U(y_{max}) = mgy_{max} = mgL(1 - \cos \theta_{max}).$$

The *loss* of energy over the ten swings is then

$$E_i - E_f = mgL(1 - \cos 31°) - mgL(1 - \cos 25°) = mgL(\cos 25° - \cos 31°)$$

$$= (10 \text{ kg})(9.8 \text{ m/s}^2)(5 \text{ m})(\cos 25° - \cos 31°) = 24 \text{ J}.$$

This is the work done by air resistance.

In this discussion and example, we used the energy loss to find the work done by nonconservative forces. But, of course, we can use this approach in a different way. If we can calculate the work done by friction—we can do this if we know its magnitude and direction and the distance over which it acts—then we can *predict* the energy loss.

Energy Conservation as One of the Underpinnings of Physics

In our discussion of energy conservation, we saw that principle as a powerful shortcut to information that could be obtained only laboriously from the equations of motion. It is important to understand that the conservation of energy is entirely equivalent to Newton's laws where ordinary mechanical systems are concerned. Indeed, energy conservation is now generally viewed as a more fundamental formulation of dynamics than is Newton's second law, even though they are equivalent for mechanical systems.

To demonstrate the equivalence, let's concentrate for simplicity on motion in one dimension. We write

$$E = \frac{1}{2}mv^2 + U(x). \tag{7-25}$$

E is a constant and both x and v are functions of time; they are the position and speed of the object in motion. Because energy is conserved, we have

$$\frac{dE}{dt} = 0. \tag{7-26}$$

We apply this to Eq. (7-25), using the chain rule when needed:

$$\frac{dE}{dt} = \frac{1}{2}m\frac{dv^2}{dt} + \frac{dU(x)}{dt} = \frac{1}{2}m\frac{dv^2}{dv}\frac{dv}{dt} + \frac{dU(x)}{dx}\frac{dx}{dt}$$

$$= \frac{1}{2}m(2v)\frac{dv}{dt} + \frac{dU(x)}{dx}v = v\left[m\frac{dv}{dt} + \frac{dU(x)}{dx}\right] = 0.$$

Unless $v = 0$ at all times, the contents of the square bracket must be zero; that is,

$$m\frac{dv}{dt} = -\frac{dU(x)}{dx} = F(x). \qquad (7\text{–}27)$$

This is just our original formulation of Newton's second law.

It is interesting to note that although physicists and mathematicians studied Newton's equations almost from the time of their discovery, the principle of energy conservation—formulated by Julius Robert Mayer (1842), Hermann von Helmholtz (1847), and James Prescott Joule (1850)—grew out of experimental research on the properties of heat and the conversion of mechanical energy into thermal energy; this principle did not develop from Newtonian mechanics.

*7–5 BARRIER TUNNELING IN QUANTUM PHYSICS

Our energy-diagram approach has shown that if the total energy of a particle is less than the potential energy in some region of space, then the particle cannot cross through that region. We must modify this statement for atoms and molecules.

For extremely small systems (in the range of 10^{-9} m or less), the description of motion by Newton's second law is no longer adequate. A proper treatment of such systems is provided by *quantum mechanics*, which predicts certain phenomena that are impossible according to Newton's laws. One such phenomenon is **barrier tunneling**, or **barrier penetration**. Let's consider a particle whose potential energy is of the form given in Fig. 7–14. Newtonian mechanics states that a particle with energy E_1 in the right-hand trough can never appear in the left-hand trough because there is a potential energy barrier between the two regions (see Section 7–1). Inside the region between points D and D' in the figure—inside the barrier—a particle with energy E_1 must have negative kinetic energy because $E_1 < U(x)$ there. The mass of a particle is positive and therefore $mv^2/2$ must be positive; thus, the motion of a particle initially in the right-side well is forever restricted to the region CD. Not so, according to quantum theory. It is possible for a particle of atomic dimensions to **tunnel** through, or cross, the barrier and thus traverse the forbidden region.

Quantum mechanics predicts quantitatively the probability of a particle crossing a barrier. This rate is proportional to a factor e^{-S}, where S depends on both the width of the barrier and the difference between the barrier height (on an energy diagram) and the energy of the tunneling particle. The greater the height and the greater the width, the larger the value of S and the less likely the particle is to cross the barrier.

The factor S also contains a controlling parameter for this process. This parameter is a fundamental physical constant determined by Max Planck that is, appropriately, known as **Planck's constant**, h, whose value is $h = 6.6 \times 10^{-34}$ J·s. This number is a small one; when it is used in the evaluation of S, it dictates that tunneling is a phenomenon that becomes important only in systems of atomic size. For example, the typical height of a barrier encountered in atomic systems is on the order of 10^{-19} J, and the typical width is 10^{-10} m. For an electron encountering a barrier of this type, S is typically on the order of unity (1). The factor e^{-S} is also on the order of unity, and the electron is likely to tunnel through the barrier. Compare this case with the case of a marble rolling with a speed of 10 cm/s and meeting a bump in the road that is 1 cm high and 1 mm wide. For this case, S takes a value on the order of 10^{28}. The factor e^{-S} is so small that we would never see the marble tunnel through the barrier. Thus quantum physics does not violate our everyday experience: Macroscopic particles do not tunnel through regions that are expected to be impassable.

Tunneling in Nuclear Fusion

Barrier penetration manifests itself in a number of phenomena at the atomic and nuclear levels. Nuclear fusion is one important example. Hydrogen comes in several forms:

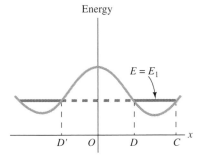

FIGURE 7–14 A potential energy barrier prevents a particle on the right with energy E_1 from entering the region on the left.

APPLICATION

Nuclear fusion

APPLICATION

Electronic devices based on
tunneling

FIGURE 7–15 Individual atoms in a solar cell, shown here in false color, are made visible by using a scanning tunneling microscope. This image is part of a project to improve solar cells.

There is the familiar hydrogen nucleus, ^1H, which is really a proton; there is also a heavy form of hydrogen, ^2H (deuterium), whose nucleus is called the deuteron, which is a proton and a neutron bound together by nuclear forces. There is another, still heavier, form of hydrogen, ^3H (tritium), whose nucleus is called the triton, which is one proton and two neutrons bound together by nuclear forces. All of these nuclei are electrically charged because the proton is electrically charged. If it were possible to bring these particles together, then a nuclear reaction could occur.

One example of such a reaction is a so-called fusion reaction, $^2H + ^2H \rightarrow ^3H + ^1H + (6.4 \times 10^{-13} \text{ J})$. The energy on the right-hand side is energy released during the reaction—energy that can be converted into electric energy, for example.

The reaction is held back because the two nuclei ^2H each carry a positive charge and, as we shall learn in Chapter 22, two positive charges repel each other. This means that when one ^2H nucleus approaches the other, there is a potential barrier that should totally prohibit contact between these nuclei. Quantum physics allows barrier penetration but, under ordinary circumstances, $U_{av} - E$ is so large (remember, the classical barrier is present when $U - E$ is positive) that the reaction cannot happen. The only way to improve the rate or, equivalently, to reduce the number S, is to make $U_{av} - E$ smaller. When the ^2H nuclei are raised to high temperatures (billions of degrees), the nuclei become agitated, move rapidly, have an increased energy E, and thus they have a smaller value of $U_{av} - E$. Effectively, by raising the energy of the nuclei, they are brought closer to the top of the barrier and both $U_{av} - E$ *and* the thickness of the barrier are reduced. Much research and development has been done since the early 1950s to bring the ^2H nuclei to these high temperatures. The technical difficulties are enormous but the reward for success would be great. The fusion reaction could yield a virtually unlimited amount of energy because seawater contains ^2H nuclei in abundance.

Other Applications

Several other manifestations of tunneling have technological application. The scanning tunneling microscope is a device that detects electrons that tunnel through a potential barrier at the surface of a piece of metal. It is used to study variations of atomic size in the surface of the metal in great detail (Fig. 7–15). Tunneling has also been used in studies of superconductivity. One of the most important applications is in the tunnel diode, a device with wide application in circuits. Quantum tunneling is not just some esoteric effect of interest to a small class of physicists. Many modern electronic devices rely in one way or another on electron tunneling. In later chapters, we will return to these applications.

SUMMARY

Conservative forces can be specified in terms of a single scalar function called the potential energy. In one-dimensional motion, the potential energy $U(x)$ depends on one coordinate only and is defined by

$$\text{for conservative forces:} \quad U(x) \equiv U(x_0) - \int_{x_0}^{x} F \, dx', \qquad (7\text{--}6)$$

where $U(x_0)$ $(= U_0)$ is an arbitrary constant and x_0 is an arbitrary point. In turn, the force is given in terms of U by

$$F(x) = -\frac{dU(x)}{dx}. \qquad (7\text{--}4)$$

Two important examples of potential energy are gravity and the spring. For gravity,

$$U(y) = mgy + U_0, \qquad (7\text{--}11)$$

where y is the height above ground level. For the spring,

$$U(x) = \frac{1}{2}kx^2, \qquad (7\text{--}17)$$

where x is the distance from the equilibrium point of the spring.

For central forces (forces aligned in the radial direction), the potential energy depends only on r, and

$$\mathbf{F} = -\frac{dU(r)}{dr}\,\hat{\mathbf{r}}. \qquad (7\text{--}21)$$

For conservative systems, the total energy—the sum of the kinetic and potential energies—does not change during the motion; thus, the total energy is a constant whose value may be determined at any time during the motion:

$$E = \frac{1}{2}mv^2 + U(x, y, z). \qquad (7\text{--}19)$$

The fact that E is constant means that any change in the kinetic energy is compensated by an equal but opposite change in the potential energy during the motion. *The total energy is conserved.* The utility of this conservation law is enormous: We can determine the speed when the position of an object is known. Energy diagrams provide us with a systematic way to approach problems that involve energy conservation. With them, we can understand such useful concepts as turning points, potential energy barriers, and stable and unstable equilibria.

Although dissipative forces are always present in the world as we see it, the law of conservation of energy holds for all the basic laws of physics. Its general validity allows us to find the energy dissipated through friction or drag forces, often without being able to calculate this energy loss directly.

Energy conservation is also a characteristic of atomic systems that must be described by quantum mechanics. There, under certain very specific circumstances, particles can act as if they have negative kinetic energy and can thereby tunnel through potential barriers. While this phenomenon may violate our intuition, it nevertheless occurs, and it conserves energy.

UNDERSTANDING KEY CONCEPTS

1. Are drag forces such as air resistance conservative?

2. Discuss why it is possible for the total mechanical energy to be negative, even if the kinetic energy cannot be negative.

3. If we add a constant term to the potential energy of an object, why doesn't this change the object's motion?

4. Explain why a rubber ball seems never to bounce back to its original height when dropped from rest. What happens to the energy? Explain why the ball can bounce back to a much greater height when it is thrown down.

5. Is energy conserved when you ingest sugar and then go out and exercise? In what ways, and in which of the various processes for converting food into muscular activity, might the energy conservation principle come into play?

6. Does a real spring exert a truly conservative force? What experiment might you perform to check the answer to this question?

7. How, if at all, does experiment demonstrate that the gravitational force is conservative?

8. A golf ball is dropped with zero initial velocity from a height of 20 m. It bounces on a concrete pad and rises to a maximum height of 19 m. Are any of the forces that the ball experiences during the motion nonconservative?

9. Consider motion with a potential energy of the shape given in Fig. 7–7 (see page 180). Let $U(x)$ describe the height of a smooth slide above the ground. In the absence of friction the motion is easily predicted to repeat with a uniform repetition time. Discuss the motion, starting from the right or the left side, when a small amount of friction is present.

10. When a force acts in two dimensions, are there two separate laws for the conservation of energy that correspond to motion in each direction?

11. The potential energy is zero at a given point. Is the force necessarily zero or nonzero at that point?

12. An object moves on a rough inclined plane. Can you still use the concept of energy conservation to relate the height and the speed? How?

13. The motion of the pendulum of a grandfather clock is slowed by air resistance and bearing friction, yet such clocks can run without stopping for years. What types of mechanisms are typically used to supply the necessary energy?

14. You are watching fireworks. You see a glowing point streak into the air. This is followed by a loud noise and, about a second later, by the appearance of many scattered dots of light, which sink in graceful curves before disappearing. Account for all of the energy supplied and spent at each stage of the spectacle.

15. A steel marble is bounced off the top step of a staircase. It rises to 1 m above the step, then falls onto the second step, bounces off, falls onto the third step, and so on for 39 steps. Assuming that there is no air resistance or other source of energy loss, how high will the marble rise after the last bounce? (This is a question, so do not do 39 calculations!)

PROBLEMS

7–1 Energy Conservation

1. (I) An unknown constant force F pushes a 10-kg body from rest on the ground vertically upward. At a height of 2 m, the velocity of the object is $\mathbf{v} = 2.4\mathbf{j}$ m/s. (a) Find the change in the potential energy associated with gravity. (b) What is the net work done, and what is the work done by the unknown force?

2. (I) Consider the expression $U = mgy$ for an object of mass m under the influence of gravity, where y is measured from the ground up. Express U in terms of z, where z is measured in a downward direction from a rooftop 30 m above the ground.

3. (I) A baseball pitcher throws a ball at 95 mi/h off a roof that is 80 m above the ground. How high will the ball be when it is traveling at 120 mi/h?

4. (I) A rock is thrown straight down into a deserted quarry from the edge, which is 18 m above the bottom. The rock has a speed of 21 m/s when it reaches the bottom. What was the rock's initial speed?

5. (I) The spring of a toy gun launches rubber-tipped projectiles with a spring constant of 5 N/m. The spring is compressed by 7 cm with the projectile in place. How much kinetic energy is imparted to the projectile?

6. (II) The potential energy of an archery bow is measured to be $U(x) = bx^2 + cx^3$, where x is the distance the bow string is pulled back from its equilibrium position. When an archer pulls the string a distance x, what force does the archer exert on the string and what force does the string exert on the archer?

7. (II) A package of mass 5 kg is subject to a constant force of 8 N pointing in the $+x$-direction. (a) Calculate the potential energy of the package as a function of its position x, defining it such that $U(x)$ at $x = 0$ m is zero. (b) Assuming that the package has a velocity of 2 m/s at $x = -1$ m, calculate the total mechanical energy of the package. (c) What is the speed of the package at $x = 3$ m?

8. (II) A spring has a spring constant of 42.38 N/m and obeys Hooke's law. How far must the spring be pulled back if its potential energy is to be 1.565 J? What is the mass of a ball at the end of the spring if the maximum speed of the ball is observed to be 2.402 m/s when the spring is released?

9. (II) The energy of a harmonic oscillator (a mass moving on the end of a spring) is given by $E = \frac{1}{2}mv^2 + \frac{1}{2}kx^2$. Plot contours of constant E on a graph in which x is measured along one axis and v is measured along the perpendicular axis. Choose the parameters $E = 16.0$ J, $m = 2.0$ kg, and $k = 8.0$ J/m^2. Such a plot is called a *phase plot*; the motion of a system is restricted to the curve corresponding to the energy E.

10. (II) Sketch the potential energy for a 5-kg mass that can move between the ground and a height of 10 m. (a) Assume that zero potential energy is at the ground, (b) at a height of 10 m, (c) at 4 m.

11. (II) A block of mass 0.528 kg slides with uniform velocity of 3.85 m/s on a horizontal frictionless surface. At some point, it strikes a horizontal spring in equilibrium. If the spring constant is $k = 26.7$ N/m, by how much will the spring be compressed by the time the block comes to rest? What is the amount of compression if the surface is rough under the spring, with coefficient of kinetic friction $\mu_k = 0.411$?

12. (II) An archery bow acts much like a spring displaced from equilibrium when the bow is drawn. Suppose that an archer displaces the string from equilibrium by 60 cm and exerts a force of 150 N. (a) What is the "spring constant"? (b) What is the speed of an arrow of mass 25 g that leaves the bow as the string reaches the equilibrium position? [*Hint*: Use the expression for energy for a mass moving under the influence of a spring.]

13. (II) A cannonball of mass 15 kg is dropped from rest from a height of 6.0 m. It falls onto a large vertically oriented spring, which is compressed from its relaxed position when the cannonball lands on it. The spring has spring constant $k = 10^4$ N/m. What is the maximum compression of the spring? How much would the spring be compressed if a man of mass 60 kg jumped onto it from a height of 1.5 m?

14. (II) A mass $m = 0.3$ kg slides along the x-axis of a horizontal frictionless surface with speed $v_x = 1.7$ m/s. It runs into a relaxed spring oriented along the x-axis. This spring had previously been observed to stretch by 0.8 cm when it was oriented vertically with the mass suspended from it. (a) What is the maximum compression of the spring when the mass runs into it? (b) The mass rebounds as a result of having compressed the spring. What is its velocity when it leaves contact with the spring?

15. (II) A spring with spring constant $k = 200$ N/m is used as a launcher for a small block whose mass is 10 g. The block is placed against the compressed spring in a horizontal arrangement on a smooth horizontal surface. The spring, with the block, is compressed 5 cm and then released. (a) Find the speed of the block just as it leaves the spring. (b) The block encounters a rough surface as it leaves the spring. How much work does friction do in bringing the block to an eventual stop? (c) The block slides a distance of 3.5 m before stopping. What is the coefficient of kinetic friction between block and surface?

7–2 Energy Conservation and Allowed Motion

16. (II) For a conservative one-dimensional force, show that the sign of the slope of the potential energy function at a position x determines the direction, positive or negative, in which the force acts.

17. (II) When two atoms on a line are far apart, there is no force between them. As they start to move closer, there is an attraction between them, which, at very close distances, turns into a strongly repulsive force. Sketch the potential energy as a function of the distance between the atoms.

18. (II) Figure 7–16 shows the force $F(x)$ that acts on a particle moving along the x-axis. (a) Plot the potential energy of the particle as a function of x. (b) The particle starts its motion at $x = -2$ m, with zero initial velocity. How far to the right will the particle travel?

FIGURE 7–16 Problem 18.

19. (II) Consider a force $\mathbf{F}(x)$, acting along the x-axis, that is opposite to the spring force (Hooke's law). In other words, the force has the single vector component $F(x) = +kx$. A mass m is under the influence of this force and of no other forces. Suppose that the mass is placed at rest just to the right of the origin. (a) Which way will the mass move, if at all? (b) Find the speed of the mass as a function of its distance from the origin. (c) Repeat the problem but assume that the mass had been placed just to the left of the origin.

FIGURE 7–17 Problem 20.

20. (II) Consider the energy diagram in Fig. 7–17. (a) What are the limits of motion for energies E_1 and E_2? Redraw the figure and label it as necessary. (b) Describe the circumstances under which the particle is always at rest. (c) Find the energies and positions for which motion within turning points is possible. (d) Find the equilibrium positions on your drawing. Are they stable or unstable?

21. (II) Draw a one-dimensional potential energy [$U(x)$] diagram with the following characteristics: (a) The particle can never reach negative x. (b) There are three regions in x where the particle can move within turning points. (c) The particle can never reach infinity. (d) The particle has unstable equilibrium positions at 1 nm and 2 nm.

22. (II) Consider the potential energy $U(x)$ shown in Fig. 7–18. (a) What is the sign of the force at the positions 1 through 6? (b) Which positions have the most positive, most negative, and zero force? (c) Find the equilibrium positions and indicate whether they are stable or unstable.

FIGURE 7–18 Problem 22.

23. (II) The potential energy of an object constrained to move in the x-direction is given by $U(x) = \alpha x^4 + \beta x^2$, where $\alpha = 26$ J/m^4 and $\beta = -3.0$ J/m^2. Find the equilibrium points and state whether they are stable or unstable.

24. (II) A particle is moving in a potential well described by the potential energy

$$U(x) = -\frac{3\,\text{J}\cdot\text{m}^2}{1\,\text{m}^2 + (x + 2\,\text{m})^2} - \frac{3\,\text{J}\cdot\text{m}^2}{1\,\text{m}^2 + (x - 2\,\text{m})^2},$$

where x is measured in meters. (a) Sketch the shape of the potential energy for -4 m $< x < 4$ m. (b) The speed of a particle of mass 0.20 kg at $x = -2$ m is 4.2 m/s. Can the particle reach $x = 3.0$ m?

7–3 Motion in Two or Three Dimensions

25. (I) A projectile fired from a gun leaves the barrel at a speed of 500 m/s. The gun is placed 180 m above a level plain. Use energy conservation to calculate the speed of the projectile when it is 16 m above the plain. Neglect all drag effects.

26. (I) What is the gravitational potential energy of the Earth–Moon system? See the appendices for the data you need.

27. (I) A cannonball is fired horizontally with an initial speed of 125 m/s from the top of a cliff that is 68 m above the sea. What is the speed of the cannonball when it hits the water? How is that changed if the cannon is inclined at a 32° angle with the horizontal without any change in initial speed? Ignore all effects of air resistance.

28. (I) A particle has potential energy that depends only on the distance r from some central point. This potential energy has the form $U(r) = U_0 - k/r^2$. What is the corresponding force law?

29. (II) A ball is thrown with initial speed v_0 in a trajectory that makes an initial angle θ with the ground. Air resistance is small. (a) By using the principle of conservation of energy, show that the speed of the ball when it reaches the height h above the

ground is $v_h = \sqrt{v_0^2 - 2gh}$, independent of the angle of the throw. (b) Use the result of part (a) to find the initial speed v_0 that is required if a vertically thrown ball is to reach height h just before turning back. (c) Use the results of part (a) to find the initial speed v_0 required so that a ball thrown at $45°$ reaches a maximum height h.

30. (II) An object of mass m is subject to two forces: One force acts only in the x-direction and is due to a spring of spring constant k; the other force acts only in the y-direction and is due to a spring with the same spring constant k. (a) What is the potential energy? (b) Use your result to show that the net force is proportional to the distance from the equilibrium point and points toward that point. (c) Using your result from part (b), show that one possible motion is uniform circular motion in the xy-plane.

31. (II) A ball of mass m is thrown with a speed v, at an angle θ from the horizontal, off the top of a tall building at height h. (a) Show that, in general, it is not possible to tell whether the angle θ is above or below the horizontal when a ball hits the ground just by a measurement of the speed of the ball when it hits the ground. (b) By knowing the horizontal distance that the ball travels, show that the angle θ can be determined.

32. (II) A skier slides down a hill, starting with zero velocity at a height of 95 m above the bottom of the hill. The shape of the terrain is shown in Fig. 7–19. What is the velocity of the skier on top of the second, intermediate hill, whose height is 32 m? What is the skier's velocity at the bottom of the hill? Neglect all frictional effects. Is this neglect reasonable?

FIGURE 7–19 Problem 32.

33. (II) Suppose that the skier of Problem 32 reaches the bottom of the hill at a speed of 23 m/s. Assuming that the skier, including equipment, has a mass of 75 kg, how much work is done by the resistive forces of friction and drag?

34. (II) A ski jumper starts from rest and follows, with its several ups and downs, the rather bumpy ski jump shown in Fig. 7–20. The jump track starts at a height $h = 20$ m above the eventual landing point, and the jumper leaves the track while he moves horizontally at a height above the landing point that is exactly one-half h. Assuming that the effects of air resistance are negligible (in real ski jumping they are not), what is the horizontal distance of the edge of the track from the landing point?

FIGURE 7–20 Problem 34.

35. (II) A particle of mass $m = 30$ g slides inside a bowl whose cross section has circular arcs at each side and a flat horizontal central portion between points a and b of length 30 cm (Fig. 7–21). The curved sides of the bowl are frictionless, and for the flat bottom the coefficient of kinetic friction $\mu_k = 0.21$. The particle is released from rest at the rim, which is 15 cm above the flat part of the bowl. (a) What is the speed of the particle at a? (b) What is the speed of the particle at b? (c) Where does the particle finally come to rest?

FIGURE 7–21 Problem 35.

36. (II) A toy car of mass M slides down a frictionless track that makes a circular loop of radius R at the bottom (Fig. 7–22). Suppose that the car starts from rest at a height H, with $H > 2R$. (a) What is the car's speed at the bottom of the circle? (b) at the top of the circle? (c) What is the force exerted by the track at the top of the circle? (d) What is the minimum

FIGURE 7–22 Problem 36.

value of H such that the car goes around the loop without falling off under the force of gravity?

37. (II) A block is constrained to move without friction along the x-axis. The block is attached to a spring of equilibrium length L. The other end of the spring is fixed at a point $x = 0$, $y = h$ (Fig. 7–23), where $h < L$. (a) What is the potential energy of the system? (b) What is the net force acting on the block? Sketch both $U(x)$ and $\mathbf{F}(x)$.

FIGURE 7–23 Problem 37.

38. (II) The potential energy of two atoms separated by a distance r may be written in the form $U(r) = U_0[(r_0/r)^{12} - 2(r_0/r)^6]$ (Fig. 7–24). Find the separation r at which there is no force between the atoms. What is the magnitude of the potential energy there?

FIGURE 7–24 Problem 38.

39. (II) The potential energy of a satellite of mass m, moving in a circular orbit of radius r about a planet of mass M, is given by $U(r) = -GMm/r$, where G is a universal constant. (a) Calculate the force on the satellite. (b) From the equation $\mathbf{F} = m\mathbf{a}$, calculate the kinetic energy for a particle in this orbit. (c) What is the total energy?

40. (II) The potential energy of a particle moving in the xy-plane is given by $U(x, y) = a_1x^2 + a_2xy + a_3y^2$, where $a_1 = 5$ J/m^2, $a_2 = -8$ J/m^2, and $a_3 = 3$ J/m^2. Calculate the force vector.

41. (III) The ski jumper of Problem 34 has survived the jump and prepares for another jump on a track of a different design: The track takes the jumper through a vertical distance H from its start to its take-off point, but the jumper can choose the angle at which he leaves the take-off point. The lip at the edge can be varied so that the jumper leaves the jump at any angle—from taking off horizontally to leaving vertically upward. Show that the angle θ with the horizontal that leads to the maximum-distance jump onto a horizontal plane a distance D below the lip is given by $\sin^2 \theta = 1/[2 + (D/H)]$. Ignore all friction and drag effects. (A real landing area is, in fact, sloped downward to allow the jumper to make a smooth landing.)

7–4 Is Energy Conservation a General Principle?

42. (I) A 10 g ping-pong ball is dropped with zero initial velocity from a height of 1.0 m, and it bounces back to a height of 0.90 m. What is the work done by the nonconservative forces in this process?

43. (I) A parachutist jumps off a training tower that is 85 m high. She starts at rest and reaches the ground with a vertical speed of 5.0 m/s. How much work was done by the drag forces acting on her, given that her mass is 75 kg?

44. (II) A track consists of a descending ramp, a straight track, and an ascending ramp. The smooth ramps both make an angle of 30° with the horizontal. The coefficient of kinetic friction on the horizontal surface is $\mu_k = 0.20$. An object starts from rest at a vertical height of 1.85 m on the descending ramp. It slides down the ramp, across the horizontal stretch, and up the ascending ramp. It reaches a vertical height of 0.75 m before coming to rest. (a) How long is the horizontal part of the track? (b) The object starts sliding back from the 1.22 m height. How far along the horizontal stretch does it slide?

*7–5 Barrier Tunneling in Quantum Physics

45. (I) Suppose that the fraction of particles that hit a potential energy barrier and tunnel through it is given by e^{-S}. What is the value of S if one particle in a million tunnels through?

46. (I) Suppose that, in Problem 45, S has the value 20. How many particles would have to strike the barrier for 10,000 of them to tunnel through?

47. (II) An interpretation of the factor S, described in Section 7–5, is that if a single particle can hit a potential energy barrier over and over again, then the fraction of times it will tunnel through is e^{-S}. Suppose that a particle is trapped between two barriers, 10^{-14} m apart, for which $S = 72$. The particle moves at a speed of 10^6 m/s back and forth between the barriers. How long, on average, will it take the particle to get out of the confined space between the barriers? This is similar to what happens to an alpha particle in a radioactive nucleus.

General Problems

48. (I) Assume that 1 kWh of electric energy costs 12 cents. Estimate the cost of lighting a three-room apartment per day.

49. (I) A furniture mover pushes a crate of mass 60 kg up a rough slope through a vertical distance of 1.0 m at a uniform speed. What is the change in the potential energy of the crate?

50. (I) A diver jumps off a rigid diving platform 6.0 m above the water with an initial upward velocity of 3.2 m/s. Assuming that his takeoff is very nearly in a vertical direction and that there are no drag forces on the diver, with what velocity will the diver hit the water? (Treat the diver as a point particle.)

51. (I) The force on an object of mass m moving along the x-axis is given by $F(x) = -ax + bx^2$, where $a = 3$ N/m and $b = 0.2$ N/m². (a) Calculate the potential energy function $U(x)$, letting $U(x) = 3$ J at $x = 0$ cm. (b) Sketch $U(x)$ as a function of x from $x = 0$ m to $x = 4.0$ m in steps of 0.5 m.

52. (II) A massless spring hangs vertically in equilibrium with no mass at its end. When a 2.0-kg mass is connected to the bottom, the new equilibrium position is 5.0 cm lower. The mass is then pulled down and released. It is observed that the speed of the mass is 2.1 m/s when the mass passes the original equilibrium position (before the mass was attached). How far down were the mass and spring pulled together when released?

53. (II) Are the following forces conservative or nonconservative? (a) The force $F(x) = ax + bx^3 + cx^4$. (b) The force $\mathbf{F}(x, y) = Ax^2\,\mathbf{i} + Bxy\,\mathbf{j}$.

54. (II) A particular spring exerts a force, as a function of displacement from equilibrium, given by $F = -(0.2$ N/m$)x - (0.4$ N/m³$)x^3$ (Fig. 7–25). This force acts on an object of mass 2 kg and displaces it from $x = 0$ m to $x = 3$ m. Consider the object being displaced along two different paths: (a) directly from $x = 0$ m to $x = 3$ m, and (b) from $x = 0$ m to $x = -2$ m, then to $x = 7$ m, and back to $x = 3$ m. Show that the work done by the force is the same for both paths.

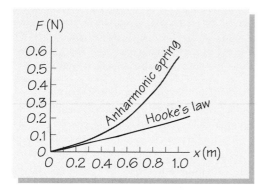

FIGURE 7–25 Problem 54.

55. (II) A conservative force does 2 J of work in moving a particle from point A to point C via path ABC (Fig. 7–26). The force does -1 J of work to move the particle from D to F, 3 J for E to B, 1 J for E to F, and 1 J for B to C. How much work does the force do as the object moves from C to A, from A to E, from D to C?

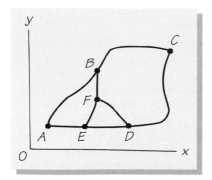

FIGURE 7–26 Problem 55.

56. (II) A particle of mass 50 g leaves, from rest, point a on a loop-the-loop. The heights of the points a, b, c, and d as measured from the table level are 10 cm, 0 cm, 8 cm, and 12 cm, respectively (Fig. 7–27). Ignore friction. (a) What are the speeds of the particle at points b, c, and d? (b) How high up on the other side does the particle rise?

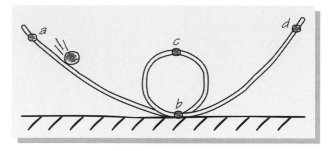

FIGURE 7–27 Problem 56.

57. (II) A skier skis from rest from a vertical height $h_1 = 18$ m over two successively lower hills of vertical heights $h_2 = 15$ m and $h_3 = 7$ m (Fig. 7–28). The summit of the third hill fits a circle of radius h_3 centered at height 0 m. Friction with the snow and air resistance are negligible. (a) Find her speeds at x_1, x_2, and x_3. (b) Does the skier leave the surface at x_3? If not, what should h_1 be so that she just leaves the surface at x_3?

FIGURE 7–28 Problem 57.

58. (II) A pile driver works by lifting a large mass and dropping it to the ground. The mass is 2500 kg, and it is raised to a height of 20 m above the pile for each stroke. The pile driver encounters resistance of a constant force of 8×10^6 N on each stroke. Use a combination of energy conservation and the work–energy theorem to determine how many strokes it takes to drive a pile 2 m into the ground. For which forces involved is it possible to use energy conservation, and for which forces is this not possible?

59. (II) The mass of a simple pendulum of length $L = 1$ m is released with the string originally in a horizontal position. (a) Calculate the speed of the mass at its lowest position. (b) What is the speed when the string makes an angle of 45° with the vertical? (c) Determine the tension in the string in both positions if the mass is 0.2 kg.

60. (II) A simple pendulum of length $L = 1.0$ m, mass $= 0.20$ kg is released from the horizontal position. When the mass is at its lowest point, the string hits a nail a distance h above the mass, so that the mass loops around the nail (Fig. 7–29). How large can h be so that the string of the pendulum remains taut even when the mass loops to a point right above the nail? [*Hint*: The string is taut as long as there is tension in the string].

FIGURE 7–29 Problem 60.

61. (II) An electron is attracted to a proton (the latter is much heavier than the former) with a central force whose magnitude is given by Coulomb's law, $F = C/r^2$, where $C = 2.3 \times 10^{-28}$ kg·m³/s². (a) Is this a conservative force? (b) Write an expression for the potential energy associated with this system. (c) An electron is very far away from a proton and starts from rest. It falls straight toward the proton under the influence of the force. What is the speed of the electron when it is 1.2×10^{-12} m from the proton?

62. (II) A batter hits a baseball and the baseball leaves the bat making a 38° angle with the ground. Air resistance has negligible effect on the trajectory of the baseball, which travels a total horizontal distance of 115 m. (a) What is the speed of the baseball just after it leaves the bat? (b) Use the conservation of the total energy of the baseball to calculate the maximum height to which the baseball rises. (c) What is the speed of the baseball when it has first risen to half its maximum height? (d) when it falls back to half its maximum height?

63. (II) A 3.0-kg block is held against a spring with spring constant $k = 25$ N/cm, compressing the spring 3 cm from its relaxed position. When the block is released, the spring expands and pushes the block upward along a rough surface inclined at a 20° angle (Fig. 7–30). The coefficient of kinetic friction between the block and the surface is $\mu_k = 0.1$. (a) What is the work done on the block by the spring as it extends from its compressed position to its equilibrium position? (b) by friction while the block moves 3 cm as in part (a)? (c) by gravity during the same motion? (d) What is the speed of the block when the spring reaches its equilibrium position? (e) If the block is not attached to the spring, how far up the incline will it slide before it comes to rest? (f) Suppose that the block is attached to the spring so that the spring is extended when the block slides past the equilibrium point. By how much will the spring be extended before the block comes to rest?

FIGURE 7–30 Problem 63.

64. (II) A block of wood of mass 3.0 kg is placed on a horizontal table and attached by a massless rope of length 6.0 cm to a vertical axis that passes through the table. The rope is initially swung around the axis, causing the wood to have a tangential velocity of 5.0 m/s. After one revolution the wood is observed to have a speed of only 3.0 m/s. (a) How much work has friction done on the block of wood during the first revolution? (b) What is the coefficient of kinetic friction between the block of wood and the table? (c) What is the potential energy of the block of wood at the beginning, after one revolution, and when the block of wood comes to rest? (d) How many revolutions does the block of wood make before stopping?

65. (II) Two blocks of mass $m_1 = 5.0$ kg and $m_2 = 2.0$ kg are supported by the system of light frictionless pulleys and massless strings shown in Fig. 7–31. Mass m_1 is at rest at a height $h = 0.8$ m above the ground when the system is released. Use the conservation of energy to determine the speed of m_1 when it hits the ground.

FIGURE 7–31 Problem 65.

66. (II) A representation of the nuclear force between two nucleons (neutron or proton) is given by the *Yukawa potential energy function*, $U(r) = -Ae^{-kr}/r$, where $1/k$ has the approximate value of 10^{-15} m and A is a constant. (a) Plot $U(r)$ versus r in steps of 0.2×10^{-15} m up to 2.4×10^{-15} m. Plot U in units of $A \times 10^{15}$. (b) At what distance is the potential energy a minimum? (c) Determine the force $F(r)$. (d) Determine the force at $r = 0.1 \times 10^{-15}$ m and 10×10^{-15} m.

67. (II) Given the relationship between the force and the potential energy, show that Newton's third law is satisfied if the potential energy has the form $U(x_1 - x_2)$ for two particles located at x_1 and x_2, respectively.

68. (III) A chain with a mass of 3.0 kg and a length of 3.0 m lies on a table, with 1.2 m hanging over the edge. How much energy is required to get all of the chain back on the table?

69. (III) Consider two-dimensional motion. You are given the components of the force **F** in the x- and y-directions, respectively, as $F_x(x, y)$ and $F_y(x, y)$. Show that if the force is conservative, then $\partial F_x/\partial y = \partial F_y/\partial x$. [*Hint*: Use the fact that the forces can be represented as derivatives of a single potential energy function $U(x, y)$ when the force is conservative.] What are the analogous conditions on the forces for three-dimensional motion?

70. (III) The surface of water in a bucket is determined by the condition that the potential energy per unit mass, as determined in the frame in which the water is at rest, is constant at all places on the surface. Assume that the bucket is set into rotation at an angular speed ω (introduced in Chapter 3) about its vertical central axis and that the water rotates with the bucket. Write a formula that gives the shape of the water surface. [*Hint*: Note that the normal force of the liquid must account for the centripetal acceleration.]

71. (III) A ball at the end of a pendulum of length L is released at rest from an initial position in which the pendulum string is horizontal (Fig. 7–32). The floor is just beneath the low point of the swing. The string is cut after the ball has passed the low point, with the string at an angle $\alpha = \theta - 90°$ to the vertical. Find the horizontal distance that the ball travels from the low point before it bounces on the floor.

FIGURE 7–32 Problem 71.

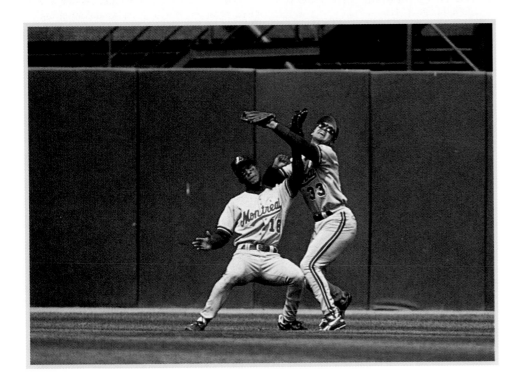

The collision between two baseball players has severely disrupted each of their trajectories. This collision is complicated by the fact that the players do not form an isolated system: Friction between their cleats and the ground also plays an important role.

Linear Momentum, Collisions, and the Center of Mass

The idea that energy is conserved is an enormously useful concept. In this chapter, we will identify another quantity that is conserved in isolated systems: momentum. The momentum of an object plays an important role in Newton's second law: A force produces a change in momentum. When a system of particles is isolated, the total momentum of the system is constant. This principle, the conservation of momentum, is particularly useful for understanding the behavior of colliding objects. We can use the conservation of momentum principle even when we know no details about the force or when the forces are very complicated.

As we study the momentum of a system, we will learn that there is a particular point of the system—the *center of mass*—which is distinguished because its motion is so simple. For an isolated system, the center of mass moves as if it were free of all forces. When external forces act on the system, the center of mass accelerates according to Newton's second law just as a point object would.

8–1   MOMENTUM AND ITS CONSERVATION

Newton's second law describes how forces change the motion of objects. We have expressed this law in terms of the mass and the acceleration of an object. A more general form of the second law ($\mathbf{F} = m\mathbf{a}$) is applicable when the mass changes, as in a

The definition of momentum

rocket. This form of the second law is

$$F = \frac{d(mv)}{dt}. \tag{8–1}$$

The combination mass times velocity, mv, is called the **linear momentum**, or just **momentum**. We denote it by $\mathbf{p}$:

$$\mathbf{p} \equiv m\mathbf{v}. \tag{8–2}$$

Newton's second law states that the force on an object equals the rate of change of the object's momentum.

The momentum of an object is a vector whose direction is that of the velocity. Its dimensions are those of a mass times a velocity, namely, $[M \cdot LT^{-1}]$; in SI, the units of momentum are kg · m/s.

In terms of momentum, the second law [Eq. (8–1)] has the general form

$$F = \frac{d\mathbf{p}}{dt}. \tag{8–3}$$

The kinetic energy of an object can also be expressed in terms of the momentum:

$$K = \frac{1}{2} mv^2 = \frac{p^2}{2m}, \tag{8–4}$$

where p is the magnitude of the momentum.

(a)

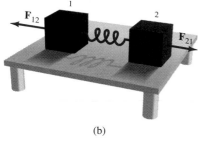

(b)

FIGURE 8–1 (a) Two billiard balls collide and (b) two masses interact because they are connected by a spring. In each case, the objects exert forces on one another. $\mathbf{F}_{12}$ is the force exerted on object 1 by object 2, and $\mathbf{F}_{21}$ is the force exerted on object 2 by object 1. In these collisions, the forces are contact forces although, in general, physical contact is not necessary for two objects to exert forces on one another.

Conservation of Momentum

When objects exert a force on one another, we say they interact. Let's consider the interaction between objects 1 and 2 in Fig. 8–1. The two objects may be in contact, as in a collision of two billiard balls; they may exert a force on each other at a distance, as in the gravitational attraction between Earth and the Moon; or they may simply be connected by a spring. Let $\mathbf{F}_{12}$ denote the force exerted on object 1 by object 2 and let $\mathbf{F}_{21}$ denote the force exerted on object 2 by object 1 (the first subscript always labels the object that is acted *upon*). Then, Newton's third law states that

$$\mathbf{F}_{12} = -\mathbf{F}_{21}. \tag{8–5}$$

When it is expressed in terms of momentum, Newton's second law [Eq. (8–3)] tells us that the rate of change of each object's momentum is the force acting on it:

$$\frac{d\mathbf{p}_1}{dt} = \mathbf{F}_{12}, \tag{8–6}$$

$$\frac{d\mathbf{p}_2}{dt} = \mathbf{F}_{21} = -\mathbf{F}_{12}. \tag{8–7}$$

Addition of these two equations leads to

$$\frac{d\mathbf{p}_1}{dt} + \frac{d\mathbf{p}_2}{dt} = \mathbf{F}_{12} + \mathbf{F}_{21};$$

$$\frac{d(\mathbf{p}_1 + \mathbf{p}_2)}{dt} = \mathbf{F}_{12} - \mathbf{F}_{12} = 0. \tag{8–8}$$

∞ Newton's third law is treated in Chapter 4.

As a consequence,

for zero external forces: $\quad \mathbf{p}_1 + \mathbf{p}_2 = $ a constant. $\tag{8–9}$

The conservation of momentum

In other words, Newton's third law implies that *the sum of the momenta of an isolated system of two objects that exert forces on one another is a constant, no matter what form the forces take.* This is called the **principle of conservation of momentum.** We will see that this result is not confined to two objects. Like the principle of the conservation of energy, it holds even when we deal with forces whose details we don't understand. The conservation laws are important both as general principles and as powerful tools for solving problems.

EXAMPLE 8–1 A billiard ball moves with a velocity of 1.20 m/s in the $+y$-direction on a billiard table and strikes an identical ball that was initially at rest (Fig. 8–2a). The rolling ball is deflected so that its velocity has a component of 0.80 m/s in the $+y$-direction and a component of 0.56 m/s in the $+x$-direction (Fig. 8–2b). What are the final velocity and final speed of the struck ball?

Solution: We will denote the initial velocity of the rolling ball as $\mathbf{v}_{1i}$ and the final velocity of that ball by $\mathbf{v}_{1f}$. Let the final velocity of the struck ball be $\mathbf{v}_{2f}$. If m is the mass of each ball, $\mathbf{p}_i = m\mathbf{v}_{1i}$ because the struck ball has an initial velocity of zero. Then

$$m\mathbf{v}_{1i} = m\mathbf{v}_{1f} + m\mathbf{v}_{2f},$$

this simplifies to $\mathbf{v}_{1i} = \mathbf{v}_{1f} + \mathbf{v}_{2f}$, or $\mathbf{v}_{2f} = \mathbf{v}_{1i} - \mathbf{v}_{1f}$. We have

$$\mathbf{v}_{1i} = (1.20 \text{ m/s})\mathbf{j} \quad \text{and} \quad \mathbf{v}_{1f} = (0.56 \text{ m/s})\mathbf{i} + (0.80 \text{ m/s})\mathbf{j},$$

so the final velocity of the struck ball is

$$\mathbf{v}_{2f} = \mathbf{v}_{1i} - \mathbf{v}_{1f} = (1.20 \text{ m/s})\mathbf{j} - [(0.56 \text{ m/s})\mathbf{i} + (0.80 \text{ m/s})\mathbf{j}]$$

$$= (-0.56 \text{ m/s})\mathbf{i} + (0.40 \text{ m/s})\mathbf{j}.$$

The final speed of the struck ball is then

$$v_{2f} = \sqrt{(v^2_{2f,x} + v^2_{2f,y})} = \sqrt{(-0.56 \text{ m/s})^2 + (0.40 \text{ m/s})^2} = 0.69 \text{ m/s}.$$

Conservation of Momentum for a System of Many Objects. The conservation of momentum is not confined to a system of two interacting objects. Suppose that there are three objects in a system on which no external forces act. Then the total force $\mathbf{F}_1$ on object 1 is given by the sum of the forces on object 1 due to objects 2 and 3:

$$\mathbf{F}_1 = \mathbf{F}_{12} + \mathbf{F}_{13}. \tag{8–10a}$$

Similarly, the total force $\mathbf{F}_2$ on object 2 is given by the sum of the forces on object 2 due to objects 1 and 3, so

$$\mathbf{F}_2 = \mathbf{F}_{21} + \mathbf{F}_{23}; \tag{8–10b}$$

and similarly

$$\mathbf{F}_3 = \mathbf{F}_{31} + \mathbf{F}_{32}. \tag{8–10c}$$

FIGURE 8–2 Example 8–1. (a) Before the collision. (b) After the collision.

Addition of these three equations, with the help of $\mathbf{F}_{12} = -\mathbf{F}_{21}$, $\mathbf{F}_{13} = -\mathbf{F}_{31}$, and $\mathbf{F}_{23} = -\mathbf{F}_{32}$, yields

$$\frac{d\mathbf{p}_1}{dt} + \frac{d\mathbf{p}_2}{dt} + \frac{d\mathbf{p}_3}{dt} = \mathbf{F}_1 + \mathbf{F}_2 + \mathbf{F}_3$$

$$= \mathbf{F}_{12} + \mathbf{F}_{13} + \mathbf{F}_{21} + \mathbf{F}_{23} + \mathbf{F}_{31} + \mathbf{F}_{32} = 0. \quad (8\text{–}11)$$

Consequently, the sum of the momenta of the three objects $\mathbf{P} = \mathbf{p}_1 + \mathbf{p}_2 + \mathbf{p}_3$ is constant throughout the motion:

$$\mathbf{P} = \mathbf{p}_1 + \mathbf{p}_2 + \mathbf{p}_3 = \text{a constant.} \quad (8\text{–}12)$$

We can extend this demonstration to N interacting objects and prove that the sum of the objects' momenta is constant throughout the motion.

PROBLEM SOLVING

A good choice of reference frames makes collision problems simpler.

At rest

(a)

Total momentum = 0

(b)

(c)

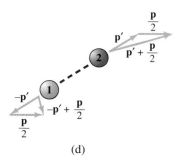

(d)

FIGURE 8–3 (a) View before collision from the frame in which ball 2 is at rest. (b) View before collision from the frame in which the total momentum is zero. (c) View after collision from the frame in which the total momentum is zero. (d) View after collision from the frame in which ball 2 was originally at rest.

Conservation of Momentum in Different Inertial Frames

We showed in Chapter 4 that Newton's laws of motion look the same to all observers in inertial frames of reference; that is, in all frames moving with uniform velocity with respect to the system being observed. It is equally true that if momentum is conserved in one inertial reference frame, it is conserved in all inertial frames. This fact may be used to great advantage because calculations involving colliding objects may be simpler in one reference frame than in another.

For example, let's consider the collision of two objects of equal mass: One of them is at rest and the other is moving with momentum $\mathbf{p} = p\,\mathbf{i}$ (Fig. 8–3a). It simplifies the problem to go to a second reference frame[†] by adding the momentum $-\mathbf{p}/2$ to the momentum of each mass. In this second frame, then, the objects have equal and opposite momentum $\mathbf{p}/2$ and $-\mathbf{p}/2$ (Fig. 8–3b). The total momentum is zero and we can immediately say that the final momenta must be $\mathbf{p}'$ and $-\mathbf{p}'$ in this frame (Fig. 8–3c). We now return to the original reference frame by adding the momentum $+\mathbf{p}/2$ to every momentum. Thus, in the original frame, the final momenta must be of the form $\mathbf{p}' + \mathbf{p}/2$ and $\mathbf{p}' - \mathbf{p}/2$, respectively (Fig. 8–3d). This way of finding the possible momenta is simpler than working entirely in the original reference frame.

8–2 IMPULSE AND ONE-DIMENSIONAL COLLISIONS

What happens when objects *collide?* The word "collision" evokes the image of a sharp contact, such as the collision between two billiard balls. In order to apply the conservation of momentum to collisions, we will first discuss these brief, sharp forces in more detail.

Impulsive Forces

During a collision, the force that alters the motion of the two objects is generally active for only a short time. This force is called an **impulsive force**. Let's see how such a force is related to the momentum change of an object on which the force acts (Fig. 8–4). We define the **impulse, J,** according to

$$\mathbf{J} \equiv \int \mathbf{F}\,dt, \quad (8\text{–}13)$$

where $\mathbf{F}$ is the (impulsive) force that acts on the object. From Newton's second law,

$$\mathbf{J} = \int_{t_1}^{t_2} \left(\frac{d\mathbf{p}}{dt} \right) dt = \mathbf{p}_{t_2} - \mathbf{p}_{t_1} = \mathbf{p}_f - \mathbf{p}_i = \Delta\mathbf{p}. \quad (8\text{–}14)$$

[†]We will see later that this is the center-of-mass reference frame.

FIGURE 8-4 The contact force that sends the football on its way is impulsive. The force acts over a limited time.

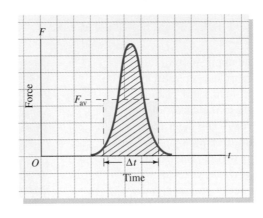

FIGURE 8-5 The momentum of an object is changed by a time-dependent force **F** that acts on it. If that force acts over a time period Δt, then the same change in momentum is produced by its average value $\mathbf{F}_{av}$ over that time period.

Here, the force acts for the time interval $\Delta t = t_2 - t_1$. Because the integral of force over time is the area under a curve of force versus time, we can also represent the impulse by the product of the time interval Δt and a quantity that we may call the *average force* $\mathbf{F}_{av}$. The average force refers to the average value of the force over the time interval Δt. In this case, the impulse or change in momentum can be written in the form

$$\Delta \mathbf{p} = \mathbf{J} = \int \mathbf{F} \, dt = \mathbf{F}_{av} \, \Delta t, \tag{8-15}$$

as depicted in Fig. 8-5. Remember that this equation is a *vector* equation. In Example 8-2, however, the vector aspect does not play a crucial role.

EXAMPLE 8-2   The average force exerted by a bat on a baseball during the time of contact, which lasts 2.00×10^{-3} s, is 6660 N. The mass of the baseball is 0.145 kg and its speed is 33.5 m/s just before the bat collides with it. What is the velocity of the ball when it leaves the bat? Assume that the ball leaves the bat along the same line of direction from which it is pitched.

Solution: Figure 8-6 illustrates this problem. To find the momentum change of the ball, we use the impulse, Eq. (8-15):

$$\Delta \mathbf{p} = \mathbf{J} = \int \mathbf{F} \, d\mathbf{t} = \mathbf{F}_{av} \, \Delta t.$$

Both the average force, $\mathbf{F}_{av}$, and the time interval are given. The motion can be described in one dimension because all motion takes place along a line; thus, we can drop the boldfaced notation. We must, however, keep track of the signs. We do so by aligning the motion along the x-axis, with the original motion of the ball—the direction of the pitch—in the $+x$-direction. The momentum change Δp is $p_f - p_i$ and the initial momentum $p_i = mv_0$, where m is the baseball's mass and v_0 is the (positive) velocity of the ball when it reaches the bat. Thus

$$p_f = p_i + F_{av} \, \Delta t = mv_0 + F_{av} \, \Delta t.$$

The final momentum is the product of m and the final velocity, v_f, and hence

$$v_f = \frac{p_f}{m} = \frac{mv_0 + F_{av} \, \Delta t}{m} = v_0 + \frac{F_{av} \, \Delta t}{m}.$$

We note that the *sign* of the impulse received by the ball is negative because the force is directed in the $-x$-direction. Then

$$v_f = +33.5 \text{ m/s} + \frac{(-6660 \text{ N})(2.00 \times 10^{-3} \text{ s})}{0.145 \text{ kg}}$$

$$= 33.5 \text{ m/s} - 91.9 \text{ m/s} = -58.4 \text{ m/s}.$$

The minus sign indicates that the ball moves in the negative direction back toward the pitcher.

Start swing

Contact (ball at rest)

Follow through

FIGURE 8-6 Example 8-2. The direction of the impulse imparted to a ball hit by a bat is to the left, in the $-x$-direction.

Momentum and impulse find a practical application in the pile driver (Fig. 8AB–1). The pile driver uses a vertical cylinder closed at the lower end and containing a piston. The upper side of this piston is attached to the center of a massive M-shaped piece of steel—a yoke. The outer legs of the M reach down beyond the bottom of the cylinder and engage with the top of the pile being driven. The whole assembly, complete with guides to allow the yoke to remain aligned with the cylinder, is hoisted by a crane and positioned on top of the pile. Compressed air and fuel is supplied to the cylinder below the piston and ignited. The resultant explosion drives the piston and the steel yoke upward. The pressure is then released from the cylinder and the yoke falls. During the fall, it gains kinetic energy and, with it, momentum. When the yoke reaches the top of the pile, it is brought to rest in an exceedingly short time (determined by the compressibility of the steel yoke and the steel pile). Thus dp/dt is very large and a large impulsive force is transmitted to the pile. This process generates forces large enough to break up small rocks in front of the pile as it is being driven into the ground. More importantly, the process we have described here does not require a fixed fulcrum such as would be needed for a static force.

FIGURE 8AB–1 The pile driver is one of the devices that enables us to build large structures.

Momentum Conservation in Collisions

We can clearly see the manifestations of the conservation of momentum in the study of collisions within an isolated system of objects. Attempts to understand collisions were carried out by Galileo and his contemporaries. The laws that describe collisions in one dimension were formulated by John Wallis, Christopher Wren,[†] and Christian Huygens in 1668.

Let's start with collisions between two objects. The two objects are free before the collision and each may be characterized by its constant momentum. During the brief interaction, their momenta change because each object experiences an impulsive force due to the other object. The impulses felt by the two objects are equal and opposite because the forces are equal and opposite. The increase in the momentum of one object is equal to the loss of momentum in the other. After the collision, the two objects are again free but have different momenta. However, *the sum of their momenta is unchanged.*

In one dimension, the vector aspect is reduced to the fact that the momenta can be negative as well as positive. Let's suppose that two objects are moving along the x-axis. Object 1 has mass m_1 and velocity v_1, whereas object 2 has mass m_2 and velocity v_2. There will be a collision if these objects approach each other from opposite directions or if one overtakes the other as they move in the same direction. The total initial momentum is given by

$$p_i = m_1v_1 + m_2v_2. \tag{8–16}$$

Several things can happen to objects during a collision:

1. The two objects can shatter into several pieces.

[†]Best remembered today as an architect.

2. Mass can be transferred from one object to the other; after the collision, then, the two objects would have masses m_6 and m_7.

3. The two masses can coalesce into one, as in the collision of two blobs of putty.

4. The masses can remain unchanged. Even in this case, though, there are different possibilities. The objects can be completely unaltered, as in the collision of two billiard balls, or they can be deformed, as in the collision of two automobiles.

Figure 8–7 shows the various final configurations of a two-object collision in one dimension. *In all cases, momentum will be conserved.* In the first case, if the objects happen to break up into three objects of masses m_3, m_4, and m_5 moving with velocities v_3, v_4, and v_5, respectively, then

$$m_3v_3 + m_4v_4 + m_5v_5 = m_1v_1 + m_2v_2. \tag{8–17}$$

In the second case, if the final velocities are v_6 and v_7, respectively, then

$$m_6v_6 = m_7v_7 = m_1v_1 + m_2v_2. \tag{8–18}$$

In the third case, if the mass of the single final object is M and its velocity is v, then

$$Mv = m_1v_1 + m_2v_2. \tag{8–19}$$

Mass conservation gives us the additional information that $M = m_1 + m_2$.[†] Finally, in the last case, if the final velocities of the two objects are v_8 and v_9, respectively, then

$$m_1v_8 + m_2v_9 = m_1v_1 + m_2v_2. \tag{8–20}$$

Only in the case of Eq. (8–19) is the final motion (meaning all final velocities) determined by this equation alone. In all the other cases, more information is required.

Energy Conservation in Collisions

We may not be able to determine *all* the unknowns in a collision—the final velocities if given the initial ones, for example—simply by using the conservation of momentum. Another important tool is the *conservation of energy*. Strictly speaking, energy is always conserved, but we cannot always take advantage of that fact in our calculations of collisions. For example, when friction is important, all of the energy does not all go into energy of motion; instead, we say that the energy is *dissipated*. For example, in a collision between two cars, some energy goes into the crumpling of metal and as a result there is less energy available for motion—the kinetic energy after the collision. If the energy decrease is known (as in Example 8–3), then we can use this information.

EXAMPLE 8–3 A 14,000-kg truck and a 2000-kg car have a head-on collision (Fig. 8–8). Despite attempts to stop, the truck has a speed of 6.6 m/s in the $+x$-direction when they collide and the car has a speed of 8.8 m/s in the $-x$-direction. If 10 percent of the initial total kinetic energy is dissipated through damage to the vehicles, what are the final velocities of the truck and the car after the collision? Assume that all motion takes place in one dimension.

[†]This need not be the case in nuclear or subnuclear collisions.

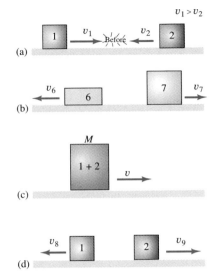

FIGURE 8–7 A collision between two objects, as shown before the collision in part (a), can have several outcomes. The two objects shatter into several pieces (not shown). (b) Mass is transferred between the objects. (c) The two masses combine as the objects coalesce into one object. (d) The original masses do not change.

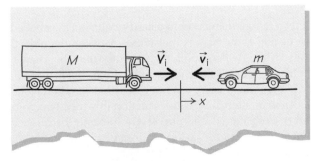

FIGURE 8–8 Example 8–3.

Solution: Let the masses of the truck and car be M and m, respectively. The initial velocities of the truck and the car are denoted V_i and v_i, respectively, and their respective final velocities are V_f and v_f. With our coordinate system, V_i, is positive and v_i is negative. By momentum conservation, we have

$$MV_f + mv_f = MV_i + mv_i \equiv p_1.$$

The initial total energy is the sum of the kinetic energies of the two vehicles, $(MV_i^2/2) + (mv_i^2/2)$. The sum of the final kinetic energies is 90 percent of this quantity, so

$$\frac{1}{2}MV_f^2 + \frac{1}{2}mv_f^2 = 0.9(\frac{1}{2}MV_i^2 + \frac{1}{2}mv_i^2) \equiv 0.9\,K_i.$$

These are two equations for the two unknowns, V_f and v_f. We have defined the initial momentum p_i and the initial kinetic energy K_i because both are known.

The first equation directly yields

$$V_f = \frac{p_i - mv_f}{M}.$$

Substituting this value into the second equation, we have a quadratic equation for vf:

$$\frac{1}{2}M\left(\frac{p_i - mv_f}{M}\right)^2 + \frac{1}{2}mv_f^2 - 0.9\,K_i = 0;$$

$$\frac{1}{2}M\frac{p_i^2}{M^2} - \frac{1}{2}M\left(\frac{2p_imv_f}{M^2}\right) + \frac{1}{2}M\frac{m^2v_f^2}{M^2} + \frac{1}{2}mv_f^2 - 0.9\,K_i = 0;$$

$$(m^2 + Mm)v_f^2 - 2mv_fp_i + p_i^2 - 2(0.9)M\,K_i = 0.$$

This equation has solutions

$$v_f = \frac{2mp_i \pm \sqrt{(2mp_i)^2 - 4(m^2 + Mm)[p_i^2 - 2(0.9)M\,K_i]}}{2(m^2 + Mm)}.$$

When numbers are inserted for the two possible solutions represented by the $\pm$ sign, the minus sign gives a negative velocity for the car and a positive velocity for the truck. This means that the car continues its motion to the left, going "through" the truck, while the truck similarly goes "through" the car. Although this scenario can happen for the collision of elementary particles, it cannot happen for motor vehicles; thus, the only physical solution corresponds to the plus sign. Inserting numbers, we find

$$v_f = 17.2 \text{ m/s} \quad \text{and} \quad V_f = 2.9 \text{ m/s}.$$

The truck continues in the $+x$-direction with a speed less than its initial speed; the car has completely reversed its direction and is moving even faster than its initial speed.

Classification of Collisions

The cases described so far provide a way to classify collisions; namely, by whether kinetic energy is lost or not. If no mass is transferred between the objects and if all the initial kinetic energy of the two colliding objects goes into the kinetic energy of the objects after the collision, we call the collision **elastic**; in other words, the kinetic energy is conserved. If the kinetic energy is not conserved in a collision, it is called an **inelastic** collision. Another alternative is the situation in which two objects collide and coalesce. This kind of collision is called **perfectly inelastic** because, as we shall see in Section 8–3, it corresponds to the maximum loss of kinetic energy.

Elastic, inelastic, and perfectly
inelastic collisions

8–3 PERFECTLY INELASTIC COLLISIONS; EXPLOSIONS

Perfectly Inelastic Collisions

We have described a range of possible collisions from elastic to inelastic to perfectly inelastic. The simplest to treat is the case of *perfectly inelastic collisions* in one dimension, which are those in which the objects coalesce as a result of the collision. These collisions are both described by Eq. (8–19). Mass conservation implies that $M = m_1 + m_2$. If we divide Eq. (8–19) by the factor $M = m_1 + m_2$, we find the velocity of the coalesced object:

$$v_1 = \frac{m_1 v_1 + m_2 v_2}{m_1 + m_2}. \tag{8–21}$$

Let's analyze this result for some special cases.

Suppose that one of the objects is at rest and the other object smashes into it (Fig. 8–9). In this case, $v_2 = 0$, and we find

$$v = \frac{m_1 v_1}{m_1 + m_2} = \left(\frac{m_1}{M}\right) v_1. \tag{8–22}$$

If $m_1 \gg m_2$, the "composite" object will move with a velocity nearly equal to that of the initially moving object; a train "catching a cow" does not slow down very much. In contrast, when $m_1 \ll m_2$, as when a stationary athlete catches a ball, we get the opposite effect; that is, the athlete will recoil with only a low velocity, just the fraction m_1/m_2 of the velocity of the ball.

Next, consider the case of a head-on collision in which the two objects have equal and opposite velocities; that is, $v_2 = -v_1$. In this case, Eq. (8–21) becomes

$$v = \frac{m_1 - m_2}{m_1 + m_2} v_1. \tag{8–23}$$

In the special case that $m_1 = m_2$, the two objects have equal and opposite *momenta* because

$$m_1 v_1 + m_2 v_2 = m_1 v_1 + m_1 v_2 = m_1(v_1 + v_2) = 0. \tag{8–24}$$

In that case, the final momentum must be zero and thus $v = 0$, as Eq. (8–23) verifies. The objects collide and stay there, as when a fullback is tackled head-on by a linebacker (Fig. 8–10).

Energy Loss in Perfectly Inelastic Collisions.

Let's find the change in energy for a collision in which the objects coalesce. Before the collision, the total energy E_i is the sum of the kinetic energies of the two objects. The final energy E_f is the kinetic energy of the composite object of mass $M = m_1 + m_2$. The change in energy $\Delta E = E_f - E_i$ is, using Eq. (8–21),

$$\Delta E = \frac{1}{2} M v^2 - \left(\frac{1}{2} m_1 v_1^2 + \frac{1}{2} m_2 v_2^2\right)$$

$$= \frac{1}{2} \frac{(m_1 + m_2)(m_1 v_1 + m_2 v_2)^2}{(m_1 + m_2)^2} - \left(\frac{1}{2} m_1 v_1^2 + \frac{1}{2} m_2 v_2^2\right)$$

$$= \frac{\frac{1}{2}\left[m_1^2 v_1^2 + 2 m_1 m_2 v_1 v_2 + m_2^2 v_2^2 - (m_1 + m_2)(m_1 v_1^2 + m_2 v_2^2)\right]}{m_1 + m_2}$$

$$= \frac{\frac{1}{2}\left[(m_1 m_2)(-v_1^2 - v_2^2 + 2 v_1 v_2)\right]}{m_1 + m_2} = -\frac{m_1 m_2}{2(m_1 + m_2)} (v_1 - v_2)^2.$$

FIGURE 8–9 An object with mass m_1 and velocity $\mathbf{v}_1$ collides with an object of mass m_2 at rest. The two objects stick together and move on together with velocity $\mathbf{v}$.

(a)

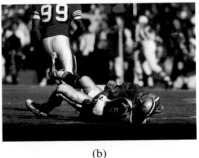

(b)

FIGURE 8–10 (a) A collision in the making. (b) When a ball carrier is tackled head-on by a defensive player, a violent collision takes place. If their masses are equal and their velocities are equal and opposite, their final velocities will be zero.

The right-hand side of Eq. (8–25) is always negative, corresponding to an energy loss. We say the collision is inelastic. When there is only one final composite object, there is a frame of reference in which the composite object is at rest. In this frame, there is no final kinetic energy; thus the collision is called *perfectly inelastic*. In this frame of reference, the total momentum is zero, and *all the energy of motion* in the initial state of the system goes into the "sticking together" of the objects.

EXAMPLE 8–4 A dog who jumps into the interior of a stationary ice boat is moving at $v_0 = 32$ km/h when he enters the boat (Fig. 8–11). The dog's mass is 14 kg and that of the boat plus boater is 160 kg. (a) Assuming that the ice surface is frictionless, what is the speed of the boat after the collision? (b) What is the ratio of the energy loss to the initial energy, and where did the energy go?

Solution: (a) The initial momentum is that of the dog because the boat is at rest. If m is the dog's mass and v_0 is the dog's speed as it enters the boat, then

$$p_i = mv_0.$$

The final momentum is given by $p_f = Mv$, where v is the unknown speed and M is the *total* mass of the boat, boater, and dog; namely, 174 kg. Equating p_f and p_i, we find

$$v = \frac{p_f}{M} = \frac{p_i}{M} = \frac{mv_0}{M} = \frac{(14\ \text{kg})(32\ \text{km/h})}{174\ \text{kg}} = 2.6\ \text{km/h} = 0.72\ \text{m/s}.$$

(b) The initial energy is the kinetic energy of the dog:

$$K_i = \frac{1}{2}mv_0^2.$$

The final energy is again all in the form of kinetic energy:

$$K_f = \frac{1}{2}Mv^2 = \frac{1}{2}M\left(\frac{mv_0}{M}\right)^2 = \frac{1}{2}\left(\frac{m}{M}\right)mv_0^2 = \frac{m}{M}K_i.$$

Thus the energy loss is given by

$$\Delta E = K_i - K_f = K_i - \frac{m}{M}K_i = K_i\left(1 - \frac{m}{M}\right),$$

and the ratio of the energy loss to the initial energy is

$$\frac{\Delta E}{K_i} = 1 - \frac{m}{M},$$

FIGURE 8–11
Example 8–4.

a number less than 1. The energy has decreased. Numerically,

$$\frac{\Delta E}{K_i} = 1 - \frac{14\ \text{kg}}{174\ \text{kg}} = 0.92.$$

Energy is lost as the boater "gives" in order to bring the dog to rest.

Explosions

Imagine that we were to film a perfectly inelastic collision in a frame of reference in which the total momentum is zero. In this reference system, the two objects approach each other and merge, leaving a composite object at rest. If we ran the film in reverse, it would look like a film of an explosion. The "initial" object of mass $M = m_1 + m_2$, at rest, breaks up into two objects, m_1 and m_2, moving with velocities such that the momentum is still zero; that is, with

$$m_1 v_1 + m_2 v_2 = 0. \tag{8–26}$$

Energy conservation suggests that an explosion is possible as long as there is an initial potential energy U such that

$$U = \frac{1}{2} m_1 v_1^2 + \frac{1}{2} m_2 v_2^2. \tag{8–27}$$

An explosive does have potential energy stored in its molecules and this potential energy is released by the detonation. Let's next take a look at a different type of "explosion," one in which an unstable atomic nucleus disintegrates.

EXAMPLE 8–5   One type of polonium nucleus (symbol ^{210}Po), with mass 3.49×10^{-25} kg, can decay into an α-particle (actually a helium nucleus), mass 6.64×10^{-27} kg, and a certain type of lead nucleus (symbol ^{206}Pb), mass 3.42×10^{-25} kg:

$$^{210}\text{Po} \rightarrow \alpha + {}^{206}\text{Pb}.$$

In this process, the decay products have a kinetic energy of 8.65×10^{-13} J above and beyond any kinetic energy possessed by the polonium nucleus itself. Consider such a decay with the polonium nucleus at rest. Find the speeds of the α-particle and the lead nucleus.

Solution: Let's denote by Q the kinetic energy of the decay products. Then conservation of momentum and the expression for the kinetic energy read, respectively,

$$M_\alpha v_\alpha = M_{\text{Pb}} v_{\text{Pb}};$$

$$Q = \frac{1}{2} M_\alpha v_\alpha^2 + \frac{1}{2} M_{\text{Pb}} v_{\text{Pb}}^2.$$

Here, v is the speed of the respective particles. These two equations can be solved for the two variables v_α and v_{Pb}. We find

$$v_\alpha = \sqrt{\frac{2Q}{M_\alpha (1 + M_\alpha/M_{\text{Pb}})}}$$

$$v_{\text{Pb}} = \sqrt{\frac{2Q}{M_{\text{Pb}} (1 + M_{\text{Pb}}/M_\alpha)}}.$$

With $Q = 8.65 \times 10^{-13}$ J, numerical evaluation gives $v_\alpha = 1.60 \times 10^7$ m/s and $v_{\text{Pb}} = 3.10 \times 10^5$ m/s. Note that the speed of the heavier of the two decay products is much less than the speed of the lighter, a result already visible in the conservation of momentum equation.

8 – 4 ELASTIC AND INELASTIC COLLISIONS

Let's continue to work in one dimension. In an *elastic collision*, there is no mass transfer from one object to another. Further, *all the kinetic energy in the initial state goes into kinetic energy in the final state.* If the final velocities of objects 1 and 2 are denoted by v_3 and v_4, then, in addition to the momentum conservation equation for one dimension,

$$m_1 v_1 + m_2 v_2 = m_1 v_3 + m_2 v_4, \qquad (8\text{--}28)$$

we have the energy conservation equation

$$\frac{1}{2} m_1 v_1^2 + \frac{1}{2} m_2 v_2^2 = \frac{1}{2} m_1 v_3^2 + \frac{1}{2} m_2 v_4^2. \qquad (8\text{--}29)$$

With this information we can find the final velocities of the colliding objects if their initial velocities are known. We rewrite the momentum conservation equation, Eq. (8–28), as

$$m_1(v_1 - v_3) = -m_2(v_2 - v_4). \qquad (8\text{--}30)$$

We use the fact that $v_1^2 - v_3^2 = (v_1 - v_3)(v_1 + v_3)$ and $v_2^2 - v_4^2 = (v_2 - v_4)(v_2 + v_4)$ to rewrite the energy conservation equation, Eq. (8–29), in the form

$$\frac{1}{2} m_1(v_1 - v_3)(v_1 + v_3) = -\frac{1}{2} m_2(v_2 - v_4)(v_2 + v_4). \qquad (8\text{--}31)$$

Dividing both sides of Eq. (8–31) by the two sides of Eq. (8–30) leads to the equation

$$v_1 + v_3 = v_2 + v_4. \qquad (8\text{--}32)$$

If we use the letter u to denote the *relative velocity* of the two colliding objects, then

$$u_i = v_1 - v_2;$$

$$u_f = v_3 - v_4.$$

Using these quantities, Eq. (8–32) can be written in the form

$$u_i = -u_f. \qquad (8\text{--}33)$$

Relative speeds remain unchanged in an elastic collision.

Equation (8–33) shows that *when the collision is elastic, the relative velocity of the colliding objects changes sign but does not change in magnitude.* A simple way to remember this result is that the relative velocity behaves like the velocity of a perfectly elastic rubber ball hitting a brick wall.

We may solve Eq. (8–32) for one of the unknown variables, v_4, for example,

$$v_4 = v_1 - v_2 + v_3,$$

and substitute this value into the momentum conservation equation, Eq. (8–28). We then have

$$m_1 v_1 + m_2 v_2 = m_1 v_3 + m_2(v_1 - v_2 + v_3),$$

which may be rewritten in the form

$$(m_1 + m_2)v_3 = (m_1 - m_2)v_1 + 2m_2 v_2;$$

$$v_3 = \frac{m_1 - m_2}{m_1 + m_2} v_1 + \frac{2m_2}{m_1 + m_2} v_2. \qquad (8\text{--}34)$$

A similar calculation leads to the formula

$$v_4 = \frac{2m_1}{m_1 + m_2} v_1 + \frac{m_2 - m_1}{m_1 + m_2} v_2. \qquad (8\text{--}35)$$

These equations are rather complicated and it is useful to consider some special cases that simplify them.

1. **Object 2 is initially at rest.** We set $v_2 = 0$, so Eqs. (8–34) and (8–35) become

$$v_3 = \frac{m_1 - m_2}{m_1 + m_2} v_1 \qquad (8\text{–}36a)$$

and

$$v_4 = \frac{2m_1}{m_1 + m_2} v_1. \qquad (8\text{–}36b)$$

Let's consider the following situations (in which object 2 is initially at rest):

1a. The objects have equal masses (Fig. 8–12a). In this case, $v_3 = 0$ and $v_4 = v_1$. The two objects in effect change roles: The moving object comes to rest and the object that was initially at rest moves with the initial velocity of the first object. This effect can be seen vividly in hard billiard shots along a line.

1b. Mass $m_1 \gg$ mass m_2 (Fig. 8–12b). In this case, Eqs. (8–36) yield $v_3 \simeq v_1$ and $v_4 \simeq 2v_1$. The velocity of the moving object decreases a little, while the object that was at rest picks up almost twice the velocity of the incoming object.

1c. Mass $m_2 \gg$ mass m_1 (Fig. 8–12c). In this case, Eqs. (8–36) yield $v_3 \simeq -v_1$ and $v_4 \simeq (2m_1/m_2)v_1$. The moving object very nearly reverses its velocity, while the object initially at rest recoils with a very small velocity. In the limit that m_2 approaches infinity, the recoil velocity can be neglected and the final velocity of the first object is equal and opposite to its incident velocity. This is what happens when a tennis ball is bounced off a wall.

2. **The initial total momentum is zero.** The two objects approach each other with velocities such that the initial total momentum is zero:

$$m_1 v_1 + m_2 v_2 = 0. \qquad (8\text{–}37)$$

Thus

$$v_2 = -\frac{m_1}{m_2} v_1.$$

When this value is substituted into Eq. (8–34), we find

$$v_3 = \frac{m_1 - m_2}{m_1 + m_2} v_1 + \left(\frac{2m_2}{m_1 + m_2}\right)\left(-\frac{m_1}{m_2}\right)v_1 = \left(\frac{m_1 - m_2 - 2m_1}{m_1 + m_2}\right)v_1 = -v_1. \quad (8\text{–}38)$$

To find the value of v_4, we can recall that the initial total momentum was zero. Thus, by momentum conservation, the final total momentum $(m_1 v_3 + m_2 v_4)$ is also zero and

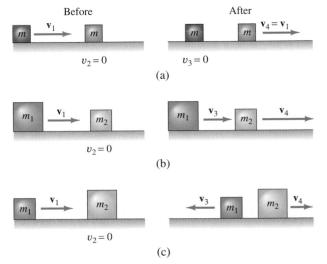

FIGURE 8–12 Two objects collide elastically in one dimension, with the second object initially at rest. (a) If the two masses are equal, the objects simply exchange velocities. (b) If $m_1 \gg m_2$, both objects move off to the right. (c) If $m_2 \gg m_1$, mass m_1 reverses its direction and m_2 moves slowly off to the right.

$$u_4 = -\frac{m_1}{m_2} v_3 = \frac{m_1}{m_2} v_1 = -v_2. \tag{8–39}$$

Therefore, in the case where the total momentum is zero, the velocities of each of the objects are unchanged in magnitude but they change sign. In effect, under these circumstances, each of the objects acts as if it hit an infinitely massive brick wall.

Examples 8–6 and 8–7 illustrate some features of collisions in one dimension.

EXAMPLE 8–6 Two spheres with masses of 1.0 kg and 1.5 kg hang at rest at the ends of strings that are both 1.5 m long. These two strings are attached to the same point on the ceiling. The lighter sphere is pulled aside so that its string makes an angle $\theta = 60°$ with the vertical (Fig. 8–13a). The lighter sphere is then released and the two spheres collide elastically. When they rebound, what is the largest angle with respect to the vertical that the string holding the lighter sphere makes?

Solution: We label the light mass m_1 and the heavy mass m_2 (Fig. 8–13). To find the velocities $\mathbf{v}_1'$ and $\mathbf{v}_2'$ resulting from the collision, we must first find the velocity $\mathbf{v}_1$ of m_1 just before the collision. Conservation of momentum and energy give us its velocity just after the collision. We can then use energy conservation to discover the height to which the lighter sphere rebounds.

We begin by calculating the initial potential energy, which is converted to kinetic energy when sphere 1 swings down to the minimum point (neglecting air resistance). From its kinetic energy, we can find its speed, v_1, just before the collision. The mass m_1 is raised a distance $L(1 - \cos \theta) = L/2$ above the minimum point. We take the potential energy to be zero when the spheres are hanging free. The initial potential energy of m_1 is then $U_i = m_1gL/2$. The kinetic energy of m_1 just before the collision is this potential energy:

$$\frac{1}{2}m_1v_1^2 = \frac{1}{2}m_1gL$$

$$v_1 = \sqrt{gL}.$$

The collision takes place at the bottom of the swing, where $\mathbf{v}_1$ is horizontal. Let the positive direction be to the right. Thus sphere 1 swinging from left to right has a velocity $+v_1$. (The collision takes place in one dimension so we drop the bold-faced notation.) The initial momentum is

$$p_i = m_1v_1 + m_2v_2 = m_1v_1.$$

To find the velocities v_1' and v_2' we use conservation of momentum, which states that

$$p_f = m_1v_1' + m_2v_2' = p_i = m_1v_1.$$

As in Eq. (8–33), energy conservation implies that the initial relative velocity, v_{rel}, and the final relative velocity are equal in magnitude but opposite in sign. With

$$v_{\text{rel}} = v_1 - v_2 = v_1,$$

FIGURE 8–13 Example 8–6. Two spheres hang from strings of equal length. (a) Sphere m_1 is pulled back to the left at an angle θ and released. It collides with sphere m_2, which is at rest. (b) The balls recoil, with sphere m_1 reaching a maximum height characterized by the angle ϕ.

(a)

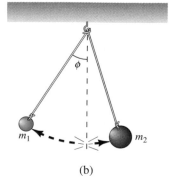

(b)

we have a final relative velocity of

$$v_1' - v_2' = -v_{\text{rel}} = -v_1.$$

We now have two simultaneous equations that can be solved for v_1' and v_2', namely, $m_1 v_1' + m_2 v_2' = m_1 v_1$ (momentum conservation) and $v_1' - v_2' = -v_1$ (energy conservation). We are interested only in the velocity v_1' of the lighter sphere, and the solution for this quantity is

$$v_1' = \frac{m_1 - m_2}{m_1 + m_2} v_1 = \frac{m_1 - m_2}{m_1 + m_2} \sqrt{gL}.$$

Because m_2 is larger than m_1, v_1' is negative, which indicates that the sphere recoils back to the left.

We next find the height to which m_1 recoils. The kinetic energy right after the collision is $m_1 (v_1')^2 / 2$. This energy will be converted into a potential energy, U_f, at the end of the swing. The expression for U_f is $m_1 g L(1 - \cos \phi)$, where ϕ is the desired angle of recoil (Fig. 8–13b). Thus the conservation of energy gives

$$m_1 g L(1 - \cos \phi) = \frac{1}{2} m_1 (v_1')^2.$$

We solve for ϕ:

$$1 - \cos \phi = \frac{(v_1')^2}{2gL} = \frac{(m_1 - m_2)^2}{(m_1 + m_2)^2} gL \frac{1}{2gL} = \frac{(m_1 - m_2)^2}{2(m_1 + m_2)^2}$$

$$= \frac{(1.0 \text{ kg} - 1.5 \text{ kg})^2}{2(1.0 \text{ kg} + 1.5 \text{ kg})^2} = 0.020,$$

or $\phi = 11°$.

∞ Friction is discussed in Chapter 5.

EXAMPLE 8–7 A 10-g bullet is fired in the $+x$-direction into a stationary block of wood that has a mass of 5 kg (Fig. 8–14). The speed of the bullet before entry into the wood block is $v_0 = 500$ m/s. What is the speed of the block just after the bullet has become embedded? What distance d will the block slide on a surface with a coefficient of friction equal to 0.50?

Solution: Let m be the mass of the bullet and v_0 be its velocity before it enters the wood block (Fig. 8–14a). The initial momentum is then $p_i = mv_0$. If M is the combined mass of the bullet plus the block and v is the velocity of the block and bullet just after the bullet enters the block, then momentum conservation implies that

$$mv_0 = Mv;$$

$$v = \frac{m}{M} v_0 = \frac{10 \text{ g}}{5010 \text{ g}} (500 \text{ m/s}) = 1 \text{ m/s}.$$

Note that we can ignore the mass of the bullet compared to that of the block and still attain good accuracy.

The force of friction is

$$\mu N = -\mu M g.$$

(The minus sign indicates that friction points to the left.) This has constant magnitude and leads to a constant acceleration a of the block, according to Newton's second law:

$$-\mu M g = Ma;$$

$$a = -\mu g.$$

The negative sign means the block slows down, traveling a distance d before it stops (Fig. 8–14b). Because the acceleration is uniform, we can use the relation $v_f^2 - v_i^2 = 2ad$ [from Eq. (2–22)]. With $v_f = 0$, we have

$$d = -\frac{v_i^2}{2a} = \frac{1}{2} \frac{v_i^2}{\mu g} = \frac{1}{2} \frac{(1 \text{ m/s})^2}{(0.5)(9.8 \text{ m/s}^2)} = 0.1 \text{ m}.$$

(a)

(b)

FIGURE 8–14 Example 8–7.

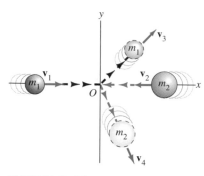

FIGURE 8–15 An object of mass m_1 and velocity $\mathbf{v}_1$ collides with another object of mass m_2 and velocity $\mathbf{v}_2$ in two dimensions.

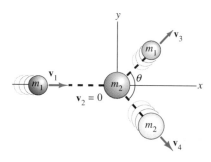

FIGURE 8–16 An object of mass m_1 and velocity $\mathbf{v}_1$ collides with an object of mass m_2 at rest. The objects move off after the collision.

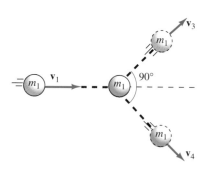

FIGURE 8–17 An object of mass m_1 and velocity $\mathbf{v}_1$ collides elastically with an object of the same mass at rest. The angle between the final velocities is 90°.

8–5 ELASTIC COLLISIONS IN SPACE

In one dimension, the possible motion of the colliding objects is severely limited. When collisions occur in space, however, as when billiard balls collide on a billiard table, there is a greater range of allowed motion and the vector nature of the mathematical equations becomes important.

The law of conservation of momentum [Eq. (8–9)] for the collision of two objects of masses m_1 and m_2, with initial velocities $\mathbf{v}_1$ and $\mathbf{v}_2$ and final velocities $\mathbf{v}_3$ and $\mathbf{v}_4$, reads:

$$m_1\mathbf{v}_1 + m_2\mathbf{v}_2 = m_1\mathbf{v}_3 + m_2\mathbf{v}_4. \tag{8–40}$$

The energy conservation law is

$$\frac{1}{2}m_1v_1^2 + \frac{1}{2}m_2v_2^2 = \frac{1}{2}m_1v_3^2 + \frac{1}{2}m_2v_4^2. \tag{8–41}$$

Let's first consider collisions in two dimensions, where everything happens in a plane, such as the xy-plane shown in Fig. 8–15. A familiar example is the collision of billiard balls on a pool table (but the masses of the colliding objects need not be equal). We want to find the magnitudes of the final velocities and the angles they make with the x-axis. Equivalently, we want the x- and y-components of the final velocities of objects 1 and 2. There are four unknowns but only three equations. [Equation (8–40) is a vector equation and actually comprises two equations: one for the x-components and one for the y-components.] Therefore, for a given set of initial velocities, there is no unique solution for the final velocities and the final objects can move in a variety of directions.

It is useful to see some of the constraints that these equations impose in the special case where one of the objects is initially at rest. Figure 8–15 shows the geometry. It is possible to solve this problem completely for the final velocities once the angles with which the objects leave the collision are specified. Let's look, though, at an interesting, if limited, result. If we choose object 2 to be at rest, then $v_2 = 0$. We take the square of Eq. (8–40) in the sense that the square of a vector equation $\mathbf{A} = \mathbf{B}$ implies $\mathbf{A} \cdot \mathbf{A} = \mathbf{B} \cdot \mathbf{B}$. With $v_2 = 0$, we find

$$m_1^2v_1^2 = m_1^2v_3^2 + 2m_1m_2\mathbf{v}_3 \cdot \mathbf{v}_4 + m_2^2v_4^2. \tag{8–42}$$

The energy conservation equation multiplied by $2m_1$ reads

$$m_1^2v_1^2 = m_1^2v_3^2 + m_1m_2v_4^2. \tag{8–43}$$

If θ is the angle between the final velocity vectors $\mathbf{v}_3$ and $\mathbf{v}_4$, so that $\mathbf{v}_3 \cdot \mathbf{v}_4 = v_3v_4 \cos \theta$, then the difference of Eqs. (8–42) and (8–43) is

$$m_1m_2v_4^2 = 2m_1m_2v_3v_4 \cos \theta + m_2^2v_4^2.$$

Dividing by m_2v_4 leads to

$$m_1v_4 - m_2v_4 = 2m_1v_3 \cos \theta. \tag{8–44}$$

In the special case where $m_1 = m_2$, Eq. (8–44) implies that $\cos \theta = 0$, or $\theta = 90°$, independent of the velocities, which is a fact known to billiards players (Fig. 8–17). The angle between a ball and the recoiling target ball is always a right angle. Energy conservation, as in Eq. (8–42), requires in this equal-mass case that

$$v_1^2 = v_3^2 + v_4^2. \tag{8–45}$$

The final velocities are constrained to form the sides of a right triangle whose hypotenuse is given by v_1.

EXAMPLE 8–8 Two billiard balls of equal mass approach each other along the x-axis; one is moving to the right with a speed of $v_1 = 10$ m/s and the other is moving to the left with a speed of $v_2 = 5$ m/s. After the collision, which is elastic, one of the balls moves in the direction of the y-axis (Fig. 8–18). What are the velocities $\mathbf{v}_3$ and $\mathbf{v}_4$ of the balls after the collision?

Solution: The fact that the masses of the two balls are equal simplifies the algebra. If we write the momentum conservation equation and divide by the factor of m that appears everywhere, we deal with a conservation of velocities.

Dividing by the mass factor in the energy conservation equation gives a conservation law for the squares of the velocities.

The initial velocities of the balls are

$$\mathbf{v}_1 = (10 \text{ m/s})\mathbf{i}$$

and

$$\mathbf{v}_2 = (-5 \text{ m/s})\mathbf{i}.$$

After the collision one ball moves in the y-direction; its velocity is given by $\mathbf{v}_3 = v_3 \mathbf{j}$. Let the velocity of the second ball after the collision be $\mathbf{v}_4 = v_{4x} \mathbf{i} + v_{4y} \mathbf{j}$. Momentum conservation now reads

$$(10 \text{ m/s})\mathbf{i} + (-5 \text{ m/s})\mathbf{i} = v_3 \mathbf{j} + v_{4x} \mathbf{i} + v_{4y} \mathbf{j}.$$

We now separately equate the coefficients of $\mathbf{i}$ and those of $\mathbf{j}$; that is, we use the fact that conservation of momentum holds in the x-direction and in the y-direction separately. We find

$$v_{4x} = (10 \text{ m/s}) + (-5 \text{ m/s}) = 5 \text{ m/s} \quad \text{and} \quad v_{4y} = -v_3.$$

Kinetic energy is conserved because the collision is elastic and this result leads to

$$\frac{1}{2}(10 \text{ m/s})^2 + \frac{1}{2}(-5 \text{ m/s})^2 = \frac{1}{2}v_3^2 + \frac{1}{2}(v_{4x}^2 + v_{4y}^2);$$
$$100 \text{ m}^2/\text{s}^2 + 25 \text{ m}^2/\text{s}^2 = v_3^2 + 25 \text{ m}^2/\text{s}^2 + v_3^2.$$

Thus we have $100 \text{ m}^2/\text{s}^2 = 2v_3^2$, or $v_3 = \sqrt{50}$ m/s. We also have $v_{4y} = -\sqrt{50}$ m/s.

What about collisions in three dimensions? Conservation of momentum allows us to show that there is always an inertial reference frame in which elastic collisions between two objects take place in a plane (Fig. 8–19). The results established for two dimensions are thus directly applicable to three dimensions.

FIGURE 8–18 Example 8–8.

FIGURE 8–19 The elastic collisions of two bodies take place in a plane, which is formed by the two momentum vectors of the incoming masses (or of the outgoing masses).

INTERIM SUMMARY

The momentum $\mathbf{p} \equiv m\mathbf{v}$ of an object of mass m moving with velocity $\mathbf{v}$ plays an important role in the interactions of objects. In particular, Newton's second law states that the net force acting on an object is the rate of change of its momentum; that is, $\mathbf{F} = d\mathbf{p}/dt$. Using Newton's second law, we find the change in momentum in terms of the impulse, $\mathbf{J}$:

$$\Delta \mathbf{p} = \mathbf{J} = \int \mathbf{F} \, dt = \mathbf{F}_{\text{av}} \, \Delta t. \tag{8–15}$$

Impulses are most useful when the force acts briefly, as in a collision.

The concept of momentum is useful for understanding isolated systems of interacting objects, which are systems on which no external force acts. The total momentum of an isolated system of objects is conserved; importantly, this can be applied to collisions of two objects. Collisions involve only interparticle forces between the colliding objects, and the effects of external forces can normally be ignored during a collision. Thus the total momentum of the system of objects is conserved whether the collisions are elastic, inelastic, or perfectly inelastic. In elastic collisions, the sum of the kinetic energies is conserved. During inelastic collisions, the sum of the kinetic energies changes. The change is generally a decrease due to the presence of dissipative forces, such as friction. In perfectly inelastic collisions, the objects that collide merge into a single object whose mass is the sum of the masses of the colliding objects.

PROBLEM-SOLVING TECHNIQUES

Collision problems involve an interaction between objects that is limited in space (and time). At some early time, the objects do not exert any forces on one another; at some later time, they again exert no forces on one another. Typically, we must find some parameters of the final (or initial) motion—the final velocity of one of the objects, for example. The following set of steps can be useful.

1. We must clearly identify the relevant interaction between the objects involved in a collision. This means that the objects are isolated or that, during the collision, external forces are small compared to the impulsive forces.

2. We must identify the objects involved before the collision and those involved after the collision. The objects that result from a collision may not be the same as the objects before the collision. Explosions or collisions in which the colliding objects coalesce are such examples.

3. Identify the quantities that are known and those unknown quantities that are to be found. In particular, a count of the number of known and unknown quantities, including the number of vector components, is helpful.

4. Remember that the conservation of momentum, which is a vector relation, is always applicable. If the mo-

tion is in one, two, or three dimensions, the conservation of momentum gives one, two, or three relations among the momenta (or velocities), respectively.

5. If the collision is known to be elastic, the conservation of energy provides an additional equation that involves the speeds of the objects. If the collision is not elastic, then an equation for the energy is available only if information about the energy loss is given.

6. The number of equations that include the unknowns must match the number of unknowns; if so, we can solve the problem by using algebra.

8-6   THE CENTER OF MASS

In a system of many objects, there is a particular point with special properties, called the *center of mass*. As we shall see, the importance of this point is that it moves under Newton's second law as if the total mass of the system were concentrated there.

To find this point, recall from Section 8–1 that, in the absence of external forces, the sum of the momenta of three objects is a constant of the motion; the procedure we used is easily generalized to an isolated N-object system. We find that the total momentum, $\mathbf{P}$, of the system does not change in the absence of external forces:

$$\mathbf{P} = m_1\mathbf{v}_1 + m_2\mathbf{v}_2 + \cdots + m_N\mathbf{v}_N. \tag{8–46}$$

The momentum of the system is a constant vector—one that *does not change in magnitude or direction as a function of time*. If we also assume that the masses do not change as a function of time, we have

$$\mathbf{P} = m_1\frac{d\mathbf{r}_1}{dt} + m_2\frac{d\mathbf{r}_2}{dt} + \cdots + m_N\frac{d\mathbf{r}_N}{dt} = \text{a constant};$$

$$\frac{d}{dt}(m_1\mathbf{r}_1 + m_2\mathbf{r}_2 + \cdots + m_N\mathbf{r}_N) = \text{a constant}. \tag{8–47}$$

The quantity in parentheses has the dimensions [ML]. We divide it by the total mass, $M = m_1 + m_2 + \cdots + m_N$, to obtain a vector that determines the position vector $\mathbf{R}$ of the **center of mass**, the average position of all of the particles in a system, weighted by their mass:

The position vector of the center of mass

$$\mathbf{R} \equiv \frac{m_1\mathbf{r}_1 + m_2\mathbf{r}_2 + \cdots + m_N\mathbf{r}_N}{M}. \tag{8–48}$$

From its definition, we see that the center-of-mass vector, $\mathbf{R}$, has components X, Y, and Z, which are

$$X = \frac{m_1x_1 + m_2x_2 + \cdots + m_Nx_N}{M}, \tag{8–49a}$$

$$Y = \frac{m_1y_1 + m_2y_2 + \cdots + m_Ny_N}{M}, \qquad (8\text{–}49b)$$

$$Z = \frac{m_1z_1 + m_2z_2 + \cdots + m_Nz_N}{M}. \qquad (8\text{–}49c)$$

Figure 8–20 illustrates several examples of the center of mass position for two and three objects.

Velocity and Acceleration of the Center of Mass

We can find the velocity, **V**, of the center of mass by taking a derivative of its position **R** [Eq. (8–48)]:

$$\mathbf{V} \equiv \frac{d\mathbf{R}}{dt} = \frac{m_1\mathbf{v}_1 + m_2\mathbf{v}_2 + \cdots + m_N\mathbf{v}_N}{M}. \qquad (8\text{–}50)$$

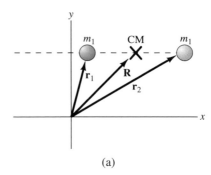

(a)

The numerator is the sum of the momenta of the individual pieces, so we may write this result as

$$\mathbf{V} = \mathbf{P}_{\text{tot}}/M. \qquad (8\text{–}51)$$

Differentiating once more with respect to time gives us the acceleration, **A**, of the center of mass:

$$\mathbf{A} \equiv \frac{d\mathbf{V}}{dt} = \frac{d^2\mathbf{R}}{dt^2}. \qquad (8\text{–}52)$$

Multiplying by M gives the relation

$$M\frac{d^2\mathbf{R}}{dt^2} = \frac{d\mathbf{p}_1}{dt} + \frac{d\mathbf{p}_2}{dt} + \cdots + \frac{d\mathbf{p}_N}{dt} = \frac{d\mathbf{P}_{\text{tot}}}{dt}. \qquad (8\text{–}53)$$

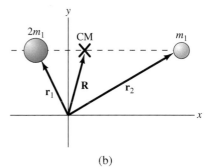

(b)

The center of mass is useful for several reasons. Equation (8–51) shows that, *in the absence of external forces, the center of mass moves with constant velocity.* No matter how complicated the motion of the constituent objects, the motion of its center of mass is simple. We shall soon see how Eq. (8–53) shows that when the system is not isolated—so that external forces act on its constituents—its *center of mass moves according to Newton's second law, with an acceleration equal to the net external force divided by the total mass.* For example, the center of mass of a diver moves with the same, simple parabolic motion that a rock with the same initial velocity would have if it were thrown from the diving board. It is important to be able to calculate the center of mass, as in Examples 8–9 and 8–10.

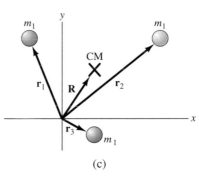

(c)

FIGURE 8–20 Several examples of the center of mass. (a) Both balls have equal masses. (b) One ball has twice the mass of the other. (c) Three balls of equal mass. We shall look below at several explicit calculations of the center of mass.

EXAMPLE 8–9 Two pointlike masses are placed on a massless rod that is 1.5 m long. The masses are placed as follows: 1.6 kg at the left end and 1.8 kg 1.2 m from the left end. (a) What is the location of the center of mass? (b) By moving the 1.8-kg mass, can you arrange to have the center of mass in the middle of the rod?

Solution: (a) Figure 8–21 illustrates this one-dimensional problem. We align the rod with length L along the x-axis, and we place its left end at the origin. We

The importance of the center of mass

FIGURE 8–21 Example 8–9. The rod is aligned with the x-axis, with its left end at the origin.

call the 1.6-kg mass m_1, placed at $x_1 = 0$ m. The 1.8-kg mass is m_2, and it is placed at $x = 1.2$ m. The location of the center of mass is given by Eq. (8–49a):

$$X = \frac{m_1 x_1 + m_2 x_2}{m_1 + m_2} = \frac{(1.6 \text{ kg})(0 \text{ m}) + (1.8 \text{ kg})(1.2 \text{ m})}{1.6 \text{ kg} + 1.8 \text{ kg}} = 0.64 \text{ m}.$$

(b) The midpoint of the rod is $x_{\text{mid}} = 0.75$ m. Let's place m_2 at a new position x_2 and ask if there is a solution for x_2 along the rod ($x_2 < L$) such that $X = x_{\text{mid}}$. We have

$$X = x_{\text{mid}} = \frac{m_1 x_1 + m_2 x_2}{m_1 + m_2} = \frac{m_1(0 \text{ m}) + m_2 x_2}{M} = \frac{m_2 x_2}{M}.$$

We can solve this equation for x_2:

$$x_2 = \frac{M x_{\text{mid}}}{m_2} = \frac{(1.6 \text{ kg} + 1.8 \text{ kg})(0.75 \text{ m})}{1.8 \text{ kg}} = 1.4 \text{ m}.$$

This value is indeed less than L, so if we place the 1.8-kg mass at this point, the center of mass is at the midpoint of the rod.

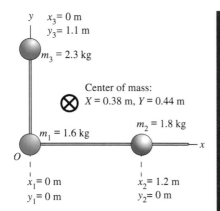

y $x_3 = 0$ m
$y_3 = 1.1$ m

$m_3 = 2.3$ kg

Center of mass:
$\otimes$ $X = 0.38$ m, $Y = 0.44$ m

$m_2 = 1.8$ kg

$m_1 = 1.6$ kg

O —— x

$x_1 = 0$ m
$y_1 = 0$ m

$x_2 = 1.2$ m
$y_2 = 0$ m

FIGURE 8–22 Example 8–10. The location of the center of mass is marked with a circled X.

EXAMPLE 8–10   Add a third mass of 2.3 kg to the masses of Example 8–9 at the point shown in Fig. 8–22. Find the center of mass.

Solution: By adding the third mass, m_3, the problem has become a two-dimensional one. We again align the first two masses along the x-axis and place the y-axis as shown in Fig. 8–22, so that the third mass is at $x_3 = 0$ m, $y_3 = 1.1$ m. We must now use Eqs. (8–49) to find both X and Y. With $M = 1.6$ kg + 1.8 kg + 2.3 kg = 5.7 kg, we have

$$X = \frac{m_1 x_1 + m_2 x_2 + m_3 x_3}{M}$$

$$= \frac{(1.6 \text{ kg})(0 \text{ m}) + (1.8 \text{ kg})(1.2 \text{ m}) + (2.3 \text{ kg})(0 \text{ m})}{5.7 \text{ kg}} = 0.38 \text{ m},$$

$$Y = \frac{m_1 y_1 + m_2 y_2 + m_3 y_3}{M}$$

$$= \frac{(1.6 \text{ kg})(0 \text{ m}) + (1.8 \text{ kg})(0 \text{ m}) + (2.3 \text{ kg})(1.1 \text{ m})}{5.7 \text{ kg}} = 0.44 \text{ m}.$$

This point is indicated in Fig. 8–22.

The Center of Mass and External Forces

What happens when there are external forces present in addition to interparticle forces? Let's again consider the three-object system shown in Fig. 8–23 and write

$$\frac{d\mathbf{p}_1}{dt} = \mathbf{F}_{12} + \mathbf{F}_{13} + \mathbf{F}_{1,\text{ext}},$$

$$\frac{d\mathbf{p}_2}{dt} = \mathbf{F}_{21} + \mathbf{F}_{23} + \mathbf{F}_{2,\text{ext}},$$

$$\frac{d\mathbf{p}_3}{dt} = \mathbf{F}_{31} + \mathbf{F}_{32} + \mathbf{F}_{3,\text{ext}}.$$

Here $\mathbf{F}_{i,\text{ext}}$ is the external force on object i. If we add these equations, use

$$\mathbf{P}_{\text{tot}} = \mathbf{p}_1 + \mathbf{p}_2 + \mathbf{p}_3,$$

and use Newton's third law—according to which $\mathbf{F}_{12} + \mathbf{F}_{21} = 0$, $\mathbf{F}_{13} + \mathbf{F}_{31} = 0$, and $\mathbf{F}_{23} + \mathbf{F}_{32} = 0$—we find that

$$\frac{d\mathbf{P}_{tot}}{dt} = \mathbf{F}_{1,ext} + \mathbf{F}_{2,ext} + \mathbf{F}_{3,ext} = \mathbf{F}_{tot,ext}. \quad (8\text{--}54)$$

This procedure can be extended to N objects, and Eqs. (8–53) and (8–54) give the general result:

$$\mathbf{F}_{tot,ext} = \frac{d\mathbf{P}_{tot}}{dt} = M\frac{d^2\mathbf{R}}{dt^2}. \quad (8\text{--}55)$$

For any system with total mass M in which there are both internal and external forces, we have shown that *the center of mass moves like a single point mass of mass M subject to the total external force on the system.* In other words, the system obeys the equation

$$\mathbf{F}_{tot,ext} = M\mathbf{A}, \quad (8\text{--}56)$$

where $\mathbf{F}_{tot,ext}$ is the total external force on the system and $\mathbf{A}$ is the acceleration of the center of mass.

This result is very useful. Consider a bowling ball rolling down an inclined plane. The bowling ball consists of some 10^{27} molecules and these interact with each other by means of electric forces. Despite this huge complexity, it is possible to treat a bowling ball as a point mass (or as a simple rigid object, as we shall see in Chapters 9 and 10) and study its motion without being aware of the underlying atomic structure of the ball or even of the laws that govern the motion of atoms. Without this possibility, the development of physics would not have occurred.

Figure 8–24 shows that the motion of a system's center of mass is simple even though there is a rather complex motion of parts of the system.

The Center of Mass of a Continuous Mass Distribution

We know that matter is composed of discrete masses; if we look closely enough, we will find evidence for the presence of atoms. But to an instrument that is not sufficiently acute to see atoms (such as our eyes), a solid object looks and behaves like a continuum of matter; that is, a *continuous mass distribution*. We may define the center of mass of a continuous distribution of mass by a simple generalization of Eq. (8–49). Consider a thin rod of mass M lying along the x-axis. We may take one end to lie at $x = a$ and the other at $x = b$. Now let's divide the interval between a and b into a large number N of tiny segments (Fig. 8–25). The segments are of equal length, Δx, but they are not necessarily of equal mass, because the rod may not have a uniform

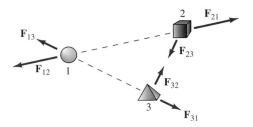

FIGURE 8–23 There are altogether six forces in a three-object system. These forces are labeled both by the object that causes them and the object on which they act. According to Newton's third law, these forces, whether they are attractive or repulsive, are equal and opposite in pairs.

The center of mass moves under the influence of external forces according to Newton's second law.

FIGURE 8–24 The center of mass of the dancer moves as a point mass under the influence of an external force. In this case, the external force is gravity so the center of mass has a parabolic trajectory.

FIGURE 8–25 A continuous mass is
divided into N tiny mass segments of
equal length Δx (where N is large).
This technique allows us to find the
center of mass. The nth segment, x_n,
has mass Δm_n.

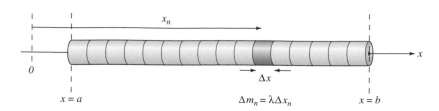

mass distribution. The position of the nth segment is labeled x_n, and we call the mass
of this tiny segment Δm_n. We may write this mass in terms of the local mass density:

$$\Delta m_n = \left(\frac{\Delta m_n}{\Delta x}\right)\Delta x. \tag{8–57}$$

The masses of all the segments add up to the total mass of the rod, so that we have

$$\Delta m_1 + \Delta m_2 + \cdots + \Delta m_n = \left(\frac{\Delta m_1}{\Delta x} + \frac{\Delta m_2}{\Delta x} + \cdots + \frac{\Delta m_n}{\Delta x}\right)\Delta x = M. \tag{8–58}$$

If we were to use infinitesimal intervals, this equation would read

$$M = \int_a^b \left(\frac{dm}{dx}\right)dx. \tag{8–59}$$

According to Eq. (8–49a), we have

$$X = \frac{\Delta m_1 x_1 + \Delta m_2 x_2 + \cdots + \Delta m_n x_n}{M}$$

$$= \frac{1}{M}\left[\left(\frac{\Delta m_1}{\Delta x}\right)x_1 + \left(\frac{\Delta m_2}{\Delta x}\right)x_2 + \cdots + \left(\frac{\Delta m_n}{\Delta x}\right)x_n\right]\Delta x.$$

As $N \to \infty$ this equation becomes an integral:

$$X = \frac{1}{M}\int_a^b \left(\frac{dm}{dx}\right)x\,dx.$$

It is usual to write the mass density dm/dx of a one-dimensional object as λ. If λ is not
a constant, it will depend on x. In this notation, the position of the center of mass is

$$X = \frac{1}{M}\int_a^b x\lambda\,dx. \tag{8–60}$$

Equation (8–60) allows us to calculate the center of mass of a one-dimensional object,
as in Example 8–11.

EXAMPLE 8–11   Consider a rod of length L, whose mass density (dm/dx) is
given by $\lambda = C(1 + ax^2)$, where x is the distance from the light end and C is a constant with dimensions of mass per length (Fig. 8–26). Calculate the center of mass
of the rod.

Solution: We first note that the total mass of the rod is given by

$$M = \int_0^L \lambda\,dx = \int_0^L C(1 + ax^2)dx = C\left(x + \frac{ax^3}{3}\right)\Big|_0^L = C\left(L + \frac{aL^3}{3}\right).$$

Now, from Eq. (8–60),

$$X = \frac{1}{M}\int_0^L xC(1 + ax^2)dx = \frac{1}{M}C\left(\frac{x^2}{2} + \frac{ax^4}{4}\right)\Big|_0^L$$

FIGURE 8–26 Example 8–11.

$$= \frac{1}{M} C\left(\frac{L^2}{2} + \frac{aL^4}{4}\right)$$

$$= \frac{C[(L^2/2) + (aL^4/4)]}{C[L + (aL^3/3)]} = \frac{(L/2) + (aL^3/4)}{1 + (aL^2/3)} = \left(\frac{L}{2}\right)\frac{1 + (aL^2/2)}{1 + (aL^2/3)}.$$

If $a = 0$, then the rod is uniform, and the center of mass is at $L/2$, as expected. In the case of nonuniformity, the center of mass is closer to the more massive end. Note that the parameter C of the mass density cancels.

Continuous Objects in Two and Three Dimensions. Suppose that instead of a one-dimensional object, we deal with a two-dimensional one such as a sheet of metal with an area A. We divide the object into tiny elements of area $\Delta A = \Delta x \, \Delta y$ and mass Δm (Fig. 8–27). Suppose that for the element located at the point (x, y) the mass density is $\sigma(x, y)$. This quantity is analogous to the one-dimensional mass density λ; σ has dimensions of mass/unit area. If we take the area element small enough so that σ is a constant within that area, then the mass of that area element is

$$\Delta m = \sigma \, \Delta x \, \Delta y.$$

Remember, the value of σ can vary from place to place, as the varying height within Fig. 8–27 indicates. We can find the total mass by summing the masses over all the elements, and, in the calculus limit, this summation takes the form[†]

$$M = \lim\sum \Delta m = \lim \sum \sigma \, \Delta x \, \Delta y = \int_{\text{Surface}} \sigma(x, y) \, dx \, dy. \tag{8–61}$$

Similar reasoning tells us that the coordinates of the center of mass are

$$MX = \lim\sum x \, \Delta m = \int_{\text{Surface}} x\sigma(x, y) \, dx \, dy; \tag{8–62a}$$

$$MY = \lim\sum y \, \Delta m = \int_{\text{Surface}} y\sigma(x, y) \, dx \, dy. \tag{8–62b}$$

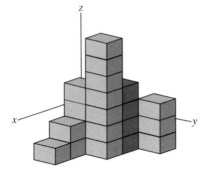

FIGURE 8–27 We have placed a two-dimensional object in the xy-plane and divided it up into squares, each of area $\Delta x \, \Delta y$. The mass of a square at a given point is the two-dimensional density σ times the area. The height of the towers in this figure represents the value of σ; we see it can vary from point to point.

When we want to learn about the center of mass of a three-dimensional object of volume V, we must know its three-dimensional mass density $\rho(x, y, z)$, measured in mass/unit volume. The dependence on x, y, z is merely an expression of the fact that the mass density can vary from point to point. The total mass is

$$M = \int_{\text{Volume}} \rho(x, y, z) \, dx \, dy \, dz, \tag{8–63}$$

while the center of mass is located at the position

$$M\mathbf{R} = \int_{\text{Volume}} \mathbf{r}\rho(x, y, z) \, dx \, dy \, dz. \tag{8–64}$$

8–7 ROCKET MOTION

A rocket is a system that undergoes a kind of continuous explosion. To understand rocket motion, imagine an astronaut who steps out of a space vehicle to do some repairs and, for some reason, finds himself or herself at rest with respect to the vehicle but separated from it. No amount of bodily contortions will bring the astronaut closer to the vehicle because the center of mass of the astronaut, if initially at rest, will remain at rest in the absence of external forces. If the astronaut is holding something that

---

[†] The subscript "Surface" on the single integral sign denotes a two-dimensional integration over the area A.

APPLICATION

Rocket propulsion

FIGURE 8–28 An astronaut can get a small push back toward the Space Shuttle by throwing a hammer in the opposite direction.

can be thrown away, then he or she is in business. For example, if the astronaut throws a hammer in a direction away from the space vehicle, he or she will start to move toward the space vehicle and safety, albeit with a small speed (Fig. 8–28). Momentum conservation lies behind this motion. If the astronaut and the hammer are initially at rest, the total momentum of the astronaut–hammer system is zero, and this must still be the case after the hammer is thrown. A rocket moves because of the same principle used by the astronaut. The ejected fuel of the rocket acts as a whole series of thrown hammers. The presence or absence of an atmosphere has nothing to do with the basic principle of operation.

A rocket is propelled by the ejection of hot gases from burning chemical fuel. Two important parameters for describing the characteristics of a rocket are the exhaust speed[†] of the hot gases, u_{ex}, and the rate at which the gases are expelled by the rocket; that is, the rate of change of mass dm/dt of the rocket (Fig. 8–29). Keep in mind that the exhaust speed involves a burning reaction within the rocket and that it is measured *with respect to the rocket*. Suppose the initial mass of the rocket with all its fuel is m_0; at some later time t, its velocity with respect to some ground-based (inertial) observer is v while its mass has been reduced to m. In the next small time interval Δt, the ground-based observer sees a small bit of exhaust gas, mass Δm, is ejected backward relative to the rocket; the ground-based observer sees this bit of gas moving with velocity $v - u_{ex}$, while the observer sees the rocket itself now traveling at velocity $v + \Delta v$ and with the decreased mass $m - \Delta m$. In the absence of external forces, the total momentum at time t equals the total momentum at time $t + \Delta t$. We have $P(t) = mv$, while $P(t + \Delta t) = (\Delta m)(v - u_{ex}) + (m - \Delta m)(v + \Delta v)$. (The last two terms are the momentum of the exhaust gas and the rocket, respectively.) Thus

$$mv = (\Delta m)(v - u_{ex}) + (m - \Delta m)(v + \Delta v)$$
$$= (\Delta m)v - (\Delta m)u_{ex} + mv + m\Delta v - (\Delta m)v - (\Delta m)(\Delta v),$$

or, neglecting the very small term $(\Delta m)(\Delta v)$,

$$m\Delta v + (\Delta m)u_{ex} = 0. \tag{8–65}$$

This equation describes the fact that the momentum given up by the rocket is taken up by the gas (Fig. 8–30). Let us divide Eq. (8–65) by the time interval Δt and take the limit $\Delta t \rightarrow 0$. We have

$$-u_{ex}\frac{dm}{dt} = m\frac{dv}{dt}. \tag{8–66}$$

This equation looks just like the usual $F = ma$ equation. The left-hand side of Eq. (8–66) is called the **thrust** of the rocket. Notice that because dm/dt is negative (the rocket loses mass), the thrust is a positive quantity. Equation (8–66) may be written in a more convenient form by dividing both sides of the equation by mu_{ex}. It then takes the form

$$\frac{1}{u_{ex}}\frac{dv}{dt} = -\frac{1}{m}\frac{dm}{dt}. \tag{8–67}$$

FIGURE 8–29 A rocket expels gases that have a speed u_{ex} with respect to the rocket. If we know the rate at which the gases are expelled, conservation of momentum allows us to find the velocity **v** of the rocket.

System "exhaust + rocket" has $\mathbf{F}_{ext} = 0$.

[†]Sometimes called the exhaust velocity. We prefer the term speed because it is always positive.

We now use

$$\frac{d(\ln m)}{dt} = \left(\frac{1}{m}\right)\left(\frac{dm}{dt}\right). \qquad (8-68)$$

[This formula follows from the calculus results in Appendix IV–7. We use

$$\frac{d(\ln x)}{dx} = \frac{1}{x}.$$

We must replace x by m, and use the chain rule of calculus.] It follows from Eqs. (8–67) and (8–68) that

$$\frac{d}{dt}\left(\ln m + \frac{v}{u_{ex}}\right) = 0.$$

This equation is equivalent to

$$\ln m = -\frac{v}{u_{ex}} + \text{a constant.} \qquad (8-68)$$

At $t = 0$, we have $m = m_0$, the initial mass of the rocket, and $v = 0$, for example. In this case, we find that the constant in Eq. (8–68) is given by $\ln m_0$. Equation (8–68) is rearranged to give

$$\frac{v}{u_{ex}} = (\ln m_0) - (\ln m). \qquad (8-69)$$

Finally, we use the fact that $(\ln a) - (\ln b) = \ln(a/b)$ to obtain

$$v = u_{ex} \ln\left(\frac{m_0}{m}\right). \qquad (8-70)$$

FIGURE 8–30 The momentum of the rocket plus the fuel it has just ejected is constant, but the rocket has a forward momentum because the fuel has a momentum toward the rear.

The speed of a rocket

This equation for the rocket speed is a fundamental one in rocket propulsion.

Let's look more closely at Eq. (8–70). First of all, we note that the speed v is proportional to u_{ex}. This dependence is to be expected because u_{ex} is the only quantity with the dimensions of speed that appears in the problem. Next, we observe that the coefficient of u_{ex} must be a dimensionless quantity, and therefore it must be a function of m/m_0, which is the fraction of the rocket mass left over after time t. Because the logarithm is a very slowly varying function of its argument, the most productive way to get a sizable speed is by making u_{ex} large. Chemical rockets can generate gas exhaust speeds up to a maximum of about 4000 m/s, and thus the typical speeds of rockets are also in the range of 10^3 m/s. Large values of m_0/m help only moderately; for example, $\ln 10 = 2.3$, so changing the mass ratio by a factor of 10 improves the speed by a factor of 2.3. Much higher speeds can be achieved with nuclear reactions, but rockets that use nuclear energy for propulsion in this way would present enormous environmental problems!

Rocket Motion in the Presence of Gravity

The previous discussion involves a rocket free of all external forces. When gravity acts on the rocket, Eq. (8–70) is modified so that, in addition to the term proportional to $u_{ex} \ln(m_0/m)$ due to the fuel exhaust, the velocity has a falling motion due to gravity superimposed on it. As we already know from Chapter 2, the velocity in free fall is $-gt$. Thus Eq. (8–70) is modified to

$$v = u_{ex} \ln\left(\frac{m_0}{m}\right) - gt. \qquad (8-71)$$

EXAMPLE 8–12 A Saturn V rocket (the vehicle that sent humans to the Moon) of mass 2.5×10^6 kg takes off from Earth in a vertical direction. It burns fuel at a uniform rate of 1.6×10^4 kg/s for a duration of 2 min. If the exhaust speed of the gas is given by 3.0 km/s, what is the speed of the rocket immediately after the combustion ceases?

Solution: We use the formula in Eq. (8–71) to find the rocket's speed. The mass m in this formula is the final mass of the rocket after the gas has been burned. We can calculate this mass from the initial mass, m_0, and the rate of combustion, dm/dt:

$$m = m_0 - \frac{dm}{dt} t = 2.5 \times 10^6 \text{ kg} - (1.6 \times 10^4 \text{ kg/s})(120 \text{ s}) = 0.58 \times 10^6 \text{ kg}.$$

We also have

$$\ln\left(\frac{m_0}{m}\right) = \ln\left(\frac{2.5 \times 10^6 \text{ kg}}{0.58 \times 10^6 \text{ kg}}\right) = \ln\left(\frac{2.5}{0.58}\right) = 1.5.$$

Thus, from Eq. (8–71),

$$v_f = (3.0 \times 10^3 \text{ m/s})(1.5) - (9.8 \text{ m/s}^2)(120 \text{ s}) = 3.2 \times 10^3 \text{ m/s}.$$

Our calculations are correct only near the surface of Earth. Earth's gravitational force on the rocket decreases as the rocket moves away from the surface, so in effect g decreases as the rocket moves away.

*8–8   MOMENTUM TRANSFER AT HIGH ENERGIES

High-energy physics is the study of the properties of the elementary particles that make up matter at the most fundamental level. These properties are studied with the aid of high-energy accelerators. There, projectile particles such as protons or electrons are accelerated to high speeds and allowed to collide with target particles of the same type. The results of such collisions are then analyzed by means of sophisticated detectors in which the energies and momenta of the particles that emerge after the collisions are precisely measured. The patterns revealed in these measurements are used to probe the underlying structure of the colliding particles.

As an example of how such studies are made, imagine a BB gun fired at two targets: One is cotton candy, the other is cotton candy with a tiny steel pellet implanted at its center. A BB shot (the projectile) will hardly be deflected by the first target. Most of the time, the same will be true for the second target. Occasionally, however, the projectile will penetrate the second target close to its center and, upon colliding with the steel pellet, will be deflected through a large angle. An analysis of the collisions in the two cases would allow us to deduce that the second target had a hard, compact center. The pattern of many projectile deflections would even allow us to say whether the steel pellet is spherical or cubical. In this example, there is no benefit in increasing the power of the BB gun. All we would need to improve the experiment is to be able to aim more accurately at the center of the cotton candy.

In the case of high-energy collisions, though, there is indeed a benefit in increasing the speed—or momentum—of the projectile. For collisions of atomic or sub-atomic particles, the rules of *quantum mechanics* apply. This is a field of study developed in the first part of the twentieth century, with some surprising consequences. One consequence of quantum mechanics is that high momenta (or high energy) are required to study the regions close to the center of a target. More precisely, the **Heisenberg uncertainty relation**, discovered by physicist Werner Heisenberg, describes a relation between the magnitude of the momentum change, $\Delta\mathbf{p}$, of a target particle in a collision and the size of the region around the center of the target that can be studied. Let us denote by r_{coll} the smallest radius of a spherical region around the center of a target that can be studied in a collision. Then the uncertainty relation is

$$|\Delta\mathbf{p}|\,r_{coll} \simeq \hbar, \tag{8-72}$$

where $\hbar \simeq 10^{-34}$ J·s is Planck's constant divided by 2π. According to Eq. (8–72), the larger the value of $\Delta\mathbf{p}$, the smaller r_{coll} can be.

Experiments with elementary particles—like the BB gun fired at cotton candy—have been carried out and have revealed much about the structure of these particles. Ernest Rutherford's experiments in the first decade of the twentieth century were the first to reveal the presence of the nucleus in atoms, which plays a role like the steel pellet in the cotton candy. More recently, experiments involving protons colliding at momenta that could probe down to distances of 10^{-16} m revealed that the proton itself contains "pellets" known as *quarks*. When a next generation of accelerators is built, we will discern even more detail. Judging by the enormous difference between the physics of atoms and the physics of nuclei, we can expect some surprises as we probe more and more deeply.

SUMMARY

The momentum (or linear momentum) of an object of mass m moving with velocity $\mathbf{v}$ is defined by

$$\mathbf{p} = m\mathbf{v}. \tag{8-2}$$

The rate of change of momentum with time is given by Newton's second law:

$$\mathbf{F} = \frac{d\mathbf{p}}{dt}, \tag{8-3}$$

where $\mathbf{F}$ is the net force acting on the object. The momentum of a collection of objects with momenta $\mathbf{p}_i$ (where $i = 1, 2, \cdots, N$) is their vector sum, and the net external force on the system determines the rate of change of the total momentum. In particular, if the only forces present are the internal forces exerted by the objects on each other, then the total momentum does not change with time; that is, momentum is conserved. In the case of two objects, this relation is

$$\mathbf{p}_1 + \mathbf{p}_2 = \text{a constant.} \tag{8-9}$$

One consequence of Eq. (8–3) is that the change in momentum of an object influenced by a force that acts over a limited time (an impulsive force) is an integral over the force and is called the impulse, $\mathbf{J}$:

$$\mathbf{J} \equiv \int \mathbf{F}\,dt = \mathbf{F}_{av}\,\Delta t. \tag{8-15}$$

The impulse does not involve the detailed time dependence of the force.

Collisions involve interparticle forces, so that the total momentum of the system of objects is conserved whether the collisions are elastic, inelastic, or perfectly inelastic. During elastic collisions, the sum of the kinetic energies is conserved. Inelastic collisions involve a change in the sum of the kinetic energies. The change is generally a decrease due to the presence of dissipative forces, such as friction. In perfectly inelastic collisions, the objects that collide merge into a single object whose mass is the sum of the masses of the colliding objects.

The center of mass of a collection of objects with masses m_i and position vectors $\mathbf{r}_i$ ($i = 1, 2, \cdots, N$) is the mean position of a system's mass, which is defined by

$$\mathbf{R} \equiv \frac{m_1\mathbf{r}_1 + m_2\mathbf{r}_2 + \cdots + m_N\mathbf{r}_N}{M}. \tag{8-48}$$

In the absence of external forces, the center of mass moves with uniform velocity $\mathbf{V}$, given by

$$\mathbf{V} = \frac{\mathbf{P}_{tot}}{M} = \frac{m_1\mathbf{v}_1 + m_2\mathbf{v}_2 + \cdots + m_N\mathbf{v}_N}{M}. \tag{8–50}$$

The total mass of the system is $M = m_1 + m_2 + \cdots + m_N$. The center of mass for a continuous object may be calculated in terms of the mass density. For a one-dimensional object with mass density λ, it is given by

$$X = \frac{1}{M} \int \lambda x \, dx, \tag{8–60}$$

where the mass is related to $\lambda \equiv dm/dx$ by

$$M = \int \left(\frac{dm}{dx} \right) dx. \tag{8–59}$$

For a two-dimensional object with mass density σ, the corresponding expressions are

$$X = \frac{1}{M} \int_A \sigma x \, dx \, dy \tag{8–62a}$$

and

$$Y = \frac{1}{M} \int_A \sigma y \, dx \, dy, \tag{8–62b}$$

where

$$M = \int_A \sigma \, dx \, dy. \tag{8–61}$$

Similar expressions hold for a three-dimensional object with mass density ρ.

When a net external force $\mathbf{F}_{net}$ acts on a system, the center of mass continues to move in a simple way, as described by Newton's second law: The position of the center of mass $\mathbf{R}$ obeys the equation

$$\mathbf{F}_{tot,ext} = M\mathbf{A} = M \frac{d^2\mathbf{R}}{dt^2}. \tag{8–55), (8–56}$$

Here $\mathbf{A}$ is the acceleration of the center of mass.

The concept of momentum conservation is useful for the description of rocket motion. For a rocket that is not influenced by external forces, whose mass is initially m_0, and whose mass at time t is m, the speed is

$$v = u_{ex} \ln \left(\frac{m_0}{m} \right). \tag{8–70}$$

Here, u_{ex} is the exhaust speed of the gas relative to the rocket.

According to the Heisenberg uncertainty relation, when collisions are used to study the structure of a target, the momentum of an incoming projectile must increase as the distance to be probed decreases.

UNDERSTANDING KEY CONCEPTS

1. Discuss momentum conservation for a comet that enters the solar system, is deflected, and then leaves the solar system in a direction different from the direction by which it entered.

2. A common toy consists of a series of five balls that touch and form a line; each ball is suspended as a pendulum. When ball 1 is pulled away and then released, it strikes the line and ball 5 rises in a motion like the reverse of ball 1. How can you explain this motion?

3. Describe a physical object for which the center of mass is not actually inside the object.

4. A diver leaps off the diving board, performing a difficult series of maneuvers. Can he cause his center of mass to perform a midair loop?

5. The center of mass of a championship-level high jumper passes below the bar even though the jumper passes above the bar. How is this possible?

6. A particle collides with another particle at rest. If there are two particles observed to come out of the collision, do the momenta of the two final state particles and the initial incoming particle have to lie in a plane? Why or why not? What if the final system consists of three particles?

7. A vase falls to the floor and shatters. Is momentum conserved in the collision? What objects need to be taken into account in describing the conservation of momentum?

8. As a tennis racket hits a tennis ball, the racket continues to move forward. Is this consistent with the conservation of momentum?

9. If a tennis player wants more power, should he or she choose a racket with more or less tension on the strings? If the player wants more control? [*Hint*: Think about how long the ball is on the strings in each case.]

10. The mass on the end of a pendulum swings due to the effects of gravity on the mass. Does momentum have to be conserved in the interaction of this mass with Earth? If so, does this mean that Earth moves back and forth along with the mass?

11. We have said that an impulsive force acts for only a short time. What decides whether a time is "short" or not?

12. Cricket players catch balls hit as hard as baseballs but they do not use padded mitts. How do they avoid injuring their hands?

13. If you have the misfortune to be in an automobile collision, you are better off in a more massive car (all other things being equal). Why?

14. If you have the misfortune to be in an automobile collision, you are better off in a car that tends to crumple on impact rather than a car that holds together stiffly. Why?

15. A closed railroad car is at rest on a flat stretch of track. A cannon located inside the car at the front end points to the rear. What is the motion of the railroad car when the cannon is fired, and the shell is absorbed by the rear wall?

16. A large, closed crate contains many pigeons that sit on the floor of the crate. A sudden noise makes them all fly up and hit the top of the crate at the same time. Will they be able to lift the crate off the ground?

17. Drag forces act when a parachute opens. The parachute and its load will slow down to a terminal velocity. Why is the drag force velocity dependent? Think of the parachute as colliding with a lot of tiny air molecules.

18. A billiard ball strikes the cushion of a billiard table at an angle of 45°. Assuming that there is no energy loss (and no effects due to the spin of the ball), what can you say about the angle at which the billiard ball bounces off? What would happen if there were some energy loss?

PROBLEMS

8–1 Momentum and Its Conservation

1. (I) Calculate the magnitudes of the momenta of (a) a 40-g arrow traveling at a speed of 110 km/h; (b) a 145-g baseball traveling at a speed of 35 m/s; (c) a 72-kg sprinter running at 22 mi/h; (d) a 95-kg tackler running 100 m in 12.5 s.

2. (I) A 0.800-kg mass falls vertically downward from a roof 36.5 m high. What is the momentum of the object after 2.0 s, given that the initial velocity is zero?

3. (I) Calculate the magnitudes of the momenta of (a) a man of mass 70 kg, running 6 m/s; (b) a freight car of 100,000 kg, moving 60 m/s; (c) a car of 1100 kg, moving 25 mi/h; (d) a proton moving at 2×10^5 m/s; (e) a feather of 10 g in an airless container that has fallen 10 cm due to gravity.

4. (II) An object of mass m is constrained to move in a circle of radius R by a central force F. (a) What is the magnitude of the momentum of the object? (b) Suppose that the force has magnitude $F = Kv$, where K is a constant and v is the speed of the object. Calculate R in terms of K and the momentum of the object. (This kind of force acts on a charged object in a uniform magnetic field.)

5. (II) A 7-kg rifle is used to fire a 10-g bullet, which travels with a speed of 700 m/s. (a) What is the speed of recoil of the rifle? (b) How much energy does it transmit to the shoulder of the person using the rifle as it stops?

6. (II) Two objects, of masses m and M, respectively, move in circular orbits that have the same center (Fig. 8–31). If the force that gives rise to this motion is a force of attraction between the two objects acting along a line joining them, using momentum conservation and Newton's second law, (a) show

that they move with the same angular speed (see Chapter 3), and (b) calculate the ratio of the radii of the two circular orbits.

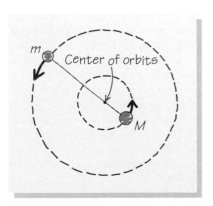

FIGURE 8–31 Problem 6.

7. (II) Two objects are moving in the xy-plane. The first, with mass 2.4 kg, has a velocity $\mathbf{v}_1 = (-2.0 \text{ m/s})\mathbf{i} + (-3.5 \text{ m/s})\mathbf{j}$; the second object, with mass 1.6 kg, has velocity $\mathbf{v}_2 = (1.8 \text{ m/s})\mathbf{i} + (-1.5 \text{ m/s})\mathbf{j}$. (a) What is the total momentum of the system? (b) If the system observed at a later time shows that the 2.4-kg object has $v'_1 = (2.5 \text{ m/s})\mathbf{i}$, what is the velocity of the 1.6-kg object? (c) Consider again the initial situation. Now suppose that there has been a mass transfer, so that the first object now has a mass of 2.1 kg. The total mass is conserved.

What is v_1' if the velocity of the second object is $(-2.5 \text{ m/s})\mathbf{j} + (1.3 \text{ m/s})\mathbf{k}$? (d) Calculate the sum of the kinetic energies in the initial configuration and in the configurations of parts (b) and (c). Compare and discuss them.

8. (III) (a) A billiard ball collides head-on with a second billiard ball that is touching a third billiard ball placed directly behind it. Describe the subsequent motion of all three billiard balls. Assume that all collisions are elastic. (b) A moving billiard ball hits the two balls at rest precisely between the two touching balls at rest. The velocity of the moving ball is perpendicular to the line between the centers of the two balls at rest. Again, all collisions are elastic. Describe the subsequent motion now.

8–2 Impulse and One-Dimensional Collisions

9. (I) A baseball of mass 0.15 kg, moving horizontally with a momentum of $4.0 \text{ kg} \cdot \text{m/s}$, is struck head-on by a baseball bat with an impulse of $10 \text{ N} \cdot \text{s}$. What is the speed of the baseball after it is struck?

10. (I) If a 25-g arrow moving at 90 km/h penetrates a block of wood suspended by a rope, what impulse is delivered to the block? What velocity will the wood block acquire if its mass is 5 kg? Ignore the mass of the arrow compared to that of the wood block.

11. (I) A 145-g baseball traveling at a speed of 36 m/s hits a bat and moves back along its incoming trajectory with a speed of 45 m/s. What is the impulse delivered to the ball by the bat? If the duration of the bat–ball collision is 7.0×10^{-4} s, what is the average force exerted by the bat on the ball during this period?

12. (I) A 120-g ball is dropped from a height of 4.8 m onto a hard floor and bounces back to exactly that height. (a) What is the impulse received by the ball? (b) What is the average force on the ball during the 0.013 s that, an independent measurement shows, it was in contact with the floor?

13. (I) A fire rescue unit uses a tightly woven net to catch an 80-kg person who jumps out of a burning building from a height of 11 m. What is the impulse transmitted to the net? If the net sinks 70 cm as it slows down the jumper, what is the average force exerted on the jumper by the net?

14. (II) (a) Estimate the possible impulse received by a nail when it is hit by a hammer. (b) Estimate the possible impulse received by a football when it is kicked in a kickoff. The ball has mass of about 250 g and travels a horizontal distance of some 50 m. Note the difference between parts (a) and (b); in the latter, you have additional information on the momentum the ball receives.

FIGURE 8–32 Problem 15.

15. (II) A high jumper of mass 55 kg clears 6′7″ (Fig. 8–32). What is the impulse transmitted to the jumper by the ground? (It has been found that the center of mass of the high jumper remains *below* the bar. In this problem, assume that the center of mass starts at 3′10″ and reaches the height 6′7″.)

16. (II) A communications satellite of mass 4000 kg can be ejected from the cargo bay of the Space Shuttle by means of springs. A particular satellite is ejected at 0.3 m/s. (a) What impulse does the spring provide? (b) If the spring operates over a time period of $\Delta t = 0.2$ s, what average force does the spring provide?

17. (II) A ball of mass 260 g is dropped from a height of 2.0 m. It hits the ground and rebounds to a height of 1.4 m. Assuming that the ball is in contact with the ground for 0.004 s, what is the average force exerted on the ball during the contact?

18. (II) A golf ball of mass 0.05 kg, placed on a tee, is struck by a golf club. The speed of the golf ball as it leaves the tee is 100 m/s. (a) What is the impulse? (b) If the time of contact between the club and the ball is 0.02 s, what is the average force? (c) If the force decreases to zero linearly with time during the 0.02 s, what is the value of the force at the beginning of the contact?

19. (II) A catcher catches a 45-m/s pitch and his glove recoils 0.25 m. The mass of the ball is 0.14 kg. Assume that the deceleration of the ball during the catching time, Δt, is constant and equal to some a_{av}. (a) Find the average force, F_{av}, that the catcher exerts. (b) How much work does he do? (c) Find Δt.

20. (II) Superman rushes to save Lois Lane, who has fallen (has been pushed?) from a window 100 m above a crowded street. Superman swoops down in the nick of time, arriving when Lois is 1.0 m above the street and stopping her just at ground level. Lois has been on a diet and has a mass of 50 kg. Ignore air resistance throughout. (a) What is the impulse that Lois receives as Superman catches her? (b) If the force that Superman supplies in stopping Lois is constant, how long does it take for Lois to come to rest? (c) What is the average force Superman applies to Lois? Compare this to the force of gravity on her. Draw your own conclusions about these last-minute rescues.

21. (II) A ball of mass 150 g is dropped from a height of 60 cm onto the first step at the top of a staircase. Each step is 20 cm above the next. The ball bounces perfectly elastically, but it has a small horizontal velocity, so it hits the second step on the next bounce, and then the third, and so on. Assume that the size of each step is such that the ball always bounces onto the next step down. What is the impulse transmitted to the nth step?

22. (II) A brother and sister live near a lake. In the middle of winter, when the lake is solidly and smoothly frozen, they play catch by sliding a rock back and forth between them across the ice. Ignore all friction. The mass of the two children is 40 kg each, and the rock has a mass of 2.0 kg. Each time the rock is given a push by the child who has just caught it, it leaves the hand with a speed of 1.0 m/s (relative to the hand). (a) The children are initially at rest, and the sister first slides the rock toward her brother. What is the initial motion of the sister after she releases the rock? (b) The brother receives the rock. What is his initial velocity after he has received it? (c) The brother then slides the rock to his sister. What is his velocity now? (d) The sister catches the rock. What is her velocity now?

(e) Will either child ever have a speed such that the rock will never reach the other player? In other words, can the game go on forever?

8–3 Perfectly Inelastic Collisions; Explosions

23. (I) A 250-g cart moves on an air track (a one-dimensional system in which motion is friction free) at 1.2 m/s. It collides with and sticks to another cart of mass 500 g, which was moving in the opposite direction at 0.80 m/s before the collision. What is the velocity of the composite cart after the collision?

24. (I) An object of mass m_0 with speed v_0 hits another object, of mass m, at rest. The two masses stick together in the process. Find the fractional change in kinetic energy.

25. (I) A snowball of mass 400 g is thrown with a speed of 10 m/s and hits the loosely placed head of a snowman. If the snowball sticks to the head, which is of mass 5 kg, with what initial speed will the now enlarged head recoil?

26. (I) A fireworks rocket is shot straight up in the air. Its "payload," mass 0.80 kg, explodes just as it reaches its peak height and separates into two fragments before eventually exploding again into spiraling colors. The first fragment, of mass 0.30 kg, heads straight down with an initial speed of 18 m/s. What is the velocity immediately after the explosion of the second fragment?

27. (I) Two objects each move with speed v in opposite directions along a line. They meet and have a completely inelastic collision. After the collision, the composite object moves along the same line with a speed of $v/2$. What is the ratio of the masses of the two objects?

28. (II) An ^{241}Am nucleus at rest emits an alpha particle (a ^{4}He nucleus). The energy released during the process is 6 MeV = 9.6×10^{-13} J. (a) What are the speeds of the alpha particle and the remaining ^{237}Np nucleus? (b) What is the kinetic energy of the ^{237}Np nucleus? (The masses of the nuclei may be taken to be $241m_0$, $4m_0$, and $237m_0$, respectively, where $m_0 = 1.66 \times 10^{-27}$ kg.)

29. (II) In a *ballistic pendulum,* a bullet of mass m and speed v embeds in a block of mass M suspended by a string (Fig. 8–33). The block and bullet, of total mass $m + M$, then moves as a pendulum and the maximum height h that it reaches can easily be measured. This apparatus can be used to measure the speed of the bullet. (a) What fraction of the bullet's kinetic energy is lost in the collision? (b) Give a formula for measuring v in terms of m, M, g, and h.

30. (II) A bullet strikes a ballistic pendulum (see Problem 29) whose target block has a mass of 5 kg. The block is observed to rise a distance of 3 cm, and the bullet is weighed and found to have a mass of 20 gm. What is the speed of the bullet when it strikes the block?

31. (III) Two persons, one of mass m_1 and the other of mass m_2, are sitting initially at rest on a bobsled of mass M on a frozen lake. Answer the following questions for a frame fixed to the lake. (a) Determine the final velocity of the bobsled if the person of mass m_1 jumps off the rear of the bobsled with horizontal velocity v, followed a few seconds later by the second person, who also jumps off in the same way with the same horizontal velocity with respect to the sled. (b) Find the velocity of the bobsled if the person of mass m_2 jumps off first. (c) What is the velocity if both jump off at the same time?

8–4 Elastic and Inelastic Collisions

32. (I) An object with velocity (1.4 m/s)$\mathbf{i}$ and mass 0.30 kg collides with an object whose velocity is $(-2.5$ m/s$)\mathbf{i}$ and whose mass is 0.15 kg. The motion takes place in one dimension. (a) What are the final velocities of the objects if the collision is elastic? (b) What is the total initial kinetic energy in the collision?

33. (I) On an air track, a 0.4-kg mass m_1 moves at 3 m/s in the positive direction. It approaches a stationary mass m_2 of 0.8 kg. They collide and, after the collision, the velocity of mass m_2 is 1.6 m/s in the positive direction. (a) What is the velocity of mass m_1 after the collision? (b) Is this collision elastic or inelastic? If the latter, what percentage of the maximum possible kinetic energy loss occurs?

34. (I) A bullet of mass 20 g is moving horizontally at a speed of 600 m/s when it strikes a 3-kg block at rest. (a) The block is made of wood; the bullet sinks in and sticks. What is the initial speed of the bullet–block combination? (b) The block is made of a very hard steel, as is the bullet, and the collision is perfectly elastic. What is the resulting velocity of the block? [*Hint*: Use the approximation that the mass of the bullet is much less than the mass of the block.]

35. (II) A block of mass 126 g is moving along the $+x$-axis with a speed of 0.875 m/s. Just ahead of it is a 9.66-kg mass, moving in the same direction with the same speed. At some point, the large mass hits a wall and bounces off the wall perfectly elastically (Fig. 8–34). What is the return speed of the small mass after its elastic collision with the large mass?

FIGURE 8–33 Problem 29.

FIGURE 8–34 Problem 35.

36. (II) A machine gun in automatic mode fires 20-g bullets with $v_{bullet} = 300$ m/s at 60 bullets/s. (a) If the bullets enter a thick wooden wall, what is the average force exerted against the wall? (b) If the bullets hit a steel wall and rebound elastically, what is the average force on the wall?

37. (II) Two spheres of mass 400 g and 600 g, respectively, are suspended from the ceiling by massless strings 1 m long. The lighter sphere is pulled aside through an angle of 80° and let go. It swings and collides elastically with the second sphere at the bottom of the swing. How high will the lighter of the spheres swing (in terms of an angle)?

38. (II) A pendulum of mass $m = 0.1$ kg and length $l = 1.5$ m is released from a horizontal position. At the bottom of its swing, it collides elastically with a mass $M = 0.5$ kg. (a) What is the velocity of mass M right after the collision? (b) To what height above its low point does mass m rebound?

39. (II) A superball has collisions that are nearly perfectly elastic. A superball of mass M is dropped from rest from a height h (where $h \gg$ the size of the superball) together with a smaller marble of mass m; the marble is initially just a little above the top of the superball and remains right over it throughout the fall. The superball hits the floor first and immediately rebounds elastically, colliding with the marble. (a) What is the speed of the superball and the marble just before the superball hits the floor? (b) Just after the superball rebounds from the floor but before it hits the marble? (c) What is the velocity of the marble after the superball hits it in the head-on collision? (d) How high does the marble go after its collision with the superball, assuming the marble has stayed in line with the superball, so that all the motion is vertical? (e) What is the answer to part (d) in the limit $M \gg m$?

40. (II) Consider a perfectly elastic collision in one dimension between a ball of mass M and another of mass m, where $M \gg m$. The light ball is initially at rest, and the heavy ball has a given initial velocity v_i and final velocity v_f. What is the fractional velocity change $(v_i - v_f)/v_i$ of the large mass? Use $M \gg m$ to find an approximate expression.

8–5 Elastic Collisions in Space

41. (I) A ball of mass 1.2 kg moves along the positive x-axis with a speed of 2.4 m/s. Another ball (mass 0.8 kg) moves along the negative x-axis with a speed of 3.6 m/s. After colliding with each other, the lighter ball moves at a speed of 1.8 m/s along a line that makes an angle of 60° with the positive x-axis. What the speed and direction of the heavier ball? Is the collision elastic?

42. (I) Two objects of equal mass approach each other with equal but opposite velocities along the x-axis. After a collision, one particle has velocity $v_1\mathbf{i} + v_2\mathbf{j}$. What is the velocity of the other particle?

43. (I) A billiard ball with velocity $\mathbf{v} = (2.50$ m/s$)\mathbf{i}$ strikes a stationary billiard ball of the same mass. After the collision, the first billiard ball has velocity $\mathbf{v}_1 = (0.50$ m/s$)\mathbf{i} + (-1.00$ m/s$)\mathbf{j}$. What is the velocity of the second ball? Is the collision elastic?

44. (I) Suppose there are two billiard balls of equal mass; one is at rest and one is moving with a speed of 1.5 m/s. They collide elastically. After the collision, one of the two balls is measured to be moving with speed 1.2 m/s. What is the speed of the other ball?

45. (I) A billiard ball moving at 3.0 m/s collides with another billiard ball at rest. The balls move off at right angles to one another. If the first ball continues with a speed of 1.5 m/s, what is the speed of the ball that was initially at rest?

46. (II) Two objects with masses 2.0 kg and 3.0 kg move toward each other, both with speeds $v_0 = 5$ m/s. They collide head on and stick together. (a) Calculate their final velocity. (b) Calculate the amount of kinetic energy lost during the process. (c) Suppose the two masses approach each other at 90° before the collision (e.g., along the x- and y-axes). What will be the kinetic energy loss in this case?

47. (II) In a target shooting game, wooden blocks are thrown into the air and shot in flight. A block of 0.80 kg has a speed of 10 m/s at the top of its trajectory when it is hit by a bullet from below at an angle of 60° from the horizontal (Fig. 8–35). The mass of the bullet is 5.0 g and its speed is 550 m/s when it hits the block. The bullet is embedded in the block. What is the velocity of the block immediately after impact?

FIGURE 8–35 Problem 47.

48. (II) An air puck with mass 0.3 kg and velocity $(-1.5$ m/s$)\mathbf{i} - (2.0$ m/s$)\mathbf{j}$ on a frictionless table collides with a second air puck, of mass 0.2 kg and velocity $(5.0$ m/s$)\mathbf{i}$ (Fig. 8–36). As a result of the collision, the first air puck comes to rest. What is the kinetic energy of the second air puck after the collision?

FIGURE 8–36 Problem 48.

49. (II) In a court hearing, a police expert reconstructs an accident in the following way (Fig. 8–37): A sports car of mass $m_1 = 1000$ kg collided with a parked pickup truck of mass $m_2 = 1500$ kg. From the skid marks, it is estimated that the speed of the pickup immediately after the collision was 21.6 m/s, at an angle of 33.7° with the direction of the road. The sports car left a mark at an angle of 60° with the road, but did not stop, leaving the scene badly damaged. Can the expert determine from these data if the sports car was speeding and its driver fully responsible?

FIGURE 8–37 Problem 49.

50. (III) A proton of mass m and kinetic energy K scatters elastically from an alpha particle (with mass four times that of the proton). The proton is deflected at 30° from its original direction. (a) At what angle does the alpha particle recoil? (b) Draw a momentum diagram of the collision. (c) What are the final energies of the two particles?

8–6 The Center of Mass

51. (I) A massless, rigid bar that is 2.24 m long connects two tiny spheres (of negligible radii compared to the length of the bar) of masses 895 g and 478 g, respectively. Locate the system's center of mass.

52. (I) Point masses of 0.2 kg, 0.1 kg, and 0.3 kg are located in the xy-plane at the origin, at $\mathbf{r}_2 = (0.3 \text{ m})\mathbf{j}$, and at $\mathbf{r}_3 = (0.2 \text{ m})\mathbf{i} + (0.2 \text{ m})\mathbf{j}$, respectively. Where is their center of mass?

53. (I) Redo Example 8–10 by placing the origin at the location of mass m_3, which was added to the original system.

54. (I) What is the velocity of the center of mass of the system in Problem 8, both before and after the collision (both parts), if the speed of the moving billiard ball is v?

55. (I) Two children of masses 25 kg and 30 kg, respectively, stand 2 m apart on skates on a smooth ice rink. The lighter of the children holds a 3-kg ball, and throws it to the heavier child. After the throw the lighter child recoils at 2.0 m/s. With what speed will the center of mass of the two children and the ball move?

56. (II) A uniform iron bar that is 0.40 m long, with mass 1.5 kg, is placed along the y-axis. From the midpoint of that bar a 0.65-m-long massless bar extends perpendicular to it in the x-direction; a 0.5-kg point mass is placed at its end. Where is the center of mass of this system? [*Hint*: First calculate the center of mass of the iron bar, $\mathbf{R}_1$, then treat that bar as a point mass of mass 1.5 kg at $\mathbf{R}_1$.]

57. (II) A mallet forms a symmetric T-shape. The top of the T is a uniform iron block of mass 4.0 kg. The wooden handle is uniform, 1.2 m long, and has a mass of 1.8 kg. Where is the mallet's center of mass? (See the hint in Problem 56.)

58. (II) Where is the center of mass of a uniform, L-shaped iron rod of sides 0.4 m and 0.7 m, respectively? (Assume that the rod is so narrow that the dimensions of the outer bend are the same as those of the inner bend of the L.)

59. (II) By using arguments based on symmetry, show that the center of mass of a uniform, thin, spherical shell made of any material of uniform density is at the center of the shell. Using this result, show that the center of mass of any sphere whose density varies arbitrarily only with distance from the center of the sphere but not with angle (in other words, a spherically symmetric density) is at the center of the sphere.

60. (II) Calculate the center of mass of the T-shaped object shown in Fig. 8–38. Assume that the mass density is uniform.

FIGURE 8–38 Problem 60.

61. (II) Find the center of mass of a flat, semicircular object of radius R. Assume that the density is uniform.

62. (III) Show that the center of mass of a planar object can be found by dividing the object into a convenient number of sections of known mass M_i, finding the center of mass of each section, and then treating each section as a collection of point masses M_i located at the center of mass of that section.

63. (III) Calculate the center of mass of a stick extending from $x = 0$ to $x = L$; assume that the density of the mass distribution is given by $\lambda(x) = K/(x^2 + a^2)$, where K is a constant with dimensions [ML].

64. (III) Find the center of mass of a flat triangular object of uniform density with sides of length a, b, c.

8–7 Rocket Motion

65. (I) The thrust of a rocket is 10 million N. If the speed of its engine's exhaust gas is 3.0 km/s, how fast is the rocket's mass changing?

66. (I) An orbiter must fire its thruster rockets to increase its speed. If the mass of the orbiter is about 80,000 kg and the exhaust gas velocity is 2800 m/s, how much mass must be discarded to increase the speed from 7900 m/s to 8100 m/s?

67. (I) After ejecting a communication satellite, the Space Shuttle must make a correction to account for the change in momentum. One of the thrusters with $u_{\text{ex}} = 10^3$ m/s is used to increase the orbital velocity by 10 m/s. What percent of the mass of the Space Shuttle must be discarded?

68. (I) A rocket that uses a fuel with an exhaust speed of 2700 m/s is shot straight up and reaches a final speed $v = 3400$ m/s when the fuel is exhausted, which happens 120 s after the launch. What fraction of the initial mass of the rocket with all its fuel is the mass of the portion of the rocket that remains? Do not ignore gravity, but assume the force of gravity to be the same throughout the burn period.

69. (II) The *burnout velocity* is the final velocity of a rocket when all the fuel is burned away. If m_0 is the total initial mass of the rocket and m_{fuel} is the mass of the fuel, show that the burnout velocity is given by

$$v = u_{ex} \ln\left(\frac{m_0}{m_0 - m_{fuel}}\right) - gt,$$

where g is assumed to be constant. If the ratio m_{fuel}/m_0 is 70 percent, $t = 90$ s, and u_{ex} is 2800 m/s, what is the burnout velocity?

70. (II) The rocket considered in Problem 69 carries human passengers. Consequently, its initial upward acceleration may not exceed $2g$. What is the largest initial burnout rate (the mass expelled over unit time) that satisfies this condition? (The mass of the rocket is 80,000 kg.)

71. (II) To test certain aspects of rocket propulsion, a student stands initially motionless on frictionless ice. He has 100 watermelon seeds that he can spit horizontally in the $-x$-direction, thereby propelling himself in the $+x$-direction. The speed he can give one seed with respect to himself is u, but if he tries to spit two seeds simultaneously, he can give them only speed $u/2$. More generally, if he tries to spit n seeds simultaneously, he can give them speed u/n. The student has mass M and each seed has mass m. (a) To maximize his speed after spitting all 100 seeds, is the student better off spitting out all 100 seeds at once, spitting out the seeds one at a time, or something in between? Assume that the ratio m/M and even $100m/M$ can be taken to be small. (b) Let $M = 50$ kg, $m = 1$ g, and $u = 3$ m/s. What is the maximum final speed of the student?

72. (III) Because the final velocity of rockets is limited by the exhaust velocity and by the mass ratio of total mass to the rest of the rocket minus fuel, engineers have developed multistage rockets in which the fuel tank of each stage is discarded before the next stage is ignited. Consider a two-stage rocket for which m_0 is the total mass of the rocket at liftoff, m_1 is the mass of the fuel in the first stage, m_1' is the mass of the first stage that is discarded, and m_2 is the mass of the second-stage fuel. Find an equation that gives the final velocity of the rocket; assume that gravity is constant.

General Problems

73. (I) A 48-kg figure skater (skater 1) moving at 6 m/s is picked up by her 82-kg partner (skater 2), who is skating in the same direction at 9 m/s. They then coast off together. Ignore all friction. (a) What is the velocity of their center of mass before he picks her up? (b) After he picks her up? (c) What were the velocities of skater 1 and 2 in the center of mass frame before he picked her up? (d) During a brief period of 0.05 s, they crash into the barrier and collapse together. What average force is exerted on the barrier?

74. (II) A railroad car, of mass 20 Mg, is released from rest in a railway switchyard and rolls to the bottom of a slope 5 m below its original height. At the low point, it collides with and sticks to another car of 10 Mg. The two cars roll off together up another slope to a position a height h above the low point, where they come to a stop. Ignoring the effects of friction, find h.

75. (II) A cannon of mass 800 kg shoots a 5-kg projectile at an angle of 35° from the horizontal. The projectile has an initial speed of 800 m/s. The cannon is on wheels. (a) Draw a dia-

gram showing the momentum vectors. (b) With what velocity does the cannon recoil along the horizontal ground? (c) What happens to the recoiling cannon's component of momentum perpendicular to the ground?

76. (II) A small bag of sand is suspended by a thread of length $L = 1.2$ m. A bullet of mass $m = 8$ g, moving with speed $v_0 = 600$ m/s in the horizontal direction, hits the bag and leaves with a speed $v = 250$ m/s (Fig. 8–39). The bag swings to a maximum angle of $\theta = 40°$. What is the mass of the sandbag?

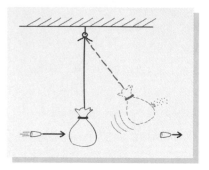

FIGURE 8–39 Problem 76.

77. (II) Two masses slide without friction down the sides of a hemispherical bowl. They each start with zero velocity at the lip, which is a height h above the bottom. They move in opposite directions and collide elastically at the bottom of the bowl. Suppose that the masses are m and M, respectively, with $m < M$. (a) By how much will the lighter mass overshoot the lip after the collision? (b) How high will the masses move if they coalesce on contact?

78. (II) A boxcar of length 8 m and height 2.4 m is at rest on frictionless rails. Inside the boxcar, whose mass is 3200 kg, a tank containing 2400 kg of water is located at one end. The tank is 1 m long and 2.4 m high. At some point, the walls of the tank start to leak and the water fills the floor of the boxcar uniformly. Assuming that all the water stays in the boxcar, describe how the breakage of the tank manifests itself to an outside observer.

79. (II) A sphere of styrofoam has radius R. A cavity of radius $R/2$, centered a distance $R/2$ directly above the center of the sphere, is hollowed out and filled with a solid material of density five times the density of styrofoam (Fig. 8–40). Where is the center of mass of the new sphere?

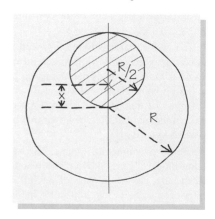

FIGURE 8–40 Problem 79.

80. (II) In the Olympic Games of the year 2020, a new sport is introduced—cyclo-shooting. A bicyclist moving in the $+x$-direction fires a gun in the y-direction in an attempt to hit a target lined up with the position $x = 0$ m. The mass of the cyclist plus the bicycle is 65 kg, and the cyclist is originally moving with speed 35 km/h. The gun shoots a projectile, with mass 15 g, at a speed of 1600 km/h. The target is a distance 30 m from the bicycle track. (a) At what x-value should the cyclist fire in order to hit the target? Ignore air resistance. (b) Immediately after she fires, the cyclist moves in a new direction due to the recoil. What is the new value of her velocity, both magnitude and direction?

81. (II) Two students sit on opposite ends of a 4-m-long sled that is initially at rest on frictionless ice. Each student has a mass of 65 kg; the sled's mass is 30 kg. The student at one end slides a 3-kg object on the sled across to the other at a uniform speed of 6 m/s relative to the sled (the object moves friction free on the sled). (a) What is the sled's speed relative to the ice before the second student catches the object? (b) After the second student catches the object? (c) Over what distance does the sled move while the object slides across? (d) Over what distance does the center of mass move while the object slides across?

82. (III) Two identical spherical billiard balls, each with a radius of 3 cm, move toward each other with velocities $(0.6$ m/s$)\mathbf{i}$ and $(-0.2$ m/s$)\mathbf{i}$. The center of one of them moves along the $+x$-axis $(y = 0$ m$)$, and the center of the other one moves in the $-x$-direction, 1.5 cm below the x-axis $(y = -1.5$ cm$)$. Assuming that the collision is perfectly elastic, what are the final velocities of the two billiard balls? Pretend that the objects slide without friction. Note that the impulse is directed along the line that joins the two centers at the moment of collision.

This water wheel provides a time-honored example of rotational motion. The weight of the water falling on the wheel supplies the dynamical means for centripetal acceleration of the wheel, and the result is motion at a constant angular speed.

Rotations of Rigid Bodies

T o this point our discussion of dynamics has treated very different objects such as crates and planets as if they were all point objects (particles). Further, our study of an extended object, or of a system of particles, has only reached the point where we can show that, under the influence of a net external force, the center of mass moves as if the entire system were concentrated at that single point. But there is more to the motion of an extended object than the motion of its center of mass. Consider the vast cloud of dust that circled the Sun and ultimately condensed into a number of planets, including Earth. Computer simulations indicate that the gravitational attractions between the dust particles led the great dust swirls to break up into isolated clusters and subsequently condense into planets orbiting the Sun and spinning on their own axes. The description of this motion is certainly more complicated than the description of the motion of the center of mass of the solar system. Even with a relatively simple object of fixed form, such as a thrown blackboard eraser, a complete description of its motion requires us to include its rotations.

In this chapter, we study the rotational motion of **rigid bodies**, which are objects that do not change volume or shape. Blackboard erasers, rocks, baseball bats, and heavy sheets of metal are all rigid bodies. We explore the rotation of rigid bodies as we did the motion of point objects: We start with a simple description (kinematics) and then continue with the causes of rotations (dynamics). Even though rotational motion in classical physics appears to encompass many new concepts, this motion is in fact described by the application of Newton's laws to collections of particles. No new physical laws are involved.

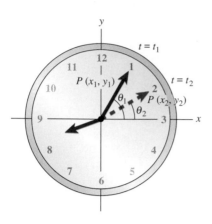

FIGURE 9–1 A flat object in the *xy*-plane rotates about a fixed axis perpendicular to the plane. Here, the object is the hand of a clock, which rotates through the angle $\theta_2 - \theta_1$. The fixed point is at the center of the clock face.

∞ Polar coordinates and the angular description of circular motion were introduced in Section 3–5.

In rigid-body rotations about an axis, a single angle is sufficient to describe the motion.

∞ In Section 3–5, we showed how angle θ acts as a convenient variable for the description of circular motion.

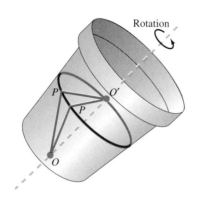

FIGURE 9–2 A three-dimensional object—a clay pot on a potter's wheel—rotates about the axis defined by the line between the two fixed points O and O'. The point P remains the same distance from these points or from any point on the axis as the pot rotates about the axis.

9–1 MOTION OF A RIGID BODY

Let's consider the possible motions of a rigid body in which one point of the object is permanently fixed in space. Any other point P will always remain at a constant distance from the fixed point; moreover, the point P will also remain at a fixed distance from any other point Q. For example, if a clock hand is moving with one end fixed at the center of the clock's face, then all the points of the hand will move in circles about the fixed point (Fig. 9–1). The motion of the hand consists of a **rotation** about an axis that passes through the fixed point and is perpendicular to the surface swept out by the hand as it moves.

Let's look more closely at Fig. 9–1. We have placed the origin at the fixed point and have drawn a set of *x*- and *y*-axes that are fixed in space. At some time t_1, point P has coordinates (x_1, y_1); at a later time, point P has moved to where its coordinates are now (x_2, y_2). *Both x and y change with time.* In terms of polar coordinates, we can describe the angle θ measured from the *x*-axis to the hand and the distance R of the hand from the origin. *The only variable that changes with time is the angle θ.* Our point P has undergone an *angular displacement*.

The special feature of a rigid body is that *when P moves through a certain angle in the rotational motion about the origin, all points in the rigid body turn through exactly the same angle.* The motion with one fixed point is thus described by the circular motion of an arbitrary line that is drawn from the origin to a point P in the object. That motion need not be uniform, but it *is* just circular motion. For the description of a two-dimensional rigid body on a flat surface, this means that we need only one variable—the angle of circular rotation θ.

In the case of a three-dimensional rigid body, once one point in the object is fixed, every other point must move in such a way that it is at a constant distance from that fixed point. However, it is sufficient for us to consider the situation in which we fix a *second* point of the object in space. The line joining the two fixed points is fixed in space; we call this the *axis*. The points lying along the axis will *always* be fixed in space. A point P that is off the axis will always be at a fixed distance from the two fixed points and at a fixed perpendicular distance from the axis (Fig. 9–2). With two points fixed, the motion of the three-dimensional object—here, a pot on a potter's wheel—consists of rotations about the axis. And as the discussion of the clock hand shows, to describe rotations about a fixed axis we need only specify a single angle. Other examples of this limited motion are provided by Earth's rotation about its axis or by a turning merry-go-round.

In considering the motion of a rigid object about a fixed axis, we will not insist that the axis pass through the center of mass; rather, we are considering circular motion about *any* axis.

9–2 ROTATIONS ABOUT AN AXIS

Angular Velocity and Angular Acceleration

Suppose that a rigid body rotates about a fixed axis; you may want to think of the pot on the potter's wheel as an example. In such motion, the angle of any point P in the object relative to the axis changes with time. We want to give a mathematical description of this motion. Let's orient the *z*-axis along the rotation axis (Fig. 9–3a). Consider the position vector **r** from the origin to point P. Take the projection of **r** on the *xy*-plane to go to point P' and let θ_0 be the angle this projection makes with the *x*-axis (Fig. 9–3b). Angle θ, which varies with time, is measured in *radians* (2π rad = 360°).[†] Figure 9–3b depicts a view down the *z*-axis, where the projection point P' has moved from angle θ_0 to angle θ in the time interval Δt.

[†]In Section 3–5, we used the notation ϕ for this angle.

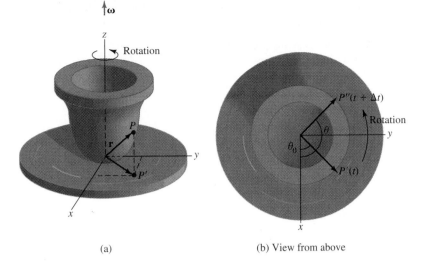

(a) (b) View from above

FIGURE 9–3 (a) A pot on a potter's wheel rotates about the z-axis. (b) The pot as viewed looking down the z-axis. The angle of rotation θ, which varies with time, is defined to be the angular displacement from the x-axis. It is positive when the pot rotates counterclockwise.

The average angular velocity, ω_{av}, over a finite time interval Δt is obtained from the change in the angle over that time interval:

$$\omega_{av} = \frac{\theta(t + \Delta t) - \theta(t)}{\Delta t}. \tag{9–1}$$

We find the *instantaneous angular velocity* (or just the *angular velocity*) $\omega(t)$, given by Eq. (3–46), by taking the limit $\Delta t \to 0$:

$$\omega = \frac{d\theta}{dt}. \tag{9–2}$$

The angular velocity and its units

The units that describe angular velocity are *radians per second* (rad/s). Because a radian is a dimensionless measure (it is the ratio of an arc length to the circumference), the dimensions of ω are $[T^{-1}]$. It is conventional to measure the angle in a counterclockwise direction from the x-axis, using a *right-hand rule: If the extended thumb of the right hand points in the +z-direction (along the axis of rotation), then the fingers curve in the positive direction of rotation,* as shown in Fig. 9–4. Thus an angle that increases in the counterclockwise direction corresponds to a *positive* angular velocity.

Suppose that the angular velocity is a *constant;* that is, $\omega(t) = \omega_0$. Then the angle θ grows linearly with time:

$$\theta = \theta_0 + \omega_0 t, \tag{9–3}$$

where θ_0 is the value in radians of the angle θ at time $t = 0$ (Fig. 9–3b). As noted in our discussion of the clock hand, any point P in the object moves in a circle. According to Section 3–5, this point makes a complete circle during a time T, which is the *period* of the motion. Because a full circle is 2π rad,

$$T = \frac{2\pi}{\omega_0}. \tag{9–4}$$

The *frequency, f,* measures the number of times per second that point P returns to its original position in uniform rotation:

$$f = \frac{1}{T} = \frac{\omega_0}{2\pi}. \tag{9–5}$$

For nonuniform motion, we also speak of an instantaneous frequency, $f(t) = \omega(t)/2\pi$.

If the angular velocity is not constant, we may define an average angular acceleration, α_{av}:

$$\alpha_{av} = \frac{\omega(t + \Delta t) - \omega(t)}{\Delta t}. \tag{9–6}$$

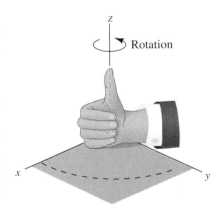

FIGURE 9–4 A right-hand rule. The thumb of the right hand points along the axis of rotation (the z-axis) when the fingers curl in the direction of rotation (from the x-axis to the y-axis).

Relations among frequency, angular velocity, and period

The angular acceleration

The instantaneous angular acceleration $\alpha(t)$ or, more simply, the **angular acceleration**, is defined as the instantaneous rate of change of the angular velocity:

$$\alpha = \frac{d\omega}{dt} = \frac{d^2\theta}{dt^2}. \tag{9–7}$$

Note that this equation has the same structure as the equation for linear motion,

$$a = \frac{dv}{dt} = \frac{d^2x}{dt^2}.$$

The angular acceleration has units of radians per second squared, and the dimension of α is $[T^{-2}]$ because radians are dimensionless.

Constant Angular Acceleration

A constant angular acceleration means that the angular velocity has a constant time derivative. Hence the angular velocity changes linearly with time; that is,

$$\omega = \omega_0 + \alpha t, \tag{9–8}$$

where ω_0 is the angular velocity at $t = 0$. The angle is then given by

The change in angle under constant angular acceleration

$$\theta(t) = \theta_0 + \omega_0 t + \frac{1}{2}\alpha t^2. \tag{9–9}$$

This solution corresponds to the following equation for linear motion:

$$x(t) = x_0 + v_0 t + \frac{1}{2}at^2.$$

We may eliminate the time from Eqs. (9–8) and (9–9) to find that

$$\omega^2 = \omega_0^2 + 2\alpha(\theta - \theta_0), \tag{9–10}$$

which is the analog of the linear motion expression

$$v^2 = v_0^2 + 2a(x - x_0).$$

EXAMPLE 9–1 A pulley is rotating at the rate of 32 rev/min. A motor speeds up the wheel so that, 30.0 s later, it is turning at 82 rev/min. (a) What is the average angular acceleration in radians per second? (b) How far will a point 0.30 m from the center of the pulley have traveled during the acceleration period, assuming that the acceleration is uniform?

Solution: (a) Each revolution consists of an angular displacement of 2π rad. Thus 32 rev/min corresponds to $(32 \text{ rev/min})(2\pi \text{ rad/rev}) = 64\pi$ rad/min, or 64π rad/60 s $= 3.35\pi$ rad/s $= \omega_0$. After 30 s, the angular velocity ω_f is

$$\omega_f = \left(\frac{82 \text{ rev}}{\text{min}}\right)\left(\frac{2\pi \text{ rad}}{\text{rev}}\right)\left(\frac{1 \text{ min}}{60 \text{ s}}\right) = 8.6\pi \text{ rad/s}.$$

Thus the average angular acceleration is

$$\alpha_{av} = \frac{\omega_f - \omega_0}{\Delta t} = \frac{(8.6\pi - 3.35\pi) \text{ rad/s}}{30.0 \text{ s}} = 0.17 \text{ rad/s}^2.$$

(b) We calculate the angular displacement $\Delta\theta$ during the period by using the for-

$$\theta = \theta_0 + \omega_0 t + \frac{1}{2}\alpha t^2,$$

mula applicable for uniform acceleration. With $\omega_0 = 3.35\pi$ rad/s and $\alpha_{av} =$

$0.17 \text{ rad/s}^2,$

$$\Delta\theta = \theta - \theta_0 = (3.35 \text{ rad/s})(30.0 \text{ s}) + \frac{1}{2}(0.17 \text{ rad/s}^2)(30.0 \text{ s})^2 = 180 \text{ rad}.$$

The distance traveled is

$$D = r\,\Delta\theta = (0.30 \text{ m})(180 \text{ rad}) = 54 \text{ m}.$$

The unit "radian" is deleted from the final value because radians are dimensionless.

Acceleration of a Rotating Rigid Body

Consider the rotating clock hand, and, more specifically, a point located at a distance r from the axis of rotation. The point moves in a circle; its velocity has a magnitude of $r\omega$ and a direction that is tangential to the circle of radius r. We learned in Chapter 3 that a point moving in a circle of radius r with speed v undergoes an acceleration in the radial direction toward the center of the circle. The magnitude of that acceleration is v^2/r. In terms of the angular speed—$\omega = v/r$—this acceleration may be expressed as

$$a = \frac{v^2}{r} = \omega^2 r. \tag{9–11}$$

If there is an angular acceleration α, then there is also a linear acceleration in the tangential direction, whose magnitude is

$$a_{\text{tan}} = \alpha r. \tag{9–12}$$

Figure 9–5 illustrates these acceleration components for a point on a potter's wheel that is undergoing an angular acceleration. Whenever a rigid body accelerates, there must be a force involved. The radial acceleration is ultimately due to the interatomic forces that fix the relative positions of the atoms of the rigid body. The tangential acceleration must involve an external force. We'll discuss this in more detail later.

Vectorial Properties of Angular Motion

We have seen that there is a parallel between position and angle, between velocity and angular velocity, and between acceleration and angular acceleration. The difference appears to be that the quantities that describe the kinematics of linear motion are vectors, whereas the notion of a vectorial description of rotations has not played any real role in our discussion. Does rotational motion have vectorial properties? In the description of rotation about an axis, there is a special direction—the direction of the axis. Can we say that the angle, angular velocity, and angular acceleration are vectors just by stating that they point in the direction of the axis? It is not quite so simple. Vectors are not just numbers with an associated direction; vectors must satisfy certain addition rules, one of them being the condition of commutativity:

$$\mathbf{A} + \mathbf{B} = \mathbf{B} + \mathbf{A}.$$

Analogous to our definition of a position vector, suppose we were to try to define a *vector angle* by associating the magnitude of the angle with the direction of the axis about which the rotation takes place. This specification includes a right-hand rule for the direction of the axis to resolve the question of which way the "vector" really points. We call this attempt at a vector angle θ_1. Let the axis for this angle, and hence its direction, be the z-axis (Fig. 9–6). We similarly define an angle θ_2 for a rotation about the y-axis. In order to check commutativity for these proposed vector angles, using the book illustrated in Fig. 9–7, we perform the operation $\theta_1 + \theta_2$ by performing the rotation θ_1 first and subsequently performing the rotation θ_2. We compare that with the operation $\theta_2 + \theta_1$, in which the rotation θ_2 is performed first (Fig. 9–8). Both angles will have magnitude 90°. If the final positions of the book are not the same for the two

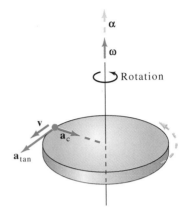

FIGURE 9–5 A rotating potter's wheel undergoes an angular acceleration. A point on its margin must have a centripetal acceleration component—this component keeps the point on its circular path—and a tangential component.

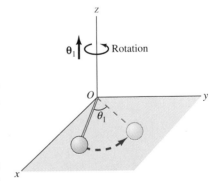

FIGURE 9–6 A proposal for defining a vector angle. When the rotation is counterclockwise in the xy-plane as seen from above, the vector angle θ_1 points in the z-direction.

The angular velocity is a vector with a direction defined by the right-hand rule.

(a)   (b)   (c)

FIGURE 9–7 (a) A book is initially oriented with its cover facing the z-axis and the book spine facing the x-axis. (b) The book first rotates from its position in part (a) by an angle 90° about the z-axis (θ_1) and then (c) 90° about the y-axis (θ_2).

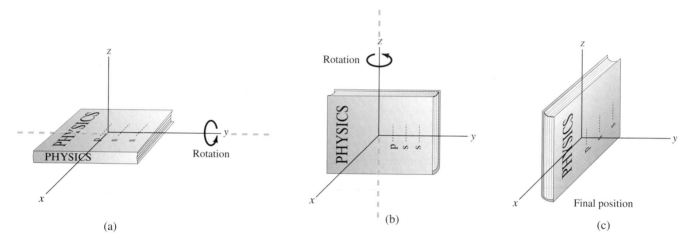

(a)   (b)   (c)

FIGURE 9–8 The order of rotations of the book is reversed from the order taken in Fig. 9–7. (a) In this case, the book starts from the position of Fig. 9–7a and (b) rotates with the rotation θ_2, followed by (c) the rotation θ_1. The final position of the book is different than the final position in Fig. 9–7c.

FIGURE 9–9 A right-hand rule is used to determine the direction of the angular velocity, $\boldsymbol{\omega}$. With the fingers of the right hand curling in the sense of the rotation, the thumb points in the direction of $\boldsymbol{\omega}$.

operations, then commutativity is not satisfied. The final positions of the book in Figs. 9–7c and 9–8c are clearly not the same. We conclude that commutativity fails and that the angles as defined above *are not vectors*!

All is not lost, however: The commutativity test *does* work with infinitesimal angles. You can check for yourself with this book by letting θ_1 and θ_2 specify rotations of only a few degrees. Although this fact does not help with large-angle rotations, it does help with angular velocities. If a small angle $\Delta\theta$ is a vector, then $\Delta\theta/\Delta t$ is a vector. This defines the instantaneous angular velocity; thus, *the angular velocity is a vector*. This vector has the magnitude ω. The direction of $\boldsymbol{\omega}$ is defined by a right-hand rule: If the fingers curl with the direction of the rotation, then the extended thumb of the right hand points in the direction of the angular velocity (Fig. 9–9). The angular acceleration vector $\boldsymbol{\alpha}$, with magnitude α, is defined as the rate of change of the angular velocity vector, exactly as the linear acceleration vector is the rate of change of the linear velocity vector.

A spinning ball consists of moving mass. This ball, or any rotating object, must have **rotational kinetic energy** (Fig. 9–10). Consider the object shown in Fig. 9–11, which rotates about the z-axis with angular velocity ω. We have chosen the origin of the co-ordinate system arbitrarily and divide the object into elements labeled by the indices i, with each element having mass Δm_i. In other words, we are treating the object as a system of discrete particles. The total kinetic energy of the object is given by

$$K = \sum_i K_i = \frac{1}{2} \sum_i \Delta m_i \, v_i^2.$$

Because each mass element is rotating about the z-axis with angular velocity ω, the speed of each element is given by $v_i = R_i \omega$. The length R_i is the *perpendicular distance* between the mass element and the rotation axis, as shown in Fig. 9–11. Substitution of this value of v_i into the expression for the kinetic energy gives

$$K = \frac{1}{2} \sum_i \Delta m_i \, R_i^2 \omega^2. \qquad (9\text{–}13)$$

The kinetic energy of our rotating object is thus proportional to ω^2. We may write

$$K = \frac{1}{2} I \omega^2, \qquad (9\text{–}14)$$

where

$$I \equiv \sum_i \Delta m_i \, R_i^2. \qquad (9\text{–}15)$$

The quantity I is a property of the rigid object called the **rotational inertia**, or the **moment of inertia**. It has dimensions of $[ML^2]$ and is measured in units of kg · m² in SI. From its definition, we see that I sums the contributions of each element of the object; for a single element or single particle, then,

$$I = mR^2, \qquad (9\text{–}16)$$

where R is the perpendicular distance from the single element to the axis of rotation. According to Eq. (9–15), the rotational inertia can be defined with respect to any axis we choose, not just the axis of rotation. Moreover, *I depends on the axis from which R_i is measured.*

The rotational inertia has a role in rotational motion that is analogous to the role of the mass in linear motion, except for the fact that it is defined in relation to a particular axis. The kinetic energy of a pointlike object is one-half its mass times the velocity squared; the kinetic energy of a rotating object is one-half its rotational inertia times its angular velocity squared; that is, $K_{\text{point}} = mv^2/2$ and $K_{\text{rotation}} = I\omega^2/2$. Just as the mass measures the resistance of an object to changes in velocity, we shall see that the rotational inertia measures the resistance of an object to changes in its angular velocity.

A *flywheel* is a rotating disk with a large rotational inertia and the ability to reach a large angular velocity; a flywheel thereby stores energy. In Zurich, the Swiss government operates a bus powered by a flywheel electrical system. A flywheel mounted in the bus powers an electrical generator; the electricity runs electric motors that drive the wheels. At each terminus of the bus run, the bus is connected to Zurich's electric power distribution net, and the bus's electrical generator is run in reverse, acting as a motor to speed up the flywheel and thereby store enough energy for the return trip.

We have defined the rotational inertia for a system of discrete elements. For an object that is a continuum of mass, we can take the limit in which the elements have

FIGURE 9–10 The flying sparks make it evident that the rotating wheel has kinetic energy.

The expression of the kinetic energy of rotation and the definition of the rotational inertia

The rotational inertia is defined with respect to an axis.

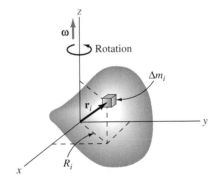

FIGURE 9–11 A rigid body rotates about the z-axis. It is divided into tiny segments labeled with the index i, of mass Δm_i, located at position $\mathbf{r}_i$. The perpendicular distance of segment i from the z-axis is R_i.

 APPLICATION

Flywheels

FIGURE 9–12 A one-dimensional object is aligned with the x-axis. A segment of width Δx is a displacement x from the origin.

infinitesimal volume and infinitesimal mass. In this case, the sum becomes an integral over the volume of the object. A symbolic way to write this continuum expression is to label the mass element by the differential notation dm, and thus

$$I = \int R^2 \, dm. \tag{9–17}$$

This formal expression is not useful as it stands because m is not a useful variable of integration. We can remedy this and the possibility that the mass element dm can vary from point to point by the use of the *mass density* function ρ. The mass density $\rho(\mathbf{r})$ at point $\mathbf{r}$, which we suppose is the point labeled by i, is defined as

$$\rho(\mathbf{r}) \equiv \lim_{\Delta V_i \to 0} \frac{\Delta m_i}{\Delta V_i} = \frac{dm}{dV}.$$

Here, both the volume of the element and the mass of that element approach zero. Now we can write the mass element as $dm = \rho \, dV$. The total mass is found from the mass density by an integration over the volume V (the single integral sign with the subscript "Volume" denotes a three-dimensional integration over the volume):

$$M = \sum_i \Delta m_i \to \int dm = \int_{\text{Volume}} \rho \, dV. \tag{9–18}$$

The rotational inertia of a continuous object is therefore

$$I = \int R^2 \, dm = \int_{\text{Volume}} \rho R^2 \, dV. \tag{9–19}$$

Don't panic when we refer to volume or surface integrals. In cases of interest in this book, we make use of symmetry to reduce all such integrals to single integrals. An equation such as Eq. (9–19) is also applicable for one-dimensional (linear) objects or for two-dimensional (flat) objects. In these cases, we may replace the mass per unit volume ρ by a mass per unit length $\lambda \equiv dm/dx$, or a mass per unit area $\sigma \equiv dm/dA$. Thus

for a one-dimensional object: $dm = \lambda \, dx$;

for a two-dimensional object: $dm = \sigma \, dA$;

for a three-dimensional object: $dm = \rho \, dV$.

The densities λ, σ, and ρ may vary from point to point of the object.

To evaluate the rotational inertia in the one-dimensional case, we can align our object with the x-axis, and

$$I = \int \lambda x^2 \, dx. \tag{9–20}$$

The distance x is measured from the origin, as in Fig. 9–12, and the integral is carried out over the length of the object. When the object is two-dimensional, we have

$$I = \int_{\text{Surface}} \sigma r^2 \, dA. \tag{9–21}$$

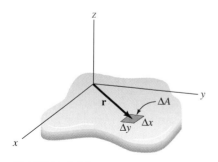

FIGURE 9–13 A two-dimensional object is aligned in the xy-plane. A segment of area ΔA is a position $\mathbf{r}$ from the origin.

We have taken the axis perpendicular to the object and integrated over its entire area (Fig. 9–13). The quantity r^2 is the square of the distance of the integration element to the axis about which I is measured.

A description of the rotational motion of an object requires us to evaluate its rotational inertia. For continuous objects, this evaluation involves integrations. For objects that have circular symmetry about the axis of rotation, it is easiest to label the points by the distance from that axis and by an angle variable rather than by Cartesian coordinates. We will see how this works in a series of examples. Table 9–1 contains a summary of the results, as well as rotational inertias of other objects for which the calculations are not carried out in this section.

Dumbbell.   The simplest rotating object that we can contemplate is a dumbbell consisting of two point masses, m_1 and m_2, connected by a massless rigid rod of length L (Fig. 9–14). The total mass is $M = m_1 + m_2$. If we put mass m_1 at the origin and mass m_2 at $x = L$, we find from Eq. (8–49a) that the center of mass is at

$$X = \frac{(m_1)(0) + (m_2)(L)}{m_1 + m_2} = \frac{m_2 L}{M}.$$

Now let's consider the case in which the axis of rotation goes through the center of mass; that is, through the point $x = m_2 L/M$. The axis is taken perpendicular to the rod. As measured from the center of mass, the coordinates of m_1 and m_2 are $-m_2 L/M$ and $L - (m_2/M) = m_1 L/M$. As in Fig. 9–14, the rotational inertia about an axis passing through the center of mass and perpendicular to the axis of the dumbbell is given by

$$I = m_1 \left(-\frac{m_2 L}{M}\right)^2 + m_2 \left(\frac{m_1 L}{M}\right)^2 = \frac{L^2(m_1 m_2^2 + m_2 m_1^2)}{M^2}$$

$$= \frac{m_1 m_2 L^2 (m_2 + m_1)}{(m_1 + m_2)^2}$$

$$= \frac{m_1 m_2}{m_1 + m_2} L^2. \tag{9–22}$$

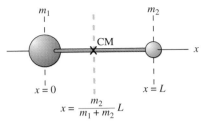

FIGURE 9–14 A dumbbell consists of two point masses, m_1 and m_2, connected by a rigid massless rod of length L. The object is aligned along the x-axis, with m_1 at the origin. The axis of rotation is perpendicular to the rod, passing through the center of mass.

TABLE 9–1   Rotational Inertias of Simple Solids

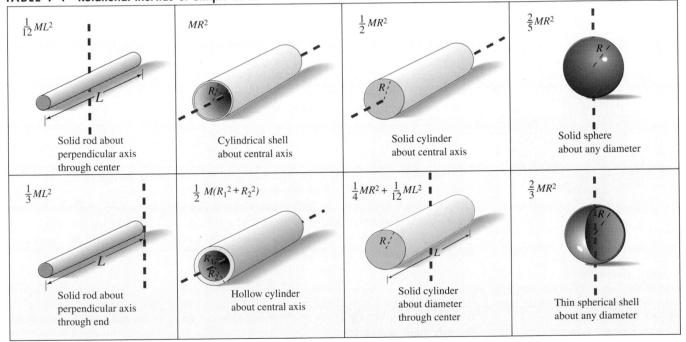

$\frac{1}{12} ML^2$	MR^2	$\frac{1}{2} MR^2$	$\frac{2}{5} MR^2$
Solid rod about perpendicular axis through center	Cylindrical shell about central axis	Solid cylinder about central axis	Solid sphere about any diameter
$\frac{1}{3} ML^2$	$\frac{1}{2} M(R_1^2 + R_2^2)$	$\frac{1}{4} MR^2 + \frac{1}{12} ML^2$	$\frac{2}{3} MR^2$
Solid rod about perpendicular axis through end	Hollow cylinder about central axis	Solid cylinder about diameter through center	Thin spherical shell about any diameter

If the masses are equal, then $m_1m_2/(m_1 + m_2) = m/2$; if $m_2 \gg m_1$, then it is a good approximation to write $m_1m_2/(m_1 + m_2) = m_1$, the mass of the light point mass. These results are easy to understand physically. In the case of equal masses m, these two point masses rotate about an axis at a distance $L/2$ from each of them, and the rotational inertia is then $I = 2m(L/2)^2 = mL^2/2$. In the case of one very light mass, $m_1 \ll m_2$, then, to a good approximation, the axis passes through the position of the heavier mass m_2, and the mass m_1 rotates at a distance L from the axis, so $I = m_1L^2$. Note that if the axis of rotation is the line connecting two point masses, then $r_1^2 = 0$, and the rotational inertia about this axis is zero.

EXAMPLE 9–2 A dumbbell consists of point masses $m_1 = 2.0$ kg and $m_2 = 1.0$ kg attached by a rigid massless rod of length $L = 0.6$ m. Calculate the rotational inertia of the dumbbell (a) about an axis perpendicular to the rod of the dumbbell and going through the center of mass; and (b) about an axis perpendicular to the rod of the dumbbell and going through the 2.0-kg mass.

Solution: (a) In Fig. 9–14, let the larger mass be at the origin, $x_1 = 0$ m, and the smaller mass at $x_2 = L$. The center of mass is located at

$$X = \frac{m_1x_1 + m_2x_2}{m_1 + m_2} = \frac{[(2.0 \text{ kg})(0 \text{ m}) + (1.0 \text{ kg})(0.6 \text{ m})]}{2.0 \text{ kg} + 1.0 \text{ kg}} = 0.2 \text{ m}.$$

Now

$$I_{cm} = m_1(x_1 - X)^2 + m_2(x_2 + X)^2$$
$$= (2.0 \text{ kg})(0 \text{ m} - 0.2 \text{ m})^2 + (1.0 \text{ kg})(0.6 \text{ m} - 0.2 \text{ m})^2 = 0.24 \text{ kg} \cdot \text{m}^2.$$

(b) Mass m_1 does not contribute to the rotational inertia because it is on the axis. Mass m_2 is a distance L from the axis. Thus

$$I = m_2L^2 = (1.0 \text{ kg})(0.6 \text{ m})^2 = 0.4 \text{ kg} \cdot \text{m}^2.$$

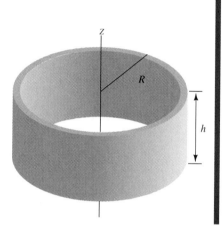

FIGURE 9–15 A thin cylinder of mass M has height h and radius R. We want to find its rotational inertia about the z-axis, which is the symmetry axis of the cylinder.

Thin Cylinder. The thin cylinder shown in Fig. 9–15 has radius R, height h, and total mass M, distributed uniformly around the ring. Let's find the rotational inertia about the central axis, which runs along the z-axis. The word "thin" means that we can consider all of the mass to be the same distance R from the axis. Thus, the rotational inertia is simply

$$I = MR^2. \tag{9–23}$$

Solid Cylinder. The rotational inertia of a uniform solid cylinder about its central axis can be calculated by using our result for the thin cylinder. The solid cylinder has exterior radius R, height h, and mass M. The mass density ρ is given by

$$\rho = \frac{M}{V} = \frac{M}{\pi R^2 h}. \tag{9–24}$$

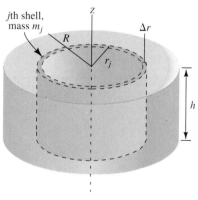

FIGURE 9–16 To find the rotational inertia of a thick cylinder of mass M, height h, and radius R about its symmetry axis, we break it into a series of thin concentric cylinders. We know the rotational inertia of the thin cylinders, and can sum these inertias to find the rotational inertia of the solid cylinder.

We can very often simplify the calculation of the rotational inertia of an object by breaking the object up into appropriate pieces. In this case, we can divide the solid cylinder into thin cylindrical shells. The jth shell is a distance r_j from the axis and has thickness Δr and mass Δm_j (Fig. 9–16). The cylindrical shell labeled j in Fig. 9–16 contributes an amount $m_jr_j^2$ to the rotational inertia, I, of the solid cylinder. We must sum these contributions to find I:

$$I = \sum_j (\Delta m_j) \, r_j^2. \tag{9–25}$$

The sum in Eq. (9–25) is a formal one. We can carry it out once we find the mass of each shell. The mass of a given volume is ρ times that volume; for the thin shell labeled j, the volume is approximately the shell thickness times the shell circumference times the height, $V_j = (\Delta r)(2\pi r_j)h$. Thus $\Delta m_j = \rho(\Delta r)(2\pi r_j)h$, and

$$I = \sum_j (\Delta m_j)r_j^2 = \sum_j \rho(\Delta r)(2\pi r_j)hr_j^2 = 2\pi h\rho \sum_j (\Delta r)r_j^3.$$

In the limit that the cylindrical shells are very thin, the sum, which includes shells from $r = 0$ to $r = R$, becomes an integral:

$$I = 2\pi h\rho \int_0^R r^3 \, dr = 2\pi h\rho \frac{R^4}{4}.$$

When we substitute Eq. (9–24) for ρ, we find

$$I = 2\pi h \frac{M}{\pi R^2 h} \frac{R^4}{4} = \frac{1}{2}MR^2.$$

We have reduced the formal three-dimensional integral for I to a one-dimensional integral. This sort of simplification is very often possible for multiple integrals. Note that the value of I is less than that of the thin cylinder of the same mass and radius. The weighted sum for the solid cylinder includes mass pieces that give a smaller contribution because they are closer to the axis. Note too that the result is independent of the height and applies for a thin disk as well as a tall cylinder.

Rod. Next we consider the rotational inertia of a uniform rod of length L about an axis that is perpendicular to it and passes through its center of mass. The center of mass is the midpoint of the rod and is the point at which we will place the origin of our coordinate system (Fig. 9–17). The mass density is the total mass divided by the rod's length, $\lambda = M/L$. Consider the mass dm contained in a slice of the rod of length dx at a distance x from the axis. We have $dm = \lambda \, dx$, so the rotational inertia about the axis is given by

$$I = \int_{-L/2}^{L/2} x^2\lambda \, dx = \lambda \int_{-L/2}^{L/2} x^2 \, dx = \lambda \frac{x^3}{3}\Big|_{-L/2}^{L/2}$$

$$= \frac{ML^2}{12}.$$

If λ is not a constant but is symmetric about the midpoint, so that the midpoint is still the center of mass, then λ is replaced by $\lambda(x)$, and the rotational inertia is the integral

$$I = \int_{-L/2}^{L/2} x^2\lambda(x) \, dx.$$

The rotational inertia for rotations about an axis perpendicular to the rod and passing through one end of the uniform rod is given by

$$I = \int_0^L x^2\lambda \, dx = \frac{\lambda}{3}x^3\Big|_0^L = \frac{\lambda L^3}{3} = \frac{1}{3}ML^2. \tag{9–26}$$

This second axis is also drawn in Fig. 9–17.

The Parallel-Axis Theorem

We have worked out the rotational inertia for several simple systems and for several specific axes. Yet how useful are such exercises? After all, there is a different answer for every axis. However, there is a natural set of axes for every object, and these are axes that go through the center of mass. It is sufficient to calculate the rotational inertia of objects about these axes because the rotational inertia about any other axis can be obtained from the **parallel-axis theorem**. This theorem, whose proof we will not detail, states that *the rotational inertia of an object about any axis is given by the sum of the rotational inertia about an axis that goes through the center of mass and is parallel to the given axis, and of the product of the total mass M of the object and the*

FIGURE 9–17 To find the rotational inertia of a massive rod of length L about an axis perpendicular to it and passing through an end point, we first find the rotational inertia about the parallel axis through the center of mass, which is at the midpoint if the rod's density is constant.

FIGURE 9–18 To find the rotational inertia of any three-dimensional object about a given axis of rotation, first find the rotational inertia about the parallel axis through the center of mass. The parallel-axis theorem gives a general relation between these two rotational inertias. Here, the object in question is a tennis racket.

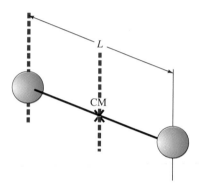

FIGURE 9–19 Two parallel axes, one through the center of mass and the other through one end, can be drawn through this dumbbell of equal masses. By using the rotational inertia for the center of mass axis and the parallel-axis theorem, we can easily find the rotational inertia about the axis through one end.

FIGURE 9–20 Example 9–3. We align the width of a rectangular plate with the *x*-axis and the length with the *y*-axis. We calculate the rotational inertia about the *z*-axis.

246

square of the perpendicular distance d between the two axes (Fig. 9–18). Mathematically, the parallel-axis theorem states that

$$I = I_{cm} + Md^2. \tag{9-27}$$

As a simple example, let's take a dumbbell with equal masses m and consider two parallel axes: one passing through the center of mass and one through one of the masses (Fig. 9–19). For the axis through the center of mass, we have $I_{cm} = m(L/2)^2 + m(L/2)^2 = mL^2/2$, while direct calculation for the axis through the end gives $I_{end} = mL^2$. What does the parallel axis theorem give for I_{end}? The distance d in Eq. (9–27) is $L/2$, while $M = 2m$, so the parallel axis theorem gives $I_{end} = I_{cm} + M(L/2)^2 = mL^2/2 + (2m)(L/2)^2 = mL^2$, the same result given by the direct calculation.

The theorem shows that the rotational inertia of any object is smallest if the axis of rotation goes through the object's center of mass ($d = 0$). This means that the kinetic energy of rotational motion of an object for fixed angular velocity ω is smallest if the rotation is about an axis containing the center of mass.

EXAMPLE 9–3 Consider a uniform, thin, rectangular sheet of metal of mass M, with width a and length b. Calculate the rotational inertia about an axis that is perpendicular to the sheet and passes through one of the corners.

Solution: Choose the corner through which the axis passes as the origin and line up the width along the *x*-axis and the length along the *y*-axis (Fig. 9–20). We have $0 \le x \le a$ and $0 \le y \le b$. The two-dimensional mass density σ is uniform and given by

$$\sigma = \frac{M}{ab}.$$

Thus

$$I_{corner} = \int_0^a dx \int_0^b dy \; \sigma \, (x^2 + y^2) = \frac{M}{ab} \int_0^a dx \int_0^b dy \; (x^2 + y^2)$$

$$= \frac{M}{ab} \left(\int_0^a x^2 \, dx \int_0^b dy + \int_0^a dx \int_0^b y^2 \, dy \right) = \frac{M}{ab} \left[\left(\frac{a^3}{3} \right) b + \left(\frac{b^3}{3} \right) a \right]$$

$$= \frac{1}{3} M(a^2 + b^2).$$

EXAMPLE 9–4 Use the result of Example 9–3 to calculate the rotational inertia of the same sheet of metal about an axis that is perpendicular to the plane of the sheet and passes through the sheet's center of mass.

Solution: The center of mass is located at the midpoint of the rectangle; that is, at $x = a/2$, $y = b/2$. The distance to the corner is given by

$$d^2 = \left(\frac{a}{2} \right)^2 + \left(\frac{b}{2} \right)^2.$$

Thus, by the parallel-axis theorem, $I_{corner} = I_{cm} + Md^2$:

$$I_{cm} = I_{corner} - Md^2 = \frac{1}{3} M(a^2 + b^2) - M \left[\left(\frac{a}{2} \right)^2 + \left(\frac{b}{2} \right)^2 \right]$$

$$= M(a^2 + b^2) \left(\frac{1}{3} - \frac{1}{4} \right) = \frac{1}{12} M(a^2 + b^2).$$

FIGURE 9–21 (a) A rod with a bob at its end is attached to a shaft and can rotate in the plane perpendicular to the shaft. The shaft acts as a rotation axis. (b) A force lying in the plane in which the rod can move is applied to the rod, causing a rotation. The effectiveness of the force in causing a rotation depends on the strength of the force, on where along the rod it is applied, and on its direction, specified by the angle θ. The lever arm is $r \sin \theta$.

9–5 TORQUE

We have described rotations; now, let's turn to the question of what causes them. The analogies we have established between rotational motion and the motion of point particles (we refer to this as *linear motion*) will be useful here. Newton's second law describes the dynamics of linear motion by the equation $\mathbf{F} = m\mathbf{a}$. We'll see here that there is in rotational motion an equation with a quality analogous to force called **torque**.

It will be helpful to keep a definite example of a rotation and its cause in mind. For example, we might think of the rotation of a rod with a bob at its end; the rod is attached at a point O to a shaft that can rotate (Fig. 9–21a). The rod and bob together have a mass M. The rod, of length R, makes a right angle with the shaft. The motion will be entirely in the plane perpendicular to the shaft. Suppose now that a force $\mathbf{F}$ acts in that plane somewhere along the rod (Fig. 9–21b). When this force causes the system to undergo rotational motion, we say that there is a torque.

We can intuitively think of a torque as a *twist,* just as a force is intuitively a push or a pull. The force will be effective in setting an extended system (such as that of Fig. 9–21) into rotational motion about a point O, depending on two features. First, a larger force will be more effective in exerting torque than would an otherwise similar force of smaller magnitude. Second, a force is most effective at producing a torque if it is applied at a point P that is *far* from point O and is at *right angles* to the line OP that connects O and the point of application of the force. In other words, the torque must be proportional to $r_\perp \equiv r \sin \theta$ in Fig. 9–21b, where θ is the angle between $\mathbf{r}$ and $\mathbf{F}$. This factor is known as the *lever arm,* or *moment arm.* Plumbers who use long wrenches to free stubborn pipes are quite familiar with the value of a large lever arm (Fig. 9–22). We can summarize these features further by writing torque, denoted by the symbol τ, as

$$\tau = rF \sin \theta. \tag{9–28}$$

FIGURE 9–22 A plumber taking advantage of a long lever arm. A long wrench is more effective than a short one for freeing a stubborn pipe joint.

The dimensions of torque are $[ML^2T^{-2}]$, and in SI it is measured in N·m. For any angle, we can interpret the factor $r \sin \theta$ as the length of the projection of $\mathbf{r}$ perpendicular to the line of action of $\mathbf{F}$. This is just the lever arm through which the force acts.

Torque and its dimensions and units

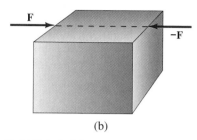

FIGURE 9–23 (a) A pair of equal and opposite forces that do not act along the same line exert a torque on an object. (b) There is no torque only if the same pair of forces act along the same line.

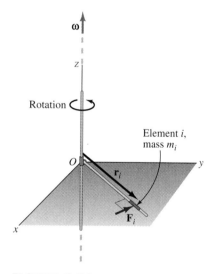

FIGURE 9–24 A force of magnitude F_i acts on an element i of a rod attached to a shaft. The force is perpendicular to the rod.

The dynamical equation of rotational motion

Equivalently, we can interpret the factor $F \sin \theta$ as the length of the projection of **F** perpendicular to the vector **r**. Only the component of **F** perpendicular to **r** is effective at producing a twist.

Dynamical Equation for Rotational Motion

To this point, our expression for torque is no more than an attempt to include the factors we know intuitively are effective at producing a twist. To make this concept a more useful one, we want to connect it *quantitatively* to a dynamical equation for rotations. We do this by applying Newton's second law.

Both internal and external forces act on a given element of the rod. In simple linear motion, internal forces do not cause any acceleration because of Newton's third law. In the same way, we can see that internal forces cannot cause a twist because they are aligned. As Fig. 9–23a shows, a pair of equal and opposite forces can produce a twist; however, they cannot if they are aligned (Fig. 9–23b). Any twist due to a force on the element labeled due to element j is canceled by an opposite twist due to a force on element j due to element i. From this point on, we will assume that the force producing torque is an external force.

Suppose, then, that a net force $\mathbf{F}_i$ acts on an element of the rod labeled by i (Fig. 9–24). We take this force perpendicular to the rod: Any component along the rod has zero lever arm and is ineffective at producing a twist. The direction of the force is *tangential*. The ith element is a distance r_i from the fixed point of attachment to the shaft and has mass m_i. The torque on element i about the origin therefore has magnitude $\tau_i = r_i F_i$. At the same time, we know from the second law that F_i produces an instantaneous tangential acceleration a_i of element i whose magnitude is $a_i = F_i/m_i$. Because the acceleration a_i is tangential, we can use Eq. (9–12), $a_i = \alpha r_i$. In other words, because the force is perpendicular to the rod, the instantaneous acceleration of the mass element is associated with an *angular* acceleration of the mass element. Because of internal forces within the (rigid) rod, *the entire rod undergoes the same angular acceleration*, so there is no subscript i on the quantity α. Combining, we have Newton's second law:

$$F_i = m_i r_i \alpha;$$

hence, the torque on the element i has magnitude

$$\tau_i = r_i F_i = m_i r_i^2 \alpha.$$

The *net* torque, τ, therefore has magnitude

$$\tau = \sum_i m_i r_i^2 \alpha = \left(\sum_i m_i r_i^2 \right) \alpha,$$

where we have used the fact that α is the same for each element. The quantity in parentheses is the rotational inertia, I, about the shaft [Eq. (9–15)], so we have found the dynamical relation we were looking for:

$$\tau = I\alpha. \tag{9–29}$$

When there is an angular acceleration, α, about an axis perpendicular to the plane of rotation, it is dynamically determined by the torque about the same point according to Eq. (9–29). *This equation is analogous to Newton's second law, which determines the linear acceleration of an object in terms of the object's inertial mass and the force acting on the object.* The major difference is that all the quantities in Eq. (9–29)—the torque, the angular acceleration, and the rotational inertia—refer to a given axis.

Newton's second law is a vector relation and we have already stated that the angular acceleration is a vector. Where is the vector aspect of the torque? Figure 9–25 shows that we can use a right-hand rule to define correctly the direction of the torque:

If the fingers of the right hand are aligned along the direction of the line perpendicular to the axis extending to the point where the force is applied (in other words, along the vector **r**) and then curled in the direction of the force, the thumb points in the direction of the torque vector τ. The procedure is illustrated in Fig. 9–26, and we can see that the torque is in the same direction as the angular acceleration. The vector equation of motion takes the form

$$\tau = I\alpha. \tag{9–30}$$

In this chapter, the vector nature of torques will not play a very important role because we are restricting ourselves to rotations about a single axis. It is important, however, to become used to thinking of the dynamics of rotational motion in terms of vectors. In the more general treatment given to rotations in Chapter 10, the vector nature of torque is critical.

EXAMPLE 9–5 A massless rod of length $L = 0.83$ m connects two small spheres of mass $m = 0.25$ kg each. The rod is constrained to rotate about an axis perpendicular to the rod and passing through its midpoint. The initial angular velocity has magnitude $\omega_0 = 2.1$ rad/s, and the rod rotates counterclockwise. A tangential force of magnitude $F = 9.6$ N is applied to one of the spheres (Fig. 9–27). If the force is applied for $t = 2.0$ s, what is the final angular velocity of the rod?

Solution: Choose the axis of rotation to be the z-axis such that the initial angular velocity ω_0 about the axis of rotation points in the $+z$-direction (Fig. 9–27). The rotational inertia of the two-sphere system about the axis is given by

$$I = m\left(\frac{L}{2}\right)^2 + m\left(\frac{L}{2}\right)^2 = \frac{mL^2}{2}.$$

The angular acceleration about the axis points in the same direction as the torque about that axis, whose magnitude is given by

$$\tau = \frac{L}{2}F.$$

Equation (9–29) gives a constant angular acceleration for a constant torque:

$$\alpha = \frac{\tau}{I} = \frac{LF/2}{mL^2/2} = \frac{F}{mL}.$$

The angular velocity then has a linear dependence on the time:

$$\omega = \omega_0 + \alpha t = \omega_0 + \frac{F}{mL}t.$$

At time $t = 2.0$ s, we have

$$\omega = 2.1 \text{ rad/s} + \frac{9.6 \text{ N}}{(0.25 \text{ kg})(0.83 \text{ m})}(2.0 \text{ s}) = 95 \text{ rad/s}.$$

Force Diagrams Revisited

The discussion of torque suggests that in the force-diagram technique, we should take into account the location at which forces act. A description of such an extended force diagram and some advice on how to approach problems that involve rigid-body motion are given in the Problem-Solving Techniques box.

Example 9–6 shows the role played by the force diagram in clarifying the forces and the torques that act on objects when rotations can occur. It also illustrates a useful energy technique.

FIGURE 9–25 The direction of the torque can be determined by a right-hand rule applied to the vectors **r** and **F**. The fingers of the right hand are aligned along **r** and then curled toward **F**, with the angle between **r** and **F** always taken to be less than 180°. The torque is then along the thumb.

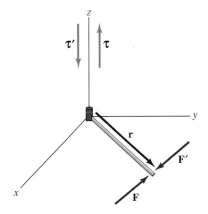

FIGURE 9–26 The direction of the angular acceleration of the rod, α, and of the torque on the rod, τ, depends on the direction of the force. When the force is in one direction (**F**), the torque τ and angular acceleration point along $+z$; when the force is in the other direction (**F'**), the torque τ' and angular acceleration point along $-z$.

249

PROBLEM-SOLVING TECHNIQUES

1. Identify and isolate the rigid body on which external forces act.

2. Identify the external forces that act on the object as well as where they act.

3. Prepare an extended force diagram: All forces should be included *plus information as to where they act* on the rigid body. Where the force acts is of paramount importance in understanding torques. Note that grav-

ity acts as if it were applied to the center of mass of the object (see Eq. (9–32) and what follows it).

4. Identify a single convenient axis, which will normally be the axis about which the rigid body rotates.

5. Find the torques about this axis that result from the acting forces. The direction of positive torque should be clearly understood.

6. Express the dynamical equations of motion that correspond to the *net* torque. This step generally requires knowledge (or calculation) of the rotational inertia about the axis of rotation. The solutions of the dynamical equations describe the rotational motion of the rigid body.

7. The net force governs the linear motion of the center of mass.

FIGURE 9–27 Example 9–5.

EXAMPLE 9–6 A bucket of water of mass $m = 12$ kg is connected to a rope of negligible mass. The rope is wrapped around a uniform solid cylindrical flywheel of mass $M = 88$ kg and radius $R = 0.50$ m (Fig. 9–28a). The flywheel is free to rotate about its axis, which is horizontal. The bucket is brought to a height $h = 28$ m and dropped, making the flywheel turn. Use the dynamical equation of motion to find the angular velocity of the flywheel after the bucket has fallen for 5 s.

Solution: We start with a force diagram drawn according to our new rules. Figure 9–28b shows a view of the flywheel from along the horizontal axis. This central axis is the one from which we measure torque. The only torque on the flywheel about this axis comes from the tension T of the rope. The angular velocity in Fig. 9–28b will point along the axis of rotation; that is, perpendicular to the page. Similarly, all forces are vertical. From now on, we will drop the boldfaced notation for these quantities.

We take the upward direction to be the $+y$-direction. The forces acting on the flywheel are the normal force, $\mathbf{F}_N$, at the axis (due to the supports that hold the rod horizontal), the force of gravity, and the tension. Because the center of mass of the flywheel does not accelerate, the net force acting on it must be zero, such that

$$\text{second law for flywheel:} \quad F_N - T - Mg = 0.$$

The forces acting on the bucket are the tension and the force of gravity. The mass will accelerate down with (as yet unknown) magnitude a, so Newton's second law gives

$$\text{second law for bucket:} \quad T - mg = -ma.$$

The torque acting on the flywheel has magnitude TR, so the dynamical torque equation is

$$TR = I\alpha.$$

Application of the right-hand rule shows that the torque is directed out of the page. Then α will also be out of the page, corresponding to an increasing angular speed in the sense shown in Fig. 9–28b.

We now have three dynamical equations and four unknowns: F_N, T, a, and α. There is, however, a fourth equation, which is a *kinematic* relation between a and α. This relation comes from the fact that if the bucket moves at a speed v, then any point on the rope moves at the same speed, and thus the outer rim of the flywheel moves at that speed. This means that the flywheel rotates about its axis at an angular speed ω such that

$$v = \omega R.$$

If we take a time derivative of this equation, we find the fourth relation:

$$a = \alpha R.$$

Now we solve our equations for the unknowns. When we substitute $\alpha = a/R$ into the dynamical equation for rotation, we find that

$$T = \frac{I\alpha}{R} = \frac{Ia}{R^2}.$$

Because the rotational inertia of a flywheel about its axis is $I = \frac{1}{2}MR^2$ [Eq. (9–26)], this becomes

$$T = \frac{1}{2}Ma.$$

We substitute this result into Newton's second law for the bucket:

$$\frac{1}{2}Ma - mg = -ma.$$

This equation can be solved for the magnitude of the bucket's acceleration:

$$a = \frac{mg}{m + M/2}.$$

In turn, the angular acceleration of the flywheel is

$$\alpha = \frac{a}{R} = \frac{mg}{[m + (M/2)]R}.$$

This angular acceleration is constant, implying that the angular velocity changes linearly with time:

$$\omega = \omega_0 + \alpha t = \omega_0 + \frac{mgt}{[m + (M/2)]R}.$$

By the initial conditions of the problem, $\omega_0 = 0$ rad/s, and after 5 s,

$$\omega = \frac{(12\ \text{kg})(9.8\ \text{m/s}^2)(5\ \text{s})}{\{(12\ \text{kg}) + [(88\ \text{kg})/2]\}(0.50\ \text{m})} = 21\ \text{rad/s}.$$

Gravity and Extended Objects

Gravity acts on every point of an object at once. Here, we demonstrate that the torque on an extended object due to gravity can be computed as if the entire mass of the object were concentrated at its center of mass. Let's consider a rod of length L and mass M fixed to a pivot at its upper end (Fig. 9–29a). We divide the rod into a series of pieces labeled by the subscript i. The pieces, which have mass Δm_i, are small enough

FIGURE 9–28 (a) Example 9–6. (b) Combined force diagram for the suspended mass m and for the cylindrical flywheel, including the locations where the various forces act. The circled dots indicate that the vectors (here, $\boldsymbol{\omega}$ and $\boldsymbol{\alpha}$) are directed out of the plane of the page.

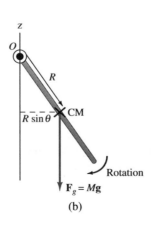

FIGURE 9–29 The force of gravity rotates a rod about the point O. (a) The differential element of force $\mathbf{F}_i$ acts on a differential mass Δm_i located a distance s_i down the rod. The circled cross indicates that the torque τ is directed into the plane of the page. (b) The net effect of the sum of the torques acting on each piece is a net torque that is the same as would be present if the entire mass of the rod were concentrated at the rod's center of mass, here marked with an $\times$.

so that we can consider each of them to be a given distance s_i down the rod. We will compute the net torque due to gravity about the pivot point, assuming the rod makes an angle θ to the vertical. The net torque, $\boldsymbol{\tau}$, is the vector sum of the torques on each piece, $\boldsymbol{\tau}_i$. All the torques are directed into the page, so this sum is the simple algebraic sum of the magnitudes τ_i. In turn, the moment arm for the force of gravity on the piece labeled by i is $s_i \sin \theta$, so the torque τ_i is

$$\tau_i = (s_i \sin \theta)\Delta m_i g.$$

The net torque then has magnitude

$$\tau = \sum_i \tau_i = \sum_i (s_i \sin \theta)(\Delta m_i g) = \left[\sum_i (s_i)(\Delta m_i)\right](\sin \theta)g. \qquad (9\text{--}31)$$

By definition (see Chapter 8), the quantity in brackets is MR, where R is the distance of the center of mass from the reference point. Thus the net torque due to gravity is (Fig. 9–29b)

$$\tau = R(Mg)\sin \theta. \qquad (9\text{--}32)$$

Compare this result with Eq. (9–28), which is our general form for the torque. *The torque due to gravity on an extended object of total mass M may be represented by the torque due to gravity acting on a particle of mass M located at the object's center of mass.* This important and very useful result is a general one that holds for any object, not just a rod (see Chapter 11 for a more general proof).

The torque due to gravity

The same reasoning can be applied to energy considerations. When an extended object moves in any way, we can find the change in its gravitational potential energy by finding the change in height, Δh, of the center of mass. The change in gravitational potential energy is then $Mg(\Delta h)$.

9–6 ANGULAR MOMENTUM AND ITS CONSERVATION

We have found a dynamical equation for rotational motion that is analogous to Newton's second law. The analogue between linear and rotational motion can be of further use to us. In our original formulation of Newton's second law, the quantity $m\,d\mathbf{v}/dt$ appears. We saw in Chapter 8 that it is useful to define the linear momentum $\mathbf{p} \equiv m\mathbf{v}$ and to formulate a more general form of Newton's second law:

$$\mathbf{F} = \frac{d\mathbf{p}}{dt}.$$

The definition of angular momentum

Similarly, it is useful to define the **angular momentum**, $\mathbf{L}$, of a symmetrical object that rotates about its symmetry axis (or an axis parallel to the symmetry axis) with angular velocity $\boldsymbol{\omega}$ by

$$\mathbf{L} \equiv I\boldsymbol{\omega}. \qquad (9\text{--}33)$$

Here, I is the rotational inertia of the object with respect to the rotation axis.

EXAMPLE 9–7 Calculate the angular momentum of Earth's motion about its axis of rotation, given that Earth's mass is 6×10^{24} kg and its radius is 6.4×10^6 m. Assume that the mass density is uniform.

Solution: Earth makes a single revolution about its axis in 24 h. Thus, the period of rotation is

$$T = (24 \text{ h})\frac{60 \text{ min}}{1 \text{ h}}\frac{60 \text{ s}}{1 \text{ min}} = 86,400 \text{ s}.$$

Hence, from the connection between ω and T [Eq. (9–4)],

$$\omega = \frac{2\pi}{T} = \frac{6.28 \text{ rad}}{86{,}400 \text{ s}} = 7.3 \times 10^{-5} \text{ rad/s}.$$

The rotational inertia of a uniform sphere is given in Table 9–1:

$$I = \frac{2}{5} MR^2 = (0.4)(6 \times 10^{24} \text{ kg})(6.4 \times 10^6 \text{ m})^2 = 10 \times 10^{37} \text{ kg} \cdot \text{m}^2.$$

Thus

$$L = I\omega = (10 \times 10^{37} \text{ kg} \cdot \text{m}^2)(7.3 \times 10^{-5} \text{ rad/s}) = 7 \times 10^{33} \text{ kg} \cdot \text{m}^2/\text{s}.$$

The assumption that Earth is a sphere of uniform density is not correct, however, which is why our calculated value of I is some 20 percent larger than the correct value of $7.9 \times 10^{37} \text{ kg} \cdot \text{m}^2$.

In terms of angular momentum, the dynamical equation for rotational motion takes the more general form

$$\boldsymbol{\tau} = \frac{d\mathbf{L}}{dt}. \qquad (9\text{–}34)$$

The torque determines the rate of change of the angular momentum.

Angular momentum for rotations of the type just described is a vector that points in the same direction as $\boldsymbol{\omega}$. For uniform rotational motion about an axis, the angular momentum does not change in either magnitude or direction. Angular momentum shares with momentum[†] the important property that it is independent of time for a system that is "left alone"; that is, a system on which there is no torque due to external forces. *Note that it is possible that the external torque is zero even when the external force is not zero, depending on where the external force is applied and on its direction.* (Similarly, it is possible to have a net torque even though the net force is zero.) When the net torque is zero, the angular momentum is independent of time and is *conserved*. For rigid bodies, the rotational inertia is constant, and the conservation of angular momentum means that the angular velocity is constant in time. When the rotational inertia can vary because the system considered can vary its shape, then the conservation of angular momentum is a very important and useful principle, as we will see in Chapter 10.

Because angular momentum is so fundamental to rotational motion, it is useful to write the energy in terms of it. We have

$$K = \frac{1}{2} I\omega^2 = \frac{1}{2} \frac{(I\omega)^2}{I} = \frac{L^2}{2I}. \qquad (9\text{–}35)$$

Parallels between Rotational and Linear Motion

Throughout this chapter, we have emphasized the similarities and the differences between rotational motion and linear motion. The major differences are that the vectors that describe the parameters of rotational motion are all measured with respect to an axis, and a right-hand rule applies to specify their direction. The similarities, or parallels, between these two types of motion are many. Although the description of rotational motion may appear to be complicated, keep in mind that no fundamentally new physical laws are involved. Everything we have discussed to this point is derived from application of Newton's laws to extended systems of point masses.

[†]The simple term "momentum" refers to linear momentum, not to angular momentum.

Angular momentum conservation plays a crucial role in drag racing. Unlike the normal operation of a car, the aim in drag racing is to have the clutch fully engaged—and therefore the rear tires spinning—before the vehicle crosses the starting line (Fig. 9AB–1). The reason for this is that when the engine turns at its maximum possible rate, it has a maximum amount of angular momentum. The transmission shafts and the rear wheels, which start at rest before the clutch begins its engagement, have a large rotational inertia. As the clutch is engaged, some of the engine's angular momentum is transferred to the shafts and wheels; thus the angular momentum of the engine is reduced. The total energy of the engine/transmission-wheel system is also reduced. (This is analogous to the collision of a moving object with a stationary one, in which momentum is conserved, but kinetic energy is reduced). If this lost energy can be replaced by burning fuel before the vehicle crosses the starting line, then the maximum energy is available to propel the automobile along the drag racing track. But if the clutch is engaged after the start of the race, then the lost energy has to be replaced *during* the race and this

FIGURE 9AB–1 The spinning rear wheels of this drag racer raise clouds of smoke from hot rubber. This is a very inefficient way to drive your car on the road.

results in a longer time to the finishing post. Note that, because the wheels spin at the start, it is kinetic rather than static friction that propels the car forward, and kinetic friction is weaker than the maximum value of static friction. Interestingly, this is less significant than the need to replace the energy lost in what we might call the "energy-transmission angular collision."

9–7 ROLLING

Kinematics of Rolling

Our discussion so far has dealt with the kinematics of pure rotation. *Rolling* is a type of motion that is more than a pure rotation. When a bicycle wheel rolls, rotation occurs *along with* linear motion. The wheel undergoes pure rotation about its axis but, to an observer on the ground, that axis is moving; as a consequence, the observer sees a point on the bicycle rim undergo motion that is more than a simple rotation (Fig. 9–30). Rolling involves a connected linear and rotational motion. Our first task is to describe rolling and discover the connection between the linear and rotational motion that it implies.

> The connection between the linear and rotational motion in rolling without slipping

Consider a wheel of radius R that rolls without slipping or skidding in a straight line on a horizontal surface (Fig. 9–31a). The center of the wheel moves at uniform speed v. In rolling without slipping, the connection between the speed v and the magnitude of the angular velocity, ω, of the wheel's rotation is

$$v = R\omega. \tag{9–36}$$

FIGURE 9–30 A rolling cylinder. A small light at the center and one at the edge show that the center of mass moves linearly, whereas a point along the edge has rotational motion and traces out a *cycloid*.

FIGURE 9–31 (a) A wheel of radius r rolls without slipping in the x-direction. Its plane lies in the xy-plane, and its center is at the position $\mathbf{r}$. (b) If the center of the wheel is at $x = 0$ at $t = 0$ and moves at speed v along the $+x$-direction, then the angular speed about the center has magnitude $\omega = v/r$.

A simple way to see this result is to imagine a chalk mark made on the rim of the wheel that rolls along a plane. The wheel leaves chalk marks on the plane as it rolls, and the distance between chalk marks made in one revolution is $2\pi R$ (the wheel's circumference). If the time taken to move this distance is T, then the center of mass will have traveled a distance vT (Fig. 9–31b). The distances are equal, so $vT = 2\pi R$, or $v = 2\pi R/T$. But $2\pi/T$ is the angular speed, ω, from which Eq. (9–36) follows.

Note that the motion of the wheel may be viewed as a rotation about the contact point with the road (Fig. 9–32a). The angular velocity about that point is determined by the fact that the wheel hub, at a distance R from the point of contact, has speed v (Fig. 9–32b). Thus this angular speed is also $\omega = v/R$. Because the top of the wheel is at a distance $2R$ from the point of contact, $v_{\text{top}} = 2R\omega = 2v$.

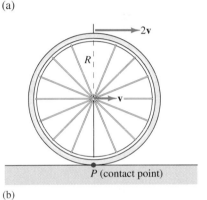

Energy in Rolling

With respect to the point of contact, the motion of the wheel is a *pure* rotation about the point of contact, and the total kinetic energy is

$$K = K_{\text{rot}} = \frac{1}{2} I_{\text{contact}} \omega^2, \qquad (9\text{–}37)$$

where I_{contact} is the rotational inertia of the wheel about an axis through the point of contact. By the parallel-axis theorem, the rotational inertia I_{contact} may be written in terms of the rotational inertia about the central axis through the center of mass:

$$I_{\text{contact}} = I_{\text{cm}} + MR^2, \qquad (9\text{–}38)$$

where M is the mass of the rolling object. Thus the total kinetic energy of the rolling object is

$$K = \frac{1}{2}(I_{\text{cm}} + MR^2)\omega^2 = \frac{1}{2} I_{\text{cm}}\omega^2 + \frac{1}{2} Mv^2, \qquad (9\text{–}39)$$

where we have used $v = R\omega$. Equation (9–39) expresses the important result that *the kinetic energy of an object rolling without slipping is the sum of the kinetic energy of rotation about its center of mass ($I_{\text{cm}}\omega^2/2$) and the kinetic energy of the linear motion of the object as if all the mass were at the center of mass.*

FIGURE 9–32 (a) A rotating bicycle wheel. The rest frame with respect to the road is the frame of reference in which the point of contact between the rolling bicycle wheel and the road is at rest. (b) In this frame, the center of the wheel moves with velocity $\mathbf{v}$ and the top of the wheel with velocity $2\mathbf{v}$. Both points rotate about the contact point.

EXAMPLE 9–8 Calculate the total kinetic energy of a solid ball of mass 2 kg and radius 10 cm that rolls without slipping on a flat surface at a speed of 0.8 m/s.

The total kinetic energy of a rolling object in terms of the center of mass

Solution: The rotational inertia of a solid ball about an axis through its center is $\frac{2}{5}MR^2$, as can be seen in Table 9–1. Thus, from Eq. (9–39), the total kinetic energy is

$$K = \frac{1}{2}(\frac{2}{5} MR^2)\omega^2 + \frac{1}{2} Mv^2 = \frac{1}{5} Mv^2 + \frac{1}{2} Mv^2 = \frac{7}{10} Mv^2$$

$$= 0.7(2 \text{ kg})(0.8 \text{ m/s})^2 = 0.9 \text{ J}.$$

The kinetic energy associated with the rolling is 40 percent of the energy associated with the linear motion.

Dynamics of Rolling

Consider a round shape of mass M and radius R that rolls without slipping down a plane making an angle θ with the horizontal (Fig. 9–33). We will insist that the object be symmetric about the central axis but it is otherwise unrestricted; for example, it could be a solid cylinder, a hollow cylinder, or a sphere. Figure 9–33 assumes that the object is a solid cylinder, but our analysis will not depend on that. We will calculate the angular velocity of the object after its center of mass has traveled a distance ℓ down the incline, starting from rest.

The forces are included in Fig. 9–33, which thereby serves as a force diagram. The force of gravity, $M\mathbf{g}$, may be separated into a component normal to the plane, $Mg \cos \theta$, and a component parallel to the plane, $Mg \sin \theta$. Both components act on the center of mass of the rolling object, as we learned in Section 9–5. There is also the contact force $\mathbf{F}_N$, normal to the plane and acting at the point of contact, which just cancels the component of gravity normal to the plane. Finally, there is the force of friction, $\mathbf{f}$, which acts at the point of contact and points back up along the plane. The existence of the frictional force is crucial if there is to be rolling rather than sliding.

To apply the equations of motion to the rolling object, we want to find the net torque acting on the object. We will use the axis through the center of the object as a reference axis, as in Fig. 9–33. Friction is the *only* force of all the forces acting on the object that exerts a torque about the center of mass; both the normal force and gravity have no lever arm through this point.

The net force along the plane determines the linear acceleration, magnitude a in this direction. According to Newton's second law,

$$Ma = Mg \sin \theta - f. \qquad (9\text{–}40)$$

Because it is static friction that is involved, its magnitude, f, is not uniquely determined until we include the dynamical equation of rotation. The torque due to friction has magnitude

$$\tau = fR. \qquad (9\text{–}41)$$

By the right-hand rule, the torque's direction is along $\boldsymbol{\omega}$ in Fig. 9-33. The dynamical equation that determines the rate of change of angular momentum about the axis of the cylinder comes from Eq. (9–29) with the torque given by Eq. (9–41):

$$I\alpha = \tau = fR. \qquad (9\text{–}42)$$

This equation implies that the friction force is given by $f = I\alpha/R$. Substituting into Newton's second law, Eq. (9–40), we get

$$Ma = Mg \sin \theta - \frac{I\alpha}{R}. \qquad (9\text{–}43)$$

FIGURE 9–33 A cylinder of radius R rolls down an inclined plane. Static friction causes the cylinder to roll without slipping. The rotational motion can be analyzed about any axis. One good one passes through the geometrical center of the cylinder and the other is formed by the line of contact between the cylinder and the plane. The forces acting are friction $\mathbf{f}$, the normal force $\mathbf{F}_N$, and gravity $M\mathbf{g}$. We have decomposed the force of gravity into components along, and perpendicular to, the inclined plane.

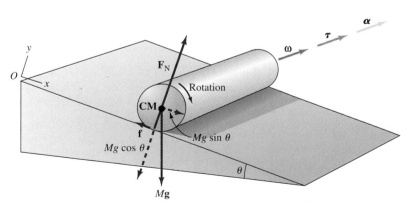

Finally, we use Eq. (9–36), $v = R\omega$, to derive a relation between a and α. The time derivative of that equation immediately gives us $a = R\alpha$. Substitution into Eq. (9–43) then gives us a single equation for the angular acceleration:

$$MaR = Mg \sin \theta - \frac{I\alpha}{R},$$

with the algebraic solution

$$\alpha = \frac{MgR \sin \theta}{MR^2 + I}. \tag{9–44}$$

Only geometric factors appear on the right, and the angular acceleration is a constant. With a constant angular acceleration, we find the usual linear dependence of angular velocity on time:

$$\omega = \omega_0 + \frac{MgR \sin \theta}{MR^2 + I} t. \tag{9–45}$$

Here, the angular velocity at $t = 0$ is ω_0. If the object starts from rest, then $\omega_0 = 0$. Similarly, the linear acceleration is constant, implying a center-of-mass velocity that increases linearly with time according to

$$v = \omega R = v_0 + \frac{(MgR^2 \sin \theta)}{(MR^2 + I)} t. \tag{9–46}$$

We can also determine the speed gained after the object has rolled a total distance ℓ down the plane. According to Eq. (2–25d), we have

$$V^2_{cm} = 2a\ell = 2 \frac{(MgR^2 \sin \theta)}{(MR^2 + I)} \ell. \tag{9–47}$$

For an object of radius R and mass M that is symmetric about a central axis, dimensional analysis tells us that the rotational inertia will equal CMR^2, where C is a pure number determined by the geometry. Thus Eq. (9–45) becomes

$$\omega = \omega_0 + \frac{(g/R) \sin \theta}{1 + C} t,$$

or, equivalently,

$$V_{cm} = v_0 + \frac{g \sin \theta}{1 + C} t. \tag{9–48}$$

Similarly,

$$V^2_{cm} = 2al = 2 \frac{(g \sin \theta)}{1 + C} \ell. \tag{9–49}$$

It is instructive to compare this last result with the speed gained by a point object that moves down the same ramp without friction: $v^2 = 2g\ell \sin \theta$. The rolling object moves less rapidly because *some of the potential energy goes into rotational motion.* Note also the striking result that *the speed of the center of mass is independent of both M and R.*

Let's look at a few special cases. The smallest possible value of C is zero, corresponding to an object with all its mass concentrated at its central axis; this is just the case of a small sliding object since there is no rotational energy in this case. The case $C = 1/2$ is the solid cylinder (see Table 9–1). The case $C = 1$ corresponds to a thin hollow cylinder with all its mass concentrated at the outside. Still larger values of C are possible if we consider yo-yo-like objects, where the rolling takes place on the inner cylinder. Thus the factor $1 + C$ increases from a minimum value of 1, and the rate at which ω, or v, increases with time for symmetric rolling objects is always less than that of a sliding object.

EXAMPLE 9–9 A ramp of length $d = 1.5$ m is set at an angle of 5.0° to the horizontal. Two objects are initially at rest at the top of the ramp and simultaneously start rolling without slipping. Object 1 is a solid cylinder of mass 0.65 kg and radius $R = 4.7$ cm. Object 2 is a hollow, thin-walled tube of the same radius R and of mass 0.13 kg. How much time does it take for each object to arrive at the bottom of the ramp?

Solution: Because there is no slipping, we can use the direct kinematic connection between linear motion and angular motion of the rolling objects; moreover, we may use the results of our discussion with different values of I for the two different objects. We therefore know the values of the linear acceleration in each case. Because we want to know the time required to travel a given straight-line distance d under constant acceleration a, we use the constant-acceleration equation $d = at^2/2$, or $t = \sqrt{2d/a}$, which applies when the initial velocity is zero. Equation (9–44) with $a = \alpha R$ tells us that

$$a = \frac{MgR^2 \sin\theta}{MR^2 + I},$$

where M is the object's mass. For the case of the solid cylinder, $I = MR^2/2$. The center of the solid cylinder then has linear acceleration

$$a_1 = \frac{g \sin\theta}{1 + \frac{1}{2}} = \frac{2}{3} g \sin\theta.$$

The hollow cylinder has a rotational inertia about its axis of $I = MR^2$; thus, its linear acceleration has magnitude

$$a_2 = \frac{g \sin\theta}{1 + 1} = \frac{1}{2} g \sin\theta.$$

In each case, the linear acceleration is independent of the mass and the radius, so this information is not needed. The only relevant difference between the solid cylinder and the hollow tube is that they have different rotational inertias because their masses are distributed differently. Numerically,

$$a_1 = \frac{2}{3} g \sin\theta = (0.67)(9.8 \text{ m/s}^2)(\sin 5.0°) = 0.57 \text{ m/s}^2;$$

$$a_2 = \frac{1}{2} g \sin\theta = (0.5)(9.8 \text{ m/s}^2)(\sin 5.0°) = 0.43 \text{ m/s}^2.$$

The time for descent of the cylinder is then

$$t_1 = \sqrt{\frac{2d}{a_1}} = \sqrt{\frac{2(1.5 \text{ m})}{0.57 \text{ m/s}^2}} = 2.3 \text{ s};$$

for the tube,

$$t_2 = \sqrt{\frac{2(1.5 \text{ m})}{0.43 \text{ m/s}^2}} = 2.6 \text{ s}.$$

It is easy to try the experiment and see that a solid cylinder reaches the bottom of a ramp faster than a hollow one. Figure 9–34 illustrates this race.

FIGURE 9–34 Example 9–9. The rate at which a rolling object accelerates depends on its rotational inertia. In a race between a solid cylinder and a hollow cylinder, both starting from rest and rolling down an inclined plane, the solid cylinder comes in first.

A rigid body can undergo rotations as well as linear motion. The linear motion can be described by specifying the motion of the center of mass. Rotations about fixed axes can be described with a single angle. Changes in this angle define the angular velocity vector, $\boldsymbol{\omega}$, and the angular acceleration vector, $\boldsymbol{\alpha}$. The directions of these vectors are determined by a right-hand rule. The period, T, and frequency, f, of rotational motion are related by

$$f = \frac{1}{T} = \frac{\omega}{2\pi}. \tag{9–5}$$

The rotational kinetic energy of a rigid body rotating about an axis is

$$K = \frac{1}{2} I\omega^2, \tag{9–14}$$

where I is the rotational inertia of the rotating object with respect to that axis. For a discrete system of point masses Δm_i that is a perpendicular distance R_i from an axis,

$$I \equiv \sum_i \Delta m_i R_i^{\,2}. \tag{9–15}$$

In particular, for a single mass m that is a perpendicular distance from an axis that can serve as a rotation axis, the rotational inertia is

$$I = mR^2. \tag{9–16}$$

The definition in Eq. (9–15) can be extended to a continuous object; thus, for an object with mass density ρ,

$$I = \int_{\text{Volume}} \rho R^2 \, dV, \tag{9–19}$$

where R^2 is the perpendicular distance of an internal point of the object to the axis.

Rotational inertia plays a role in rotational motion analogous to the role played by mass in linear motion. A useful tool for its evaluation is the parallel-axis theorem. This theorem states that the rotational inertia of an object about a given axis is

$$I = I_{\text{cm}} + Md^2, \tag{9–27}$$

In this equation, I_{cm} is the rotational inertia about an axis that goes through the center of mass and is parallel to the given axis, M is the mass of the object, and d is the perpendicular distance between the axes.

Torque, $\boldsymbol{\tau}$, is the cause of rotational motion and is analogous to force. For a rigid body rotating about a fixed axis with rotational inertia I about that axis, the equation of motion is analogous to Newton's second law:

$$\boldsymbol{\tau}_{\text{net}} = I\boldsymbol{\alpha}, \tag{9–30}$$

where $\boldsymbol{\alpha}$ is the angular acceleration. If the torque is constant, the angular velocity grows linearly with time.

The net torque is expressed in terms of the net force by the expression

$$\tau = rF \sin\theta, \tag{9–28}$$

where r is the distance from the point of application of the force to the axis of rotation and θ is the angle between the net force and r. The combination $r \sin\theta$ is the lever arm, or perpendicular distance from the axis to the line along which the force acts. The direction of the torque is specified by a right-hand rule. The torque due to gravity acts as if it were applied to the center of mass.

The technique of force diagrams for solving linear motion can be extended to rotational motion. Note that the point of application of the force must be specified.

The equation of motion also takes the form

$$\tau = \frac{d\mathbf{L}}{dt},$$ (9–34)

where $\mathbf{L}$ is the angular momentum—a quantity analogous to the linear momentum in linear motion. For rotations of a symmetric, rigid body about a symmetry axis, the angular momentum is

$$\mathbf{L} = I\boldsymbol{\omega}.$$ (9–33)

When the net torque is zero, the angular momentum is constant; thus, the angular momentum is conserved for isolated systems.

In rolling, linear motion is combined with rotational motion. Objects with radius r that roll without slipping have an angular velocity about their axes, ω, that is related to the speed v of the center of mass of the object:

$$v = r\omega.$$ (9–36)

The kinetic energy of a rolling object is the sum of its rotational kinetic energy about its axis of rotation, $I_{cm}\omega^2/2$, and the kinetic energy of its linear motion:

$$K = \frac{1}{2}(I_{cm} + Mr^2)\omega^2 = \frac{1}{2}I_{cm}\omega^2 + \frac{1}{2}Mv^2.$$ (9–39)

UNDERSTANDING KEY CONCEPTS

1. A record turntable rotates in the clockwise sense when seen from above. In what direction is the angular velocity of the turntable? If you had used a left-hand rule rather than a right-hand rule, which direction would you have chosen for the direction of the angular velocity vector? Is the choice between a right-hand or a left-hand rule purely conventional?

2. In the discussion of rolling without slipping, we saw that the kinetic energy of the rolling object is greater than the kinetic energy associated with the linear motion alone. Are there cases of rolling, with or without slipping, for which this is not true?

3. Suppose that you have a set of spherical objects with the same total mass but different radial distributions of mass. Which objects have the larger rotational inertias about an axis through the center: the spheres with more mass at the center or the spheres with more mass toward the outer surface?

4. A paddle wheel that propels a Mississippi River boat dips down into the water. Discuss the direction of motion of the portion of the paddle wheel under the water, according to an observer on the shore, then according to an observer on the boat.

5. A ball starts from rest down the inside of a parabolic bowl and rolls without slipping. At the bottom, the surface is a frictionless surface. The ball then moves up the other side of the bowl to a certain height. Is this height higher than, lower than, or the same as the height from which the ball started?

6. Solid and hollow cylinders both roll from rest down an inclined plane. Explain the difference in speeds at the bottom.

7. Devise a method to determine the rotational inertia of a sphere whose density can vary with the distance from its center.

8. Two skaters holding on to opposite ends of a rope circle a common point between them. In order to double their angular velocity about that point, should they lengthen or shorten the distance between them, and by how much?

9. You have a flat outline of the continental United States cut out in $\frac{1}{4}$-in plywood. Devise a method to find the center of mass of this outline.

10. When canoeing in rapidly flowing water, it is possible for the opposite ends of the canoe to become pressed against rocks, such that the canoe is aligned across the flow of the stream. Why is this bad?

11. Why can't flywheels be made to store arbitrarily large amounts of energy in their rotational motion?

12. A solid and a hollow cylinder are rolled from rest down an inclined plane. At the bottom, the center of the solid cylinder is moving more rapidly, even if the two cylinders have the same radius and mass. Is this a violation of the conservation of energy?

13. Why did the wheels used to steer large sailing ships have large radii?

14. When you switch to a longer wrench, you have not suddenly become stronger. So why is it easier to loosen a pipe with a long wrench than with a short wrench?

15. Attach one end of a given spring to a fixed pivot point on a frictionless table. Attach a mass to the other end, and start the mass in circular motion about the pivot point (Fig. 9–35). What happens when the revolution time is decreased by some external means, and why?

16. When a wheel of radius R rolls without slipping, the relationship between the speed of the center of the wheel, v, and the angular speed, ω, is $v = R\omega$. How would this relationship change if there were some slipping?

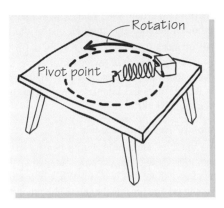

FIGURE 9–35 Question 15.

17. Suppose that you swing a rock at the end of a string in a circle in a vertical plane. What happens to the rock if the string suddenly breaks?

18. You have two wheels with rotational inertias in the ratio 2:1. Initially, the wheels are at rest. If the same torque is applied to both of them for the same length of time, what will be the ratio of their angular momenta at the end of that time interval?

19. You have two wheels with rotational inertias in the ratio 2:1. Initially, the wheels are at rest. If the same torque is applied to both of them for the same length of time, what will be the ratio of their rotational kinetic energies at the end of that time interval?

20. A wheel rotates about a central axis. When the angular speed increases beyond a certain value, the wheel breaks. Why?

PROBLEMS

9–2 Rotations about an Axis

1. (I) When a phonograph turntable is switched off, it comes to rest from its original $33\frac{1}{3}$ rev/min in 5.15 s. What is the magnitude of the angular deceleration, assuming that it is constant while the turntable comes to rest?

2. (I) How many revolutions does the turntable of Problem 1 make between the moment it is switched off and the moment it stops?

3. (I) A skater does a pirouette at the rate of 2.0 rev/s and then stops within 3/4 rev. Assume that the angular deceleration is constant, and calculate its magnitude.

4. (I) A turntable slows down from 2.0 rev/s to rest in 350 s. What is the average angular acceleration of the turntable?

5. (I) What is the angular velocity of Earth's rotation about its axis? What is the angular velocity of Earth in its orbital motion around the Sun?

6. (I) A carousel has a 7-m radius and requires 8 s for a single revolution at full speed. A carousel pig sits at a distance of 3 m from the axis, and a carousel horse at a distance of 6 m. (a) What is the period T for a single revolution of the pig? (b) of the horse? (c) What is the angular frequency of the motion of the pig? (d) of the horse? (e) What is the velocity of the pig? (f) of the horse? (g) What is the centripetal acceleration of the pig in its motion around the axis? (h) of the horse?

7. (II) A centrifuge whose maximum rotation rate is 10,000 rev/min can be brought to rest in 4.00 s. (a) What is the average angular acceleration of the centrifuge? (b) What is the distance that a point on the rim travels during the deceleration time, assuming that the radius of the centrifuge is 8 cm and that the acceleration is uniform?

8. (II) A phonograph turntable is rotating at 33.33 rev/min in a clockwise direction, viewed from above. (a) What is its angular velocity, ω, both direction and magnitude, in rad/s? (b) The turntable is switched off and comes to rest in 4.791 s. What is the average value of the angular acceleration, direction, and magnitude, during that period?

9. (II) A carousel initially at rest has an angular acceleration of 0.4 rad/s^2 and accelerates for 5 s. It then rotates at a constant angular velocity for 30 s before slowing down at the same rate with which it accelerated. (a) What is the average acceleration during the first 20 s? (b) How many total revolutions does it make? (c) How far does a child sitting on a horse 3 m from the center travel?

10. (II) A thread is wrapped around a cylindrical spool, of radius 1.5 cm, whose central axis is fixed on a support (Fig. 9–36). The thread is pulled off at a constant rate, causing the spool to spin at a constant rate; it takes 2 s to pull off 3 m of thread. What is the angular velocity of the spool while the thread is pulled off?

FIGURE 9–36 Problem 10.

11. (II) A more careful measurement of the unwrapping of the thread in Problem 10 shows that the spool accelerates from rest at a steady rate in the 2 s it takes to pull off 3 m of thread. (a) Give a formula for the position of the hand that pulls the thread as a function of time. What is the value of the constant (linear) acceleration? (b) What is the value of the angular acceleration of the spool? (c) Give a formula for the magnitude of the angular velocity as a function of time.

12. (II) A vacuum pump is connected to its electric motor by a belt drive (Fig. 9–37). The motor rotates at a rate of 4200 rev/min and the diameter of the motor shaft is 4 cm. How large should the pulley be if it is designed for a speed of 15 rev/s?

FIGURE 9–37 Problem 12.

13. (II) The path of the tip of a needle on a phonograph record may be described by the formula $r = r_0 - (\theta\rho/2\pi)$, where r_0 is the outer starting radius, θ is the angle (in radians) that a fixed radial line in the record makes with the needle arm, and ρ is the spacing between grooves. Assuming that the record turns at a rate of $33\frac{1}{3}$ rev/min and that the radial distance traveled by the needle is 9 cm in 20 min, what is ρ?

14. (II) The angular acceleration of a wheel starting from rest has magnitude $C_1 t + C_2 t^3$, where $C_1 = 48$ rad/s^3 and $C_2 = -9.5$ rad/s^5. (a) What is its angular velocity at 3.0 s? (b) How many revolutions has the wheel made after 2.0 s? (c) When will the wheel be at rest again?

15. (III) A rigid solid undergoes rotational motion about an axis. Its angular velocity has magnitude $\omega = \alpha t$, where α is constant. As the angular speed, ω, increases with time, the period, T, decreases. (a) Show that the rate of change of the period is described by the equation $dT/dt = -2\pi/\alpha t^2$. (The period is infinite at $t = 0$ because the rotation has not yet started, and a measurement of the time for 1 rev at that rate would be infinite!) (b) Show that the change in the period between $t = t_1$ and $t = t_2$ is given by

$$T(t_2) - T(t_1) = \frac{2\pi}{\alpha}\left(\frac{1}{t_2} - \frac{1}{t_1}\right).$$

9–3 Rotational Kinetic Energy

16. (I) A metal ball of mass 5.0 kg at the end of a 0.92-m-long wire rotates with an angular speed of 125 rev/min ($125 \times 2\pi$ rad/min). What is the rotational kinetic energy of the ball?

17. (I) Two identical balls are spinning on a flat surface. Ball #1 has three times the angular speed of ball #2. What is the ratio of their kinetic energies?

18. (I) Measurements of the amount of energy used by an electric motor to speed up a wheel from rest show that to bring the wheel from rest to an angular speed of 3.7 rad/s, the motor expends 7600 J. In a second use of the motor and wheel, the motor expends 9200 J to bring the wheel up from rest to an unmeasured angular speed. The motor and wheel have practically negligible amounts of friction or other type of damping. What is the unmeasured angular speed?

19. (II) A string is wrapped around a cylindrical spool of radius 1 cm. The axis of the spool is fixed. A length of string of 0.8 m is pulled off in 1.5 s at a constant tension of 20 N. What is the rotational inertia of the spool?

20. (II) What is the rotational kinetic energy of a dumbbell consisting of two equal (compact) masses of 2.5 kg each, connected by a massless rod of length 0.80 m, when the dumbbell rotates about an axis through the center of, and perpendicular to, the rod at 400 rev/min? What is the rotational

kinetic energy if the dumbbell rotates with the same angular velocity about a parallel axis through one of the masses?

21. (II) A ball of mass 0.75 kg is attached by a 1.5-m-long rope to the top of a rod. The ball swings in a circle at the rate of 25 rad/s with the rope making an angle of 30° with the vertical. What is the rotational kinetic energy of the ball? What is it when the angle is 60°?

9–4 Evaluation of Rotational Inertia

22. (I) A pipe made of aluminum with a density of 2.7 g/cm^3 is a right cylinder 20 cm long, whose outer diameter is 10 cm and whose inner diameter is 7 cm. What is the rotational inertia about the central axis of the pipe? Note that the rotational inertia of the thick cylinder can be expressed as the rotational inertia for a solid cylinder of radius R_2 minus the rotational inertia of the solid cylinder of radius R_1.

23. (I) What is the rotational inertia of a uniform 4.0-kg iron rod, 0.25 m long, about (a) an axis through its center point and perpendicular to the rod? (b) an axis through an end point and perpendicular to the rod?

24. (I) *Estimate* the rotational inertia of a tennis ball about an axis through its center.

25. (II) A neutron star has a constant density of 6×10^{17} kg/m^3 and a mass five times that of our Sun. Compare its rotational inertia with that of Earth (assume constant density). In both cases the reference axis is an axis through the center of the sphere; Table 9–1 gives the rotational inertia for such an axis.

26. (II) Mass m_1 sits at the point $(x, y, z) = (0$ m, 0 m, 0 m$)$, and mass m_2 at point $(0$ m, 1 m, 0 m$)$. (a) Where is the center of mass? (b) What is the rotational inertia about an axis through the center of mass and parallel to the z-axis? (c) parallel to the y-axis? (d) parallel to the x-axis? (e) What is the rotational inertia about an axis through the origin and along the z-axis? (f) Verify the parallel-axis theorem for this system, using the results of parts (b) and (e).

27. (II) Find the rotational inertia about the symmetry axis of a thick cylinder of mass M. Take the inner radius to be R_1 and outer radius to be R_2.

28. (II) Calculate the rotational inertia of a section of a right circular cylinder of radius R that subtends an angle θ_0 at the origin (Fig. 9–38), when the reference axis is at the origin and perpendicular to the section.

FIGURE 9–38 Problem 28.

29. (II) Calculate the rotational inertia about the central axis of the solid cone of mass M illustrated in Fig. 9–39, of opening half-angle α and height H.

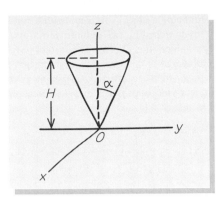

FIGURE 9–39 Problem 29.

30. (II) A thin stick of length $L = 2$ m is denser at one end than at the other: Its mass density is $\lambda = (0.42 \text{ kg/m}) - (0.15 \text{ kg/m}^2)x$, where x measures the distance from the heavier end of the stick. The stick rotates with period $T = 0.3$ s about an axis perpendicular to the stick through the heavy end. Determine the rotational kinetic energy of the stick.

31. (II) Calculate the rotational inertia of a sphere of radius R and mass M about an axis through the center of the sphere; assume that the density is not uniform but is given by ρ_1 for $0 \le r \le R_1$ and by ρ_2 for $R_1 \le r \le R$ (Fig. 9–40).

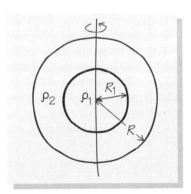

FIGURE 9–40 Problem 31.

32. (III) Use Eq. (9–19) to show that the rotational inertia about an axis through the center of a uniform, constant-density sphere of mass M and radius R is $\frac{2}{5}MR^2$.

33. (III) In Problem 79 of Chapter 8, we described a styrofoam sphere of radius R. A cavity of radius $R/2$, centered a distance $R/2$ directly above the center of the sphere, was hollowed out and filled with a solid material of density five times the density of styrofoam. In that problem, the location of the center of mass of the composite sphere was determined. What is the rotational inertia of the composite sphere about a horizontal axis through the center of mass? Express your result in terms of the total mass M of the composite sphere and its radius R. [*Hint*: You may view the mass as consisting of a large sphere of radius R and density ρ and a small sphere, off-center, of radius $R/2$ and density 4ρ. Use the parallel-axis theorem.]

34. (III) A square of mass m is made of thin wire with sides of length a. Calculate the rotational inertia about the axes shown in Fig. 9–41.

FIGURE 9–41 Problem 34.

9–5 Torque

35. (I) An instruction manual calls for a bolt to be tightened to 20 ft·lb. All you have is a 10-in wrench. How much force do you need to apply to the end of the wrench to tighten the bolt as required?

36. (I) A plumber of mass 74 kg just loosens a rusted-in bolt with the help of a 45-cm-long wrench. He places the wrench in a horizontal position and applies torque by hanging from the end of the wrench. What is the torque applied?

37. (I) The assistant of the plumber in Problem 36 meets a similar situation but aligns the wrench at an angle of 50° from the vertical (Fig. 9–42). What must his mass be in order for him to use the same technique to loosen the same bolt?

FIGURE 9–42 Problem 37.

38. (I) A flywheel of rotational inertia $I = 25$ kg·m² rotates with angular speed 6.0 rad/s. A tangential force of 10.0 N is applied at a distance of 0.30 m from the center in such a way that the angular speed decreases. How long will it take for the wheel to stop?

39. (II) A uniform rod of length L lies along the x-axis. A force F_{1y} is applied to one end of it, and a force $-F_{2y}$ is applied to the other end of it. How large is the torque on the rod about its center of mass?

40. (II) A uniform rod 1 m long with mass 0.6 kg is pivoted at one end, as shown in Fig. 9–43, and released from a horizon-

FIGURE 9–43 Problem 40.

tal position. Find the torque about the pivot exerted by the force of gravity as a function of the angle that the rod makes with the horizontal direction.

41. (II) A seesaw pivots as shown in Fig. 9–44. (a) What is the net torque about the pivot point? (b) Give an example for which the application of three different forces and their points of application will balance the seesaw. Two of the forces must point down and the other one up.

FIGURE 9–44 Problem 41.

42. (II) A two-dimensional object placed in the *xy*-plane has several forces acting on it. Find the torques about points A and B in Fig. 9–45.

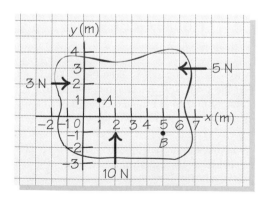

FIGURE 9–45 Problem 42.

43. (II) An aluminum casting is finished with a band sander. The sandpaper is stretched by two cylindrical rollers, one of them driven by a motor (Fig. 9–46). The rollers have diameter of 16 cm. How large is the torque that has to be applied to the driven cylinder, if the coefficient of kinetic friction between aluminum and sandpaper is 1.2, and the force applied to push the casting against the sandpaper is 4 N?

FIGURE 9–46 Problem 43.

44. (II) A massless rope is wrapped around a hollow cylinder of radius 25 cm whose central axis is fixed in a horizontal position. A mass of 10 kg hangs from the rope and, starting from rest, moves 80 cm in 2.0 s. What is the mass of the cylinder?

45. (II) A wheel of radius 24.6 cm whose axis is fixed starts from rest and reaches an angular velocity of 4.15 rad/s in 2.68 s due to a force of 13.4 N acting tangentially on the rim. (a) What is the rotational inertia of the wheel? (b) What is the change in the angular momentum during the 2.68 s? (c) How many revolutions does the wheel make? (d) How much rotational kinetic energy does the wheel have after 2.68 s?

46. (III) A motorcycle has a mass of 500 kg, the wheels have a diameter of 60 cm, and the centers of the wheels are separated by 1.5 m. Assuming that the weight is distributed uniformly over the wheels, that the wheels roll without sliding, and that the coefficient of friction between the wheels and the road is $\mu = 0.5$, calculate the torque about the center of the front wheel exerted by the forces between the road and the wheels when there is a maximum braking.

9–6 Angular Momentum and Its Conservation

47. (I) A student sits on a piano stool and holds the axle of a bicycle wheel that rotates with angular velocity ω, of magnitude 4π rad/s, pointing upward. The wheel's axis of rotation goes through the axis of the stool, which is at rest. The rotational inertia of the wheel about its axis is 1.2 kg·m², and the rotational inertia of the student and stool about the stool's axis is 8 kg·m². The student suddenly flips the shaft of the wheel, so that its angular velocity points down. How fast and in what direction will the student and stool rotate? Ignore friction.

48. (I) Two identical tops spin with angular velocities 10π rad/s up and 12π rad/s down, respectively, about vertical axes on a table. The tops bump into one another and separate. After the

collision, one of the tops has an angular velocity of 5π rad/s in its original direction. What is the angular velocity of the other top?

49. (I) A spherically symmetric celestial object rotates at 3.24592 rev/s. Through some mechanism, which does not involve an application of external torque or a change in the spherical symmetry, its radius decreases rapidly and uniformly. As a consequence, the rate of revolution changes to 3.24608 rev/s. What is the fractional change in the radius?

50. (II) A child of mass 25 kg stands at the edge of a rotating platform of mass 150 kg and radius 4.0 m. The platform with the child on it rotates with an angular speed of 6.2 rad/s. The child jumps off in a radial direction. (a) What happens to the angular speed of the platform? (b) What happens to the platform if, a little later, the child, starting at rest, jumps back onto the platform? (Treat the platform as a uniform disk.)

51. (II) Compact disks and long-playing records are made from the same material. The former have a diameter of about 12 cm; the latter, about 32 cm. When in use, records spin at $33\frac{1}{3}$ rev/min, and compact disks spin at, say, 400 rev/min. What is the ratio of the angular momentum of a compact disk in use to that of a record? Assume that a compact disk has half the thickness of a record.

9–7 Rolling

52. (I) Bicycle racers sometimes use solid wheels in order to cut down the drag force between the air and the spokes of an ordinary wheel. This can be an important effect because drag forces rise quite rapidly as the speed of an object through the air increases. If the radius of a wheel is 35 cm and the speed of the bicycle relative to the ground is 30 mi/h (13.4 m/s), what is the speed relative to the ground of the end of the spoke closest to the rim for (a) a spoke leading to the contact point with the ground; that is, a spoke pointing vertically down? (b) a spoke pointing vertically up? (c) a spoke that is horizontal and points forward? (d) a spoke that is horizontal and points backward?

53. (II) A cylindrical shell starting from rest rolls down an inclined plane that makes an angle of 20° with the horizontal. How far will the shell travel in 4 s? How far would a solid cylinder travel in the same time?

54. (II) The following objects all roll without slipping, have uniform density, mass M, and radius R, and the speed of the center of mass in each case is v. Find the ratio of the rotational kinetic energy to the total kinetic energy for (a) a solid cylinder, (b) a hollow cylinder, and (c) a sphere.

55. (II) Figure 9–47 shows a disk attached to an axle placed on an incline made of two parallel bars. The radius of the disk is 0.12 cm and its mass is 0.8 kg; the radius of the axle is 0.020 m and its mass is 0.10 kg, not including the part inside the disk. Calculate the acceleration of the system if the incline mades an angle of 5° with the horizontal and the axle rolls without slipping.

56. (II) A hollow cylinder moves down an inclined plane of length ℓ and angle θ. The cylinder has uniform density, mass M, and radius R. It is initially at rest at the top of the plane. Calculate and compare the times taken to reach the bottom if the cylinder rolls without slipping as opposed to the case in which the cylinder slips all the way down the plane without rolling. In the latter case, the coefficient of friction must be zero.

FIGURE 9–47 Problem 55.

General Problems

57. (I) A basketball player shoots a desperate last shot, spinning the ball at an angular speed of 15 rad/s to give it "action." The ball is shot with an initial velocity of 4.1 m/s at an elevation angle of 45° and leaves the player's hands 1.7 m off the floor. Unfortunately, the shot misses the backboard, the rim, and the net. How many revolutions has the ball made when it hits the floor? Ignore air resistance.

58. (I) *Estimate* the angular momentum of a spinning ice skater.

59. (II) Earth's radius is 6.4×10^6 m. Assume that its density is not uniform; that is, the inner core has a density of 8.0×10^3 kg/m^3 and the outer mantle has a density of 3.0×10^3 kg/m^3. Given that the rotational inertia of Earth is 8.3×10^{37} kg·m^2, calculate the radius R at which the density changes. [*Hint:* Go back to Problem 31.]

60. (II) A solid cylinder of mass 20 kg and radius 50 cm rotates at 300 rev/min about its central axis. What is the rotational kinetic energy of this motion? Suppose that a 0.2-kg mass is attached at one point on the rim of the cylinder. If the additional mass can be treated as a point mass, and the rotational speed is unchanged, what is the percentage change in the rotational kinetic energy?

61. (II) A solid uniform cylinder of mass M and radius R is projected up an incline of angle θ. It rolls without slipping from an initial speed v_0 of the center of mass. What distance s does the center of the cylinder travel before it starts to fall back?

62. (II) When a bicycle rider accelerates, he must accelerate his own and his bicycle's linear motion as well as the angular motion of the wheels. Suppose that the cyclist has mass 55 kg; the bicycle (not counting wheels), 8 kg; and both wheels together, 1.8 kg. Assume that the wheels, each of radius 30 cm, have all their mass concentrated in the (thin) rim. (a) At 25 km/h, what fraction of the kinetic energy of the rider plus his bicycle is in linear motion, and what fraction is in rotational motion? (b) Suppose that the cyclist loses 3 kg on a diet. What percentage of the original force is required to accelerate the system uniformly from 0 km/h to 25 km/h in 10 s? (c) Suppose that instead of going on a diet, the cyclist replaces his wheels with ones of total mass 1.2 kg. Now what is the percentage of the original force required to accelerate the system uniformly from 0 km/h to 25 km/h in 10 s?

63. (II) A hollow cylinder of radius 15 cm and mass 3.0 kg rolls without slipping at a constant speed of 1.6 m/s. (a) What is its angular momentum about its symmetry axis? (b) What is its rotational kinetic energy? (c) What is its total kinetic energy?

64. (II) A pulsar (the remnant of a star after a supernova explosion) has a mass of 4×10^{30} kg and a radius of 10 km. It rotates with a period of 0.03 s. (a) Assuming a spherical shape and constant density, what is the kinetic rotational energy of the pulsar? (b) If the period changes by 1 part in 10^8 in 1 yr, what is the rate of energy loss of the pulsar?

65. (II) A yo-yo has mass M and external radius R. The central stem has negligible mass and a radius r. The string is pulled horizontally on the lower side with a constant force F, while the yo-yo rests on a rough horizontal surface (Fig. 9–48). What is the maximum value of F for which the yo-yo will roll without slipping, assuming that the coefficient of friction between the yo-yo and the surface is μ?

FIGURE 9–48 Problem 65.

66. (II) A cylinder of known mass M, radius R, and rotational inertia I is placed on an inclined plane with angle θ (Fig. 9–49). A string is wound around the cylinder and pulled up with a tension T parallel to the inclined plane. The coefficient of static friction is large enough to prevent slipping. (a) Find the tension T_0 needed to keep the cylinder in equilibrium. (b) Find the acceleration of the cylinder if the tension is known and is different from T_0.

FIGURE 9–49 Problem 66.

67. (II) Suppose that the yo-yo described in Problem 65 is released from rest while the upper end of the cord is held steady. What is the tension in the cord during the yo-yo's downward motion?

68. (II) A thin rod of mass M and length ℓ is lying on a frictionless table. It is given an impulse of 1 N·s by a force **F** on one end, at an angle of 45° to the rod. How far will the center of mass of the rod travel when the rod has completed 2 rev?

The large torque acting on the sailboat by wind forces on the sail tends to rotate the sailboat out of the water; the torque due to gravity tends to rotate the sailboat back to a normal position. The forces and torques on a sailboat such as this one are rather complex.

More on Angular Momentum and Torque

We studied the rotations of rigid bodies in Chapter 9. Many real physical systems, such as the collections of stars that make up galaxies or a figure skater executing a complex maneuver, are not rigid, however. We will see that angular momentum and its conservation are far more useful concepts than might have been evident in Chapter 9—even for systems of particles that do not form rigid bodies and are not in pure rotational motion. We will extend to these systems the notions of angular momentum and torque. We will also explore situations in which the vector nature of these quantities helps in understanding more complicated motions. In doing so, we will be able to extend the correspondence between the dynamics of linear motion and rotational motion.

10-1 GENERALIZATION OF ANGULAR MOMENTUM

We have attributed an angular momentum to rigid systems in pure rotation. *But even a point mass moving in space has an angular momentum with respect to a fixed origin.* To assign that angular momentum, we will consider a free particle, which moves along a straight line with constant momentum $\mathbf{p}$, together with a point that we label as the origin O. A point and a line form a plane and we will label this plane the xy-plane (Fig. 10–1a). We will insist that the angular momentum, $\mathbf{L}$, that we assign to the par-

(a)

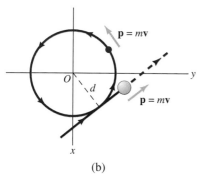

(b)

FIGURE 10-1 (a) A mass m moves freely in a plane with velocity **v**. (b) To find its angular momentum with respect to a point O we match its motion when it is closest to O to a particle of the same mass moving in a circle about O.

The angular momentum of a point mass moving in the xy-plane

There is angular momentum even if there is no circular motion.

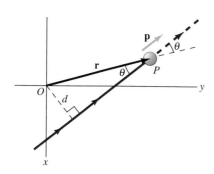

FIGURE 10-2 At a later time mass m has moved to point P. Its angular momentum, however, is the same as it was in Fig. 10-1.

ticle is constant because there are no forces—thus no torques—acting on the particle. As the particle moves, its point of closest approach to O is a distance d from the origin. At this point d, the line of motion is perpendicular to the line from O to the origin; further, the particle's motion at point d is *instantaneously* identical to that of a second particle of the same mass moving in a circular path in the xy-plane about point O (Fig. 10-1b). We already know how to find the angular momentum of the second particle: Its angular momentum is in the $+z$-direction and has magnitude

$$L = I\omega = (md^2)(v/d) = mvd = pd.$$

Therefore this is the angular momentum at all times that we assign to the freely moving particle:

$$\mathbf{L} = pd\,\mathbf{k}. \tag{10-1}$$

As Fig 10-2 illustrates, the freely moving particle may at some time be at point P, a distance r from the origin. Here, the momentum makes an angle θ with respect to the line from O. Geometry shows that $d = r\sin\theta$ so that, in term of these variables,

$$\mathbf{L} = pd\,\mathbf{k} = pr\sin\theta\,\mathbf{k}. \tag{10-2}$$

We can now simply take Eq. (10-2) as the basis for a more general definition of the angular momentum with respect to a point O of a particle with momentum p that is a distance r from O, with the angle θ between r and p. It no longer matters that the particle be free. The difference between a free particle and one on which a torque acts about point O is that for the free particle, Eq. (10-2) is the angular momentum *for all time*; whereas for the particle on which a torque acts, Eq. (10-2) is only the value of **L** at the moment picked out in Fig. 10-2. It does not matter if the particle is moving in a circle, in a straight line, or on any other trajectory (Fig. 10-3). Note that a force may act, but if that force exerts no torque on the particle about point O, Eq. (10-2) remains the constant value of **L**. Notice that we now determine angular momentum with respect to a point rather than an axis.

Reduction of Angular Momentum to Earlier Form

How does our result reduce to a symmetric rigid body rotating about point O with angular velocity ω in the $+z$-direction—a case we studied in Chapter 9? To find the net angular momentum of any system, we sum the angular momenta of the individual pieces of the system. In particular, let's look at the angular momentum about the origin of a dumbbell rotating around the z-axis, where the dumbbell is off the xy-plane (Fig. 10-4). Once we understand this case, the generalization to any three-dimensional symmetric object rotating about the z-axis follows by decomposing the object into pairs of tiny masses forming dumbbells like the one we consider here. At the moment shown in Fig. 10-4, the masses are above the y-axis. $\mathbf{L}_1$ and $\mathbf{L}_2$ have magnitudes r_1mv_1 and r_2mv_2, and are perpendicular to r_1 and r_2, respectively; $\mathbf{L}_1$ and $\mathbf{L}_2$ are also shown in that figure. The horizontal components of the $\mathbf{L}_i$ (the subscript i stands for 1 and 2 here) cancel each other, while the z-components add. To find the z-component, we observe from the figure that the angle ϕ equals the angle θ. Thus the mass labeled with the subscript i has z-component

$$(L_i)_z = L_i\sin\phi = r_imv_i\sin\phi = r_imv_i\sin\theta.$$

The combination $r_i\sin\theta$ is equal to d, which is the radius of the dumbbell, so that $(L_i)_z = mv_id$. Finally, the velocity of the mass labeled i has magnitude $v_i = \omega d$. Thus the z-component of the angular momentum of each of the two masses is

$$(L_i)_z = mvd = m\omega d^2.$$

The net angular momentum of the dumbbell is then in the $+z$-direction, with magnitude

$$L = (L_1)_z + (L_2)_z = 2(m\omega d^2).$$

We can break a more general symmetric rigid body into pairs of masses in this way; the subscript i now runs over many values, and we must retain the subscript on each mass m and each distance d. However, the angular speed ω is the same for each piece of the object, so it would have no index. Thus

$$L = \left[\sum_i (m_i d_i^2)\right]\omega = I\omega. \tag{10-3}$$

We have recognized the sum in Eq. (10–3) as the rotational inertia about the z-axis. We can conclude that our new definition of the angular momentum about a point—Eq. (10–2)—reduces to the earlier form, Eq. (9–33), for the angular momentum about an axis in the case of a symmetric rigid body.

Angular Momentum and Central Forces

Central forces—forces directed along a line from a given source—provide examples of forces with no torque about a special point. In this case, the point is the source of the force itself. The Sun is responsible for a central force on the planets, and this force is directed toward the Sun itself. Thus the gravitational force due to the Sun exerts no torque on the planets, and the angular momentum of the planets as they circle the Sun is constant. Because both gravitation and electrostatic forces are central, the case of central forces is of particular importance. Example 10–1 illustrates angular momentum for an object under the influence of a central force.

EXAMPLE 10–1 A comet, of mass $M = 10^{15}$ kg, moves in a highly eccentric orbit about the Sun, as shown in Fig. 10–5. At its closest approach to the Sun, $d = 10^6$ km, the comet is measured to be moving at a speed of 6×10^6 m/s. What is the angular momentum of the comet with respect to the Sun if the comet is treated as a point mass?

Solution: The momentum of the comet has magnitude $p = Mv$, where v is the speed of the comet. The angular momentum is given by the momentum times the distance of closest approach. Therefore it has magnitude

$$L = Mvd = (10^{15}\text{ kg})(6 \times 10^6\text{ m/s})(10^6\text{ km})\frac{10^3\text{ m}}{1\text{ km}} = 6 \times 10^{30}\text{ kg} \cdot \text{m}^2/\text{s}.$$

The direction of the angular momentum is given by a right-hand rule. In Fig. 10–5, this direction is up out of the plane of the orbit, in what is labeled the $+z$-direction.

FIGURE 10–5 Example 10–1. The direction of the angular momentum is found from a right-hand rule to be up out of the plane of the orbit, in the $+z$-direction.

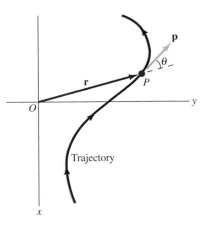

FIGURE 10–3 A particle of the same mass and velocity as that of the free particle in Fig. 10–2 at the same point P. Its angular momentum at that point is the same as that of the freely moving particle.

FIGURE 10–4 To find the angular momentum of a symmetric rigid body about the symmetry axis in terms of our results for the angular momentum of point masses, we break up the rigid body into pairs of symmetric mass elements.

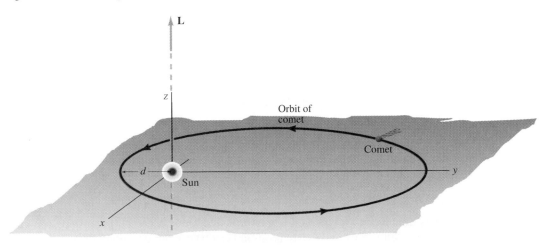

PROBLEM–SOLVING TECHNIQUES

The Vector Product

In addition to the scalar product discussed in Chapter 6, we can define a product of two vectors—**A** and **B**—that is itself a vector. The **vector product** (or **cross product**) of **A** and **B** is defined to be perpendicular to both **A** and **B**; it is denoted **A** × **B**. Any two vectors form a plane; for example, we can choose the x-axis along **A** and then define the xy-axis so that **B** has only x and y components. The direction of the vector product of **A** and **B** is then perpendicular to this plane. A right-hand rule is used to define the direction of the vector product because there is an ambiguity as to whether the direction is into or out of the plane. The direction of **A** × **B** is specified as follows (Fig. 10B1–1): Point the fingers of your right hand in the direction of **A** and curl toward **B** according to the rule that the angle through which the fingers are curled to reach **B** must always be less than 180° (it will be hard to get the fin-

gers to follow the opposite rule!). The direction of your thumb then indicates the direction of the vector product **A** × **B**. The magnitude of the vector product is given by

$$|\mathbf{A} \times \mathbf{B}| = AB \sin\theta, \quad \text{(B1–1)}$$

where θ is the angle (less than 180°, so that the sine is positive) between **A** and **B**.

An immediate consequence of the right-hand rule is that

$$\mathbf{A} \times \mathbf{B} = -\mathbf{B} \times \mathbf{A}. \quad \text{(B1–2)}$$

Thus the vector product is not commutative. It also follows from Eq. (B1–1) that *the vector product of two vectors that are parallel (or antiparallel) to each other is zero*: When $\theta = 0°$ or 180°, $\sin\theta = 0$. The vector product attains its maximum magnitude when **A** and **B** are perpendicular to each other because then $\sin\theta = 1$. Contrast this to the scalar product, which is zero when the two vectors are perpendicular and

has a maximum magnitude when they are parallel.

The unit vectors **i**, **j**, and **k** along the x-, y-, and z-axes, respectively, obey the relations

$$\mathbf{i} \times \mathbf{j} = -\mathbf{j} \times \mathbf{i} = \mathbf{k}, \quad \text{(B1–3a)}$$

$$\mathbf{j} \times \mathbf{k} = -\mathbf{k} \times \mathbf{j} = \mathbf{i}, \quad \text{(B1–3b)}$$

$$\mathbf{k} \times \mathbf{i} = -\mathbf{i} \times \mathbf{k} = \mathbf{j}. \quad \text{(B1–3c)}$$

Also $\mathbf{i} \times \mathbf{i} = \mathbf{j} \times \mathbf{j} = \mathbf{k} \times \mathbf{k} = 0$. We can expand the vectors **A** and **B** into their components and find

$$\mathbf{A} \times \mathbf{B} = (A_x\mathbf{i} + A_y\mathbf{j} + A_z\mathbf{k})$$
$$\times (B_x\mathbf{i} + B_y\mathbf{j} + B_z\mathbf{k})$$

$$= A_xB_y(\mathbf{i} \times \mathbf{j}) + A_zB_z(\mathbf{i} \times \mathbf{k})$$
$$+ A_yB_x(\mathbf{j} \times \mathbf{i}) + A_yB_z(\mathbf{j} \times \mathbf{k})$$
$$+ A_zB_x(\mathbf{k} \times \mathbf{i}) + A_zB_y(\mathbf{k} \times \mathbf{j}).$$

With the results of Eqs. (B1–3), we find that

$$\mathbf{A} \times \mathbf{B} = (A_yB_z - A_zB_y)\mathbf{i}$$
$$+ (A_zB_x - A_xB_z)\mathbf{j}$$
$$+ (A_xB_y - A_yB_x)\mathbf{k}. \quad \text{(B1–4)}$$

FIGURE 10B1–1 A right-hand rule specifies the direction of the vector product **A** × **B** between two vectors **A** and **B**. In particular, $\mathbf{k} = \mathbf{i} \times \mathbf{j}$.

Angular Momentum as a Vector Product

The angular momentum of a point mass about an origin

Vector products (see "The Vector Product" box) are useful for many relationships involving rotation. One of the most important examples is provided by the angular momentum *about some origin O* of a point mass with momentum **p**, which is given by

$$\mathbf{L} \equiv \mathbf{r} \times \mathbf{p}, \quad \text{(10–4)}$$

where **r** is the position vector of the point mass with respect to O (Fig. 10–6). We can reduce this result to that previously given in Eq. (10–2) and described there. The magnitude of **L** is $rp \sin \theta$, where θ is the angle between **r** and **p**, and $r \sin \theta$ is the perpendicular distance d of Eq. (10–2). By the property of the vector product, the vector **L** is perpendicular to *both* **r** and **p**, with the direction given by the right-hand rule. Note that **L** is zero when **r** and **p** are parallel; that is, when the straight-line extension of the vector **p** passes through the point of reference, O. Examples 10–2 and 10–3 illustrate the use of Eq. (10–4).

EXAMPLE 10–2 In an engineering design, a light but stiff rod of length R is attached at an angle θ to a shaft along the z-axis; it is used to rotate a mass M about the shaft (Fig. 10–7). The mass moves with speed v. Describe the angular momentum, **L**, of the mass with respect to the attachment point of the rod.

Solution: The angular momentum vector is perpendicular both to the vector from point O to the mass, which runs along the rod, and to the mass's momentum. Thus **L** has a component both in the z-direction and in the horizontal plane, pointing to the shaft—that is, radial. **L** has magnitude Rmv. To find its components, we can refer directly to Fig. 10–4, which treated a similar case with a pair of masses; here we are interested in the angular momentum for one of the pair. Replacing r_1 in Fig. 10–4 with R, we have

FIGURE 10–6 Vector relation for the angular momentum **L** of a point particle about a point O.

$$L_z = L \sin \theta = L \frac{d}{R} = (RMv) \frac{d}{R} = dMv;$$

$$L_{\text{radial}} = -L \cos \theta = -L \frac{\sqrt{R^2 - d^2}}{R} = -\sqrt{R^2 - d^2} \, Mv.$$

Note that because the angular momentum is not constant (the horizontal component rotates so as to point to the shaft), there must be a torque about point O. This classic engineering problem tells us that the attachment to the shaft must be appropriately constructed to supply the torque.

EXAMPLE 10–3 A tether ball is tied by a rope to a central pole. As the ball whirls around the pole in a (nearly) horizontal plane, the rope gradually winds around (or unwinds from) the pole, shortening (or lengthening) the amount of rope between the ball and the pole (Fig. 10–8). The motion is spiral rather than circular: Near the time t, when the ball is at a distance nearly equal to R and rotating around the pole with angular speed nearly equal to ω, the distance is more precisely $r = R - A\omega t$, where A is a constant with dimensions of length and $A\omega t$ is small compared to R. Find the ball's angular momentum with respect to the pole at time t.

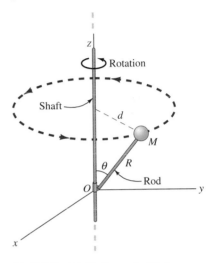

FIGURE 10–7 Example 10–2.

Solution: Because the tether ball does not follow a circular path, we must use the more general form $\mathbf{L} = \mathbf{r} \times \mathbf{p}$ for the angular momentum. We take the horizontal plane to be the xy-plane. We first write out the x- and y-coordinates of the ball as a function of time, so as to be able to calculate the velocities and thus the momentum. We have

$$x = r \cos \theta = (R - A\omega t) \cos(\omega t);$$

$$y = r \sin \theta = (R - A\omega t) \sin(\omega t).$$

From this expression, we can also calculate the momentum components of the ball:

$$p_x = Mv_x = M\frac{dx}{dt} = M[(R - A\omega t)(-\omega \sin(\omega t)) + (-A\omega)\cos(\omega t)];$$

$$p_y = Mv_y = M\frac{dy}{dt} = M[(R - A\omega t)(\omega \cos(\omega t)) + (-A\omega)\sin(\omega t)].$$

We can now use Eq. (B1–4) in "The Vector Product" box on page 270 to find **L**:

$$\mathbf{L} = (xp_y - yp_x)\mathbf{k}$$

$$= M\left[\begin{array}{c} (R - A\omega t)^2\omega \cos^2(\omega t) + (R - A\omega t)(-A\omega) \cos(\omega t) \sin(\omega t) \\ -(R - A\omega t)^2(-\omega) \sin^2(\omega t) - (R - A\omega t)(-A\omega) \sin(\omega t) \cos(\omega t) \end{array} \right]\mathbf{k}$$

$$= M(R - A\omega t)^2\omega[\cos^2(\omega t) + \sin^2(\omega t)]\mathbf{k} = M(R - A\omega t)^2\omega\mathbf{k}.$$

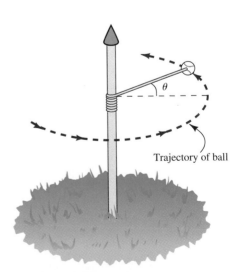

FIGURE 10–8 Example 10–3. As the rope winds around the pole, it shortens.

FIGURE 10–9 Vector relation for the torque about a point O. The force is applied at a specific point on an extended object.

L always lies in the z-direction because the motion is, by assumption, always in the xy-plane. But the angular momentum changes with time—albeit slowly ($R \gg A\omega t$). As a check, note that if $A = 0$ (no change in length), the angular momentum has the familiar magnitude $MR^2\omega = MvR$.

FIGURE 10–10 The tugboat can rotate the large ship by exerting a force on it. Notice that the force is perpendicular to the large ship and acts at the bow; this maximizes the torque.

The torque on an object about a reference point

The rate of change of angular momentum is given by the torque.

10-2 GENERALIZATION OF TORQUE

Torque, introduced in Chapter 9, has a magnitude that is given by a lever arm times the magnitude of force; it has a direction specified by a right-hand rule. The torque, $\boldsymbol{\tau}$, with respect to any reference point O due to a force $\mathbf{F}$ applied at some point P of an object is therefore compactly written by using a vector product:

$$\boldsymbol{\tau} = \mathbf{r} \times \mathbf{F}, \tag{10-5}$$

where $\mathbf{r}$ is the position vector from O to the point of application (P) of the force (Fig. 10–9). The object in question can be an extended system (Fig. 10–10) or it can even be a point mass located at P. The reference point O about which the torque is defined can be located inside or outside the object. Aligning the fingers of the right hand along $\mathbf{r}$ and curling them toward $\mathbf{F}$, the torque is in the direction of the thumb.

The test as to whether this is the proper definition of torque is whether the dynamical relation between torque and angular momentum, Eq. (9–34), continues to be satisfied:

$$\boldsymbol{\tau} = \frac{d\mathbf{L}}{dt}. \tag{10-6}$$

We can check that Eq. (10–6) is indeed satisfied if we take Eq. (10–4), which is our generalized expression for the angular momentum of a point mass, and compute its rate of change. Keep in mind that we can always use Newton's second law. The rate of change of angular momentum can be found by using the chain rule of calculus (Appendix IV–7):

$$\frac{d\mathbf{L}}{dt} = \left(\frac{d\mathbf{r}}{dt} \times \mathbf{p}\right) + \left(\mathbf{r} \times \frac{d\mathbf{p}}{dt}\right). \tag{10-7}$$

The first part of Eq. (10–7) drops out because it is of the form $\mathbf{v} \times (m\mathbf{v})$, and the vector product of two parallel vectors is zero. For the second part of Eq. (10–7) we use Newton's second law in the form $d\mathbf{p}/dt = \mathbf{F}$ and obtain the equation

$$\frac{d\mathbf{L}}{dt} = \mathbf{r} \times \mathbf{F} = \boldsymbol{\tau},$$

which is just the desired Eq. (10–6).

Having shown that the general form $\mathbf{L} = \mathbf{r} \times \mathbf{p}$ reduces to the form $I\omega$ for symmetric rigid bodies, we can use the fact that $\boldsymbol{\tau} = d\mathbf{L}/dt$ to see that the general form $\boldsymbol{\tau} = \mathbf{r} \times \mathbf{F}$ similarly reduces to the form $I\alpha$.

We may use the properties of the unit vectors **i** and **j** in Eqs. (B1–3) to express torque in terms of the components of **r** and **F**.

If **r** is written as $\mathbf{r} = x\mathbf{i} + y\mathbf{j} + z\mathbf{k}$, Eq. (B1–4) shows that torque is given in terms of components by

$$\boldsymbol{\tau} = (yF_z - zF_y)\mathbf{i} + (zF_x - xF_z)\mathbf{j} + (xF_y - yF_x)\mathbf{k}. \qquad (10\text{–}8)$$

EXAMPLE 10–4 A stone of mass $m = 1$ kg is dropped from an outstretched arm of length $\ell = 0.8$ m (Fig. 10–11). For the time when the stone has dropped exactly $d = 1$ m, find the net torque on the stone about the shoulder, labeled point O. Ignore air resistance.

Solution: The net torque is determined by the net force, and in this case the only force acting is gravity, $\mathbf{F} = -mg\mathbf{k}$. We are using the coordinate system with the shoulder at the origin. Although the distance of the rock from point O increases with time, the lever arm for the force of gravity about point O is constant, given by the arm length ℓ. The magnitude of the torque will therefore be ℓmg, and its direction will be given by the right-hand rule to be in the $+y$-direction. Equivalently, we can use the vector product definition in terms of components. We have, from Eq. (10–8),

$$\boldsymbol{\tau} = (\ell\,\mathbf{i}) \times (-F_g\mathbf{k}) = \ell F_g\,\mathbf{j} = \ell mg\,\mathbf{j}.$$

Note that the torque is constant over time and independent of the distance the stone has fallen.

The magnitude of the torque is

$$\tau = (0.8 \text{ m})(1 \text{ kg})(9.8 \text{ m/s}^2) = 8 \text{ N} \cdot \text{m}.$$

FIGURE 10–11 Example 10–4. The origin of the coordinate system is placed at the shoulder. $\mathbf{F}_g$ is the force of gravity on the falling stone.

10–3 THE DYNAMICS OF ROTATION

In Section 10–2, we saw that the dynamical equation of motion [Eq. (10–6)] holds for a point mass whose angular momentum is defined with respect to a particular point. To see that the same kind of equation of motion applies to extended objects, we will now consider a system that is a collection of point particles. Our more general expressions for the angular momentum and torque will provide some new and powerful ways to think about rotations and the motions of extended objects, both rigid and nonrigid.

To find our new expressions, let's consider a collection of particles with masses m, located at positions $\mathbf{r}_i$ with respect to a point O, as in Fig. 10–12. If the momenta of the particles are given by $\mathbf{p}_i$, then the particle with the label i has an angular momentum $\mathbf{L}_i = \mathbf{r}_i \times \mathbf{p}_i$, and if the force acting on the ith particle is $\mathbf{F}_i$, then

$$\boldsymbol{\tau}_i = \mathbf{r}_i \times \mathbf{F}_i = \frac{d\mathbf{L}_i}{dt}. \qquad (10\text{–}9)$$

The total angular momentum of the collection of particles is

$$\mathbf{L} = \sum_i \mathbf{L}_i. \qquad (10\text{–}10)$$

Similarly, the total torque is the sum of the individual torques on the particles:

$$\boldsymbol{\tau} = \sum_i \boldsymbol{\tau}_i. \qquad (10\text{–}11)$$

The rate of change of the total angular momentum is obtained by summing Eq. (10–9) over the component particles, and we obtain Eq. (10–6):

$$\boldsymbol{\tau} = \frac{d\mathbf{L}}{dt}. \qquad (10\text{–}12)$$

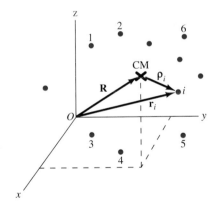

FIGURE 10–12 A system of many particles labeled by the index i. The position of particle i with respect to an origin O is $\mathbf{r}_i$, and the displacement of particle i from the center of mass is $\boldsymbol{\rho}_i$.

The rate of change of the total angular momentum of a system is given by the total torque.

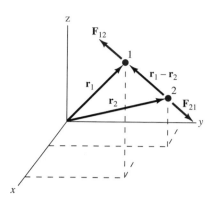

FIGURE 10-13 Two mass points, labeled 1 and 2, within an extended object, and the internal forces they exert on one another.

Interparticle forces play no role in the torque on a system.

Features of rotations about any axis

Equation (10–12) expands the rotational analogue of Newton's second law to any extended system.

The Role of Interparticle Forces in Torque

What role do the forces between the individual particles (or between different regions of a continuous object) play in the dynamics of rotations? Let's consider just two particles (or two small mass regions), with labels 1 and 2; this result can be generalized. For the torque on particle 1, we have (Fig. 10–13)

$$\boldsymbol{\tau}_1 = \mathbf{r}_1 \times (\mathbf{F}_1^{\text{ext}} + \mathbf{F}_{12}), \tag{10–13a}$$

where $\mathbf{F}_{12}$ is the (internal) force exerted on particle 1 due to particle 2 and $\mathbf{F}_1^{\text{ext}}$ is the external force on particle 1. We also have

$$\boldsymbol{\tau}_2 = \mathbf{r}_2 \times (\mathbf{F}_2^{\text{ext}} + \mathbf{F}_{21}). \tag{10–13b}$$

We must assume that the forces between 1 and 2 are directed along a line between the two particles.

The interparticle forces contribute to the total torque through the terms

$$(\mathbf{r}_1 \times \mathbf{F}_{12}) + (\mathbf{r}_2 \times \mathbf{F}_{21}) = (\mathbf{r}_1 \times \mathbf{F}_{12}) - (\mathbf{r}_2 \times \mathbf{F}_{12})$$

$$= (\mathbf{r}_1 - \mathbf{r}_2) \times \mathbf{F}_{12}. \tag{10–14}$$

We have used Newton's third law. Observe that $(\mathbf{r}_1 - \mathbf{r}_2)$ is the vector leading from particle 2 to particle 1; further, observe that the interparticle force $\mathbf{F}_{12} = -\mathbf{F}_{21}$ is a vector *along the direction between the two particles*. Thus the vector product that appears in Eq. (10–14) is zero, and *only external forces appear in the total torque* $\boldsymbol{\tau}$ that is to be used in Eq. (10–12). This result is similar to the result derived in Chapter 8, where we showed that interparticle forces do not contribute to the change in linear momentum of a system composed of many interacting particles. As in the case of the momentum equation, this allows us to discuss the motion of extended bodies without knowing the details of how their individual parts interact with each other. Only the torques due to external forces enter.

Dynamical Equations of Motion with Reference to Any Axis

We showed in Eq. (9–33) that, for a symmetric rigid body with some rotational inertia I about its center of mass axis, the dynamical equation for rotation takes the form

$$\boldsymbol{\tau} = \frac{d\mathbf{L}}{dt} = I\frac{d\boldsymbol{\omega}}{dt} = I\boldsymbol{\alpha}. \tag{10–15}$$

In solving equations that describe the angular motion of an extended object, however, it is often useful to consider rotations about a second axis A, parallel to the axis through the center of mass. We can derive three important results concerning the rotations of a rigid body about any axis A:

1. The angular momentum about A is the sum of two terms: (a) the angular momentum $\mathbf{R} \times \mathbf{P}$ about A of a point mass that is carrying the whole mass (M) of the object, as though it were located at the center of mass and moving with the velocity ($\mathbf{V}$) of the center of mass; and (b) the angular momentum $\mathbf{L}_{\text{cm}}$ of the object about the center of mass:

$$\mathbf{L} = (\mathbf{R} \times \mathbf{P}) + \mathbf{L}_{\text{cm}}. \tag{10–16}$$

Here, $\mathbf{P} = M\mathbf{V}$.

2. The total torque about A is the sum of two terms: (a) the torque about A due to the total external force applied to the center of mass; and (b) the torque about the center of mass:

$$\boldsymbol{\tau} = (\mathbf{R} \times \mathbf{F}_{\text{tot}}) + \boldsymbol{\tau}_{\text{cm}}. \tag{10–17}$$

3. The rate of change in each term of the total angular momentum in Eq. (10–16) is equal to the corresponding term of the total torque; that is,

$$\mathbf{R} \times \mathbf{F}_{\text{tot}} = \frac{d}{dt}(\mathbf{R} \times \mathbf{P}) \quad \text{and} \quad \boldsymbol{\tau}_{\text{cm}} = \frac{d}{dt}\mathbf{L}_{\text{cm}}. \qquad (10\text{–}18)$$

How to Get Equations (10–16) through (10–18) for Angular Momentum and Torque

We break our extended object into a set of discrete pieces labeled with the subscript i. The center of mass position $\mathbf{R}$ with respect to an origin O was defined in Eq. (8–48) by

$$\mathbf{R} = \frac{1}{M}\sum_{i}^{N} m_i\mathbf{r}_i,$$

where $M = \sum_i^N m_i$ is the total mass, N is the number of pieces, and $\mathbf{r}_i$ is the position of the ith piece. We introduce the vector $\boldsymbol{\rho}_i$ (Fig. 10–12), the position vector of particle i as measured from the center of mass:

$$\mathbf{r}_i = \mathbf{R} + \boldsymbol{\rho}_i.$$

The angular momentum with respect to point O may now be written in the form

$$\mathbf{L} = \sum_i \mathbf{r}_i \times \mathbf{p}_i = \sum_i (\mathbf{R} + \boldsymbol{\rho}_i) \times \mathbf{p}_i = \left(\mathbf{R} \times \sum_i \mathbf{p}_i\right) + \left(\sum_i \boldsymbol{\rho}_i \times \mathbf{p}_i\right)$$

$$= (\mathbf{R} \times \mathbf{P}) + \left(\sum_i \boldsymbol{\rho}_i \times \mathbf{p}_i\right),$$

where $\mathbf{P} = \sum \mathbf{p}_i$ is the total momentum of the system. The term $\sum \boldsymbol{\rho}_i \times \mathbf{p}_i$ is the angular momentum $\mathbf{L}_{\text{cm}}$ about the center of mass. We have thus proven Eq. (10–16).

There is a similar decomposition of the total torque:

$$\boldsymbol{\tau} = \sum_i \mathbf{r}_i \times \mathbf{F}_i = \sum_i (\mathbf{R} + \boldsymbol{\rho}_i) \times \mathbf{F}_i = \left(\mathbf{R} \times \sum_i \mathbf{F}_i\right) + \left(\sum_i \boldsymbol{\rho}_i \times \mathbf{F}_i\right)$$

$$= (\mathbf{R} \times \mathbf{F}_{\text{tot}}) + \left(\sum_i \boldsymbol{\rho}_i \times \mathbf{F}_i\right).$$

The second term in the final equality is the torque about the center of mass, and so we have demonstrated Eq. (10–17).

To derive the third result, we take the rate of change of the $\mathbf{R} \times \mathbf{P}$ term in Eq. (10–18). We find that it is the same as the first term of Eq. (10–17):

$$\frac{d(\mathbf{R} \times \mathbf{P})}{dt} = \left(\frac{d\mathbf{R}}{dt} \times \mathbf{P}\right) + \left(\mathbf{R} \times \frac{d\mathbf{P}}{dt}\right) = \left[\left(\frac{1}{M}\right)\mathbf{P} \times \mathbf{P}\right] + (\mathbf{R} \times \mathbf{F}_{\text{tot}})$$

$$= \mathbf{R} \times \mathbf{F}_{\text{tot}}.$$

This, in turn, implies that the rate of change of the total angular momentum about the center of mass is equal to the torque about the center of mass due to all the forces:

$$\frac{d\mathbf{L}_{\text{cm}}}{dt} = \boldsymbol{\tau}_{\text{cm}}.$$

Why It Is Useful to Be Able to Choose the Reference Point

The results just summarized are of practical importance because they provide us with alternative ways to approach problems with both linear and rotational motion. We can choose points of reference for torque, angular momentum, and rotational inertia for convenience. This is illustrated by taking a second look at a cylinder rolling down an incline, which is a problem first examined in Chapter 9 (Fig. 10–14a). There we used a point on the symmetry axis of the cylinder as a reference. Let's now choose our ref-

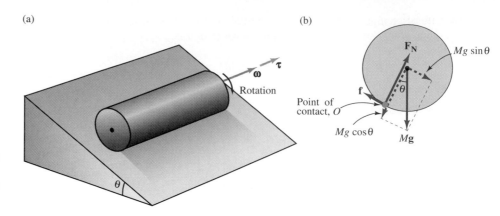

FIGURE 10–14 (a) A cylinder rolls without slipping down an inclined plane. (b) Extended force diagram for the rolling cylinder.

erence point to be point O, the point of contact between the cylinder and the plane. It is most efficient to choose as a reference point the point through which the largest number of individual forces act. For such a point the lever arm is zero; thus, these forces do not contribute to the torque (Fig. 10–14b). In this case, only gravity produces a torque about the contact point; moreover, only the component of gravity that is parallel to the plane—of magnitude $mg \sin \theta$—contributes. Using the right-hand rule, we see that the corresponding torque about O points into the plane of the page and has magnitude

$$\tau = (Mg \sin \theta)R. \qquad (10\text{–}19)$$

Having calculated the torque about the contact point, let's look at the angular momentum about this same point. As indicated in Eq. (10–16), there are two terms. First, there is the angular momentum of the total mass as though that mass were placed at the object's center of mass, $\mathbf{R} \times \mathbf{P} = \mathbf{R} \times (M\mathbf{V})$. $\mathbf{R}$ and $\mathbf{V}$ are perpendicular to each other, so this term has magnitude

$$MVR = M(\omega R)R = MR^2\omega. \qquad (10\text{–}20)$$

Second, there is the angular momentum about the symmetry axis passing through the center of mass, $L = I_{cm}\omega$. Thus the total angular momentum about the point of contact is the sum of Eq. (10–20) and $I_{cm}\omega$:

$$L = (MR^2 + I_{cm})\omega. \qquad (10\text{–}21)$$

We have chosen the same reference point for both the torque and the angular momentum. The dynamical equation $dL/dt = \tau$ is therefore applicable and, from Eqs. (10–19) and (10–21), it takes the form

$$(MR^2 + I_{cm}) \frac{d\omega}{dt} = MgR \sin \theta;$$

$$\frac{d\omega}{dt} = \frac{MgR \sin \theta}{MR^2 + I_{cm}}.$$

This equation states that the angular acceleration $d\omega/dt$ is constant, so the angular velocity increases linearly with time:

$$\omega(t) = \omega_0 + \frac{MgR \sin \theta}{MR^2 + I_{cm}} t. \qquad (10\text{–}22)$$

Here, the angular velocity at $t = 0$ is ω_0. For a symmetric rolling object, I_{cm} will take the form $I_{cm} = CMR^2$, where C is a numerical constant. Then the velocity $v = \omega R$ of the center axis becomes

$$v(t) = v_0 + \frac{g \sin \theta}{1 + C} t. \qquad (10\text{–}23)$$

PROBLEM SOLVING

Choose a convenient axis. Physical results won't depend on that choice.

This result is the same one found in Eq. (9–48), yet we have used a different reference point. The choice of reference point does not affect the result and can be chosen for convenience.

The angular impulse is as useful for the motion of extended systems as the "ordinary," or linear, impulse is for the motion of point masses. The linear impulse, $\mathbf{J} = \mathbf{F}\,\Delta t$, describes the change $\Delta \mathbf{p}$ in momentum when a force $\mathbf{F}$ acts for a short duration Δt. Similarly, the dynamical equation for rotational motion tells us that the change in the angular momentum of a system, $\Delta \mathbf{L}$, when a torque, $\boldsymbol{\tau}$ acts for a duration Δt is given by the angular impulse $\mathbf{J}_\tau$:

The definition of angular impulse

$$\Delta \mathbf{L} = \boldsymbol{\tau}\,\Delta t \equiv \mathbf{J}_\tau. \qquad (10\text{–}24)$$

Here, both the torque and angular momentum are measured with respect to the same reference point.

Let the reference point for torque and angular momentum be O. The force, $\mathbf{F}$, is applied briefly at $\mathbf{r}$. The torque is then $\boldsymbol{\tau} = \mathbf{r} \times \mathbf{F}$; multiplying by Δt, we see that the angular impulse is given by

$$\mathbf{J}_\tau = \boldsymbol{\tau}\,\Delta t = \mathbf{r} \times (\mathbf{F}\,\Delta t) = \mathbf{r} \times \mathbf{J}. \qquad (10\text{–}25)$$

EXAMPLE 10–5 A court expert models a collision in which a stationary automobile is struck from the side by treating the automobile as a rod of mass M and length ℓ at rest on a frictionless surface. An impulse $F\,\Delta t$ is applied at right angles at a distance $\ell/3$ from one end of the rod (Fig. 10–15). Describe the subsequent motion of the rod.

Solution: The motion may be decomposed into motion of the center of mass and rotation about the center of mass. The initial momentum of the center of mass is zero, and therefore its final momentum is given by the impulse:

$$\Delta P = P_f - P_i = P_f = MV_f = F\,\Delta t.$$

Similarly, the angular impulse is equal to the change in angular momentum in time Δt. The initial angular momentum is zero, and the distance from the point of application of the force to the center of mass is $(\ell/2) - (\ell/3)$. Thus, if ω_f is the final angular velocity of the rod about the center of mass,

$$\Delta L = L_f - L_i = I\omega_f = (F\,\Delta t)\left(\frac{\ell}{2} - \frac{\ell}{3}\right) = \frac{(F\,\Delta t)\ell}{6}.$$

Because $I = M\ell^2/12$ (see Table 9–1), this relation reads

$$\frac{M\ell^2\,\omega_f}{12} = \frac{(F\,\Delta t)\ell}{6};$$

$$\omega_f = \frac{2}{M\ell}(F\,\Delta t) = \frac{2}{\ell}V_f.$$

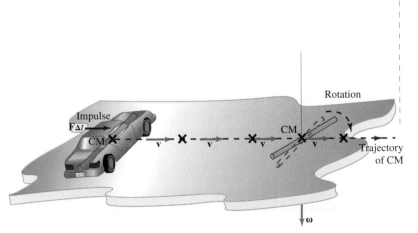

FIGURE 10–15 Example 10–5. The automobile is modeled by a uniform rod. (a) Before: An impulse is given to the rod a distance $\ell/3$ from the top. (b) After: The rod moves with a combination of a linear center-of-mass motion and a rotation about the center of mass.

FIGURE 10–16 (a) A rapidly rotating bicycle wheel is spinning with its angular velocity, **ω**, and its initial angular momentum, **L**$_i$, aligned as shown. One end of the axis is fixed on a pivot point, and the other end is given an impulse **J** = $-J$**k**. (b) When a corresponding angular momentum change, Δ**L**, is added to the initial angular momentum, the result is that the free end of the wheel tends to rotate toward the *y*-axis.

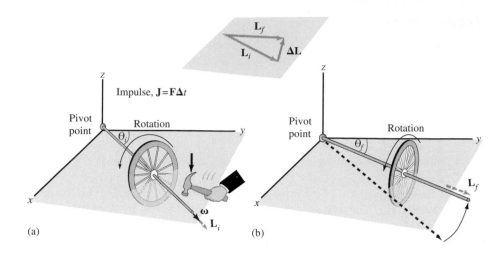

As in Fig. 10–15, the motion of the rod consists of a linear center-of-mass motion, with velocity $V = F \, \Delta t/m$ along the direction of the impulse, plus a rotation about the center of mass, with angular velocity $\omega = 2V/\ell$.

The Vectorial Nature of Angular Impulse. The angular impulse illustrates one of the most surprising aspects of the dynamics of rotation, one in which the *vector* nature of angular momentum and torque plays a crucial role. Consider an object such as a bicycle wheel rotating about its axle, with its angular velocity **ω** and initial angular momentum **L**$_i$, aligned horizontally in the *xy*-plane (Fig. 10–16a). The wheel is pivoted about one end of the axle and the other end is struck by a hammer blow downward, receiving an impulse **J** = $-J$**k**. What is the immediate response of the wheel as a result of the impulse?

Let's compute the angular impulse, **J**$_\tau$, using the fixed end (pivot point) of the axle as a reference point. If the vector from the fixed end to the point where the impulse is applied is $\boldsymbol{\ell}$, then the angular impulse is

$$\mathbf{J}_\tau = \boldsymbol{\ell} \times \mathbf{J}.$$

If we use a right-hand rule on this vector product, we see that its direction, and hence the direction of the instantaneous angular momentum change Δ**L**, points in the direction indicated in Fig. 10–16b. Even though the blow comes from above, *the free end of the wheel tends to rotate back toward the y-axis*! The wheel moves in a direction quite different from the direction of the blow. This is a surprising consequence of the vector nature of angular momentum and torque that will be echoed by a similar phenomenon in the motion of rotating objects (gyroscopes) under the effect of gravity (Section 10–7).

10–4 CONSERVATION OF ANGULAR MOMENTUM

The general dynamical equation for rotational motion is $\boldsymbol{\tau}_{\text{net}} = d\mathbf{L}/dt$, where the torque always refers to *external* forces. If there is no net external torque on a system—rigid or otherwise—the angular momentum of the system is constant or *conserved*,

$$\frac{d\mathbf{L}}{dt} = 0, \tag{10–26}$$

so that the angular momentum is constant during the motion of the system. For a non-rigid system, the rotational inertia may change, and the conservation of angular momentum is a powerful tool for study of the motion.

The rotation axis of a rotating object that has no torques acting on it will point in a fixed direction. A device such as this with a built-in fixed direction is a self-contained tool for navigation. An interesting version consists of a small metal sphere, engraved with black and white marks and levitated by a magnetic field (Fig. 10AB–1). The whole assembly is placed inside a vacuum enclosure mounted on gimbals. The sphere can then be sped up to a very high angular velocity with the aid of special magnetic fields. The rotating sphere is held in a horizontally stable position by an electrostatic field; any drifting up or down in the enclosure is corrected by servomechanisms. Once the rotation is established, the sphere's rotation axis remains aligned in a constant spatial direction. As there are no mechanical bearings, there are no vestigial torques that can cause the sphere's axis to rotate. The whole assembly can then be placed in an airplane. With the aid of photocells, the markings on the sphere can be read; the position of these marks relative to the casing, which is attached to the airplane, then measures the exact rotation of the sphere relative to the plane. This allows for precise control of the

FIGURE 10AB–1 Two versions of the heart of a navigating device based on conservation of momentum for a rotating sphere.

flight path. The particular device described here provided an accuracy of 200 m on a transatlantic crossing. Note that this device does not require any communication from the ground in order to work.

Central Forces

The torque on a particle is zero if the force and the displacement vector **r** from the point of reference (the origin) to the point of application of the force are *parallel* (or antiparallel) as for central forces; this is because the vector product between **r** and **F** is zero. Here, the angular momentum about the origin is conserved, as illustrated in Example 10–1.

Consider an object moving under the influence of a central force, and take the origin of the coordinate system to be at the source of this force. Let the initial velocity of the object be $\mathbf{v}_0$. The initial direction of the angular momentum is given by $\mathbf{r} \times \mathbf{v}_0$, that is, perpendicular to the plane formed by **r** and $\mathbf{v}_0$. Because the angular momentum is constant and its direction does not change, *the motion is always confined to the initial plane formed by* **r** *and* $\mathbf{v}_0$. To obtain the magnitude of the angular momentum, consider the path shown in Fig. 10–17a. At any given moment t, the particle is at position **r** and is moving with a velocity **v** tangent to the trajectory. The magnitude of the angular momentum is given by

$$L = \left| \mathbf{r} \times \mathbf{p} \right| = \left| \mathbf{r} \times m\mathbf{v} \right| = rmv \sin \alpha, \tag{10–27}$$

where α is the the angle between **r** and **v**. The velocity $\mathbf{v} = d\mathbf{r}/dt$ so that, if the position changes from **r** to $\mathbf{r} + \Delta\mathbf{r}$ in a small time interval Δt as in Fig. 10–17b, and if we replace **v** by $\Delta\mathbf{r}/\Delta t$, we find

$$L = mr\frac{\left| \Delta\mathbf{r} \right| \sin \alpha}{\Delta t} = mr\frac{r \, \Delta\theta}{\Delta t}. \tag{10–28}$$

We have used the geometric relation; $\left| \Delta\mathbf{r} \right| \sin \alpha = r \, \Delta\theta$. In the limit $\Delta t \to 0$, Eq. (10–28) becomes

$$L = mr^2 \frac{d\theta}{dt} = mr^2\omega, \tag{10–29}$$

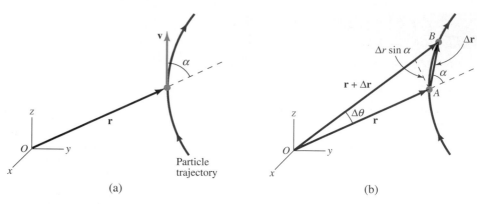

(a)

(b)

FIGURE 10–17 The trajectory of a particle with constant angular momentum. (a) The velocity at any given time is tangent to the trajectory and makes an angle α with the position vector. (b) In a small time interval, the particle moves from point A to point B and is displaced by $\Delta\mathbf{r}$. The angle $\Delta\theta$ is approximately $\Delta\theta = (|\Delta\mathbf{r}| \sin\alpha)/r$.

FIGURE 10–18 If a particle moves with constant angular momentum, then its trajectory sweeps out equal areas in equal times. If the motion of a comet from A to B takes time Δt and the motion from C to D takes the same amount of time, then the particle must move more quickly in traveling from C to D.

Kepler's second law

v = a constant

FIGURE 10–19 The triangles that a uniformly moving particle sweeps out with respect to any point O have equal areas.

where ω is the instantaneous angular velocity about the origin. Thus $r^2\omega$ is constant throughout the motion when L is constant. For uniform circular motion, both r and ω are constant and there is nothing new in Eq. (10–29). If r is not constant, then Eq. (10–29) can be interpreted as follows: Consider the time interval Δt and the geometry of Fig. 10–17b. For infinitesimal displacements, the area of the triangle OAB is given by

$$\Delta A = \frac{1}{2}(r\,\Delta\theta)r = \frac{1}{2}r^2\,\Delta\theta, \qquad (10\text{–}30)$$

from which it follows that

$$dA = \frac{1}{2}r^2\frac{d\theta}{dt}\,dt = \frac{1}{2}r^2\omega\,dt.$$

From Eq. (10–29),

$$\frac{dA}{dt} = \frac{1}{2}r^2\omega = \frac{L_0}{2m}, \qquad (10\text{–}31)$$

where L_0 is the constant value of the angular momentum. Whatever the trajectory of the particle (which depends on the detailed form of the central force), a consequence of angular momentum conservation is that *the rate at which the radius vector sweeps out an area is constant.* In the context of the gravitational force (Chapter 12), this is known as **Kepler's second law**. For example, a comet, whose motion about the Sun is an ellipse with the Sun located at one of the focal points, sweeps out equal areas in equal lengths of time. This means that it must move very rapidly when it is near the Sun compared with its speed at large distances from the Sun (Fig. 10–18).

Note that a freely moving particle experiences no torque. Thus it too satisfies the condition of Eq. (10–31) that its position vector, as measured from any point, sweeps out each area at a uniform rate, as shown in Fig. 10–19.

Nonrigid Objects

When an object is not rigid and its angular momentum is constant, the rotational inertia and the angular velocity can each change in such a way that their product remains constant. This possibility is commonly realized by divers or figure skaters (Fig. 10–20). In another example, water going down the drain holds its angular momentum constant by swirling more quickly as it approaches the drain.

EXAMPLE 10–6 In a common classroom experiment, a student sits on a spinning stool with weights in his hands. As an idealized version of this experiment, consider a solid cylinder of diameter $d = 0.5$ m and mass $M = 50$ kg (Fig. 10–21). The cylinder is oriented vertically and spins freely about its axis with a period of 3 s. Two massless rods are attached horizontally to the cylinder, with their ends $L = 1$ m from

(a)   (b)

FIGURE 10–20 (a) Angular momentum is conserved even for nonrigid objects. The cat is not a rigid body. By rotating the central part of the body one way and the outer part of the body (legs and head) the opposite way, the cat can land on its feet even though its angular momentum must remain constant at zero. (b) Although a figure skater spins more slowly when her arms are extended than when her arms are close to her body, her angular momentum remains constant.

the surface, and there is a mass $m = 2$ kg at the end of each rod (Fig. 10–21a). The rods are drawn into the cylinder by an internal mechanism until the 2-kg masses are at the surface of the cylinder (Fig. 10–21b). What are the initial and final angular velocities?

Solution: The angular momentum, L, about the symmetry axis of the cylinder is constant because there are no external torques. (The force pulling in the "arms" is *internal* and *radial*.) Thus the product $I\omega$ is constant. As the masses are pulled in, I decreases; therefore, ω must increase. If I_0 and ω_0 refer to the initial conditions and I and ω refer to the final conditions, we find

$$I_0\omega_0 = I\omega;$$

$$\omega = \omega_0 \frac{I_0}{I}.$$

(a)　(b)　(c)　(d)

FIGURE 10–21 (a) Example 10–6. (b) When the masses are brought in closer to the rotation axis, the conservation of angular momentum requires the rotation to speed up. (c), (d) A demonstration of this.

I_0 is the rotational inertia of a solid cylinder ($MR^2/2$) of mass $M = 50$ kg and radius $R = d/2 = 0.25$ m, plus the rotational inertia of the two masses of 2 kg a distance $L + R = 1.25$ m from the axis:

$$I_0 = \frac{MR^2}{2} + 2m(L + R)^2.$$

I is found the same way, except the two masses are a distance R from the axis:

$$I = \frac{MR^2}{2} + 2mR^2 = \left(\frac{M}{2} + 2m\right)R^2.$$

The initial angular velocity is found from the period:

$$\omega_0 = \frac{2\pi \text{ rad}}{T} = \frac{6.28 \text{ rad}}{3 \text{ s}} = 2 \text{ rad/s}.$$

(a)　(b)　(c)

FIGURE 10–22 (a) A student on a stool that can rotate holds a freely rotating wheel. (b) When the wheel is inverted, the stool begins to turn in order to hold the total angular momentum constant. (c) To the now negative angular momentum of the wheel must be added a positive angular momentum from the student. Only in this way can the total angular momentum be conserved.

Thus

$$\omega = \omega_0 \frac{(MR^2/2) + 2m(L + R)^2}{[(M/2) + 2m]R^2}$$

$$= (2 \text{ rad/s}) \frac{(0.5)(50 \text{ kg})(0.25 \text{ m})^2 + 2(2 \text{ kg})(1.25 \text{ m})^2}{(0.5)(50 \text{ kg})(0.25 \text{ m})^2 + 2(2 \text{ kg})(0.25 \text{ m})^2}$$

$$= (2 \text{ rad/s}) \frac{7.8 \text{ kg} \cdot \text{m}^2}{1.8 \text{ kg} \cdot \text{m}^2} = 9 \text{ rad/s}.$$

The angular velocity has increased by a factor of 4.5. The movement of the end weights into the axis is remarkably effective at decreasing the rotational inertia! You can perform this demonstration in class (Figs. 10–21c and 10–21d).

The conservation of angular momentum is the conservation of a vector quantity. Suppose that a spinning bicycle wheel with its angular momentum pointing up (the wheel spins counterclockwise as seen from above) is handed to a student perched on a stool that can itself rotate (Fig. 10–22a). If the student now turns the wheel upside down, so that its angular velocity points downward (negative), he or she will begin to spin counterclockwise—the original direction of rotation of the wheel (Fig. 10–22b). This new spin direction is generated to conserve the total angular momentum (Fig. 10–22c).

An Astrophysical Example. A supernova provides a dramatic astrophysical illustration of the conservation of angular momentum. When a massive star stops burning because it has used up its fuel, it undergoes a collapse due to gravitational forces. There is a stupendous explosion and some of the stellar material is blown off in a radial direction. What is left behind is a *neutron star*: an extremely dense sphere of matter with a mass of several solar masses and a diameter of perhaps 10 km (compared with 10^6 km for the original star). This large ratio leads to very fast rates of rotation for neutron stars, some of which have periods on the order of milliseconds, compared to days for the original star. Many such rapidly rotating neutron stars have been observed.

10–5   WORK AND ENERGY IN ANGULAR MOTION

Rotating systems of particles—even those that are nonrigid—must have energy, and work must be done on them to change their energy. Now we want to explore the energy of rotating systems of many objects. In Chapter 9, we studied the rotational kinetic energy of a symmetrical rigid body whose rotational inertia about the axis of rotation is I and whose angular velocity is ω. We found that this energy is

$$K = \frac{1}{2} I \omega^2. \tag{10–32}$$

The angular velocity is a vector: if it is given by

$$\boldsymbol{\omega} = \omega_x \mathbf{i} + \omega_y \mathbf{j} + \omega_z \mathbf{k}, \tag{10–33}$$

then

$$\omega^2 = \boldsymbol{\omega} \cdot \boldsymbol{\omega} = \omega_x^2 + \omega_y^2 + \omega_z^2. \tag{10–34}$$

If a torque is present, the rotational kinetic energy will change because the angular speed changes. We calculate the rate of change of energy; that is, the *instantaneous power*

$$\frac{dK}{dt} = \frac{1}{2} I \frac{d}{dt} (\omega^2) = \frac{I \, d\boldsymbol{\omega}}{dt} \cdot \boldsymbol{\omega} = I \boldsymbol{\alpha} \cdot \boldsymbol{\omega} = \boldsymbol{\tau} \cdot \boldsymbol{\omega}, \tag{10–35}$$

Power in rotating systems

where we have used Eq. (10–15), $I\boldsymbol{\alpha} = \boldsymbol{\tau}$. Equation (10–35) is the analogue of the linear motion equation for power that we saw in Chapter 6: $dK/dt = \mathbf{F} \cdot \mathbf{v}$.

The Work–Energy Theorem for Rotations

It is fairly straightforward to derive a work–energy theorem for rotational motion about a fixed axis. To define the work, we follow the analogue with linear motion that has proven so useful. Just as the infinitesimal work in linear motion is defined as $dW = \mathbf{F} \cdot d\mathbf{x}$, the infinitesimal work done in rotating a rigid body through an infinitesimal angle $d\boldsymbol{\theta}$ about the axis is defined to be

$$dW \equiv \boldsymbol{\tau} \cdot d\boldsymbol{\theta}. \qquad (10\text{--}36)$$

Note that an infinitesimal angle *is* a vector. When the torque, $\boldsymbol{\tau}$, points along the axis of rotation, this reduces to the form

$$dW = \tau \, d\theta. \qquad (10\text{--}37)$$

This will be a reasonable definition for the work if it leads to an appropriate work–energy theorem. We have, using $\tau = I \, d\omega/dt$ and $d\theta = \omega \, dt$.

$$W = \int_{\theta_0}^{\theta} \tau \, d\theta = \int_0^t I \frac{d\omega}{dt} \omega \, dt = \int_0^t I \left(\frac{1}{2} \right) \frac{d\omega^2}{dt} \, dt$$

$$= \frac{1}{2} I \int_{\omega_0}^{\omega} d\omega^2 = \frac{1}{2} I(\omega^2 - \omega_0^2) = K - K_0. \qquad (10\text{--}38)$$

Thus the work–energy theorem for rotational motion about a fixed axis is that *the work done on an object with rotational inertia I by a torque about an axis of rotation is equal to the increase in the rotational kinetic energy of the object.*

The Energy of an Extended Object in Motion

We next consider the energy of an extended object—rigid or not. We do this as we did in the subhead in which we derived Eqs. (10–16) and (10–17). First, we break up the object into a number of individual pieces with masses m_i located at $\mathbf{r}_i$ relative to some origin. We introduce the center of mass position $\mathbf{R}$ and the coordinates $\boldsymbol{\rho}_i$ relative to the center of mass according to $\mathbf{r}_i = \boldsymbol{\rho}_i + \mathbf{R}$, as in Fig. 10–12. When we take a derivative with respect to time, we have

$$\mathbf{v}_i = \frac{d\mathbf{r}_i}{dt} = \frac{d\mathbf{R}}{dt} + \frac{d\boldsymbol{\rho}_i}{dt} = \mathbf{V} + \mathbf{u}_i, \qquad (10\text{--}39)$$

where $\mathbf{u}_i \equiv d\boldsymbol{\rho}_i/dt$ are the velocities of the masses m_i with respect to the center of mass and $\mathbf{V}$ is the velocity of the center of mass. We can then show that the kinetic energy of the system takes the form

$$K = \sum_i \frac{1}{2} m_i \mathbf{v}_i^2 = \frac{1}{2} MV^2 + \frac{1}{2} \sum_i m_i \mathbf{u}_i^2. \qquad (10\text{--}40)$$

(We do not prove this result here.) The total kinetic energy consists of two parts: the kinetic energy of the total mass of the object moving with the velocity of the center of mass, and the kinetic energy of the motion relative to the center of mass.

The kinetic energy of Eq. (10–40) takes on a familiar form for a *rigid* body. In that case, the vectors $\boldsymbol{\rho}_i$ have a fixed magnitude, and only the angular variable changes. For rotations with angular speed ω about an axis through the center of mass, the speed of any point that is a radial distance r from the axis is given by $v = \omega r$. Thus $u_i = \omega \rho_i$, and $u_i^2 = \omega^2 \rho_i^2$. We can substitute this expression into Eq. (10–40) to find that

$$K = \frac{1}{2} MV^2 + \frac{1}{2} I_{cm} \omega^2. \qquad (10\text{--}41)$$

The total kinetic energy of a rotating
rigid body consists of a center-of-mass
term and a term for rotations about the
center of mass.

Equation (10–41) shows that *the total kinetic energy of a rigid body consists of the kinetic energy of the total mass moving with the velocity of the center of mass together with the rotational kinetic energy of the object rotating about an axis passing through the center of mass.* We had already derived this result [see Eq. (9–39)] for the special case of rolling objects. Now we see that it is more general. This result is useful in calculations involving a combination of rotational and linear motion, as Example 10–7 shows.

EXAMPLE 10–7 A spool of mass M, rotational inertia I about its axis, and radius R unwinds under the force of gravity (Fig. 10–23). By using energy considerations, find the speed of the spool's center of mass after it has unwound a length h of thread.

Solution: The total kinetic energy of the spool is the sum of the kinetic energy of motion of the center of mass and the kinetic energy of rotation about the center of mass. Thus

$$K = \frac{1}{2} M v^2 + \frac{1}{2} I \omega^2.$$

The relation between the angular velocity and the velocity of the center of mass is identical to that of a cylinder rolling without sliding, so that

$$v = R\omega.$$

This implies that

$$K = \frac{1}{2} M R^2 \omega^2 + \frac{1}{2} I \omega^2 = \frac{1}{2} (M R^2 + I)\omega^2.$$

Note that a simple application of the parallel-axis theorem shows that the quantity in parentheses is just the rotational inertia about the point at which the thread separates from the surface.

Another contribution to the energy is the potential energy of gravity. We can measure the potential energy from the initial height of the spool so that if the spool falls a distance h, the potential energy is $-Mgh$ and the spool has gained a kinetic energy Mgh. With a zero initial angular velocity, the kinetic energy after the spool has fallen a distance h is

$$K = Mgh.$$

Thus

$$\frac{1}{2} (M R^2 + I)\omega^2 = Mgh,$$

so

$$v = R\omega = R \sqrt{\frac{2Mgh}{MR^2 + I}}.$$

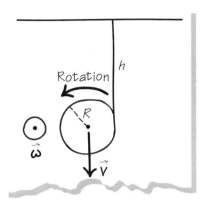

FIGURE 10–23 Example 10–7.

More Parallels Between Rotational and Linear Motion

Despite all the subtleties that appear when the full vector nature of angular momentum is involved, it is important to stress that there are no new laws of physics that describe rotational motion on the classical (the non-quantum-mechanical) level. *Every result on classical physics obtained in these last two chapters is a direct consequence of the application of Newton's laws of motion to aggregates of particles.* The rotational motion of both rigid and nonrigid objects has so many simplifying features that it can be described in terms of derived quantities such as angular velocity, angular momentum, and torque, but the basic laws are not new. Thus the analogies between quantities that appear in classical linear motion and those that appear in classical rotational motion are not accidental; instead, they are consequences of the fact that aggregate systems are made up of individual pieces that separately obey the laws of linear motion discovered by Newton (Table 10–1).

*10–6 QUANTIZATION OF ANGULAR MOMENTUM

When systems with atomic dimensions or smaller are involved, the effects of quantum physics become important. One such effect is that *angular momentum can have only certain discrete values*—we say that it is *quantized.* Consider an electron that is or-

TABLE 10-1   Analogies Between Linear and Rotational Motion

Linear Motion	*Rotational Motion*
Infinitesimal linear displacement: $\mathbf{dr}$	Infinitesimal angular displacement: $d\theta$
Velocity: $\mathbf{v}$	Angular velocity: $\boldsymbol{\omega}$
Acceleration: $\mathbf{a} = \dfrac{d\mathbf{v}}{dt}$	Angular acceleration: $\boldsymbol{\alpha} = \dfrac{d\boldsymbol{\omega}}{dt}$
Momentum: $\mathbf{p} = m\mathbf{v}$	Angular momentum: $\mathbf{L} = \mathbf{r} \times \mathbf{p}$
Force: $\mathbf{F} = \dfrac{d\mathbf{p}}{dt}$	Torque: $\boldsymbol{\tau} = \mathbf{r} \times \mathbf{F} = \dfrac{d\mathbf{L}}{dt}$
Impulse: $\Delta\mathbf{p} = \mathbf{F}\,\Delta t$	Angular impulse: $\Delta\mathbf{L} = \boldsymbol{\tau}\,\Delta t$
Kinetic energy: $\frac{1}{2}mv^2$	Kinetic energy: $\frac{1}{2}I\omega^2$
Work: $\displaystyle\int \mathbf{F} \cdot \mathbf{dr}$	Work: $\displaystyle\int \boldsymbol{\tau} \cdot d\boldsymbol{\theta}$
Power: $\mathbf{v} \cdot \mathbf{F}$	Power: $\boldsymbol{\omega} \cdot \boldsymbol{\tau}$

*The quantization of angular
momentum*

biting the nucleus of an atom, under the influence of the central force due to the nucleus. The electron has an angular momentum, $\mathbf{L}$, about an axis perpendicular to the plane in which it orbits, much like a planet orbiting the Sun. In Newtonian physics, angular momentum can have any magnitude at all. Niels Bohr proposed in 1913 that the angular momentum can only have a component in the direction perpendicular to the plane of motion given by

$$L_z = n\hbar \qquad\qquad (10\text{--}42)$$

(in the z-direction, here). In this quantum mechanical expression, n has *only* the integer values $0, \pm 1, \pm 2, \ldots$, and $\hbar$ is Planck's constant, h, divided by 2π, with value $\hbar \simeq 10^{-34}$ J·s.

Equation (10–42) must apply to macroscopic as well as microscopic systems. However, it is not in conflict with our experience about macroscopic systems because $\hbar$ is so small. A particle of mass 0.001 mg rotating about an axis at a distance of 1 μm with angular velocity 1 rad/s has a value of L_z given by

$$L_z = mvr = m\omega r^2 = (10^{-9}\,\text{kg})(1\,\text{rad/s})(10^{-6}\,\text{m})^2 = 10^{-21}\,\text{J·s}.$$

This value corresponds to 10^{13} units of $\hbar$. The quantization is not observable because its observation would require a determination of the radius or the angular velocity to an accuracy of one part in 10^{13}. In molecular or atomic systems, however, the masses and distances are very tiny, and the values of the classical angular momentum are of the same order of magnitude as $\hbar$. For example, consider a molecule of nitrogen, N_2. Its classical angular momentum is given by $I\omega$. For the purpose of determining I, we can consider the molecule to form a tiny dumbbell made of two nuclei of mass M a distance d apart.[†] The rotational inertia for rotations about an axis perpendicular to, and halfway down, the line connecting the nuclei is $I = Md^2/2$. The mass of a nitrogen nucleus is the mass of 14 protons, $M = 14(1.6 \times 10^{-27}\,\text{kg}) = 2.2 \times 10^{-26}\,\text{kg}$. The typical distance between the nuclei of any molecule is about 10^{-10} m, so $I \simeq 10^{-46}$ kg·m^2. To determine L, we use the expression for the kinetic energy of the system, from Eq. (9–35): $K = L^2/2I$. The typical kinetic energy of a molecule is in the range of 10^{-2} eV $\simeq (10^{-2}\,\text{eV})(1.6 \times 10^{-19}\,\text{J/eV}) \simeq 1.6 \times 10^{-21}$ J. Thus we estimate a classical angular momentum

$$L = \sqrt{2IK} \simeq \sqrt{2(10^{-46}\,\text{kg·m}^2)(1.6 \times 10^{-21}\,\text{J})} \simeq 5.6 \times 10^{-34}\,\text{J·s}.$$

This value of the angular momentum is between 5 and 6 units of $\hbar$. The difference between 5 and 6 units of $\hbar$ corresponds to a 10- to 20-percent difference in energy and the fact that the angular momentum is quantized—it has only the discrete values dic-

---

[†]The electrons are some 10,000 times lighter than the nuclei and can be ignored.

tated by quantum mechanics—is easily detectable in experiments that measure the energies of the molecule, such as those described next.

Quantization of Energy

Equation (9–35) shows that if the angular momentum is quantized, then *the allowed values of energy of a system such as the N_2 molecule or of an electron orbiting an atomic nucleus are quantized as well.* That is because the energy of an orbiting electron can be expressed in terms of L.

When the system has one of its possible values of energy, corresponding to different values of n, it is said to be in an **allowed state**. The possible values of energy can then be labeled by the particular value of n—as E_n—and we then refer to a particular **energy level**. When a molecule or an orbiting electron absorbs or emits energy as a result of collisions or other interactions, then this change can occur only in discrete jumps corresponding to changes from one state to another. For example, the electron of an atom may lose energy by jumping from a state labeled n to a state labeled $n - 1$. When it does so, the conservation of energy requires that the energy loss be compensated by an energy gain somewhere else. This energy appears as radiation (light), with a frequency (color) given by

$$f_{rad} \sim E_n - E_{n-1}. \tag{10–43}$$

Thus if the allowed energies are discrete, the frequencies of Eq. (10–43) take on only discrete values. Equation (10–43) then provides us with the means to detect the effects of the quantization of angular momentum. Atoms or molecules initially excited (moved into an allowed state with an increased energy) by collisions or by impinging radiation will jump spontaneously to lower energy levels, one step at a time, emitting radiation (light) in the process. Thus light is emitted with characteristic frequencies determined by equations such as Eq. (10–43). The frequency of this radiation can be measured. It was the observation that atoms and molecules emit light with discrete frequencies that provided one of the keys to the development of quantum mechanics.

The consequences of the discreteness of atomic and molecular energies are momentous. For example, there is a minimum energy—an *energy gap*—required to move an atomic system such as a hydrogen atom up the "quantum ladder" from the lowest state, characterized by $n = 1$, to states with higher n. It is the relatively large size of the energy gap that explains the stability of atoms. It is difficult to excite a hydrogen atom. The fact that even large molecular systems have energy gaps is crucial to the stability of biological systems. If there were no energy gaps, even the slightest perturbation would change molecular systems in important ways.

*10–7 PRECESSION

Recall the motion of a top: There is a rapid rotation about the symmetry axis: **precession**, which is a rotation of the symmetry axis itself around a vertical axis through the point of the top; and *nutation*, the name given to the up-and-down bobbing motion of the symmetry axis while it rotates about the vertical. The general motion of tops and of gyroscopes is beyond the scope of this book, but we can indicate with an example how precession arises.

In linear motion, there are two rather different ways that Newton's second law can be satisfied. First, the momentum can change its magnitude; this leads to changes in speed. Second, if the applied force is not parallel to the momentum, the momentum can change *direction* rather than magnitude, as in circular motion. The same is true for rotational motion. If the applied torque is parallel to the angular momentum, then we have changes in the magnitude of the angular momentum, and this situation represents the majority of the cases we have studied. But if the torque is in some other direction, then we can have changes in the direction of the angular momentum.

EXAMPLE 10–8 In making a science-fiction movie about an asteroid that collides with Earth, a studio constructs a model Earth attached firmly to a support fixed at the South Pole and spinning rapidly with ω in the $+z$-direction (Fig. 10–24a). A model asteroid approaches horizontally from the $+y$-direction and strikes a grazing blow with the model Earth just at the North Pole. Describe the new motion of the model Earth.

Solution: Because the model Earth is fixed at its pivot at the South Pole, we calculate the torque on the model Earth about this point. The collision delivers an impulse $\mathbf{J} = -J\mathbf{j}$ to the model Earth in the given coordinate system. According to

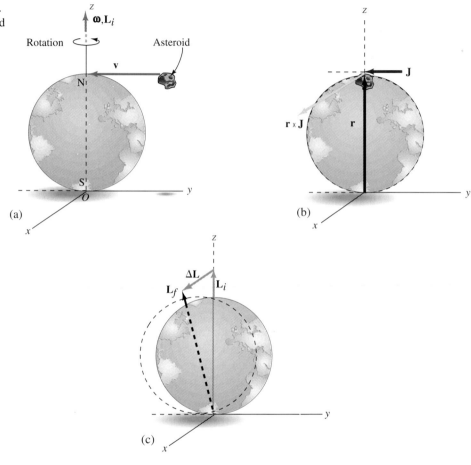

FIGURE 10–24 (a) Example 10–8. (b) The collision of the model asteroid along the $-y$-direction causes the spinning model Earth to tilt toward the $+x$-direction.

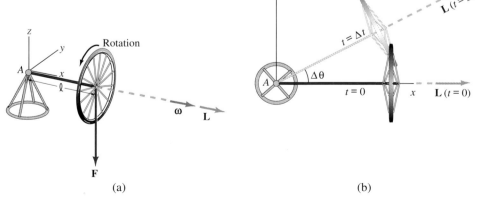

FIGURE 10–25 (a) A wheel rotates with angular velocity $\boldsymbol{\omega}$ about a massless horizontal shaft that can pivot about point A. The force of gravity, $\mathbf{F}_g = m\mathbf{g}$, acts at the center of the wheel. (b) As seen from above, the shaft has rotated an angle $\Delta\theta$ after a time Δt. The angular momentum vector has therefore changed direction.

Eq. (10–25), the angular impulse about the support point is $\mathbf{J}_\tau = \mathbf{r} \times \mathbf{J}$, where $\mathbf{r}$ is the vector from the fixed point to the point where the impulse is delivered. Because $\mathbf{r}$ is in the $+z$-direction, the vector product $\mathbf{r} \times \mathbf{J}$ is in the $+x$-direction. But this describes the change in angular momentum, $\Delta\mathbf{L}$, produced by the angular impulse. The initial angular momentum $\mathbf{L}_i$ is in the $+z$-direction, and when $\Delta\mathbf{L} = (+\Delta L)\mathbf{i}$ is added to this, the new angular momentum $\mathbf{L}_f$ corresponds to Earth having tilted toward the $+x$-axis (Fig. 10–24b). Note that this is *not* the direction from which the model asteroid comes.

Torque on a Spinning Wheel

As another example of the consequence of the vector nature of the dynamical equations of motion, consider a wheel of rotational inertia I rotating with angular velocity $\boldsymbol\omega$ about a horizontal shaft of length ℓ, which has one end pivoted at point A, and the other end free. The system is shown at $t = 0$ in Fig. 10–25a. If the sense of rotation of the wheel is as shown, then the angular momentum about the pivot initially has magnitude $L = I\omega$ and points in the x-direction. Now recall that a gravitational force of magnitude Mg, pointing in the $-z$-direction, acts on the center of mass of the wheel. The torque about point A due to this force has magnitude

$$\tau = Mg\ell, \tag{10–44}$$

and it points in *the $+y$-direction* by the right-hand rule. Because

$$\Delta L = \tau\,\Delta t,$$

the angular momentum after a short time interval Δt is given by

$$\mathbf{L} = I\omega\mathbf{i} + Mg\ell(\Delta t)\mathbf{j}. \tag{10–45}$$

Thus the angular momentum vector has rotated slightly in a counterclockwise direction, as seen from above. As Fig. 10–25b shows, the new angular momentum makes an angle

$$\Delta\theta = \frac{Mg\ell\,\Delta t}{I\omega} \tag{10–46}$$

with the original direction. This is the angle of the shaft, and so we have seen that the direction of the shaft changes with time in a steady way. When the shaft of the spinning wheel steadily rotates with the pivot point fixed, we say that the shaft *precesses* (Fig. 10–26a). The angular velocity of the precession of the angular velocity vector determined from Eq. (10–46) is

$$\omega_p = \frac{\Delta\theta}{\Delta t} = \frac{Mg\ell}{I\omega} \tag{10–47}$$

and is known as the *angular frequency of precession*. The precession frequency is, in fact, a vector that points in the $+z$-direction due to the right-hand rule. The precession frequency turns out to be independent of the angle the initial value of $\mathbf{L}$ makes with the vertical (see Problem 43). This formula applies for any top—not just a rotating wheel—if ℓ is the distance of the top's center of mass from the support point (Fig. 10–26b).

Our description of the precession of the wheel's shaft is only approximately correct. Energy considerations give rise to a perturbation to the simple precession just described, which consists of an up-and-down bobbing motion of the shaft below the xy-plane: a nutation (Fig. 10–27). The effect is small if the angular velocity of the wheel is large, or more precisely, if $\omega_p \ll \omega$. In this section, we have been able to point out only the simplest features of a top's motion. Nevertheless, the laws of rotational motion described in this chapter are sufficient to account for all these features.

(a)

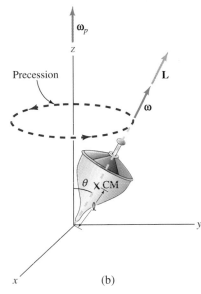

(b)

FIGURE 10–26 (a) Simple precession. (b) The precession of a top is governed by Eq. (10–45), where ℓ is the distance between the pivot point and the top's center of mass.

FIGURE 10–27 The interplay of gravity, angular momentum, and torque lead to perturbations on the precession of a top, resulting in the up-and-down bobbing motion called nutation.

For a particle moving with momentum **p**, the general form of the angular momentum about a point is

$$\mathbf{L} = \mathbf{r} \times \mathbf{p},\qquad(10\text{--}4)$$

where **r** is the position vector from the point to the particle. This result involves the vector product of the vectors **r** and **p** with angle θ between them. The vector product has magnitude $rp \sin \theta$, and its direction—perpendicular to both **r** and **p**—is determined by a right-hand rule. The angular momentum of a system of particles, rigid or otherwise, about some point is a sum over terms such as Eq. (10–4).

A dynamical equation obeyed by extended systems follows from consideration of Newton's second law. It states that the rate of change of angular momentum about some point O is given by

$$\boldsymbol{\tau} = \frac{d\mathbf{L}}{dt},\qquad(10\text{--}12)$$

where $\boldsymbol{\tau}$ is the torque applied to the object. The torque is generally defined by

$$\boldsymbol{\tau} = \mathbf{r} \times \mathbf{F},\qquad(10\text{--}5)$$

where **r** is the vector from the point of reference to the point where the force, **F**, is applied.

If a rigid body rotates about an axis A that does *not* go through its center of mass, then (1) the angular momentum about A is the sum of the angular momentum about A of a point mass carrying the whole mass of the object, as if it is located at the center of mass and moves with the velocity of the center of mass, plus the angular momentum of the object about the center of mass,

$$\mathbf{L} = (\mathbf{R} \times \mathbf{P}) + \mathbf{L}_{\text{cm}};\qquad(10\text{--}16)$$

(2) the total torque about A is the sum of the torque about A due to the total external force applied to the center of mass and the torque about the center of mass,

$$\boldsymbol{\tau} = (\mathbf{R} \times \mathbf{F}_{\text{tot}}) + \boldsymbol{\tau}_{\text{cm}};\qquad(10\text{--}17)$$

and (3) the rate of change of each part of the total angular momentum is equal to the corresponding part of the total torque.

In the absence of torque, angular momentum is constant throughout the motion, and this is an important tool in solving problems of motion of extended systems, especially nonrigid ones. It also helps us understand the motion of particles. The torque will naturally be zero in the absence of external forces, but it also is zero for central forces. As a consequence, in the orbit of an object moving under the influence of a central force, the areas swept out by position vectors from the source of the force to the moving object in equal times are equal.

The kinetic energy of rotational motion is given by $K = I\omega^2/2$, and the change in K equals the work done—just as in linear motion. For rotational motion about an axis, this relation (the work–energy theorem) reads

$$W = \int_{\theta_0}^{\theta} \tau \, d\theta = K - K_0,\qquad(10\text{--}38)$$

where K_0 is the initial rotational energy at the angle θ_0. In many problems both rotational and linear motion are present. The total energy of a rigid body may be written as the sum of the linear kinetic energy, calculated as if all the mass of the object were concentrated at the center of mass, and the rotational energy of the object about the center of mass:

$$K = \frac{1}{2} MV^2 + \frac{1}{2} I_{\text{cm}}\omega^2.\qquad(10\text{--}41)$$

Angular momentum in atomic systems is quantized in the form

$$L_z = n\hbar = n\frac{h}{2\pi}, \qquad (10\text{--}42)$$

where h is Planck's constant and $n = 0, \pm 1, \pm 2, \ldots$. The consequences of this quantization include the result that not all energy values are permitted for atoms and molecules.

Many subtle effects are associated with the effect of torque on angular momentum, and we illustrated one of them: the precession of angular momentum for a top. If the top is spinning while inclined from the vertical, then gravity produces a torque on the top that causes its axis to precess about the vertical with a precession frequency

$$\omega_p = \frac{Mg\ell}{I\omega}. \qquad (10\text{--}47)$$

UNDERSTANDING KEY CONCEPTS

1. Angular momentum is sometimes given in units of joule-seconds ($J \cdot s$). Is this a correct SI unit for angular momentum?

2. A comet is heading at high speed straight into the center of the Sun. Why is the angular momentum of the comet with respect to the Sun zero?

3. When a quarterback throws a football, he tries to put quite a bit of spin its axis. Since some of the energy put into the ball then goes into rotational motion, this means that there is less energy for translational motion. Why is it nevertheless done?

4. In taking a fast corner on a bicycle, it is safer if you crouch as low as possible. Why is that?

5. Films from Skylab show the astronauts reorienting themselves as they float weightlessly by spinning their arms. Is this consistent with the conservation of angular momentum?

6. A long, flexible, heavy bar can be very useful to a tightrope walker. Why?

7. Why is it easier on your back for you to lift heavy objects by bending your knees, keeping your back straight, and straightening your knees, rather than by picking up the object as you bend over from your waist with your legs straight?

8. An astronaut floating in a space station holds the axle of a rotating wheel. When the astronaut is vertical and the axle points away from her, she suddenly rotates the axle to point up. What happens to the astronaut?

9. A diver prepares for a complex set of midair maneuvers. In springing off an elastic diving board, the diver wishes to acquire angular momentum relative to her center of mass. What should she strive for in the take-off?

10. In Example 10–3, we studied a tether ball and found that as the rope tying it to the pole changes, so does the angular momentum. Where does the corresponding torque come from?

11. It is possible to tell whether an egg is hardboiled without cracking it open by setting it in rotation on a table. Discuss what you would find, and why. Compare the behavior of the hardboiled egg to that of a raw one.

12. If you were to tie a rock to a rope and swing it in a horizontal circle above your head, you could rather easily do so without spinning around yourself. Is this a violation of the conservation of angular momentum?

13. The center of mass of an object accelerates as the result of an impulse (a brief force). If the object is extended, the impulse may also be an angular impulse. Is it possible to have an angular impulse without a linear impulse?

14. A diver executes a series of midair maneuvers. To do so, is it necessary for the diver to give herself some angular momentum about her center of mass?

15. We saw that the torque–angular momentum relation is a consequence of Newton's laws. Is it also true in Newtonian mechanics that the conservation of angular momentum follows from Newton's laws?

16. You are given a stool of known rotational inertia that can rotate with minimal friction. You are also given a stopwatch, a very light meter stick, and two known masses that can slide along the meter stick. How would you use this apparatus to measure the rotational inertia about the rotational axis of a person sitting on the stool?

17. The propeller of a single-engine propeller airplane rotates clockwise, as seen from the cockpit. The plane makes a slow turn to the right. What else happens?

18. If only one set of hand brakes on your bicycle works when you descend a steep downhill slope, which set would you prefer?

19. A woman stands on the edge of a freely rotating platform. She walks toward the center along a radius. Will the speed of rotation of the platform change? If so, in what way, and what is the source of the torque?

20. A horizontal platform is rotating at a certain speed. A boy jumps onto the platform from an overhanging tree branch. He lands with both feet straddling the center and remains standing. Will the platform speed up, slow down, or neither?

21. A boat can be stabilized against rolling by attaching a large flywheel to the sides (Fig. 10–28). The attachment point is above the waterline. Both ω and the rotational inertia, I, about the axis are large. (a) A wave hits the boat on the side. The wave would tend to rock the boat or rotate it about its longitudinal axis without the flywheel. With the flywheel installed, how does the boat react to this wave? (b) A wave comes straight at the bow (the front). It would lift the bow of the boat without the flywheel. With the flywheel, what happens?

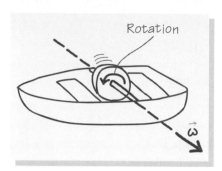

FIGURE 10–28 Question 21.

22. When jugglers perform juggling acts, the props they use are frequently put into a spinning motion when they are tossed into the air. Why is this a good idea?

23. A baseball is thrown horizontally, hits a rough floor, and bounces. The ball is spinning in the direction of its motion (like a rolling bicycle wheel). Which of the following is true: When the ball leaves the floor, it does so at (a) the same angle at which it struck the floor; (b) a larger angle with respect to the floor; (c) a smaller angle with respect to the floor?

24. Think about the demonstration in which someone spinning on a stool pulls in his or her arms and speeds up because of angular momentum conservation. In the process, does the energy of rotation decrease, remain constant, or increase? Explain.

PROBLEMS

10–1 Generalization of Angular Momentum

1. (I) An airplane of mass 2000 kg located 100 km north of New York City is flying 200 km/h in an easterly direction. (a) What is its angular momentum with respect to New York City? (b) What if it is flying in a northeasterly direction?

2. (I) What is the angular momentum about the origin of a particle of mass 200 g at position $\mathbf{r} = (2\mathbf{i} - 3\mathbf{j} + \mathbf{k})$ m, moving with a velocity of $\mathbf{v} = (5\mathbf{i} + 15\mathbf{j} - 10\mathbf{k})$ m/s?

3. (I) You are standing on the corner of Main Street and Elm, watching the cars on Main pass at a steady 10 m/s. You stand 5 m from the line of traffic. You watch a red convertible, whose mass is about 1000 kg, from the moment it is one block away, a distance of 200 m. Treat the car as pointlike, and assume that Main Street is straight. (a) What are the car's angular momentum, magnitude, and direction, with respect to you when it is one block away? (b) What is its angular momentum with respect to you when it passes your position on the corner?

4. (I) Assuming that each object is pointlike, how large is the angular momentum (a) of Earth about the Sun; (b) of the Moon about Earth? (c) Compare these results to the angular momentum of Earth (no longer pointlike) about its own axis; assume constant density.

5. (I) A bicycle travels east. The mass of the wheel, 1.8 kg, is uniformly distributed along the rim, with the mass of the hub and spokes negligible. If the radius of the wheel is 18 cm and the wheel rotates at the rate of 4.2 rev/s, what is the direction and the magnitude of the angular momentum of the wheel about its axis? (Neglect the width of the rim.)

6. (II) Calculate the angular momentum about the origin of a particle of mass m moving along the trajectory $y = ax + b$ with uniform speed v.

7. (II) A rock of mass 60 g is thrown with initial horizontal speed $v_x = 25$ m/s off a building from a height of 30 m. Calculate the angular momentum of the rock about the line along the edge of the roof as a function of time.

8. (II) A unicycle has a wheel of mass 3.5 kg, rotational inertia 0.45 kg·m² about the axle, and radius 0.4 m. What is its angular momentum with respect to a point on the road if the wheel rolls without slipping with an angular velocity of 8 rad/s?

9. (II) The position vector of an object of mass m subject to two constant forces that act at right angles is given by $\mathbf{r} = (\frac{1}{2}at^2)\mathbf{i} +$ $(vt)\mathbf{j} + (\frac{1}{2}bt^2 - wt)\mathbf{k}$. Calculate the angular momentum of this object about the origin.

10. (II) An object of mass M moves in a path given by $\mathbf{r} = (x_0 + \rho \cos [\omega t])\mathbf{i} + (y_0 + \rho \sin [\omega t])\mathbf{j}$. What is the angular momentum of the object about the origin?

11. (II) Consider two objects whose position vectors are given by $\mathbf{r}_1$ and $\mathbf{r}_2$ and whose momenta are given by $\mathbf{p}_1 = m_1\mathbf{v}_1$ and $\mathbf{p}_2 = m_2\mathbf{v}_2$, respectively (Fig. 10–29). Show that in the special case that the center of mass of the two bodies is at rest at the origin (that is, $\mathbf{P} = \mathbf{p}_1 + \mathbf{p}_2 = 0$, and the position of the center of mass is $\mathbf{R} = 0$), the sum of the angular momenta of the two objects about the center of mass equals the angular momentum of a single object of mass $\mu = m_1m_2/(m_1 + m_2)$, rotating in circular motion about the origin at a distance $r = r_2 - r_1$. The quantity μ is called the *reduced mass*. [*Hint*: Introduce $\mathbf{r} \equiv \mathbf{r}_2 - \mathbf{r}_1$, and express $\mathbf{p}_1$ in terms of $d\mathbf{r}/dt$.]

FIGURE 10–29 Problem 11.

12. (II) Earth is not a point object but is a sphere with a rotational inertia of 9.8×10^{37} kg·m² about its axis. Assume that Earth's axis of rotation is parallel to the axis of the orbital motion of Earth around the Sun and that the Sun is so massive that it can be considered to be fixed. Calculate (a) the rotational inertia of Earth about the axis of its orbital motion around the Sun; (b) the total angular momentum of Earth about that same axis. (c) Calculate the fractional difference between your result for part (b) and the angular momentum you would find for part (b) if Earth were pointlike; explain why the difference is small.

13. (II) A square, 20 cm on the side, is made of very light sticks. Four identical masses of $m = 0.1$ kg form the corners of the

square. The square rotates with an angular velocity of 8 rad/s about an axis perpendicular to its plane through the center of the square. (a) Calculate the rotational inertia of the system about the rotation axis and use it to find the angular momentum about this axis. (b) Use the general definition of angular momentum to calculate the angular momentum of each mass with respect to the center of the square, and add these up. Compare the results of (a) and (b).

14. (II) Consider the square studied in Problem 13. Calculate the angular momentum of each particle about a point on the axis of rotation 25 cm below the plane of the square. Compare the total angular momentum calculated in this way with the results of Problem 13.

15. (II) Three identical masses m are attached to the corners of an equilateral triangle of sides d. Calculate the angular momentum (a) if the triangle rotates at an angular velocity ω about the center of mass around an axis perpendicular to the plane of the triangle; (b) if the triangle rotates with angular velocity ω about one of its sides; (c) if the triangle rotates about an axis going through one of its vertices to the midpoint of the opposite side.

10–2 Generalization of Torque

16. (I) A construction worker of mass 72 kg stands at the end of a 3.4-m-long (massless) horizontal mast attached to a building. What is the magnitude of the torque exerted on the hinge that holds the mast fixed in position?

17. (I) What is the torque about the origin on a particle positioned at $\mathbf{r} = (3\mathbf{i} - \mathbf{j} - 5\mathbf{k})$ m, exerted by a force of $\mathbf{F} = (2\mathbf{i} + 4\mathbf{j} + 3\mathbf{k})$ N?

18. (I) A flagpole 2.4 m long is attached to a building. It makes an angle of 30° with the horizontal. A mass of 65 kg is suspended from the end. What is the torque acting on the point of attachment to the building due to the suspended mass?

19. (I) A 1.0-m-long massless stick lies on a table. A force of 200 N is applied to one end of the stick for 0.1 s at an angle of 45° to the stick, and in a direction pointing away from the center of the stick. What is the torque about the other end of the stick during the brief period that the force is acting?

20. (II) Consider two forces $\mathbf{F}$ and $-\mathbf{F}$ that act at different points on an extended object. Show that the net force of this combination is zero and that the torque about any point P is independent of the location of P, and depends only on the separation of the two points at which the forces act.

10–3 The Dynamics of Rotation

21. (I) A point mass M is attached to a turntable at a distance R from the center (Fig. 10–30). The turntable rotates with constant angular speed ω about its axis. If the axis is horizontal, so that the turntable rotates in a vertical plane, what is the torque that the force of gravity on the mass exerts about the axis as a function of time? Assume that the mass is at the topmost position at $t = 0$.

22. (II) The position of a ball of mass m thrown from a building is given by $\mathbf{r} = (v\cos\theta t)\mathbf{i} + (v\sin\theta t - gt^2/2)\mathbf{j}$, measured from the point from which the ball was thrown. What is the torque about the origin that the force of gravity exerts on the ball?

FIGURE 10–30 Problem 21.

23. (II) A ball of mass m slides at speed v on a frictionless horizontal surface and bounces elastically from a wall. The initial path of the ball makes an angle θ with the wall (Fig. 10–31). Find the initial and final angular momenta of the ball about the point A. What causes the change in angular momentum?

FIGURE 10–31 Problem 23.

24. (II) A pulley system is used to lift a heavy mass. How much force must be applied to lift the object in Fig. 10–32 at a steady speed? Neglect friction at the axle.

FIGURE 10–32 Problem 24.

25. (II) What is the vector product of $\mathbf{A} = 2\mathbf{i} - 4\mathbf{j} + 5\mathbf{k}$ and $\mathbf{B} = \mathbf{i} + 3\mathbf{j} - 2\mathbf{k}$?

26. (II) Show that the magnitude of the vector product of two vectors is the area of the parallelogram for which the two vectors form adjacent sides.

27. (I) A playground merry-go-round of diameter 3.4 m and rotational inertia 120 kg·m² is pushed with no one on it by three children to an angular speed of 2.5 rad/s. Two of the children, of mass 25 kg each, jump on the edge of the merry-go-round, coming radially in. What is the new angular speed?

28. (I) A uniform disk rotating without friction about its (vertical) central axis, with total mass 138 kg and radius 1.5 m, acts as a turntable. Its angular speed is $\omega = 1.2$ rad/s. A person of mass 74 kg jumps straight down onto the rotating turntable. The person lands 1.2 m from the axis. What is the new angular speed of the turntable?

29. (I) A firetruck, mass 6000 kg, passes a parked car on a straight street at $t = 0$ with speed 15 m/s. A physics graduate student finds that, 10 s later, the angular momentum of the firetruck with respect to the parked car is twice its value at $t = 0$. What is the speed of the truck at $t = 10$ s?

30. (I) A skater twirls at 1.0 rev/s with her arms extended and holds a 5-kg mass in each hand; each mass is 0.9 m from the axis of rotation. She pulls the masses in along the radial direction until they are 0.3 m from the axis of rotation. Assuming that the rotational inertia of the arms is negligible and that the rotational inertia of the skater without the masses is 2.9 kg·m², what is the speed of rotation after the masses have been pulled in?

31. (II) A bug of mass $m = 2.0$ g walks around a horizontal turntable, which may be viewed as a uniform cylinder of mass $M = 0.24$ kg. If both the turntable and the bug are initially at rest, how much does the turntable rotate relative to the ground while the bug makes one full circle relative to the turntable?

32. (II) A small mass of 0.5 kg slides down a frictionless slope starting from rest at 1.6 m above the ground level. When the slope reaches the bottom, it levels off, and the mass strikes the bottom of a vertical uniform bar of mass 3.2 kg and length 1.0 m, pivoted at its midpoint, and sticks to it (Fig. 10–33). With what angular speed will the bar start its rotation?

1.6 m

FIGURE 10–33 Problem 32.

10–5 Work and Energy in Angular Motion

33. (I) An airplane engine develops 240 hp while turning the propeller at 3200 rev/min. What is the torque exerted on the propeller axis by the engine?

34. (I) A flywheel has radius 1.2 m and mass 680 kg, almost all of which is concentrated on the rim. It is spinning at $\omega = 4.5$ rad/s about its axis when a torque is applied along the axis,

producing an angular acceleration of 0.3 rad/s². How much time does it take for ω to increase to 6 rad/s, and how much work is done by the torque during this time?

35. (I) The work done to slow down a spool used in a manufacturing process during a certain period is 1200 J. The spool is fixed on an axis, and has rotational inertia of 0.033 kg·m² about that axis. The spool is rotating at an angular speed of 490 rad/s before the specified period. What is the spool's angular speed at the end of the period?

36. (II) Repeat the problem of the cylinder rolling down the inclined plane treated in Section 9–7, but this time use energy techniques to learn the speed of the cylinder as a function of distance traveled.

37. (II) A cylinder of mass 0.2 kg rolls without slipping down an inclined plane of 15°. What is its rotational kinetic energy after it rolls 80 cm?

38. (II) A particular flywheel used for the storage of energy is a solid steel cylinder of density 8 g/cm³. The cylinder has a radius of 1.4 m, is 30 cm thick, and spins about its axis at a frequency of 320 rev/min. (a) What is the rotational inertia of the flywheel? (b) What is the work the flywheel can do in being brought to a halt, if there is no energy loss to frictional forces?

*10–6 Quantization of Angular Momentum

39. (II) Consider an object of mass m that moves in a circular orbit caused by a central force given by $F = -kr$. Suppose that the Bohr quantization condition is applied to this motion. What are the allowed quantized radii, velocities, and kinetic energy values? [*Hint*: The acceleration for circular motion is v^2/r, and $E = (mv^2/2) + U(r)$.]

40. (II) Repeat the calculation of Problem 39 for an object that moves in a circular orbit caused by a central constant attractive force, obtained from the potential energy $U(r) = Cr$.

41. (II) The energy of the hydrogen atom, when quantized, is given by $E_n = -(13.6 \text{ eV})/n^2$, where $n = 1, 2, 3, \ldots$. Particles of energy 2.0 eV repeatedly pass through a gas of hydrogen atoms in the $n = 1$ state but never excite them out of this lowest-energy state. Explain. What energies would excite the hydrogen atoms?

42. (II) A proton has mass 1.67×10^{-27} kg and "radius" 1.3×10^{-15} m. (We put the radius in quotation marks because the proton is not a classical object with a radius, like a baseball.) It also can be thought of as having angular momentum with respect to an internal axis of 0.5×10^{-34} kg·m²/s. Take as a model that the proton is a uniform sphere. (a) What, according to the model, is the angular frequency of the proton's rotational motion? (b) What is the speed with which the outermost portion of the proton, on its equator, moves? Compare this to the speed of light, 3×10^8 m/s. (c) What is the energy associated with the rotational motion of the proton?

*10–7 Precession

43. (II) By repeating the derivation of the precession frequency of a top, but with the top making an angle ϕ with respect to the vertical, show that the precession frequency is *independent* of the angle ϕ.

44. (II) A student sits on a piano stool that is not rotating. She holds a vertical shaft on which a bicycle wheel of rotational inertia I is mounted and rotates with an angular speed ω oriented upward. She wants to tilt the wheel away from herself in a radial direction. What is the direction of the torque that she must exert? Suppose she succeeds in reversing the direction of the shaft by 180°. What will the speed and direction of rotation of the student and the piano stool be, assuming that the rotational inertia of the student and stool together is I^*?

45. (II) A wheel with massless spokes has mass 1 kg and radius 10 cm and is mounted on one end of a massless axle (Fig. 10–34). The axle rests on a pivot at a point 16 cm from the mounting point and 10 cm from the wheel. At the other end a mass of 0.8 kg is attached. The wheel spins at an angular frequency of 10 rad/s. What is the rate of precession?

FIGURE 10–34 Problem 45.

General Problems

46. (I) An 80-g ball is thrown out of a second-story window 8.0 m above the ground. The initial velocity of the ball is horizontal, with magnitude 2.4 m/s. What is the angular momentum of the ball, as a function of time, about the point on the ground directly below the window?

47. (II) A solid cylinder of mass 0.85 kg and radius 4.2 cm initially at rest rolls down a plane inclined at 28° with the horizontal and 1.5 m long. Use energy conservation to calculate the angular velocity of the cylinder at the bottom of the ramp; assume that all the kinetic energy of the cylinder is in rolling motion (that is, there is no sliding).

48. (II) A constant-density cylinder of mass 0.5 kg and radius 4 cm can rotate freely about an axis through its center. It has thread wound around an attached axle of radius 0.5 cm that also runs through its center (Fig. 10–35). The thread is attached to a mass of 1 kg, which slides down an inclined plane with an acceleration of 0.1 m/s². What is the coefficient of kinetic friction between the block and plane?

FIGURE 10–35 Problem 48.

49. (II) A child of mass 32 kg stands at the center of a platform of radius 2 m and rotational inertia 450 kg·m². The circular platform rotates about a frictionless shaft with angular speed of 0.8 rad/s. The child walks in a radial direction until he reaches the rim. What will the angular velocity of the platform be when that happens? What is the change in energy of the platform plus child? Identify the source of the work responsible for the change in rotational kinetic energy.

50. (II) A bullet of mass 5.0 g and velocity 330 m/s passes through a wheel at rest (Fig. 10–36). The wheel is a solid disk of mass 2.0 kg and radius 18 cm. The bullet passes through the wheel at a perpendicular distance of 15 cm from the center, and the bullet's final velocity is 220 m/s. What are the wheel's angular velocity, angular momentum, and kinetic energy? Is energy of motion conserved?

FIGURE 10–36 Problem 50.

51. (II) Figure 10–37 shows the Atwood's machine treated in the problems of Chapter 5, with a rope of negligible mass 2 m long. Earlier the pulley was treated as massless. Now suppose that the pulley can be approximated by a solid disk of radius 0.1 m and mass 2 kg. The system is released from rest with the 4-kg mass 1.5 m from the floor and the 1-kg mass on the floor. (a) What is the speed of either block just before the 4-kg mass hits the floor? (b) How long does it take the 4-kg mass to reach the floor?

FIGURE 10–37 Problem 51.

52. (II) Two weights of mass 1.2 kg and 0.85 kg, respectively, are connected by a massless string that passes over a pulley. The pulley is a hollow cylinder of radius 18 cm and mass 0.45 kg. What are (a) the acceleration of the system and (b) the time that it takes for the larger mass to descend a distance of 1.6 m, if the weights start from rest?

53. (II) A ball of mass M, radius R, and uniform density falls from rest from the top of a hemispherical bowl (Fig. 10–38). The left side of the bowl is frictionless, but the right side has a large coefficient of friction with the ball, and for all practical purposes the ball immediately rolls without slipping. How far up the right side of the bowl does the ball reach? Explain your answer in light of our assumptions about the very short distance in which sliding changes to rolling.

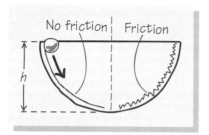

FIGURE 10–38 Problem 53.

54. (II) A solid door of mass $M = 20$ kg and width $\ell = 1$ m is hit at a right angle by a mud ball of mass $m = 0.5$ kg, which, as Fig. 10–39 shows, hits the door at the edge with speed $v = 10$ m/s and sticks. (a) What is the rotational inertia of the door about the hinges? (b) What is the angular velocity of the door after having been struck? (c) What fraction of the initial energy does the moving door–mud ball system retain?

FIGURE 10–39 Problem 54.

55. (II) A 75-kg bank robber is escaping on a 450-kg motorcycle, runs out of gas, and thereafter coasts (friction free) at 80 km/h. As he passes under an overpass, a 75-kg policeman drops vertically onto the back of the seat and hangs on. Ignore the impulse due to the sudden change of the force of friction on the road on the translational and rotational motion. (a) Find the final velocity of the motorcycle. (b) What fraction of the initial kinetic energy (motorcycle plus robber) is lost? Ignore the rotational energy in the wheels. (c) Redo part (b) but include the effect of the rotational energy of the wheels, which each have a rotational inertia about their axes of 3 kg·m² and a radius $r = 0.5$ m.

56. (II) Electric power is used to speed up a centrifuge whose rotational inertia is 0.2 kg·m²; 2.7 kW of power were used to make the centrifuge accelerate at a steady rate from rest to 24,000 rev/min. If the electricity use were 100 percent efficient, how much time is required to speed up the centrifuge?

57. (II) A point mass $m = 0.2$ kg is attached to a string, which passes through a hole in a table and rotates in a circle of radius $r = 0.8$ m with an angular velocity of 40 rad/s. What mass M must be attached to the end of the string under the table to maintain this motion? Suppose that mass M is slowly increased by an amount that makes it descend a distance 0.1 m. What is the amount of the increase of M? What will the new angular velocity of the point mass be? (*Hint*: Use angular momentum conservation.)

58. (II) Show that $(\mathbf{r} \times \mathbf{p}) \cdot (\mathbf{r} \times \mathbf{p}) = r^2 p^2 - (\mathbf{r} \cdot \mathbf{p})^2$. (*Hint*: It is convenient, without any loss of generality, to assume that both $\mathbf{r}$ and $\mathbf{p}$ lie in the xy-plane.) Use this result to express the kinetic energy of a particle in terms of the momentum in the radial direction and of the square of the angular momentum.

59. (II) A cylindrical shaft of radius 5 cm is connected by a band to a solid cylindrical flywheel of mass 300 kg and of radius 0.35 m (Fig. 10–40). A motor brings the shaft up to a rotational rate of 1400 rev/min. Calculate the amount of work done by the motor, neglecting the rotational inertia of the shaft.

FIGURE 10–40 Problem 59.

60. (II) A thin rod of mass M, length ℓ, and constant density is standing on end on a rough table that forms the xy-plane. The rod begins to fall, with its top moving in the $+x$-direction, but as it falls, its point of contact does not move. As the rod hits the table, what are its (a) angular velocity, (b) angular momentum, and (c) kinetic energy?

61. (II) An object of mass M moves in a circular planar orbit about a center of gravitational attraction. The force of attraction has magnitude $F = K/r^2$, where r is the radius of the circle, and it is directed toward the center. Calculate (a) the velocity, (b) the radius, (c) the period, T, and (d) the acceleration of the object, all in terms of the angular momentum L, M, and K.

62. (III) A putty ball of mass $m = M/3$ is thrown with velocity $\mathbf{v} = v\mathbf{i}$ and hits the top of the thin rod in Problem 60 as the rod stands vertically. If the putty ball makes a completely inelastic collision, and if again the point of contact between the rod and the table does not move, what are the angular velocity, angular momentum, and kinetic energy of the system as it hits the table?

63. (III) A particular top can be approximated as a solid cylinder of mass 100 g and radius 2 cm. A string of negligible mass and length 1 m is wound around the top, which is started by pulling horizontally on the string with a constant force of magnitude 0.6 N. The top starts from rest at point O, and the string is pulled off. Ignore all friction between the top and the table on which it moves. (a) What is the final velocity of the center of mass of the top? (b) the final angular velocity of the top about its center of mass?

64. (III) A uniform solid cylinder of radius R and mass M rests against a vertical curb of height h, where $h < R$ (Fig. 10–41). The cylinder is mounted through its axis on a frictionless horizontal axle. You exert a horizontal force of magnitude F on the axle, pushing the cylinder against the curb. What is the minimum value of F that will cause the cylinder to roll up over the curb?

FIGURE 10–41 Problem 64.

The tightrope artist Philippe Petit, here seen in Frankfort, uses the large bar to control the position of his center of mass and minimize the torque acting on the system formed by him and the bar.

Statics

As Chapters 9 and 10 show, the motions of extended objects present a rich array of possibilities because both linear and rotational motion are possible. In the area of physics called **statics**, we are interested in the conditions that leave rigid bodies motionless. The problems presented by statics for the motion of extended objects must include the possibility that there are torques acting as well as forces. Statics has its most notable use in the field of civil engineering, as in the design of bridges or buildings. Applications extend to the analysis of the role of muscle, tendon, and bone in living systems.

11-1 STATIC CONDITIONS FOR RIGID BODIES

In principle, it is enough to say that a building or a bridge will be static if the net force on every component is zero. A full calculation of the net force on any one component must include forces due to other components, which are sometimes very difficult to include. For example, it is not an easy problem to calculate the bending of a beam supported at two ends. However, a discussion of the conditions under which the building is static is greatly simplified if the building is *rigid*. The external forces acting on such an object have two effects: First, no matter where they are actually applied on the object, their vectorial sum produces a linear acceleration of the center of mass. Second, depending on where they are applied, they may produce torques that act to rotate the object. In the previous two chapters, we were interested in the resulting motion. In this chapter, we will look at the conditions for *no center-of-mass acceleration and no angular acceleration about any axis*; in most practical applications, there is also *no center-of-mass velocity and no angular velocity about any axis*. This area of study is known as statics (Fig. 11–1).

Suppose that we have a set of external forces $\mathbf{F}_i$, labeled by the index i, that act at various points described by the position vectors $\mathbf{R}_i$ on a rigid body of mass M. Then the full dynamical equations for the object are

$$\mathbf{F}_{\text{net}} = \sum_i \mathbf{F}_i = M\mathbf{A} \tag{11-1}$$

and

$$\boldsymbol{\tau}_{\text{net}} = \sum_i (\mathbf{R}_i \times \mathbf{F}_i) = I\boldsymbol{\alpha}. \tag{11-2}$$

The points at which the forces act have position vecters $\mathbf{R}_i$ measured with respect to any convenient origin. Here, $\mathbf{A}$ is the acceleration of the center of mass due to the net force, $\mathbf{F}_{\text{net}}$; $\boldsymbol{\alpha}$ is the angular acceleration of the object body due to the net torque, $\boldsymbol{\tau}_{\text{net}}$, *about the point chosen as the origin*. (Ordinarily, the rotation about a point also corresponds to rotation about an axis, which is why we have included the form $I\boldsymbol{\alpha}$ in Eq. (11–2). In that case, the rotational inertia, I, must be calculated about the axis of rotation.)

We can now write down the conditions for statics. The condition under which the linear acceleration drops out is contained in Eq. (11–1) with $\mathbf{A} = 0$, which states that the net force is zero:

$$\sum_i \mathbf{F}_i = 0. \tag{11-3}$$

The condition that the net torque drops out is obtained from Eq. (11–2) with the angular acceleration $\boldsymbol{\alpha} = 0$:

$$\sum_i (\mathbf{R}_i \times \mathbf{F}_i) = 0. \tag{11-4}$$

Equations (11–3) and (11–4) are all we need to study the statics of rigid bodies.

Recall from Chapter 9 that a torque has a magnitude given by the product of the force and a length known as the *moment arm*. To find the moment arm about a certain point O, extend the line of the force. The closest, or perpendicular, distance between the extension line and point O is the moment arm.

The Condition of No Torque Is Independent of the Choice of Axis

When we say that there is no angular acceleration [Eq. (11–4)], to which axis are we referring? We can show that if an object has no angular acceleration about any one axis, then, assuming that the object is not in linear acceleration, it is not in angular acceleration about another axis. Equation (11–4) applies to *any* choice of origin when Eq. (11–3) holds.

In order to prove this important result, we consider a new origin displaced by $\mathbf{D}$ from the old origin and compute the angular acceleration about this new origin. Any angular acceleration is due to a series of forces, labeled by an index i, applied to the extended object. If the force labeled i is applied at point $\mathbf{R}_i$ with respect to the original origin, then it is applied at the point $\mathbf{R}_i'$ in the new system:

$$\mathbf{R}_i' = \mathbf{R}_i + \mathbf{D}.$$

The condition for no rotational acceleration is

$$0 = \sum_i (\mathbf{R}_i \times \mathbf{F}_i) = \sum_i [(\mathbf{R}_i' - \mathbf{D}) \times \mathbf{F}_i] = \sum_i (\mathbf{R}_i' \times \mathbf{F}_i) - \sum_i (\mathbf{D} \times \mathbf{F}_i)$$

$$= \sum_i (\mathbf{R}_i' \times \mathbf{F}_i) - \left(\mathbf{D} \times \sum_i \mathbf{F}_i\right).$$

FIGURE 11–1 Among the forces and torques acting on this rock are those due to contact forces and to gravity. This static marvel has remained balanced for a very long time, suggesting that the equilibrium is stable.

∞ The calculation of torque is treated in Chapter 10.

The conditions under which a rigid body is static

"No torque about any one axis" means "No torque about any axis."

From Eq. (11–3), the sum over the forces in the second term of the right-hand side is zero. Thus

$$\sum_i (\mathbf{R}_i \times \mathbf{F}_i) = \sum_i (\mathbf{R}_i' \times \mathbf{F}_i) = 0,$$

which shows that if there is no torque about one origin, then there is no torque about another origin.

Because a static object has no torque about any axis, we can choose the axis about which we find the torques that compose the net torque. A good choice leads to considerable simplification; the calculation of the torques about some points may be especially simple. In particular, the torque about the point where a force is applied, or anywhere along the line of that force, is zero because the moment arm is zero for that force. For example, let's consider a hard ball being pushed against a curb by a horizontal force **F** that is applied at precisely curb height (Fig. 11–2a). The ball itself has a radius equal to the curb height. A horizontal force can make the ball climb the curb if it causes the ball to rotate about the point P. This occurs when there is a net torque about P. But the contact force, $\mathbf{F}_C$, and the external force, **F**, are both directed so that extensions of them pass through point P. These forces therefore produce no torque about P, as the extended force diagram Fig. 11–2b shows. Thus the external force cannot make the ball climb the curb and the ball is static.

PROBLEM SOLVING

The calculation of torque can be simplified by a good choice of axis.

(a)

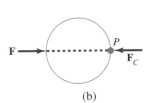

(b)

FIGURE 11–2 (a) The ball has a radius equal to the height of the curb, and a horizontal force **F** is applied at exactly curb height. (b) Extended force diagram (we have ignored the vertical forces). The forces shown can have no torque about point P.

11–2 GRAVITY AND RIGID BODIES

We often apply the static conditions to mechanical structures such as bridges, where gravity is an external force. Unless the structure is so huge that the acceleration of gravity differs over its length or height (an unusual situation, to say the least), gravity acts equally on all its different portions. In Section 9–5, however, we showed that there is a considerable simplification: *Gravity acts as though it were applied to a concentrated (point) mass of the total mass M at the center of mass of the extended object.* Thus, gravity produces a torque as though the gravitational force acts on the center of mass (which is sometimes called the *center of gravity* in this context).

EXAMPLE 11–1 Consider a rectangular book with a uniform mass density and a width L. The book is lying on a table with one side parallel to the table edge, but the book is hanging off the edge by an amount ℓ (Fig. 11–3a). What is the largest possible value of ℓ for which the book does not rotate off the edge?

Solution: Figure 11–3b shows a force diagram with the force of gravity, $m\mathbf{g}$, acting on the book's center of mass at a point midway along its width. If the book falls off the table, it will do so by rotating about the table edge. Thus we calculate the torque about this axis and place the origin at the edge of the table. We can satisfy the first equilibrium condition, Eq. (11–3), if

$$\mathbf{F}_N = -m\mathbf{g}.$$

∞ Torque due to gravity was first discussed in Chapter 9.

Gravity produces a torque as though it acts on the center of mass.

(a)   (b)   (c)

FIGURE 11–3 (a) Example 11–1. (b) The center of mass of the book is over the table surface. (c) The center of mass of the book is beyond the table edge.

FIGURE 11–4 In an extension of Example 11–1, it is possible to show that, with as few as four books, the topmost book in a stable pile can lie entirely over the edge of the table.

APPLICATION

Stable construction

FIGURE 11–5 The arch is a stable architectural element that has been used since ancient times.

When the center of mass of the book lies above the table surface, we can satisfy the second equilibrium condition—Eq. (11–4)—for rotations about the edge of the table if $\mathbf{F}_N$ acts at the same point as the force of gravity, as shown in Fig. 11–3b. Note that at the point where the book is ready to rotate off the table, $\mathbf{F}_N$ does indeed act only at the edge of the table.

If the book's center of mass is beyond the edge of the table, then the second equilibrium condition can no longer be satisfied. The normal force will be acting at the edge of the table; at this point, the lever arm for the torque of the normal force is zero. The force of gravity, however, does have a torque for rotations about this point. Thus the equilibrium condition can no longer be satisfied, and the book falls (Fig. 11–3c). The equilibrium conditions *can* be satisfied if the center of mass of the book lies above the table surface and the largest possible value of ℓ is $L/2$.

Structures Without Mortar

An interesting generalization of Example 11–1 involves the static conditions for a stack of several books of mass m on a table. In Problem 13, we consider the conditions under which one book overhangs a second, identical book on the edge of a table. If you stack enough books, it is possible to make a stable pile with the topmost book lying as far off the table as desired. Figure 11–4 shows such a stack. Even with a stack of only four books, the topmost book can be made to lie completely off the table edge. These results can be applied to other constructions; for example, they suggest that if building blocks are used judiciously, it is possible to build stable structures without worrying about mortar. A crucial added ingredient is the keystone of an arch, which, in effect, allows the joining of the topmost edges of two stacks of books, such as the one in Fig. 11–5. With the keystone, neither pile need be stable by itself. Mortar often represents the weak point of a structure and many ancient building structures that have lasted through the ages have made good use of the possibility of equilibrium without mortar. Among these we may list Egyptian, Greek, and Roman monuments, as well as the mighty cathedrals erected during the middle ages in Europe. Modern buildings, too, use principles like these; Frank Lloyd Wright's Falling Water is a splendid example.

An Experimental Method to Find the Center of Mass

The fact that gravity effectively acts on the center of mass shows that a suspended object will be in stable equilibrium when the center of mass lies directly below the suspension point. Figure 11–6a shows a suspended object with its center of mass indicated. The forces acting on the object are the tension, T, of the suspending rope, acting at the suspension point, and the object's weight, acting at the center of mass. The equi-

FIGURE 11–6 (a) If a suspended object does not line up with its center of mass directly below the pivot point, there will be a net torque about some point, and the object cannot be in static equilibrium. (b) With the center of mass directly below the pivot point, there is no net torque about any point, and the object is in static equilibrium.

(a)   (b)

(a)

(b)

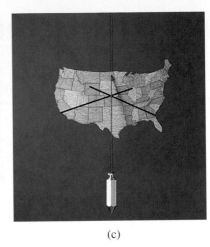
(c)

FIGURE 11–7 Here's a technique to find the center of mass of this cutout map of the United States. (a) Suspend the map from any point, and drop a plumb line from that point. The center of mass lies somewhere along the line. (b) The same is true for a second plumb line dropped from a second point, and the intersection of the two lines is the location of the center of mass. (c) The plumb line dropped from a third point does indeed pass through the center of mass.

librium condition for the force is that T is equal and opposite to the weight. However, if the center of mass does not lie directly below the suspension point as in Fig. 11–6a, there will be a net torque and the object cannot be in equilibrium. For example, take the origin at the center of mass; in this case, the tension will have a torque about this point. If the center of mass lies directly below the suspension point, then neither force exerts a torque about the center of mass (Fig. 11–6b).

The center of mass of a flat object can be determined by using a plumb bob. The object is suspended three times in succession, each time from a different point on the object. A plumb bob is dropped from each suspension point and a vertical line is drawn on the object. In each case, the center of mass lies along these lines, and thus the intersection of these lines is the center of mass. Figure 11–7 illustrates the method.

APPLICATION

The limits of safe construction

11–3   APPLICATIONS OF STATICS

How many beams of a given size are required to support a roof of a given weight? Can the beams be reduced in size? The equations for statics are often used to determine the forces ("loads") on such components, which in turn are used to find the size of components needed for construction on various scales, or to determine the limits of safe construction (Fig. 11–8). In Examples 11–2 and 11–3, we treat some of these aspects.

EXAMPLE 11–2   In order to handle a very hot object of mass 5 kg, a worker hooks it with a light but stiff pole of length $L = 3$ m, which he then carries horizontally with two hands (Fig. 11–9). His right hand is at one end of the pole (point A), while his left hand is at point B, a distance $\ell = 0.75$ m farther down the pole, and the mass is at the other end. What forces must the worker's hands exert on the pole?

Solution: We want to use the static conditions to determine the forces exerted by the hands on the pole. To do so, we first note that the extended object on which the forces act is the pole along with the hot object at its end; the external forces acting on it are gravity and contact forces from each hand. In the force diagram of Fig. 11–9b, we have used our intuition that the hands will exert upward forces but, as is usual with force diagrams, it is not necessary that the forces drawn correspond precisely in direction and magnitude to the forces to be determined; the static conditions will themselves determine these forces.

FIGURE 11–8 The Statue of Liberty forms a complex system whose successful construction required knowledge of many static forces. Interestingly, the raised arm was originally designed to lie along the black dashed line. The additional torque caused by the misplaced arm required an extensive repair, completed in 1986.

303

PROBLEM–SOLVING TECHNIQUES

Suggestions for Statics Problems

1. Prepare an extended force diagram that includes all the forces and where they act. Follow the first three steps described in the Problem-Solving Techniques box in Chapter 9.

2. Choose a single convenient origin and think about the choice of axes. In particular, the torques due to the forces may be simpler to compute about some origins than about others; further, the more forces are aligned with coordinate axes, the better.

3. Be clear about the direction you have chosen for positive torque.

4. Write down the conditions for statics that follow from steps 1 through 3—Eqs. (11–3) and (11–4).

5. Count the resulting equations as well as the unknown forces to be determined to make sure that the number of equations corresponds to the number of unknowns and the problem does indeed have a unique solution.

6. In solving the equations for statics, carry through an algebraic rather than a numerical solution as far as possible. In this way, you can make checks in various limits or in special cases and you can see if the results are reasonable.

0.75 m   2.25 m

(a)

(b)

FIGURE 11–9 (a) Example 11–2. (b) Extended force diagram. The object on which the forces act is the massless rod with the hot, massive object at the end. The problem solution shows that the actual direction of $\mathbf{F}_A$ points down.

Let's choose point A to be the origin. Equation (11–3) reduces to one equation for the vertical components of the force; Eq. (11–4) is also one equation because the torques about point A are all into or out of the page. Two is the right number of equations to determine the two unknown forces exerted by the hands. Then our static conditions are

$$F_A + F_B - mg = 0$$

and

$$\ell F_B - Lmg = 0.$$

Note that F_B and mg must exert torques in opposite directions and that there is no torque due to F_A with our choice of A as origin.

These equations can be solved for the unknown forces:

$$F_B = mg\left(\frac{L}{\ell}\right);$$

$$F_A = mg - mg\frac{L}{\ell} = mg\left(\frac{\ell - L}{\ell}\right).$$

Because $\ell < L$, F_A is negative: Hand A must exert a downward force. In effect, point B, which lies between the two ends of the pole, acts as a pivot point about which the pole may rotate. The force at A must point in the *same* direction as the force of gravity on the mass to allow the torque to be zero. $\mathbf{F}_B$ must then counteract mg and $\mathbf{F}_A$. Note the magnification factor in the determination of the unknown forces. For example, F_B is much larger than mg if L is much larger than ℓ. It is difficult to hold a mass at the end of a horizontal pole!

Inserting numbers, we find

$$F_A = (5 \text{ kg})(9.8 \text{ m/s}^2)\frac{0.75 \text{ m} - 3 \text{ m}}{0.75 \text{ m}} = -150 \text{ N};$$

$$F_B = (5 \text{ kg})(9.8 \text{ m/s}^2)\frac{3 \text{ m}}{0.75 \text{ m}} = 200 \text{ N}.$$

EXAMPLE 11–3 A crane whose cabin and engine are effectively fixed to Earth is used to lift a mass m of 5300 kg (Fig. 11–10a). The arm of the crane is supported at its base, at point B, by a strong but friction-free pivot, and at its top, at point A,

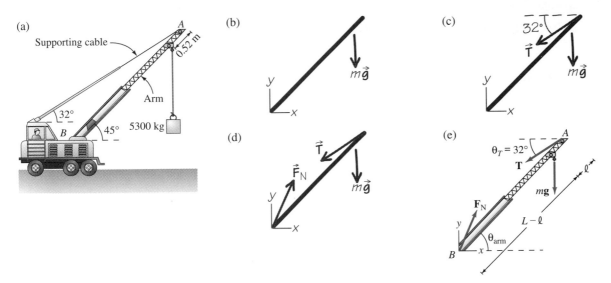

FIGURE 11–10 (a) Example 11–3. (b–d) Sketches illustrating the preparation of the extended force diagram. (e) Extended force diagram. The object on which the forces act is the arm of the crane. The relevant angle between the gravitational force and the position vector from point B along the crane's arm is given by $\theta_{arm} + 90°$. The relevant angle between the tension and the position vector from point B along the crane's arm is given by $-\theta_{arm} + 180° + \theta_T$. The sines of these angles enter into the expression for the torque about point B due to the forces of gravity and tension, respectively.

by a cable. The arm and the cable make angles of 45° and 32°, respectively, with the horizontal, and the arm is 10.0 m long. The mass is lifted from a point on the arm 0.52 m from point A. Ignoring the mass of the arm, compute the tension in the cable.

Solution: We want to find the magnitude of one of the forces (the cable tension) that acts on the crane arm and therefore we take the arm to be the object for which a force diagram is drawn (Fig. 11–10b). The forces acting on the arm are: (1) the tension, $\mathbf{T}$, acting at A along the cable; (2) the weight of the 5300-kg mass; and (3) an unknown contact force, $\mathbf{F}_N$, at the bottom pivot point (Fig. 11–10c, d). The contact force points in an arbitrary direction to be determined. We must choose an origin to write the equations of motion. If the origin is at the application point of any of the forces, then there will be no torque due to that force. Aside from this consideration, there is no special advantage to choosing any one of the three candidate points, and we choose point B as shown (Fig. 11–10c). There are three equations: two for the x- and y-components of the net force and one for the torque. There are three unknowns: the two components of $\mathbf{F}_N$ and the magnitude of $\mathbf{T}$. The number of equations will match the unknowns, hence the problem has a solution.

The net force equation has the two vector components:

$$F_{Nx} - T \cos \theta_T = 0, \qquad (11–5)$$

$$F_{Ny} - T \sin \theta_T - mg = 0. \qquad (11–6)$$

The torques are perpendicular to the page, so the net torque equation has one vector component:

$$-(L - \ell)mg \sin(\theta_{arm} + 90°) + LT \sin(-\theta_{arm} + 180° + \theta_T) = 0, \quad (11–7)$$

where L is the arm length (AB) and ℓ is the distance away from point A that the mass hangs. Equation (11–7) is sufficient to solve for T because the choice of B as the origin removes F_N from the net torque equation:

FIGURE 11–11 A crane is held rigid by internal forces. External forces, such as those provided by cables and bolts, hold the crane static.

$$T = \left(\frac{L-\ell}{L}\right) \frac{\sin(\theta_{arm} + 90°)}{\sin(-\theta_{arm} + 180° + \theta_T)} mg$$

$$= \left(\frac{10.0\text{ m} - 0.52\text{ m}}{10.0\text{ m}}\right) \frac{\sin(45° + 90°)}{\sin(-45° + 180° + 32°)} (5300\text{ kg})(9.8\text{ m/s}^2)$$

$$= 1.5 \times 10^5 \text{ N}.$$

A useful check on this result is that if $\ell = L$, the mass is hung directly from point B and, from physical considerations, we expect no tension to be required; this is indeed the case. It is also true that if $\theta_T = \theta_{arm}$—so that the cable is also attached at the pivot point at B—then there is no way that T can provide a torque of the opposite sign from the torque coming from the weight, and the arm cannot be stabilized. Indeed, we cannot solve for T when the angles are equal.

Example 11–3 is one you might use if you were going to design a crane and had to choose the size of the supporting cable (Fig. 11–11). Notice that in this design, the tension in the cable must be much greater than the weight itself. It may also be necessary to solve for the contact force to ensure that the pivot itself is sufficiently strong.

It is a good exercise to choose some other point—say, point A—to be the origin and verify that the answer is independent of the choice of origin.

EXAMPLE 11–4 A ladder of length 3.0 m is leaning against a wall at an angle of 58° as in Fig. 11–12a. Its eight rungs are spaced 0.33 m apart. The ladder's mass is insignificant compared to the mass $m = 85$ kg of a window washer who is climbing the ladder. The coefficient of static friction between the rubber feet of the ladder and the floor is $\mu_s = 0.51$, but assume that the wall is smooth and there is no friction between the top of the ladder and the wall. Is the ladder safe from slipping if the washer climbs to the seventh rung?

Solution: The extended object on which the forces act is the ladder; the question of interest concerns the magnitude of the friction force, and hence of the normal force at the floor. The normal forces are labeled $\mathbf{F}_{floor}$ and $\mathbf{F}_{wall}$ in Fig. 11-12b, which is an extended force diagram. The friction force at the floor acts to the left. To find out if the ladder slips, it is necessary to note that the static friction force takes on its *maximum value* just before slipping. The magnitude of this force is

$$f_{floor} = \mu_s F_{floor}. \tag{11–8}$$

The angle of the ladder is θ and the ladder length is L. The weight, $m\mathbf{g}$, acts at a distance d up the ladder, and points directly downward.

There will be three equations: two for the cancellation of forces and one for the torque. The three unknowns will be the two contact forces and the friction force. If the friction force that we determine exceeds the maximum static value, Eq. (11–8), then the ladder will slip. The bottom of the ladder is a useful choice for the origin

FIGURE 11–12 (a) Example 11–4. The window washer is a thin person standing vertically, so his center of mass is over the rung on which he stands. (b) Extended force diagram for the ladder. (c) The lever arm for the force of gravity about the origin is $d \cos \theta$, and the lever arm for the normal force of the wall is $L \sin \theta$.

(a)

(b)

(c)

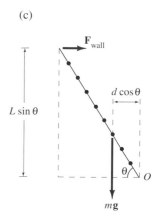

because two forces act at this point. We let $+x$ be to the right and $+y$ be upward. The statics equations for the forces are

$$F_{\text{floor}} - mg = 0, \tag{11–9}$$

$$F_{\text{wall}} - f_{\text{floor}} = 0. \tag{11–10}$$

For the torque, we see in Fig. 11–12c that the lever arm for the force of gravity about the origin is $d \cos \theta$, whereas the lever arm for the normal force of the wall is $L \sin \theta$. Thus

$$mgd \cos \theta - F_{\text{wall}}L \sin \theta = 0. \tag{11–11}$$

The contact force at the wall tends to rotate the ladder clockwise, while the weight tends to rotate the ladder counterclockwise.

Equations (11–10) and (11–11) can be solved for the contact force at the wall and the friction, and we find that

$$f_{\text{floor}} = \frac{mgd \cot \theta}{L}. \tag{11–12}$$

Equation (11–9) gives the contact force at the floor directly. The condition that the friction force at the floor not exceed the maximum value of static friction is

$$\frac{mgd \cot \theta}{L} \leq \mu_s F_{\text{floor}} = \mu_s mg,$$

$$d \leq \mu_s L \tan \theta.$$

The higher the washer climbs, the greater the danger! For the seventh rung, $d = 2.33$ m, and with $\theta = 58°$, the condition becomes

$$2.33 \text{ m} \leq 0.51(3.0 \text{ m})(\tan 58°) = 2.4 \text{ m}.$$

The condition is satisfied, but barely. One more rung and the ladder would slip.

The reason the ladder becomes more likely to slip as the washer climbs is that, as the washer moves up the ladder, the lever arm of the washer's weight with respect to the contact point with the ground increases. The torque due to the washer's weight about the contact point with the ground increases, until the torque about this point due to the normal force with the wall can no longer counteract the effect of the torque due to the weight.

EXAMPLE 11–5 The biceps muscle is responsible for bending your arm. It acts through a kind of lever system, as in Fig. 11–13a. Some typical values for a, the elbow–hand distance, and x, the distance from the biceps attachment point to the

FIGURE 11–13 (a) Example 11–5. (b) Extended force diagram for example.

(a)

(b)

FIGURE 11–14 Biological systems can change the relative orientations of their different parts to allow them to be stable. (a) At least five different extenal forces act on the professor. (b) These forces add to a net force of zero.

(a)

(b)

FIGURE 11–15 (a) A uniform four-legged table. (b) Extended force diagram for the table.

elbow, are $a = 30$ cm and $x = 4$ cm. If a mass M is held in your hand with your forearm horizontal and your upper arm vertical, what upward force does the biceps have to exert on the forearm bones? Ignore the weight of the forearm bone itself, and assume the hand-forearm forms a single system.

Solution: If the forearm weight is ignored, there are three vertical forces on the forearm: the contact force from the upper arm acting at the elbow; the force $\mathbf{F_B}$ of the biceps on the forearm; and the weight of the mass. These are shown in the extended force diagram in Fig. 11–13b. The condition that there is no net force on the forearm determines the contact force from the upper arm. The condition that there is no net torque about the elbow determines F_B in terms of the distances and the mass's weight. Note that the contact force from the upper arm does not enter because that force exerts no torque about the elbow. We have

$$0 = (Mg)a - (F_B)x,$$

$$F_B = Mg\,\frac{a}{x}.$$

The force exerted by the contracting muscle must be a large factor greater than the weight itself. For typical arm dimensions, this factor is (30 cm)/(4 cm) = 7.5.

The physiological example described here is representative of the adaptable structures that occur throughout the kingdom of living organisms. For the considerations of this chapter, we must think of such structures as a collection of rigid parts—the bones—with variable orientation (Fig. 11–14).

Underdetermined Systems

We sometimes face problems in static systems where the forces cannot be uniquely determined by the conditions of force and torque equilibrium. There are fewer equations for the forces to be determined than there are forces. Such systems are said to be *underdetermined*. When a system is underdetermined, small incidental effects can become important.

As an explicit example, consider a uniform square table of mass M with four light legs on a horizontal surface (Fig. 11–15a). The top has sides of length L and, if we take the axes shown with the origin at leg 1, legs 1, 2, 3, and 4 make contact with the ground at the respective points $(x, y) = (0, 0)$, $(L, 0)$, $(0, L)$, and (L, L). The center of mass of the table is at its geometric center, the point $(x, y) = (L/2, L/2)$.

The forces acting on the table are the four contact forces $\mathbf{F_{N_1}}, ..., \mathbf{F_{N_4}}$ acting upward on the legs, and the force of gravity, $M\mathbf{g}$, acting on the center of mass. Figure 11–15b is the force diagram. The forces all act in the vertical, or z-, direction. We calculate the torque about the origin. The torque vectors due to the forces are perpendicular to these forces and therefore lie in the xy-plane. Our static conditions thus consist of one force equation (for the z-component) and two torque equations (for the x- and y-components). These equations are, respectively,

$$F_{N_1} + F_{N_2} + F_{N_3} + F_{N_4} - Mg = 0, \tag{11–13}$$

$$LF_{N_3} - \frac{L}{\sqrt{2}}\,Mg\,\sin 45° + \sqrt{2}LF_{N_4}\sin 45° = 0,$$

$$-LF_{N_2} + \frac{L}{\sqrt{2}}\,Mg\,\sin 45° - \sqrt{2}LF_{N_4}\sin 45° = 0.$$

Sin $45° = 1/\sqrt{2}$, so the last two equations simplify to

$$F_{N_3} - \frac{1}{2}\,Mg + F_{N_4} = 0, \tag{11–14}$$

$$F_{N_2} - \frac{1}{2}\,Mg + F_{N_4} = 0. \tag{11–15}$$

The tower crane used at construction sites (Fig. 11AB–1) consists of a very flimsy tower constructed from three planes arranged in a triangle. Each plane consists of a lattice of triangular bars (this is called a geodetic construction). At the top of this tower is a normal crane boom, which can lift large masses, such as furnaces, air conditioners, and steel girders. The tower can rotate to place the mass at the correct angle, while the crane can move the mass radially inward or outward to position the object at the correct location; the object is finally set down. The strength of the tower is completely insufficient to withstand the bending moment of a large mass at the end of the boom, so there is a balance weight at the far side of the crane that moves in an opposite direction from the lifting equipment and the object being moved. The motion of the counterweight is controlled by strain sensors at the top of the tower and ensures that the crane boom produces no net torque at the top of the tower.

A similar idea is applied to the portable cranes that are carried by trucks. These cranes are provided with sensors that measure the downward force of the crane and truck body on each wheel. If this force reaches a lower limit, indicating that a wheel is lifting off the ground, movement of the crane boom is halted, with operator controls being overridden. Thus the crane remains in static equilibrium and does not tip over.

FIGURE 11AB–1 This type of construction crane has become a familiar feature of the urban landscape.

We have only three equations to determine the four unknown forces $\mathbf{F}_{Ni}$. Indeed, any one leg could be removed completely. In this case, Eqs. (11–13) through (11–15) are three equations for the three remaining unknown contact forces. Here we see how a small change could make a big difference in the situation. If a teacup were placed near the location of the missing leg, just off the center of the table, the table would tip over toward the missing leg.

Finite-Element Methods

In this chapter, we have treated a rigid body as a collection of perfectly rigid parts. In some of the examples given, we can alter the relation between the orientations of the various rigid components of an extended object—like the crane of Example 11–3.

In reality, the components of extended objects are more complicated than such idealizations. Beams stretch, compress, bend, or twist. It is important to try to understand this behavior because the behavior of construction components is of paramount importance in the beauty, cost, and safety of a structure. The behavior of a beam depends not only on the external forces that act on it, but also on the forces within it. Many of these internal forces are the result of the action of external forces. These internal forces are in general referred to as **stresses**, and the response of extended objects—their stretching or bending or deforming—to stresses are generally called **strains**.[†] Because of the complexity of a real system, we do not have a mathematically exact way of computing, for example, the way in which a real beam will bend under the influence of a suspended weight.

[†]A more complete treatment of stresses and strains is given in Chapter 21.

Model

FIGURE 11–16 A thin beam is not truly rigid and may stretch or compress when it is subjected to tension or compression forces. It may be modeled by a series of masses connected by springs.

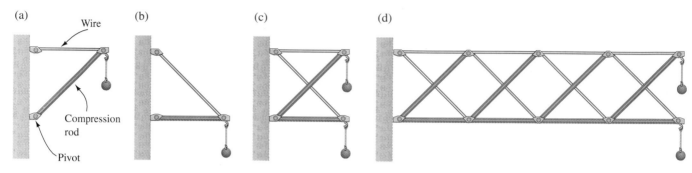

(a)

Wire

Compression rod

Pivot

(b)

(c)

(d)

FIGURE 11–17 A finite-element model of a cantilever beam. (a), (b) Two basic parts are combined to provide (c) the primitive elements for the beam. (d) Together these elements make a model for a beam. (After J. E. Gordon, *The New Science of Strong Materials*, Princeton University Press, Princeton, New Jersey, 1976.)

APPLICATION

Applying finite-element methods

$m\mathbf{g}$

FIGURE 11–18 By using the model described in Fig. 11–17, it is possible to calculate various characteristics of a cantilever beam, such as how it will strain (bend) under stress.

There is, however, a numerical approximation technique that springs directly from the methods we developed earlier. We can imagine breaking a continuous object into tiny elements, each of which can be thought of as an idealized rigid component. The accuracy of this method, known as the **finite-element method**, improves systematically as the number of elements grows larger. Consider as an example a one-dimensional system: a thin beam subject to forces in this one direction. We idealize the beam as a series of N rigid and massive components connected by springs (Fig. 11–16). The springs model the internal forces in the beam by resisting compression and stretching. In the model, the components are spaced and assigned masses so that the average mass density of the model matches that of the real beam. The strength of the individual springs is determined by matching some measured properties of the real beam, such as how the beam stretches when a known tension is applied. The model can now be used to see how the beam reacts to some external longitudinal force. There will be a static condition for each of the components and an array of N simultaneous equations will have to be solved. In other words, in solving problems with finite-element methods, we repeat the same relatively simple steps for many elements, a method that is well suited for numerical solution on a computer.

A more realistic type of example treats a *cantilever*, which is a projecting beam of finite thickness supported at one end only, as a succession of repeated components. The components pictured in Figs. 11–17a and 11–17b can each act as cantilevers and, when they are combined (Fig. 11–17c) and strung together in succession (Fig. 11–17d), they form a finite-element model of the cantilever beam. An element to the right of a second element acts as a weight for that second element; the succession of elements is treated as a succession of coupled statics problems. In this way, we can determine the size of beam necessary to support a certain weight at the end with a certain tolerated amount of bending (Fig. 11–18) and the amount of bending that results.

Finite-element methods, in which a continuous system is approximated by numerous small elements, have uses in many areas of the physical sciences. Among numerous examples are supercomputer studies of the turbulent behavior of fluids, thermonuclear fusion research, and the quantum mechanics of atomic and molecular systems.

SUMMARY

An extended object is static when the net external force and the net external torque about any convenient origin are both zero:

$$\sum_i \mathbf{F}_i = 0 \qquad\qquad (11\text{–}3)$$

and

$$\sum_i (\mathbf{R}_i \times \mathbf{F}_i) = 0. \qquad\qquad (11\text{–}4)$$

In order to apply these conditions, it is necessary to know both what the external forces acting on an object are and *where* they act. Gravity acts as though it were applied to the center of mass of a rigid body. We can say that the weight of an object acts on its center of mass.

The static conditions are most often used to learn under what conditions a structure can be held static or to learn what forces various components of a structure must endure. Some structures are underdetermined, which means that the static conditions do not by themselves determine the forces acting.

UNDERSTANDING KEY CONCEPTS

1. It is a common (and quite useful) piece of advice for those who are learning mountaineering techniques to "stand away" from the mountain on steep slopes rather than to follow one's instincts and "hug" the mountainside. Explain why this is so, using your knowledge of torques and forces on a climber standing on the steep slope of a mountain.

2. A baby pulls straight down with all his might on the flush handles of the closed drawer of a bureau (Fig. 11–19). Can he cause the bureau to tip over?

FIGURE 11–19 Question 2.

3. Why does a rope from which an object is suspended line up with the vertical?

4. Give at least three examples for which an object is not in equilibrium even though the net force on the object is zero.

5. You are sitting quietly in a porch rocker that is suspended by chains from the ceiling. Are you in a stable or unstable equilibrium?

6. Bridges and buildings are not really rigid. Does this mean that nothing we have said in this chapter is relevant to them? How can an object be "approximately" rigid?

7. You have a flat map of the continental United States cut out of plywood. Devise two separate methods for determining the map's center of mass, one of them using only one finger.

8. Is the height of the four-legged table discussed in Section 11–3 irrelevant to the calculation of the equilibrium conditions for the table?

9. A pendulum suspended from the roof of an accelerating rail car makes a nonzero angle with the vertical. The pendulum is not swinging. Is this a case of stable equilibrium?

10. For some objects (U-shaped objects, for example), the center of mass is outside the object itself. For such objects, can we still think of gravity as though it acts on the center of mass?

11. The doughnut-shaped space station in Stanley Kubrick's film *2001, A Space Odyssey* rotates about its axis of symmetry (see Fig. 5–25a). Is the rotating station in stable equilibrium, unstable equilibrium, or neither?

12. Does the method of finding the center of mass of a flat object, as discussed in this chapter, work even if the density of the object is not constant?

13. You want to hold a beam with its end against a frictionless wall. Is it possible to do so by running a rope from the far end of the beam to any attachment point on the wall?

14. Will a ladder placed against a rough vertical wall remain standing when the floor is so smooth that there is no friction between it and the feet of the ladder?

15. Identical twins are placed at opposite ends of a seesaw pivoted about its midpoint. No forces other than those due to the twins, the pivot, and gravity act on the seesaw. What determines the inclination to the horizontal made by the seesaw when it is balanced? Is the equilibrium stable, unstable, or neutral?

16. Suppose that the seesaw of Question 15, still pivoted about its midpoint, has a sharp downward bend at the midpoint. If there is an equilibrium, will it be stable, unstable, or neutral?

11–2 Gravity and Rigid Bodies

1. (I) A uniform board of mass 80 kg and length 3.6 m is placed on top of a pivot 1.2 m from one end. What mass must be put at that end to allow the board to balance?

2. (I) A 30-kg board 4 m long is supported in a horizontal position at the two ends. A 70-kg worker stands 1.5 m from one end. What forces are exerted by the board on the two support points?

3. (I) Two workmen each carry one end of a 2.2-m-long ladder of mass 24 kg. The ladder is tapered so that its center of mass is 0.9 m from the wider end. What are the forces exerted by the ladder on the two workmen?

4. (I) A rail of length 2.0 m and mass 5.0 kg runs horizontally between two scales; a bowling ball of mass 5.5 kg is allowed to roll at a steady speed of 0.2 m/s from the left scale to the right scale (Fig. 11–20). During the 10 s the ball is moving, how do the readings on the two scales change?

5. (I) A projectile has broken into two parts. At time t_0, the two parts are located at the following points: m_1 at $(x, y, z) = (3, 0, 0)$ and m_2 at $(0, 0, 3)$, where all distances are measured in meters. The mass m_1 is twice the mass m_2. What is the location at time t_0 of the point that follows a parabolic trajectory (assuming that there is such a point)?

FIGURE 11–21 Problem 8.

9. (II) In order to hang a load of mass $M_1 = 30$ kg from the horizontal, flat roof of a building, a plank of length $L = 2.4$ m is placed on the roof (Fig. 11–22). One end is held in place with a chunk of concrete of mass $M_2 = 15$ kg, and the other supports the load M_1 with a light rope. How far can the end of the plank reach without tipping over? Neglect the mass of the plank.

FIGURE 11–20 Problem 4.

FIGURE 11–22 Problem 9.

6. (II) Two people of unequal strength must carry a uniform beam of length L while holding it horizontal. The weaker of the two holds the beam at one end. (a) How far from the other end must the stronger person hold the beam in order to support three-quarters of the weight? (b) Is there a way in which the stronger person can carry the beam at one end and still support more than half the weight of the beam?

7. (II) A playground seesaw is balanced at its midpoint. Two children, weighing 25 kg and 40 kg, respectively, want to balance on the seesaw. If the children are separated by a distance of 2.8 m, how far from the pivot point will the lighter child sit?

8. (II) Consider a seesaw whose total mass is 20 kg and total length is 4.0 m. Suppose the seesaw is placed off center on the pivot point so that the pivot point is 30 cm from the center of the seesaw (Fig. 11–21). How far from the center will the children of Problem 7 have to sit if the lighter child sits on the longer part of the seesaw? (Their separation is still 2.8 m.)

10. (II) A uniform beam of mass 80 kg and length 3 m rests on two pivots, one at the left edge and one 2.4 m from the left edge. How far to the right of the right pivot can a mass of 150 kg be placed without the beam tipping?

11. (II) A rectangular piece of plywood (60 cm × 120 cm) lies in the horizontal xy-plane. The surface mass density of the plywood is 3 kg/m². Calculate the torque about one of the corners due to the force of gravity.

12. (II) A book of mass 1 kg is placed such that 2/3 of the book is hanging over the edge of a table. A paperweight of mass m is placed centrally on top of the 1/3 of the book that is on the table. What is the minimum mass of the paperweight such that the book doesn't fall off the table?

13. (II) Two books are stacked at the edge of the table, with their lengths perpendicular to the table edge. If the width of each book is L, how far out from the table edge can the top book's extreme edge be placed without the books falling down? How does this generalize to three books?

14. (I) A door 90 cm by 195 cm of mass 14 kg hangs on two hinges: One is attached to the bottom of one side of the door, and the other to the top of the same side. What are the horizontal forces exerted on the door by the hinges?

15. (I) A football player is at the top of a pushup. The angle that the (rigid) torso makes with the floor is 25°. His arms are perpendicular to his torso and his center of mass is located at a point 3/8 of the distance from the shoulders to the feet. Assuming that the mass of his head can be neglected, what is the force, in terms of the player's weight, along his arms?

16. (I) A student wants to place a flower pot on a board that juts out from a window so it will get more sunlight (Fig. 11–23). The flower pot has a mass of 2.0 kg and needs to be 1.0 m from the window sill. The student can only place a nail into the sill 4 cm from the edge. Neglect the mass of the board and find out how much force the nail must exert to hold the board in place.

FIGURE 11–23 Problem 16.

17. (I) A uniform rod, mass 12 kg and length 1.5 m, rests on two points, one at its left end and one at the center point. What are the contact forces on the rod at these points? Comment on the stability of the situation.

18. (II) An 8.5-m extension ladder of mass 26 kg is propped up against a wall, touching at a point 8.0 m above the level ground. A man of mass 75 kg climbs 7 m up the ladder to repair a window. The ladder rests against a frictionless wall, but the ground has friction. Determine all the forces on the ladder.

19. (II) Consider a ladder of mass 10 kg and length 4 m, leaning against a vertical wall at an angle of 30° with the vertical. The

FIGURE 11–24 Problem 19.

coefficient of friction between the ladder and the floor is $\mu_s = 0.4$, and there is no friction between the wall and the ladder. A man of mass 80 kg climbs up the ladder. (a) How high can he climb before the ladder begins to slip? (b) Work out your calculation by taking the torques about the three points A, B, and O in Fig. 11–24, and show that the resulting equations are independent of the choice of point.

20. (II) Consider the ladder in Example 11–4 (see Fig. 11–12a). The 3.0-m-long ladder is placed against a frictionless wall, making the same 58° angle to a different horizontal surface; this time, much to his dismay, the same 85-kg window washer finds himself starting to slip when he steps to the second rung. What is the coefficient of static friction between the ladder and floor?

21. (II) Using a uniform strut, a rigid brace hinged at the floor, a person holds a 30-kg engine in equilibrium while it is being repaired. The strut has a mass of 12.5 kg. A smooth rope passes over a pulley at the end of the strut (Fig. 11–25). (a) What is the force exerted on the rope by the person? (Specify the direction of this force by calculating the angle θ that the rope makes with the horizontal.) (b) What forces are exerted by the strut?

FIGURE 11–25 Problem 21.

22. (II) A piece of plywood leans against a wooden wall with which it has a coefficient of static friction of $\mu_s = 0.3$. (a) If the coefficient of static friction between the board and the floor is also 0.3, what is the minimum angle that the board makes with the floor? (b) What happens if the coefficient of friction between the board and the floor is zero?

23. (II) A stepladder consisting of two ladders of mass M and length L is held together by a crossbar attached to the midpoints of the two ladders (Fig. 11–26). What force is exerted on the crossbar by each ladder if the length of the crossbar is $L/2$? Assume that any friction between the ladders and the floor is negligible.

FIGURE 11–26 Problem 23.

24. (II) A flagpole of mass 6 kg and length 2.4 m is hinged at a wall and supported in a horizontal position by a cable attached to the free end (Fig. 11–27). The cable makes an angle of 25° with the horizontal. What is the tension in the cable? What is the vector force exerted on the hinge at the wall?

FIGURE 11–27 Problem 24.

25. (II) A desk of height 0.82 m, length 1.54 m, and mass 43 kg is pushed across a horizontal floor at a steady speed with a horizontal force **F** applied at the top (Fig. 11–28). The coefficient of kinetic friction between the legs and the floor is $\mu_k = 0.45$. What is the friction force at each leg, and what is **F**? Assume that the two right legs each support the same forces, as do the left legs.

FIGURE 11–28 Problem 25.

26. (II) Repeat Problem 25, but this time assume that the desk is being pushed down a slope of 5.5°. The force applied is parallel to the sloping floor. The center of mass of the desk is in the middle of its long dimension and 0.28 m down from its top surface.

27. (II) Figure 11–29 is a side view of a seat used for babies. It enables a baby to sit at the edge of a table by means of four points of contact; in the side view shown, two of these points are visible—point A at the top of the table, and point B beneath the table. The other two points are aligned directly behind. The center of mass of the baby plus the seat can be approximated by a mass m at point C. (a) Taking into account only forces in the xy-plane, the plane of the figure, calculate the forces *on the table* at points A and B. (Remember that there are four contact points, not two. Assume that the symmetric legs share the force equally.) (b) What happens for $\ell_2 \rightarrow 0$? for $\ell_1 \rightarrow 0$? (c) Work out numerical values for the forces for $m = 10$ kg, $\ell_1 = 20$ cm, and $\ell_2 = 30$ cm.

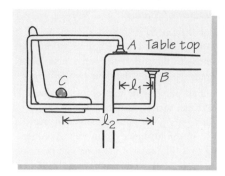

FIGURE 11–29 Problem 27.

28. (II) Consider the baby seat of Problem 27. The table has a total mass $M = 18$ kg, all concentrated in the (uniform) top, and the center of the table is 70 cm from the vertical extension of point P. For the geometry of the baby seat and the table top shown in Fig. 11–30, what mass m of the baby plus the seat will cause the table to tip over by a rotation at point P? Can the possible movements of a baby seriously destabilize the situation?

FIGURE 11–30 Problem 28.

29. (II) A trap door 1.8 m square with mass 20 kg is hinged at one edge and is attached to a rope at the opposite edge (Fig. 11–31). The trap door makes an angle of 55° with the horizontal, and the rope is perpendicular to the door. (a) What is the tension in the rope? (b) What is the force vector acting on the door at the hinge?

FIGURE 11–31 Problem 29.

30. (II) A beam of mass 10 kg is hinged at the top, and its bottom end is pulled to the side by a horizontal rope. The beam makes an angle of 30° with the vertical. What is the tension in the rope, and what is the force exerted on the beam at the suspension point?

31. (II) A variant of the crane in Example 11–3 is shown in Fig. 11–32. The pivot at point B is frictionless, and the beam, of length 3 m, has mass 100 kg. The rope makes an angle of 30° with the horizontal and can withstand a tension of 10,000 N before breaking. If this arrangement is used to lift masses from the point shown on the beam, what is the maximum mass that can be so raised?

FIGURE 11–32 Problem 31.

32. (II) A uniform door is attached to the wall by two hinges. Draw a force diagram for the door and describe the various forces that can act. Describe the conditions for static equilibrium, and find the forces that are solvable. Is this an underdetermined system? If so, which forces are underdetermined?

33. (II) Consider the four-legged table discussed in Section 11–4, with leg 1 slightly shorter than the others. Suppose that a teacup were placed on the diagonal between legs 1 and 4, at location

$$(x, y) = \left(\frac{L}{2} + x, \frac{L}{2} + x \right).$$

(a) Find a satisfactory solution to the magnitude of the contact forces on the legs if $x > 0$. Neglect the masses of the legs. (b) Repeat part (a) for $x < 0$.

34. (II) A large, spherical satellite is held in place in the bay of a space station with six ropes: four are attached and equally spaced around the equator and the other two are attached at the poles. (a) What equations describe the situation in which the satellite is held motionless with respect to the station? (b) Is this an underdetermined system? (c) Is the system underdetermined when all the ropes are under identical tension?

35. (III) A seaside tower is supported by a cable (Fig. 11–33). A horizontal wind that increases with height often blows from the left, exerting a horizontal force to the right on the tower.

FIGURE 11–33 Problem 35.

The horizontal force on a unit length of the tower increases with the height h according to force/unit length = αh, where h is the height from ground level, and, if h is measured in meters, $\alpha = 50 \text{ N/m}^2$. The total height of the tower is 20 m. What is the tension in the cable?

36. (III) The mythical Greek king Sisyphus, pushing a large, round rock up a mountain, wishes to take a rest and supports the rock with a horizontal rope attached to the top of the rock (Fig. 11–34). The coefficient of static friction between the rock and the slope is $\mu_s = 0.6$. (a) What is the largest value of θ for which this method of support is possible? (b) If the mass of the rock is 288 kg, what tension must the rope be able to support on this maximum slope? (c) How does the tension vary as a function of θ for angles less than the maximum angle? [*Hint*: It is easiest to take moments about the point of contact.]

FIGURE 11–34 Problem 36.

General Problems

37. (II) A 30-cm-wide shelf is supported at the wall by an L-shaped bracket and a cable at each end of the shelf placed at 45° to the wall (Fig. 11–35). A 20-kg sack of potatoes is placed and centered on the shelf. What is the tension of the cable if we ignore the mass of the shelf?

FIGURE 11–35 Problem 37.

38. (II) The cross section of an A-frame house is shown in Fig. 11–36. The total height of the apex is 5.0 m, and the 1.5-m-long crossbeam is two-thirds of the way up the roof line. The crossbeam must support two roof beams, each of mass 3000 kg. (a) Does the crossbeam push the roof beams out or does it pull them in? (b) What is the force exerted by the roof beams on the crossbeam?

FIGURE 11–36 Problem 38.

39. (II) A uniform beam of length L and mass M is freely pivoted at one end about an attachment point in a wall. The other end is supported by a horizontal cable also attached to the wall, so that the beam makes an angle θ_0 with the horizontal (Fig. 11–37). (a) What is the tension in the cable? (b) The cable snaps. What is the angular acceleration of the beam about its pivot point immediately afterward? (c) What is the angular velocity of the beam as it falls through the horizontal position?

Length L, mass M

θ_0

FIGURE 11–37 Problem 39.

40. (II) A uniform board of length 2.4 m and weight 47 N has one end on the ground. With the aid of a horizontal force applied at the upper end by means of an attached horizontal rope, the

2.4 m

47 N

θ

$\mu_s = 0.32$

FIGURE 11–38 Problem 40.

board is held at an angle θ with respect to the vertical (Fig. 11–38). The coefficient of static friction between the end of the board and the ground is $\mu_s = 0.32$. What is the range of angles the board can make with the vertical and still be in static equilibrium? How does the tension in the rope vary with the angle within the angle's possible range?

41. (II) A sign is to be constructed from a piece of plywood in the shape of a 30° right triangle of mass 15 kg. It is to be attached to a wall, as shown in Fig. 11–39. The lower attachment point is a frictionless pivot, and the upper point is a rope that can be reeled in or out to make the bottom of the sign and the rope itself horizontal. What is the tension in the rope?

TRIANGLE
SIGN COMPANY
30°

FIGURE 11–39 Problem 41.

42. (II) A car is lifted vertically by a jack placed at the car's rear end 40 cm off the central axis, so that the weight of the car is supported by the jack and the two front wheels (Fig. 11–40). The distance between the front wheels is 1.60 m, the distance from the axis connecting the two wheels to the center of mass of the car is 80 cm, and the distance from the rear of the car to the center of mass is 2.10 m. What fraction of the car's weight is carried by each of the wheels, and what fraction is carried by the jack? (Note that the weight on the wheels will *not* be symmetrically distributed.)

FIGURE 11–40 Problem 42.

43. (II) A roller of radius 30 cm and mass 80 kg is pulled by a force F applied horizontally to the axle. How large must F be in order for the roller to climb a step 15 cm high?

44. (II) Consider the roller of Problem 43. Assume that the force F can be applied in any direction. What direction minimizes the magnitude of F necessary to make the roller climb the step?

45. (II) A chest of drawers 58 cm wide, 1.6 m long, and 1 m high is pulled by a horizontal rope attached at a height of 58 cm to the midpoint of the long side. The force is such that the chest moves with uniform velocity. (a) Express the force on the legs on the side with the rope, and the force on the legs on the opposite side, in terms of the coefficient of kinetic friction μ_k. (b) For what value of μ_k will the chest topple over?

46. (II) A box 0.8 m long and 1.2 m tall is placed on a flat-bed truck. The truck accelerates at a rate of 2 m/s². Will the box topple over?

47. (II) A cylinder of mass M and radius R rests on an inclined plane (Fig. 11–41). It is held in place by a horizontal string that is attached to the edge of the cylinder. If the angle that the plane makes with the horizontal is θ, and the coefficient of static friction between the cylinder and the plane is μ_s, what is the smallest value of μ_s that will maintain this position as an equilibrium position?

FIGURE 11–41 Problem 47.

48. (II) A centrifugal governor consists of light rods pivoted at points A, B and C, as shown in Fig. 11–42. They are loaded with masses M, m, and m respectively, and the whole apparatus rotates about the vertical axis. Depending on the angular velocity, the mass M slides up or down, and its position can be used to control the flow of steam in a steam engine. Derive a relationship between the angular velocity and the angle the arms make with the vertical.

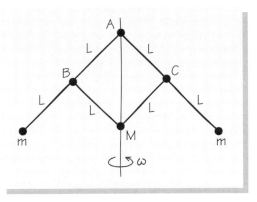

FIGURE 11–42 Problem 48.

49. (III) Two steel ball bearings of mass m and radius 2 cm are placed in a vertical tube of diameter 6 cm that is closed on the bottom. Find all the forces on the two balls.

50. (III) Three steel ball bearings of mass m and radius 2 cm are placed in a closed vertical tube of diameter 6 cm that is closed on the bottom. Find the forces on the middle ball bearing.

This montage of photos from the Voyager I satellite shows four of Jupiter's largest moons. The force that is largely responsible for their motions—gravitation—is the same force that governs the motion of a falling apple.

Gravitation

Most of the forces that we see every day are disguised forms of the more basic forces that govern the interactions of matter. Examples include friction, tension, and contact forces, which are all ultimately explained by the forces holding the components of matter together. There is, however, one important exception, and that is gravity. Gravity is simply a special limit of the force of universal gravitation; as far as we know today, gravitation is one of the truly fundamental forces. Gravitation governs not only the motion of a falling apple but also the majestic orbital motions of the Moon around Earth, the planets around the Sun, and the stars in their voyage around the galaxy. In 1687, Isaac Newton published the realization in his *Principia* that the same force governs both the falling apple and the orbiting Moon; this was one of the great intellectual leaps in human history.

Happily, the law of gravitation is a force law for which the equations of motion can be solved. We can thus find a full description of the motion of astronomical bodies. Like the other fundamental forces, gravitation is conservative and lends itself to a description in terms of potential energy. As we will see in this chapter, the variety of motions possible when gravitation acts presents a rich and beautiful design.

12–1 FUNDAMENTAL FORCES

The forces that act on an object determine its motion. We can distinguish between fundamental (or basic) forces and derived forces. Fundamental forces are those for which we cannot find (or have not yet found) an underlying force from which they are derived. Derived forces result from the operation of some underlying fundamental force.

Gravitation is a fundamental force.

For example, friction and the spring force are both derived forces. In the final analysis, these forces result from the forces between molecules, and *these forces turn out to be electric forces*, which *are* fundamental. The fundamental forces are all conservative, and that is why energy conservation is such an important principle.

In the realm of classical physics and in most parts of atomic physics, there are only two domains in which the fundamental forces appear undisguised. For both the gravitational force between masses and the electromagnetic forces between charges, *the force that is observed is the basic force*.

The development of the theory of gravitation is one of the important achievements of physics, both in its Newtonian form and in its later elaboration by Einstein. It is the Newtonian form of the law, combined with advances in computing technology, that allows us to plan interplanetary journeys for probes that can pass the planets Jupiter, Uranus, and Neptune on one long trajectory—even with our present chemical sources of fuel (Fig. 12–1).

12-2 EARLY OBSERVATIONS OF PLANETARY MOTION

The earliest astronomical observations of the night sky led ancient peoples to divide the points of light they saw into two classes: the so-called fixed stars, which move in seemingly perfect circles around Earth, and the planets (in Greek, the "wanderers"), which appear to move in complicated, erratic patterns in the night sky. The first interpretations of these observations placed Earth at the center of the universe, in a **geocentric** frame. Earth was pictured as being surrounded by a rotating spherical shell in which the fixed stars were embedded. The planets, the Sun, and the Moon seemed to have more complicated motions—motions explained by placing the planets upon moving, transparent shells within the outer, fixed-star shell. These shells had to move in complicated ways if the observations were to be consistent with this frame. In the second century A.D., Claudius Ptolemy made the most detailed formulation of the notions just outlined. To explain the various paths taken by planets, he had to depart from the simple spherical shell description and construct the paths of motion to form circles that themselves lay on circles around Earth (Fig. 12–2). His theory was limited by a culturally imposed belief that perfect, circular motion describes celestial motion. His *epicy-*

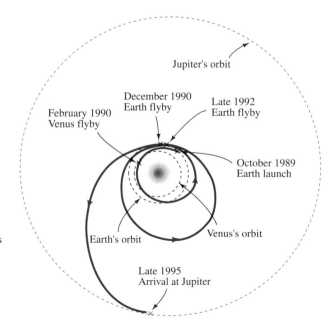

FIGURE 12–1 An understanding of the inverse-square force law permits us to calculate numerically rather complicated satellite trajectories. The satellite *Galileo* was launched in 1989 by NASA to study Jupiter and its atmosphere.

cles (the circles on circles used to describe the movements of the planets) gave a reasonable description of the motion of the planets, and the idea proved serviceable for over 1400 years.

The Copernican Picture

In 1543, Nicolaus Copernicus introduced the **heliocentric** frame, with the Sun at the center of the solar system. Unfortunately, Copernicus continued to insist on describing all motions with circles and, because the true motions of the planets about the Sun are not circles, epicycles continued to be needed in the Copernican description to accommodate the observations of planetary motion (Fig. 12–3). The apparent immobility of the stars was ascribed to their great distance. Partly because the epicycles meant that the Copernican description was not really more economical than the Ptolemaic description, the iconoclastic Copernican hypothesis was not accepted for almost a century. During this transitional period, the construction of more refined instruments—there were still no telescopes—allowed Tycho Brahe, toward the end of the sixteenth century, to improve the knowledge of planetary **orbits**, or paths, to an accuracy of less than half a minute of arc.

Kepler's Laws

At Brahe's death in 1601, his assistant Johannes Kepler inherited the data that Brahe had accumulated. Kepler spent some 20 years analyzing these data, looking for mathematical regularities. He came to the conclusion that the idea of circular orbits should be discarded and replaced with elliptical orbits. This was one of the crucial breakthroughs in the data analysis. **Kepler's laws** summarize Kepler's most important conclusions:

1. Planets move in planar elliptical paths with the Sun at one focus of the ellipse (Fig. 12–4a).

2. During equal time intervals, the radius vector from the Sun to a planet sweeps out equal areas (Fig. 12–4b).

3. If T is the time that it takes for a planet to make one full revolution around the Sun, and if R is half the major axis of the ellipse (R reduces to the radius of the planet's orbit if that orbit is circular), then

$$\frac{T^2}{R^3} = C, \tag{12–1}$$

where C is a constant *whose value is the same for all planets.*
Kepler's laws were so simple that it was no longer possible for scientists to cling to the pre-Copernican ideas (Fig. 12–5).

We have already found the origin of Kepler's second law in Chapter 10: Eq. (10–31). The law follows from the conservation of angular momentum and it is a consequence of the fact that *the gravitational force between the Sun and each planet is central*; that is, the force acts along the line between the Sun and the planet. In fact, Kepler's second law can be taken as evidence that the gravitational force is central. Conservation of angular momentum also implies that the trajectories of planets must lie in a plane—the plane that is perpendicular to the direction of the fixed angular momentum vector.

12–3 NEWTON'S INVERSE-SQUARE LAW

Newton understood the importance of Kepler's laws. Because the planets do not move in straight lines, Newton realized that they must be subject to a net force. He concluded from Kepler's second law that the net force on a planet must point from the planet to

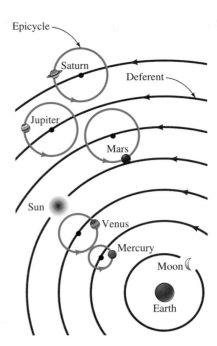

FIGURE 12–2 Diagram of planetary motion based on the Ptolemaic system. According to Ptolemy, Earth is the center of the world system. Planets move in small circular paths called epicycles, and the centers of the epicycles move around Earth on large circles called deferents. (After D. Scott Birney, *Modern Astronomy,* Allyn and Bacon, 1969.)

Kepler's laws of planetary motion

Kepler's second law (the equal-area law) follows from the conservation of angular momentum.

FIGURE 12–3 The Ptolemaic universe has a difficult time explaining the retrograde motion of Mars shown here. The Sun and Moon move west to east relative to fixed stars from night to night, but planets do a retrograde "loop" from east to west every once in a while. Only the Keplerian description explains these observations.

Kepler's first law (elliptical orbits) follows from a central inverse-square force law.

(a)

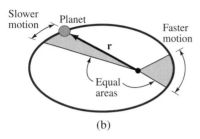

(b)

FIGURE 12–4 (a) Kepler's first law specifies that the planets move in elliptical orbits with the Sun at one focus. The perihelion is the distance of closest approach and the aphelion is the farthest distance between a planet and the Sun. (b) Kepler's second law specifies that the radius vector from the Sun to a planet sweeps out equal areas in equal times. Thus the two shaded areas shown are equal if the same time is taken to sweep out each of them. This means that the planet must move more quickly over a part of the orbit closer to the Sun than over a more distant part.

FIGURE 12–5 Galileo's early improvements to the telescope allowed him to make critical observations— including observing the moons of Jupiter and the phases of Venus—that helped destroy the pre-Copernican point of view. Unfortunately, the Church treated Galileo harshly for propagating his findings.

322

the Sun. He also found that the elliptical paths described by Kepler are the consequence of an *inverse-square force law:* If the source of the force is at the origin O and the mass on which the force acts is at position **r**, then the force has magnitude k/r^2, where k is a constant.

More generally, Newton showed that the paths of the mass under the influence of this force had to be conic sections. **Conic sections** are the shapes obtained if a cone is sliced by a plane. As can be seen in Fig. 12–6a, if a cone is sliced by a plane parallel to the cone's base, the conic section is a **circle**. If the plane is tilted a little, the conic section is an **ellipse**. If the slice is cut parallel to the slope of the cone, the conic section turns out to be a **parabola**. If the slice is steeper than the slope of the cone, the conic section is a **hyperbola**. The planets from which Kepler discovered his first law move in closed elliptical orbits, but there are astronomical bodies that move along trajectories described by all these mathematical curves (Fig. 12–6b).

By postulating a central inverse-square force law, Newton was able to explain not only Kepler's first two laws, but Kepler's third law as well. To deal with Kepler's third law, let's consider the very special case of a circular orbit of radius R. The force has magnitude[†] k/R^2. We know from Chapter 3 that the acceleration for an object moving with speed v in a circle of radius R is v^2/R. We can now apply Newton's law of motion $F = ma$ for the case of a planet of mass m, so that F is replaced by k/R^2 and ma is replaced by mv^2/R:

$$\frac{k}{R^2} = \frac{mv^2}{R}. \tag{12–2}$$

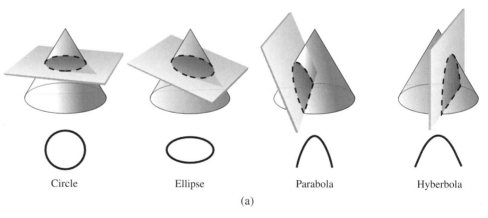

Circle   Ellipse   Parabola   Hyberbola

(a)

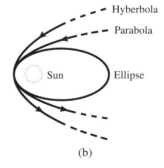

(b)

FIGURE 12–6 (a) Slicing a cone with a plane produces conic sections. Several familiar geometric curves are shown here. (b) The orbits of astronomical objects such as planets can follow all the curves drawn in (a).

[†]We are assuming that the Sun is (nearly) fixed, an assumption justified by the Sun's large mass.

In terms of the *period*, T, of the planet's orbital rotation, we have $v = $ circumference$/T$, so that $T = 2\pi R/v$. Squaring, we have

$$T^2 = \frac{4\pi^2 R^2}{v^2}.$$

If we now substitute $v^2 = kR/mR^2 = k/mR$ from Eq. (12–2), we find

$$T^2 = \frac{4\pi^2 R^2}{k/mR} = \left(\frac{4\pi^2 m}{k}\right)R^3. \tag{12–3}$$

This is just the statement of Kepler's third law as it applies to a circle, with the additional condition that the constant $C = (4\pi^2 m/k)$ in Eq. (12–1) should be the same for all planets—as it will be if k is proportional to the mass of the planet.

Kepler's third law follows from a central inverse-square force law.

The Law of Universal Gravitation

The constant k just described involves the force between two objects; by Newton's third law, there is a symmetry between these objects as they appear in the force law. If k is proportional to one of the masses, it must also be proportional to the other one. Then k must have the form GmM, where m is the mass of one object, M is the mass of the other, and G is a proportionality constant called the **gravitational constant**. Newton put all this together into what is now known as **Newton's law of universal gravitation**: The force of gravitation acting on a point mass of mass m due to another point mass of mass M is an attractive force with an inverse-square form,

Newton's law of universal gravitation

$$\mathbf{F} = -\left(\frac{GmM}{r^2}\right)\hat{\mathbf{r}}. \tag{12–4}$$

Here $\hat{\mathbf{r}}$ is the unit vector in the direction from the mass M to the mass m, and r is the distance separating them. The minus sign indicates that the force is attractive. Of course, Newton's third law states that if Eq. (12–4) is the force exerted on m by M, then its negative is the force exerted on M by m. Mass M is attracted to m with the same magnitude of force that m is attracted to M.

We can now return briefly to Kepler's third law. According to Newton, the constant C in Kepler's third law is given by

$$C = \frac{4\pi^2 m}{k} = \frac{4\pi^2 m}{GmM} = \frac{4\pi^2}{GM}. \tag{12–5}$$

Kepler's third law, Eq. (12–3), thus takes the form

$$T^2 = \frac{4\pi^2 R^3}{GM}. \tag{12–6}$$

The Gravitational Constant. The constant G must be determined from experimental data. It characterizes the strength of the gravitational force. Its dimensions are $[L^3 M^{-1} T^{-2}]$, and its units in SI are $N \cdot m^2/kg^2$.

The constant G in Eq. (12–4) cannot be measured independently unless the two masses involved are known; for this reason, G cannot be measured for astronomical objects such as Earth, the Moon, or the Sun. The value of G was first determined by Henry Cavendish in 1798. The basic idea of his experiment is shown in Fig. 12–7. Two masses m at the end of a rod of negligible mass and length $2L$ are suspended by a quartz fiber. Large masses M are placed near the masses m, with their centers of mass separated by a distance d. The magnitude of the torque about the point of suspension due to the force of gravitation on the masses is counterbalanced by a second torque due to the resistance of the fiber to twisting and proportional to the angle of deflection

The gravitational constant

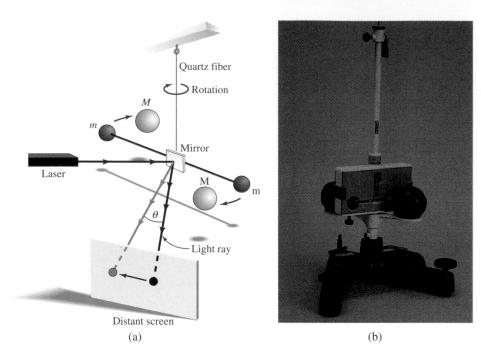

FIGURE 12–7 (a) Schematic diagram of Cavendish's experiment. The attraction of the smaller masses to the larger ones twists the fiber holding the rod. A light shining on a mirror on the rod indicates the small rotation by reflection on a distant screen. Switching the positions of the large masses reverses the rotation. (b) A Cavendish experiment apparatus.

θ. When this resistance is known, the torque due to the gravitational force, and hence G, can be determined. The most recent measurements lead to the value

$$G = 6.673 \times 10^{-11} \text{ N} \cdot \text{m}^2/\text{kg}^2. \tag{12–7}$$

with an uncertainty of about 0.06 percent. The Cavendish experiment is not a simple one. The magnitude of the gravitational force between two 10-kg masses separated by 10 cm is only about 7×10^{-7} N.

The Potential Energy Associated with the Gravitational Force

∞ Potential energy is discussed in Chapter 7.

The gravitational force is central and depends only on the distance of the influenced object from the force center. It is therefore *conservative* and can be derived from a potential energy function. We show here that the potential energy of a system of two point masses interacting with each other through the gravitational force (Fig. 12–8) is

The potential energy of a system of two masses that interact with each other through the gravitational force

$$U(r) = -\frac{GmM}{r}. \tag{12–8}$$

We have chosen the potential energy to be zero at infinity. To demonstrate this result, we recall the definition of potential energy from Chapter 7, Eq. (7–9):

$$U(r) - U(\infty) = -\int_{\infty}^{r} \mathbf{F}(\mathbf{r}') \cdot d\mathbf{r}',$$

where we have chosen the point x_0 in Eq. (7–9) at ∞. The force points from the location of mass m to the origin (the location of M), and we choose the path to go directly along a radial direction so that $\hat{\mathbf{r}} \cdot d\mathbf{r}' = dr'$. (Any path gives the same result: The force is conservative.) We thus obtain

$$U(r) - U(\infty) = -\int_{\infty}^{r} \frac{-GmM}{r'^2} \, dr' = -\frac{GmM}{r'} \bigg|_{\infty}^{r} = -\frac{GmM}{r}.$$

By choosing $U(\infty) = 0$, Eq. (12–8) follows.

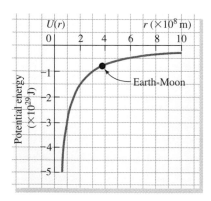

FIGURE 12–8 The potential energy of the gravitational force is negative and has a $1/r$ dependence. We define it to be zero at $r = \infty$.

The Superposition Principle

The gravitational force obeys the **principle of superposition**: When several particles of varying masses exert gravitational forces on a point mass m, the net force on m is the (vectorial) sum of the individual forces. The same idea holds for potential energies.

The potential energy of a point mass m at position $\mathbf{r}$ due to a point mass M_1 at position $\mathbf{r}_1$ is $U_1 = -GmM_1/|\mathbf{r} - \mathbf{r}_1|$, and the potential energy of m due to a point mass M_2 at position $\mathbf{r}_2$ is $U_2 = -GmM_2/|\mathbf{r} - \mathbf{r}_2|$. Then, if both M_1 and M_2 are present in their original positions, the potential energy U of m in the presence of the two masses is the *sum* of the potential energies U_1 and U_2. If the source of the gravitational force or potential energy for a point mass is an extended continuous object, we can apply the principle of superposition through integration.

∞ The superposition principle applied to gravitation. See also Section 4–1.

PROBLEM SOLVING

Use superposition for a continuous object.

EXAMPLE 12–1 A satellite is to be sent to the position between the Moon and Earth where there is no net gravitational force on an object due to those two bodies. Locate that point.

Solution: Figure 12–9 shows us that the force on the satellite due to the Moon, $\mathbf{F}_M$, is to the right, while the corresponding force due to Earth, $\mathbf{F}_E$, is to the left. Let's write the Earth–Moon distance as d, while the Earth–satellite distance is x, and the Moon–satellite distance is $(d - x)$. The net force on the satellite then has magnitude

$$F_{net} = F_M - F_E = Gm\left[\frac{M_E}{(x)^2} - \frac{M_M}{(d - x)^2}\right].$$

The condition that this equal zero determines x:

$$\left[\frac{M_E}{(x)^2} - \frac{M_M}{(d - x)^2}\right] = 0,$$

or

$$(d - x)^2 = x^2 R,$$

where we have defined the ratio $R = M_M/M_E = 0.0123$ (see Appendix III–1.1). This quadratic equation for x has the following solution with x between 0 and d:

$$x = \frac{1 - \sqrt{R}}{1 - R}\,d = \frac{1 - \sqrt{0.0123}}{1 - 0.0123}\,d = 0.900\ d.$$

The desired position is thus 9/10ths of the way to the Moon. The result is independent of the mass of the satellite.

Measuring Earth's Mass

Local gravity is just a manifestation of the general law of gravitation. Later, we will discuss the fact that the gravitational force due to a spherically symmetric object—Earth is a good approximation of one—acts as if the mass of the object were con-

FIGURE 12–9 Example 12–1. This view of the Moon from some 17,000 km was taken by the Apollo 11 astronauts on their way home. The whole Earth view was taken by the Apollo 17 astronauts on their way to the Moon. The satellite feels gravitational forces from both the Moon and Earth.

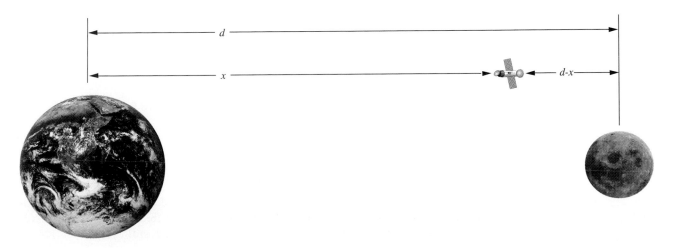

centrated at its center. Given this statement, we can determine Earth's mass, M_E, from the size of the acceleration due to gravity, g, and Earth's radius, R. The force exerted on an object of mass m on Earth's surface by Earth's gravitational attraction is given by

$$F = \frac{GmM_E}{R^2} = mg. \tag{12–9}$$

We thus obtain

$$M_E = \frac{gR^2}{G} = \frac{(9.80 \text{ m/s}^2)(6.37 \times 10^6 \text{ m})^2}{6.673 \times 10^{-11} \text{ N} \cdot \text{m}^2/\text{kg}^2} = 5.96 \times 10^{24} \text{ kg,}$$

where we have used the measured values of g, R, and G. This mass determination shows that Earth has an average density of 5.5 g/cm^3.

In making this calculation, we are using what might be called the *central dogma* of physics: The laws of nature are the same everywhere, be it Cavendish's laboratory, the environs of Earth, the solar system, or extragalactic space.

12–4 PLANETS AND SATELLITES

The force of gravitation [Eq. (12–4)] determines the trajectories (or orbits) followed by astronomical bodies and other properties of their motion. Newton's second law for the orbit of a single object subject to the gravitational force law due to a single source can be solved to give an *analytic form for the orbit*. (This is certainly not true of *every* force law!)

Escape Speed

It is simplest to approach the orbit of a single object—we refer to such objects as satellites—from the point of view of energy. The energy is the sum of the kinetic and potential energies and is thus given by

$$E = K + U = \frac{1}{2}mv^2 - \frac{GmM}{r}. \tag{12–10}$$

The **escape speed**, v_{esc}, of a projectile launched from Earth's surface is the minimum speed with which the projectile must leave the surface in order to leave the vicinity of Earth forever; that is, travel an infinite distance from Earth. In other words, the escape speed is the speed at $r = R_E$ that gives zero speed at infinite r. The object will have zero kinetic energy at $r = \infty$, and the potential energy is defined to be zero at that point, so the total energy will be $E = 0$ at infinity. Because the total energy is conserved, we can set $E = 0$ everywhere; then, the condition that $E = 0$ at the surface is the condition that determines v_{esc}. Let the mass of the projectile be m and Earth's mass be M. At Earth's surface, $r = R_E$, and

$$E = \frac{1}{2}mv^2_{\text{esc}} - \frac{GMm}{R_E} = 0.$$

If we solve this equation for v_{esc}, the factor m cancels, and we find that

$$v_{\text{esc}} = \sqrt{\frac{2GM}{R_E}}. \tag{12–11}$$

With the known values of G, M, and R_E, we find that the escape speed from Earth's surface is $v_{\text{esc}} = 1.12 \times 10^4$ m/s = 11.2 km/s.

It is interesting that the escape speed is independent of the direction in which the object leaves the surface. (Of course, the object cannot pass through Earth itself!) Even if the object leaves in an initially horizontal direction, it will still escape.

Types of Orbits

We can classify orbits according to the sign of the energy in Eq. (12–10). If the energy is positive, then as r becomes infinite, $|v| > 0$. This corresponds to an orbit that never closes and it is, in fact, a hyperbolic orbit. When the energy of the object is exactly zero, we have the special case of a parabolic orbit. The object starts at infinity with a speed of zero, swings past the Sun in a parabolic orbit, and slows down as it moves away from the Sun, ending up with zero velocity when r becomes infinitely large. For negative energies, r cannot become too large because then v^2 would become negative, which is impossible. In that case, the orbit is limited and it is elliptical (Fig. 12–10a) or circular. Halley's comet, like other comets, follows a very elongated elliptical orbit (Fig. 12–10b). Circular orbits are the simplest of all, as Examples 2 through 4 illustrate.

EXAMPLE 12–2 A particle of mass m moves in a circular orbit of radius r under the influence of the gravitational force due to an object of mass $M \gg m$ (Fig. 12–11). Calculate the total energy of the particle as a function of r.

Solution: Equation (12–10) gives the total energy, but it is a function of both v and r. We want to eliminate the speed. This can be done by using Newton's second law, $\mathbf{F} = m\mathbf{a}$. For a circular orbit, the acceleration is centripetal and has the form $a = v^2/r$, directed toward the center. The force has the magnitude GmM/r^2 and is also directed toward the center. The second law therefore has the form

$$\frac{GmM}{r^2} = \frac{mv^2}{r},$$

or, equivalently,

$$v^2 = \frac{GM}{r}.$$

This equation can be used to eliminate the speed. The total energy is

$$E = \frac{1}{2}mv^2 - \frac{GmM}{r} = \frac{1}{2}m\frac{GM}{r} - \frac{GmM}{r} = -\frac{1}{2}\frac{GmM}{r}.$$

The total energy is just one-half the potential energy for a circular orbit. It is negative, as is appropriate for a closed orbit.

EXAMPLE 12–3 Calculate the mass of the Sun, assuming that Earth's orbit around the Sun is circular, with radius $r = 1.5 \times 10^8$ km.

Solution: Kepler's third law relates the distance of Earth from the Sun, the period of Earth's orbit, and the mass of the Sun. This law is expressed by Eq. (12–6), which may be rewritten as

$$M = \frac{4\pi^2 r^3}{GT^2}.$$

Given the value of G, the radius of the orbit, r, and the period, T, which is 365 days = (365 d)(24 h/d)(60 min/h)(60 s/min) = 3.15×10^7 s, we find that

$$M = \frac{4(3.14)^2(1.5 \times 10^{11} \text{ m})^3}{(6.67 \times 10^{-11} \text{ N} \cdot \text{m}^2/\text{kg}^2)(3.15 \times 10^7 \text{ s})^2} = 2.0 \times 10^{30} \text{ kg}.$$

This is a factor of 3×10^5 larger than the mass of Earth.

EXAMPLE 12–4 A satellite moves in a circular orbit around Earth, taking 90 min to complete 1 rev (Fig. 12–12). We are given the following information: The distance from the Moon to Earth is $d_{ME} = 3.84 \times 10^8$ m; the Moon's orbit is circular; the period of the Moon's rotation about Earth is $T_M = 27.32$ d; Earth's radius is

(a)

(b)

FIGURE 12–10 (a) For negative total energies, the value of r is limited and the orbit is elliptical (or circular). (b) Halley's comet has an elongated elliptical orbit with a period of 76 yr.

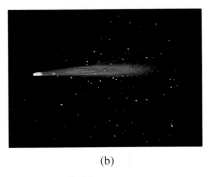

For Example 12-2: $F = \dfrac{mv^2}{r} = \dfrac{GmM}{r^2}$

For Example 12-3: $M = \dfrac{4\pi^2 r^3}{GT^2}$

For Example 12-4: $r^3 \propto T^2$

FIGURE 12–11 This simple sketch can be used to help with Examples 12–2, 3, and 4.

FIGURE 12–12 Example 12–4. A satellite in orbit about Earth.

$R_E = 6.37 \times 10^6$ m; and Earth's gravitational force acts as if all of Earth's mass were concentrated at its center. With this information, calculate the height of the satellite above Earth.

Solution: A satellite in circular orbit around Earth behaves like a planet in circular orbit around the Sun. We can thus use Kepler's third law in the simplifying case that the ellipse reduces to a circle: The cube of the radius of the orbit is proportional to the square of the orbital period. To make use of this result, the period of the satellite must be in the same units as that of the Moon. We thus have $T_S = 90$ min $= 1.5$ h $= (1.5$ h$)/(1$ d$/24$ h$) = 0.0625$ d. If we call the distance from the satellite to the center of Earth D, then Kepler's third law [Eq. (12–1)] as applied to the satellite is

$$\left(\frac{D}{d_{ME}}\right)^3 = \left(\frac{T_S}{T_M}\right)^2;$$

$$\left(\frac{D}{3.84 \times 10^8 \text{ m}}\right)^3 = \left(\frac{0.0625 \text{ d}}{27.32 \text{ d}}\right)^2;$$

that is,

$$D = (3.84 \times 10^8 \text{ m})(0.00229)^{2/3} = 6.67 \times 10^6 \text{ m}.$$

To find the height above the surface, we must subtract Earth's radius,

$$(6.67 - 6.37) \times 10^6 \text{ m} = 3 \times 10^5 \text{ m} = 300 \text{ km}.$$

Many satellites use this orbit, which is one that lies just above Earth's atmosphere.

(a)

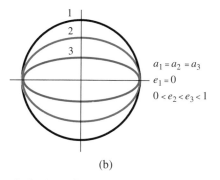

(b)

FIGURE 12–13 Two ellipses with (a) the same eccentricity and different semimajor axes; and (b) the same semimajor axes but different eccentricities.

Properties of Noncircular Orbits. The hyperbolic, parabolic, and elliptical orbits corresponding to different signs of total energy are closely related; as stated in Section 12–3, they are all *conic sections* (see Fig. 12–6). Let's focus our attention on the solar system, although we should keep in mind that our results will be applicable to, say, satellites orbiting Earth as well. The Sun, which is very heavy in comparison to the planets, plays a special role in the orbits. For closed orbits in particular, the Sun sits at one *focus* of the ellipse, a fact noted by Kepler in his first law and illustrated in Fig. 12–4a. The point where the planet makes its closest approach to the Sun ($r = r_{min}$) is called the **perihelion**; the point where the farthest distance between a planet and the Sun is attained ($r = r_{max}$) is the **aphelion**.[†] The **semimajor axis**, a, is a length that sets the overall size of the orbit (Fig. 12–13). From Fig. 12–4a, we have

$$\text{semimajor axis} = a = \frac{1}{2}(r_{min} + r_{max}). \quad (12\text{–}12)$$

The **eccentricity**, e, of the orbit is a dimensionless measurement of the elongation of the orbit and is proportional to the difference between r_{max} and r_{min}:

$$e = \frac{r_{max} - r_{min}}{2a}. \quad (12\text{–}13)$$

Note that when $r_{max} = r_{min}$, the orbit reduces to a circular one of radius a and the eccentricity is zero. The most extreme possible orbit corresponds to $r_{min} = 0$, when $r_{max} = 2a$, and

$$e_{extreme} = \frac{2a}{2a} = 1.$$

The eccentricity therefore varies between zero and one.

[†]For Earth satellites, the corresponding words are perigee and apogee.

EXAMPLE 12–5 Consider the motion of a comet in an elliptical orbit around a star. The eccentricity of the orbit is given by $e = 0.20$ and the distance between the perihelion and the aphelion is 1.0×10^8 km. (a) Find the distances of nearest and farthest approaches of the comet. (b) If the speed of the comet is 81 km/s at perihelion, what is its speed at aphelion?

Solution: (a) We are given e, and $2a$ is the sum of the aphelion and perihelion distances. Equations (12–12) and (12–13) are two equations that can be solved for r_{max} and r_{min} in terms of a and e. If we add Eq. (12–12) to a times Eq. (12–13), r_{min} cancels, and

$$r_{max} = a + ae = a(1 + e). \qquad (12\text{–}14)$$

In turn,

$$r_{min} = 2\left(a - \frac{r_{max}}{2}\right) = 2\left[a - \frac{a(1 + e)}{2}\right] = a(1 - e). \qquad (12\text{–}15)$$

Numerically, with $e = 0.20$ and $2a = 1.0 \times 10^8$ km,

$$r_{max} = \frac{(1.0 \times 10^8 \text{ km})(1.20)}{2} = 6.0 \times 10^7 \text{ km};$$

$$r_{min} = \frac{(1.0 \times 10^8 \text{ km})(0.80)}{2} = 4.0 \times 10^7 \text{ km}.$$

This would be an unusually circular orbit for a solar comet.

(b) At both perihelion and aphelion, the orbit is at right angles to the vector **r** from the star to the comet. The angular momentum is therefore just the product of r with mv, where v is the speed. The angular momentum is conserved in motion under the influence of gravitation, so we find

$$mv_{perihelion}r_{min} = mv_{aphelion}r_{max}.$$

We can solve to find the speed at aphelion:

$$v_{aphelion} = v_{perihelion}\frac{r_{min}}{r_{max}} = (81 \text{ km/s})\left(\frac{4.0 \times 10^7 \text{ km}}{6.0 \times 10^7 \text{ km}}\right) = 54 \text{ km/s}.$$

We used the conservation of angular momentum to relate the speed at perihelion to the speed at aphelion in this example. The use of energy conservation would also allow us to find the star's mass (see Problem 36).

12-5 GRAVITATION AND EXTENDED OBJECTS

We have viewed the planets and the Sun on the scale of the entire solar system and have treated them as point masses. But they are not really pointlike. This problem is especially evident if we think about how the gravitational force from Earth acts on an apple. To an apple, Earth most certainly is an extended object! We can study how extended objects behave by using the principle of superposition (Sec. 12–3).

The Gravitational Force Due to a Spherically Symmetric Object

We suggested in Section 12–3 that when an extended object is spherically symmetric, the force it exerts on a point mass outside the extended object is the same as the force that would be exerted if the entire mass of the extended object were concentrated at its center. It is this assertion that allowed us to relate the acceleration due to gravity at Earth's surface, g, to Earth's mass and radius, as in Eq. (12–9). In fact, for any spher ically symmetric object[†] of mass M and radius R, the acceleration due to gravity at its surface is

$$g = \frac{GM}{R^2}. \qquad (12\text{–}16)$$

Thus we would expect the acceleration of gravity to be different at the surfaces of the different planets (Fig. 12–14).

Many mass distributions—such as Earth—are at least approximately spherically symmetric. For such systems, the mass density depends only on the distance from the center of the distribution. The mass density at the center could be larger or smaller than that at the surface but, at a given radial distance from the center, it must be the same. Earth conforms well to these requirements. Earth's core is denser than the outer layers—it most likely consists primarily of iron. The small deviations from spherical symmetry come from the fact that, because of its rotation, Earth bulges a little in the equatorial region; there is an additional distortion that makes the planet look a little pear-shaped. In addition, there are small local regions of greater or lesser density. Here, we will summarize the principal conclusions about the gravitational force due to spherical systems. The same results hold for the gravitational potential energy.

1. Suppose that a point mass is *outside* a spherically symmetric object. The gravitational force experienced by the point mass is identical to the force that would arise if the whole mass of the spherical object were concentrated at its center. In other words, the force exerted on the point mass is the same as the force that would be exerted on it if the uniform spherical object (total mass M) were a point mass M located at the center of the sphere.

2. Suppose that a point mass is *inside* a thin spherical shell of constant density. Then there is no gravitational force on the mass. This conclusion holds for a point mass inside an arbitrarily thick shell as well, as long as the mass density of the shell depends *only* on the distance from the geometric center of the shell.

These two results show that the gravitational force on a point mass m within Earth at a distance r from the center, for example, would be due to a mass M' concentrated at Earth's center; M' is only the mass contained within Earth up to the radius r (Fig. 12–15a).

EXAMPLE 12–6 Suppose that a tunnel is drilled through our planet along a diameter. Assume that Earth's mass density is uniform and is given by ρ. Describe the force on a point mass m dropped into the hole as a function of the distance of the mass from the center.

Solution: The gravitational force on the point mass m is due only to the mass of the material contained within a radius r, where r is the distance from the point mass m to the center of Earth. The force is attractive, toward Earth's center, and it is given by

$$F = -\frac{GmM'}{r^2},$$

where the mass M' that attracts the point mass is the total mass *inside* radius r (Fig. 12–15a). Mass M' is given by (volume inside) × (density):

$$M' = \left(\frac{4\pi r^3}{3}\right)\rho.$$

PROBLEM SOLVING

The gravitational effects of spherically symmetric objects

FIGURE 12–14 The force of gravity—the weight—is less on the Moon's surface than on Earth's surface. Astronaut John Young, on the Apollo 16 mission, was able to jump to quite a height despite the massive spacesuit.

[†]This means its density varies only with distance from the center.

Thus, as in Fig. 12–15b,

$$F = -\left(\frac{4\pi Gm\rho}{3}\right)r.$$

The result that F is proportional to r shows that, inside Earth, the point mass acts as if it were moving under the influence of a spring with spring constant $k = 4\pi Gm\rho/3$. This motion is oscillatory, and the point mass moves from one end of the tunnel to the other and back (Chapter 13).

The results for spherically symmetric objects will not be derived mathematically until Chapter 24, where we give a simple but indirect proof. Direct proof, which would follow by integrating the contributions of the force from different pieces of our extended object, would take us too far afield now. However, we can give a physical picture of just how it is that the mass *outside* a point mass exerts no force on our mass.[†] Figure 12–16 shows our point mass placed within a shell. The point mass is taken to be off-center. Consider a double cone making an opening angle θ at the point mass and, in particular, the gravitational effects of the two circular sections that the cone cuts on the shell. The point mass is a distance r_1 from each point in the area on the right, A_1, and a different distance r_2 from each point in the area on the left, A_2. The shell is so thin that these areas do not change as we go from the inner to the outer surface of the shell. If the density of the shell is ρ and its thickness is τ, then the force F_1 that attracts the point mass to the right-hand area is

$$F_1 = \frac{Gm}{r_1^2}\rho\tau A_1,$$

and the force F_2 that attracts the point mass to the left-hand side is

$$F_2 = \frac{Gm}{r_2^2}\rho\tau A_2.$$

Now we note that areas A_1 and A_2 are proportional to r_1^2 and r_2^2, respectively, so that

$$\frac{A_1}{r_1^2} = \frac{A_2}{r_2^2}.$$

This means that the two forces are independent of the distance of the point mass to the circular sections and the forces cancel because they pull in opposite directions. In effect, as the point mass moves closer to one side, the inverse-distance-squared factor in the force increases, but the amount of mass seen decreases by the same factor; thus, the pull of every sector of the shell is the same and cancels.

This curious effect is very special to the inverse-square law—a law that also happens to hold for electric charges. Indeed, it was Benjamin Franklin who noticed that an electric charge surrounded by a shell of the opposite charge feels no net force, and it was on the basis of this observation that Joseph Priestley in the eighteenth century first suggested that electric forces obey the inverse-square law.

How g Varies with Altitude

Equation (12–16), together with the result that a symmetric, spherical Earth behaves like a centered point mass, implies that g varies with altitude. Suppose that we measure altitude, h, from sea level, and that Earth's radius at sea level is R_E. From Eq. (12–16), we find

$$g(h) = \frac{GM}{(R_E + h)^2} = \frac{GM}{R_E^2[1 + (h/R_E)^2]}. \qquad (12–17)$$

[†]Strictly speaking, our arguments hold only for small values of θ. But generalization is possible.

(a)

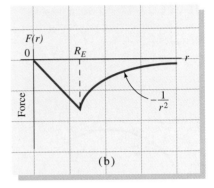

(b)

FIGURE 12–15 (a) The gravitational force on an object with mass m located inside Earth depends on the total amount of mass M' inside the sphere whose density is ρ and whose radius r is the distance of mass m from the center. (b) A graph of the variation of the force with r.

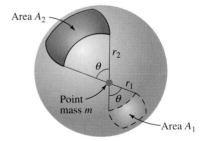

FIGURE 12–16 The total attraction that a point mass m experiences inside a spherical shell of uniform density is zero, because the force due to the mass in area A_1 is equal to that from the area A_2.

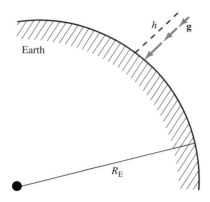

FIGURE 12–17 As we move away from the surface, the value of g decreases.

Now, if the ratio $h/R_E \ll 1$ (even for the top of Mount Everest, $h/R_E \simeq 10^{-3}$), then we can use the approximation $1/(1 + x)^2 \simeq 1 - 2x$ (for $x \ll 1$; see Appendix IV–9), so $g(h) \simeq (GM/R_E^2)[1 - (2h/R_E)]$, or

$$\frac{g(h)}{g(0)} \simeq \frac{(GM/R_E^2)[1 - (2h/R_E)]}{(GM/R_E^2)} = 1 - \frac{2h}{R_E}. \qquad (12\text{–}18)$$

To estimate the size of the effect, we can see from Eq. (12–18) that, at the top of the Sears Tower in Chicago, with $h = 443$ m, g is 99.99 percent of its value at sea level; even at the top of Mount Everest, with $h = 8848$ m, g is 99.74 percent of its value at sea level (Fig. 12–17).

Other effects modify Eq. (12–16). In Chapter 5, we mentioned the effect on g of Earth's rotation. This effect is largest at the equator, where g is 99.57 percent of what it would be if Earth did not rotate. This effect decreases with latitude and it is not present at the poles. Finally, Earth is not a perfectly uniform sphere. Not only is it not spherical, but it contains lumps of higher or lower density. Measurements of g are routinely made to many significant figures, more than enough to reveal all the effects we have discussed.

*12–6 A CLOSER LOOK AT GRAVITATION

When we have applied the law of gravitation, it has always been in the context of the attraction exerted on a "small" object by a "large" object. We have considered planets or comets moving around the Sun, tennis balls falling on Earth's surface, and so on. We assumed that the massive object was at rest and that the orbits being studied (for example, the ellipses) were those of the light object. We know that, just as Earth exerts a force on the Moon, the Moon exerts a force on Earth. The Moon moves in an orbit around Earth, but does Earth move around the Moon? The answer to this question is yes. When two objects move under the influence of the gravitational forces they mutually exert on each other, they move about their common center of mass. This is a property we first observed in the context of momentum. If we regard the two objects as an isolated system, then there is no external force on that system, and its center of mass remains unaccelerated. When one of the two objects is much heavier than the other, the center of mass is close to the heavier object and may even be within that object. In the Earth–Sun system, the Sun's motion is a very small orbit about a point very close to the center of (and well within) the Sun. Cases in which the center of mass is well separated from the two objects in question occur in double star systems with stars of comparable masses. These stars each move in a circular orbit around a point between them (Fig. 12–18).

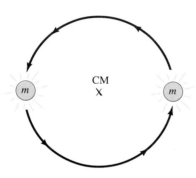

FIGURE 12–18 In a double star system with stars of equal mass, the center of mass is midway between the stars. Each star orbits around this point.

Ocean Tides

In the absence of a moon, the oceans would form a concentric envelope of water around Earth (Fig. 12–19a). With the Moon added to this picture, a superficial view of the system has the Moon attracting water on the near side of Earth more strongly than water on the far side. This would lead to a bulge in the water distribution which would peak

FIGURE 12–19 (a) If there were no Moon, the ocean could surround Earth in a uniform manner. (b) The Moon's gravitational attraction is stronger on the portion of the ocean nearer the Moon. If this were the only effect acting, there would be only one tide a day. Newton correctly pointed out that the orbiting of Earth about the center of mass of the Earth–Moon system leads to two bulges and two high tides per day.

(a)   (b)

There is a great deal of interest in mapping the exact details of Earth's gravity for prospecting purposes. Although Earth's average density is 5.5×10^3 kg/m³, the density near the surface varies significantly. Water has a density of 1.0×10^3 kg/m³, oil less than this; iron-containing rocks have a higher density than the average. Thus an accurate map of Earth's gravity would provide a global mineral prospecting tool. This is relatively easy to do because of the inverse-square law, which implies that irregularities close to the region being measured have a large effect on the size of *g*. NASA proposed a global project that involved a satellite in the form of an evacuated hollow sphere, containing a smaller dense sphere. Once in flight, the position of the smaller sphere within the larger would be sensed optically. This sensing would control gas jet thrusters on the large sphere and keep the spheres concentric. Normally a satellite is not in exact free fall because there are weak nongravitational forces acting on it. The most significant of these are the frictional drag due to the very dilute atmosphere at the satellite's altitude and momentum transfer produced by gases escaping from the satellite. (That is why the Space Shuttle is described as a microgravity rather than zero gravity environment.) The interior gravitation-sensing sphere is not acted on by any forces except the radiation pressure of the light from the optical sensors, and this is negligible. By following the precise path of the interior satellite, variations in the strength of the gravitational force acting on it can be measured.

in the direction of the Moon (Fig. 12–19b). Earth would rotate within this stationary, distorted water envelope, and thus the water level would be seen to rise once every 24 h. This argument, due to Galileo, fails to account for the fact that there are *two* high tides per day, not one.

Newton showed that, just as the Moon attracts the water nearest it more strongly than Earth as a whole, making a bulge toward the Moon, so the Moon also attracts Earth as a whole more strongly than the water on the *far* side of Earth, leaving behind a second bulge on the far side. The result is two high tides per day (Fig. 12–20).

Let's look at this argument in more detail. In addition to experiencing a gravitational force due to Earth itself, the surface points are attracted to the Moon. Because of Earth's finite size, the Moon exerts slightly different forces on different parts of the surface; points nearer the Moon experience stronger forces than points farther from the Moon. In particular, we want to see how these forces *differ* from the "average" force due to the Moon, the force that the Moon exerts on Earth treated as a point at its own center. A mass of water on the surface moves differently from the overall motion according to how much the force on it *differs* from the average force. Water nearest the Moon experiences *a little bit more* than the average force due to the Moon. Thus the force there differs from the average force by a vector pointing *toward* the Moon. Similarly, water farthest from the Moon experiences *a little bit less* than the average force. There, the force differs from the average force by a vector pointing *away from* the Moon. The force differences form what are called *tidal forces* (Fig. 12–21a). Earth's shell of water is accordingly distorted by the tidal forces to form the two bulges shown in Fig. 12–21b.

The Sun also contributes to tide formation, although less than half as much as does the Moon. This is not because the gravitational force of the Sun is smaller than that of the Moon. On the contrary, it is about 175 times larger. However, the *difference* in the Sun's gravitational force from one side of Earth to the other is smaller due to the fact that Earth's radius is so much less than the Earth–Sun distance. When the Moon and

(a)

(b)

FIGURE 12–20 Alma beach in the Fundy National Park, New Brunswick, Canada, at (a) high and (b) low tide. The Bay of Fundy has some of the largest tidal differences in the world.

Moon

North
+
Pole

Earth

(a)

Moon

Earth

(b)

FIGURE 12–21 (a) The arrows show how the forces from the Moon over Earth differ from the average value of the force on Earth. (b) The resulting water distribution has two bulges.

the Sun are lined up, at a new or full Moon, an especially large tide results; when they are at right angles, the tidal effects partially cancel.

The friction between the water and Earth moving beneath the bulges leads to a dissipation of energy, and Earth's rotation about its axis slows down. Calculations show that the day is thereby lengthened by about 10^{-3} s every century, a prediction verified by experiment.

Effects of Other Objects

The previous discussion involved the gravitational effects of more than one object. This has implications for planetary orbits. The superposition principle tells us that the orbit of any planet is affected not only by the Sun, but also by the presence of all other planets—although to a much smaller extent. The largest effects are due to the most massive planets, Jupiter and Saturn. One consequence of this is that the potential energy of a planet of mass m will have the form

$$U = -Gm\left(\frac{M_S}{r_S} + \frac{M_1}{r_1} + \frac{M_2}{r_2} + \cdots\right),$$ (12–19)

where the terms in parentheses represent the contribution of the Sun and those of the other planets. Thus the net force on a planet is no longer a pure $1/r^2$ force directed exactly at the Sun; there are small corrections to it. As a consequence, the orbits are no longer exact ellipses that close on themselves, but instead, the orbits *precess* (Fig. 12–22). Almost all major figures in nineteenth-century mathematics worked on the problem of computing orbits subject to perturbations from additional masses; in fact, accurate orbits were computed even before the advent of large-scale computing machines. Two interesting historical events are worth mentioning in this connection. First, the calculation of the orbit of Uranus, with the inclusion of all the perturbations, did not fit the observed orbit. In 1845, both John Adams (an undergraduate at Cambridge University) and Urbain Le Verrier in France calculated the potential effects of a hypothetical new planet and published their results. Adams's work was ignored, whereas Le Verrier was more successful in mounting a search for the new planet, which culminated in the discovery of the planet Neptune in 1846. Second, the calculation of the precession of the perihelion of the planet Mercury was also carried out to great accuracy by Le Verrier. The result of the comparison of observation and theory left a discrepancy in the rate of precession of the perihelion that amounted to only 43″ of arc

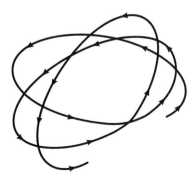
FIGURE 12–22 An orbit that does not close on itself. The orbit is said to precess.

per century (out of an observed total of some 5600″ of arc per century). Both theory and experiment were so good that there was no doubt of the existence of this discrepancy, whose explanation had to wait for Einstein's theory of gravitation.

Equality of Inertial and Gravitational Masses

The parameter m in $\mathbf{F} = m\mathbf{a}$ describes a property of an object that is properly called the *inertial mass* [see Eq. (5–1)]. It is a constant that characterizes the object, and it appears as a coefficient of the acceleration in response to *any* force. The parameter m that appears in the expression for the gravitational force that is exerted on that object is the *gravitational mass*, and there is no a priori reason why the inertial and gravitational masses should be equal. Their equality to one part in 10^{11} has been demonstrated by Robert Dicke and Vladimir Borisovitch Braginsky. Newton had already measured the equality of these quantities to an accuracy of one part in 10^3, and Loránd von Eötvös carried out measurements in the period from 1890 to 1922 to an accuracy of one part in 10^9. The equality of inertial and gravitational masses made a great impression on Einstein and led toward his formulation of the equivalence principle, a cornerstone of the general theory of relativity (see Section 12–7).

How Accurate Is Newton's Law of Gravitation?

With what accuracy do we know that the inverse-square law suggested by Newton holds? It is well verified that celestial phenomena obey an inverse-square law, but these phenomena occur on scales ranging from hundreds of kilometers to much greater distances. The Cavendish experiment and its modern versions verify the inverse-square law at distances ranging from centimeters to meters. But because there is no precise, direct confirmation of the law on distances ranging from meters to hundreds of kilometers, tests of the force of gravitation in this range are still proceeding. It is useful to remember that Newton's predictions are subject to the same experimental verification as are those of any other scientist.

*12–7 EINSTEIN'S THEORY OF GRAVITATION

As remarkably accurate as it is, Newton's theory of gravitation was superseded in 1915 by a still more accurate description: Albert Einstein's theory of gravitation. The Einstein theory, also known as the **general theory of relativity**, reduces to Newton's theory for objects that move with a speed v that is small compared to the speed of light, c ($v \ll c$), and when the gravitational potential energies are small compared to mc^2. These conditions are satisfied except in regions close to very large masses; this is why Newton's theory was so very successful in virtually all its applications.

Einstein's theory arose out of his attempts in 1915 to combine Newton's theory with the special theory of relativity, which will be discussed in Chapter 40. While working on this problem, Einstein had what he described as "the happiest moment of my life": the idea that a freely falling person does not feel (and has no way to measure) his or her own weight. This idea was generalized in 1907 to form the **equivalence principle**. According to this principle, no experiment can distinguish between the following two situations: (1) a physical system at rest that is subject to a uniform gravitational force; and (2) a physical system that is uniformly accelerating in the absence of gravity. A simple example will impart the flavor of what is implied. Suppose an observer stands in an elevator and experiences a force on his feet. He can interpret this effect as being due to an upward acceleration of the elevator. Because the observer, of mass M, experiences an acceleration, a, he must be subject to an upward force of magnitude $F = Ma$. In fact, this force is the contact force that the elevator floor exerts on his feet. A second interpretation is that the elevator is at rest,

The equivalence principle

but there is a uniform gravitational force acting to pull him downward. He again feels the upward normal force, F_N, of the floor and, because the elevator and thus the observer are at rest, that force must just cancel the gravitational force. Thus $F_N = Mg$, where g is the acceleration due to gravity. Notice that we have used the same mass M in both descriptions. Strictly speaking, the mass M in $F = Ma$ is the *inertial* mass, whereas it is the *gravitational* mass in the force-identity equation $F_N = Mg$. The gravitational and inertial masses were first observed by Newton to be identical, but he was unable to draw deep conclusions from this observation. Einstein's equivalence principle states that if g has the same numerical value as a, there is no way for the observer to distinguish the two cases; thus, the inertial mass and the gravitational mass are *required* to be equal.

Predictions of the Equivalence Principle

Light Falls under the Influence of Gravity. Light falls just like matter does. To see this, let's consider the elevator again. A beam of light shines across the elevator in a horizontal direction from one side just as the elevator accelerates upward (Fig. 12–23). Because it takes light some time to travel across the elevator, it will hit the opposite wall closer to the floor than a horizontal line parallel to the floor would indicate. More importantly, for constant acceleration, the amount by which the elevator moves upward is proportional to the square of the time. That means that a *series* of measurements made by an occupant of the elevator that located the light beam relative to the floor would show that the light beam follows a parabolic path, just as a falling object would under the effect of gravity. If we are to interpret this observation from the point of view that the elevator and its contents feel a gravitational force, then the observed deflection of the light must be due to the gravitational force. Light, in this sense, does not behave any differently than matter, except that it moves faster. The first observation of the deflection of light due to gravitation was made during a solar eclipse in 1919. At such a time, pairs of stars whose light passes very close to the Sun become visible. The angular spread between the stars when their light comes around the two sides of the Sun can be compared with their observed angular spread when they are seen away from the Sun (Fig. 12–24). The confirmation of the predicted effect did more than anything else to bring Einstein's theory to the public's attention.

Gravitational Lenses. Recently, astronomers have observed two identical *quasars* (extraordinarily bright sources of light) very close to one another as seen from Earth (Fig. 12–25). Because there is good evidence that quasars are billions of light-years

FIGURE 12–23 (a) A beam of light shines across an elevator at a time t when the elevator starts to rise. (b) At time $t + \Delta t$ the light hits the opposite side of the elevator at a point lower than it would have hit had the elevator not been moving. (c) The observer in the elevator cannot know whether the elevator is accelerating upward at **a** or whether gravity **g** is present. We conclude from thought experiments like this that light must be bent when gravitational forces are present.

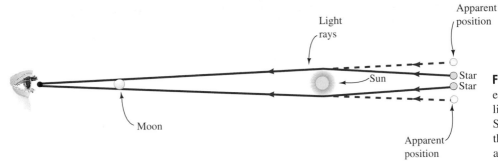

FIGURE 12–24 During a solar eclipse, it is possible to observe the light of two stars on either side of the Sun. Because the Sun's gravity bends the light, the stars appear to be farther apart than they actually are.

away, any two quasars are literally quite disconnected from one another; thus, it is highly improbable that any two could be as close to identical as the observed pair appear to be. The generally accepted interpretation is that there is only one quasar, but its light passes close to an extremely massive object, which bends light going around it just as a lens would; this gives rise to a double image; the effect is said to be due to a *gravitational lens*. An even more spectacular demonstration of this effect is seen in a second discovery, where we can detect a ring image of the source (Fig. 12–26). (Can you see how the ring is produced?) To produce such a large effect, the light must be bent by a galaxy whose mass equals that of about 300 billion suns.

Black Holes. The fact that light falls when it passes near masses implies that it is possible to imagine a large enough mass, localized in a small enough region, for which the speed of light is smaller than the escape speed from the surface of the mass. Such a mass would not be directly visible because light or matter could not escape from it. It would manifest itself only through the gravitational force it exerts. Such a mass forms a *black hole*. Recent observations have provided indirect evidence for black holes. There exist pairs of stars—only one of which can be seen directly. The other is invisible, but its presence and mass can be deduced from the motion of the visible star. In addition, some such pairs emit X-rays copiously. The characteristics of the X-rays demand the interpretation that they come from matter falling into a black hole. The strong gravitational force necessary to produce such dramatic effects can come only from a black hole.

Precession of Planetary Orbits. The equivalence principle forms the foundation of the general theory of relativity, which is the full theory of gravitation. The mathematical application of the theory leads to subtle corrections to Newton's gravitational force law. The corrections to the Newtonian form predict, among other things, that the perihelion of Mercury should precess by 43″ of arc per century, an amount that is within observational error when the effects of other planets on the motion of Mercury are taken into account. Other predictions of the general theory of relativity are being confirmed by recent experiments.[†]

_____

[†]See Clifford M. Will, *Was Einstein Right?*, Basic Books, New York, 1986.

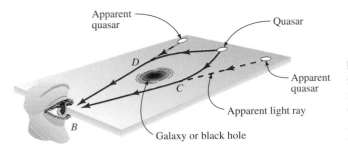

FIGURE 12–25 A double image results when light from a single quasar is bent as it passes on either side of an object with strong gravity, such as a galaxy or a large black hole. The effect is greatly exaggerated here.

FIGURE 12–26 This photograph, taken by the Hubble Space Telescope, of a massive, compact galactic cluster illustrates gravitational lensing. The lensing effects are evident in the arc-like pattern. They are caused when light from an object far beyond the cluster passes near the cluster, which magnifies, distorts, and brightens that light on its way to our eyes. The image tells us a great deal both about the matter far beyond the cluster and about the cluster itself.

Astronomical observations led Kepler to three laws for planetary motion:

1. Planets move in planar elliptical paths with the Sun at one focus of the ellipse.

2. During equal time intervals, the radius vector from the Sun to a planet sweeps out equal areas.

3. If T is the time that it takes for a planet to make one full revolution around the Sun, and if R is half the major axis of the ellipse (R reduces to the radius of the orbit of the planet if that orbit is circular), then

$$\frac{T^2}{R^3} = C, \tag{12–1}$$

where C is a constant whose value is the same for all planets.

Newton showed that these laws are a consequence of a law of universal gravitation, which states that any two point masses m and M, separated by a distance r, will attract each other with a force that is along the radius vector connecting the masses. The force on object 2 due to object 1 has the inverse-square form

$$\mathbf{F}_{21} = -\left(\frac{GmM}{r^2}\right)\hat{\mathbf{r}}, \tag{12–4}$$

where $\hat{\mathbf{r}}$ is the unit vector pointing from object 1 to object 2. The constant G has been measured, and its value is $G = 6.673 \times 10^{-11}$ N·m²/kg². The gravitational force is conservative and may be derived from a potential energy:

$$U(r) = -\frac{GmM}{r}. \tag{12–8}$$

Masses that exert gravitational forces are not always pointlike. The most important case occurs when there is a spherically symmetric mass distribution, such as Earth or the Sun. In this case, the gravitational force is the same as if all the mass of the extended object were concentrated at the center of the spherical distribution. Another consequence of the formula given in Eq. (12–8) is that the gravitational force exerted on an object anywhere inside an arbitrarily thick spherical shell, with a mass distribution that depends only on the distance from the center, is zero.

The gravitational force acts on both objects; for example, not only does Earth revolve around the Sun, but the Sun also revolves around Earth. More precisely, both objects revolve around their center of mass. When this phenomenon is taken into account in the Earth–Moon system, the tides can be understood.

The Newtonian theory of gravity is a limiting case of a more accurate and fundamental theory of gravity: Einstein's general theory of relativity. That theory is based on the equivalence principle, which states that no experiment can distinguish between a uniform acceleration and the effects of a uniform gravitational force. One consequence is the equality between the inertial mass (m_i in $F = m_i a$) and the gravitational mass (the masses that appear in the law of universal gravitation). Another consequence is the fact that light falls just like ordinary matter when it is subject to gravity.

UNDERSTANDING KEY CONCEPTS

1. What are some forces, other than those mentioned in this chapter, that are not fundamental? From what fundamental forces are they derived?

2. Any projectile fired with enough initial speed will eventually escape Earth, regardless of the direction in which it is fired. How do you reconcile this statement with the fact that the height above Earth to which a cannonball will rise depends on the angle at which it is fired?

3. To a good approximation, the Sun and the Moon both move with respect to Earth in the plane of Earth's equator. Does this mean there would be little or no tide at the North Pole if there were surface water at the North Pole?

4. The European satellite launch area is in French Guiana, because less energy is required to launch rockets into orbit from there than from a point in Europe. Explain why this is so.

5. How can Earth's rotational motion be used to minimize the fuel needed to boost a satellite into a given orbit around the planet? How can Earth's orbital motion be used to minimize the fuel needed to boost a satellite into a given orbit around the Sun?

6. The satellites of Jupiter follow Kepler's third law: The square of their periods, divided by the radius of their orbits cubed, is a constant. Is this the same constant as for the planets moving around the Sun?

7. Newton performed a thought experiment when he described an artillery shell fired horizontally from the top of a high mountain. Ignoring air resistance, the shell will land farther and farther out as the speed with which the shell is fired increases. Eventually, the curvature of Earth becomes significant, and the surface falls away from beneath the shell. At last, for a sufficiently high initial speed, Earth's surface falls away from the path of the shell as the shell circles Earth, and this is a circular orbit. How is this result consistent with the statement that an artillery shell follows a parabolic path rather than the arc of a circle?

8. Describe the path of a celestial object whose angular momentum with respect to the Sun is zero.

9. If the gravitational force were a central force proportional to $1/r^3$ rather than $1/r^2$, the planetary orbits would no longer be closed (unless they are circular). Would the planets still sweep out equal areas in equal times?

10. We often hear that Earth satellites burn up when they leave their orbit and return to Earth. Why don't satellites burn up as they go up into orbit?

11. Communication satellites are placed in *geosynchronous* orbit, which means that they are always above a given point on Earth. Is it possible to place such a satellite above the North Pole? Above what points on Earth is it possible to have a geosynchronous orbit?

12. When astronauts float in the bay of their spacecraft, they are in outer space, above Earth's atmosphere. Is gravity acting on them?

13. Very careful measurements of the orbits of satellites can help teach us about Earth's internal structure. How can we use such measurements to study regions with a mass density that is higher or lower than that of Earth's average mass density?

14. If Earth were a perfect sphere, would you weigh more or less at the equator than at the poles?

15. Earth is not a perfect sphere. What types of observations might we use to learn this fact?

16. When a satellite is in circular orbit around Earth, there is a direct relationship between its angular momentum and the radius of the orbit. Suppose that such a satellite collides elastically with a meteor that was heading directly toward the center of Earth. Because the impulse is in a radial direction, the angular impulse is zero. What happens? Draw some diagrams, using the conservation laws that you know. (If you get stuck, review Kepler's first law.)

17. In the text, we spoke of friction between Earth and its shell of water as slowing down the rotation frequency of Earth. Would this still occur if the Moon were not present?

18. Is it possible for friction from the tides to slow down Earth's rotation and still conserve Earth's angular momentum (with its oceans)? The origin of the tides involves the gravitational force due to the Moon on Earth's oceans, so Earth and its oceans do not form an isolated system.

19. A satellite is in circular orbit around Earth. How much work is done on the satellite by the gravitational force of Earth during one orbit?

20. Assume that Earth is perfectly spherical and that its density depends only on the distance from its center. A large asteroid makes a close pass to Earth. Could such an asteroid (which is nevertheless small compared to Earth itself) change the rate of Earth's rotation without actually colliding?

21. You are in a spaceship very far away (say, 100 Earth radii) from Earth. Could you move to a region where the gravitational force due to Earth is less by ejecting some material from the spaceship in the proper direction?

22. Edgar Rice Burroughs's character Tarzan discovers that Earth is hollow. Tarzan finds an entry and discovers a whole new civilization, with modernistic buildings, people walking around or driving modernistic vehicles, and so on, in Earth's interior. What is wrong with this picture?

PROBLEMS

12–2 Early Observations of Planetary Motion

1. (I) Use the data listed in the appendices to calculate C in Eq. (12–1).

2. (I) In planetary tables, we find that Jupiter's satellites Io and Europa each follow nearly circular orbits: Io's orbit has a mean radius of 422,000 km and Europa's orbit has a mean radius of 671,400 km. The period of Io is 152,854 s. What is Europa's period?

3. (II) Consider an object of mass m, moving in a circular orbit, subject to a central attractive force whose magnitude is given by $F(r) = h/r^3$. (a) What are the dimensions of h? (b) Show that the angular momentum for the motion is uniquely determined by h and m. (c) What is the resulting relation between period and radius analogous to Kepler's third law for this force?

4. (III) Angular momentum is conserved for a radial, or central, force. Show that the orbits due to a radial force lie in a plane.

12–3 Newton's Inverse-Square Law

5. (I) A man of mass 95 kg is dancing with his wife, who has a mass of 68 kg. Assume that each person's mass is concentrated at their respective centers of mass, which are separated by 48 cm. (a) What is the gravitational attraction between them? (b) Which person has the greater gravitational attraction toward the other?

6. (I) Calculate the gravitational attraction between a proton and an electron in a hydrogen atom if the radius of the atom is 0.6×10^{-10} m. The masses can be found in Appendix V.

7. (I) A Cavendish experiment involves the force between two spheres of 1 kg each whose centers are separated by 40 cm. Using the known value of G, find the gravitational force between these spheres. Compare this force to the weight of a fly.

8. (I) What is the acceleration due to gravity on the surface of (a) the Moon ($R = 1.74 \times 10^3$ km, $m = 7.35 \times 10^{22}$ kg);

(b) Mars ($R = 3.40 \times 10^3$ km, $m = 6.42 \times 10^{23}$ kg); (c) Jupiter ($R = 7.14 \times 10^4$ km, $m = 1.90 \times 10^{27}$ kg); (d) the Sun ($R = 6.96 \times 10^5$ km, $m = 1.99 \times 10^{30}$ kg)?

9. (I) A satellite orbits Earth in 90 minutes. What is the radius of its motion around the center of Earth?

10. (I) What is the period of a satellite circling the Moon at a height of 300 km above the Moon's surface?

11. (I) What is the period of a satellite circling Earth at a height of 300 km above Earth's surface?

12. (I) What is the surface gravity (the value of g) on an asteroid of diameter 150 km and density 4800 kg/m³?

13. (I) Two identical satellites move in circular orbits around Earth. One has twice the kinetic energy of the other. The radius of the faster one's orbit is three Earth radii. What is the radius of the slower one's orbit?

14. (II) A weight lifter can lift 115 kg on Earth. What mass could the same weight lifter lift on (a) the Moon, (b) the Sun (use the data in Problem 8)?

15. (II) The height achieved in a jump is determined by the initial vertical velocity that the jumper is able to achieve. Assuming that this is a fixed number, how high can an athlete jump on Mars if she can clear 1.85 m on Earth?

16. (II) There is a point on the line joining two astronomical bodies where there is no gravitational force on a rocket. Find this point for (a) the Earth–Sun system, (b) the Earth–Moon system.

12–4 Planets and Satellites

17. (I) Pluto has the most eccentric orbit of all the planets, with $e = 0.25$. Its semimajor axis is 39.5 AU. What is ($r_{max} - r_{min}$)/($r_{max} + r_{min}$) for Pluto's orbit?

18. (I) The semimajor axis of Earth's orbit is 149.6×10^6 km, while its eccentricity is 0.017. What is the maximum distance between Earth and the Sun as Earth traces out its orbit?

19. (I) Determine the escape speed of an object from the Sun's surface.

20. (I) What are the escape speeds on the surface of (a) the Moon, (b) Mars, and (c) Jupiter? (Use the data in Problem 8).

21. (I) Consider the asteroid of Problem 12. What is the escape speed from the surface of that satellite?

22. (I) The radius of a neutron star is 600 times smaller than Earth's radius, and its mass is 4×10^5 times larger than Earth's mass. What is the escape velocity from the surface of a neutron star? (Ignore the fact that, at high speeds, one should not really use $mv^2/2$ for the kinetic energy.)

23. (II) If the asteroid of Problem 12 rotates with an angular speed ω about an axis, material on the equator will have a tendency to be thrown off. How slowly must it rotate so that material that is not attached permanently just barely stays on the surface?

24. (II) A rocket is sent vertically upward from Earth's surface with an initial speed of 7 km/s. How far above Earth's surface will it go before falling back? Ignore atmospheric friction.

25. (II) Astronomers discover a meteorite at a distance of 80,000 km from the center of Earth. The meteorite is moving directly toward Earth with a velocity of 2000 m/s. What will be the velocity of the meteorite when it hits Earth's surface? Ignore all drag effects.

26. (II) Determine the minimum energy needed to allow an unmanned rocket of mass 1200 kg to leave the Moon's surface and arrive at a point very far away. Ignore the effect of Earth.

27. (II) Calculate the distance from Earth's center to a satellite in circular orbit with a period that is (a) one-third the Moon's period; (b) three times the Moon's period; (c) one-thousandth of the Moon's period.

28. (II) Using the information in Example 12–3, determine a satellite's time of revolution about Earth if it is in a circular orbit 250 km above Earth. How would your answer change if the mass of the satellite were to double?

29. (II) Determine the velocity with respect to Earth of a satellite of mass 500 kg in a circular orbit 200 km above Earth's surface. What is its kinetic energy? What is its angular momentum?

30. (II) A geosynchronous communications satellite orbits Earth, always positioned above the same point on the equator. (a) What is the period and angular velocity of the satellite? (b) What is the radius of the orbit? (c) Show that Kepler's third law applies to the orbits of the satellite and the Moon.

31. (II) The mass of Mars is 6.42×10^{23} kg, and its radius is 3393 km. What is the period of a satellite in a circular orbit 95 km above the surface of Mars?

32. (II) The Moon goes around Earth once in 27.3 d. What is the distance between the Moon and Earth?

33. (II) Use the mass of the Sun, given in Problem 8, to estimate the distance from Earth to the Sun. Assume that the orbit is circular.

34. (II) A satellite is fired off horizontally with an initial speed $v_0 = 9.3$ km/s from the North Pole. Ignore air resistance. (a) What is the maximum distance from Earth's center attained by the satellite? Use both the conservation of energy and the conservation of angular momentum about the center of Earth. (b) What is the maximum distance attained if the satellite is fired vertically with the same speed? (c) if the satellite is fired at an angle of 30° with the same speed? (d) Sketch the motion of the satellite in the three cases.

35. (II) A small package is fired off Earth's surface with a speed v at a 45° angle. It reaches a maximum height h above the surface at $h = 6370$ km, a value equal to Earth's radius itself (Fig. 12–27). What is its speed when it reaches this height? Ignore any effects that might come from Earth's rotation.

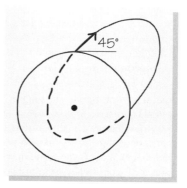

FIGURE 12–27 Problem 35.

36. (II) In Example 12–4, we used the conservation of angular momentum to find the speed of a comet at perihelion and aphelion around a star. Use the fact that the energy at these two points must be equal to find the mass of the star.

37. (II) An object of mass 3×10^{15} kg approaches the solar system (Fig. 12–28). When it is very far away—where the gravitational potential energy can be neglected in comparison with its kinetic energy—the object moves with a velocity of 12 km/s in a straight line. By straight-line extrapolation, the closest this line would come to the Sun is 3×10^8 km. The point of the object's nearest approach to the Sun is characterized by the fact that the radius vector from the object to the Sun is perpendicular to the tangent to the path at that point. (a) Sketch the orbit of the object. (b) Use conservation of energy and of angular momentum to calculate the velocity of the object at the point of nearest approach. (c) Calculate the distance of nearest approach.

FIGURE 12–28 Problem 37.

38. (II) A satellite is in a circular orbit of radius R_1 around Earth. Small rockets aboard the satellite change its direction so that it has an elliptical orbit. The change causes the satellite to lose half its orbital angular momentum, but the total energy remains constant. In terms of R_1, what are the perigee and apogee distances of the new orbit with respect to the center of Earth?

39. (II) A satellite of mass 300 kg is in circular orbit 2000 km above Earth's surface ($M_E = 6 \times 10^{24}$ kg, $r_E = 6370$ km). (a) What is the orbital speed of the satellite? (b) What is the angular momentum of the satellite? (c) A rocket engine is fired, reducing the speed of the satellite to half its initial value, but leaving the direction of motion unchanged. What is the new angular momentum? (d) Does the satellite crash as a result of the maneuvers in part (c)? Explain your answer.

40. (III) The distance of closest approach of Halley's comet to the Sun is 8.9×10^{10} m. Its period is 76 yr. What is the nature of its orbit? Calculate the following: (a) semimajor axes; (b) eccentricity; (c) aphelion distance (farthest distance from Sun).

41. (III) A satellite of mass 2000 kg is in circular orbit about Earth at a distance of 300 km above the surface. (a) What is the speed of the satellite in its orbit? (b) What is the angular momentum of the satellite about the center of Earth? (c) What is the total energy of the satellite in its orbit? (d) Controllers back on Earth wish to move the satellite to a new orbit 500 km above the surface. They propose to do this by briefly firing a rocket engine on the satellite for several seconds in the direction of the cen-

ter of Earth; that is, the force on the satellite is directly away from Earth's center. What is the torque on the satellite about the center of its orbit? (e) Can the new orbit be circular?

12–5 Gravitation and Extended Objects

42. (I) What is the approximate difference between the value of g at sea level and the value on top of a 10,000-ft-high mountain? Assume that Earth has a constant density, and take Earth's radius as the radius at sea level.

43. (I) Equation (12–19) is an approximate form. The corresponding exact form can be derived from Eq. (12–18): $g(h)/g(0) = (R_E^2)/(R_E + h)^2$. Verify the accuracy of the approximation by calculating the ratio $g(h)/g(0)$ for $h = 10,000$ m (the altitude of a cruising passenger jet) according to the approximate and the exact forms.

44. (II) A deep hole in Earth reaches a depth of one-half of Earth's radius (Fig. 12–29). How much work is done when a 1-kg mass is slowly lifted from the bottom of the hole to Earth's surface?

FIGURE 12–29 Problem 44.

45. (II) What is the speed of the mass in the tunnel through Earth in Example 12–5 as it passes through the center of Earth?

46. (III) Rather than a tunnel through Earth's center, as in Example 12–5, consider a tunnel drilled along a chord of the earth, meaning that it passes a perpendicular distance d away from the center of Earth (Fig. 12–30). Find the potential energy of a mass placed in such a tunnel as a function of (a) its distance r from the center of Earth and (b) its distance x from the midpoint of the tunnel.

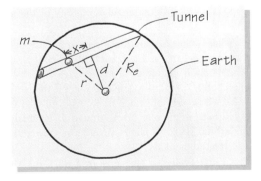

FIGURE 12–30 Problem 46.

*12–6 A Closer Look at Gravitation

47. (I) The mean Earth–Sun separation is 1.50×10^8 km, Earth's radius is 6.37×10^3 km, the mass of the Sun is 1.99×10^{30}

kg, and Earth's mass is 5.98×10^{24} kg. Given these data, locate the center of mass of the Earth–Sun system.

48. (I) At a certain moment, the Sun, Jupiter, and Saturn are lined up, with Jupiter between the Sun and Saturn. Suppose the orbits of Jupiter and Saturn were circular, with mean distances from the Sun of 7.78×10^8 km and 1.42×10^9 km, respectively. The mass of Saturn is 0.029 percent of that of the Sun. What is the ratio of the gravitational force on Jupiter due to Saturn to the gravitational force on Jupiter due to the Sun, at the specified moment?

49. (II) A binary star system consists of two stars, each of mass M, orbiting around their common center of mass with radii R from the center of mass. Determine the period of revolution. [*Hint*: See Problem 11 of Chapter 10.]

*12–7 Einstein's Theory of Gravitation

50. (I) Fighter pilots are able to withstand accelerations up to $7g$ for a short period. A jet dives toward Earth and pulls up at the last second in a parabolic orbit. Draw a force diagram showing the various forces on the pilot at the bottom of the orbit. If the dive was at night, could the pilot tell the difference between an increased value of gravitational force and the effect of the contact forces on him at the bottom of the dive?

51. (II) A very sharply defined laser beam, directed horizontally, enters a hotel room at height h. At what height does the light beam hit the opposite wall, which is 8 m from the first wall? Compare the difference in heights to the size of an atom. Is this a feasible experiment with which to test Einstein's theory of gravitation?

52. (II) An elevator of width w in free space is accelerated upward with acceleration g. A ray of light, traveling with speed c, enters through a pinhole on one side of the elevator, at right angles to the side at the moment the elevator starts to accelerate. It will strike the opposite wall at a somewhat lower height. What is the angle of deflection of the light? According to the equivalence principle, a passenger in the elevator could not distinguish this bending of light from a bending due to the effects of gravity.

General Problems

53. (I) A spaceship of the future is cylindrical in shape, with a radius of 60 m. In order to simulate terrestrial gravity on the inside surface of the cylinder, the spaceship is made to rotate about its axis. What is the angular velocity of the spaceship about its axis?

54. (I) (a) What is the acceleration g due to gravity on a planet with the same density as Earth but with twice the radius? (b) The orbital period and radius of Jupiter's Moon Ganymede are 7.16 d and 660,000 mi, respectively. What is the period of the Moon Io, whose orbital radius is 262,000 mi? (c) Planets A and B are both in circular orbits around a star. Planet A has half the orbital speed of planet B. What is the radius of A in terms of the radius of B?

55. (II) The Little Prince (a character in a book by Antoine de Saint-Exupery) lives on the spherically symmetric asteroid B-612 (Fig. 12–31). The density of asteroids, including B-612, is 5.2×10^3 kg/m³. Assume that the asteroid does not rotate. The Little Prince noticed that he felt lighter whenever he walked quickly around his asteroid. In fact, he found that he became weightless and started to orbit the asteroid like a satel-

lite whenever he speeded up to 2 m/s. (a) Estimate the radius of the asteroid from these data. (b) What is the escape speed for the asteroid? (c) Suppose that B-612 does rotate about an axis such that the length of the day there is 12 h. Can the Little Prince take advantage of this rotation when he wants to orbit his asteroid?

FIGURE 12–31 Problem 55.

56. (II) A neutron star has a mass of 4×10^{30} kg and a radius of 10 km. (a) Calculate the acceleration due to gravity at the surface of the neutron star. (b) What is the difference between the gravitational forces acting on the top and the bottom of a dumbbell held vertically on the surface (that is, with one end on the surface of the neutron star and the other 1 cm above the surface)? The dumbbell consists of two 1-g point masses connected by a massless connector of length 1 cm.

57. (II) Suppose that, instead of a $1/r^2$ dependence, an attractive central force varied with distance as $1/r^n$. (a) Would such forces support a circular orbit? (b) Find the resulting relation between period and radius analogous to Kepler's third law for this force.

58. (II) Consider a cluster of galaxies that fills a sphere of radius R and average mass density ρ. (There are so many galaxies that you can assume uniform density.) There is a galaxy of mass M at the edge of this sphere (Fig. 12–32). (a) Write an expression for the energy of the galaxy. (b) In the big-bang model of the origin of the universe, the velocity of the galaxy is directed radially outward from the center of the sphere; the galaxy's speed is $v = HR$, where $H = (15$ km/s$)/(10^6$ ly$)$ is the Hubble parameter. For what critical density ρ_c of the large cluster will the galaxy be able to escape to infinity with a final velocity of zero?

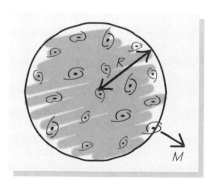

FIGURE 12–32 Problem 58.

59. (II) A satellite in low circular Earth orbit is subject to a very small constant friction force, f, due to the thin atmosphere. As it spirals in, it slowly decreases its radius. Find the decrease in radius per revolution under the assumption that the orbit is approximately circular with radius r. Find the changes in potential energy, total energy, and kinetic energy per orbit. [*Hint:* If you find the kinetic energy increasing, you are on the right track!]

60. (II) The electric force between an electron and a (much more massive) proton is attractive and of magnitude e^2/r^2, where $e^2 = 2.3 \times 10^{-28} \text{ N} \cdot \text{m}^2$. The electron circles the proton in a circular orbit of total energy $E = -2.2 \times 10^{-18} \text{ J}$. (We have taken zero potential energy to be at infinite separation.) (a) What is the radius of the orbit? (b) What is the period of the orbit? (c) If another electron were in another circular orbit around the proton, with an orbit radius four times as large as the first, what would its total energy and period be?

61. (II) An astronaut of mass 115 kg (including equipment) finds himself drifting away from his orbiting space ship at 0.05 m/s. He throws a 3-kg wrench in the direction of his drift and comes to rest relative to the ship 1 m from its surface. The ship is a sphere of radius 12 m and mass 10^5 kg. (a) At what speed does he throw the wrench? (b) How many hours must he wait for the gravitational attraction of the ship to pull him to its surface, assuming that the force of gravity is approximately constant in the region of interest?

62. (II) Suppose that an object of mass m is placed at the point at which the gravitational attraction of the Moon is just canceled by that of Earth; further, suppose that the object is displaced by a small distance x along a line perpendicular to the line connecting the centers of Earth and the Moon (Fig. 12–33). What are the magnitude and direction of the net force on the object as a function of x? Calculate your answer by using the approximation $(r^2 + x^2)^n = r^{2n}[1 + (nx^2/r^2) + \cdots]$, valid for $x^2/r^2 \ll 1$.

63. (III) Three stars, each the mass of the Sun, form an equilaterial triangle. Each moves in a circular orbit about the center of mass of the system because of the gravitational force exerted by the other two stars. (a) Is such an arrangement possible? (b) If so, what is the period of the motion, assuming that the side of the triangle is an Earth–Sun distance? (c) Is the system stable?

64. (III) Astrologers claim that a person's life is influenced by the position of the planets at the moment of that person's birth. To check whether this influence could be due to gravity, compare the following two quantities: the change in the gravitational force on a baby in a hospital due to the change in the position of Jupiter from one day to the next, and the change in the gravitational force due to the presence or absence of a 2-ton truck parked near the hospital at a distance of 100 m. Jupiter has a mass of 1.90×10^{27} kg; its mean distance from the Sun is 0.78×10^9 km, and its period is 11.9 yr. Assume a circular orbit for Jupiter, and a circular orbit of radius 1.5×10^8 km for Earth. Choose the region of closest approach of the two planets for convenience.

65. (III) When the first nuclear weapons were detonated, concern was expressed in some quarters that a huge nuclear chain reaction would be set up, blowing Earth to pieces. Show that the energy that would be required to disassemble Earth completely into pieces totally separate from each other is $\frac{3}{5}GM^2_E/R_E$. [*Hint:* Imagine that layers of Earth are peeled off one by one like layers of an onion.]

FIGURE 12–33 Problem 62.

An earthquake in Oakland, California caused this highway bridge to collapse under the influence of harmonic resonant forces. By studying the behavior of systems—such as the soil underlying the highway—under harmonic forces, we can learn to build structures that can resist damage.

Simple Harmonic Motion

R hythmic motion—also known as periodic motion—is an important feature of the physical world. The very concept of time may have arisen from the observation that certain motions, such as the human heartbeat and the cycling of the seasons, repeat themselves in a reliable and regular way. The most basic type of rhythmic motion appears over and over again: simple harmonic motion. Examples are provided by the motion of a mass on the end of a spring and by the motion of a pendulum. *Simple harmonic motion is motion in which the position of a point varies with time as a sine or a cosine.* Such motion occurs when we have restoring forces—forces that tend to bring an object back to a point—that vary linearly with a position variable. While the spring force is an example that we will use repeatedly, simple harmonic motion is of universal importance because *virtually any stable equilibrium situation involves a linear restoring force.* We will discover that simple harmonic motion is a fundamental part of the physical world.

13-1 THE KINEMATICS OF SIMPLE HARMONIC MOTION

Simple harmonic motion—the motion followed by a mass on the end of a spring or a pendulum—is a basic form of **periodic**, or **oscillatory**, motion: motion that repeats. The word "harmonic," signifying agreement and accord, reveals that we have always seen beauty in this motion. In the back-and-forth of simple harmonic motion, the position $x(t)$ of an object is of the form $\sin(\omega t)$ or $\cos(\omega t)$. Both sines and cosines repeat

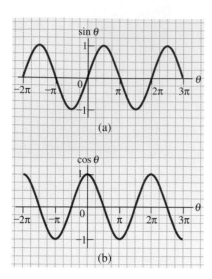

FIGURE 13–1 Plots of (a) $\sin\theta$ and (b) $\cos\theta$, both as a function of θ.

themselves periodically as time passes. The trigonometric functions are functions of a dimensionless argument, an angle measured in degrees or radians. Thus the coefficient ω of the time must have the dimensions $[T^{-1}]$. We'll see later that ω is a fundamental property of the motion, determined by the inertia of the moving objects and the restoring forces acting on them.

How do we decide if the motion of a mass on the end of a spring has a sine or a cosine dependence? Let's look at a graph of $\sin\theta$ versus θ next to a graph of $\cos\theta$ versus θ (Fig. 13–1). Both functions repeat every time the angle θ changes by 2π rad. When $\theta = 0$, the sine function is zero, whereas the cosine function is $+1$, but this is only a matter of placing the axis. Indeed, the functions are *identical* if the origin of the θ axis is shifted; if θ is shifted by an angle we call the **phase**, δ, the curves drawn in Fig. 13–1 shift along the horizontal axis and move between the sine and the cosine functions. Thus, if δ is chosen properly, the function $\sin(\omega t + \delta)$ can represent $\sin(\omega t)$, $\cos(\omega t)$, or anything in between.

The sine function is dimensionless and, moreover, varies between -1 and $+1$. But $x(t)$ has dimensions of length. We therefore multiply the sine function by an **amplitude**, A, which has the proper dimensions of length. The amplitude is defined to be positive and describes the magnitude of the maximum excursion away from the point of zero displacement (Fig. 13–2). The resulting expression for the position of an object in simple harmonic motion is

$$x(t) = A\sin(\omega t + \delta). \tag{13–1}$$

Properties of Simple Harmonic Motion

The mathematical definition of simple harmonic motion, amplitude, and phase

Three independent parameters appear in simple harmonic motion: the amplitude, A; the phase, δ; and the constant ω. The amplitude and the phase are determined by specifying the value of position $x(t)$ and velocity $v(t) = dx/dt$ at an initial time $t = 0$. We say that A and δ are determined by the **initial conditions** for the motion. It follows from Eq. (13–1) that, at $t = 0$,

$$x = A\sin\delta$$

and

$$\frac{dx}{dt} = \omega A\cos\delta.$$

These equations are enough to specify both A and δ once ω is known.

The period of simple harmonic motion

The constant ω has to do with the repetition time for the motion. The sine function repeats itself either when the angle increases by 2π rad (see Fig. 13–1) or, because

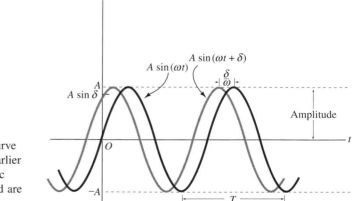

FIGURE 13–2 The phase, δ, corresponds to a sliding of the curve of displacement versus time to earlier or later times for simple harmonic motion. The amplitude and period are also indicated.

δ is a constant, when ωt increases by 2π. The time that is required for this is by definition T, so that $\omega T = 2\pi$. We can solve for T, the **period** of simple harmonic motion:

$$T = \frac{2\pi}{\omega}. \qquad (13\text{--}2)$$

This equation provides us with an interpretation of ω, whose value specifies the value of T. In Chapter 3, where we described uniform circular motion, we defined the **frequency**, f, as the inverse of the period. A period of 5 s means a frequency of one every five seconds, while a period of 0.5 s means a frequency of two per second, and so forth:

The frequency of simple harmonic motion

$$f = \frac{1}{T}. \qquad (13\text{--}3)$$

If the period is measured in seconds, the frequency is measured in s^{-1}. In SI, the unit s^{-1} is the **hertz** (Hz), after the physicist Heinrich Hertz:

The SI unit of frequency

$$1 \text{ Hz} = 1 \text{ s}^{-1}. \qquad (13\text{--}4)$$

By comparing Eqs. (13–2) and (13–3), we find that

$$f = \frac{\omega}{2\pi}. \qquad (13\text{--}5)$$

The constant ω is referred to as the **angular frequency**. We will see in Section 13–2 that ω can be identified with the angular speed, a quantity defined and used in Chapters 3 and 9. Inversion of Eq. (13–2) or (13–5) gives

The angular frequency

$$\omega = \frac{2\pi}{T} = 2\pi f. \qquad (13\text{--}6)$$

When the position is specified as a function of time, the velocity and the acceleration are determined by taking successive derivatives. As a consequence of Eq. (13–1), we have (see Appendix IV–7)

$$v = \frac{dx}{dt} = \omega A \cos(\omega t + \delta). \qquad (13\text{--}7)$$

One further derivative gives the acceleration as a function of time:

$$a = \frac{dv}{dt} = -\omega^2 A \sin(\omega t + \delta) = -\omega^2 x. \qquad (13\text{--}8)$$

The acceleration is proportional to the displacement, with a crucial minus sign. We will argue that virtually all stable equilibrium situations are associated with simple harmonic motion; thus, the proportionality of the acceleration and the displacement is a universal property of motion near equilibrium.

In simple harmonic motion, the acceleration is proportional to the position.

Relations among Position, Velocity, and Acceleration in Simple Harmonic Motion

In Fig. 13–3, we plot the position, velocity, and acceleration of an object in simple harmonic motion over two full periods, starting with $x(t) = A \sin(\omega t)$. (For convenience, the phase has been taken to be zero. The relations discussed here are not affected by the phase.) Here, we take the $+x$-direction to the right and the $-x$-direction to the left.

Let's follow Fig. 13–3. The object starts at the origin at $t = 0$. Here, the velocity is maximum in magnitude and is positive. The acceleration is zero, so that the velocity is not changing. After one-quarter of the period ($\omega t = \pi/2$), the object has moved

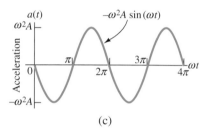

FIGURE 13–3 Starting with (a) position $x(t) = A\sin(\omega t)$, a single derivative gives (b) the velocity $v(t) = \omega A\sin(\omega t)$. (c) One further time derivative gives the acceleration $a(t) = -\omega^2 A\sin(\omega t)$. We have in each case plotted two cycles, or periods, for zero phase; the curves repeat after every period.

to the right-hand extreme and is ready to turn around. The velocity is zero at this turnaround point, but the acceleration has actually reached a maximum in magnitude and is negative, indicating that the velocity will be turning to the left and will become negative. (Think of a ball thrown in the air; at the maximum height—the turnaround point—the velocity is zero even if the acceleration is nonzero and directed toward Earth.) After one-half the period ($\omega t = \pi$), the object once again passes through the origin, this time moving to the left. The acceleration is again zero. The three-quarter mark ($\omega t = 3\pi/2$) is at another turnaround, characterized by a maximum negative value of x—the object is at its left-hand extreme—and zero velocity. The acceleration is maximum and positive, meaning that the velocity is becoming positive, and the object will subsequently move back to the right. Finally, after one full period ($\omega t = 2\pi$), the object has come back to its starting point, moving to the right through the origin with its largest positive velocity and zero acceleration. The situation at $t = 2\pi/\omega$ is identical to what it was at $t = 0$.

EXAMPLE 13–1 A cork floating on a pond moves in simple harmonic motion, bobbing up and down over a range of 4 cm (Fig.13–4). The period of the motion is $T = 1.0$ s, and a clock is started at $t = 0$ s when the cork is at its minimum height. What are the height and velocity of the cork at $t = 10.5$ s?

Solution: We suppose that the cork moves along the z-axis, and we take the origin $z = 0$ to be the midpoint of the motion. Thus the maximum value of z is $z_{max} = 2$ cm, and the minimum value is $z_{min} = -2$ cm. The motion takes the general form $z(t) = A\sin(\omega t + \delta)$. We know the period, T, and from Eq. (13–6), $\omega = 2\pi/T$. The constants A and δ must be determined from other information, namely, the initial conditions. The amplitude, A, is the maximum excursion from equilibrium, and is given by $A = z_{max} = |z_{min}|$. The phase, δ, is then determined by the initial condition that the height is a minimum when $t = 0$ s. Thus the equation determining δ is

$$z_{min} = A\sin(\omega t + \delta)\big|_{t=0} = A\sin\delta.$$

When we substitute $A = |z_{min}|$, this equation becomes $z_{min} = |z_{min}|\sin\delta$. Because z_{min} is negative, this result implies

$$\sin\delta = -1.$$

When the sine function is -1, its argument is $-\pi/2$ or $3\pi/2$. In fact, any integer multiple of 2π can be added to or subtracted from $-\pi/2$, and it is just a matter of convenience to choose the phase to be $-\pi/2$. When a simple phase such as this occurs, it is often worthwhile to expand the sine function with trigonometric identities. In particular, from Appendix IV–4,

$$\sin(\omega t + \delta) = \sin\left(\omega t - \frac{\pi}{2}\right) = \sin(\omega t)\cos\left(\frac{\pi}{2}\right) - \cos(\omega t)\sin\left(\frac{\pi}{2}\right) = -\cos\omega t.$$

We have used the fact that $\cos(\pi/2) = 0$ and $\sin(\pi/2) = 1$. Then, instead of $\sin(\omega t + \delta)$, we have $-\cos(\omega t)$ appearing in the expression for $z(t)$. We gather our results:

$$z = -A\cos\left(\frac{2\pi t}{T}\right),$$

where $A = 2$ cm and $T = 1$ s.

The velocity is the time derivative of this expression:

$$v = \frac{dz}{dt} = -A\left(\frac{-2\pi}{T}\right)\sin\left(\frac{2\pi t}{T}\right) = \frac{2\pi A}{T}\sin\left(\frac{2\pi t}{T}\right).$$

We now want to evaluate z and v at $t = 10.5$ s, or 10.5 periods. Both z and v repeat themselves every period, so the values of z and v at 10.5 s are the same as at

0.5 s (0.5 period):

$$\cos\left[\frac{(2\pi)(10.5\text{ s})}{1.0\text{ s}}\right] = \cos[(2\pi)(10.5)] = \cos[2\pi(10) + 2\pi(0.5)]$$

$$= \cos[2\pi(0.5)] = -1;$$

$$\sin\left[\frac{(2\pi)(10.5\text{ s})}{1.0\text{ s}}\right] = \sin[(2\pi)(10.5)] = \sin[2\pi(10) + 2\pi(0.5)]$$

$$= \sin[2\pi(0.5)] = 0.$$

Thus, for $t = 10.5$ s,

$$z = -A(-1) = A = +2\text{ cm}$$

and

$$v = \frac{2\pi A}{T}(0) = 0\text{ cm/s}.$$

It is simple to deduce these results from physical reasoning. We are interested in where the cork is after exactly one-half a period. Because the cork starts at its minimum height, half a period later it is at its maximum height, +2 cm in this case. That is a point where the cork stops and starts back down, so the velocity is zero there.

FIGURE 13–4 Example 13–1.

13–2 A CONNECTION WITH CIRCULAR MOTION

In Chapter 3, we discussed another kind of periodic motion: uniform circular motion. We now demonstrate that this motion is intimately related to simple harmonic motion (Fig. 13–5). Figure 13–6 recalls uniform circular motion for a point moving in the xy-plane a constant distance R from the origin. The motion is described by an angle θ, measured from the x-axis, that varies linearly with time:

$$\theta = \omega t + \delta. \tag{13–9}$$

The phase, δ, is just the value of θ at time $t = 0$.

If we were to look at a side view of the uniform circular motion of a thumbtack stuck on a rotating turntable, we would see the thumbtack oscillate in simple harmonic motion. Figure 13–6 indicates the *projection* of the circular motion on the x- and y-axes. Simple trigonometry gives us these projections:

$$x = R\cos\theta = R\cos(\omega t + \delta); \tag{13–10}$$

$$y = R\sin\theta = R\sin(\omega t + \delta). \tag{13–11}$$

Thus uniform circular motion corresponds to simple harmonic motion in both of the x- and y-directions. A cosine rather than sine appears in x but, as we have already discussed, this is just the standard form with a different phase. We can use the trigonometric identity $\sin[\theta + (\pi/2)] = \sin\theta\cos(\pi/2) + \cos\theta\sin(\pi/2) = \cos\theta$ to replace the cosine in Eq. (13–10) with a sine function, and we thereby obtain

$$x = R\sin\left(\omega t + \delta + \frac{\pi}{2}\right). \tag{13–12}$$

Both the x- and y-motions are now in the standard form of Eq. (13–1). The two motions have a phase that differs by exactly $\pi/2$ (90°), and the sign of this extra phase specifies the direction—clockwise or counterclockwise—of the corresponding uniform circular motion (see Problem 22).

PROBLEM SOLVING

Each period, the motion repeats.

FIGURE 13–5 The relation between uniform circular motion and simple harmonic motion is evident in the piston-linkage connection on the wheel.

Projections of uniform circular motion are simple harmonic motion.

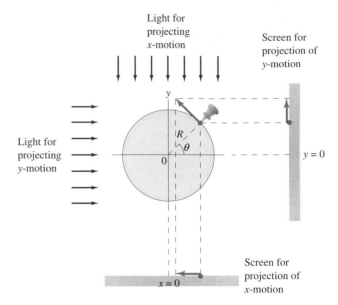

FIGURE 13–6 Uniform circular motion in the *xy*-plane, and its projection onto the *x*- and *y*-axes. These projections are simple harmonic motion.

13-3   SPRINGS AND SIMPLE HARMONIC MOTION

Springs give rise to simple harmonic motion. Let's restrict ourselves to one-dimensional motion and dispense with vector notation. If the spring is not overly stretched or compressed, the spring force on a mass displaced by *x* from the equilibrium position of the spring has the form known as *Hooke's law*:

$$F = -kx. \tag{13–13}$$

∞ The spring force. Hooke's law is discussed in Chapter 6.

Here *k* is the *spring constant*. Thus the force is linear in the displacement. The minus sign in Eq. (13–13) makes the force a *restoring force*. A displacement in the $+x$-direction gives rise to a force that acts in the $-x$-direction and vice versa (Fig. 13–7). Newton's second law provides us with the connection between the force and the acceleration; namely, $F = -kx = ma$. Thus the acceleration of a mass on the end of a spring is proportional to its displacement:

$$a = -\frac{k}{m}x. \tag{13–14}$$

The angular frequency of simple harmonic motion

This is indeed simple harmonic motion because the acceleration is proportional to the position. Comparison of Eqs. (13–8) and (13–14) yields the important result that the angular frequency is determined by the mass and the spring constant:

$$\omega^2 = \frac{k}{m}; \tag{13–15}$$

$$\omega = \sqrt{\frac{k}{m}}. \tag{13–16}$$

The period of simple harmonic motion does not depend on the amplitude.

In turn, Eqs. (13–2) and (13–3) give the period and the frequency of the oscillations:

$$T = 2\pi\sqrt{\frac{m}{k}} \quad \text{and} \quad f = \frac{1}{2\pi}\sqrt{\frac{k}{m}}. \tag{13–17}$$

Remarkably, *the period of the motion is independent of the amplitude.*

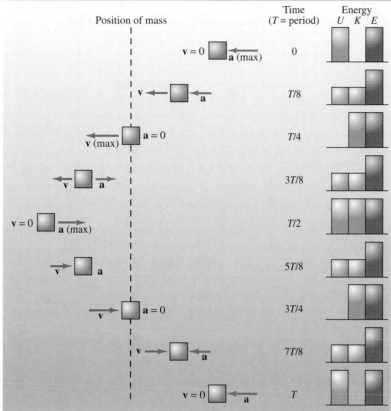

FIGURE 13-7 We can follow the velocity and acceleration of a mass that moves in simple harmonic motion on the end of a spring. The speed is lowest (and the acceleration is highest) when the displacement from equilibrium is a maximum, and the speed is highest (and the acceleration is lowest) when the displacement is a minimum. We can also see the play between kinetic and potential energy; one is large where the other is small.

EXAMPLE 13-2 A mass of 0.50 kg moves along the *x*-direction under the influence of a spring with spring constant $k = 2.0$ N/m. The origin of the *x*-axis is at the equilibrium point of the mass. At $t = 0$ s, the mass is at the origin and moving with a speed of 0.5 m/s in the $+x$-direction. (a) At what time t_1 does the mass first arrive at its maximum extension? (b) What is this maximum extension?

Solution: (a) The information given includes k, m, the position at $t = 0$ s, and the velocity at $t = 0$ s. This is enough information to calculate ω, the period, the amplitude, and the phase. From Eq. (13-16), the angular frequency, ω, is

$$\omega = \sqrt{\frac{k}{m}} = \sqrt{\frac{2.0 \text{ N/m}}{0.50 \text{ kg}}} = \sqrt{4.0 \text{ s}^{-2}} = 2.0 \text{ rad/s}.$$

The period T is

$$T = \frac{2\pi}{\omega} = \frac{2\pi \text{ rad}}{2.0 \text{ rad/s}} = 3.1 \text{ s}.$$

The mass travels from the equilibrium position to its maximum extension in precisely one-quarter period, so

$$t_1 = \frac{T}{4} = 0.79 \text{ s}.$$

(b) We use the information about x and v at $t = 0$ s to find the amplitude. Writing $x(t) = A \sin(\omega t + \delta)$, we get

$$t = 0 \text{ s}, \qquad x = A \sin \delta = 0.$$

This implies that $\delta = 0$. We use this, in turn, for the value of v:

$$t = 0 \text{ s}, \qquad v = A\omega \cos(0) = A\omega.$$

The argument of the cosine is zero because both t and δ are zero. Thus

$$A = \frac{v|_{t=0}}{\omega} = \frac{0.50 \text{ m/s}}{2.0 \text{ rad/s}} = 0.25 \text{ m},$$

which is the maximum excursion of the mass from the origin.

The spring is the prototype of dynamic systems that have simple harmonic motion. Many other dynamic systems also exhibit simple harmonic motion. All such systems are reducible in one way or another to a spring, meaning that the *form* of the force is the same as that of the spring: a restoring force linear in some variable.

The Hanging Spring: How Constant Forces Affect Simple Harmonic Motion

When a constant force such as gravity influences a mass that is already attached to a spring, remarkably little changes. The motion of this mass is exactly as it would be with no gravity, except that the equilibrium point changes. To show this, suppose that a mass is suspended vertically from a spring (Fig. 13–8). The mass would be at height $y = y_0$ if there were no gravity, where y is measured up from ground level. In the presence of gravity, the mass hangs at a new, lower equilibrium position, $y_1 = y_0 - \Delta y$. The positive value Δy is determined by a cancellation between the downward (negative) force of gravity, whose magnitude is $F_g = mg$, and the upward (positive) force exerted by the spring,

$$F_{\text{spring}} - F_g = 0.$$

Because $F_{\text{spring}} = -k(-\Delta y) = k \, \Delta y = k(y_0 - y_1)$, we have

$$k(y_0 - y_1) - mg = 0. \qquad (13\text{–}18)$$

Suppose that we now move the mass to a new position, y. The stretch of the spring from its original equilibrium position (no gravity) is $y - y_0$, and the spring force is

(a) (b) (c)

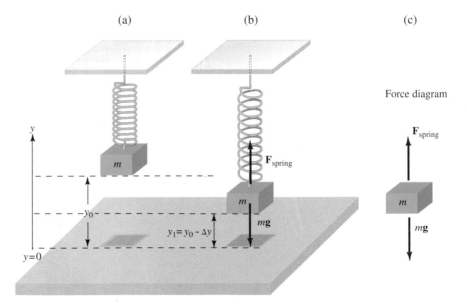

Force diagram

FIGURE 13–8 A mass on the end of a spring is suspended vertically. (a) If its equilibrium length would place it at height y_0 in the absence of gravity, then (b) it will be stretched an additional amount, Δy, to a new equilibrium position, y_1, under the influence of gravity. (c) Force diagram for the mass.

$F_{spring} = -k(y - y_0)$. The signs are satisfactory because if y is higher than y_0, then $y - y_0$ is positive and the spring force is negative (or downward). If y is below y_0, then $y - y_0$ is negative and the spring force is positive (or upward). The force on the mass is then

$$F_{net} = F_{spring} - F_g = -k(y - y_0) - mg. \qquad (13\text{-}19)$$

According to Eq. (13–18), we can replace the quantity mg by $k(y_0 - y_1)$, and the net force is

$$F_{net} = -k(y - y_0) - k(y_0 - y_1) = -k(y - y_1). \qquad (13\text{-}20)$$

Equation (13–20) is again the equation for a spring force with exactly the same spring constant, and hence *the motion of a mass on the end of a vertical spring has exactly the same frequency as it does when it is on the end of a horizontal spring.* The only change is that the displacement is measured from the new equilibrium position, y_1, rather than the old one, y_0.

13-4 ENERGY AND SIMPLE HARMONIC MOTION

We extensively examined questions involving energy as they apply to the spring force in Chapter 7, where the topic was the conservation of energy. In particular, we found that the work done by a spring force in moving a mass from one position to another is independent of the path taken by the mass. This means that *the spring force is conservative.* A conservative force depending on a position variable x has a potential energy function $U(x)$ associated with it; the total energy, E (the sum of kinetic energy, K, and potential energy), is *conserved* throughout any motion.

The computation of the potential energy $U(x)$ of an object attached to a spring was treated in Section 7–2. We found

$$U(x) = \frac{1}{2} kx^2. \qquad (13\text{-}21)$$

In Eq. (13–21), zero potential energy has been chosen at the equilibrium position of the spring, $x = 0$. The kinetic energy is simply

$$K = \frac{1}{2} mv^2. \qquad (13\text{-}22)$$

The potential energy associated with a spring

Because both x and v are known for simple harmonic motion from Eqs. (13–1) and (13–7), the variation in time of U and K can be plotted. If we write the argument of $\omega t + \delta$ as θ, we have

$$U = \frac{1}{2} kA^2 \sin^2 \theta, \qquad (13\text{-}23)$$

$$K = \frac{1}{2} mA^2 \omega^2 \cos^2 \theta = \frac{1}{2} kA^2 \cos^2 \theta. \qquad (13\text{-}24)$$

In writing this form for K, we have used $\omega^2 = k/m$, Eq. (13–15). Figure 13–9 is a plot of the potential and kinetic energy functions as θ varies between 0 and 2π, which corresponds to a complete cycle. Both $\sin^2 \theta$ and $\cos^2 \theta$ vary between 0 and 1; when $\sin^2 \theta$ is a minimum, $\cos^2 \theta$ is a maximum and vice versa. Thus U and K each vary between 0 and $kA^2/2$. Suppose that an object attached to a spring starts at the origin and moves to the right. At the origin, the potential energy is zero and K is a maximum. As the mass moves to the right, it slows until it has reached its turnaround point at one-quarter cycle, where the velocity and hence K are zero. Because x is at its maximum here, U is also a maximum. The mass now moves to the left, gaining speed until the speed is a maximum as it passes through the origin once more. Here, after one-half cycle, K is a maximum and U is a minimum. Finally, at the left-hand turnaround point, K is a minimum and U is a maximum. *The energy flows back and forth between U and K.*

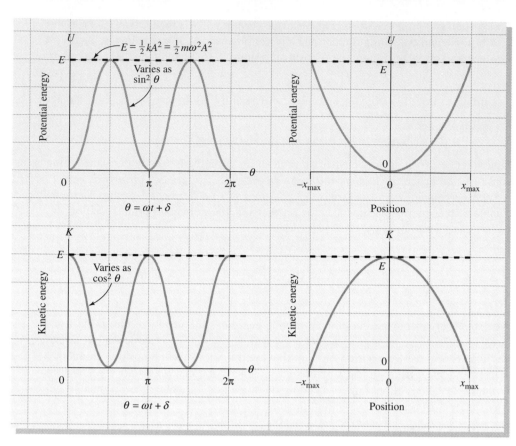

FIGURE 13-9 (a) The potential energy and (b) the kinetic energy of a mass in simple harmonic motion plotted over one cycle. When one is a maximum, the other is a minimum, and their sum, the total energy, is conserved.

The Total Energy

The total energy, $E = U + K$, must be constant. We have

$$E = \frac{1}{2} kA^2 \sin^2(\omega t + \delta) + \frac{1}{2} kA^2 \cos^2(\omega t + \delta)$$

$$= \frac{1}{2} kA^2 [\sin^2(\omega t + \delta) + \cos^2(\omega t + \delta)]. \qquad (13\text{-}25)$$

The total energy in simple harmonic motion

Because the sum of $\sin^2\theta$ and $\cos^2\theta$ is unity for any θ, E is indeed constant in time:

$$E = \frac{1}{2} kA^2. \qquad (13\text{-}26)$$

The quadratic dependence of energy on amplitude is typical of simple harmonic motion.

EXAMPLE 13-3 A mass m attached to a spring of spring constant k is stretched a length X from its equilibrium position and released with no initial motion. (a) What is the maximum speed attained by the mass in the subsequent motion? (b) At what time is this speed first attained?

Solution: (a) The conservation of energy is a useful tool in this type of problem. Just before the mass is released from rest, all of its energy is potential energy; that is, the total energy is

$$E = \frac{1}{2} kX^2.$$

This agrees with Eq. (13–26) because the maximum excursion in space of the motion is, by definition, the amplitude of the motion. E is the value of the energy for all time. When the maximum speed is attained, all the energy is in the form of kinetic energy:

$$\frac{1}{2} m v_{\max}^2 = E = \frac{1}{2} k X^2.$$

We solve for $v_{\max}$:

$$v_{\max} = \sqrt{\frac{k}{m}} X = \omega X.$$

(b) The conservation of energy is not very helpful when we want to learn about time dependence. To answer the second question, we must make use of our knowledge about the time dependence of simple harmonic motion. The maximum speed is attained here when $x = 0$. Because the mass is released at the maximum value of x, the first time the mass passes through the origin is one-quarter period later:

$$t = \frac{T}{4} = \frac{1}{4} 2\pi \sqrt{\frac{m}{k}} = \frac{\pi}{2} \sqrt{\frac{m}{k}}.$$

PROBLEM SOLVING

For time dependence, energy is less useful.

13–5 THE SIMPLE PENDULUM

Simple harmonic motion occurs throughout nature. A particularly important example is the ordinary pendulum, such as the swinging pendulum in some clocks. For centuries, such clocks were the most accurate way to measure time. When we idealize the form of the pendulum to a point mass suspended from a massless string of length ℓ, as in Fig. 13–10, we have the **simple pendulum**. The mass moves along the arc of a circle traced out by the end of the taut string (Fig. 13–11a).

Let's first look at how Newton's second law applies. Suppose that the string makes an angle θ with the vertical. The force diagram for the mass, Fig. 13–11b, includes the force of gravity, $m\mathbf{g}$, and the tension of the string, $\mathbf{T}$. The tension is perpendicular to the path of the motion. Its only role in the motion is to constrain that motion to lie along the arc of a circle of radius ℓ. The position s of the mass along the arc of the circle is given by

$$s = \ell\theta, \tag{13–27}$$

where s is measured from $\theta = 0$. The angle θ varies with time and is the quantity we wish to determine dynamically. Accordingly, we are interested only in components of forces along the arc of the path.

To obtain the velocity along the arc of the circle, we differentiate s with respect to time. Because ℓ is a constant, we find

$$v = \frac{ds}{dt} = \ell \frac{d\theta}{dt}. \tag{13–28}$$

The simple pendulum

APPLICATION

Pendulum clocks

FIGURE 13–10 The simple pendulum, strobed at equal time intervals. The pendulum bob moves faster near the bottom of its swing and more slowly near the ends.

(a)

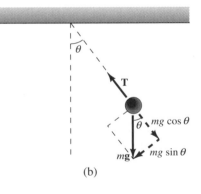

(b)

FIGURE 13–11 (a) Geometry of the simple pendulum. (b) Force diagram for the simple pendulum. The tension, $\mathbf{T}$, is along the radial direction, but $m\mathbf{g}$ has components along both the radial direction and tangent to the arc of the circle traced out by the motion of the mass.

The tangential acceleration, which is the component of the total acceleration along the arc of the circle, is, in turn, associated with changes in the magnitude of this velocity:

$$a = \frac{dv}{dt} = \ell \frac{d^2\theta}{dt^2}. \tag{13–29}$$

This component of the total acceleration (which should *not* be confused with the centripetal component of the acceleration) is due to the tangential force component

$$F_t = -mg \sin\theta. \tag{13–30}$$

The sign of this force component is important. It is in the negative direction when θ is itself positive—when the mass is on the right side of the vertical—and in the positive direction when the mass is on the left side of the vertical. This means that *the force of gravity always acts to bring the mass back to the vertical.* This is enough to ensure that the motion will be oscillatory but not enough to guarantee simple harmonic motion: In simple harmonic motion, the force must be *linear* in the dynamical variable itself—in this case, the angle.

Using Eq. (13–29) for the acceleration and Eq. (13–30) for the force, we see that Newton's second law takes the form

$$m\ell \frac{d^2\theta}{dt^2} = -mg \sin\theta.$$

Canceling the mass from this equation, we get

$$\ell \frac{d^2\theta}{dt^2} = -g \sin\theta. \tag{13–31}$$

Equation (13–31) would satisfy our requirement for simple harmonic motion if, instead of $\sin\theta$, θ itself appeared on the right-hand side. But this will indeed happen when the angle θ is small. To see this, refer to Fig. 13–12a for $\sin\theta$ and θ, which gives

$$\sin\theta = \frac{x}{\ell} \quad \text{and} \quad \theta = \frac{s}{\ell}, \tag{13–32}$$

where s is the arc length corresponding to angle θ. In Fig. 13–12b, note that x approaches s more and more closely as θ becomes smaller. Thus, for small θ, $\sin\approx\theta$. This can also be seen from the expansion for the sine function:

$$\sin\theta = \theta - \frac{\theta^3}{3!} + \frac{\theta^5}{5!} - \cdots. \tag{13–33}$$

When θ is small, the terms of order θ^3, θ^5, and so forth can be ignored, justifying our approximation. For example, when $\theta = 0.2$ rad (about 11°), the difference between $\sin\theta$ and θ is about 1 percent, but when $\theta = 0.1$ rad (about 6°), the difference is only about 0.1 percent.

The small-θ approximation means that a pendulum will have true simple harmonic motion only for small excursions. This is because, with this approximation, Eq. (13–31) becomes

$$\text{for small } \theta: \quad \ell \frac{d^2\theta}{dt^2} = -g\theta. \tag{13–34}$$

Equation (13–34) is *precisely* the equation for simple harmonic motion, as shown by comparison with Newton's second law for the spring, $m\, d^2x/dt^2 = -kx$. We must change only some variable names and constants to go from spring to pendulum:

$$x \to \theta, \quad k \to g, \quad \text{and} \quad m \to \ell. \tag{13–35}$$

(a)

(b)

FIGURE 13–12 Comparing the horizontal displacement x of the mass of a simple pendulum with the arc length s traced out by the motion. These two lengths differ by a larger percentage (a) when θ is large than (b) when θ is small.

(Here, the horizontal arrow simply indicates a name substitution.) The solution to the motion of the simple pendulum for small angles is then taken directly from the solution for the motion of the spring:

$$\theta = \theta_0 \sin(\omega t + \delta) \quad \text{with} \quad \omega = \sqrt{\frac{g}{\ell}}. \qquad (13\text{--}36)$$

Again, dynamics has determined the frequency. Here, θ_0 is the amplitude of the angular motion—the maximum angle attained. The period and frequency of the pendulum's motion come from Eqs. (13–5) and (13–6):

$$T = 2\pi \sqrt{\frac{\ell}{g}} \quad \text{and} \quad f = \frac{1}{2\pi} \sqrt{\frac{g}{\ell}}. \qquad (13\text{--}37)$$

As for the mass on the spring, *the period of the pendulum is independent of the amplitude.* This explains why a pendulum clock can be accurate: The amplitude does not enter into the period of the swing!

The period and frequency of the simple pendulum

EXAMPLE 13–4 A simple pendulum 2.00 m long is suspended in a region where $g = 9.81$ m/s^2. The point mass at the end is displaced from the vertical and given a small push, so its maximum speed is 0.11 m/s. What is the maximum horizontal displacement of the mass from the vertical line it makes when at rest? Assume that all the motion takes place at small angles.

Solution: The angle that the string makes with the vertical varies harmonically, $\theta = \theta_0 \cos(\omega t + \delta)$, where ω is the angular frequency. The horizontal displacement from the vertical is $x = \ell\theta$ (where ℓ is the string's length), as long as θ remains small (see Fig. 13–12b). Thus x also varies harmonically: $x = A \cos(\omega t + \delta)$, where $A = \ell\theta_0$. This is the quantity we want to find. Another good small-angle approximation is that the vertical component of the velocity is small, so $v \simeq dx/dt$. Thus we have

$$v \simeq \frac{d}{dt}[A \cos(\omega t + \delta)] = A\frac{d}{dt}[\cos(\omega t + \delta)] = -A\omega \sin(\omega t + \delta).$$

From this expression, we see that v varies harmonically, with amplitude $A\omega$. The maximum value of v occurs at $\theta = 0$ and is given by $v_{max} = A\omega = 0.11$ m/s. Finally, we can find ω from Eq. (13–36):

$$\omega = \sqrt{\frac{g}{\ell}} = \sqrt{\frac{9.81 \text{ m/s}^2}{2.00 \text{ m}}} = 2.21 \text{ rad/s}.$$

Thus, from $v_{max} = A\omega$,

$$A = \frac{v_{max}}{\omega} = \frac{0.11 \text{ m/s}}{2.21 \text{ rad/s}} = 0.05 \text{ m}.$$

This horizontal displacement of 5.0 cm is indeed small compared to the length, so our small-angle approximations are good. Figure 13–13 illustrates the motion.

The Energy of the Simple Pendulum

Let's now consider the simple pendulum from the point of view of energy. The kinetic energy, K, which is a function of θ, is found by expressing the speed as a function of θ. Equation (13–28) gives us what we require, and

$$K(\theta) = \frac{1}{2}mv^2 = \frac{1}{2}m\ell^2\left(\frac{d\theta}{dt}\right)^2. \qquad (13\text{--}38)$$

The only force with a component along the motion of the mass is gravity. This force is conservative and to find the associated potential energy function, U, express the height h gained by the mass in terms of θ. From Fig. 13–11a, we see that $h = \ell - \ell \cos\theta$, so that

$$U = mgh = mg\ell (1 - \cos\theta). \qquad (13\text{--}39)$$

FIGURE 13–13 Example 13–4.

Here, we have taken zero potential energy to be at the bottom of the swing, at $\theta = 0$. In the small-angle approximation, $\cos\theta \simeq 1 - (\theta^2/2)$, and

$$U(\theta) = mg\ell(1 - 1 + \frac{1}{2}\theta^2)$$

$$= \frac{1}{2}mg\ell\,\theta^2. \qquad (13\text{–}40)$$

The potential energy of the simple pendulum

Comparing Eqs. (13–38) and (13–40) for the kinetic and potential energies with their counterparts for the spring, Eqs. (13–22) and (13–21), we see that they are formally the same—with the name changes listed in Eq. (13–35).

The period of oscillation of a simple pendulum provides a highly accurate method for the measurement of g because the time of many oscillations gives a very precise measure of the period of one oscillation. The period of a pendulum has played an important historical role in the definition of time. The grandfather clock, now more a curiosity or a work of art than a scientific instrument, was crucial in setting standards for time (Fig. 13–14).

FIGURE 13–14 The pendulum clock was the best timekeeper available for many years.

Simple harmonic motion nearly always occurs near points of stable equilibrium.

∞ The concept of the potential energy well is treated in Chapter 7.

13–6 HARMONIC MOTION AND EQUILIBRIUM

The universal importance of the motion described in this chapter stems from the fact that *almost all small departures from stable equilibrium exhibit simple harmonic motion.* Everyday experience bears this out, at least in a qualitative way. A marble nudged a little from its stable equilibrium at the bottom of a bowl rolls back and forth, as does a mass hanging from a string or an automobile rocking on its worn shock absorbers. It is for this reason that simple harmonic motion occurs so often in physics. Table 13–1 gives a sampling of the range of periods of mechanical systems that move in simple harmonic motion.

In the spring, the pendulum, and indeed every case of stable equilibrium, some mass is trapped in a *potential energy well.* This term refers to a *minimum* in the curve of potential energy versus some position variable (Fig. 13–15). In the spring case, the position variable is the stretch of the spring and the minimum is at zero stretch. For the spring, we have seen that the potential energy forms a parabola. For a pendulum, the appropriate variable is the pendulum's angle θ; and the potential energy curve is also a parabola in θ near $\theta = 0$.

The Taylor expansion (App. IV–8) is a very general mathematical result that allows us to see why the minimum of a well forms a parabola and hence why simple harmonic motion is universal near equilibrium. Suppose we apply the Taylor expansion to a potential energy function near a minimum, which characterizes a stable equilibrium point. Let's label the position of the minimum as the origin, $x = 0$. Then, with each differentiation with respect to x labeled with a prime, we have

$$U(x) = U(0) + xU'(x)|_{x=0} + (x^2/2)U''(x)|_{x=0} + \cdots.$$

TABLE 13–1 Periods of Mechanical Systems in Simple Harmonic Motion

Mechanical System	*Period (s)*
Sloshing of water in a tidal basin or large lake	10^2 to 10^4
Large structures (bridges, buildings)	> 1
Strings or air columns of musical instruments	5×10^{-2} to 10^{-4}
Fluid machinery	10^{-4} to 10^{-5}
Piezoelectric crystals, ultrasound generators	10^{-5} to 5×10^{-1}
Vibrations in molecules	10^{-14}

The constant $U(0)$ plays no physical role and, as we know, we can always replace it by 0. The first derivative of U at $x = 0$ is zero because that is a minimum point. Thus, if we keep the first nonzero term in the Taylor expansion, we find

$$U(x) \simeq (x^2/2)U''(0). \qquad (13\text{--}41)$$

(Here, we have used the notation $U''(x)|_{x=0} = U''(0)$.) This is indeed in the form of a spring force, with $U''(0)$ playing the role of the spring constant. Thus the force takes the general form near the equilibrium point

$$F(x) = -\frac{dU}{dx} \simeq -U''(0)x, \qquad (13\text{--}42)$$

the familiar linear restoring force.

The Physical Pendulum

A pendulum need not consist of a massless string with a pointlike mass at the end of it. When the suspended, swinging object has some other form, we call it a *physical pendulum.* Any object can be suspended from any point on the object and act as a physical pendulum. This illustrates the point that simple harmonic motion is a general characteristic of motion about a stable equilibrium. The appropriate approach to such problems is to recast them as a spring problem, through analysis of either the potential energy or the dynamical equation, and then use the results we have already found for the spring.

We have the necessary tools to handle the physical pendulum: We can study the torque, τ, on it and the corresponding angular acceleration. Figure 13–16 illustrates a physical pendulum allowed to pivot through some horizontal axis, called y, which defines the vertical plane of the swinging. We take the y-axis to be into the page and through the oscillation point. We need to know the rotational inertia, I, about the pivot axis, the total mass M of the object, and the distance r from the center of mass to the pivot axis. The stable equilibrium point for this object is $\theta = 0$, when the center of mass hangs directly below the axis. When $\theta \neq 0$, only the force of gravity, which acts on the center of mass, exerts a (restoring) torque. From Eq. (9–28), the equation of motion governing the behavior of the object is

$$\tau = I\alpha, \qquad (13\text{--}43)$$

where α is the angular acceleration about the pivot axis. The magnitude of α is $d^2\theta/dt^2$ and, because it acts to *decrease* θ, its sign is negative.

The torque is given by $\boldsymbol{\tau} = \mathbf{r} \times \mathbf{F}$, where $\mathbf{r}$ is the vector from the pivot axis to the center of mass and $\mathbf{F}$ is the force of gravity acting on the center of mass. According to the right-hand rule, this torque is along the $+y$-axis (when θ is positive, as shown); α must therefore also be along the $+y$-axis, which corresponds in Fig. 13–16 to an angular acceleration that brings the object back to equilibrium. (Both $\boldsymbol{\tau}$ and $\boldsymbol{\alpha}$ would change signs if the object were drawn with θ on the other side of the vertical axis.) The magnitude of $\boldsymbol{\tau}$ is

$$\tau = Fr \sin \theta = Mgr \sin \theta, \qquad (13\text{--}44)$$

so the equation of motion, Eq. (13–44), becomes

$$Mgr \sin \theta = -I\frac{d^2\theta}{dt^2}. \qquad (13\text{--}45)$$

Equation (13–45) has exactly the same form as the equation governing the motion of the simple pendulum, which is Eq. (13–31). For small angles, the sine of θ can be replaced with θ itself and, as for the simple pendulum, simple harmonic motion fol-

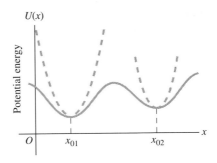

FIGURE 13–15 A potential well, in which potential energy has a minimum at $x = x_0$. This point is a point of stable equilibrium. The dashed line is a parabola that matches the minimum of the well.

PROBLEM SOLVING

Try to restate stable equilibrium situations to resemble the spring problem.

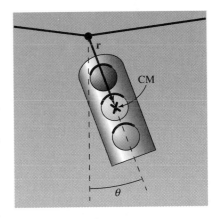

FIGURE 13–16 A swinging traffic light is an example of a physical pendulum. The center of mass of the oscillating light executes simple harmonic motion about the stable equilibrium position, $\theta = 0$.

lows: $\theta = \theta_0 \sin(\omega t + \delta)$. To find ω, it is necessary only to replace ℓ and g for the simple pendulum with I and Mgr, respectively, for the physical pendulum. Thus, from Eqs. (13–36) and (13–37),

$$\omega = \sqrt{\frac{Mgr}{I}}, \tag{13–46a}$$

$$T = 2\pi \sqrt{\frac{I}{Mgr}}. \tag{13–46b}$$

Because I is always M times some length squared, M will cancel from Eqs. (13–46). *The mass of a physical pendulum does not enter into its period.* One check on the period of the physical pendulum as expressed in Eq. (13–46b) is to verify that it does reduce to the simple pendulum when the swinging object is a point mass m on a massless string of length ℓ. In that limit, $I \rightarrow m\ell^2$, $r \rightarrow \ell$, $M \rightarrow m$, and T reduces to the appropriate value,

$$T \rightarrow 2\pi \sqrt{\frac{m\ell^2}{mg\ell}} = 2\pi \sqrt{\frac{\ell}{g}}.$$

For a physical pendulum of a given shape, the period of the small amplitude oscillations are independent of the mass.

FIGURE 13–17 Example 13–5.

EXAMPLE 13–5 A thin, uniform rod of mass M and length L swings from its end as a physical pendulum (Fig. 13–17). What is the period of the oscillatory motion for small angles? Find the length ℓ of the simple pendulum that has the same period as the swinging rod.

Solution: Equation (13–46b) gives the period of a physical pendulum swinging in small angles and thus is directly applicable here. We need to find the rotational inertia of a rod about an axis through its end. This quantity was calculated in Eq. (9–24) and is listed in Table 9–1. It is

$$I = \frac{1}{3}ML^2.$$

The distance of the center of mass from the end is $r = L/2$, and Eq. (13–46b) then gives the period of the motion as

$$T_{\text{physical pendulum}} = 2\pi \sqrt{\frac{I}{Mgr}} = 2\pi \sqrt{\frac{ML^2/3}{MgL/2}} = 2\pi \sqrt{\frac{2}{3}\left(\frac{L}{g}\right)}.$$

The length ℓ of a simple pendulum of the same period is determined by

$$T_{\text{simple pendulum}} = 2\pi \sqrt{\frac{\ell}{g}} = 2\pi \sqrt{\frac{2}{3}\left(\frac{L}{g}\right)};$$

$$\ell = \frac{2}{3}L.$$

13-7 DAMPED HARMONIC MOTION

Almost all physical systems, including springs and other oscillating systems, are affected by friction or drag (resistive) forces (Fig. 13–18). These forces tend to remove energy from a moving system and thereby slow it down, or damp its motion. Their universality and importance are evident in the world around us. The pendulum of a clock does not swing forever. There is an energy loss due to damping; as can be seen from Eq. (13–26), a decrease in energy implies a decrease in amplitude. For the pendulum to maintain a given amplitude, energy must be supplied from, say, a wound spring or falling weights.

What are the *quantitative* effects of these nonconservative, or dissipative, forces? We can answer in the case of a drag force, $\mathbf{F}_d$, proportional to velocity:

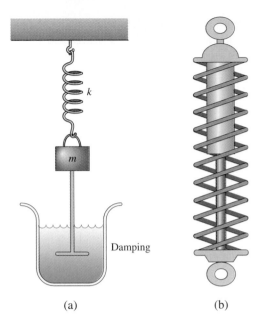

FIGURE 13–18 (a) A damped oscillator. (b) An automobile shock absorber is a damped harmonic oscillator.

$$\mathbf{F}_d = -b\mathbf{v} = -b\frac{d\mathbf{x}}{dt}, \qquad (13\text{–}47)$$

where b is the *damping coefficient* (or *damping parameter*). The minus sign indicates that this force is always opposite to the direction of motion. For this case the equations of motion are *solvable* in analytic form, and we can at least take these solutions as a guide to the behavior of damped systems. The equation of motion is

$$-kx - b\frac{dx}{dt} = m\frac{d^2x}{dt^2}. \qquad (13\text{–}48)$$

Just how does the damping modify simple harmonic motion? To solve differential equations such as Eq. (13–48), we attempt trial solutions that are educated guesses and see if they work. As a guess to the solution of Eq. (13–48), we keep a sinusoidal component but also incorporate a decreasing term due to damping. Our trial solution is

$$x = Ae^{-\alpha t}\sin(\omega' t + \delta), \qquad (13\text{–}49)$$

where A, δ, and α are constants to be determined. We use a frequency ω' rather than ω_0, the frequency in the absence of drag, because we want to allow for the possibility that the frequency is changed by the damping. We check whether the trial solution is a good one by seeing if the equation of motion, in this case, Eq. (13–48), is satisfied when the trial solution is plugged into it. This procedure is left to Problem 71. The result is that the trial solution is satisfactory provided that

$$\alpha = \frac{b}{2m} \qquad (13\text{–}50)$$

and that

$$\omega' = \sqrt{\frac{k}{m} - \frac{b^2}{4m^2}} = \sqrt{\omega_0^2 - \frac{b^2}{4m^2}}. \qquad (13\text{–}51)$$

Thus *the damping factor α and the modified angular frequency ω' are determined by the equations of motion.* (A and δ remain undetermined.) It is easy to check that Eq. (13–49) reduces to standard simple harmonic motion when $b = 0$.

The falling exponential function in Eq. (13–49) is a kind of *envelope* (shown by the dashed lines in Fig. 13–19a) that modulates what would otherwise be simple harmonic motion; this causes the amplitude of the motion to decrease as time goes on.

PROBLEM SOLVING

Solve differential equations with trial solutions.

The mass thereby settles back to rest at the equilibrium point. The argument of the exponential is directly proportional to b; that is, to the size of the drag force. Figure 13–19 shows the motion of Eq. (13–49) for two values of b, one larger than the other. Here, we have taken $k = 1$ N/m, $m = 1$ kg, and $A = 1$ m. In Fig. 13–19a, $b = 0.1$ kg/s, and in Fig. 13–19b, $b = 0.8$ kg/s. In each case, the decrease of the amplitude with time is clearly visible. The difference between ω' and the angular frequency $\omega_0 = \sqrt{k/m}$ of the spring without drag is harder to see.

As b increases, the angular frequency decreases (and hence the period of the motion increases) until, from Eq. (13–51), $\omega' = 0$ when $b^2 = 4mk$. We refer to this value of b as the *critical* value b_c,

$$b_c = \sqrt{4mk} = \sqrt{4m^2\omega_0^2};\qquad(13\text{--}52)$$

Different degrees of damping

the system is said to be **critically damped**. The system is **underdamped** for $b < b_c$, when it still exhibits oscillatory behavior—albeit with decreasing amplitude. When $b > b_c$, the system is **overdamped**. Then, the strength of the damping causes the system to come to rest at equilibrium without oscillation; it crosses the equilibrium point at most once following an initial displacement.

The quantity $b/2m$, which appears multiplying t in the exponential factor in Eq. (13–49), has dimensions of inverse time because the argument of an exponential must be dimensionless. We therefore define the **lifetime** (or the *mean life*), τ, of the damped oscillator by

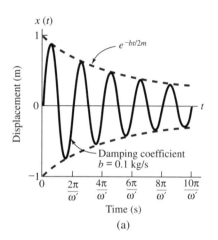

$$\tau \equiv \frac{m}{b}.\qquad(13\text{--}53)$$

In terms of τ, the exponential envelope has the form $e^{-t/2\tau}$. The larger the value of τ, the slower the exponential falloff. Still another nomenclature employs the dimensionless **Q factor**, defined by

$$Q \equiv \omega_0\tau.\qquad(13\text{--}54)$$

Because Q is proportional to τ, it too measures the amount of damping. The less the damping, the larger are the values of τ and Q.

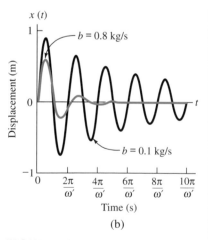

EXAMPLE 13–6 A struck gong can be modeled as a damped harmonic system, with its frequency of oscillation measured by the musical tone emitted and the "loudness" of its sound measured by the amplitude squared (we will see this in Chapter 14). A certain gong is struck. After 9.0 s, the loudness has dropped to 0.85 times the original loudness. How much more time will have elapsed before the loudness is 0.25 times the original loudness?

Solution: In the damped systems studied in this section, the amplitude decrease is governed by the exponential factor $e^{-t/2\tau}$, where τ is the lifetime characteristic of the system. The loudness drop-off is proportional to the square of this factor, or $e^{-t/\tau}$. When we denote the loudness as $L(t)$, we have $L(t) = L(0)\,e^{-t/\tau}$. At 9.0 s, $L = L(0)(0.85)$, and therefore

$$e^{-(9.0\text{ s})/\tau} = 0.85.$$

FIGURE 13–19 Damping of simple harmonic motion by a drag force $\mathbf{F} = -b\mathbf{v}$. Plotted is the function $x(t) = A\,e^{-bt/2m}\sin(\omega't)$ versus t, with $\omega' = \sqrt{k/m - b^2/4m^2}$. We have taken $k = 1$ N/m, $m = 1$ kg, and $A = 1$ m. In (a), $b = 0.1$ kg/s, and in (b), $b = 0.8$ kg/s.

This equation allows us to find τ. The inverse of the exponential is the natural logarithm, ln (see Appendix IV–6). If we take the natural log of both sides of this equation, we have

$$\ln[e^{-(9.0\text{ s})/\tau}] = -\frac{9.0\text{ s}}{\tau} = \ln(0.85) = -0.16,$$

$\tau = (-9.0\text{ s})/(-0.16) = 55$ s. We want to find the time for which $L(t)/L(0) = 0.25$, or the time for which $e^{-t/\tau} = 0.25$. Using the value of τ we have found, we take the natural log of both sides:

$$\ln[e^{-t/(55 \text{ s})}] = -\frac{t}{55 \text{ s}} = \ln(0.25) = -1.4,$$

or $t = (1.4)(55 \text{ s}) = 76 \text{ s}$.

This gong has a large Q value; it undergoes many oscillations before the damping factor has reduced the amplitude substantially.

13–8 DRIVEN HARMONIC MOTION

At times, systems such as a mass on the end of a spring are subject to external forces. We saw the effect of a constant force (gravity) on an oscillator in Section 13–5. Another common situation involves a *harmonic driving force*, in which the driving force varies sinusoidally with time (Fig. 13–20). The resulting motion of the mass is called **driven harmonic motion**. (Damped harmonic motion is undriven.)

In **resonance** phenomena, the amplitude of simple harmonic motion grows enormously when the frequency of the driving force matches the natural frequency of the oscillating system. The Tacoma Narrows Bridge collapse of 1940 is a good example (Fig. 13–21). Strong winds drove the bridge into oscillatory motion and the amplitude of the oscillations increased to the point where the bridge actually collapsed. You may have read about the possibility of a bridge being driven to large oscillations and collapse if soldiers march over it in step with the right rhythm. Less spectacular examples are familiar: the trampoline jumper who, by timing her jumps, can make them much more effective or the coffee that sloshes out over the edge of the cup when you walk at just the wrong pace. Resonance phenomena are even more important in microscopic situations, as in the action of a microwave oven, where microwave radiation drives the electrons of water molecules with a natural frequency of the molecular system. Resonances also occur in acoustical phenomena, and many musical instruments are driven in harmonic resonance in order to produce their notes. Finally, electric circuits often behave like driven harmonic systems even though they are not mechanical.

It is easy to demonstrate the resonance phenomenon for yourself on a swing (Fig. 13–22) or with the simple aid of a loose rubber band tied to a mass such as a kitchen utensil. By extending the suspended band and releasing it, you can get a good idea of the natural frequency of the system and observe the amplitude steadily reduced by the damping. If you then tie the band to one hand, you can drive the system by moving your hand up and down in an approximation of harmonic motion—first with a frequency less than, then greater than, and finally equal to, the natural frequency. At resonance, when the hand frequency equals the natural frequency of the system, the oscillation amplitude is dramatically large.

Equations of Motion for Driven Harmonic Motion

Let's consider a mass subject to a spring force, a drag force proportional to the speed, and an external harmonic driving force F whose time variation is determined by the angular frequency ω, where $F(t) = F_0 \sin(\omega t)$. F_0 is a type of amplitude for the driving force. In this problem, it is useful to recall the natural frequency of the unencumbered mass on the spring, $\omega_0 = \sqrt{k/m}$. All motion is in the x-direction, so the x-component of Newton's second law, $\mathbf{F} = m\mathbf{a}$, is

$$-kx - b\frac{dx}{dt} + F_0 \sin(\omega t) = m\frac{d^2x}{dt^2}. \qquad (13\text{–}55)$$

Equation (13–55) appears to be even more complicated than the equation of motion with a drag force alone, but a moment's reflection can help us. By definition, the

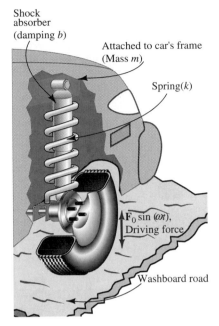

Shock
absorber
(damping b)

Attached to car's frame
(Mass m)

Spring(k)

$F_0 \sin(\omega t)$,
Driving force

Washboard road

FIGURE 13–20 A car passing over a rough road is an example of driven, damped harmonic motion. The washboard road provides the force, the shock absorber provides the damping, and the spring provides the harmonic motion.

APPLICATION

Applications of resonance

In driven harmonic motion, the mass on the spring moves with the frequency of the driving force.

(a)

(b)

FIGURE 13–21 The Tacoma Narrows Bridge in Tacoma, Washington. (a) In November 1940, strong winds drove the bridge into oscillatory motion. (b) It may have been a resonant phenomenon that increased the amplitude of the oscillations to the point of collapse.

(a)

(b)

FIGURE 13–22 An example of resonant driven periodic motion. (a) A push in rhythm with the swing's natural frequency drives the swing to large amplitudes. (b) The child can rhythmically "pump," a more complex way of bringing in the resonant effect.

driving force has gone on and will go on forever—we mean, of course, for a long time—and the mathematical solution must show that *the motion of the mass is simple harmonic motion with the frequency of the driving force.* In other words, the mass will follow the time dependence of the driving force. Any exponential effects due to friction will have long since died out—exponentially, of course. Imagine, for example, a spring with a mass on its end suspended from a harmonically moving hand. In time, the mass will move with the frequency of your hand motion, even if the motion of the mass and your hand are not in phase. After long times, the solution to the equation of motion is then

$$x = A \sin(\omega t + \delta). \tag{13–56}$$

This solution can be verified by substitution into Eq. (13–55), and the amplitude A and phase δ are determined in this substitution. This is a complicated exercise in algebra, which we forego. The physically interesting amplitude of oscillation is in this case determined by Eq. (13–55):

$$A = \frac{F_0}{\sqrt{m^2(\omega^2 - \omega_0^2)^2 + b^2\omega^2}}. \tag{13–57}$$

This amplitude has the remarkable property of *resonance: It is peaked when the driving frequency, ω, nears the natural frequency, ω_0.* In fact, if there were no damping ($b = 0$), the amplitude would become infinite when $\omega = \omega_0$. This is not a realistic physical situation because it corresponds to the spring being stretched to infinite length. A real spring will snap rather than accept an infinite stretch; in other words, some form of damping will ultimately occur. But it does illustrate that, at resonance, the response of a harmonic system to a driving force can be catastrophically large.

Properties of Resonance

We plot amplitude as a function of ω in Fig. 13–23 for several values of b. We choose $F_0 = 0.01$ N, the natural frequency $\omega_0 = 1$ rad/s, and $m = 1$ kg. The damping coefficient b is given the values 0.01, 0.05, and 0.2 kg/s. From the plot, we can see that the position of the amplitude shifts slightly from $\omega_0 = 1$ rad/s when b increases from a small value. From Eq. (13–57), the peak amplitude generally occurs at

$$\omega_{max}^2 = \omega_0^2 - \frac{1}{2}\left(\frac{b^2}{m^2}\right). \tag{13–58}$$

That the peak is less sharp for larger b is sometimes expressed as the fact that the *width* of the peak is a measure of b. To be more precise, it is possible to compute the *total width at half-maximum* $\Delta\omega$ of the peak, found by evaluating the amplitude where it is one-half its peak value and measuring the spread of frequencies to which this corresponds, $\Delta\omega$ (Fig. 13–24). For small b, this width can be shown to be

$$\Delta\omega \simeq \frac{2b}{m}. \tag{13–59}$$

Thus the sharpness $\Delta\omega$ of the resonance peak is a direct measure of the damping coefficient, b, divided by m: The smaller the damping, the sharper the resonance peak.

*Resonance and Uncertainty

We have seen that the amplitude for a driven, damped harmonic oscillator has a resonance peak. We have also seen that the undriven damped harmonic oscillator has a characteristic decay time. These phenomena are closely related, and this relation has important consequences for our ability to construct systems with resonances, such as radio tuners or filters for screening electronic noise.

Equation (13–49) can be used to find the rate at which the energy is dissipated in an undriven damped oscillator. The energy is proportional to the amplitude *squared*

The most obvious application of simple harmonic motion can be seen in clocks. The pendulum clock is the simplest example. A more modern example is the quartz crystal that provides the time standard in a wristwatch (Fig. 13AB–1). The crystal takes the form of a bar, end mounted and cut in relation to the "piezo axis" of the crystal. The "piezo" property of a crystal is an electrical property; in this case, it refers to the fact that an applied electric field causes the crystal to flex, and a harmonically varying electric field leads to harmonic flexing. The amplitude of flexure at the resonant frequency of the crystal is very much larger than at other frequencies, and the energy loss due to damping passes through a maximum at the resonant frequency. The electric circuitry that provides the driving frequency senses where the energy loss is greatest, so there is a feedback to the fixed resonant frequency. In a cheap watch, the crystal is driven only at its resonant frequency, and the accuracy of the period is on the order of one part in 10^5 to 10^6. Better watches can adjust the driving frequency, so that the quartz bar can be operated slightly away from resonance. Such

FIGURE 13AB–1 The quartz watch has put accurate timekeeping within the reach of everyone.

watches have accuracies of one part in 10^7. Time standards that can keep the oscillating bar at a constant temperature achieve stabilities of better than one part in 10^{10}, not as good as atomic clocks, but much more convenient.

and therefore falls as $e^{-t/\tau}$, where $\tau = m/b$ is the lifetime. The width in frequency of the driven harmonic oscillator is given in Eq. (13–59), and we see that it is inversely proportional to τ. From Eqs. (13–53) and (13–59), we have

$$\tau\Delta\omega \text{ is on the order of 1.} \qquad (13\text{--}60)$$

This equation is known as the **uncertainty principle**: It expresses the impossibility of measurements that are arbitrarily precise in both time and frequency. Strictly speak-

The uncertainty principle

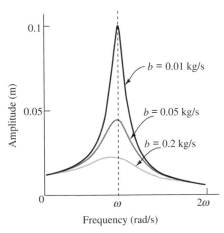

FIGURE 13–23 The amplitude of a harmonic system under the influence of a harmonic driving force with frequency ω. The natural frequency of the harmonic system is $\omega_0 = 1$ rad/s. There is a drag force with damping coefficient b. Resonance occurs at the peak (near $\omega = \omega_0$) and is strongest when b is smallest.

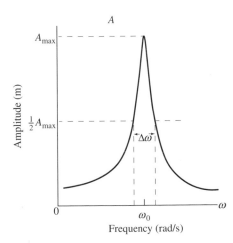

FIGURE 13–24 The width at half-maximum of a resonant peak near ω_0 is $\Delta\omega$.

ing, we have derived it only for a special damping force. But it actually represents a very general property. It states that if the damping time of a harmonic oscillator is large, then the resonance width is small, and vice versa. The weaker the damping of a harmonic oscillator, the more sharply it responds to, or selects, a harmonic driving force of the right frequency. What is the significance of this result? One of the important uses of the resonance phenomenon, both in mechanical systems and in electric circuits, is that it allows us to select, or filter, certain frequencies by letting these frequencies act as a driving force on our selector. For example, the tuner on a radio selects a given station by letting in only frequencies near the central frequency of the station. Equation (13–60) sets strong limits on our ability to design filters that can respond to only a narrow range of driving frequencies. The transient effects of such filters die away more slowly with time as the frequencies they select are more and more limited. This is a very general result that cannot be avoided with clever design.

SUMMARY

Simple harmonic motion is characterized by a particular type of periodic behavior; namely, sinusoidal time dependence of some position variable of a moving object. In one dimension, this dependence can be written as

$$x(t) = A \sin(\omega t + \delta). \tag{13–1}$$

The amplitude, A, measures the maximum displacement from equilibrium. The phase, δ, is the angle by which the motion is shifted from $x = 0$ at $t = 0$; the angular frequency, ω, measures how rapidly the motion repeats. The period, T, and frequency, f, of the simple harmonic motion are related to ω by

$$T = \frac{2\pi}{\omega} \tag{13–2}$$

and by

$$f = \frac{1}{T}. \tag{13–3}$$

In simple harmonic motion, the velocity and the acceleration are also sinusoidal. In particular, the acceleration is proportional to the displacement but opposite in sign:

$$a = -\omega^2 x. \tag{13–8}$$

Uniform circular motion is closely related to simple harmonic motion. It is just the result of simultaneous simple harmonic motion of equal amplitudes and frequencies in the x- and y-directions, 90° out of phase.

Because acceleration is proportional to displacement in simple harmonic motion, the force that leads to this motion is a restoring force proportional to the displacement. The prototype of this is the force exerted on a mass by a spring, Hooke's law:

$$F = -kx, \tag{13–13}$$

where k is the spring constant and x is the displacement of the mass from the equilibrium position of the spring. The resulting angular frequency of the motion of a mass m subject to this force is

$$\omega = \sqrt{\frac{k}{m}}. \tag{13–16}$$

The period of the motion is independent of the amplitude.

The potential energy of a mass subject to a spring force is

$$U(x) = \frac{1}{2}kx^2. \tag{13–21}$$

It oscillates with time as $\sin^2(\omega t + \delta)$, whereas the kinetic energy oscillates with time as $\cos^2(\omega t + \delta)$, and the total energy, E, which is their sum, remains constant with time:

$$E = \frac{1}{2}kA^2. \qquad (13\text{–}26)$$

The simple pendulum, consisting of a mass on the end of a light string of length ℓ, also undergoes simple harmonic motion when it is not allowed to swing too far. In this case, it is the angle with respect to the vertical that varies sinusoidally, and the period of the motion is independent of the mass as well as the amplitude:

$$T = 2\pi\sqrt{\frac{\ell}{g}}. \qquad (13\text{–}37)$$

Almost all small departures from stable equilibrium situations exhibit simple harmonic motion. Two examples are a bead rolling back and forth across the bottom of a bowl and a physical pendulum. The period of a physical pendulum depends on the geometric factor contained in its rotational inertia about the axis through which it swings, but its total mass cancels from the period.

The presence of additional velocity-dependent drag, or resistive, forces causes the amplitude of a particle that moves under the influence of springlike forces to decrease. When a system that, by itself, would move in simple harmonic motion is driven by a force with sinusoidal time dependence, the system moves with the frequency of the driving force. The amplitude of the resulting motion of the system shows resonant behavior when the frequency of the driving force equals the natural frequency of the system. The width of the resonance peak is inversely related to the exponential rate of falloff of the undriven system due to damping—a result known as the uncertainty principle.

UNDERSTANDING KEY CONCEPTS

1. Describe some examples of simple harmonic motion that are not discussed in the chapter.
2. Describe an experiment to determine the spring constant of a spring.
3. If the amplitude of simple harmonic motion is doubled, what happens to the maximum kinetic energy?
4. Will a pendulum clock lose time or gain time when it is taken from sea level to the top of a mountain? Ignore all damping effects.
5. In the previous question, we suggested that all damping effects be ignored. Suppose that you could not ignore damping effects due to the air, which becomes less dense as you gain in altitude. Why might you find it difficult to predict whether a pendulum clock gains or loses time when it is taken from sea level to the top of a mountain?
6. A damped harmonic oscillator driven by a harmonic external force maintains a steady oscillatory motion. Is energy lost to friction in the motion? If so, what keeps the oscillator moving?
7. The length of a simple pendulum is doubled, and the mass at the end is halved. What happens to the period?
8. What is the phenomenon that allows you to increase the amplitude of your motion when you swing on a swing?
9. Suppose you stand on a swing instead of sitting on it. Will your frequency of oscillation increase or decrease?
10. It can be shown that if the frequency, the energy, or the amplitude of a harmonic oscillator changes very slowly then, although the energy E is no longer constant, the ratio E/ω, where ω is the angular frequency, does not change. Suppose you have a pendulum whose maximum angle of deflection is θ_0. If the bob at the end of the pendulum consists of fine sand in a spherical container, and the sand leaks very slowly out the bottom, what will happen to the motion?
11. Make use of the assertion in Question 10 to discuss what happens to the period and amplitude of a pendulum in which the length is very slowly decreased by a factor of 1.5.
12. Discuss if and why each of the following is an example of simple harmonic motion, damped harmonic motion, and/or driven harmonic motion: (a) leaves blowing on a tree limb; (b) children seesawing; (c) a child swinging on a playground swing; (d) a car bouncing up and down after hitting a large pothole in the road; and (e) water sloshing back and forth in a tub.
13. A mass is suspended from two ropes of equal length, attached to different points on a ceiling. Would small oscillations about the point of stable equilibrium represent simple harmonic motion?
14. A simple pendulum acts as a *conical pendulum* when the mass moves in a horizontal circle (Fig. 13–25). What force keeps it moving in a circle? Does the total energy consist of potential and kinetic energies that each vary in this case?

FIGURE 13-25 Question 14.

15. Suppose that there were a harmonic driving force on a mass attached to a horizontal spring and that very little damping occurs. What could supply the energy that would allow the amplitude to become enormously large near resonance?

16. What are you doing when you adjust your bounce up and down on a diving board to get a big boost for your dive?

17. A large sled is sliding down a snowy hill. Resting on the sled are a mass and a spring; one end of the spring is attached firmly to the sled and the other end is attached to the mass. There is little friction between the mass and the sled and the mass is set into oscillation. Will the sled accelerate smoothly down the hill?

PROBLEMS

13-1 The Kinematics of Simple Harmonic Motion

1. (I) What phase, δ, is necessary if $\sin(\theta + \delta) = \cos(\theta)$? This can be shown graphically, with Fig. 13-1, or by using a trigonometric identity for the sine of the sum of two angles.

2. (I) A simple harmonic motion along the x-direction has the following properties: The maximum amplitude = 0.15 m, the time between the maximum and minimum x-values = 0.4 s, and $x = 0.1$ m for $t = 0.1$ s. Find the period, the angular frequency, and the general equation of motion.

3. (I) A spring has the speed $v = 0.4 \sin(\omega t + \pi)$ m/s, where $\omega = 2.00$ rad/s. Plot x, v, and a as functions of time for three periods of motion.

4. (I) The angular frequency of a mass on the end of a spring in simple harmonic motion is 3.827 rad/s. What is the period of the motion?

5. (I) The amplitude of the motion of a mass attached to a spring is $A = 2.84$ m, while the maximum speed of the mass is 4.36 m/s. What is the period of the motion?

6. (I) The maximum speed of a mass attached to a spring is $v_{max} = 0.486$ m/s, while the maximum acceleration is 1.66 m/s^2. What is the maximum displacement of the mass?

7. (I) What is the position of the mass in Problem 6 at time $t = 1$ s, given that the mass is precisely at the origin ($x = 0$ m) at $t = 0$ s?

8. (II) The expression for the position of a simple harmonic oscillator is given by $x(t) = B \cos(\omega t) + C \sin(\omega t)$. (a) Show that this can be written in the form given in Eq. (13-1) and express the constants B and C in terms of the constants A and δ. (b) Express B and C in terms of the position x_0 and the speed v_0 at time $t = t_0$. (c) What is the maximum speed in terms of B and C?

9. (II) A harmonic oscillator operates at a frequency of 813.52 Hz. What is the amplitude for which the maximum acceleration is 183.25 m/s^2?

10. (II) A small object is placed on a horizontal platform, which vibrates vertically with an amplitude of 1 cm. The frequency of the vibration is slowly increased. At what frequency will the object start bouncing on the platform?

11. (II) The motion of a mass can be described by the function $x(t) = A \sin(\omega t + \delta)$, where $\omega = 2.0$ rad/s and $\delta = 0.40$ rad. Express the motion as a cosine function.

12. (II) A professor pacing back and forth is observed by the class

to move back and forth along the x-axis in a rough approximation to simple harmonic motion. Relative to the center of the classroom, the motion is between the two extremes $+3$ m and -3 m. The professor was at $x = -0.3$ m at $t = 0$ s, and the motion is observed to repeat itself 6 times in 90 s. What are the amplitude, phase, period, frequency, and angular frequency of the motion?

13. (II) A spring has spring constant 0.5 N/m and a 0.2-kg mass on its end, which has a maximum speed of 2.0 m/s. (a) What are the angular frequency and period of the system? (b) What is the amplitude of the motion?

14. (II) A sailing ship rolls sideways in simple harmonic motion, with the period given by $T = 8$ s (Fig. 13-26). The tip of a 25-m mast travels a maximum of 3 m from the vertical position. What is the speed of the tip of the mast at the instant it is in a vertical position?

FIGURE 13-26 Problem 14.

15. (II) A particle undergoing simple harmonic motion travels a total distance of 6.98 cm during one cycle of 1.71 s. (a) What is the average speed of the particle? (b) What are its maximum speed and acceleration?

16. (II) A mass $m = 0.3$ kg is attached to an ideal spring of spring constant $k = 15$ N/m. All motion takes place in the (horizontal) x-direction, and the equilibrium position of the mass is defined to be $x = 0$ m. The mass is then displaced to $x = 0.050$ m and released from rest. (a) Write an expression for the function $x(t)$ that describes the subsequent motion. (b) At what po-

sition x does the maximum positive acceleration of the mass occur? (c) What is the magnitude of the maximum acceleration? (d) What is the maximum speed of the mass?

17. (II) An object oscillates with an angular frequency of 3.0 rad/s. Its initial displacement from equilibrium is $+3.0$ cm, and the initial velocity is 5 cm/s in the direction of the equilibrium point. (a) Find the displacement as a function of time. (b) How soon will the displacement be $+3$ cm again? (c) At what times will be object move with a speed of 5 cm/s? (d) Sketch x and v as functions of t.

13–2 A Connection with Circular Motion

18. (I) Find the phase angle δ for uniform circular motion when $x = -R$ and $y = 0$ at $t = 0$.

19. (I) A particle moves in the xy-plane so that its x and y coordinates are described by Eq. (13–10) and Eq. (13–11); namely,

$$x = R\cos(\omega t + \delta), \qquad y = R\sin(\omega t + \delta).$$

Show that the distance of the particle from the origin is a constant, and find that constant.

20. (I) When a certain uniform circular motion in the xy-plane is projected onto the x- and y-axes, this projection gives $x(t) = R\sin[\omega t + \delta - (\pi/2)]$ and $y(t) = R\sin(\omega t + \delta)$. (a) Show that the uniform circular motion is clockwise. (b) What would happen if instead of an extra phase $-\pi/2$ in $x(t)$, there was an extra phase $+\pi/2$ in $y(t)$?

21. (I) Write down a formula that describes the motion of Earth around the Sun as seen by a distant observer in the plane of Earth's orbit. Assume a circular orbit.

22. (II) A small object is placed at the outer edge of an LP record of diameter 12 in. The record rotates clockwise at $33\frac{1}{3}$ rev/min. (a) What is the projection of the object's motion on the x-axis, assuming that at $t = 0$ the projection of the motion places the object at $x = 0$? (The point $x = 0$ corresponds to the center of the record.) (b) Give the amplitude, angular frequency, and largest speed of the projection of the motion.

23. (II) Show that if the motion along the x-and y-axes of a mass moving in the xy-plane is $x(t) = R_1\cos(\omega t + \delta)$ and $y(t) = R_2\sin(\omega t + \delta)$, then this motion traces out an ellipse of axis lengths R_1 and R_2. (An ellipse is the curve described by $[x^2/a^2] + [y^2/b^2] = 1$, where a and b are the axis lengths.)

13–3 Springs and Simple Harmonic Motion

24. (I) A 10-kg ball is suspended motionless under the influence of gravity from a spring with a force constant of $k = 2 \times 10^3$ N/m. (a) How much is the spring stretched? (b) The same spring is sent to the Moon and suspended in the same way. By how much is the spring stretched in this case?

25. (I) A student has a spring with $k = 200$ N/m and wants to build a horizontal mass–spring system with period 1.0 s. What mass should the student use at the end of the spring?

26. (I) The spring on a scale is compressed by 2.45 cm when a 60-kg man stands on it. What is the spring constant of the spring?

27. (I) The period of a 45-g mass attached to the end of a spring is 3.1 s. Find the spring constant of the spring.

28. (I) A mass moves along the x-axis under the influence of a spring whose equilibrium position is at the origin. The mass moves between limits of -2 cm and $+2$ cm, and the period of its motion is 0.5 s. (a) What is the period if the mass is dou-

bled? halved? (b) What is the period if the amplitude is doubled to 4 cm? halved to 1 cm?

29. (II) The plate at the base of a floor sander moves back and forth in simple harmonic motion at a frequency of 20 oscillations/s, and the amplitude of the motion is 0.8 cm. If the mass of the oscillating plate is 1200 g, what is the maximum value of the driving force?

30. (II) A spring is placed in a vertical position by suspending it from a hook at its top. A similar hook on the bottom of the spring is measured to be 45 cm above a table top. A mass of 180 g, of negligible size, is then suspended from the bottom hook, which is now found to be 20 cm above the table top. The mass is then pulled down a distance of 15 cm and released (Fig. 13–27). Find (a) the spring constant, k; (b) the angular frequency, ω; (c) the position of the bottom hook after 5 s.

FIGURE 13–27 Problem 30.

31. (II) A spring is suspended from the ceiling. A mass of 25 g is attached to it, and the spring stretches by 12 cm. Ignore damping. What is the period of the oscillation of a 75-g mass attached to the spring?

32. (II) Two identical springs of spring constant k are attached end to end to make one longer spring. Show that this new spring has spring constant $k/2$. The springs are said to be attached *in series*. The case of n springs attached in series gives a spring n times as long, with a spring constant $k_n = k/n$.

33. (II) A small object of mass $m = 0.06$ kg is held in place by two springs (Fig. 13–28). The one acting on the left has spring constant $k_1 = 100$ N/m; the one acting on the right has spring constant $k_2 = 200$ N/m. The object is moved away from its equilibrium position by 1 cm to the right, and released at time $t = 0$. What is the displacement of the object as a function of time?

FIGURE 13–28 Problem 33.

34. (I) What is the energy of a 0.5-kg mass that moves with amplitude 5 cm on a flat, frictionless table and is attached to a spring of spring constant 860 N/m?

35. (I) A trampoline acts like a spring of spring constant 1200 N/m. A mass is placed on the trampoline. The trampoline surface is depressed by a total of 30 cm. What is the period of oscillation of the system if this mass is pushed beyond the equilibrium displacement?

36. (I) A mass of 1.2 kg, attached to a spring, is in simple harmonic motion along the x-axis, and its period is $T = 2.5$ s. If the total energy of the spring and mass is 2.7 J, what is the amplitude of the oscillation?

37. (I) A spring with a 1-kg fish at its end and with spring constant $k = 2$ N/m is compressed 3 cm from equilibrium and then released. Use the conservation of energy to find the maximum speed of the fish.

38. (I) A puck of mass 582 g moves horizontally with speed $v = 2.5$ m/s on a frictionless surface toward the end of a relaxed spring for which $k = 840$ N/m, and compresses the spring. By how much is the spring compressed? How would your answer change if the mass of the puck were increased by 100 percent?

39. (I) A mass at the end of a string moves in a circle in a vertical plane. The only forces acting on it are the tension of the string, which is central, and the force of gravity. If the length of the string is 1.2 m and the angular velocity at the top of the circle is $\omega = 2.2$ rad/s, what is the angular velocity at the bottom of the circle?

40. (II) Consider a mass m, moving along the x-axis, with potential energy given by $U(x) = \frac{1}{2}m\omega^2 x^2$. Show that the motion of this mass is simple harmonic motion with angular frequency ω by using $dE/dt = 0$, where E is the (constant) total energy of the object.

41. (II) A point mass m on a turntable that rotates with angular frequency ω is located a distance d from the center of the turntable. Show that the energy of the point mass is the sum of the energies of the harmonic motions in the x- and y-directions.

42. (II) A mass m of 0.2 kg is attached to the end of a spring and released from rest at $t = 0$ s from an extended position $x_{max} = 6$ cm. All other forces acting on the spring cancel, so that the motion of the mass is due entirely to the effect of the spring. After its release, the mass's speed drops back to zero for the first time after 0.3 s. Find the maximum speed of the mass.

43. (II) A mass m attached to the end of a spring is released from rest at $t = 0$ s from an extended position x_{max}. The mass $m = 0.2$ kg, and $k = 1$ N/m. After 0.5 s, the speed of the mass is measured to be 1.5 m/s. Calculate x_{max}, the maximum speed of the motion, and the total energy.

13–5 The Simple Pendulum

44. (I) A simple pendulum 1.20 m long is suspended in a location where g is 9.82 m/s^2. What is the period of the pendulum?

45. (I) A simple pendulum has a frequency of 0.342 Hz. The length of the pendulum string is 2.12 m. What is the local value of g?

46. (I) A thin wire 7.58 m long is attached to the ceiling of a large lecture room. A lead ball is attached at the bottom. The wire

is displaced by an angle of 0.075 rad and released. Express the angular displacement of the lead ball as a function of time, given that $g = 9.80$ m/s^2.

47. (I) You need to measure the height of a room. You have a watch but no meter stick. You also have your human brain, not to mention an opposable thumb. A long pendulum with a point mass at the end extends from the ceiling to the floor and has a period $T = 3$ s. (a) What is the height of the room? (b) You take the same pendulum to the top floor of a skyscraper and measure a period $T = 3.0002$ s. What is the height of the skyscraper?

48. (II) The difference in temperature between summer and winter causes the length of the pendulum in a clock to change by one part in 30,000. What time-difference error will this make in one week?

49. (II) A small lead ball of mass 2 kg is suspended at the end of a light string 1 m in length. A small peg, 0.5 m below the suspension point, catches the string in its swing (Fig. 13–29). The ball is set swinging through small angles. (a) What is the period of this pendulum? (b) The ball is started swinging on the side that does not catch the peg, at an initial height 0.05 m above the low point. How high does it rise on the side where the peg restricts the pendulum length to 0.5 m?

FIGURE 13–29 Problem 49.

13–6 Harmonic Motion and Equilibrium

50. (I) A small door to allow a dog to pass in and out of the house has a mass of 0.6 kg and is 25 cm wide and 35 cm tall. The door is hinged along the top of the 25-cm width. What is the frequency of oscillation of the door?

51. (I) A student wants to build a pendulum out of a circle of plywood as shown in Fig. 13–30. The circle has a radius of $R = 10$ cm and the plywood has a mass of 200 g. What is the period of the motion?

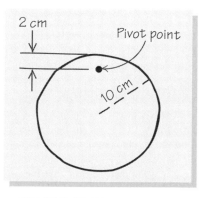

FIGURE 13–30 Problem 51.

52. (II) A uniformly dense book of dimensions 5 cm × 12 cm × 20 cm has a mass of 3 kg. You pick it up close to one corner with a pair of calipers and let it swing. What is the angular frequency? You can do this experiment with your own textbook.

53. (II) A thin, uniform rod of mass M and length L oscillates with small amplitude in a vertical plane about a pin that pierces the rod a distance y from the upper end. Calculate the period of the motion as a function of y. [*Hint*: You need to use the parallel-axis theorem to calculate the rotational inertia to be used in Eq. (13–46b).]

54. (II) An object of mass m has potential energy given by $U(x) = U_0(x^2 - a^2)^2$. (a) What are the stable equilibrium positions of the object? (b) What is the angular frequency of its motion about a stable equilibrium point, when the object is displaced by an amount that is small compared with a? [*Hint*: Let $x = x_{eq} + z$, and keep only up to z^2 terms in $U(z)$.]

55. (II) A straight wire 80 cm long is bent in the middle into an L-shape and balanced with the two ends down on a knife edge. With what frequency will it oscillate about its equilibrium position?

56. (II) Imagine a tunnel that has been drilled through Earth: a smooth, straight tunnel with a frictionless interior (Fig. 13–31). The deepest point of the tunnel is at depth d, and the coordinate x measures the distance along the tunnel from its deepest point to an arbitrary point P a distance ℓ from Earth's center. The known parameters are the mass and radius of Earth, M_E and R_E, respectively, as well as G. Earth is assumed to have uniform density. (a) What is the total mass of that portion of Earth that lies within the distance ℓ in terms of ℓ, M_E, R_E, and G? (b) What is the gravitational force, in direction and magnitude, acting on a ball of mass m at point P? (c) What is the *total* force, in direction and magnitude, acting on the ball *as a function of x* and of the constants of the problem? Why is there no net force acting perpendicular to the tunnel? (d) What is the period of the motion if the ball is released at rest at an entrance to the tunnel? Ignore air resistance. (e) What is the period of a satellite in circular orbit around Earth at a radius equal to Earth's radius? Ignore air resistance.

FIGURE 13–31 Problem 56.

57. (II) A thin uniform disk of mass M and radius R hangs from a nail driven straight into the disk at a distance ℓ from the center. (a) What is the rotational inertia of the disk about the nail? (b) What is the equation of motion for small oscillations of this pendulum about the point where the nail is driven in?

(*Hint*: Use Newton's law for the torque about the point in question.) (c) What is the period, T, of oscillations about the point of suspension? (d) What is T in the limit where ℓ goes to zero?

58. (II) A uniform circular disk of radius R and mass M is attached at its center to one end of a massless, rigid rod of length L. The other end of the rod is attached to the ceiling and pivots freely about that point. The system thus makes a physical pendulum. What is the period of small oscillations of the pendulum?

59. (II) A *torsion pendulum* consists of a dumbbell suspended from its center by a wire that resists being twisted (Fig. 13–32). The dumbbell has length ℓ and has masses m on each end. It remains in the horizontal plane as it twists back and forth about the equilibrium position, $\theta = 0$ rad. For small oscillations about this equilibrium position, the torque exerted by the wire has the form $\tau = -\gamma\theta$, where γ is the *torsion constant*. (a) For $m = 80$ g, $\ell = 30$ cm, and $\gamma = 2 \times 10^5$ g·cm²/s², what is the period of the oscillation? (b) If the system is started from rest at $\theta = 0.1$ rad, what is the total energy of the system? (c) What is the maximum speed of either mass, given the initial conditions of part (b)?

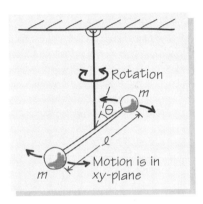

FIGURE 13–32 Problem 59.

60. (II) A slender, uniform rod of mass m and length ℓ can be pivoted on a frictionless horizontal support at any point along the rod's length. The rod then moves as a physical pendulum for small oscillations about the vertical equilibrium position. Suppose that the pivot point is at a point $z\ell$ from the end of the rod, where z is a fraction between 0 and 1. Using an equation for the angular acceleration of the rod as a function of the angle θ, which measures the departure from the vertical, find the period of small-angle oscillations. Check the answer by specifying $z = 0.5$, where the period should become large.

61. (III) A small metal block of mass m is placed on a smooth horizontal table and constrained to move along a frictionless, rectilinear groove. The block is attached to one end of a spring (of spring constant k) whose other end is fastened to a pin P (Fig. 13–33). Let length ℓ be the equilibrium length of the spring and the perpendicular distance to the groove. The spring is now pulled a distance x_0 from the equilibrium position and released. Show that, if the displacement along the groove $x \ll \ell$, the restoring force on the block is proportional to x^3, so the motion is not simple harmonic (although it will still be periodic). [*Hint*: For $x \ll \ell$, $\sqrt{x^2 + \ell^2} \simeq \ell + (x^2/2\ell)$.]

FIGURE 13–33 Problem 61.

62. (III) A bead of mass $m = 50$ g slides without friction at the bottom of a bowl whose bottom traces out the arc of a circle of radius $R = 20$ cm. If the bead moves in a plane that passes through the bottom of the bowl, its position can be described by an angle θ measuring its angular displacement from the bottom. (a) Express the potential energy as a function of θ for very small angles θ. [*Hint*: The potential energy is a linear function of height. Reexpress the energy in terms of θ, and use an approximation suitable for small θ.]

At $t = 0$ s the mass is released from rest from a very small angle $\theta_0 = 0.1$ rad. Parts (b)–(e) refer to this initial condition. (b) Give an expression $\theta(t)$ describing the subsequent motion. In particular, what is the frequency of small oscillations at the bottom of the bowl? (c) What is the total energy of the system? (d) the velocity of the bead at $t = 0.1$ s? (e) the acceleration of the bead at $t = 0.2$ s?

13–7 Damped Harmonic Motion

63. (I) Consider the gong struck in Example 13–6. What is the loudness of the sound (as a fraction of the original loudness) after 2 min? How much time does it take for the sound to have a loudness 1/10,000 of the original loudness? Your ear is a very good detector of loudness, and you would have no trouble hearing the gong for this entire time.

64. (I) A mass on a spring with a natural angular frequency $\omega_0 = 2.5$ rad/s is placed in an environment in which there is a damping force proportional to the speed of the mass. If the amplitude is reduced to 0.65 times its initial value in 5.4 s, what is the angular frequency of the damped motion?

65. (I) The damping coefficient of a damped harmonic oscillator can be adjusted. Two measurements are made: First, when the damping coefficient is zero, the angular frequency of motion is 3880 rad/s. Second, a static measurement shows that the effective spring constant of the system is 184 N/m. To what value should the damping coefficient be set in order to have critical damping?

66. (II) A mass m is attached to a horizontal spring, with spring constant k, and the mass can move under the spring's influence along the x-axis on a frictionless tabletop. It is initially at rest at the equilibrium position $x = 0$. At $t = 0$, the mass is struck a brief blow that gives it a speed v_0 in the $-x$-direction. (a) What is the position x of the mass as a function of time? (b) Suppose that there is a small drag force $f_D = -bv$, where b is a constant and v is the speed of the mass. What is the position x of the mass as a function of time? (c) Suppose that, rather than a drag force proportional to speed, there is a small force f of kinetic friction, constant in magnitude but always opposing the motion, acting on the object. How, qualitatively, does the motion differ from that described in part (b)?

67. (II) A harmonic oscillator with natural period $T = 2.0$ s is placed in an environment where its motion is damped, with a damping force proportional to its speed. The amplitude of the oscillation drops to 70 percent of its original value in 5.0 s. What is the period of the oscillator in the new environment?

68. (II) Consider a damped harmonic oscillator. The damping, proportional to the speed, is sufficiently weak so that it is a good approximation to view the amplitude as constant over the duration of a cycle. What is the energy of the oscillator at time t if its original energy at time $t = 0$ is E_0? [*Hint*: For one cycle, the usual energy formula involving the square of the amplitude can be used.]

69. (II) A spring with $k = 12.0$ N/m and an attached bob oscillates in a viscous medium (Fig. 13–34). A given maximum, of $+6.0$ cm from the equilibrium position, is observed at $t = 1.5$ s, and the next maximum, of $+5.6$ cm, occurs at $t = 2.5$ s. What will the position of the bob be at 3.0 s and at 4.8 s? What was its position at $t = 0$ s?

FIGURE 13–34 Problem 69.

70. (II) For what mass m of the bob in Problem 69 will the spring system have its critical value? What are the lifetime and Q factor in this case?

71. (II) Show by direct substitution that Eq. (13–49) is a solution to the equations of motion of the damped harmonic oscillator, Eq. (13–48), provided that Eqs. (13–50) and (13–51) hold. [*Hint*: You will need to show that the sine and cosine terms *separately* vanish.]

13–8 Driven Harmonic Motion

72. (I) A long, flexible, and very light strip of metal, such as a saw blade, is clamped to the edge of a table, extending horizontally over the edge like a miniature diving board. When a gob of putty is placed on the end, the strip sags and comes to a new equilibrium position with its tip a distance 1.0 cm below its original position. If the end of the strip is then lightly tapped, its tip will oscillate in simple harmonic motion. With what frequency should the tip be tapped to make the strip oscillate with maximum amplitude?

73. (I) A mass of 0.5 kg is suspended from a spring, which stretches by 8 cm. The support from which the spring is suspended is set into sinusoidal motion. At what frequency would you expect resonant behavior?

74. (II) Show that resonance occurs for driven harmonic motion when $\omega = \sqrt{\omega_0^2 - (b^2/2m^2)}$. The resonant frequency occurs when the amplitude has a maximum as a function of frequency.

75. (II) A particular spring has a spring constant of 86 N/m and a mass of 0.548 kg at its end. When the spring is driven in a viscous medium, the resonant motion occurs at an angular frequency of 12.2 rad/s. What is (a) the damping parameter due to the viscous medium? (b) the lifetime of the system? (c) the sharpness of the resonance peak?

76. (II) Consider the driven, damped harmonic oscillator discussed in Section 13–8. (a) Calculate the power dissipated by the damping force. (b) Calculate the average power loss, using the fact that the average of $\sin^2(\omega t + \delta)$ over a cycle is one-half. [*Hint*: Recall that $P = Fv$.]

General Problems

77. (II) When you bounce up and down on a diving board of negligible mass, you find yourself and the board bouncing with the maximum amplitude when your bounce frequency is once every 1.2 s. How much will the board deflect vertically when you stand on the end without bouncing? Assume that the damping that must be present is small.

78. (II) Consider simple harmonic motion of a mass on the end of a spring, $x = (0.5 \text{ m}) \sin(\omega t + \delta)$. At $t = 0$ s the position is -0.1 m, and the velocity is 1 m/s in the $-x$-direction. The total energy of the motion is 5 J. What is the value of the (a) phase, δ; (b) frequency, f; (c) acceleration at $t = 0$ s; (d) spring constant, k; (e) mass, m?

79. (II) A mass m on the end of a spring oscillates with angular frequency ω. The mass is removed, the spring is cut in two, and the mass is reattached. What is the new angular frequency?

80. (II) A spring of equilibrium length 30 cm has one end anchored, and a mass is attached to the other end. The mass is set in uniform circular motion in a plane. The angular frequency of the rotational motion is two-thirds the natural angular frequency of the spring, ω_0. (a) By how much is the spring extended by the motion? (b) Derive the general result for $\omega = \alpha\omega_0$.

81. (II) In the spectacular sport of bungee jumping, a light elastic cord (a bungee cord) is tied tightly around the ankles of someone who jumps from a bridge of height H (the other end of the cord is attached to the bridge). The length of the cord is calculated so that the jumper, of mass m, will not quite reach the surface of the water below the bridge before he or she springs back up. Suppose that the cord behaves like a spring of spring constant $10mg/H$, where g is the acceleration due to gravity. (a) How long must the cord be so that a jumper just touches the water before being pulled back up? Neglect the height of the jumper and any effects due to friction. (b) Friction damps the up-and-down motion of the jumper that results after the initial jump. How far above the water would the jumper be when the oscillations have ceased? Express your answer as a fraction of H.

82. (II) A block of mass $m = 0.3$ kg is dropped onto a horizontal platform supported by a spring with spring constant $k = 5$ N/m. The mass of the platform and spring are negligible compared with m. The speed of the block when it hits the platform is 2 m/s. (a) Find the maximum compression of the spring. (b) How long does it take to reach the lowest point after the block hits the platform?

83. (II) The following systems, illustrated in Fig. 13–35, exhibit simple harmonic motion. What is the period of each motion? (a) A toy tightrope walker, with geometry as shown, whose body has mass much less than that of the barbell weights, each of mass 50 g. The toy sways from side to side in harmonic motion. (b) A mass m attached to two parallel springs, each of spring constant k. Ignore gravity. (c) A spring with spring constant k and a mass m attached to each end. (d) A mass m hanging vertically from a spring, of spring constant k, under the influence of gravity.

FIGURE 13–35 Problem 83.

84. (II) A simple pendulum of mass $m = 0.5$ kg and length $L = 0.8$ m is attached to a cart of mass $M = 1$ kg (Fig. 13–36). The mass of the pendulum support is negligible. The cart can roll freely on a horizontal surface. At $t = 0$, the pendulum bob is released from rest when the string makes an angle of $10°$ with the vertical. Assume that the resulting motion of the cart relative to the ground is simple harmonic motion. Determine the amplitude of the motion of the cart. Why would you expect the cart to move in simple harmonic motion?

FIGURE 13–36 Problem 84.

85. (II) A bar 5.0 m long with a mass of 12 kg has a sharp bend (totaling 14°) at its midpoint (Fig. 13–37). The bar rests on a

pivot placed under the bend. Twins of mass 32 kg each are seated at opposite ends of the bar. What is the period of small oscillations of this modified seesaw?

FIGURE 13–37 Problem 85.

86. (II) When a rope of length L is stretched by an amount x, the rope acts like an imperfect spring that exerts the force $F = -f_1(x/L) - f_2(x/L)^2$, where $f_1 > 0$, $f_2 > 0$, and $f_1 \gg f_2$. The motion is restricted to the x-direction. (a) Describe qualitatively what happens when a large bucket hanging vertically at the end of the rope is slowly filled with water. (b) Find the potential energy function for the rope and discuss what happens when the bucket, at rest in equilibrium, is set in motion with an initial vertical velocity $\mathbf{v}_0$. What happens when the initial velocity is increased?

87. (II) The potential energy of a diatomic molecule whose two atoms have the same mass, m, and are separated by a distance r is given by the formula $U(r) = (A/r^2) - (e^2/r)$, where A and e^2 are positive constants. (a) Find the equilibrium separation r_0 of the two atoms. (b) Show that if the atoms are slightly displaced, so that their separation is $r_0(1 + x)$, then they will undergo simple harmonic motion about the equilibrium position. Calculate the angular frequency of the harmonic motion. Use $(1 + x)^{-2} = 1 - 2x + 3x^2 - \cdots$ and $(1 + x)^{-1} = 1 - x + x^2 - \cdots$ for $x \ll 1$. (Remember the reduced mass.)

88. (II) A small mass m is placed midway on the line between two large masses M. The large masses are separated by a distance L. (a) What gravitational force does the small mass experience? (Assume that the system is isolated from other masses.) (b) The small mass is displaced a distance y in a direction perpendicular to the line between the masses. What is the direction of the net force on that mass? (c) Evaluate the magnitude of the net force for $y \ll L$. Give an expression that incorporates the inequality. (d) What is the motion of m when it is displaced from the midway point?

89. (II) A thin, circular hoop of mass m and radius R hangs on a small peg (Fig. 13–38). What is the frequency of small oscil-

lations with the center of mass moving back and forth along the x-direction?

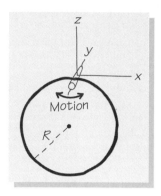

FIGURE 13–38 Problem 89.

90. (II) A machine gun fires bullets in a direction perpendicular to the plane of a target 250 m away from the gun. The bullets leave the muzzle at a speed of 410 m/s. The machine gun is set up on a platform that is vibrating with a frequency of 5.0 Hz and an amplitude of 10.0 cm in a horizontal direction parallel to the plane of the target. Describe the distribution of bullet impact points on the target. The bullets are fired with a frequency much greater than 5 Hz.

91. (III) A solid, uniform cylinder of mass m and radius R is fitted with a frictionless axle along the cylinder's long axis. A horizontal spring (spring constant k) is attached to this axle (Fig. 13–39). Under the influence of the spring, the cylinder rolls back and forth without slipping on a horizontal surface. What is the frequency of this motion?

FIGURE 13–39 Problem 91.

These steel band musicians in Trinidad are making complex disturbances in the air. We perceive these disturbances as sound. Sound is only one example of the widespread set of phenomena that we classify under the heading of mechanical waves.

Waves

W aves are familiar to all of us. We can see ocean waves by means of light waves. Music may originate as waves on piano strings, be transported as waves of electrical current in a stereo system, form waves on the cone of a loudspeaker, and reach our ears as sound waves in air, which, in turn, form waves within the mechanical structure of our ears and are carried as waves of electrical impulses to our brains. Waves may appear as traveling waves, which move in some direction, as ocean surf moves toward a beach. They may appear as standing waves, like the vibrations of a guitar string. They may appear in one-dimensional systems, such as a Slinky spring; two-dimensional surfaces, such as the surface of a pond; or three-dimensional media, such as the air around us.

The point masses and perfectly rigid bodies that we have considered up to now are an idealization: We assumed that the internal forces for a rigid body are so strong that its different parts cannot move relative to each other. Waves form because continuous media *can* distort in response to internal forces. In Chapters 14 and 15, we study waves on a guitar string, on the sea, or in the air, all of which involve continuous media. Eventually, we shall encounter (in Chapter 35) electromagnetic waves such as light waves, in which no medium at all is involved!

14-1 ORGANIZED OSCILLATIONS

A guitar string under tension oscillates, or vibrates, in an organized way when it is plucked. The atoms comprising the guitar string are bound together by interatomic forces. Undisturbed, the atoms are in stable equilibrium positions and the whole string is in equilibrium. When the string is plucked, a group of atoms is displaced from its equilibrium position, exerting forces on the neighboring atoms and displacing them. This displacement will, in turn, lead to a displacement of the next neighboring set of atoms, and so on. Thus the disturbance propagates through the length of the entire string, distorting the string. It is the whole string that moves *in an organized way* to form what we call **waves**. We talked earlier about the motion of either point masses (one or several) or rigid extended bodies, in which the relative motion of different parts of the body is unimportant. In wave motion, however, it is the relative motion of different parts of an extended body that interests us.

To understand better the origin of wave motion, think of a model in which an extended medium such as a solid is represented by a collection of point masses connected by springs. Such a medium can be distorted, but if the distortion is not too great, the medium can return to its original shape, and it will support waves. We call such a medium an **elastic medium**. A stretched guitar string is formed from an elastic medium. Elastic media include materials that we might not normally think of as elastic. The ocean surface responds elastically to distortions, and the surface supports waves. Molecules connected by springs make a poor model of air, yet air responds elastically when a small region of it is abruptly compressed; the result is sound. In all of these cases, the elastic response of the medium leads to an organized and collective motion: waves that move and spread. If waves require matter to propagate, as is true of waves on strings, water waves, or sound waves, we call them **mechanical waves**. Such waves are the subject of this chapter and the next.

Although the microscopic model of a guitar string as a succession of point masses connected by springs can be used for a quantitative treatment of waves, we will take a more descriptive approach. For example, we employ the tension in the string without worrying about the microscopic explanation of tension—namely, that tension comes from the forces holding the molecules of the string together.

Transverse Motion and Longitudinal Motion

Let's look a little more closely at the kinds of waves that occur on a string under tension. A mechanical model of this system, shown in Fig. 14–1, is a series of point masses connected by springs and attached at the ends. Such a system exhibits the organized oscillatory motion of waves. This model illustrates a way to place wave motion into two important classes, whose significance goes beyond the stretched string.

The first class of wave motion is illustrated in Fig. 14–2a: In **transverse motion**, all the point masses move *perpendicular* to the original, equilibrium line of masses and springs, with a maximum displacement called the **amplitude**. This describes one possible **mode of vibration** of the string. This type of wave is easily seen on a Slinky toy

FIGURE 14–1 A series of point masses connected by springs in a line. In equilibrium, the entire system is at rest or is moving uniformly as a whole.

Transverse waves versus longitudinal waves

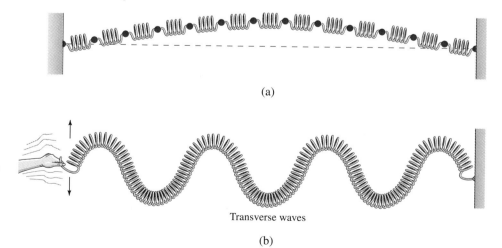

(a)

Transverse waves

(b)

FIGURE 14–2 (a) The point masses of the system illustrated in Fig. 14–1 are displaced transverse to the equilibrium line (dashed). All the springs are stretched and act to bring the system back to equilibrium. (b) Transverse waves on a Slinky toy.

Stretched   Stretched

Compressed   Compressed

(a)

← Motion →   Longitudinal waves

(b)

FIGURE 14–3 (a) The point masses of the system illustrated in Fig. 14–1 are displaced longitudinally along the equilibrium line. Some springs are stretched, and some are compressed. (b) Longitudinal waves on a Slinky toy.

or a rope. If one end of the Slinky is whipped from side to side, transverse waves will be generated (Fig. 14–2b).

The second class of wave motion—**longitudinal motion**—is illustrated in Fig. 14–3a and consists of displacements *along* the line of the springs. Some of the springs are stretched and some are compressed, but the spring forces ultimately act to bring each point back to its equilibrium position. The Slinky also shows this type of wave. If the end is pumped back and forth along the direction of the Slinky, longitudinal waves are generated (Fig. 14–3b).

Some waves include both transverse and longitudinal motion. An example is provided by shore waves. If a cork is put into the ocean to mark the passage of a wave, it will move partly up and down and partly toward and away from the shore. In doing so, it traces out a nearly circular path (Fig. 14–4). This is because water is incompressible; it cannot be squeezed. Therefore the trough of a water wave can be formed only by pushing water forward or backward, toward a neighboring crest. This is the origin of the horizontal (longitudinal) motion that accompanies the up-and-down (transverse) motion.

Standing Waves and Traveling Waves

In addition to the transverse or longitudinal classification of waves, a string under tension also illustrates a different classification: standing (or stationary) waves and traveling (or progressive) waves. In **standing waves**, all the different parts of the medium move together with the same time dependence (Fig. 14–5). A vibrating guitar string is an example of a standing wave. Standing waves are typically present when the ends of a string are fixed or otherwise constrained.

In **traveling waves**, energy is propagated in a definite direction and with a definite speed. If we whip one end of a rope back and forth, a traveling wave moves down the rope (Fig. 14–6). The intriguing feature of this wave is that *all the actual motion is transverse to the overall alignment of the rope, even though the disturbance—the wave—moves along the rope.* Think of a wave formed by a crowd in a stadium, in which each spectator successively moves his or her arms up and down in response to a neighbor's motion, propagating a "wave" around the stadium perpendicular to the direction of the arm motion. The wave travels from one side of the stadium to the other much faster than any spectator could.

We have illustrated the concepts of standing waves and traveling waves with transverse waves, but longitudinal waves can also form both types. Sound waves, which are longitudinal, can form both standing waves and traveling waves. For example, standing sound waves are generated within an organ pipe, whereas a Swiss yodel carrying from one mountain top to another is a traveling sound wave.

Multidimensional Waves

Waves on a string, a one-dimensional system, will continue to be our primary example, whether they are standing or traveling waves. But a two-dimensional system, such as the surface of the ocean, can also support waves. Just as standing waves can be gen-

Standing waves versus traveling waves

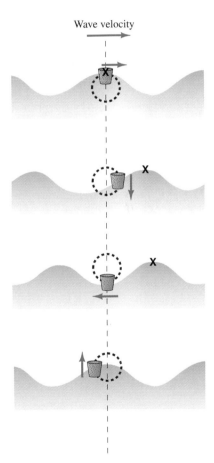

Wave velocity

Wave velocity

x

x

x

FIGURE 14–4 A cork floats on the ocean surface, where the water is not too deep, as a wave passes. The x marks the crest of the passing wave; the dashed circle marks the path followed by the cork; and the arrow, the direction of motion of the cork at the particular point in the wave's passage.

FIGURE 14–5 Vibrating strings of a guitar are examples of standing waves, which can be established on strings fixed at both ends.

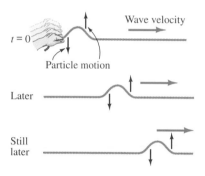

FIGURE 14–6 A traveling wave on a string. At time $t = 0$, a hand generates a wave that travels down the taut string.

FIGURE 14–7 Standing waves on a drumhead. The *nodes,* points where the drumhead is not in motion, form lines rather than individual points as they did for the string. The lines shown are not necessarily nodes. Can you determine the nodes for each pair of vibrations?

erated on a string, they can also be generated on a drumhead. These waves consist of an organized harmonic motion transverse to the drumhead itself (Fig. 14–7). Traveling waves in two-dimensional systems form **wave fronts**, which are lines perpendicular to the direction of motion of the wave. This is a familiar phenomenon, illustrated by the spreading ripples generated by throwing a rock into a pond, as in Fig. 14–8. The ripples move radially away from a central point and the wave front is an expanding circle. For a three-dimensional system, such as open air, the wave fronts formed by traveling waves move out as spheres from a central point.

14-2   STANDING WAVES

As we have seen, a guitar string with two fixed ends supports standing waves. The fact that the two ends are fixed is referred to as a set of *boundary conditions* for the wave motion. The variables needed to understand standing waves on a string with fixed ends are the tension, T, the length L, and the mass density, μ.[†] The mass density, or mass per unit length, is such that the mass of any segment of length ℓ is $m = \mu\ell$. These three variables can be changed. For example, we tighten or loosen a guitar string by turning its peg, we change the length of the string when we press it against a fret, and we change the mass density when we replace a light string with a heavy one.

If we pluck the guitar string, it vibrates in a transverse standing wave. Some possible vibrations are shown in Fig. 14–9. The transverse displacement of a standing wave may differ at different points along the string; however, the *time dependence* of the transverse motion is the same everywhere along the string. For example, the center point on the string might move up and down with an amplitude of 1 cm, while a point halfway between one end and the center point might move up and down with an amplitude of only 0.5 cm; but the two points move together.

Mathematical Description of Standing Waves

To describe a standing wave of a guitar string mathematically, suppose that the string is aligned along the x-direction, with one end at $x = 0$ and the other end at $x = L$. The position along the string is labeled x, and the transverse displacement of a point on the string is labeled $z(x, t)$. Then every point along the string moves with the same time dependence if

$$z(x, t) = f(x)g(t). \tag{14-1}$$

In view of the close relation between waves and springlike forces discussed in Section 14–1, we expect harmonic time dependence,

$$g(t) = \cos(\omega t + \phi), \tag{14-2}$$

where ω is just like the angular frequency of simple harmonic motion and ϕ is a phase. Thus any particular point on the string is in simple harmonic motion. For the space dependence, a simple function that fits the kind of situation we are discussing is

$$f(x) = z_0 \sin(kx + \delta). \tag{14-3}$$

The constant z_0 is the amplitude, which measures the maximum displacement; δ is another phase. The constant k, which is positive, should not be confused with a spring constant. It is just a constant analogous to ω, whose dimensions make the argument of the sine dimensionless. Substituting for the functions $f(x)$ and $g(t)$ in Eq. (14–1), we get a standing wave of the form

$$z(x, t) = z_0 \sin(kx + \delta) \cos(\omega t + \phi). \tag{14-4}$$

We can call this a *harmonic standing wave.*

[†] The mythical massless strings we have so often used could not support waves.

Properties of Harmonic Standing Waves

Let's study the various features of the standing wave expressed by Eq. (14–4). The *angular frequency, ω*, is familiar from Chapter 13. The *frequency* of the repetition is $f = \omega/2\pi$, and the *period* of the repetition is $\tau = 1/f = 2\pi/\omega$.[†] The function $\sin(kx + \delta)$ repeats itself in space, and the distance over which it repeats is $2\pi/k$. We refer to this distance as the **wavelength**, λ, and k is called the **wave number**. These quantities are related by

$$\lambda = \frac{2\pi}{k}. \qquad (14\text{–}5)$$

Boundary Conditions Determine the Wavelength. The possible values of k are determined by boundary conditions. For this particular case, the boundary conditions are that the ends of the string are fixed: $z = 0$ at $x = 0$ and at $x = L$ for all times t. In order to satisfy $z(0, t) = 0$, we take $\delta = 0$. In order to satisfy $z(L, t) = 0$, we require that

$$\sin(kL) = 0. \qquad (14\text{–}6)$$

Equation (14–6) can be satisfied if the argument of the sine function is any positive integer multiple of π (negative integer multiples only reverse overall sign of the wave), or if k takes on the discrete values

$$k_n = \frac{n\pi}{L} \qquad \text{for } n = 1, 2, 3, \dots. \qquad (14\text{–}7)$$

(The case $n = 0$ gives just a flat or undistorted string—we say the string is *unexcited.*) Equation (14–5) then implies that the allowed wavelengths also carry an index n:

$$\lambda_n = \frac{2\pi}{k_n} = \frac{2}{n} L \qquad \text{for } n = 1, 2, 3, \dots. \qquad (14\text{–}8)$$

When the string is set into motion with values of k that correspond to a certain standing wave, we say the string is *excited* into those waves. Equation (14–8) can be rewritten as $L = \frac{1}{2} n\lambda_n$, providing a convenient way to recall the possible values of the wavelength: *The total length of a string can be divided into an integral number of half-wavelengths.* The value of n labels the **modes**, or possible vibration patterns, of the possible standing waves; $n = 1$ labels the *fundamental mode* (see Fig. 14–9), also called the *first harmonic. $n = 2, 3, \dots$* are called the second, third, … harmonics. Note that for odd values of n, the number of wavelengths that can fit on the string is a whole number plus one-half, while for even values of n, a whole number of wavelengths fits into the string. Note also that, for $n > 1$, there are *nodes* where the string rests motionless with no transverse displacement. Although these nodes are not caused by a physical clamping of the string as at the end points, they might as well be because these points do not move. Not counting the end points, there are $n - 1$ of these nodes— for example, one node in the $n = 2$ case of Fig. 14–9. We can also see the presence of *antinodes,* locations where the displacement is a maximum.

EXAMPLE 14–1 Standing waves can be excited on a string of length $L = 1.00$ m whose ends are fixed at $x = 0$ and $x = L$. Find the two smallest values of the wave number and the two longest wavelengths allowed for standing waves. Sketch the space dependence of these waves.

Solution: The possible wave numbers, k_n, are proportional to n, whereas the possible wavelengths, λ_n, are proportional to n^{-1}. Thus the smallest values of k and the largest values of λ are the two cases $n = 1$ and $n = 2$ of Eqs. (14–7) and (14–8),

[†]We use τ for the period because we do not want to confuse it with the tension, T, of the string.

FIGURE 14–8 The spreading waves resulting from a disturbance of a water surface form circles centered on the disturbance.

A harmonic standing wave on a string

The wavelength and wave number

Boundary conditions fix the possible values of the wavelengths of standing waves.

PROBLEM SOLVING

A way to remember the possible values of the wavelength.

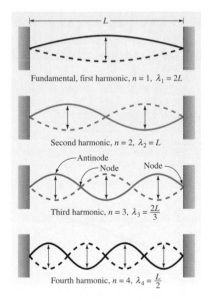

Fundamental, first harmonic, $n = 1$, $\lambda_1 = 2L$

Second harmonic, $n = 2$, $\lambda_2 = L$

Antinode, Node, Node

Third harmonic, $n = 3$, $\lambda_3 = \frac{2L}{3}$

Fourth harmonic, $n = 4$, $\lambda_4 = \frac{L}{2}$

FIGURE 14–9 Some transverse standing waves of a vibrating string with fixed ends. We see some of the possible modes, together with nodes and antinodes. The string moves back and forth between the extremes shown here, as the arrows indicate.

respectively. These are

$$k_1 = \frac{\pi}{L} = 3.14 \text{ m}^{-1}, \qquad k_2 = \frac{2\pi}{L} = 6.28 \text{ m}^{-1}$$

and

$$\lambda_1 = 2L = 2.00 \text{ m}, \qquad \lambda_2 = \frac{2L}{2} = 1.00 \text{ m}.$$

The space dependence of the waves is given by the function $\sin(k_n x)$ times an arbitrary amplitude z_0. Figure 14–9 shows the possibilities $n = 1$ through $n = 4$, with z_0 left unspecified.

Dynamics Determines the Frequency. We have seen how boundary conditions determine k. Initial conditions, which are a type of boundary condition in time, determine the phase ϕ. But *boundary conditions do not determine the angular frequency ω.* We will next show that, as in Chapter 13, ω *is determined by Newton's second law, not by the boundary conditions.*

The Wave Equation

Just how do we apply $F = ma$ to the string? In other words, what is the dynamical equation of motion? We find the equation of motion by studying the forces on a small segment of the string as it is stretched and distorted (see the next subsection "How To Get the Wave Equation"). One crucial approximation leads to a simplified equation of motion. This is the *assumption of small distortion*, in which the stretch of the string is kept small. As a result of the small-distortion approximation, we find the dynamical equation of motion for transverse displacements of the string to be

$$T\frac{\partial^2 z}{\partial x^2} = \mu \frac{\partial^2 z}{\partial t^2}. \qquad (14\text{–}9)$$

A wave equation

Relations such as this are important in many areas of science and engineering and are known as **wave equations**. The symbol ∂ is used for partial differentiation. The transverse displacement z is a function of *two* variables (x and t). The partial-derivative notation simply means that for $\partial z/\partial x$, the variable t is held fixed while the derivative is taken with respect to x. The wave equation (or its generalization to two or three dimensions) describes not only transverse and longitudinal mechanical waves—both standing and traveling—but also electromagnetic waves, such as light (see Chapter 35).

PROBLEM SOLVING

Partial derivatives

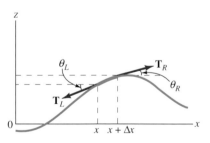

FIGURE 14–10 A string has been displaced. We have isolated a segment of width Δx, as well as the tension forces that act on it. Due to the curvature, the direction through which the tension acts differs on each side. The magnitudes of $\mathbf{T}_L$ and $\mathbf{T}_R$ are equal, and we call them T.

*How To Get the Wave Equation

Consider a small segment of length Δx and mass $\mu\Delta x$ of a uniform string, where μ is the mass density of the string. The ends of the segment are at x and $x + \Delta x$. We suppose the string has been displaced in the transverse direction (the z-direction) to take the shape shown in Fig. 14–10. We look for the tranverse forces acting on this segment.[†] We assume the distortion of the string is small, meaning that (a) the tension is the same throughout the string, and (b) all angles are small.

The forces acting on the segment are the tensions on the left and right sides. The *directions* of the tensions are slightly different because the string is curved, and that is what gives a net vertical force on the segment. On the left side the vertical component of the tension is downward and on the right it is upward. For small angles, the magnitudes of the vertical components are $T\sin\theta \simeq T\tan\theta$; that is, the tension times the

---

[†]For waves other than transverse waves on a string, only the details would change.

slope of the string at the respective points. In turn, this slope is the derivative of z with respect to x. Thus the vertical component of the tension on the left side is

$$F_L = -T\frac{\partial}{\partial x} z(x, t).$$

The slope is usually denoted by dz/dx but, because z depends on two variables, x and t, we use the partial-derivative notation described in the text. The force on the right end of the segment points upward. This time, the slope involved is the slope at $x + \Delta x$:

$$F_R = T\frac{\partial}{\partial x} z(x + \Delta x, t).$$

Combining, the net vertical force on our element is

$$F = F_R + F_L = T\frac{\partial}{\partial x} z(x + \Delta x, t) - T\frac{\partial}{\partial x} z(x, t)$$

$$= T\left[\frac{\partial}{\partial x} z(x + \Delta x, t) - \frac{\partial}{\partial x} z(x, t)\right].$$

(14-10)

Now recall that by definition of the derivative, in the limit that Δx approaches zero,

$$\frac{f(x + \Delta x) - f(x)}{\Delta x} = \frac{df}{dx}$$

for any function $f(x)$. We apply this to Eq. (14-10) for $f(x) = \partial z/\partial x$:

$$F = T(\Delta x)\frac{\partial}{\partial x}\left(\frac{\partial z}{\partial x}\right) = T(\Delta x)\frac{\partial^2 z}{\partial x^2}.$$

This equation is the net force on our small segment.

The mass times the vertical acceleration of the segment is simply given by

$$ma = \mu(\Delta x)\frac{\partial^2 z}{\partial t^2}.$$

(14-11)

The dynamical equation of motion results from the equality $F = ma$. In equating F and ma, the factor Δx cancels, and we are left with Eq. (14-9).

How the Wave Equation Determines the Frequency

We want to verify that our trial standing-wave solution, Eq. (14-4), is indeed a solution of the wave equation. In so doing, we determine the angular frequency, ω. We test the solution by taking the necessary derivatives of Eq. (14-4) and substituting them into the wave equation, Eq. (14-9). The first time derivative of Eq. (14-4) gives

$$\frac{\partial z}{\partial t} = z_0 \sin(kx + \delta)\omega \sin(\omega t + \phi)$$

and hence

$$\frac{\partial^2 z}{\partial t^2} = z_0 \sin(kx + \delta)(-\omega^2) \cos(\omega t + \phi).$$

For the left-hand side of Eq. (14-9) with Eq. (14-4) substituted, the derivatives with respect to x leave the time dependence unchanged. Therefore the $\cos(\omega t + \phi)$ term cancels from both sides of Eq. (14-9) with Eq. (14-4) substituted, and we find that

$$T\frac{\partial^2[z_0 \sin(kx + \delta)]}{\partial x^2} = -\mu\omega^2 z_0 \sin(kx + \delta);$$

(14-12)

$$-Tk^2 z_0 \sin(kx + \delta) = -\mu\omega^2 z_0 \sin(kx + \delta).$$

The angular frequency is determined by the dynamics.

Finally, we can cancel the factor $z_0 \sin(kx + \delta)$ from both sides, leaving

$$Tk^2 = \mu\omega^2. \tag{14–13}$$

Recall that the wave number, k, is related to the wavelength, λ, by $k = 2\pi/\lambda$. Thus, from Eq. (14–13), the angular frequency is given by

$$\omega = \frac{2\pi}{\lambda}\sqrt{\frac{T}{\mu}}. \tag{14–14}$$

When we substitute $\omega = 2\pi/\tau$, Eq. (14–14) becomes a relation between the period of the oscillations of the string, τ, and the wavelength of the standing waves:

$$\tau = \lambda\sqrt{\frac{\mu}{T}}. \tag{14–15}$$

What can we learn about the motion of the string from Eqs. (14–14) and (14–15)?

1. As the wavelength of possible standing waves on a particular string decreases, so does the period of the motion.

2. A less dense string vibrates more rapidly than a dense one.

3. A string under great tension vibrates more rapidly than a string under less tension.

Harmonics

For the boundary conditions previously discussed, $\lambda_n = (2/n)L$, and the frequency, angular frequency, and period can have the discrete values

$$f_n = \frac{n}{2L}\sqrt{\frac{T}{\mu}}, \tag{14–16a}$$

$$\omega_n = \frac{n\pi}{L}\sqrt{\frac{T}{\mu}}, \tag{14–16b}$$

$$\tau_n = \frac{2L}{n}\sqrt{\frac{\mu}{T}}. \tag{14–16c}$$

The frequency f_1, the lowest frequency, is known as the **fundamental frequency**, and $\lambda_1(= 2L)$ is the **fundamental wavelength**. The series of frequencies that are integral multiples of the lowest possible frequency are called the **harmonic series**. The frequency $f_n = nf_1$ is called the nth *harmonic* of the string; Fig. 14–9 depicts some harmonics. (We will return to these topics in Section 14–6.)

EXAMPLE 14–2 The time for 100 full vibrations (100 periods) in the fundamental mode of a piano wire (Fig. 14–11) is 0.5 s. The wire length is $L = 2$ m and the total mass of the wire is 25 g. What is the tension on the wire? By how much must the tension be increased in order to halve this time?

Solution: The fundamental mode is labeled $n = 1$. For the period of one vibration, Eq. (14–16c) gives

$$\tau_1 = 2L\sqrt{\frac{\mu}{T}}.$$

This may be solved for the tension, T:

$$T = \frac{4\mu L^2}{\tau_1^2}.$$

FIGURE 14–11 This piano tuner adjusts the natural frequencies of the strings by changing their tensions.

We know that $L = 2$ m, $\tau_1 = (0.5 \text{ s})/100 = 5 \times 10^{-3}$ s, and the mass density, μ, is $(0.025 \text{ kg})/(2 \text{ m}) = 0.0125$ kg/m. Thus the tension is

$$T = \frac{4(0.0125 \text{ kg/m})(4 \text{ m}^2)}{2.5 \times 10^{-5} \text{ s}^2} = 800 \text{ N},$$

which is equivalent to the weight of an average man.

From Eq. (14–16c), it can be seen directly that the period will be halved if the square root of the tension is doubled or if the tension is increased by a factor of 4. The new tension will be 3200 N.

Standing Waves in More than One Dimension

The boundary conditions determining the allowed values of k and the dynamics determining the allowed values of ω are far more complicated when the system is a surface or a volume rather than a one-dimensional system like the string. For example, the boundary conditions for a drumhead require that the vibrations vanish on the circular rim of the drum. Figure 14–7 showed a few of the modes possible on a circular drumhead. The modes on an elliptical drumhead differ from those on a circular drumhead. The common feature of all these systems is that the wavelengths and frequencies are discrete, so they can be labeled by integers.

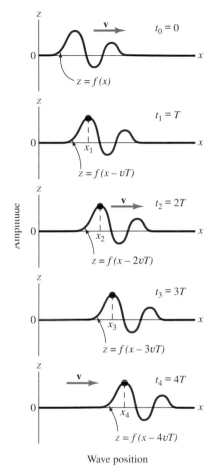

14–3 TRAVELING WAVES

Suppose that a hand shakes one end of a string and disturbs it. If the string is so long that we do not need to worry about what is happening at the other end, then we can have traveling waves on the string. Traveling waves can move over large distances even though the material that comprises the propagating medium moves in a very limited way.

FIGURE 14–12 The curve is some function f of the variable $x - vt$. As time goes on ($t_2 > t_1$), the curve moves to the right with speed v.

Mathematical Description of Traveling Waves

Picture a string oriented along the x-axis, with mass density μ and tension T. We want to characterize the transverse displacement $z(x, t)$ of a traveling wave. We require a mathematical form that describes a shape moving _as a whole_ with speed v to the right, for example. We will now show that this can be done with _any function_ of the quantity $x - vt$:

$$z(x, t) = f(x - vt), \tag{14–17}$$

where the choice of function f corresponds to a particular moving shape.

To see why this works, consider a function such as that in Fig. 14–12. We single out a particular point on the function, such as a peak, and mark this point with a black dot. At time t_1, this point is at an x-value of x_1. If t_1 increases to t_2, the argument of f will have the same value—the point will still be at the black dot—only if x increases to compensate for the decreased value of the term $-vt$. In order for the point to stay at the same argument, we must have

$$x_2 - vt_2 = x_1 - vt_1;$$

$$x_2 = vt_2 + x_1 - vt_1 = x_1 + v(t_2 - t_1). \tag{14–18}$$

Because Eq. (14–18) describes a point moving to the right with speed v, $f(x - vt)$ describes some form moving in the same way. Similarly, $f(x + vt)$ represents a shape moving to the left. In summary, $f(x - vt)$—or $f(x + vt)$—_represents a shape that moves to the right—or left—with speed v._

Any function of $x - vt$ (or $x + vt$) describes a traveling wave moving in the positive (negative) x-direction with speed v.

A harmonic traveling wave on a string

(a) $t = 0$

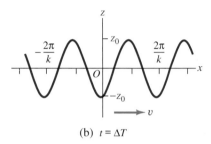

(b) $t = \Delta T$

FIGURE 14–13 (a) The harmonic traveling wave of Eq. (14–19) is plotted at the fixed time $t = 0$. (b) As time passes, this pattern will move to the right.

The relations between frequency, wavelength, wave number, angular frequency, and wave speed

Again, let's be guided by our experience in simple harmonic motion and use a sinusoidal form for f. We cannot, however, use the simple form $\sin(x - vt)$ because the argument of the sine must be dimensionless. To make a dimensionless argument, we multiply $x - vt$ by the wave number, k, as in Eq. (14–3). Thus, the following equation represents a harmonic traveling wave on a string:

$$z(x, t) = z_0 \sin[k(x - vt)] = z_0 \sin(kx - kvt). \tag{14–19}$$

Properties of Harmonic Traveling Waves

If we were to take a snapshot of $z(x, t)$ in Eq. (14–19) at $t = 0$ and plot $z(x, t) = z_0 \sin(kx)$, we would get Fig. 14–13a. The sine curve is zero at $x = 0$ and at every value of x equal to an integer multiple of π/k. It repeats every time x increases by $2\pi/k$. The repetition length is the wavelength, λ; *as before, the relation between λ and k is $\lambda = 2\pi/k$.* The maxima of the sine curve, $z = z_0$, are known as **crests** of the wave, and the minima, $z = -z_0$, are **troughs**.

We have already shown that the wave described by Eq. (14–19) moves to the right with speed v (Fig. 14–13b). How much time does this wave take to move by one wavelength? The wave has the same configuration when kvt changes by 2π,

$$kv\tau = 2\pi;$$

$$\tau = \frac{2\pi}{kv}. \tag{14–20}$$

The time τ is the period of the wave. We have a frequency f and an angular frequency ω with the usual relation to the period:

$$f = \frac{1}{\tau}, \qquad \omega = 2\pi f = \frac{2\pi}{\tau}. \tag{14–21}$$

Equation (14–21) states *exactly the same relations as those that hold in simple harmonic motion.* Comparison of Eqs. (14–20) and (14–21) shows that the wave speed is related to the wave number, wavelength, frequency, and angular frequency by

$$v = \lambda f = \frac{\lambda}{\tau} = \frac{\omega}{k}. \tag{14–22}$$

The important relation $v = \lambda f$ among speed, wavelength, and frequency is easily understood if we think about a passing train possessing cars of equal length. The frequency is measured by counting the number of cars that pass in a given time and the wavelength is the length of one car. The speed of the train is then the length of a car times the number of cars that pass per unit time. If we measure three cars passing every 5 s, then $f = 3/(5\ \text{s}) = 0.6$ Hz, and if a car has a length of $\lambda = 30$ m, the speed is $v = \lambda f = (30\ \text{m})(0.6\ \text{Hz}), = 18$ m/s.

Equation (14–22) suggests alternative useful ways to write Eq. (14–19) for the transverse displacement:

$$z(x, t) = z_0 \sin(kx - \omega t) \tag{14–23}$$

$$= z_0 \left[\sin 2\pi \left(\frac{x}{\lambda} - \frac{t}{\tau} \right) \right]. \tag{14–24}$$

If we examine the traveling wave at some fixed position x_0, then the kx_0 term in Eq. (14–23) acts as a phase, and the z-motion at x_0 is simple harmonic motion.

EXAMPLE 14–3 A person standing at the narrow entrance to a harbor sees sinusoidal water waves moving into the harbor. He counts 50 wave crests in 1.0 min and he estimates the distance between the crests to be 3.0 m (with the help of an an-chored boat of known size). Write an expression for the form of the wave height where the person is standing. What are the wavelength, wave number, frequency, an-gular frequency, and speed of these waves?

Solution: If $+x$ is taken as the direction into the harbor, and if h is the height of the waves above the average water level, then an expression for h is

$$h(x, t) = h_0 \sin(kx - \omega t).$$

The amplitude, or maximum height h_0, is not specified in the problem, but the wave number and the angular frequency are specified. The wavelength, λ, is just the distance between crests, $\lambda = 3.0$ m. The wave number, k, is given in terms of λ by Eq. (14–5):

$$k = \frac{2\pi}{\lambda} = \frac{2\pi}{3.0} \text{ m}^{-1} = 2.1 \text{ m}^{-1}.$$

The frequency, f, of the waves is 50 min^{-1}, or

$$f = (50 \text{ min}^{-1}) \frac{1 \text{ min}}{60 \text{ s}} = 0.83 \text{ s}^{-1}.$$

The angular frequency, ω, and the frequency, f, are related by Eq. (14–21):

$$\omega = 2\pi f = (2\pi \text{ rad})(0.83 \text{ s}^{-1}) = 5.2 \text{ rad/s}.$$

Finally, the speed v is given by Eq. (14–22),

$$v = \lambda f = (3.0 \text{ m})(0.83 \text{ s}^{-1}) = 2.5 \text{ m/s}.$$

Dynamics Determines the Speed. Unlike the standing waves of Section 14–2, trav-eling waves have no boundary conditions for determining wavelength—even though the wavelength is related to the frequency and the wave speed by $v = \lambda f$. In fact, *the dynamical equations determine the speed of a traveling wave.* We showed in Section 14–2 that the wave equation, or dynamical equation of motion, takes the form of Eq. (14–9): $T \partial^2 z/\partial x^2 = \mu \partial^2 z/\partial t^2$. It is not difficult to check that our traveling wave satis-fies this equation by calculating the derivatives. We leave this exercise for the prob-lems; here, we simply state that if we use one of the forms for $z(x, t)$, such as Eq. (14–23), then we find an algebraic relation that must be satisfied after taking the ap-propriate derivatives:

$$Tk^2 = \mu\omega^2. \tag{14–25}$$

The quantities k and ω are already kinematically related by the wave speed, $v = \omega/k$. Thus Eq. (14–25) gives the speed of the wave in terms of the mass density and ten-sion of the string:

$$v = \sqrt{\frac{T}{\mu}}. \tag{14–26}$$

The speed of a traveling wave on a string

Equation (14–25) is exactly the same as Eq. (14–13) of the standing wave problem. The only difference is that we did not interpret the ratio ω/k as a speed for the case of the standing wave.

EXAMPLE 14–4 A very long wire is held under tension by suspending weights from an end that passes over a pulley. The speed of transverse traveling waves on the wire is 51 m/s. The wire is replaced by another with three times the mass den-sity and twice as much weight is suspended from the end of the new wire. What is the speed of the traveling waves now?

Solution: Equation (14–26) gives the original speed v_1 in terms of the original tension T_1 and mass density μ_1:

$$v_1 = \sqrt{\frac{T_1}{\mu_1}}.$$

When Newton's second law is applied to the wire in equilibrium—with zero acceleration—it implies that the tension is given precisely by the suspended weight. Thus T has been doubled and the new tension is $T_2 = 2T_1$. The new mass density is $\mu_2 = 3\mu_1$. Then the new speed is

$$v_2 = \sqrt{\frac{T_2}{\mu_2}} = \sqrt{\frac{2T_1}{3\mu_1}} = \sqrt{\frac{2}{3}} v_1.$$

Thus the speed has been decreased to

$$v_2 = (\sqrt{0.67})(51 \text{ m/s}) = 42 \text{ m/s}.$$

Increasing the tension tends to increase the speed, whereas increasing the mass density tends to decrease the speed. In this case, the mass density is increased by a larger factor than is the tension, so the speed decreases.

14–4 ESTIMATING THE SPEEDS OF TRAVELING WAVES

The speed of a traveling wave is one of its most important features. If the time interval between a lightning flash and the thunderclap is 5 s, how far away was the lightning strike? How much time does it take for a tsunami (a tidal wave) to travel from the point at which an earthquake generates it to a distant shore? Here we shall examine some techniques for estimating wave speeds.

A General Form of the Wave Equation

A string of mass density μ under tension T obeys the wave equation [Eq. (14–9)]. Using the fact that $v = \sqrt{T/\mu}$, we can rewrite the wave equation as

$$\frac{\partial^2 z}{\partial x^2} = \frac{1}{v^2} \frac{\partial^2 z}{\partial t^2}. \tag{14–27}$$

In this form, the wave equation takes on a more universal aspect than the special situation of transverse waves on a string. The variable z now represents any physical quantity whose variation describes a wave; for example, the height of a water wave, the back-and-forth motion of the coils of a Slinky, the compression of a metal, or the variation of air pressure in a sound wave. The direction of motion of the wave is represented by x and v is the speed with which the wave moves.

All of these waves can take the very general form $z(x, t) = f(x \pm vt)$ and this form will automatically satisfy the wave equation. To demonstrate this, we need only take the derivatives of $f(x \pm vt)$. For example, denote $y = x - vt$ so that $f(x - vt)$ is written $f(y)$. Then use the chain rule for derivatives (Appendix IV–7):

$$\frac{\partial f(y)}{\partial x} = \frac{df(y)}{dy} \frac{\partial y}{\partial x} = \frac{df(y)}{dy}.$$

In the last step, we have used $\partial y/\partial x = \partial(x - vt)/\partial x = (\partial x/\partial x) - [\partial(vt)/\partial x] = 1$. By following this procedure and inserting the appropriate derivatives in the wave equation—Eq. (14–27)—we can verify that the wave equation is satisfied (see Problem 17).

Surface waves, like small amplitude water waves, are used in an interesting fashion in the preparation of the printed circuit boards that form the central parts of computers, radios, televisions, and VCRs. A printed circuit is a fiberglass plastic board with many holes punched into it. The electronic components of the circuit are placed at the correct location on one side of the board, and wire legs that enable electrical contact to these components to be made are placed through the appropriate holes. The attachment of the component wires is by soldering; i.e., by melting a metal alloy at the appropriate intersections. Previously, this was done by hand—one joint at a time. Today, the board is suspended over a rectangular container of molten solder. A traveling surface wave of solder is then started at one end of the container. The height of this wave is very carefully controlled. When the wave reaches the board, it passes just below the surface of the board, touching each of the joints, and depositing solder at each. The bath temperature is chosen so that a properly fused connection occurs at every appropriate location. This method of assembly, which is called "wave soldering," is fast, and it does not "forget" any connections. In addition, and very importantly, wave soldering subjects the component wires of the board to a controlled temperature rise because the solder contacts the joint for a brief controlled period; excessive heating of the components is thereby avoided.

How the Wave Speed Depends on the Medium

Once we have established that a medium is elastic and obeys a wave equation, we can try to determine the wave speed, which depends on the nature of the medium. For the string ($v = \sqrt{T/\mu}$), the tension, T, acts as a restoring force that tends to bring the string back to its equilibrium position. The mass density, μ, describes the reaction of the string to this restoring force. Generally, this suggests that the speed of mechanical waves is a function of an internal force factor (an elastic restoring force) divided by a mass factor characteristic of the system.

A dimensional analysis helps us push further. The tension is a force and has dimensions $[MLT^{-2}]$. The mass density (mass per unit length) has dimensions $[ML^{-1}]$. Therefore

∞ Dimensional analysis is described in Chapter 1.

$$\left[\frac{\text{tension}}{\text{mass density}} \right] = \left[\frac{MLT^{-2}}{ML^{-1}} \right] = [L^2 T^{-2}].$$

Thus, if v is a function of the ratio of the restoring force factor to the mass factor, *the particular function must be a square root*:

$$\text{wave speed} = \sqrt{\frac{\text{restoring force factor}}{\text{mass factor}}}. \qquad (14\text{--}28)$$

A general form for the wave speed

If we restrict ourselves to bulk (macroscopic) properties of the string, Eq. (14–26) is the *only* dimensionally correct relation for wave speed. Indeed, the *only* bulk properties of the string that we know are the tension, the total mass, the length, and the thickness; the first three of these are figured into the wave speed. (For the true mass density, or mass per unit *volume*, the thickness of the string would also enter.) As we discussed in Chapter 1, this analysis does not permit a calculation of constant factors such as 2 or π, but it will give the system's dependence on the bulk properties, and it will give a correct estimate of orders of magnitude.

The Speed of Sound in Solids

To illustrate the dimensional approach, let's estimate a longitudinal wave speed in solids. To do so, we need to know that the *Young's modulus*, Y, of a solid determines the (small) stretch of the solid under the influence of an external force (see Chapter 21). In particular, Young's modulus describes the *fractional* longitudinal stretch, $\Delta L/L$, of a given material of length L when a force per unit area, F/A, is applied to its end: $(\Delta L/L)Y = F/A$. Young's modulus therefore has dimensions of force per area.

Let's now suppose that one end of a metal bar is vibrated rhythmically. Waves will form and travel along the bar. As with the string, we expect that the speed of these waves is independent of the exact dimensions of the bar. Then only two mechanical properties of the bar can determine the wave speed: Young's modulus and the mass density (mass per unit volume), ρ. The first factor is a measure of restoring force and the second is a mass factor. In addition, the ratio Y/ρ has dimensions of force/area divided by mass/volume, or $[(MLT^{-2}L^{-2})/(ML^{-3})] = [L^2T^{-2}]$. The dimensions are indeed those of speed squared. We have thus estimated the speed of mechanical waves in a metal bar to be

$$v = (\text{a constant})\sqrt{\frac{Y}{\rho}}. \qquad (14\text{--}29)$$

(We expect the constant to be on the order of 1.) The waves whose speed we have estimated here are longitudinal waves because Young's modulus measures the resistance of the bar to longitudinal compression or longitudinal stretching. These waves are known as **sound waves**.

A more careful analysis, based on the wave equation, shows that the simple form $v = \sqrt{Y/\rho}$ is *exact* (the constant = 1); this result is confirmed by experiment. Young's moduli are given in Table 21–1 for several solids. For silver, for example, whose density is roughly 10^4 kg/m^3 and whose Young's modulus is 7.5×10^{10} N/m^2, the quantity $\sqrt{Y/\rho} = 2700$ m/s, which is a very good approximation of the measured value of the speed of sound in silver.

INTERIM SUMMARY

Mechanical waves are organized motions within elastic media. Standing waves are waves for which all the motion within the medium has the same time dependence. For sinusoidal standing waves, the displacement of any point in the medium has the form

$$z(x, t) = z_0 \sin(kx + \delta) \cos(\omega t + \phi). \qquad (14\text{--}4)$$

The important parameters of this wave are the amplitude z_0, the angular frequency, ω, and the wave number, k. The wavelength, λ (the length over which the wave repeats itself) is related to the wave number by $\lambda = 2\pi/k$. We also use the frequency, f, and the period, τ. These parameters are related by $\lambda f = \lambda/\tau = \omega/k$. Possible values of the wavelength are restricted by the boundary conditions of the medium.

Traveling waves are characterized by a series of internal motions that propagate progressively with some definite speed through the medium. These waves can travel over large distances even though the material that makes up the medium moves in only a limited way. The wavelength, wave number, frequency, and period are also useful parameters for traveling waves. In addition, traveling waves move at a definite speed v, which is related to the parameters already introduced according to $v = \lambda f = \lambda/\tau = \omega/k$. When the particles of the medium move parallel to the direction of wave propagation, longitudinal waves form; when the particles move perpendicular to the wave direction, transverse waves form. Standing waves can also be classified into longitudinal and transverse types.

The angular frequency of waves is determined by the dynamical wave equation, obtained by the application of Newton's second law to an elastic medium:

$$\frac{\partial^2 z}{\partial x^2} = \frac{1}{v^2}\frac{\partial^2 z}{\partial t^2}. \qquad (14\text{--}27)$$

This equation applies to all types of mechanical waves. The wave speed v is a property of the medium:

$$v = \sqrt{\frac{\text{restoring force factor}}{\text{mass factor}}}. \qquad (14\text{--}28)$$

In the case of a string of mass density μ under tension T, $v = \sqrt{T/\mu}$.

In waves, masses are in motion, so energy must be present (Fig. 14–14). In traveling waves, energy can be delivered from one end of the system to the other even though no mass is transferred in the motion. It is an important observation that *energy can be transmitted without the transport of mass.*

The Power Delivered by Waves

Let's think about right-moving transverse traveling waves on a uniform string that has a mass per unit length μ and tension T. The power P delivered at a point x along the string is given by $P = \mathbf{F} \cdot \mathbf{v}_{\text{transverse}}$, where $\mathbf{v}_{\text{transverse}}$ is the (transverse) velocity of the string at that point and $\mathbf{F}$ is the force acting on the portion of the string to the right of x. Only the transverse component of the force enters into the scalar product. To find the transverse component of the force, refer to Fig. 14–15, which shows that the transverse component of force at x acts downward (negative sign) with magnitude

$$T \sin \theta \simeq T \tan \theta = T \frac{\partial z}{\partial x}.$$

We have used the small distortion approximation here—the angle θ is small—and we have remembered to use the partial-derivative notation again, as a reminder that z depends on two variables, x and t. The velocity of the string at x is given by

$$v_{\text{transverse}} = \frac{\partial z}{\partial t}. \tag{14–30}$$

Thus the power delivered to the string at point x is

$$P = \mathbf{F} \cdot \mathbf{v}_{\text{transverse}} = -T \frac{\partial z}{\partial x} \frac{\partial z}{\partial t}. \tag{14–31}$$

It is simple to calculate the derivatives involved here for a harmonic traveling wave, $z = z_0 \sin(kx - \omega t)$. We find

$$P = T k \omega z_0^2 \cos^2(kx - \omega t) = \mu v \omega^2 z_0^2 \cos^2(kx - \omega t). \tag{14–32}$$

To find the last form, we have used Eq. (14–26), which gives $T = \mu v^2 = \mu(\omega/k)v$, where v is the wave velocity. (Don't confuse v with the transverse velocity of a portion of the string.)

It is often useful to think about the time average of the power delivered. Let us denote this average with angular brackets. To find the time average, we need the average value taken by the cosine squared over a long time period, namely 1/2. Thus

$$\langle P \rangle = \frac{1}{2} \mu \omega^2 z_0^2 v. \tag{14–33}$$

Energy Transport

There is no mechanism for energy loss in a string such as we have described. Thus the power supplied by whatever generates the waves in the first place is transmitted from one piece of the string to another. This power is thus the rate at which energy is carried by the system, or the rate of energy transport. We can also think of the vibrating string as having an energy density u—an energy per unit length. This energy density must also take the form of a traveling wave. How is the energy density u related to the rate P at which energy is delivered? Suppose we knew the energy density. Then note that the quantity $v \Delta t$ is the length of wave that moves past a given point in a brief time interval Δt. All the energy in that length, namely the energy density times the length, or $uv \Delta t$, will be delivered; the rate of energy delivery P is then found by dividing this quantity by Δt. In other words,

$$u(x, t) = P/v = \mu \omega^2 z_0^2 \cos^2(kx - \omega t). \tag{14–34}$$

Traveling waves transmit energy.

FIGURE 14–14 That waves carry energy is a fact well known to anyone who has spent time at the seashore.

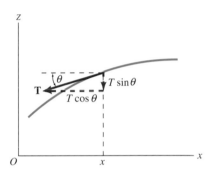

FIGURE 14–15 A string that is distorted is subject to a tension force from the left whose direction depends on the string distortion. We are interested in this case in the vertical component.

The average power delivered by a harmonic wave

A traveling harmonic wave transports energy at the wave speed.

The power transmitted by a wave is proportional to the squares of both the amplitude and the frequency.

We have shown that the power delivered and the energy density on the string are each traveling waves moving with speed $v = \omega/k = \sqrt{(T/\mu)}$. These quantities are always positive. Moreover, *they are proportional to the squares of both the amplitude and the frequency.* Although we have derived our results for traveling waves on a string, the quadratic dependence of the rate of energy transport, of power, or of energy density on amplitude and frequency is a general feature of *all* waves.

FIGURE 14–16 Example 14–5.

EXAMPLE 14–5   A long rope of mass density 150 g/m is tied at one end. A person holds the other end with a horizontal force of 95 N so that the equilibrium position of the rope is horizontal. The person then shakes the held end of the rope up and down (Fig. 14–16) such that sinusoidal traveling waves are generated with a frequency of 1.0 s^{-1} and an amplitude of 5.0 cm. What power must be delivered by the hand that is shaking the rope?

Solution: The power delivered to the rope is the power carried by the rope in its wave motion. Thus Eq. (14–33) can be used to calculate the power. We are given the tension $T = 95$ N, the mass density $\mu = 0.15$ kg/m, the frequency $f = 1.0$ Hz, and the amplitude $z_0 = 0.050$ m. To apply Eq. (14–33), we need $\omega = 2\pi f = 6.28$ rad/s, and $v = \sqrt{T/\mu} = \sqrt{(95 \text{ N})/(0.15 \text{ kg/m})} = 25.2$ m/s. Then Eq. (14–33) gives

$$P = \frac{1}{2}(0.15 \text{ kg/m})(6.28 \text{ s}^{-1})^2(0.050 \text{ m})^2(25.2 \text{ m/s}) = 0.19 \text{ W}.$$

This is a small fraction of 1 hp and is easily delivered by the hand holding the rope.

The energy density of standing waves has the same dependence on amplitude and frequency as the total energy density of traveling waves. However, no power is delivered by standing waves because such waves do not propagate through space.

Consequences of Energy Conservation

We can use our results for the energy density, together with the conservation of energy, to determine how the amplitude (wave height) of a periodic wave originating at a point varies with distance from that point. For example, a toe periodically dipped into a pond is the source of the energy contained in the ripples that spread outward in circles from the toe. If a given ripple has radius r, the energy in the ripple is spread evenly over the circle, which has length $2\pi r$. Thus the total energy density falls off as $1/r$; further, the amplitude falls off as $1/\sqrt{r}$ because the energy density is proportional to the square of the wave's amplitude. Similarly, a horn sounding a note in open air is the source of a wave that spreads outward from the source as a sphere. Sound energy is spread over the spherical surface of area $4\pi r^2$ when a particular part of the wave has spread to a radius r. Thus the energy density at a point decreases as $1/r^2$, and the amplitude of the sound waves from the horn decreases as $1/r$.

14–6   SOUND

In Section 14–4, we referred to the compressional waves traveling down a metal bar as *sound*. Another type of sound wave is generated in air by the mechanical motion of some object in contact with the air—the vibration of vocal cords, an oscillating violin string, or the collapsing hot air column formed by a lightning stroke. A traveling sound wave transports energy, which can then be picked up by another mechanical system such as an eardrum.

The Nature of Sound Waves

Let's look at sound in air on a microscopic level. The molecules in air are widely spaced so the intermolecular forces that act as springs in solid or liquid bodies play a very limited role here. Except for brief, random collisions with each other, air molecules move freely. How, then, can a wave propagate? Consider a pipe oriented along the x-axis as in Fig. 14–17. At one end of the pipe is a speaker. When the speaker diaphragm moves to the right, the air molecules in a thin slice to the right of the diaphragm within the pipe are subject to a force that acts along the +x-direction. These air molecules acquire a net momentum in the +x-direction in addition to their random motion. Some of the molecules in the original slice will then move to an adjacent region, increasing the number of molecules there and creating a region of *compression*, where the molecular density is higher. When the diaphragm moves back to the left, fewer molecules remain in this region, which corresponds to a region of *rarefaction* where the molecular density is lower. We thus have adjacent regions of compression and rarefaction, as in Fig. 14–17. In the region of compression, there are more molecules and therefore more molecular collisions. The air molecules that recoil from this region of more collisions will fill in the rarefied region. The molecules to which the net momentum had been transferred will move in the +x-direction, so the region of compression moves to the right in the form of a pulse. If the diaphragm undergoes periodic motion, regions of compression and rarefaction are propagated in the +x-direction as a periodic wave in air. Figure 14–17 shows the molecular pattern of a sound wave in air. The displacement of the molecules is aligned with the propagation direction of the wave. Sound waves in air are therefore *longitudinal* waves.

Sound waves in air are longitudinal waves.

Note that there is no net movement of molecules in the direction of the wave. It is the alteration of the density from the average (or equilibrium) density that propagates. If ρ is the density of air (or any other gas), then sound waves may be described in terms of $\Delta\rho = \rho - \rho_0$, the deviation from the equilibrium density ρ_0. A harmonic sound wave traveling in the +x-direction would be described by

$$\Delta\rho = A_\rho \sin(kx - \omega t), \qquad (14\text{--}35)$$

where A_ρ is the maximum deviation of the density from ρ_0. Generally, $A_\rho << \rho_0$, so we may speak of a nearly constant mass density, even in the presence of sound waves.

The Speed of Sound

We can use the techniques of Section 14–4—in particular Eq. (14–28)—to estimate the speed of sound waves in a gas (such as air). We must find a restoring force factor and a mass factor for the gas; these factors should not depend on the total amount of gas because the speed is not dependent on the total amount of gas. The mass factor is

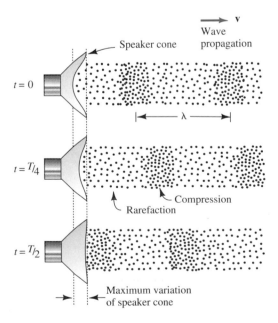

FIGURE 14–17 When a piston—here, a speaker diaphragm—at one end of a pipe filled with air moves along the +x-direction, there is a force on the air molecules to the right of the piston. If the piston moves back and forth, the result is alternate regions of higher and lower density propagating along the pipe.

easy: The only quantity that fits is the mass density ρ_0. The restoring force factor is the pressure P_0 (see Chapter 16). If we were to insert a tiny membrane into the gas and observe one side of the membrane, we would see it being bombarded by a stream of molecules. When these molecules collide elastically with the membrane, they transfer momentum to it. A transfer of momentum is equivalent to a force. Therefore, we may characterize the gas by a force per unit area, which is the **pressure**. Pressure has dimensions of force per area, or $[MLT^{-2}][L^{-2}] = [ML^{-1}T^{-2}]$, whereas density has dimensions of mass per volume, or $[ML^{-3}]$. The ratio P_0/ρ_0 has dimensions $[L^2 T^{-2}]$, a speed squared. We thus estimate for the speed of sound in a gas

$$v_{\text{sound}} \simeq \sqrt{\frac{P_0}{\rho_0}}. \tag{14–36}$$

This result was first found by Newton, and it is appropriately called Newton's formula. A more detailed calculation gives a better value for the speed of sound,

The speed of sound

$$v_{\text{sound}} = \sqrt{\frac{\gamma P_0}{\rho_0}}. \tag{14–37}$$

Here γ is a constant with the approximate value 1.4 for air. The numerical correction to our estimate due to the factor γ is small. For normal atmospheric conditions at sea level, $P_0 \simeq 10^5$ N/m^2, $\rho_0 \simeq 1.3$ kg/m^3,

$$v_{\text{sound}} \simeq 330 \text{ m/s} = 740 \text{ mi/h}. \tag{14–38}$$

In air, sound travels 1 mi in about 5 s. Sound waves move at different speeds in different materials. Table 14–1 gives the speed of longitudinal sound waves in various media.

Standing Sound Waves

Standing sound waves are generated if the necessary boundary conditions exist. A pipe much narrower than the wavelength of standing waves set up in it provides a suitable one-dimensional environment. If the pipe is closed at an end, then there can be no displacement of gas molecules across the end. The closed end of the pipe is then like the fixed end of a string. If both ends are closed, we have the equivalent of a string fixed at both ends. As Fig. 14–18 shows, the wavelength for these standing waves takes on only the discrete values allowed by Eqs. (14–7) and (14–8). The frequencies are determined in terms of the wavelengths by Eq. (14–22),

$$v_{\text{sound}} = \lambda f. \tag{14–39}$$

As for the vibrating string, the discrete integer n labels the modes [Eqs. (14–16)]. When $n = 1$, we have the fundamental frequency (discussed in Section 14–2); when $n > 1$, we speak of the nth harmonic.

TABLE 14–1
The Speed of Sound Waves in Various Media†

Medium	Speed of Sound (m/s)
Hydrogen	1284
Air	330
Liquid mercury (20°C)	1450
Methyl alcohol	1189
Water	1402
Polyethylene	920
Lead	1210
Silver	2700
Aluminum	5000
Beryllium	12,870

†Longitudinal waves. Temperature = 0°C, pressure = 1 atm unless otherwise stated.

EXAMPLE 14–6 A pipe closed at both ends has length $L = 4.0$ m. Compute the fundamental frequency for standing sound waves as well as the frequencies of the next two modes.

Solution: As Eq. (14–8) states, the fundamental wavelength, λ_1, for a system with two closed ends is twice the length of the system: $\lambda_1 = 2L$. Figure 14–18 shows how the standing wave fits within the pipe with a node at each end. Higher modes have shorter wavelengths, enumerated in this case by $n = 2, 3$: $\lambda_2 = \lambda_1/2$, $\lambda_3 = \lambda_1/3$. The relation between frequency and wavelength is given by Eq. (14–39). Thus the frequencies in question are determined by

$$f_n = \frac{v_{\text{sound}}}{\lambda_n} = \frac{v_{\text{sound}} n}{2L}.$$

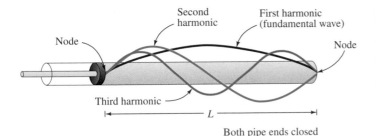

Second harmonic

First harmonic (fundamental wave)

Node

Node

Third harmonic

L

Both pipe ends closed

FIGURE 14–18 Example 14–6. There must be a node of molecular displacement at each end of a closed pipe. Thus one-half of the standing sound wave with the longest possible wavelength fits in the pipe. The longest wavelength represents the fundamental frequency, which corresponds to the first harmonic. The next two harmonics are also shown. The curves represent the amplitudes (not to scale) of the density wave.

Taking v_{sound} from Eq. (14–38), we find

$$f_1 = \frac{330 \text{ m/s}}{2(4.0 \text{ m})} = 41 \text{ Hz},$$

$$f_2 = 2f_1 = 83 \text{ Hz},$$

$$f_3 = 3f_1 = 124 \text{ Hz}.$$

Figure 14–18 also shows these second and third harmonics.

Other boundary conditions are possible for standing sound waves. When one end of a pipe is open, the pressure at that end takes on the value of the pressure of the exterior. There is no variation of the pressure there, which means that there is a pressure node. This can be shown to mean that the displacement has an amplitude maximum (or *antinode*) rather than a node at the open end (Fig. 14–19).

EXAMPLE 14–7 An organ pipe with the dimensions of the pipe in Example 14–6 is closed at one end and open at the other. Find the fundamental frequency and next harmonic in this case.

Solution: When the organ pipe is closed at one end, then the longest possible standing wave has a wavelength four times the length of the pipe, as shown in Fig. 14–19. This is because the wave at the open end is a maximum, or an antinode. Thus $\lambda_1 = 4L$, and the fundamental frequency is

$$f_1 = \frac{v_{\text{sound}}}{\lambda_1} = \frac{330 \text{ m/s}}{4(4.0 \text{ m})} = 21 \text{ Hz}.$$

This is just half of the fundamental frequency when both ends are closed.

The next harmonic is not the harmonic with a frequency twice the fundamental frequency; such a wave would have a node at the open end. Rather, it is the harmonic with three times the fundamental frequency, also illustrated in Fig. 14–19. For this wave, three-quarters of one wavelength fits into the pipe.

Figure 14–20 shows the first few harmonics for an organ pipe with both ends open. The condition that there is an antinode at each end produces a different harmonic series.

Excitation of Standing Waves. We can excite standing waves in an organ pipe if we vibrate the air within at the standing wave frequency; for example, with externally generated sound waves of that frequency. In addition, these standing waves must be able to generate traveling sound waves of the same frequency if the distant members of an audience are to perceive the standing waves within the organ pipe as sound. The couplings between these different types of waves normally involve rather complicated mechanisms. The many different ways that standing waves in musical instruments are excited (bow, single and double reed, mouthpiece, and so forth) indicates the rich variety of physical phenomena involved.

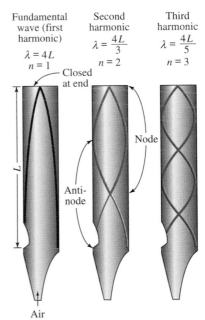

Fundamental wave (first harmonic)
$\lambda = 4L$
$n = 1$

Second harmonic
$\lambda = \frac{4L}{3}$
$n = 2$

Third harmonic
$\lambda = \frac{4L}{5}$
$n = 3$

Closed at end

Node

Antinode

L

Air

FIGURE 14–19 The amplitude of a standing sound wave in an organ pipe open at one end has a maximum at the open end. In this situation, one-fourth of the standing wave with the longest possible wavelength fits within the pipe. The next harmonic is also shown. The curves represent the amplitudes (not to scale) of the density wave.

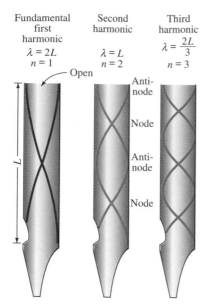

Fundamental
first
harmonic
$\lambda = 2L$
$n = 1$
Open

Second
harmonic
$\lambda = L$
$n = 2$

Third
harmonic
$\lambda = \dfrac{2L}{3}$
$n = 3$

Anti-node
Node
Anti-node
Node

FIGURE 14–20 An organ pipe open at each end has standing waves with antinodes at each end, hence a different harmonic series than the organ pipe closed at one end.

PROBLEM SOLVING

Exciting sound waves

The intensity is the average power per unit area.

FIGURE 14–21 The goblet breaks because a resonance phenomenon acts that allows energy to be transferred easily from sound waves in air to the glass. The sound frequency corresponds to a natural frequency of the goblet.

In each case, the excitation of waves involves a transfer of energy. This transfer of energy, from the energy of an external source—such as the air forced across a clarinet mouthpiece—to the energy of the standing wave within the tube of a clarinet, takes place in ways closely related to the *resonance* phenomenon discussed in Chapter 13. Thus, for example, if a tuning fork is placed near the end of an organ pipe and if the frequency of the tuning fork is the same as the fundamental frequency of the pipe, strong standing waves will rapidly be established within the pipe. Perhaps you have seen a resonance phenomenon act in the opposite direction: When a trumpet plays a note whose frequency matches the natural frequency of a delicate goblet, standing waves can be induced in the goblet strong enough to make it break (Fig. 14–21).

Hearing Sounds

The combination of ear and brain acts as a very sensitive instrument with many interesting properties. For present purposes, it suffices to note that the combination can detect pitch and loudness. *Pitch* is a measure of frequency. The human ear can hear frequencies ranging from about 20 Hz to 20,000 Hz. *Loudness* is a measure of the power carried in a wave.

Because sound waves in air form two-dimensional fronts, it is appropriate to define the average power per unit area rather than the total power in the wave. This average power per unit area is called the **intensity**, I. The intensity measures energy per unit time per unit area, and has SI units of watts per meters squared (W/m^2). Intensity, like power, is proportional to the square of both the amplitude and the frequency. Intensity is not a particularly convenient measure physiologically; a measure better suited to the human ear is the *decibel scale*, in which the measure of intensity is a dimensionless quantity, β, with units of decibels (dB). It is defined by

$$\beta \equiv 10 \log_{10}\left(\frac{I}{I_0}\right), \qquad (14\text{–}40)$$

where

$$I_0 = 10^{-12} \text{ W/m}^2.$$

A logarithmic scale such as this is designed to cover a large range. The smallest detectable intensity is I_0 itself, for which $\beta = 0$, and the largest intensity still perceived as sound is about $I = 1$ W/m^2, or $\beta = 120$ dB. Higher intensities are just painful. The range of intensities detectable as sound by the ear–brain combination is thus an impressively large factor of 10^{12} (Fig. 14–22). Curiously enough, the range of sound intensities detectable by the human ear is about the same as the range of light intensities detectable by the human eye.

14–7 THE DOPPLER EFFECT

Traveling waves move with a finite velocity. It is therefore possible for an observer who measures the waves (the receiver) to move relative to them or for the generator of traveling waves (the source) to move relative to these waves. The movement of the source or of the receiver affects the measured frequencies. An extreme case that illustrates this is a receiver who moves along with the wave crests. The receiver measures a wavelength, just as if he or she were stationary with respect to the medium, by extending a yardstick from one crest to another. But the receiver measures no frequency at all, for the frequency measures the rate at which crests pass the observer.

The perception of a sudden change in the pitch of an automobile horn (a source) as a vehicle passes is a common experience. Related effects occur when the listener (a receiver) moves. In 1842, Christian Doppler gave an explanation for this phenomenon,

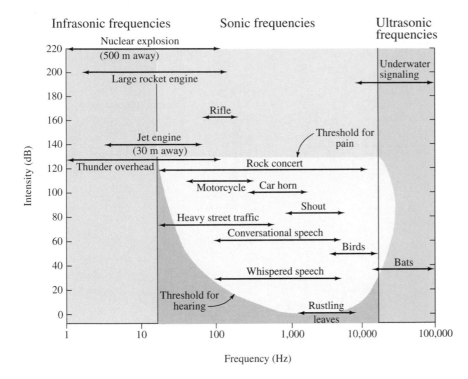

FIGURE 14-22 Typical noise levels at the source: frequency versus β, the intensity measure on the decibel scale. In acoustics—the study of sound—frequency ranges of sound are categorized as infrasonic (low frequency), sonic (or audible, medium frequency), or ultrasonic (high frequency).

as well as a corresponding phenomenon for light. The **Doppler effect** is the shift in frequency and wavelength of waves that results from a source moving with respect to the medium, receiver moving with respect to the medium, or even a moving medium. We'll treat the various possibilities in order.

Case A: A Moving Source. Consider a point source of traveling waves, such as a siren wailing at some frequency f_0 (period $\tau_0 = 1/f_0$). These waves spread away from the source at a wave speed v that is characteristic of the medium alone and is unaffected by the source's motion. If the source is stationary with respect to the medium, these waves are symmetric on all sides of the source, and the wavelength is $\lambda = v/f_0$ [from Eq. (14-22)]. But if the source moves with velocity $\mathbf{v}_s$ with respect to the medium, then the wave fronts will be squeezed together in the direction of $\mathbf{v}_s$ and spread apart in the opposite direction (Figs. 14-23a and b). This is because larger circles correspond to emission at some earlier time, when the source was farther back in the direction $-\mathbf{v}_s$. Thus the wavelengths are shorter along the direction $+\mathbf{v}_s$ and longer in the opposite direction.

For simplicity, consider the source to be moving either directly toward or directly away from the observer (Fig. 14-23c). If the source is moving toward the observer, then during one period τ_0 it moves a distance $v_s\tau_0 = v_s/f_0$. The wavelength is then decreased by this amount:

$$\lambda' = \lambda - \frac{v_s}{f_0} = \frac{v - v_s}{f_0}. \tag{14-41}$$

Because the wave speed remains unaffected, the frequency with which the observer receives the waves becomes

$$f' = \frac{v}{\lambda'} = f_0\left(\frac{v}{v - v_s}\right)$$

$$= \frac{f_0}{1 - (v_s/v)}. \tag{14-42}$$

When the source moves toward the observer, the wavelength *decreases*, and the frequency *increases*—the pitch will be higher in the case of sound. To find the result when the source moves away from the observer, simply reverse the sign of v_s:

(a)

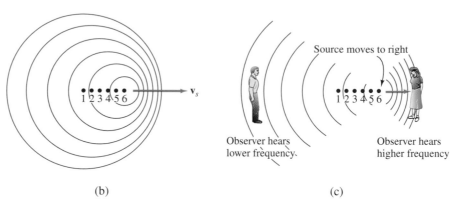

(b)   (c)

FIGURE 14–23 (a) The Doppler effect as seen in the waves of a ripple tank. (b) A moving source of waves. The dots along the direction of the velocity represent the position of the source when the wave centered on them was emitted. The dot at the farthest left corresponds with the outermost wave, and so forth. (c) Here we show two observers at rest, on either side of the source.

$$\lambda' = \lambda + \frac{v_s}{f_0} = \frac{v + v_s}{f_0}; \qquad (14\text{–}43)$$

$$f' = \frac{f_0}{1 + (v_s/v)}. \qquad (14\text{–}44)$$

When the source moves away from the observer, the wavelength *increases*, and the frequency *decreases*—for sound, the pitch drops. A manifestation of this phenomenon is familiar to anyone who has heard a siren drop to a lower pitch as the vehicle with the siren approaches, then passes.

Case B: A Moving Observer. Figure 14–24 depicts an observer moving with velocity $\mathbf{v}_r$ with respect to a wave-carrying medium, within which a stationary source is exciting waves with frequency f_0. We have illustrated the case of an observer moving directly *toward* the source. Such an observer will measure exactly the same wavelength λ as if he or she were at rest but the observer sees the wave crests pass by more quickly. The observer measures a *modified wave speed*

$$v' = v + |v_r|. \qquad (14\text{–}45)$$

The wavelength and wave speed provide the modified wave frequency that the moving observer measures:

$$f' = \frac{v'}{\lambda} = \frac{v + |v_r|}{\lambda} = f_0 + \frac{|v_r|}{\lambda} = f_0\left(1 + \frac{|v_r|}{v}\right). \qquad (14\text{–}46)$$

When an observer moves *toward* the source of traveling waves, the wavelength is unchanged, the wave speed *increases*, and the frequency *increases*. We reverse the sign of v_r and repeat the derivation to learn that when an observer moves *away from* a wave source, the wavelength is again unchanged and the wave speed and the frequency both *decrease*:

$$v' = v - |v_r|; \qquad (14\text{–}47)$$

$$f' = \frac{v - |v_r|}{\lambda} = f_0 - \frac{|v_r|}{\lambda} = f_0\left(1 - \frac{|v_r|}{v}\right). \qquad (14\text{–}48)$$

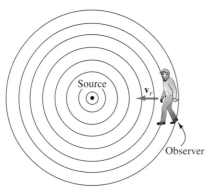

FIGURE 14–24 An observer moving with velocity $\mathbf{v}_r$ receives waves from a fixed source. In this situation, the observer hears an increased frequency.

EXAMPLE 14–8 While moving along a straight road, you first approach, and then pass, a stationary car just at the side of the road whose one-note horn is stuck on. Having perfect pitch, you recognize that the frequency you hear approaching the car is 853 Hz, while you hear a note of 741 Hz receding from the car. Use the value 330 m/s for the speed of sound in air to find the frequency the unfortunate occupant of the car hears and your own speed.

Solution: As you move toward the car, the frequency you hear, f_+, is increased over the at-rest frequency f_0 according to Eq. (14–46), $f_+ = f_0\left(1 + \dfrac{v_r}{v}\right)$, where v_r is your speed and v is the speed of sound. Similarly, as you move away, you hear a decreased frequency $f_- = f_0\left(1 - \dfrac{v_r}{v}\right)$. These are two equations for the two unknowns f_0 and v_r. We solve them by taking their sum and difference:

$$f_+ + f_- = 2f_0$$
$$f_+ - f_- = 2f_0(v_r/v) = (f_+ + f_-)(v_r/v).$$

The first equation immediately gives us f_0 as the average of the two frequencies,

$$f_0 = \frac{1}{2}(f_+ + f_-) = \frac{1}{2}(853 \text{ Hz} + 741 \text{ Hz}) = 797 \text{ Hz}.$$

This is the frequency the car's occupant hears. The second equation gives us v_r:

$$v_r = v\frac{f_+ - f_-}{f_+ + f_-} = (330 \text{ m/s})\frac{853 \text{ Hz} - 741 \text{ Hz}}{853 \text{ Hz} + 741 \text{ Hz}} = 23.2 \text{ m/s}.$$

Case C: A Moving Source and a Moving Observer.   If all motion is kept to the same line, which we label the x-axis, then we can simply combine the results of cases A and B. *Any velocity, including that of the wave itself, is positive if it is to the right and negative if it is to the left.* With this reminder about the sign of the velocity, we shall drop the absolute-value signs. These velocities refer to motion with respect to the medium. The effect of the moving source is to change the wavelength but not the wave speed, and the effect of the moving observer is to change the wave speed but not the wavelength. This can be expressed as

$$\lambda' = \frac{v - v_s}{f_0}; \qquad (14\text{–}49)$$

$$v' = v - v_r.$$

The modified frequency f' is then

$$f' = \frac{v'}{\lambda'} = \left(\frac{v - v_r}{v - v_s}\right)f_0. \qquad (14\text{–}50)$$

The Doppler effect for a moving source and a moving receiver

Source and observer approach one another when v_s and v have the same sign and v_r has the opposite sign. In this case, f' is *increased* over f_0. When the source and observer are moving away from each other, v_r and v have the same sign and v_s has the opposite sign. The perceived frequency is *decreased*. Equations (14–49) and (14–50) include cases A and B as special cases. Note that Eq. (14–50) is *not* symmetric between the source and the observer. If we know the relative speed, we can tell what is moving—the source or the observer—by measuring the at-rest and in-motion frequencies.

When the speeds of the observer and the source are *small* compared to v, as is often the case when the Doppler effect applies to sound waves, then Eq. (14–50) can be shown to be approximately

$$f' \simeq \left(1 + \frac{v_s - v_r}{v}\right)f_0. \qquad (14\text{–}51)$$

In this approximation, the Doppler shift (the change in frequency) depends only on the relative velocity between source and observer, and it is no longer possible to say who is moving—the source or the observer.

Case D: A Moving Medium.   We can consider a fourth possibility: that the medium is itself in motion with respect to some fixed reference frame. Waves generated in a stream provide an example. The source is fixed with respect to the bank, and the stream

FIGURE 14–25 Traveling waves generated at a fixed point, marked by a cross, in a moving medium. The centers of the circles move downstream.

is moving with speed v_m (Fig. 14–25). Let's take a look at how an observer on the bank downstream from the source sees the waves. There are two effects: First, the observer sees the crests moving by at an increased speed $v' = v + v_m$. Second, the wavelength is also increased because a downstream crest moves an additional distance before a second crest is emitted. From Fig. 14–25, it is easy to deduce that the wavelength increases from λ to $\lambda' = \lambda[1 + (v_m/v)]$. The frequency measured by the observer is $f' = v'/\lambda' = (v + v_m)/\lambda[1 + (v_m/v)] = (v/\lambda)[(v + v_m)/(v + v_m)] = v/\lambda = f$.

Similarly, an observer on the bank upstream from the source will see both the wave speed and the wavelength decreased, and in the same ratio, $[1 + (v_m/v)]$, so the frequency $f' = v'/\lambda'$ is unchanged. In other words, because both the speed and the wavelength are changed in the same way by the movement of the medium, *the frequency is unchanged by the movement of the medium*. The fact that the frequency is unchanged is the reason why it is possible to give outdoor concerts even under windy conditions. The intensity of the sound may be decreased if the wind is blowing in the wrong direction, but at least the pitch remains true.

*14–8 SHOCK WAVES

When a source of traveling waves moves with a speed exceeding the wave speed, we encounter a new situation in which *shock waves* can form. This can be seen from Eqs. (14–42) or (14–50), which show that when $v_s = v$, the observed frequency becomes infinite. When $v_s = v$, the waves emitted directly in front of the source pile up on top of one another, so the wavelength is zero. Thus the time separation between the waves is zero, which corresponds to an infinite frequency. When v_s exceeds v, the source outruns the motion of the waves themselves; thus, instead of the squeezed wave fronts that occur for $v_s < v$ (see Fig. 14–23), the wave fronts are actually left behind.

The geometry of this situation can be analyzed as follows. Suppose that the wave source is at point x_0 at $t = 0$ and emits a crest (Fig. 14–26). In a time t_2, the source has traveled to x_2, a distance $v_s t_2$ from x_0. In this same time, the crest has moved out to a circular wave front a distance $v t_2$ from x_0. The perpendicular to the line drawn from x_2 tangent to this circle makes an angle θ_0 with respect to the line perpendicular to the motion of the source, and

$$\cos \theta_0 = \frac{v t_2}{v_s t_2} = \frac{v}{v_s}. \tag{14–52}$$

What is special about this line is that *all* the wave fronts emitted as the source moves

FIGURE 14–26 When v, the wave speed, is less than v_s, the speed of the source, a linear wave front forms along the red lines. This front is called a shock wave. The source continuously emits disturbances that generate circular waves. At $t = 0$, the source is at the center of the larger circle. At $t = t_1$, the source is at the center of the smaller circle, and at $t = t_2$, the source is about to emit another wave front. Successive circles form the continuous line of the shock wave.

FIGURE 14–27 A bullet passing through the hot gases above a candle flame produces a shock wave because its speed is faster than the speed of sound in air. The ratio of these speeds determines the angle of the shock wave.

FIGURE 14–28 The blue glow is Cerenkov radiation that is produced when a charged particle moves through the water at a speed greater than the speed of light in water. In this case, the particles are the result of radioactivity in the reactor elements suspended in the water.

lie along this tangent. For instance, consider the crest emitted at t_1, when the source is at x_1, a distance $v_s t_1$ from x_0. The geometry of the figure gives

$$\cos \theta_1 = \frac{v(t_2 - t_1)}{v_s(t_2 - t_1)} = \frac{v}{v_s}.$$

Thus $\theta_1 = \theta_0$ and the tangent is formed by all the waves emitted during the voyage of the source. This line—a linear wave front called a **shock wave**—moves out with a speed v, maintaining the angle given by Eq. (14–52). When the wave is propagating in a three-dimensional medium, such as air, a cone is formed instead. It must be stressed that *there is only one such wave front*. A cone whose apex is at any location on the trajectory of the source represents just an earlier or later location of the cone that appears in Fig. 14–26. The wave front will pass an observer located on the ground just once (Fig. 14–27).

When $v < v_s$, Eq. (14–52) determines without ambiguity the angle of the shock wave. The ratio v_s/v is called the *Mach number;* the larger the Mach number, the closer θ_0 comes to 90°; that is, the more acute is the angle of the line or cone. The sonic boom that occurs whenever an airplane moves faster than the speed of sound is a manifestation of the arrival of a shock wave of sound at an observer's ear. A second type of shock wave is associated with the movement of an electrically charged particle through some medium such as a gas or Plexiglas. The speed of light in this medium can be less than the speed of light in a vacuum, and a charged particle, whose speed cannot exceed that of light in a vacuum, can nevertheless move through the medium faster than the speed of light in that medium. Light waves then propagate along a shock wave front whose angle is given by Eq. (14–52). These waves are called *Cerenkov radiation,* after Pavel Cerenkov (Fig. 14–28). When the speed of light in a given medium is known, measurement of the angle of the shock wave gives the speed of the charged particle. This is a useful measurement tool.

Wakes

The *wake,* or track, that extends out behind an ocean liner, a row boat, or a paddling duck (Fig. 14–29) is unlike the shock wave that is responsible for a sonic boom (see Fig. 14–26). Upon observation, we see that the wake forms the same angle for *all* these

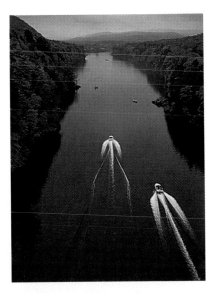

FIGURE 14–29 The wakes of these motorboats on the Connecticut River are a consequence of the fact that the speed of water waves depends on their wavelengths.

examples, even though the ocean liner moves much more swiftly than the duck. This wake forms a vee with an angle of 39° between the linear wave fronts. A wave in deep water (the depth is much greater than the wavelength) behaves differently from all the other waves we have studied in this chapter, because *its speed depends on its wavelength*. The wake of a ship, and other differences between that wake and the shock waves discussed here, depend on the variation of speed with wavelength, a phenomenon called *dispersion*.

SUMMARY

Extended bodies that respond elastically to external and internal forces support organized motions called waves. Waves can be classified into standing waves, in which the motion of the entire medium has the same sinusoidal time dependence, and traveling waves, in which limited internal motion propagates progressively with some definite speed and in some definite direction through the medium. Both traveling and standing waves are further classified into either longitudinal waves or transverse waves. In transverse waves, the motion within the medium is perpendicular to the direction in which the waves are organized. In longitudinal waves, the motion is parallel to the wave direction.

For harmonic standing waves, the motion of any point in the medium is described by the equation

$$z(x, t) = z_0 \sin(kx + \delta) \cos(\omega t + \phi). \tag{14–4}$$

The parameters of this wave are the amplitude z_0, the angular frequency ω, and the wave number k. The wavelength, λ, the length over which a standing wave repeats itself, is related to the wave number by Eq. (14–5), $\lambda = 2\pi/k$. We also use the frequency, f, and the period, τ. These parameters are all related by $\lambda f = \lambda/\tau = \omega/k$, from Eq. (14–22).

For standing waves, possible values of k and of λ are determined by the boundary conditions of the medium and are generally discrete. The angular frequency, however, is determined by the dynamics of the internal motion of the medium. For example, for transverse waves on a string, the dynamical equation of motion is a form of the wave equation:

$$T\frac{\partial^2 z}{\partial x^2} = \mu \frac{\partial^2 z}{\partial t^2}, \tag{14–9}$$

where T is the string's tension and μ is its mass density. The frequency is determined from this equation to be

$$\omega = \frac{2\pi}{\lambda} \sqrt{\frac{T}{\mu}}. \tag{14–14}$$

More generally, the dynamical wave equation takes the form

$$\frac{\partial^2 z}{\partial x^2} = \frac{1}{v^2} \frac{\partial^2 z}{\partial t^2}. \tag{14–27}$$

Here, the quantity v is given by

$$\text{wave speed} = \sqrt{\frac{\text{restoring force factor}}{\text{mass factor}}}. \tag{14–28}$$

For a sinusoidal traveling wave along the x-direction, the displacement z of a point in the medium has the form

$$z(x, t) = z_0 \sin(kx - \omega t). \tag{14–23}$$

The speed of the traveling wave is

$$v = \lambda f = \frac{\lambda}{\tau} = \frac{\omega}{k}. \qquad (14\text{–}22)$$

This speed is determined by the wave equation. In the case of a string of mass density μ under tension T, the speed is

$$v = \sqrt{\frac{T}{\mu}}. \qquad (14\text{–}26)$$

Waves carry energy, and traveling waves carry energy and momentum along the direction of their motion, even if the medium has no net motion in this direction. For both standing waves and traveling waves, the total energy density is proportional to the amplitude squared. The total energy density of a traveling wave in particular is

$$\frac{\mathrm{d}E}{\mathrm{d}x} = \mu\omega^2 z_0^2 \cos^2(kx - \omega t). \qquad (14\text{–}32)$$

This energy density moves along with the wave, delivering an average power

$$\langle P \rangle = \frac{1}{2}\mu\omega^2 z_0^2 v. \qquad (14\text{–}33)$$

Sound waves are longitudinal waves in solids, liquids, or gases. The speed of sound in air is given by

$$v_{\text{sound}} = \sqrt{\frac{\gamma P_0}{\rho_0}}, \qquad (14\text{–}37)$$

where γ is a constant close to 1, and P_0 and ρ_0 are the pressure and the density of the medium, respectively. For normal atmospheric conditions at sea level, the speed of sound in air is roughly 330 m/s.

The Doppler effect describes what happens to the frequency and wavelength of traveling waves emitted by a source moving with velocity $\mathbf{v}_s$ and detected by a receiver moving with velocity $\mathbf{v}_r$. If v is the speed of a wave within the medium, then the frequency perceived by the receiver is

$$f' = \left(\frac{v - v_r}{v - v_s}\right) f_0. \qquad (14\text{–}50)$$

Here, v_r and v_s are positive (negative) when the corresponding motion is along (against) the motion of the wave.

UNDERSTANDING KEY CONCEPTS

1. Give some examples of transverse waves and longitudinal waves and discuss the nature of the mechanical motion in each wave.

2. When a gong is sounded, it produces sound waves in air. If the same gong were struck in a vacuum, would it continue to vibrate for less time, for the same amount of time, or for more time than it does when struck in the same way in air? Explain your reasoning.

3. In old cowboy movies, you have probably seen someone place his ear on railroad rails to tell if a train is coming. Why is this done, and how does it work?

4. Would you expect the speed of sound (a) in hydrogen and (b) in argon to be greater than or less than the speed of sound in oxygen if all gases are under the same pressure?

5. When a lecturer in a classroom demonstration takes a lungful of helium and speaks, his or her voice is comically higher in pitch. Why is that?

6. Captain Kirk is somewhere between galaxies. When a gong sounds in a neighboring spaceship, Kirk reacts to the sound. What is wrong with this scenario?

7. Suppose that a string supports a standing wave with a node at a particular location. How would things change if the point of the node were clamped physically?

8. What prevents sound waves from being transverse waves in a dilute gas such as air? In solids, transverse "sound" waves are generally possible. What can we learn about the nature of the intermolecular forces in such a solid?

9. Are the nodes actually motionless when a longitudinal standing wave is established on a Slinky?

10. The speed of deep water waves of wavelength λ is given by the formula $v = \sqrt{g\lambda/2\pi}$. When there is a storm somewhere on the ocean, it manifests itself by large breakers (breaking waves) at the shore. Initially, you observe that the breakers arrive every 20 s at the shore. Somewhat later, you observe that they are coming every 10 s. How do you interpret this?

11. A child concocts a telephone with two cans that are open at one end and connected by a string attached through a small hole to the closed ends. When the string is made taut by pulling on the cans at opposite ends, two children talking into the open ends can carry out a soft conversation over rather large distances. How does this happen?

12. A common toy consists of a series of five balls that touch and form a line; each ball is suspended as a pendulum. When ball 1 is pulled away and then released, it strikes the line and ball 5 rises in a motion like the reverse of ball 1. If there is a small gap between the balls at equilibrium, then a series of collisions occurs in motion, each with an exchange of velocities, until the last ball swings up. The phenomenon does not change even if the balls touch when at rest. If we describe the phenomenon as a collision of the first ball with an object four times its mass (this is how we describe the four touching balls), we do not get the observed motion of the balls. Why is the last description inadequate?

PROBLEMS

14–2 Standing Waves

1. (I) A guitar string has end points 54 cm apart. What are the three largest wavelengths of standing waves this string can support? Sketch the standing waves for these wavelengths.

2. (I) The third harmonic of a certain piano wire ($\mu = 0.01$ kg/m, $T = 500$ N) has a frequency of 600 Hz. What is the wavelength of the fundamental?

3. (I) A standing wave is formed on a string for which three and one-half wavelengths fit into the total length of the string, which is 2.7 m. The wave has a period of 0.10 s. If the string has a mass density of 220 g/m, what is the tension in the string?

4. (I) A string that is fastened at the two end points, $x = \pm 60$ cm, oscillates in its fundamental mode (no nodes except at the walls). The period of the oscillation is 1 s, the displacement at $x = 0$ cm at $t = 0$ s is 2 cm, and at $t = 0.1$ s, 3 cm. Express $z(x, t)$ for the displacement.

5. (I) Two wires of the same length and under the same tension have fundamental frequencies that are in the ratio 3:1. What is the ratio of their masses?

6. (I) The tension on a wire is 2840 N. The wire is 2.6 m long and the fundamental frequency is 310 Hz. What is the mass of the wire?

7. (II) A steel string under tension vibrates in its lowest mode at a frequency of 1200 Hz. A string of the same material and length but three times the thickness vibrates in its lowest mode at a frequency of 400 Hz. What is the ratio of the tensions of the two strings?

8. (II) A string with fixed ends is in a standing wave that vibrates in five segments (Fig. 14–30). The frequency of this mode is 120 Hz. (a) Find the fundamental frequency of the string. (b) The tension in the string is reduced by a factor of 9. What is the new fundamental frequency?

9. (II) A rope has a mass density of 40 g/cm and is under tension of 200 N. The rope is fixed at one end and is connected to a rod at the other end that sets up standing waves by moving slightly. The rope is 1 m long. Determine the three lowest possible frequencies and the corresponding angular frequencies and periods.

10. (II) Figure 14–31 shows a pulse at $t = 0$ traveling to the right with a speed of 0.5 m/s. (a) Sketch, properly scaled, the shape of the rope at $t = 4$ s. (b) Plot the displacement of the point at $x = 5$ m as a function of time.

FIGURE 14–31 Problem 10.

11. (II) The mass density of copper is 8.92×10^3 kg/m^3. What is the tension on a copper wire 1.00 mm in diameter and 60.0 cm long, fastened at each end, when the frequency of vibration in the $n = 3$ mode is 870 Hz?

12. (II) A string with length L and linear density μ is under tension T. One end is fixed; the other end remains horizontal while it moves periodically up and down. Standing waves with an antinode (a point of maximum displacement) can be induced in this way, as in Fig. 14–32. Find the possible wavelengths and frequencies.

FIGURE 14–30 Problem 8.

FIGURE 14–32 Problem 12.

13. (I) A wave is described by a displacement function $y = A \sin(ax - bt)$, where $a = 0.3$ m^{-1} and $b = 0.02$ s^{-1}. Plot the displacement versus t at both $x = 1$ m and $x = 12$ m.

14. (I) A wave is described by $y = A \sin(kx + \omega t)$, where $k = 3.0$ m^{-1}, $\omega = 2.0$ s^{-1}, and $A = 0.40$ m. Determine (a) the wavelength, (b) the frequency, (c) the maximum amplitude, (d) the period, and (e) the wave number.

15. (II) Use Eq. (14–23) for the traveling wave $z(x, t)$ and the wave equation to show that $Tk^2 = \mu\omega^2$.

16. (II) A perfect triangular pulse travels along a string with a velocity of 4 m/s. The shape of the pulse at $t = 0$ is shown in Fig. 14–33. Plot the displacement and velocity of the segment at $x = 3$ m as a function of time. Is such a pulse physically possible?

FIGURE 14–33 Problem 16.

17. (II) Show by substitution that the general form of a traveling wave, $f(x - vt)$, satisfies the wave equation.

18. (II) Between $x = 0$ cm and $x = 10$ cm, a very long wire (so long that the boundary effects do not matter) is distorted from its straight-line equilibrium form into a triangular shape like that of Fig. 14–33 with the peak 4 cm above the original position. The tension and density of the wire are such that waves travel along the wire at a speed of 400 m/s. (a) What is the shape of the wire at the initial time $t = 0$? (b) What is the shape of the wire at an arbitrary time t, assuming that the wave travels in the $-x$-direction?

19. (III) Suppose you make a little hump in a hall rug with linear mass density 1.5 kg/m and that the hump propagates with speed 2 m/s to the right. In this case, the segments of the rug do not return to their original position; rather, the pulse results in the displacement of the whole rug. Assuming that the hump involves 0.6 m of the rug and extends 0.5 m along the floor, calculate the momentum carried by the pulse. Similar disturbances, called dislocations, play an important role in the plastic deformation of solids. [*Hint*: Consider the global effect of the pulse on a long rug.]

14–4 Estimating the Speeds of Traveling Waves

20. (I) The French train known as the TGV travels at 300 km/hr. What is the frequency with which the cars of this train pass a given point at that speed, given that the cars are 30 m in length?

21. (I) Consider a long length of the wire used in Example 14–2. The wire is taut under a tension of 500 N. If the wire is plucked, what will be the speed of the traveling wave?

22. (I) A rope of density μ_1 is attached to another rope, of density μ_2 (Fig. 14–34). A traveling wave of speed v_1 is sent down the first rope. What will the speed in the second rope be?

FIGURE 14–34 Problem 22.

23. (I) A harmonic wave with wavelength 0.27 m moves with speed 13 m/s. Find the wave number k and the frequency f of this wave.

24. (I) A mountain climber of mass 68 kg hangs from a nylon rope 28 m below the top of a cliff. The total mass of the rope is 1.3 kg; assume that this mass is small enough so that the variation in equilibrium tension along the rope can be ignored. How long does it take a wave pulse to travel up the rope from the man to the top of the cliff?

25. (I) The values of Young's modulus, Y, for aluminum and copper are $Y_{Al} = 7.0 \times 10^{10}$ N/m^2 and $Y_{Cu} = 11 \times 10^{10}$ N/m^2. Their mass densities are 2.70×10^3 kg/m^3 and 8.96×10^3 kg/m^3, respectively. Calculate the speed of elastic wave propagation in these two media.

26. (II) Assume that experiment has shown that the velocity of waves in shallow water is independent of the wavelength of these waves. Use dimensional analysis to show that the speed of these waves is proportional to the square root of the depth D. Express the speed in terms of the relevant constants.

27. (II) Assume that the expression for the velocity of shallow water waves in terms of g and depth D that you derived in Problem 26 is exact. A speed boat that travels close to the speed of such waves tends to build up a large wave just at its bow and thus will have trouble going any faster. At what speed in kilometers per hour will this happen in a harbor in which the water is 5 m deep?

28. (II) The considerations of Problem 26 apply to all water waves for which the wavelength is much larger than the depth. For tidal waves (or *tsunami*: waves caused by underwater earthquakes), the wavelengths are enormous, many hundreds of kilometers. What is the speed of a tidal wave in water that is 3 km deep?

29. (III) When n springs, each of spring constant k and equilibrium length ℓ, are attached in a line, the resulting system is a spring with spring constant $k_n = k/n$ and length $L = n\ell$. If there is a mass m at the connection point between each spring (the springs themselves are assumed to be massless), then the resulting system can support longitudinal, or compressional, waves (see Fig. 14–3b). By imagining an arbitrarily long series, you can see that the speed of these waves is independent of the total length of the spring, and hence of n. Use the available constants k, ℓ, and m to estimate the speed of longitudinal waves. This system is an excellent model for compressional waves on a real spring. [*Hint*: Dimensional analysis can be useful here.]

30. (III) A set of tiny beads is connected in a line by springs, as in Problem 29. The equilibrium length of the springs is 3 cm. The beads have mass 5 g each, and the spring constant of the springs is 20 N/m. Longitudinal waves can be generated on the string of beads by moving the end of the string along the direction of the string. Estimate the speed of such longitudinal pulses. [*Hint*: Dimensional analysis can be useful here.]

14–5 Energy and Power in Waves

31. (I) A wave of wavelength 0.7 m is sent moving down a rope of density 12 g/cm, under a tension of 800 N. The amplitude of the wave is 5 cm. What is the average power transported by the rope? Suppose that the wavelength is doubled; what is the average power transported by the rope now?

32. (I) Calculate the kinetic energy and potential energy contained in one wavelength of a harmonic traveling wave on a string whose mass density is μ and whose tension is T.

33. (I) The power delivered per unit area by a sound wave is given by an expression similar to that of Eq. (14–33). In this case, μ is replaced by the density, ρ, of the gas in which the sound propagates: $P/\text{area} = \text{intensity} = \frac{1}{2}\rho\omega^2 A^2 v$, where A is the amplitude of the displacement of air molecules from their average positions. Given an 8600-Hz sound wave with an intensity of 5.0×10^{-7} W/m^2, what is A? The density of air is 1.30 kg/m^3, and the speed of sound in air is 330 m/s.

34. (II) A sound wave has an intensity of 1.7×10^{-6} W/m^2. Given that the speed of sound in air is 330 m/s, what is the energy density of the sound wave?

35. (II) Obtain an expression analogous to Eq. (14–32) for the rate of energy transport in a left-moving wave. In particular, show that the power remains positive.

36. (II) Harmonic water waves of a given amplitude are generated on a water surface by rhythmic dipping of a pointer at one spot. The waves thus generated spread in circular wave fronts. If R is the distance from the generation point, show that conservation of energy implies that the amplitude z_0 must vary with R, and that this variation is of the form $z_0 \approx 1/\sqrt{R}$.

37. (III) We derived the power or rate of energy transport for a sinusoidal wave [Eq. (14–32)]. Obtain an expression for the power delivered by an arbitrary right-moving wave of the form $f(x - vt)$.

14–6 Sound

38. (I) Sound waves in air travel at 330 m/s, whereas light travels at 3×10^8 m/s. You can determine how far away a lightning strike is by counting the number of seconds between the time you see the lightning flash and the time you hear the thunderclap. How far are you from the lightning bolt for each 1-s difference between the flash and the thunderclap?

39. (I) Using the information in Table 14–1, compute the densities of air and hydrogen.

40. (I) A telephone receiver is at your left ear, with the sound of the conversation at a level of 58 dB. At the same time, your little sister's screams, at a level of 93 dB, enter your right ear. What is the ratio of the sound intensities that enter your two ears?

41. (I) A series of noises have intensities 8.3×10^{-4} W/m^2, 3.5×10^{-6} W/m^2, and 7.2×10^{-9} W/m^2, respectively. What is the sound level of each in decibels?

42. (I) A sound suddenly appears to the ear to be twice as intense as usual. How has the amplitude of the pressure wave (the magnitude of the maximum pressure in the wave) changed?

43. (II) A sound source emits sound with equal intensity in all directions. What is the power output of the source if the intensity of the sound measured at 5 m from the source is 83 dB?

44. (II) Two sounds have levels of 35 dB and 53 dB, respectively. What is the ratio of the power of the two noises? What is the ratio of the amplitudes of the pressure waves they make?

45. (II) In an ideal fluid, the speed of sound is given by $v = \sqrt{B/\rho}$, where B is the *bulk modulus*, a measure of how easy it is to compress the fluid. Sound waves in a fluid are only longitudinal because an ideal fluid cannot bounce back from the application of transverse forces, sometimes called shearing forces. Using the information given in Table 14–1, calculate the bulk modulus of water.

46. (II) A source of sound waves is operating in midair, far from any reflecting surfaces. If, 1 m from the source, the sound intensity level is 120 dB, what is the intensity 1000 m from the source? What is the total power output of the source?

14–7 The Doppler Effect

47. (I) A horn emits sound at a frequency of 160 Hz. What frequency is heard by an observer moving away from the source at 26.0 m/s?

48. (I) A musician with perfect pitch—the capacity to identify pitch by ear—has run a red light and is being pursued by a police car whose siren emits at the single frequency $f = 400$ Hz. The musician's car is moving east at 60 mi/h (27 m/s), and the police car follows at 80 mi/h. The speed of sound is 330 m/s, and the air is still. What does the musician say the frequency of the siren is?

49. (I) A whistle emits sound with a frequency of 333 Hz. The whistle is mounted on a train that travels at 140 km/h in the $-x$-direction toward an observer at rest. (a) What frequency is heard by the observer? (b) Suppose that the whistle is at rest, and the observer moves in the $+x$-direction at 140 km/h. What frequency is heard by the observer?

50. (I) A child is blowing a whistle while riding her bicycle toward you on the sidewalk. You have set up an accurate audio receiver that measures the frequency of the whistle sound to be 793 Hz. You estimate the bicycle to be moving at 7 mph (3.1 m/s). What would you measure the frequency of the whistle to be at rest?

51. (II) The frequency of light emitted by a star moving away from us (the observer) with speed v will be *red shifted* (its frequency will be lowered) by an amount properly described by the treatment in Section 14–7, provided $v/c \ll 1$, where c is the speed of light ($c = 3 \times 10^8$ m/s). How fast is a galaxy receding from us if the observed frequency is 0.95 times the natural frequency of the light?

52. (II) A truck is traveling at 60 mi/h (27 m/s) down the interstate highway where you are changing a flat tire. As the truck approaches, the truck driver sounds an air horn, which has a frequency of 185 Hz. (a) What frequency do you hear? (b) Suppose that you were driving your car at 50 mi/h while the truck driver blows his air horn after he passes you. What frequency would you hear?

53. (II) An ambulance siren is the source of sound emitted with a frequency of 1600 Hz. The ambulance is driving toward the base of a large cliff at 95 km/h (Fig. 14–35). What is the frequency of the reflected sound wave heard at the ambulance?

FIGURE 14–35 Problem 53.

54. (II) You are moving southward on a divided highway and, heading northward on the other side, a fire engine passes with its (one-note) siren screaming. You are close enough to each other as you pass to assume that you and the fire engine are moving on the same line. Your speedometer reads 45 mi/h, and, being a trained musician, you recognize that the frequency you hear before you pass the fire engine is 910 Hz and 680 Hz after you pass. Use the value 330 m/s for the speed of sound in air to find the frequency emitted by the siren and the speed of the fire engine.

55. (III) A tuning fork with frequency 440 Hz is dropped from a tower of height 100 m by an enterprising student. She can hear the fork sounding as it falls. Given that the speed of sound in air is 330 m/s and that $g = 10$ m/s^2, what frequency does the student hear as a function of time? Do not forget that it takes time for sound to travel back to the student.

56. (III) Show that, for $v \ll c$, where c is the speed of a wave in a certain medium, the frequency shift for an observer at rest and a source moving away at speed v is the same as that for a source at rest and the observer moving at speed v in the opposite direction.

*14–8 Shock Waves

57. (I) The angle made by the shock wave associated with the sonic boom of a low-flying airplane is $\theta_0 = 70°$. What is the speed of the airplane?

58. (I) An airplane moving at three times the speed of sound (Mach 3) flies horizontally and at a very low altitude. It emits sound continuously with a frequency that the pilot hears as 300 Hz. (a) What angle does the plane's shock wave cone make with the vertical? (b) What is the frequency of the sound heard by a stationary observer after the plane has passed? (Assume that the plane is moving directly away from the observer.)

General Problems

59. (II) A string is fastened to a wall at $x = 0$ cm and is free to move on another wall at $x = 60$ cm, with the constraint that it is always perpendicular to the wall there (Fig. 14–36). What are the permissible values of the wavelength?

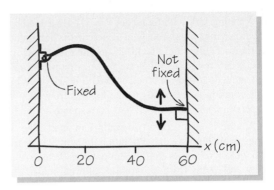

FIGURE 14–36 Problem 59.

60. (II) A string is stretched between a pulley and a wave generator consisting of a plate vibrating up and down with small amplitude and frequency 120 Hz (Fig. 14–37). A standing wave with two intermediate nodes is generated when the string has a mass of 215 g attached to it. How large a load is needed to generate standing waves with one and with four nodes?

FIGURE 14–37 Problem 60.

61. (II) The sound intensity at a rock concert reaches the pain threshold of 120 dB. Normal conversation has an intensity of 55 dB. What is the ratio of the power delivered to the ear at a rock concert compared to that delivered by a typical conversation?

62. (II) Transverse traveling waves are generated on a uniform wire of constant tension and a given mass. By what factor will the required power be increased or decreased if (a) the given mass of wire is doubled in length and the angular frequency remains constant? (b) the amplitude is doubled and the angular frequency is halved? (c) both the wavelength and the amplitude are doubled? (d) the given mass of wire is doubled in length and the wavelength is halved?

63. (II) The density of neon is approximately 0.9 kg/m^3 at standard atmospheric pressure and temperature. The value of γ for neon is 5/3. A loudspeaker emits a note with a frequency of 400 Hz in air. What will the frequency and wavelength of that note be if the loudspeaker is put into a chamber filled with neon?

64. (II) The left-hand end of a long, taut string is moved harmonically up and down with amplitude 0.14 m. This motion produces a wave of wavelength $\lambda = 2.5$ m that travels in the $+x$-direction at speed $v = 245$ m/s. (a) What is the frequency, f, of the traveling wave? (b) What is the maximum transverse velocity of a point along the string? (c) What is the maximum transverse acceleration of a point along the string?

65. (II) A tuning fork with frequency $f = 512$ Hz is held by someone who swings it vigorously in a circle in the horizontal plane. The radius of the circle is 1.0 m, and the frequency of revolution is 3.0 rev/s. (a) What are the maximum and minimum frequencies that a second person would hear? (b) Which part of the rotation corresponds to the highest frequency the second person hears, and which part corresponds to the lowest?

66. (II) A wire is attached to two walls 1 m apart. The wire oscillates in such a way that a point on the wire 25 cm from one end is held fixed at its equilibrium position. What are the possible wavelengths for the first three allowed modes?

67. (II) A tuning fork with a frequency of 440 Hz is held just above the top of a uniform tube containing water (Fig. 14–38). The tube can excite the column of air above the water, whose level can be changed by a spigot at the bottom of the tube. As the water is drained out, the sound intensity of the fork is enhanced when the air column has a length of 0.6 m and again when the air column has a length of 1 m. Use these data to find the speed of sound in air.

FIGURE 14–38 Problem 67.

68. (III) A square sheet of metal with sides of length L is clamped on all four sides. It is convenient to put these sides at $x = 0$, $x = L$, and $y = 0$, $y = L$ (Fig. 14–39). The vertical displacement at any point on the sheet is given by $z(x, y, t)$, and it satisfies the wave equation

$$\frac{\partial^2 z}{\partial x^2} + \frac{\partial^2 z}{\partial y^2} = \frac{1}{v^2} \frac{\partial^2 z}{\partial t^2}.$$

Show that a standing wave of the form $z(x, y, t) = A \sin(kx) \sin(qy) \cos(\omega t)$ satisfies the wave equation, and find the values of k and q that satisfy the boundary conditions and the constraints of the wave equation.

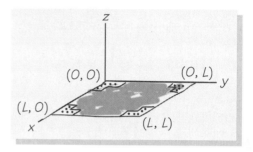

FIGURE 14–39 Problem 68.

69. (III) The relation between angular frequency and wave number for a stiff wire takes the form $\omega^2 = (T/\mu)k^2 + \alpha k^4$, where α is a measure of the stiffness. Suppose that such a stiff wire is clamped at $x = 0$ and $x = L$. Express the frequency for the two lowest modes of the wire. For which frequency is the stiffness a more important effect?

As waves enter the protected harbor through a narrow opening, they spread in opening circles. This is a diffraction phenomenon that can be most easily understood in terms of Huygens' principle.

Superposition and Interference of Waves

The harmonic waves discussed in Chapter 14 are idealizations. Waves observed in nature, such as water waves, sound waves, waves on strings, or light waves, are rarely as simple as harmonic waves. The surface of the ocean is a chaotic welter of waves moving in various directions. A guitar string is set into a very complicated motion when it is plucked. A single shake of the end of a rope sends a single pulse rather than a sine function traveling down the rope.

Nevertheless, we can understand more complicated waves entirely in terms of simple harmonic waves: Harmonic waves can be combined, or superposed, to produce the full variety of waves observed in nature. Superposed waves, which sometimes reinforce and sometimes cancel each other, are said to interfere. This interference produces regular patterns of reinforcement and cancellation; such patterns are seen most easily in light waves as the colors in oil spots or in the pattern of ocean waves as they enter a harbor.

Our goal in this chapter is to understand the different ways in which waves can combine. We will see how standing waves can be formed by superpositions of traveling waves, how beats arise, how there can be well-organized interference patterns in space, and how waves of any shape can be formed by the superposition of harmonic waves.

The superposition principle

15-1   THE SUPERPOSITION PRINCIPLE

The waves that appear in nature are often complicated ones. Nevertheless, our work on harmonic waves is relevant because elastic media that undergo small displacements exhibit **linearity**. Linearity means that if two waves can travel through a medium, then a third wave whose form is the algebraic sum of these waves is possible. Stated mathematically, *the sum of two solutions of the equations of motion of these systems is also a solution.* When linearity applies, the sum of any number of waves can add to form an allowed wave, and a wave multiplied by any constant is also an allowed wave. This fact is known as the **superposition principle**.

To understand how a sum of two solutions of an equation can also be a solution, let's reconsider the dynamical equations that determine the solutions. Newton's second law, applied to large systems with elastic restoring forces, led us to the wave equation in Chapter 14. For transverse waves on a string, Eq. (14–27) took the form

$$\frac{\partial^2 z}{\partial x^2} = \frac{1}{v^2} \frac{\partial^2 z}{\partial t^2}, \qquad (15\text{-}1)$$

where x is the distance along the string and $z(x, t)$ is the transverse displacement of the string at point x and time t. Equation (15–1) is *linear* because the function z appears linearly; that is, as z rather than as z^2 or as z to some other power. If $z_1(x, t)$ and $z_2(x, t)$ are two different solutions satisfying some linear equation such as the wave equation, then the sum $z_1(x, t) + z_2(x, t)$ also satisfies that equation. Superposition works only for waves of small amplitude because when a medium is stretched or distorted too far, the restoring force is no longer purely springlike, and the dyamical equation becomes nonlinear.

Interference

The superposition principle allows us to add sinusoidal waves of different amplitudes, wavelengths, and frequencies, each of them moving in different directions. The sums considered are *algebraic* sums, complete with the possibilities of algebraic cancellations (Fig. 15–1). The results of adding different waves present a rich palette of patterns called **interference phenomena**.

FIGURE 15-1 The superposition of (a) two waves of the same frequency, with a slight phase difference between them; (b) two waves of the same amplitude and frequency, with a phase difference of nearly 180° between them; (c) two waves of the same frequency and phase but different amplitudes; (d) two waves of the same amplitude but frequencies that differ by a factor of 3; (e) two waves of the same amplitude but slightly different frequencies, resulting in beats; (f) three waves of different amplitudes and frequencies.

(a)

(b)

(c)

(d)

(e)

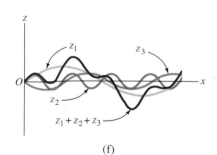

(f)

A simple and striking example of interference is the case of two transverse waves that travel to the right on a string. Let the waves have the same amplitude and be 180° (π rad) out of phase. The two waves have the form

$$z_1(x, t) = z_0 \sin(kx - \omega t) \tag{15-2}$$

and

$$z_2(x, t) = z_0 \sin(kx - \omega t + \pi) = -z_0 \sin(kx - \omega t). \tag{15-3}$$

The superposition of these two transverse waves adds to zero! This is an extreme case of **destructive interference**, which is interference that occurs at any point where two waves have the opposite sign. Although interference may be *fully* destructive (waves adding to zero at a given point), keep in mind that intermediate levels of cancellation are possible. **Constructive interference** occurs when the waves add in such a way that they reinforce. If, for example, our two waves z_1 and z_2 had exactly the same phase, we would find a summed solution with the same x- and t-dependence of either wave, but with amplitude $2z_0$.

Destructive and constructive interference

Coherence

What conditions are necessary for waves from two different sources (for example, the sound waves coming from two loudspeakers) to exhibit an interference pattern? If the two sources put out a series of waves with different frequencies and with phases that change randomly, then there is no consistent interference pattern. Any pattern would change so rapidly that we could not recognize it; *an interference pattern will only appear if the waves have a definite, stable relation between their frequencies and phases.* The waves are then said to be **coherent**. When we have waves without a definite, stable relation between their frequencies and phases, the waves are **incoherent**. Sources that are not point sources, such as piano strings or light bulbs or the human voice, are generally incoherent with other sources because different parts of the waves produced by those sources generally have different phases.

Coherence. Only coherent waves can exhibit stable interference patterns.

15-2 STANDING WAVES THROUGH INTERFERENCE

Standing waves can be viewed as the result of the superposition of traveling waves moving in opposite directions. Imagine a very long string with two waves that move in opposite directions but are otherwise identical, each a solution of the wave equation:

for the wave moving to the left: $z_l(x, t) = z_0 \sin(kx + \omega t)$;   (15-4a)

for the wave moving to the right: $z_r(x, t) = z_0 \sin(kx - \omega t)$.   (15-4b)

We have taken the wavelengths, frequencies, phases, and amplitudes to be equal. According to the principle of superposition, the resultant displacement is

$$z(x, t) = z_r(x, t) + z_l(x, t) = z_0[\sin(kx - \omega t) + \sin(kx + \omega t)]. \tag{15-5}$$

If we use the trigonometric identity 5.3 of Appendix IV-4, with $a = kx - \omega t$ and $b = kx + \omega t$, then we obtain

$$\sin(kx - \omega t) + \sin(kx + \omega t)$$

$$= 2 \sin\left[\frac{(kx - \omega t) + (kx + \omega t)}{2}\right] \cos\left[\frac{(kx - \omega t) - (kx + \omega t)}{2}\right]$$

$$= 2 \sin(kx) \cos(\omega t).$$

In the last line, we have used the fact that $\cos(-a) = \cos(a)$. Thus,

$$z(x, t) = 2z_0 \sin(kx) \cos(\omega t). \tag{15-6}$$

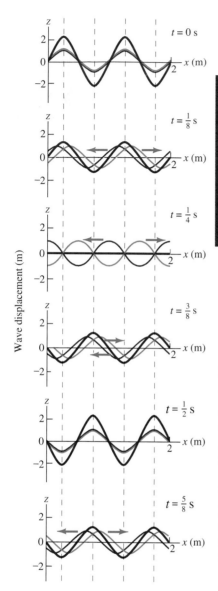

FIGURE 15–2 Superposition of two traveling waves produces a standing wave, plotted in red. Shown are the waves $\sin(kx - \omega t)$, moving rightward, and $\sin(kx + \omega t)$, moving leftward. The amplitude of each wave is 1 m, the wavelength is 1 m, and the period is 1 s. Thus, $k = 2\pi$ m^{-1} and $\omega = 2\pi$ rad/s. Arrows indicate the direction of motion of the two traveling waves. Time progresses in $\frac{1}{8}$-s intervals starting at $t = 0$ s and ending at $t = \frac{5}{8}$ s. Note that the peaks of the standing wave remain at exactly the same x-values.

This expression is identical to our expression for standing waves, Eq. (14–4), except for an (unimportant) phase difference and a factor of 2 that can be taken as part of the amplitude. It is now easier to understand why the wave speed v appeared in the equation for a standing wave: It is the speed of the two traveling waves that are superposed to form the standing wave. In Fig. 15–2, we plot two traveling waves and their superposition. The superposition produces a standing wave whose period and wavelength are the same as those of the traveling waves that compose it.

EXAMPLE 15–1 Two traveling waves move in opposite directions in a one-dimensional medium. They have the form given by Eqs. (15–4a, b). For each wave, the wave speed is 1.8 m/s and the angular frequencies of the traveling waves are 29 rad/s. Find the wavelength of the resulting standing wave.

Solution: We have seen that the wave number k, and hence the wavelength λ, of the standing wave formed by two waves traveling in opposite directions such as those of this example are the *same* as the wave number and wavelength of the traveling waves. This result is expressed in Eq. (15–6). In turn, we can find the wavelength of the traveling waves from the wave speed and the frequency:

$$\lambda = v/f = v/(\omega/2\pi) = 2\pi v/\omega = (6.28)(1.8 \text{ m/s})/(29 \text{ rad/s}) = 0.39 \text{ m}.$$

Interference of Incident and Reflected Waves

Let's look at an *incident,* or incoming, traveling wave that encounters a wall; this will help us to understand the production of standing waves (Fig. 15–3). We can attach the two ends of a string to two walls so that the displacement at the walls ($x = 0$ and $x = L$) is always zero. An incident wave that moves to the right and arrives at $x = L$ will be *reflected* from the wall. The energy carried to the right by the traveling wave bounces back and travels to the left as a reflected wave. The reflected wave has the same frequency, f, as the incoming wave because collisions with the wall of any part of the incoming wave, such as the crests, also have frequency f, and it is these collisions that generate the reflected wave. The reflected wave moves with the same speed as the original traveling wave because the wave speed v is just a property of the string. Figure 15–4a shows the incoming wave; Fig. 15–4b, the reflected wave. The reflected wave travels to the left and can now interfere with the rightward-moving incident wave to produce a standing wave (Fig. 15–4c) with the same frequency and wavelength—exactly as in the derivation of Eq. (15–6). It is thus correct to think of standing waves on a string fixed at both ends as a superposition of traveling waves and of their reflections at the two ends.

One feature of the reflection deserves special attention. The standing wave formed on the string must have nodes at the two fixed ends. This means that the incident wave and its reflection must *cancel* at each end (and hence form a node). This requires that *the sinusoidal wave that reflects from a fixed end must have a displacement with a sign opposite to that of the incoming wave and must therefore be inverted* (see also Section 15–6).

15–3 BEATS

Beats occur when two waves of nearly the same frequency are superposed; a long wavelength (or small frequency) interference pattern is produced. Musicians use beats to tune their instruments, and the phenomenon is of scientific importance because it provides a sensitive technique for measuring frequency (or period) differences. A simple visual analogue to beats can be realized with the help of two combs with slightly different tooth spacings; the slightly different spacings are analogous to slightly different frequencies. In Fig. 15–5, two combs are superposed. Where the teeth lie directly on top of one another, the spaces are visible. Some distance away, the teeth will lie out of phase and no spaces are visible. Thus, the superposed combs exhibit a pattern that repeats over a much greater distance than the separation of the teeth. The very large spacing of the interference pattern observed when the two combs are superposed is like the very large period of the interference pattern observed when waves with two slightly different periods are superposed.

Consider two traveling waves of slightly different frequencies, as in Figs. 15–6a and 15–6b. The two waves have equal amplitudes of 1 m. The first wave has $\omega_1 = 1$ Hz and $k_1 = 1.0$ m^{-1}, and the second has $\omega_2 = 0.9$ Hz and $k_2 = (1/0.9)$ m^{-1} = 1.1 m^{-1}. To the eye, the two waves are barely distinguishable from each other. Yet their superposition has a striking visual signature: *A very different wave—one with a very long wavelength—is formed* (Fig. 15–6c).

What has happened is the following: Both waves start out together at $x = 0$ m, where their similarity causes them to interfere constructively. The intensity of the superposed wave is large there. Because their wavelengths differ slightly, the two waves begin to differ from one another as we move along x; at some point along x (at about 30 m), they are almost perfectly out of phase. At this point, they interfere destructively. The intensity of the superposed wave is small there. The cycle repeats itself as the waves slowly come together again, at about 55 m, as the combs of Fig. 15–5 demonstrated.

Calculation of the Beat Frequency

To analyze this situation mathematically, start with the two waves

$$z_1(x, t) = z_0 \sin(k_1 x - \omega_1 t) \qquad \text{and} \qquad z_2(x, t) = z_0 \sin(k_2 x - \omega_2 t). \quad (15\text{–}7)$$

A convenient way to account for the fact that the frequencies of the two waves are not very different is to deal with the sum and the difference of the frequencies. The frequency difference will be a small quantity. If $\omega_1 > \omega_2$, and $k_1 > k_2$, we define

$$\delta\omega \equiv (\omega_1 - \omega_2) \qquad \text{and} \qquad \Omega \equiv \frac{1}{2}(\omega_1 + \omega_2), \quad (15\text{–}8)$$

and similarly,

$$\delta k \equiv (k_1 - k_2) \qquad \text{and} \qquad K \equiv \frac{1}{2}(k_1 + k_2). \quad (15\text{–}9)$$

It is easy to show that the ratios $\delta\omega/\delta k$ and Ω/K are given by the wave speed $v = \omega_1/k_1 = \omega_2/k_2$ in the medium. Both $\delta\omega$ and δk are positive and small.

We can easily reexpress the original frequencies and wave numbers in terms of their sums and differences:

$$\omega_1 = \Omega + \frac{1}{2}\delta\omega \qquad \text{and} \qquad \omega_2 = \Omega - \frac{1}{2}\delta\omega, \quad (15\text{–}10)$$

$$k_1 = K + \frac{1}{2}\delta k \qquad \text{and} \qquad k_2 = K - \frac{1}{2}\delta k. \quad (15\text{–}11)$$

When Eqs. (15–10) and (15–11) are substituted into the waves of Eq. (15–7), we find that

$$z_1(x, t) = z_0 \sin\left[\left(K + \frac{\delta k}{2}\right)x - \left(\Omega + \frac{\delta\omega}{2}\right)t\right]$$

$$= z_0 \sin\left\{(Kx - \Omega t) + \frac{1}{2}[(\delta k)x - (\delta\omega)t]\right\}, \quad (15\text{–}12)$$

$$z_2(x, t) = z_0 \sin\left[\left(K - \frac{\delta k}{2}\right)x - \left(\Omega - \frac{\delta\omega}{2}\right)t\right]$$

$$= z_0 \sin\left\{(Kx - \Omega t) - \frac{1}{2}[(\delta k)x - (\delta\omega)t]\right\}. \quad (15\text{–}13)$$

APPLICATION

Beats are used to measure frequency differences.

FIGURE 15–3 The parallel waves generated at the bottom of this photograph are reflected from the 45° barrier and move to the right.

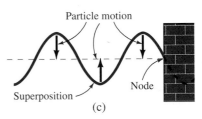

FIGURE 15–4 The reflection of a transverse traveling wave on a string from a wall (the second wall is far to the left). (a) The incident wave $\sin(kx - \omega t)$ moves to the right. (b) The reflection of that wave has the form $-\sin(kx + \omega t)$ and moves to the left. (c) The superposition of the traveling wave and its reflection is a standing wave with nodes at the walls. The arrows in parts (a) and (b) represent the direction of the waveform; in (c) the arrows indicate the actual motion of the string itself.

411

FIGURE 15–5 Two superposed combs whose tooth spacing is slightly different illustrate beats by analogy. They produce what is known as a moiré pattern.

The beat frequency

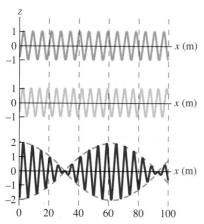

FIGURE 15–6 Two traveling waves of identical amplitude, $z_0 = 1$ m, and nearly the same frequency and wavelength, at $t = 1$ s. (a) The first traveling wave has $\omega = 1$ rad/s and $k_1 = 1$ m^{-1}. (b) The second traveling wave has $\omega_2 = 0.9$ rad/s and $k_1 = 1.1$ m^{-1}. (c) Beats are visible in their superposition, formed by the envelope shown as a dashed line. The vertical lines trace the regions of constructive and destructive interference.

To sum these two terms and find the resultant superposition, we expand the trigonometric functions with identity 5.3 of Appendix IV–4. The result is

$$z(x, t) = z_1(x, t) + z_2(x, t)$$

$$= 2z_0 \sin(Kx - \Omega t) \cos\left[\left(\frac{\delta k}{2}\right)x - \left(\frac{\delta \omega}{2}\right)t\right]. \quad (15\text{–}14)$$

We have used the fact that $\cos(-a) = \cos(a)$ to arrive at this result.

Equation (15–14) states that the sum of the original two waves can be reformulated as a *product* of two traveling waves. The wavelength and frequency of the first traveling wave, which is the sine term in Eq. (15–14), are given by the *average* of corresponding quantities for the original two waves. The envelope formed by the second wave is the dashed curve in Fig. 15–6c. This envelope is a wave traveling with the same speed as the sine wave but with a much longer wavelength. The envelope wave is called a *beat* of the original two waves. The coefficient of x in the cosine factor in Eq. (15–14), $\delta k/2$, gives the wave number of the beats, whereas the coefficient of the time, $\delta\omega/2$, gives the angular frequency. In turn, we can use Eqs. (15–8) and (15–9) to find the wavelength and frequency of the beats:

$$\frac{1}{\lambda_{\text{beat}}} = \frac{k_{\text{beat}}}{2\pi} = \frac{k_1 - k_2}{4\pi} = \frac{1}{2}\left(\frac{1}{\lambda_1} - \frac{1}{\lambda_2}\right); \quad (15\text{–}15)$$

$$f_{\text{beat}} = 2\pi\omega_{\text{beat}} = \pi(\omega_1 - \omega_2) = \frac{1}{2}(f_1 - f_2). \quad (15\text{–}16)$$

The beat has a wave number and frequency that are 1/2 the *differences* of the original wave numbers and frequencies. Because this difference is small by assumption, the wave number of the envelope is small and the wavelength is *large*. Similarly, the frequency of the beat is *small*. In deriving this result, we tacitly assumed $f_1 > f_2$, so that f_{beat} was positive. Note, however, that the beat factor in Eq. (15–14) is a cosine factor, and the cosine is unchanged if its argument changes sign. This means that if $f_2 > f_1$, f_{beat} would be given by $(f_2 - f_1)/2$ rather than $(f_1 - f_2)/2$: the beat frequency is always positive.

Figure 15–7 is an enlarged version of the superposed waves plotted in Fig. 15–6c. Recall that the original two waves have $k_1 = 1.0$ m^{-1} and $k_2 = 1.1$ m^{-1}. From Eq. (15–15), the beat wavelength is then

$$\lambda_{\text{beat}} = \frac{4\pi}{|(k_1 - k_2)|} = \frac{4\pi}{|(1 \text{ m}^{-1} - 1.1 \text{ m}^{-1})|} = 110 \text{ m}.$$

Only a little over half a wavelength is plotted in Fig. 15–7, and we can indeed see that $\lambda_{\text{beat}}/2 \approx 55$ m. The frequency of the beat is, in this case,

$$f_{\text{beat}} = \pi(\omega_1 - \omega_2) = \pi(1.0 \text{ Hz} - 0.9 \text{ Hz}) = 0.31 \text{ Hz}.$$

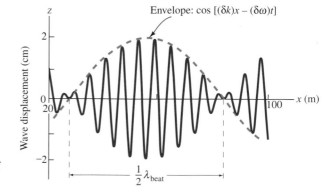

FIGURE 15–7 An enlarged look at the superposition of Fig. 15–6c. The dashed envelope is a traveling wave of wavelength $\lambda_{\text{beat}} \approx 110$ m.

Beats in Sound. When we hear beats of sound, the combination of ear and brain has a definite physiological response. In addition to the note of average frequency, we hear a second note with the pitch of the beat frequency if the beat frequency exceeds 25 Hz or so. If the beat frequency is less than this, however, we hear regular pulses in amplitude. Because the ear interprets loudness as proportional to the wave intensity, which is in turn proportional to the amplitude squared, the ear hears a pulse every time the beat passes through either a maximum or minimum. Thus *the frequency of the perceived pulse is twice the beat frequency.*

PROBLEM SOLVING

The pulse frequency differs from the beat frequency.

EXAMPLE 15–2 Two tuning forks are each labeled 440 Hz. Yet when they are struck simultaneously, there is a 1-Hz pulsing in the strength of the sound. What is the difference in the frequencies of the two tuning forks? Is it possible to tell which one has the higher pitch from the information given?

Solution: The frequency of the *perceived* pulses occurs with twice the beat frequency:

$$f_{pulse} = 2f_{beat} = 2\pi(\omega_1 - \omega_2) = f_1 - f_2.$$

Thus $f_1 - f_2 = 1$ Hz, and this is the difference in frequencies of the forks. The frequency of the beat depends only on the absolute value of the frequency differences, so there is no way to tell which fork has the higher pitch from the information given.

FIGURE 15–8 Two sources that produce waves in a ripple tank form an interference pattern, visible as regions of constructive and destructive interference radiating out from the sources.

15–4 SPATIAL INTERFERENCE PHENOMENA

Sound waves, light waves, and water waves all exhibit *interference patterns* in space.[†] For example, if a laser beam is separated into two beams by passing it through two holes, it is possible to produce a regular series of intense spots where the two beams interfere constructively. A ripple tank is another useful tool for demonstrating interference patterns (Fig. 15–8). *Any* wave that propagates in two or three dimensions can show such spatial interference phenomena. Spatial interference phenomena are the basis of many precision measurements of distance, as we shall see when we study the interference of light waves in Chapters 38 and 39.

In order to see how spatial interference phenomena arise, imagine a single source of harmonic traveling waves with wavelength λ and frequency f, such as a harmonically driven loudspeaker and two pipes that carry the sound waves away from the loudspeaker. The pipes fix the path that the waves can follow, and the two paths do not necessarily have the same length even if they end up at the same place (Fig. 15–9). (This is what is special about two or three dimensions.) At the beginning of the respective paths, the waves start in phase. For example, if there is a wave crest at the start of path 1, then the same crest is at the start of path 2. These two crests each travel down their paths at the same speed. If the paths are the same length, $L_1 = L_2$, then they will both arrive at the observation point P at the same time. They will still be in phase at P, and they will interfere constructively. Under these circumstances, the adjacent troughs will arrive together, as will the next crests, and so forth. The interference will always be constructive.

Conversely, if the path lengths are different, then these same crests will not arrive at the same time. Whether the interference is constructive or destructive depends on the *difference* in path length, $\Delta L = L_2 - L_1$. Suppose path 2 is longer, so that ΔL is positive. If ΔL is one full wavelength, $\Delta L = \lambda$, then the second crest will arrive at P along path 1 just when the first crest arrives at P along path 2. The two waves will be

FIGURE 15–9 Unless they propagate on a string or on other one-dimensional media, waves from a source S may follow different paths to get to a given observation point P. Here two pipes of different lengths determine two different paths between S and P.

PROBLEM SOLVING

Determining the interference pattern

[†]Not to be confused with beats, in which time dependence is primary.

The conditions for constructive and destructive interference

in phase and the interference will be constructive. If $\Delta L = \lambda/2$, then when the crest arrives at P from path 2, it meets a trough from path 1. In this case, the interference is destructive, and no disturbance is recorded at P. The situation can be summed up as follows:

$$\text{for constructive interference:} \quad \Delta L = n\lambda, \, n = 0, \pm1, \pm2, \dots; \qquad (15\text{--}17)$$

$$\text{for destructive interference:} \quad \Delta L = (n + \tfrac{1}{2})\lambda, \, n = 0, \pm1, \pm2, \dots. \quad (15\text{--}18)$$

The effect just described also occurs with two sources at different distances from an observation point. Imagine two separate sources emitting harmonic waves of the same wavelength in a given medium, as though we were making a rhythmic disturbance at two different points on a water surface. At each point, a series of spreading ripples that interfere with one another is started. In Fig. 15–10, we locate two separated sources, S_1 and S_2, each producing waves with wavelength λ and frequency f. We assume that these waves are *in phase*; that is, a crest emerges from S_1 at the same moment that a crest emerges from S_2. These waves are *coherent*. An observation point P is located a distance L_i from source S_i; the path difference is $\Delta L = L_2 - L_1$. Then the conditions for constructive interference are simply that the crests, or the troughs, of the wave arrive at P together. For this two-source situation, the same conditions respectively apply for constructive and destructive interference as before; namely, Eqs. (15–17) and (15–18).

Locations of Maxima and Minima

The observation point P can be anywhere that the waves propagate. Some locations will have constructive interference, some destructive, and some points will have an intermediate level of interference, depending only on the distances from the sources. The locations where the interference is constructive are said to exhibit interference *maxima*, and the locations where the interference is destructive exhibit interference *minima*. Finding the spatial pattern of the maxima and minima is an exercise in geometry—of the type sketched out in Fig. 15–11.

FIGURE 15–10 (a) The water waves generated by two coherent sources in phase spread away in circles from these sources. Whether the interference is constructive or destructive at observation point P depends on the difference in path lengths. The circles are (arbitrarily) drawn on the wave crests. (b) The geometry necessary to determine the path difference $\Delta L = L_2 - L_1$ and hence for the pattern of interference maxima and minima. The observation point P is far away compared to the separation of the two sources, d.

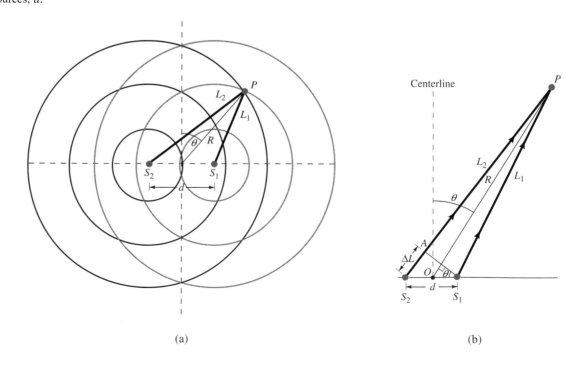

(a)

(b)

We can get a feeling for patterns of this type with a visual analogue to wave interference. Figure 15–12a shows a set of gray concentric circles around a point representing a wave source. Think of the gray circles as simplified crests and the clear areas between the circles as simplified troughs. When two such wave sources are present, as in Fig. 15–12b, we can think of constructive interference as an overlap of the gray areas (crests) with themselves or of the clear areas (troughs) with themselves; we can think of destructive interference as a uniform gray where the gray areas and the clear areas overlap. In fact, bands of maxima and minima are clearly visible in the figure.

To find the patterns of maxima or of minima defined by Eqs. (15–17) and (15–18), the necessary geometric information is the wavelength, λ, and the separation of the two sources, d. It is convenient to put an origin, O, midway between the sources and to measure the distance R from point P to the origin, as well as the angle θ between the centerline and the line from the origin to P. The centerline is the line perpendicular to and bisecting the line of the sources. The geometry is shown in Fig. 15–10b. Note that at the angle $\theta = 0$, where $L_1 = L_2$, we have $\Delta L = 0$; *there is a maximum everywhere along the centerline.* The regions of the other maxima and minima are simple to see when $R \gg d$; that is, when P is very far from the pair of sources. This case is illustrated in Fig. 15–10b, where, to a good approximation, all three lines—the line from S_1 to P (S_1P), the line from the origin to P, and the line from S_2 to P (S_2P)—make the same angle θ with the centerline. Then ΔL can be measured by dropping a line from S_1 perpendicular to S_2P. This line meets S_2P at point A in the figure. S_2S_1 is perpendicular to the centerline, and AS_1 is perpendicular to S_2P, so the angle formed by lines AS_1 and S_2S_1 is θ. Because $AP = S_1P$, we have $\Delta L = L_2 - L_1 = S_2A$, and $\sin \theta = \Delta L/d$, or

$$\Delta L = d \sin \theta. \tag{15–19}$$

For $R \gg d$, we can restate Eqs. (15–17) and (15–18) for the maxima and minima in terms of the angle θ by using Eq. (15–19):

for maxima: $\quad \sin \theta = n\dfrac{\lambda}{d}, \; n = 0, \pm 1, \pm 2, \dots ; \tag{15–20}$

for minima: $\quad \sin \theta = \left(n + \dfrac{1}{2}\right)\dfrac{\lambda}{d}, \; n = 0, \pm 1, \pm 2, \dots . \tag{15–21}$

These equations determine a set of angles for which there are interference maxima and minima. Positive θ values are to the right of the centerline and negative values are to the left. The array of angles is symmetric about the centerline, so we can concentrate on the right side. The case $n = 0$ is a maximum at $\theta = 0$, which is along the centerline itself. As we increase θ, the first minimum occurs at $\sin \theta = \lambda/2d$. The second maximum is at $\sin \theta = \lambda/d$, the second minimum at $3\lambda/2d$, and so forth. In Fig. 15–12b, these angles are marked on the pattern made by overlaying two sets of concentric circles, one set slightly displaced from the other. In the original pattern from which these figures were made, $\lambda = 5$ mm and $d = 16$ mm. The first three minima should then be

(a)

(b)

FIGURE 15–11 The conditions for maxima and minima in the interference of coherent waves of the same wavelength emitted by two sources. (a) Maxima correspond to the difference in path length equal to an integer multiple of λ, while (b) minima occur when the path difference is an integral multiple of λ plus $\lambda/2$.

The angle locations of the maxima and minima in the interference pattern of two sources in phase

FIGURE 15–12 (a) Concentric circles representing waves propagating out from a source. (b) Overlay of the circles that spread from two adjacent sources. Regions of constructive and destructive interference, marked as max (maxima) and min (minima), are clearly visible. We have measured $\lambda = 5$ mm and $d = 16$ mm on the original of this reproduction, so the ratio $\lambda/d = 0.32$. The first minimum should occur at $\sin \theta_1 = \lambda/2d = 0.16$, or $\theta_1 = 9°$, and this is the value of the angle found in the figure.

(a)

(b)

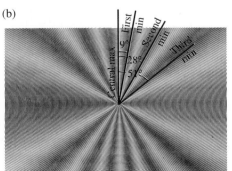

at $\sin\theta_1 = 0.16$, $\sin\theta_2 = 0.47$, and $\sin\theta_3 = 0.79$, or $\theta_1 = 9°$, $\theta_2 = 28°$, and $\theta_3 = 51°$, in agreement with the figures.

Any constant phase difference between two harmonic waves—coherent waves—will lead to a definite interference pattern. Example 15–3 describes the pattern that occurs when two waves of the same wavelength have phases differing by π.

FIGURE 15–13 Example 15–3. The geometry is the same as that illustrated in Fig. 15–10b.

EXAMPLE 15–3 Two loudspeakers, S_1 and S_2, are separated by $d = 0.50$ m. These speakers form sound waves with the same amplitude at a frequency of precisely $f = 4400$ Hz. The amplifier emits the two waves 180° out of phase. A set of chairs is arranged in a semicircle 30.0 m from the speakers. Find the amplitude of the wave at the chair on the perpendicular bisector (centerline) of the line between the speakers. At what distance D to the right of this central chair is there a first maximum in sound intensity?

Solution: As long as we are not too far from the centerline, the distance D along the semicircle is close to the horizontal distance, and the geometry used for the discussion of the in-phase sources can be used here (Fig. 15–13). The fact that the sources have a phase difference of 180° means that when one speaker emits a wave crest, the other emits a wave trough. Think again about the centerline, for which the difference ΔL of the path lengths from the speaker to the observation point is zero. Along this line, the 180° phase difference implies that the wave generated by S_1 is the negative of the wave generated by S_2, and the waves cancel along the centerline. The same reasoning implies that all the results given by Eqs. (15–20) and (15–21)—for sources far from the observation point—are switched: *The maxima become minima, and vice versa*:

$$\text{for minima:} \quad \sin\theta = n\frac{\lambda}{d}, n = 0, \pm1, \pm2, \ldots ;$$

$$\text{for maxima:} \quad \sin\theta = \left(n + \frac{1}{2}\right)\frac{\lambda}{d}, n = 0, \pm1, \pm2, \ldots .$$

The first maximum ($n = 0$) is at an angle given by

$$\sin\theta = \frac{\lambda}{2d}.$$

If the semicircle of chairs is a distance L from the speakers, then $\theta = D/L$. For small θ, $\sin\theta \simeq \theta$, and the value of D for which the first maximum occurs is

$$\sin\theta = \frac{D}{L} = \frac{\lambda}{2d};$$

$$D = \frac{\lambda}{2d}L = \frac{v}{2fd}L.$$

We have used the relation $\lambda f = v$, where v is the wave speed. The speed of sound is roughly 330 m/s, so

$$D = \frac{(330 \text{ m/s})}{2(4400 \text{ Hz})(0.50 \text{ m})}(30.0 \text{ m}) = 2.2 \text{ m}.$$

We see that $D \ll L$, so our small-angle approximation is justified.

Broadening and Narrowing

Interference patterns broaden as wavelengths increase and/or source separations decrease.

Note that as λ increases and/or d decreases, λ/d increases, and the angles given by Eqs. (15–20) and (15–21) increase. The pattern *spreads*, or is *broadened*. The pattern *shrinks*, or *narrows*, as λ decreases and/or d increases. For example, as the separation of our two loudspeakers decreases, the pattern of interference maxima and minima spreads in angle.

One modern use of the interference between waves is in the design of noise-canceling headphones in aviation (Fig. 15AB–1). Aircraft cockpits are noisy environments; even with the type of headphones that fully cover the ears, sufficient noise penetrates to cause hearing discomfort. This is particularly true for aircraft such as helicopters, which have a high power-to-weight ratio. The most recent headsets have a speaker at the outside of the headphone and a microphone located close to the entrance of the ear. Sound received at the microphone is amplified and supplied to the speaker with a phase difference, such that the sound from the speaker cancels the original signal as completely as possible just beyond the entrance to the ear, nominally at the surface of the eardrum. True communication from the radio or intercom of the aircraft is not subject to cancellation and is therefore much more audible to the crew members.

FIGURE 15AB–1 An electronic noise cancellation head set such as this one can reduce noise in the low frequencey range by some 10–15 dB over conventional head sets.

15–5 PULSES

Real waves normally appear as **pulses**, which are disturbances with a definite beginning and a definite end—like a train passing in the night. Pure harmonic waves (cosines or sines) are an idealization in that they are infinitely long. Any truly periodic wave is also an idealization. However, pulses do not present any special difficulty as far as the wave equation is concerned. We saw in Chapter 14 that *any* function of the form $f(x \pm vt)$ satisfies the wave equation. The function f can describe a traveling pulse as easily as a sine or cosine.

Pulses are of interest to us because they occur throughout the physical world. Our messages, of whatever type, are communicated through pulses. The sound of a drum, a flash of light, and a ripple caused by a pebble dropped into a pond are all pulses. We will see in Section 15–7 that pulses are themselves formed from the superposition of harmonic waves. As we will now see, the superposition of pulses also provides us with insight into wave interference.

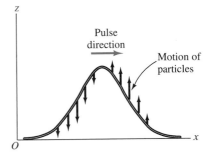

FIGURE 15–14 A pulse moving to the right (horizontal arrow) on a string. The vertical arrows indicate the velocities of elements of the string.

Collisions of Pulses

Suppose that there is a pulse that moves to the right (Fig. 15–14). There is upward transverse motion along the leading edge of the illustrated pulse. The transverse motion vanishes at the peak, and there is downward transverse motion along the trailing edge until the string again comes to rest after the pulse has passed. The velocity of the transverse motion is indicated in the figure by vertical arrows.

Consider the superposition of two pulses of identical shape on the string: Pulse 1 is positive and moves to the right; pulse 2 is negative and moves to the left. Figure 15–15 illustrates a sequence in which the pulses approach each other, superpose, then continue on. At the moment when their respective peaks would be at exactly the same position, $t = 0$ s, the superposition of the two pulses gives a flat string. This situation is in fact no more peculiar than that of a ball that has been thrown vertically and is

417

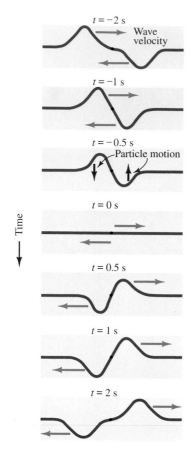

$t = -2$ s — Wave velocity

$t = -1$ s

$t = -0.5$ s — Particle motion

$t = 0$ s

Time

$t = 0.5$ s

$t = 1$ s

$t = 2$ s

FIGURE 15–15 The superposition of two pulses, one moving to the right and one to the left. The center point, at the dot, remains motionless throughout.

A wave is inverted when it reflects from a boundary where the medium is fixed.

momentarily at rest at the peak of its flight. Even if the ball is momentarily at rest, its acceleration remains constant. Similarly, even though the string is flat momentarily, it may still be moving. Not only do the displacements of the pulses superpose, but so do their time derivatives, which measure their transverse velocities. The velocities of the string are indicated in sequence in Fig. 15–15. When the peaks superpose, the leading edge of pulse 1 is superposed on the trailing edge of pulse 2. Both edges are moving upward, so there is a region just right of the center of the superposed peaks moving upward rapidly. Similarly, the region just left of the center of the superposed peaks, consisting of the trailing edge of pulse 1 and the leading edge of pulse 2, is moving downward rapidly. Even though the string is flat for a brief moment, it is not at rest!

15–6 REFLECTION AND TRANSMISSION

When an incident ocean wave meets a wall, the wave rebounds and forms a reflected wave that moves back toward the sea. If the wall is a low one, part of the incident wave can pass over it and continue as a transmitted wave. We are familiar with the fact that waves that meet obstacles can reflect from, and be transmitted across, boundaries.

Reflection

In our discussion of two pulses of opposite sign that meet and interfere, the central point between the two pulses—marked on Fig. 15–15 by a dot—never moves at all. At this point, the displacements of the two waves have equal magnitude but are opposite in sign. This depiction helps us to think about what happens when an incident wave or pulse reflects from the fixed end of a string (or rope). The point that is stationary in the collision of two pulses can be considered to correspond to the fixed end of the string. The two colliding pulses must be of opposite sign to leave a central point stationary. Similarly, as we remarked in Section 15–2, the sinusoidal wave that reflects from a fixed wall must be inverted or else there would be a displacement at the wall. Thus, we can think of reflection as the meeting of two waves in which one wave is coming from the other side of the wall. The string and the inverted pulse or wave on the other side of the wall are purely imaginary, just an aid to picturing the reflected wave, which is not at all imaginary. Figure 15–16 shows a sequence of a pulse reflecting from the fixed end of a string. The pulse moving away from the boundary is the *reflected* pulse.

Suppose that, instead of being fixed at the wall, the end of a rope under tension were *free to move in the vertical direction*. This could be arranged by fixing the rope to a support that allows it to slide freely, as in Fig. 15–17a. The analogous sequence of reflection is shown in Fig. 15–17b. As before, the reflected pulse matches the incident pulse in shape. This time, however, the reflected pulse must be right-side up rather than inverted because the end of the string has been thrown upward by the incident pulse. In this case, we can also imagine a traveling wave on the other side of the slide support moving to the left. The two pulses meet and combine at the slide support, and the pulse height at the support is twice as high as the original amplitude. What has happened here is that the free end gets a vertical impulse when the pulse arrives. Because there is no mass to be accelerated farther down the string (there is no string to the right of the spot!), the end can move an additional amount. That the reflected wave is right-side up is confirmed by experiment.

EXAMPLE 15–4 A harmonic sound wave travels in the $+x$-direction. We can describe this wave by using the quantity z that measures the longitudinal displacement of a layer of air from its equilibrium position. z will be a function of the longitudinal position x and will take the form $z_i = A_i \cos(kx - \omega t + \phi_i)$. The wave encounters a brick wall at $x = 0$ and, because no air can cross the brick wall, the boundary condition $z = 0$ at $x = 0$ holds. What is the form of the reflected wave, and what is the form of the superposition of the incident and reflected waves?

Solution: The reflected wave will travel to the left and thus it takes the general form $z_r = A_r \cos(kx + \omega t + \phi_r)$. The boundary condition stipulates that the superposition of the two waves (the net wave) must be zero at $x = 0$: $z_i + z_r = 0$ at $x = 0$ or, for all t,

$$A_i \cos[k(0) - \omega t + \phi_i] + A_r \cos[k(0) + \omega t + \phi_r]$$

$$= A_i \cos(-\omega t + \phi_i) + A_r \cos(\omega t + \phi_r) = 0.$$

We can expand the cosine functions by using the identities in Appendix IV–4—such as $\cos(\omega t + \phi) = \cos(\omega t) \cos \phi - \sin(\omega t) \sin \phi$—and separate the coefficients of $\sin(\omega t)$ and $\cos(\omega t)$ that result:

$$(A_i \cos \phi_i + A_r \cos \phi_r) \cos(\omega t) + (A_i \sin \phi_i - A_r \sin \phi_r) \sin(\omega t) = 0.$$

If this equation is to hold for all t, the coefficients of both $\cos(\omega t)$ and $\sin(\omega t)$ must be zero. For example, if the expression is to be zero at $t = 0$, where $\sin(\omega t)$ is zero automatically, the coefficient of $\cos(\omega t)$ must be zero; if the expression is to be zero at $t = \pi/2\omega$, where $\cos(\omega t)$ is zero automatically, the coefficient of $\sin(\omega t)$ must be zero. Thus

$$A_i \cos \phi_i = -A_r \cos \phi_r;$$

$$A_i \sin \phi_i = A_r \sin \phi_r.$$

We suppose that all signs are taken care of by phases; that is, the amplitudes are positive. If we square these two equations, add them (use $\sin^2 \phi + \cos^2 \phi = 1$), and take the square root, we find that

$$A_i = A_r.$$

FIGURE 15–16 An incident pulse reflecting from the fixed end of a string is inverted. If the pulse is symmetric, then at some point the string is flat, because the reflected pulse cancels the incoming pulse by destructive interference.

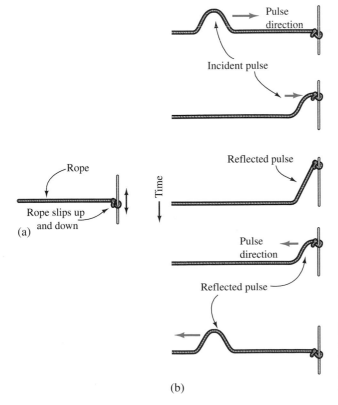

FIGURE 15–17 (a) The right-hand end of a string that is free to slide vertically. (b) The reflection of an incident pulse is upright when the end of the string is free.

This gives us the amplitude of the reflected wave. To learn about the phase, note that we can choose $\phi_i = 0$, because this is only a matter of setting the clock. In that case, $\cos \phi_i = 1$, and we have $A_i = -A_r \cos \phi_r = -A_i \cos \phi_r$, or $-1 = \cos \phi_r$. Thus $\phi_r = \pi$, and the reflected wave takes the form

$$z_r = A_i \cos(kx + \omega t + \pi) = -A_i \cos(kx + \omega t),$$

where we have again used the same identity from Appendix IV–4.

Finally, with $\phi_i = 0$, the total (superposed) wave is the algebraic sum of the incident and reflected waves:

$$z_{\text{total}} = z_i + z_r = A_i[\cos(kx - \omega t) - \cos(kx + \omega t)]$$
$$= 2A_i \sin(kx) \sin(\omega t).$$

This is a standing wave, a result we have already derived for waves on a string.

FIGURE 15–18 A string is attached to a heavier string (top) and to a lighter string (bottom).

Transmission

In addition to being reflected at boundaries, waves and pulses can also pass through boundaries separating two media. This is the process of *transmission*. Up to this point, we have supposed that one end of a string was either tightly fixed or completely free. Imagine instead that our string is attached to a new string that is heavier or lighter than the original one (Fig. 15–18). If the new string is sufficiently heavy, then it acts just like the wall; if the new string is sufficiently light, then it serves as a device that allows the end of the original string to move freely. This arrangement gives us a new feature: a *transmitted wave*, a wave carried by the new string.

Properties of Reflected Waves and Transmitted Waves

Figure 15–19a shows what happens when a pulse is incident on a heavier string and Fig. 15–19b shows what happens when a pulse is incident on a lighter string. Consider first the reflected pulse. In view of our discussion of the string with fixed or free ends,

FIGURE 15–19 (a) A pulse is incident on a heavy string from a lighter one. The reflected pulse is reduced in amplitude and inverted. The transmitted pulse has an amplitude less than that of the incident pulse and moves more slowly than either the incident or reflected pulse. (b) A pulse is incident on a light string from a heavier one. The reflected pulse is reduced in amplitude and right-side up. The transmitted pulse has an amplitude less than that of the incident pulse and moves more rapidly than either the incident or reflected pulses.

it is not surprising that when the new string is heavier than the original string, the reflected pulse is inverted; when the new string is lighter, the reflected pulse is right-side up. The amplitude of the reflected pulse decreases to zero in the limit where the new string has a mass density equivalent to that of the original string because, in that case, there is no boundary. This behavior is both smooth and reasonable.

Consequences of the Conservation of Energy. If we have an incident pulse that is partly reflected and partly transmitted, then the sum of the energies in these two pulses equals the energy of the incident pulse. If there is no transmitted pulse, as is the case when the new string is either infinitely heavy or infinitely light, then the energy in the reflected pulse matches the energy in the incoming pulse. Otherwise, both the transmitted and reflected waves must share the total amplitude of the incident wave and each therefore must have an amplitude reduced from that of the incident wave.

We can state all this in a slightly different way. If the wave speed in the medium containing the incident wave is greater than the wave speed in the medium containing the transmitted wave, then the reflected wave is inverted but is otherwise similar in shape to the incident wave. The transmitted wave is always right-side up, but its shape will be stretched if the wave speed is higher in the second medium and compressed if the wave speed is lower in the second medium. This information is further summarized in Table 15–1.

EXAMPLE 15–5 A string with mass density μ_1 is attached at $x = 0$ to a second string, with mass density μ_2. The strings are under tension T. An incident harmonic traveling wave of amplitude A_i arrives on the first string. A reflected harmonic wave of amplitude A_r results on the first string, and a transmitted harmonic wave of amplitude A_t results on the second string (Fig. 15–20). The strings behave elastically so there is no mechanism for energy loss. Use the fact that energy is conserved to relate the three amplitudes.

Solution: For traveling waves, energy conservation dictates that the rate of energy delivery that leaves the boundary in the form of reflected and transmitted waves must on average equal the rate at which energy reaches the boundary; if not, energy would be lost somewhere. This is equivalent to saying that the sum of the power delivered by the reflected and transmitted waves must equal the power delivered by the incident wave. The power delivered by traveling waves was discussed in Chapter 14; Eq. (14–33) gives the average power in a harmonic wave of amplitude A (equivalent to z_0): $P = \frac{1}{2}\mu\omega^2 A^2 v$. The speed of both the incident and reflected waves is v_1, whereas the speed of the transmitted wave is v_2. Thus, energy conservation gives

$$\frac{\mu_1\omega_i^2 A_i^2 v_1}{s} = \frac{\mu_1\omega_r^2 A_r^2 v_1}{2} + \frac{\mu_2\omega_t^2 A_t^2 v_2}{2},$$

where ω_i, ω_r, and ω_t are the frequencies of the three waves. These frequencies must all be the same because they simply represent a counting of the rate at which crests reach and leave the boundary. (If the frequency changed, wave crests would be lost or would pile up somewhere.) Finally, we can substitute the wave speeds $v_1 = \sqrt{T/\mu_1}$

A_i incident pulse amplitude
A_r reflected pulse amplitude
A_t transmitted pulse amplitude

FIGURE 15–20 Example 15–5.

TABLE 15–1 **Properties of Reflected Waves and Transmitted Waves[†]**

A Wave in Medium 1 Is Incident on a Boundary that Separates Medium 1 from Medium 2		
Wave	*Reflected Wave*	*Transmitted*
Medium 1 is less dense than medium 2 (wave speed in 1 is higher than in 2)	Inverted	Right-side up, Compressed
Medium 1 is denser than medium 2 (wave speed in 1 is lower than in 2)	Right-side up	Right-side up, Stretched

†See Fig. 15–19.

and $v_2 = \sqrt{T/\mu_2}$. The frequencies and the factors of $\sqrt{T}$ and of $\frac{1}{2}$ cancel, and we are left with the relation between the three amplitudes:

$$\sqrt{\mu_1}A_i^2 = \sqrt{\mu_1}A_r^2 + \sqrt{\mu_2}A_t^2.$$

Note that when there is no transmitted amplitude—either because there is a wall at $x = 0$ or because the end of the first string is free—then the relation is $A_i^2 = A_r^2$. Because this relation depends on the *squares* of the amplitudes, it tells us nothing about whether the reflected wave is right-side up (positive A_r) or inverted (negative A_r).

Uses of Reflected and Transmitted Waves

We use reflected light waves constantly: This is how we see the world around us. Modern technology uses the phenomena of reflection and transmission in a variety of different ways. Geophysicists use sound waves to explore Earth's interior. On a small scale, explosions are set near Earth's surface; the sounds from these explosions travel outward from the source. Their arrival times are monitored at a variety of locations after they have been reflected and attenuated by their passage through different materials and across boundaries. Analysis of these measurements can reveal the presence of ore bodies or of geologic formations that may contain oil. On a large scale, earthquakes initiate waves of such intensity that they can travel, by a variety of different paths, all the way through Earth. Seismographs measure and record the arrival time and amplitude of such waves. These waves, called seismic waves, are reflected, transmitted, or both, during their travel; thus, they carry information both about the boundaries, or layers, they encounter within Earth's interior and about the nature of the material between these boundaries. Fourier analysis of the detected waves (a wave-decomposition technique that we will discuss in Section 15–7) helps geophysicists form an image from these signals.

Doctors use ultrasound (sound of very high frequencies) to explore the human body (Fig. 15–21). These same techniques are also used for investigating the imperfections in materials. Radar and sonar devices emit electromagnetic waves or sound waves, then detect the arrival of their reflections, allowing us to learn about the reflecting object. Finally, Doppler shifts of the reflected wave tell us about the speed of the reflecting object.

FIGURE 15–21 The colors in this ultrasound image of a human fetus are the result of computer processing of the image.

*15–7 FOURIER DECOMPOSITION OF WAVES

We have now studied two types of waves: single pulses, which are apparently not associated with periodic phenomena, and harmonic (or sinusoidal) waves, which are periodic. Another important category of waves are periodic waves that are not harmonic (Fig. 15–22). In fact, the superposition principle, together with a remarkable mathematical result called *Fourier's theorem*, shows that waves that fall into these three categories are very closely related. The substance of the theorem is that *any wave can be approximated by a superposition of purely harmonic (or sinusoidal) waves with different frequencies.*

Fourier's theorem

The theorem states rules for finding just what harmonic components are necessary to make up a wave—what we call the *decomposition* of the wave. Moreover, it states how only a few harmonic components can form an approximation to the wave. Suppose that the wave we wish to fit with harmonic forms has period T or, equivalently, frequency $f = 1/T$, and angular frequency $\omega = 2\pi f$ (where f and ω are the *fundamental* frequency and *fundamental* angular frequency of the wave). For such waves, the period in time is related by properties of the medium (in particular, the wave speed) to a period in space—the wavelength. Thus, for a periodic wave, we could refer either to the periodic time dependence at a particular point or to a snapshot in time of a wave in space. Let's look at the periodic time dependence at a particular point, $f(t)$, of the wave we want to fit. This fit is known as a *Fourier expansion* and takes the general form

$$f(t) = \sum_{n=1}^{\infty} A_n \sin(n\omega t + \phi_n). \tag{15–22}$$

The only frequencies that enter into the sum are integer multiples of the fundamental frequency, $f_n = nf$. These frequencies are called harmonics of the fundamental frequency, with second harmonic referring to $n = 2$, third harmonic referring to $n = 3$, and so on.

If, instead of concentrating on the time dependence, we take a snapshot of a periodic wave $g(x)$ in space, then that wave, however complex in form, has a fundamental wavelength λ or a fundamental wave number $k = 2\pi/\lambda$. The Fourier expansion of our wave takes the form

$$g(x) = \sum_{n=1}^{\infty} A'_n \sin(nkx + \theta_n). \tag{15–23}$$

Here, the higher harmonics have wave numbers k_n that are integer multiples of the fundamental wave number. The Fourier expansion specifies rules for the calculation of the constants in Eqs. (15–22) and (15–23).

If the infinite sum in Eq. (15–22) or (15–23) is truncated (cut off), the resulting finite sum forms an approximation to the original waveform. Such approximations are quite useful. We may want to learn about the important frequencies in a particular waveform so that we can suppress or enhance certain modes. For example, in the construction of bridges or buildings, it is important to suppress certain types of movement by mechanical reinforcements, so that frequencies that are resonant with external forces are eliminated and the system cannot be driven to large amplitude oscillations.

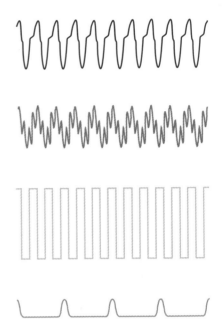

FIGURE 15–22 Some nonharmonic periodic waves.

APPLICATION

Using the Fourier expansion

Fourier Decomposition of a Triangular Wave

To demonstrate Fourier decomposition, let's try to fit the triangular wave illustrated in Fig. 15–23 with harmonic functions. (Such waves might be used in electronics applications where they are generated precisely by the superposition of harmonic waves.) The triangular wave has a period T and an angular frequency $\omega = 2\pi/T$. The phases ϕ_n in the Fourier expansion for the triangular wave in the figure are equal to $\pi/2$, which means that only cosines appear in the expansion. The triangular waveform is symmetric about $t = 0$, and the cosine also has this symmetry. The cosine is thus a "natural" expansion function for the triangular wave. Cosines that reach negative maxima at $t = \pm T/2$ would make it easy to build the sharp negative trough of the triangular wave at these points. This feature is shared by $\cos(\omega t)$, $\cos(3\omega t)$, $\cos(5\omega t)$, and so forth, but *not* by $\cos(2\omega t)$, $\cos(4\omega t)$, and so on. The calculation of A_n bears this out: A_2, A_4, ... all turn out to be zero. When the expansion rules are applied, the exact Fourier expansion of our wave is

$$z(t) = \frac{8}{\pi^2} z_0 \sum_{n=\text{odd}} \frac{1}{n^2} \cos(n\omega t)$$

$$= (0.81)z_0 \left[\cos(\omega t) + \frac{1}{9} \cos(3\omega t) + \frac{1}{25} \cos(5\omega t) + \cdots \right]. \tag{15–24}$$

Figure 15–24 illustrates this Fourier expansion. Figure 15–24a shows the individual terms $z_1 = 0.81 z_0 \cos(\omega t)$ and $z_2 = (0.81/9)z_0 \cos(3\omega t)$. Figure 15–24b shows how

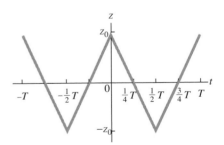

FIGURE 15–23 A triangular wave of period T. The time dependence of the transverse displacement z of a string is plotted at some fixed point in space.

(a)

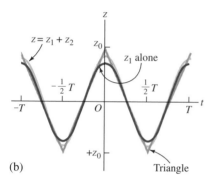

(b)

FIGURE 15–24 Fourier analysis of the triangular waveform of Fig. 15–22. (a) The first two contributing harmonics, $z_1 = 0.81z_0 \cos(\omega t)$ and $z_2 = (0.81/9)z_0 \cos(3\omega t)$. (b) The triangular waveform, the expression z_1 alone, and an approximation containing the first two terms of Eq. (15–24). The approximation with z_1 alone is already a good one, and the two-term expression is an even better one.

∞ We saw a different version of the uncertainty principle in Chapter 13.

first one term then two terms of Eq. (15–24) match the triangular wave. The $\cos(\omega t)$ term alone does quite well in approximating the triangular wave. By the time the second term is included, the approximation is excellent.

The Fourier theorem is of great importance in engineering and physics because of its relevance to the solution of many different problems. In addition to the applications in vibrations and wave motion that are simply sketched out here, heat flow problems and the solutions of a great class of problems in electricity and magnetism require the Fourier expansion.

Fourier Analysis and the Ear

Our ears work very well as Fourier analysis devices. When the note A is struck on a tuning fork designated A, a pure harmonic wave of period $T = 1/(440 \text{ Hz}) = 2.27 \times 10^{-3}$ s emerges. The same note played on a piano, a violin, a trombone, an oboe, or a saxophone contains a harmonic wave of the same fundamental frequency. But each instrument emits its own complicated waveform (Fig. 15–25). In the language of Fourier analysis, this means that a note played by any of those instruments contains vast numbers of differing harmonic components. The ear is capable of picking out the fundamental frequency; we can all agree when the same note is played by a piano and an electric guitar. It is the higher components by which the ear differentiates between different instruments or between different voices. Even different violins have very different higher harmonic contents, a very subtle matter as far as the construction of violins is concerned. Although the physical differences between Stradivarius violins and others of lesser value are so slight that attempts to duplicate "Strads" fail, the sounds of Strads are generally more appealing than those of ordinary violins. A Fourier analysis shows that better violins have a far greater component of higher harmonics. Our ears can even distinguish between individual violins made by the same master!

---

* 15 – 8   PULSES AND THE UNCERTAINTY PRINCIPLE

When we treat pulses with Fourier's theorem, we discover a fundamental relation between the width of the pulse and the frequencies of the periodic waves of which it is composed. This relation is known as the **uncertainty principle**. In Section 15–7, we used Fourier's theorem to treat general periodic waves. At first sight, pulses would seem to have little to do with such periodic waves. Pulses are not periodic. Actually, however, there is a way to think of pulses as periodic, and this is illustrated in the last drawing in Fig. 15–22. Here we see a periodic wave that is a series of short, repeated pulses, with a very long period, T, between the pulses, or, equivalently, separated by a very long wavelength, λ. This means that the fundamental angular frequency, ω (or the fundamental wave number, k) is very small, because $\omega = 2\pi/T$ (or $k = 2\pi/\lambda$).

A true pulse in time is like the periodic sequence shown in the limit that $T \to \infty$, or $\omega \to 0$. Similarly, a true pulse in space is like the periodic sequence shown in the limit that $\lambda \to \infty$, or $k \to 0$. In the Fourier expansion, the sum is taken over integer multiples of angular frequencies (wave numbers). Therefore, *all* frequencies enter into the sum and *the sum in the Fourier expansion is converted into an integral.*

Fourier Decomposition of Pulses

The range of frequencies that enter into the Fourier expansion of a pulse is limited.

What frequencies are involved in the Fourier expansion of a pulse? Although the exact answer to this question depends on the shape of the pulse, we can state an important principle. *If a pulse is strictly limited in time (or space), then it is strictly limited* in the range of angular frequencies (or wave numbers) that enter into it. In other words, the angular frequencies that contribute are in a narrow range, and we can estimate that range by the following argument. The different frequencies that contribute to a pulse interfere constructively in the region where the pulse is strong and interfere destructively in the region where the pulse drops to zero. Suppose that a pulse lasts a time Δt; we say it has a **width** in time of Δt. Suppose also that the contributing frequencies go from a minimum angular frequency ω_{min} to a maximum ω_{max}, with $\Delta\omega = \omega_{max} - \omega_{min}$. We want these waves to be out of phase at the beginning of the pulse, to be in phase at the center of the pulse, and then to be out of phase once more at the end of the pulse. If, in time span Δt, the number of periods of the waves of frequency ω_{max} differs by just one from the number of periods of the waves of frequency ω_{min}, then waves that are out of phase at the beginning of the pulse will be in phase at the center and out of phase at the end. Now, the number of periods of a wave of angular frequency ω contained in a time span Δt is

$$\frac{\Delta t}{T} = \Delta t \frac{\omega}{2\pi}.$$

Our condition is then

$$\frac{\Delta t}{2\pi}(\omega_{max} - \omega_{min}) \simeq 1;$$

$$(\Delta t)(\Delta\omega) \simeq 2\pi. \tag{15–25}$$

This equation is a **reciprocal relation**; if Δt is large, $\Delta\omega$ is small—and vice versa—because 2π is a constant. A more detailed mathematical study shows that this equation holds for all pulses, even if the value of the constant on the right-hand side of Eq. (15–25) might be slightly different for different mathematical pulse forms. A very sharp pulse in time contains a very broad range of harmonic components. Similarly, a pulse that is very broad in time will have only a limited range of angular frequencies.

A more commonly used reciprocal relation follows if we use the frequency, $f = \omega/2\pi$, in place of the angular frequency, ω:

$$(\Delta t)(\Delta f) \simeq 1. \tag{15–26}$$

It is also worthwhile to see how the reciprocal relation works when we study a pulse in space rather than time. Here, the breakdown of the pulse is in terms of a continuous range of wavelengths rather than angular frequencies. Suppose that the pulse has width Δx in space. Then the same method of estimating the range of wavelengths, $\Delta\lambda$, of the contributing harmonic waves can be used. The condition would then be

$$\Delta x\left(\frac{1}{\lambda_{min}} - \frac{1}{\lambda_{max}}\right) \simeq 1,$$

or, in terms of the wave number, $k = 2\pi/\lambda$,

$$(\Delta x)(\Delta k) \simeq 2\pi. \tag{15–27}$$

Here, $\Delta k = k_{max} - k_{min}$ is the range of wave numbers that must enter into the construction of a pulse whose width in space is Δx. Again, the narrower Δx is, the larger the range of wave numbers must be.

There is a very real place where the reciprocal relation is relevant: communication by signal construction. Pure harmonic waves cannot convey information because they have neither a beginning nor an end. Signals and information transfer are necessarily in the form of pulses. For example, all movement of information within and between

(a) Trumpet

(b) Synthesized trumpet

(c) Guitar

(d) Synthesized guitar

FIGURE 15–25 The characteristic sounds of different instruments are associated with frequencies above that of the sound of the "note" being played. These higher frequencies are known as overtones, and their presence makes the waveform of the sound different than the purely harmonic sine or cosine form. A synthesizer electronically generates the characteristic pattern of overtones to trick the ear. Shown in Figs. (a) to (d) are the note of F, played, respectively, by a trumpet, a synthesized trumpet, a guitar, and a synthesized guitar.

The reciprocal relation for frequency and time

The reciprocal relation for width in space and for wave number

APPLICATION

The costs of communication

computers is done with sequences of pulses in binary form. Rapid communication of information requires that the pulses carrying the information be very narrow. If not, and, in particular, if pulses overlap to make wider pulses, information is lost in the confusion between the pulses. According to the reciprocal relation, this means that a very broad spectrum of frequencies must go into the construction of these narrow pulses: The narrower the pulse, the broader the range of frequencies must be. But this costs money—in a very literal sense. The energy density and power in harmonic waves of frequency ω are proportional to ω^2, as discussed in Section 14–5. The reciprocal relation forces a trade-off: A sharper pulse, which is more effective in transferring information, is more expensive.

Uncertainty Relations

In the context of the microscopic world of atoms, the reciprocal relations we have discussed are called **uncertainty relations**; the fact that such relations come into play is the *uncertainty principle*. In atomic physics, quantum mechanics is important and the key idea in quantum mechanics is that all matter has wavelike characteristics. Although the words "uncertainty principle" are often used only in connection with quantum mechanics, the principle really applies to any wavelike phenomenon. The trade-off in the context of quantum mechanics involves the position and speed of microscopic particles. If you measure the position of a particle at any given time with high precision, then you can know the speed—or the momentum—only with poor precision, and vice versa. The momentum is closely related to a wave number in quantum mechanics. We will discuss this further in Chapter 41.

SUMMARY

The algebraic sum, or superposition, of two solutions of the wave equation is also a solution. The collection of physical phenomena that follow from this fact fall under the general heading of wave interference. When two waves that are superposed have opposite algebraic signs at a particular point or a particular moment of time, they interfere destructively. When they have the same sign, they interfere constructively.

Standing waves can be considered to be the result of the superposition of traveling waves moving in opposite directions. Standing waves on a string can be decomposed into the superposition of traveling waves with their own reflections from the ends.

When two waves with nearly equal frequencies, f_1 and f_2, are superposed, the result is that the original wave is modulated by a wave of much smaller frequency and larger wavelength called a beat. The beat frequency is

$$f_{\text{beat}} = \frac{1}{2}(f_1 - f_2). \tag{15–16}$$

Two waves of the same wavelength are coherent if there is a constant phase difference between them. If the distances that the waves travel differ by ΔL, an interference pattern that depends only on ΔL and on the phase difference is traced out. If the phase difference at the source is zero, the interference pattern is

for constructive interference: $\quad \Delta L = n\lambda, \; n = 0, \pm 1, \pm 2, \dots;$ (15–17)

for destructive interference: $\quad \Delta L = (n + \frac{1}{2})\lambda, \; n = 0, \pm 1, \pm 2, \dots.$ (15–18)

Suppose that two coherent wave sources of wavelength λ and zero phase difference are separated by a distance d that is much smaller than the distance to the observation point. Then there is an interference pattern that is a function of the angle θ made with the perpendicular bisector of the line separating the wave sources:

$$\text{for maxima:} \quad \sin\theta = n\frac{\lambda}{d}, \quad n = 0, \pm1, \pm2, \ldots; \quad (15\text{–}20)$$

$$\text{for minima:} \quad \sin\theta = \left(n + \frac{1}{2}\right)\frac{\lambda}{d}, \quad n = 0, \pm1, \pm2, \ldots. \quad (15\text{–}21)$$

The pattern described by these equations broadens (the angle θ corresponding to a particular minimum or maximum increases) as the wavelengths involved increase and/or the separation between the sources decreases. This is a very general feature of interference patterns.

Waves reflect from boundaries and are transmitted past boundaries that separate one wave-carrying medium from another. A reflected wave is inverted or right-side up, compared to the original wave, according to whether the medium from which the wave reflects is denser or less dense than the original medium.

Fourier's theorem shows that any wave, both periodic and nonperiodic (pulses), can be arbitrarily approximated as well as we like by a superposition of purely harmonic waves with different frequencies. This result allows for the construction of arbitrary waveforms from harmonic waves and for the analysis of physical systems in terms of harmonic waves.

A pulse can also be treated by Fourier analysis. If a pulse has a time span of Δt, then the range of frequencies, Δf, that comprise that pulse is limited:

$$(\Delta t)(\Delta f) \simeq 1. \quad (15\text{–}26)$$

There is a similar relation between the range in space of the pulse, Δx, and the range in wave numbers, Δk, that enter into the pulse's decomposition:

$$(\Delta x)(\Delta k) \simeq 2\pi. \quad (15\text{–}27)$$

These relations hold whenever wave phenomena are present and explain why sharp information-carrying signals are expensive to generate.

UNDERSTANDING KEY CONCEPTS

1. A standing wave on a string can be constructed from two traveling waves that move in opposite directions. Can a traveling wave be constructed from the sum of two standing waves?

2. When transverse positive and negative pulses that have the same symmetric shape and size but travel in opposite directions meet, is it necessary that there be a moment when the string or wire on which they move is flat? If so, how do the pulses "know" to continue moving on the string?

3. Two people on each end of a long rope send off a wave pulse. If both wave pulses are on the same side of the rope, describe what happens when the pulses meet. What about when the pulses are on opposite sides? What happens when one pulse is oriented at a 90° angle to the second pulse?

4. Listening to the beat frequency between an unknown and a standard (known) tuning fork tells you only about the difference between the two frequencies. The unknown frequency could be larger or smaller than the known frequency. With a piece of chewing gum at your disposal, can you devise a method to determine the unknown frequency?

5. A wave with truly sharp edges, such as the triangular wave discussed in Section 15–7, can never exist on a string in nature. What might limit how sharp the changes can be?

6. Consider a sinusoidal traveling wave that moves down a sequence of successively lighter strings tied together end to end. The wave starts at the end containing the heaviest string. Qualitatively describe what happens to the frequency and wavelength of the wave by the time it reaches the lightest end of the system.

7. A pebble dropped into still water produces a pulse. The speed of water waves of different wavelengths differ from one another. Why does a dropped pebble produce a series of spreading concentric waves rather than a single wave front?

8. How do you reconcile the principle of conservation of energy with the observation that, in one region, the wave medium is not in motion when waves interfere destructively?

9. In Section 15–1, we referred to two waves that move to the right and cancel; they give no wave at all. Yet each wave alone has an energy density. What happens to the energy? [*Hint*: What are the implications of the superposition principle for the energy density?]

10. Musicians often use beats to tune to some standard, such as an oboe. Can a trumpet player tell from the beat frequency whether one particular note is sharp (at a higher frequency) or flat (at a lower frequency) compared to the same note played on the oboe?

PROBLEMS

15–1 The Superposition Principle

1. (I) Show that the superposition of the two traveling waves $\sin(kx - \omega t)$ and $\cos(kx - \omega t)$ can be written in the form $A \sin(kx - \omega t + \phi)$, and find A and ϕ.

2. (I) Two traveling waves have the forms $z_1 = z_0 \sin(kx - \omega t)$ and $z_2 = z_0 \sin(kx - \omega t + \pi/2)$, respectively. Sketch the superposition of these two waves.

3. (I) Work out the result of the superposition of the two waves of Problem 2 in the form $z = z_1 + z_2 = A \sin(kx - \omega t + \delta)$.

4. (II) Consider the superposition of two harmonic waves, $\psi(x, t) = A \sin(kx - \omega t) + 2A \cos(kx - \omega t + \theta)$. (a) For what values of θ will this superposition result in the maximum constructive interference? (b) the maximum destructive interference? (c) If this superposition is written in the form $B \sin(kx - \omega t + \alpha)$, express B and α in terms of A and θ.

5. (II) Two sinusoidal waves with the same frequency travel down a long rope (Fig. 15–26). The waves have amplitudes 3 cm and 2 cm, respectively, but the second wave trails the first wave by a phase of $\pi/2$. Determine the amplitude of the resultant motion.

FIGURE 15–26 Problem 5.

6. (II) Four sinusoidal waves have the same frequency, but the first two waves have amplitudes twice that of the other two waves. The phases of the four waves are 0, $\pi/2$, π, and $3\pi/2$, respectively. Plot the resulting superposed wave, and describe its motion.

7. (II) Two sine waves, $A \sin(\omega t + kx)$ and $A \sin(\omega t + kx + \delta)$, combine to form a sine wave with the same amplitude A. For what values of δ of the second wave is this possible?

15–2 Standing Waves through Interference

8. (I) Two waves are described by $\psi_1 = (5.0 \text{ cm}) \sin(kx - \omega t)$ and $\psi_2 = (5.0 \text{ cm}) \sin(kx + \omega t)$, where $k = \pi/2 \text{ cm}^{-1}$ and $\omega = 3\pi \text{ s}^{-1}$. (a) Show that the addition of the two waves produces a standing wave. (b) Determine the amplitude and frequency of the resulting standing wave.

9. (I) Consider traveling waves moving in opposite directions: one has the form $z_r(x, t) = z_0 \sin(kx - \omega t)$ and the other has the form $z_l(x, t) = z_0 \cos(kx + \omega t)$. Do these waves add to a standing wave? If so, what is the expression for the standing wave, and what are the locations of the nodes? Note that $\cos \theta$ can be written as $\sin[\theta + (\pi/2)]$.

10. (II) Show by a specific construction that two standing waves can add up to give a traveling wave. What standing wave would you have to add to $A(\cos kx)(\cos \omega t)$ to get a wave traveling to the left?

11. (II) Two traveling waves that move in opposite directions along the x-axis interfere to produce a standing wave of the form $6 \sin(kx) \cos(\omega t)$, where $k = 3\pi \text{ m}^{-1}$ and $\omega = 22\pi \text{ s}^{-1}$. (a) What are the frequencies and speeds of the traveling waves? (b) What are the forms of the traveling waves?

12. (II) One harmonic of an organ pipe has a single node at 2/3 of the length of the pipe and an antinode at the open end (see Fig. 14–20). Write an expression to describe this mode as a standing wave. Find two traveling waves that produce the same result. The length of the pipe is 1.2 m and the speed of sound is 330 m/s.

13. (II) Show that the sum of a wave that travels to the right and a wave that travels to the left with the same amplitude, wave number, and frequency is a standing wave, independent of phase. [*Hint*: Consider $\psi(x, t) = A \sin(kx - \omega t) + A \sin(kx + \omega t + \theta)$; define $u \equiv kx + (\theta/2)$ and $v \equiv \omega t + (\theta/2)$, and use the appropriate identities.]

14. (II) The most general forms of two traveling waves that move in opposite directions along the x-axis with the same amplitude, wavelength, and frequency are $z_r(x, t) = z_0 \sin(kx - \omega t + \phi_1)$ and $z_l(x, t) = z_0 \sin(kx + \omega t + \phi_2)$. What is the form of the standing wave that results from their superposition?

15. (II) A harmonic wave traveling to the right along the x-axis has its maximum amplitude at the time $t = 0$ s at the point $x = 0$ m. Superimposed on this is a wave traveling to the left with equal amplitude, wavelength, and frequency, such that the net standing wave has nodes at $x = 3$ m and at $x = 6$ m. What is the largest possible value of the wavelength?

15–3 Beats

16. (I) You have a tuning fork of unknown frequency and a tuning fork of frequency 264 Hz. Combined, the two produce a pulse frequency ($f_{\text{pulse}} = 2f_{\text{beat}}$) of 2 Hz. What are the possible values of the frequency emitted by the first tuning fork?

17. (I) You have a tuning fork of frequency 512 Hz and look for another tuning fork of smaller frequency so that there is a perceived beat frequency of 6 Hz between them. What frequency should the other tuning fork have?

18. (I) Two violinists play their (identical) A-strings, nominally 440 Hz. A beat is heard at a pulsing of 2 Hz. What is the magnitude of the fractional difference in wave speed, $(v_1 - v_2)/v_1$, for the two strings? [*Hint*: The length L of a violin string is the same from violin to violin.]

19. (I) Two strings of the same material on a pair of guitars are in tune with a frequency of 414 Hz. The string of one of them is tightened slightly, and you hear a beat frequency of 8 Hz. What is the frequency of the tightened string?

20. (I) Two speakers are wired so that they emit sound coherently. Because of a manufacturing error, the same electrical impulse that leads one to emit sound of wavelength 80.10 cm leads the other one to emit sound of wavelength 79.71 cm. What is the pulse frequency ($f_{\text{pulse}} = 2f_{\text{beat}}$)?

21. (II) Two identical piano wires have a frequency of 512 Hz when under the same tension. As a consequence of slippage, the tension of one of the wires changes slightly, so that its frequency decreases slightly. If 4 beats/s are heard when the wires both vibrate, what is the fractional change in the tension of the lower-frequency string?

15–4 Spatial Interference Phenomena

22. (I) Two speakers that emit waves with the same wavelength λ are placed at $x = L$ and $x = -L$, respectively. Assuming that the speakers emit waves with equal amplitudes A and that the phase of the wave on emission is zero at $t = 0$ in both cases, what is the form of the traveling waves that travel toward a microphone located between them at $x = 0$?

23. (I) Find the waveform picked up by the microphone in Problem 22; that is, the superposition of the two waves arriving at the microphone.

24. (I) Two speakers separated by 30 cm are fed with the same pure sinusoidal wave of 1500 Hz. A microphone is placed 6.0 m away along the centerline between the speakers (Fig. 15–27). It picks up an intensity maximum. It is then moved slowly out to the side. (a) How far out to the side, in meters, does the microphone have to be moved to pick up a first minimum? (b) How much farther does the microphone now have to be moved to pick up a second maximum? (c) How do the answers to parts (a) and (b) change if the speakers are now separated by just 10 cm? The approximation $\sin\theta \simeq \theta$ is not valid here, although the angles of the lines from the speakers to the microphone are nearly equal.

FIGURE 15–27 Problem 24.

25. (I) Two coherent sources are 3.0 cm apart and make harmonic ripples of the same frequency (see Fig. 15–8). Consider the ripples along a straight line L_1 parallel to the line L_2 that connects the sources and 40 cm away from line L_2. If the distance between the central maximum at L_1 and the next maximum on L_1 is 8.0 cm, what is the wavelength of the ripples?

26. (I) Sound of frequency 600 Hz is emitted in phase from two speakers 1.4 m apart (Fig. 15–28). A microphone can move along a line parallel to the line that joins the two speakers and is 10 m from the base of the speakers; it measures sound intensity, which can then be plotted versus distance. The interference produced is constructive in the center. Where is the first minimum in the sound intensity curve along the track of the microphone?

27. (II) Two coherent wave sources 5 cm apart make ripples in a water tank (see Fig. 15–8). The wavelength of the waves is 3 cm. (a) Determine the angle between the perpendicular bisector of the line segment connecting the two wave sources and

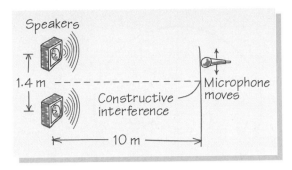

FIGURE 15–28 Problem 26.

the direction of the first noncentral maximum. (b) What is the direction of the first maximum if the two sources operate with opposite phases? (c) How will the answers to (a) and (b) change if the wavelength is changed to 6 cm?

28. (II) Harmonic light waves are emitted by a source equidistant between two narrow vertical slits 0.2 mm apart in a screen (Fig. 15–29). The light waves detected at a screen parallel to the screen containing the slits but 25 cm away have the usual maximum at the point opposite the midpoint of the two slits, and the first minimum occurs at a point 0.4 mm from the maximum. What is the wavelength of the light?

FIGURE 15–29 Problem 28.

29. (II) Coherent waves emerge from two slits separated by a distance $2a$. A screen is placed at a distance L away from the slits, and the first interference minimum is found at a distance d from the center point. Use the small-angle approximation to show that the relationship between the wavelength (λ) and the other lengths is given by $\lambda = 4ad/L$.

30. (II) Two tuning forks vibrating in phase with frequency 440 Hz are placed 2 m apart. A detector is moved in a circle of radius 40 m around the center of the tuning forks (Fig. 15–30). Plot the approximate variation of the square of the net amplitude as a function of angle. Neglect the small variation in the magnitude of the amplitude that results from the sources not being at the same distance from the detector. Take the velocity of sound to be 330 m/s.

31. (II) An oscillator, located at $(x, y) = (-2 \text{ cm}, 0 \text{ cm})$ that dips in and out of water in a tank produces a water wave of the form $\psi_1 = (3 \text{ cm}) \sin(\pi r_1 - \pi v t)$, where r_1 is the distance from the oscillator; a second oscillator, located at $(x,$

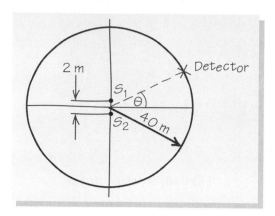

FIGURE 15–30 Problem 30.

$y) = (+2$ cm, 0 cm), produces a water wave of the form $\psi_2 = (3$ cm$) \sin(\pi r_2 - \pi v t)$, where r_2 is the distance from the second oscillator. (a) Where on the line between the sources (the x-axis) are the interference maxima located? (b) What is the form of the wave along the line $x = 0$ cm? (Take all times in seconds.)

32. (II) Two tuning forks that vibrate coherently (in phase) at 512 Hz are placed 0.5 m apart on a horizontal surface. A circular wall of radius 10 m, centered on the midpoint between the sources, is placed around them. Describe the interference pattern of maxima and minima at the wall.

33. (II) Suppose that two in-phase sources of surface waves in water generate water waves that spread out in circles with equal amplitudes. (a) Show that the intensity of the disturbance, or the energy generated in the water, is, at a maximum, four times the intensity for each wave taken separately. (b) Similarly, show that there is no intensity at a minimum. (c) Show that the intensity at any given point if there were no wave interference is twice the intensity from each wave. (d) Do the different answers to parts (a), (b), and (c) mean that the conservation of energy principle must be abandoned for waves? Explain your answer.

15–5 Pulses

34. (I) A student pulls a rope taut and shakes it once, sending a wave pulse down the rope. The pulse is described by a *Gaussian function*,

$$z(x, t) = z_0 \exp[-(x - (0.5 \text{ m/s})t)^2/(0.1 \text{ m})^2].$$

Where is the center of the pulse at time t?

35. (I) At $t = 0$, a sawtoothed pulse has the shape $z = 0$ for $x < -a$ and for $x > 0$; $z = k(x + a)$ for $-a < x < 0$. Sketch this pulse. Suppose that the pulse moves at speed v in the $+x$-direction. What is the algebraic expression for the moving pulse?

36. (II) Consider a train of pulses of the type described in Problem 35. Here, at $t = 0$ the displacement is $z = 0$ for $x < -a + nb$ and for $x > nb$; $z = k(x + a - nb)$ for $-a + nb < x < nb$; n is any integer—positive, negative, or zero—and $b > a$. (a) Sketch this wave train of pulses. (b) Suppose that the entire train of pulses moves at speed v in the $+x$-direction. What is the algebraic expression for the train? (c) What are the period and wavelength of this nonharmonic wave?

37. (II) Consider a transverse wave that moves to the right on a string, in the form of a *Gaussian function*, namely

$$z(x, t) = z_0 e^{[-(x-vt)^2/\alpha^2]}.$$

By taking a time derivative, compute the transverse velocity of this pulse, and sketch it as a function of $(x - vt)/\alpha$. Do the same for the acceleration. Note that any time derivative is automatically a function of $x - vt$ and is itself therefore a quantity that forms a traveling wave.

38. (II) By using the results of Problem 37, *estimate* the energy in a Gaussian pulse that moves to the right. What would the total energy in a harmonic wave be? One of the reasons that real waves have beginnings and ends and are therefore, strictly speaking, pulses, is that pulses involve only a finite amount of energy.

39. (II) Two transverse pulses travel along a string at speed v in opposite directions. The pulse that moves to the right approximates a square with sides of length ℓ projecting above the equilibrium level of the string. The pulse that moves to the left approximates a square with sides of length $\ell/2$ projecting below the equilibrium level. Assume that the superposition principle holds, and describe the approximate shape of the string as the pulses collide.

15–6 Reflection and Transmission

40. (I) A pulse shaped like an upright two-dimensional pyramid, or an isosceles triangle, travels to the left along a string under tension (Fig. 15–31). The left-hand end of the string is free to move up and down. Sketch a sequence of drawings that show how the pulse reflects from the free end.

FIGURE 15–31 Problem 40.

41. (I) A string with linear mass density μ_1 is attached to another string, with linear mass density μ_2. A wave with amplitude A that travels along the μ_1 string will at the attachment point give rise to a reflected wave with amplitude B and a transmitted wave with amplitude C, where

$$B = \frac{\sqrt{\mu_2} - \sqrt{\mu_1}}{\sqrt{\mu_2} + \sqrt{\mu_1}} A, \qquad C = \frac{2\sqrt{\mu_1}}{\sqrt{\mu_2} + \sqrt{\mu_1}} A.$$

For $\mu_1 = 5.0 \times 10^{-2}$ kg/m and $\mu_2 = 3.8 \times 10^{-2}$ kg/m, what are B and C in terms of the amplitude of the incoming wave?

42. (I) Use the formulas in Problem 41 to show that the average power of the reflected wave plus the average power of the transmitted wave add to the average power of the incident wave.

43. (I) A string with mass density μ_1 is attached to a string with mass density μ_2. The amplitude of a reflected pulse is 85 percent of the amplitude of the incident pulse and is inverted. What is the ratio of μ_1 to μ_2? (See Problem 41.)

44. (II) When a symmetric pulse is incident on the fixed end of a rope, the reflected and incident pulse interfere, and there is a moment when the rope is completely flat. Is this a violation of the conservation of energy? (The energy of waves is discussed in Chapter 14.) Where is the energy at the moment the rope is flat?

45. (III) A string with mass density μ_1 is attached at the point $x = 0$ to a string with mass density μ_2. A tension T is maintained in both strings. A transverse wave of the form $A \cos(kx - \omega t)$, moving to the right, is generated far to the left in the first string. Use the fact that the string is unbroken at $x = 0$ and that no sharp kinks can develop there to obtain the forms of the reflected and transmitted waves. [*Hint*: The stated boundary conditions are sufficient to determine all the parameters of the reflected and transmitted waves. These conditions are (a) that the sum of the incident and reflected waves at $x = 0$ on string 1 equals the transmitted wave at $x = 0$ (unbroken string), and (b) that the derivative with respect to x of the sum of the incident and reflected waves at $x = 0$ on string 1 equals the derivative with respect to x of the transmitted wave at $x = 0$ (no kinks).]

*15–7   Fourier Decomposition of Waves

46. (III) Figure 15–32 shows a square wave $f(t)$ with period $T = 2\pi$ (or angular frequency $\omega = 1$) and amplitude = 1 unit. The wave's Fourier expansion is

$$f(t) = \frac{4}{\pi}\left[\sin(\omega t) + \frac{1}{3}\sin(3\omega t) + \frac{1}{5}\sin(5\omega t) + \cdots \right].$$

Plot an approximation to the square wave that corresponds to the first, the first two, and the first three terms in this series. (It helps in this to have access to computer software or to a calculator that can plot graphs.)

FIGURE 15–32 Problem 46.

*15–8   Pulses and the Uncertainty Principle

47. (I) A synthesizer, which is an electronic device for generating and combining harmonic waves, generates an electromagnetic pulse of width 10^{-6} s . Use Eq. (15–26) to find the range of frequencies that the synthesizer must utilize to generate the pulse.

48. (II) Waves are usually not infinite, but come in wave pulses caused by the turning on and turning off of the source. Suppose a laser with wavelength 650 nm is switched on for one second. What is the average frequency, and the width of the range of frequencies in this pulse?

General Problems

49. (I) Two harmonic waves are given by $y_1 = A \cos(kx - \omega t)$ and $y_2 = A \sin[kx - \omega t + (\pi/3)]$, where $k = 5\pi$ m^{-1}, $\omega = 800\pi$ s^{-1}, and $A = 4$ cm. (a) What is the frequency of each wave?

What is the wavelength of each wave? What is the speed of each wave? (b) What is the amplitude of the wave that is the superposition, $y_1 + y_2$, of these two waves?

50. (I) Two rectangular pulses 6 cm long and of amplitude 4 cm but opposite sign are traveling down a string from opposite directions. At $t = 0$ s, the pulses approach each other with their centers 20 cm apart. The wave speed on the string is 1 cm/s. (a) Describe and sketch the shape of the string at $t = 0$, 2, 5, 8, 10, and 15 s. (b) What has happened at $t = 10$ s? What is the energy at this time?

51. (I) Two violinists play their (identical) A-strings, nominally 440 Hz. A beat is heard at a pulsing of 2 Hz. What is the fractional difference in tensions $(T_1 - T_2)/T_1$ of the two strings? [*Hint*: The length of a violin string is the same from violin to violin; when a violinist tunes his or her instrument, the string tension is changed.]

52. (II) The police are busy on the day that you hear two police cars sounding their one-note sirens as you wait at a light (Fig. 15–33). The police cars are moving at 67.1 mi/h (30.0 m/s) and their drivers hear their own sirens emitting at a frequency of 650 Hz. One police car passes you while the other is still approaching. (a) What frequency do you hear for the siren that has passed? (b) for the siren that is approaching?

FIGURE 15–33 Problem 52.

53. (II) Consider two sources of sound that both emit waves with a frequency of 1800 Hz. These sources are placed on a turntable of radius 50 cm that rotates with an angular speed of 120 rev/min. Calculate the pulse frequency ($f_{\text{pulse}} = 2f_{\text{beat}}$) that results from the opposite Doppler shifts of the two sources, heard at a large distance from the turntable as a function of time. Ignore the fact that the sources are not quite equidistant from the detector of the sound. The speed of sound in air is 330 m/s.

54. (II) A guitar string has a fundamental frequency of 400 Hz. Another apparently identical guitar string is stretched beside the first one. (a) During successive stretching, each time tightening the second guitar string, the beat frequencies are 10, 5, and 1 Hz, consecutively. During the tightening, the beat frequency never passes through zero. What are the fundamental frequencies of the second guitar string for each case? (b) By what percentage should the tension be increased for the final adjustment to make the guitar strings have identical frequencies?

55. (II) A loudspeaker (source) that emits sound waves uniformly in all directions is 10 m from a microphone (receiver). Both the source and the receiver are 20 m from a smooth, rigid wall that reflects the sound waves from the source (Fig. 15–34). As the frequency of the waves from the source is smoothly increased from zero, there is a first interference maximum detected by the receiver and subsequently a series of further maxima. List the frequencies of the three maxima that occur above 80 Hz. Assume that sound reflects from the walls as would a ball in an elastic collision.

FIGURE 15–34 Problem 55.

56. (II) Consider two wave sources on a water surface. They generate identical waves of $\lambda = 1$ cm. What possible values of the separation of the sources, d, can you take to ensure that there is no disturbance to the side of the sources (that is, on the extension of the line joining the sources)? When this happens, is there a disturbance on that same line but *between* the sources? If so, describe it; if not, explain why not. Neglect the change in amplitude as a function of distance from the source; that is, treat this as a one-dimensional problem.

57. (II) The speed of the propagation of light in different media depends on the properties of the media. The formula is given by $v = c/n$, where $c = 3 \times 10^8$ m/s and n is the *index of refraction* ($n > 1$ for all media). Consider light of frequency 0.5×10^{15} Hz entering from empty space, with $n = 1$, into glass, with $n = 1.52$. What is the wavelength of the light in empty space? What is the wavelength of the light in glass?

58. (II) Waves are generated in phase in a water tank. The waves move along the x- and y-axes, respectively, with equal amplitude, so that the total wave is given by $\psi(x, y, t) = \sin(kx - \omega t) + \sin(ky - \omega t)$. Show that the pattern of interference maxima is given by the line $x - y = 2n\pi/k$ (where $n = 0, \pm 1, \pm 2,...$).

59. (II) Which of the following equations are linear? Explain your reasoning:

(a) $\dfrac{\partial^2 \psi}{\partial t^2} = v^2 \dfrac{\partial^2 \psi}{\partial x^2}$.

(b) $\dfrac{\partial \psi}{\partial t} = -K \dfrac{\partial^2 \psi}{\partial x^2}$.

(c) $\dfrac{\partial^2 \psi}{\partial t^2} = a\psi^2 + b \dfrac{\partial \psi}{\partial x}$.

(d) $\dfrac{\partial \psi}{\partial t} = 2\psi \dfrac{\partial \psi}{\partial x} + 3\psi$.

60. (II) Two sources emit sound coherently and in phase at a frequency of 845 Hz with the same intensity. They are 35.0 cm apart on a horizontal surface. A circular wall of radius 5.3 m (centered on the midpoint between the sources) is placed around them. Describe systematically the interference pattern of maxima and minima at the wall.

61. (III) Two waves of the form $z_1 = z_0 \sin(k_1 x - \omega_1 t)$ and $z_2 = z_0 \cos(k_2 x - \omega_2 t)$, respectively, propagate at speed v in a nondispersive medium. The frequency difference $\Delta f = f_1 - f_2$ is small. What function describes the envelope of the resulting superposition of the two waves? Express this envelope as a wave, and give its frequency and wavelength.

62. (III) Sinusoidal waves are generated on an infinitely long rope. One wave, $y_1(x, t)$, moves to the left and has amplitude y_0, wave number k_1, and angular frequency ω_1; the other, $y_2(x, t)$, moves to the right with amplitude y_0, wave number k_2, and angular frequency ω_2. Each wave has the same phase. (a) Express $y_i(x, t)$ for each of the waves. (b) Assuming that the rope is a nondispersive medium, what is ω_2 in terms of k_1, k_2, and ω_1? In parts (c) and (d), assume that $k_2 = k_1 + \delta k$ and $\omega_2 = \omega_1 + \delta \omega$. Take the speed of traveling waves on the rope to be v. (c) What is $\delta \omega$ in terms of δk and v? (d) Superpose these waves, assuming that δk and $\delta \omega$ are small. Use the notation $k_{av} = (k_1 + k_2)/2$ and $\omega_{av} = (\omega_1 + \omega_2)/2$. Show that the result of the superposition is two waves, one moving at the low speed $v_l = v \, \delta k/2k_{av}$ and wavelength $2\pi/k_{av}$, and one moving at the high speed $v_r = 2vk_{av}/\delta k$ and the long wavelength $4\pi/\delta k$.

63. (III) Imagine that sinusoidal waves are generated on an infinitely long rope. One wave, for which the displacement is $y_1(x, t)$, moves to the left and has amplitude y_0, wave number k_1, and angular frequency ω_1. The other wave, with displacement $y_2(x, t)$, moves to the right with amplitude y_0, wave number k_2, and angular frequency ω_2. Each has the same phase, which you may take to be zero. (a) Write the appropriate displacement function $y_i(x, t)$ for each of these two waves. (b) What is ω_2 in terms of k_1, k_2, and ω_1?

Suppose now that $k_2 = k_1 + \delta k$ and $\omega_2 = \omega_1 + \delta \omega$, where δk and $\delta \omega$ are small. Take the speed of traveling waves on the rope to be v. (c) What is $\delta \omega$ in terms of δk and of the speed v of waves on the rope? (d) Consider the superposition of these waves. In the limit of $\delta k \to 0$ and $\delta \omega \to 0$, show that the result of the superposition is a standing wave. In particular, what are the wavelength and angular frequency of the standing wave? (e) Superpose these waves again, but this time assume only that δk and $\delta \omega$ are small. Use the notation $k_{av} = (k_1 + k_2)/2$ and $\omega_{av} = (\omega_1 + \omega_2)/2$. Show that the result of the superposition is the product of two waves, one moving at the low speed $v_l = v \, \delta k/2k_{av}$ and short wavelength $2\pi/k_{av}$, and one moving at the high speed $v_r = 2vk_{av}/\delta k$ and long wavelength $4\pi/\delta k$.

Rafting down a highly turbulent section of a river. The motion of the water is very complex and depends on many properties that vary from fluid to fluid.

Properties of Fluids

Waves on the sea, smoke rings, and a whirlpool of water flowing down a drain are all part of the fascinating variety of responses that fluids have to forces. The motion of individual particles and of rigid, extended bodies is completely predictable and relatively simple if the forces are known. As soon as we leave these idealized forms of matter, however, the description of motion becomes more complicated. We have learned that forces that induce only small departures from stable-equilibrium configurations lead universally to simple harmonic motion and, in extended systems, to waves. But the response of fluids to forces goes well beyond wave motion.

In this chapter, we discuss density and pressure, general properties of fluids, which leads us to the notion of buoyancy. We also study the consequences of the conservation of mass and energy for fluids in motion. A good deal of idealization is necessary here; however, even with idealization, we can still arrive at a useful description of the behavior of real fluids.

16-1 STATES OF MATTER

In between rigid solids and dilute gases, matter comes in liquid form. The particular form that matter takes is determined ultimately by the forces between its molecules. Figure 16–1 is a curve that depicts the typical force between two molecules as a function of their separation, r. At very small separations, the force is large and positive; it is strongly repulsive. It becomes attractive at a separation r_0, whose value is approximately 10^{-10} m, and then tapers off to zero at a distance of a few times 10^{-10} m. Close

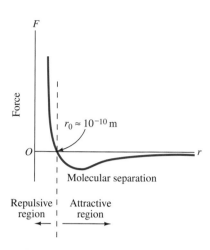

F

Force

$r_0 \approx 10^{-10}$ m

O   r

Molecular separation

Repulsive | Attractive
region | region
←——— |

FIGURE 16–1 Qualitative behavior of the force between a pair of molecules as a function of their separation, r. Near $r = r_0$, the point at which the force is zero, the force is springlike (restoring). For $r < r_0$, the force is strongly repulsive; for $r > r_0$, it is attractive.

The mass density

FIGURE 16–2 A lattice structure characteristic of solids. For cubic lattices, the molecules or atoms reside at the corners. The forces between the molecules are springlike for small displacements from the equilibrium positions, as shown here schematically.

to the distance $r = r_0$, the curve is almost a straight line with a negative slope, and $F = -k(r - r_0)$ is just the typical restoring spring force that leads to oscillatory behavior, as discussed in Chapters 13 and 14.

Gases

If molecules are, on average, sufficiently far apart, then they experience the force shown in Fig. 16–1 only at large values of r, where the force is small. In effect, their only interaction occurs when the molecules occasionally collide with each other. The molecules behave as a **gas**. We will study gases further in Chapters 17 through 19 and only note here that gases are easily compressed and that the shape of a body of gas is easily deformed.

Solids

If molecules are sufficiently close together that the forces due to neighboring molecules are significant, matter will take on a form determined by the nature of the molecules and by the detailed properties of the forces. In some cases, the forces act to organize the molecules into a **lattice**, which is a three-dimensional structure in which the molecules or atoms appear to be connected by rigid bonds and have little freedom to move, except for small displacements (Fig. 16–2). This is a typical **solid**: a substance whose structure resists forces that would deform its shape.

Liquids

In liquids, the molecules are closely spaced and the intermolecular forces between the molecules are strong, but they do not lead to an organized lattice structure. For such aggregates of molecules, there is no resistance to deformation. Nevertheless, unlike the molecules of gases, the molecules of a liquid are close to each other, and there are repulsive forces that tend to resist compression.

Density

A quantity of matter has a mass, M, and a volume, V. The average **mass density**, ρ, of the matter is given by

$$\rho = \frac{M}{V}. \qquad (16–1)$$

Table 16–1 gives the densities of some typical substances. Densities vary with temperature; the density of water at sea level is 0.9998×10^3 kg/m³ at 0°C and 0.9584×10^3 kg/m³ at 100°C. The densities of liquids also can change under compression, although the effect is often small. The fractional density change of water in a cylinder is only about 1.5×10^{-8} when a force per unit area of 3000 N/m² acts on a piston. This is equivalent to the force per unit area produced by a 75-kg person standing on a piston of area 0.25 m². It is therefore a very good approximation to treat water as incompressible. Finally, we can define the **specific gravity** of a substance as the ratio of the mass density of that substance to the mass density of water; thus, for example, the specific gravity of glycerin is 1.26.

EXAMPLE 16–1 A neutron star is a star that is much smaller than our Sun and has the density of an atomic nucleus. A typical neutron star has a radius of 10 km and a mass of 2×10^{30} kg (the mass of the Sun). How much would a 1-cm³ volume of such a star weigh under the influence of Earth's gravity?

Solution: To find the weight of 1 cm³ of a neutron star on Earth's surface, we must find the mass of this volume. To determine this, we need the density, which we

TABLE 16-1   Mass Densities

Phase	Material	Density (kg/m³)
Gas	Hydrogen (0°C, 1 atm)	0.090
Gas	Air at sea level	1.29
Gas	Chlorine (0°C, 1 atm)	3.21
Liquid	Ethyl alcohol	0.79×10^3
Liquid	A good olive oil	0.92×10^3
Liquid	Water	1.00×10^3
Liquid	Glycerin	1.26×10^3
Liquid	Earth's core	9.5×10^3
Liquid	Liquid mercury	13.6×10^3
Liquid	Sun's core	1.6×10^5
Solid	Cork	0.25×10^3
Solid	Applewood	0.745×10^3
Solid	Ice	0.917×10^3
Solid	Quartz	2.65×10^3
Solid	Steel alloys	$7.6 \times 10^3 - 8.9 \times 10^3$
Solid	Iridium	22.4×10^3

can calculate from the given information. The star's radius is $10 \text{ km} = 10^4$ m. Thus the density is

$$\rho = \frac{\text{star mass}}{\text{star volume}} = \frac{M}{4\pi R^3/3} = \frac{2 \times 10^{30} \text{ kg}}{4\pi(10^4 \text{ m})^3/3} = 0.5 \times 10^{18} \text{ kg/m}^3$$

$$= 0.5 \times 10^{12} \text{ kg/cm}^3.$$

Thus 1 cm³ of the material of a neutron star has a mass 0.5×10^{12} kg, and it weighs $(0.5 \times 10^{12} \text{ kg})g = (0.5 \times 10^{12} \text{ kg})(9.8 \text{ m/s}^2) = 0.5 \times 10^{13}$ N. For comparison, 1 cm³ of water on Earth's surface has a mass of 1 g and weighs 0.01 N. A neutron star is very dense indeed!

We stress that the distinction among the three forms of matter is not absolute. If the forces are large enough, even rock will flow. Gases share many properties with liquids; one major difference between gases and liquids is in the compressibility. A collection of molecules can form a solid, a liquid, or a gas, depending on external circumstances. Water is solid at low temperatures, but it can become a liquid at 0°C and a gas at 100°C. We say that matter appears in different **phases**; the particular phase depends on external conditions. At very high temperatures, the molecules themselves may dissociate into negatively charged electrons and positively charged ions. That kind of matter has very different properties because electrical forces between the constituents play an important role. Matter that is electrically dissociated into positive ions and negatively charged electrons is called a **plasma**.

We shall define a **fluid** to be a substance that is not resistant to deformation. This description involves how a material responds to external forces and applies to both liquids and gases. Therefore, except for phenomena in which compressibility is important, our description of the behavior of fluids applies to gases as well as to liquids.

16-2   PRESSURE

The first part of this chapter concerns **hydrostatics**—the study of fluids at rest. Density and pressure are the major variables of interest for such fluids. We have already discussed density. **Pressure** is a measure of the force transmitted by fluids, even fluids at rest. Swimmers know from experience that liquids can transmit forces. If you dive too

The definition of pressure

FIGURE 16–3 A device that would allow the measurement of pressure in a liquid. The fit between the piston and cylinder walls allows the piston to move but is tight enough so that a near vacuum can be maintained on the side with the spring. The compression of the spring provides a measurement of the force, from which the pressure is determined.

The pressure is the same in all directions.

Units of pressure

deeply into water, your eardrum may hurt or even rupture. This is due to the water pressure on the eardrum. But a sheet of glass does not shatter under water, because there is as much pressure on one side of it as on the other. One way to determine the force exerted by a fluid is to balance that force against a calibrated spring (Fig. 16–3). The plate on the spring balance is in contact with the fluid, while the spring is inside a fluid-free enclosure, so that there is no force on the plate from the inside due to the fluid. The average pressure, p, on any surface of area A is defined as the *force per unit area* acting perpendicular to that surface:

$$p = \frac{F}{A}. \tag{16-2}$$

Under the influence of gravity, pressure will vary as a function of depth. Thus, for a more precise definition, we consider a small area ΔA located at point $\mathbf{r}$, and take the limit $\Delta A \to 0$. Denoting the force perpendicular to this area as ΔF, we find

$$p = \lim_{\Delta A \to 0} \frac{\Delta F}{\Delta A} = \frac{dF}{dA}. \tag{16-3}$$

It is important to note that *when a fluid is at rest, the pressure at any given point must be the same in all directions*. The microscopic origin of pressure is the bombardment of the surface on which the pressure acts by the individual molecules of the substance (Chapter 19). The pressure is the same in all directions because the molecular movements are oriented randomly. For fluids at rest, the pressure is simply a positive scalar quantity. The pressure at a static interface between two fluids is the same on both sides of the interface; if it were not, there would be a net force at the interface and a resulting motion of the interface.

In SI, pressure is measured in *pascals* (Pa), named after Blaise Pascal, who studied pressure in the mid-seventeenth century:[†]

$$1 \text{ Pa} = 1 \text{ N/m}^2. \tag{16-4}$$

16–3 VARIATION OF PRESSURE IN A FLUID AT REST

The gravitational force causes the pressure of fluids at rest to be higher at ocean depths than at Earth's surface. In order to establish exactly how pressure varies with depth for an incompressible fluid, consider a liquid of fixed density ρ at rest in a container. Imagine isolating a very thin cylinder of that liquid (Fig. 16–4). (The top, base, and walls of the cylinder are purely imaginary; their sole purpose is to help us visualize the forces acting on an element of the fluid.) The cylinder's top, of area A, is horizontal and lies a distance y below the surface of the liquid. The bottom of the cylinder is at a depth $y + \Delta y$. Note that we measure the depth y downward from the surface, so it increases as we go down. We denote the pressure at the top of our cylinder by p and the pressure at the bottom by $p + \Delta p$. We want to find an expression for Δp.

The downward force on the cylinder, which is applied at the top surface by pressure from the surrounding fluid, is pA; the upward force, which is applied at the bottom surface by the pressure there, is $(p + \Delta p)A$. Additionally, there is a downward force on the cylinder due to gravity; namely, the weight of the liquid inside the cylinder. That force has a magnitude of $(\Delta m)g$, and the mass of the cylinder, Δm, is the density times the volume of the cylinder. Thus, the downward force of gravity is $\rho(A \Delta y)g$.

[†]Other common units for pressure are pounds per square inch (lb/in^2): 1 lb/in^2 = 6.9 × 10^3 Pa; the atmosphere (atm), which is the atmospheric pressure at sea level: 1 atm = 1.01 × 10^5 Pa = 14.7 lb/in^2; the bar: 1 bar = 10^5 Pa; and the torr: 1 torr = 133.32 Pa.

FIGURE 16–4 An imaginary thin cylinder of fluid is isolated to show the forces that act to keep it in equilibrium.

All together, we find

$$F_{up} = (p + \Delta p)A,$$

$$F_{down} = pA + (\Delta m)g = pA + \rho(A \Delta y)g.$$

Because the cylinder does not accelerate, the net force must cancel:

$$F_{net} = F_{up} - F_{down} = (p + \Delta p)A - pA - \rho(A \Delta y)g = 0.$$

We cancel the pA terms, divide by A, and rearrange terms to find that

$$\frac{\Delta p}{\Delta y} = \rho g.$$

In the limit $\Delta y \to 0$, this result becomes

$$\frac{dp}{dy} = \rho g. \tag{16–5}$$

According to Eq. (16–5), the pressure change is positive for positive Δy. Pressure therefore increases with increasing depth (Fig. 16–5). In physical terms, this phenomenon is a measure of the increased amount, and hence increased weight, of the liquid above any given point as the depth of the point increases. There is a greater weight of water above a submarine at an ocean depth of 500 m than above a submarine at a depth of 20 m.

Equation (16–5) looks just like the formula for the constant acceleration g of an object acted on by the constant force of gravity, $dv/dt = g$. When we studied projectile motion in Chapters 2 and 3, we saw that this equation has a solution $v = v_0 + gt$. By comparison, we can write the solution to Eq. (16–5), which gives the pressure as a function of depth, as

$$p = p_0 + \rho g y. \tag{16–6}$$

Alternatively, we can integrate Eq. (16–5) to find this result. The constant p_0 is the value of the pressure at the surface, where $y = 0$; for a body of fluid such as the ocean, this is just atmospheric pressure. The pressure increase with depth y is equal to the weight per unit area of the overlying fluid. The total mass of the fluid above the cylinder is $\rho V = \rho A y$, so its weight is $\rho A y g$, and thus the weight per unit area is $W/A = \rho g y$.

Note that our derivation is completely independent of the horizontal location of the cylinder within the liquid. The pressure is *independent* of the horizontal position and is dependent only on depth. Thus the pressure of a liquid in a series of open, connected vessels at the same depth is the same. The pressure at the fluid surfaces is equal to atmospheric pressure (the surface is in contact with the atmosphere and is therefore at the same pressure), and so the heights of the fluid surfaces in these vessels must be the same (Fig. 16–6).

An immediate consequence of Eq. (16–6) is that if p_0 is changed as a result of some external effect, p is changed by exactly the same amount. In other words, *the same change in pressure applied to any point in a fluid at rest is transmitted to every part of the fluid.* This statement is commonly known as **Pascal's principle**.

EXAMPLE 16–2 A hydraulic press (Fig. 16–7) is used to lift an object of mass 400 kg. The supporting cylindrical column has a diameter of 20 cm. The object is lifted by an operator using foot power to push on a piston. The height of the opera-

FIGURE 16–5 To equalize the pressures of the surrounding water and the sinus cavities, a diver blows air into his or her sinuses.

The pressure of an incompressible fluid under the influence of gravity

Pressure varies only with depth.

FIGURE 16–6 The fluid level in a series of connected vessels is the same because the pressure at the surface is the same.

Pascal's principle

FIGURE 16–7 Example 16–2.

Area A_1

tor's foot and the height at which the supporting column contacts the fluid are the same. Assuming that the force applied by the operator is 150 N, what is the diameter of the piston to which the foot is applied?

Solution: Pascal's principle tells us that the pressure p at the foot piston equals the pressure at the supporting column. We can find the force at the supporting column by multiplying p by the area A_1 of the column; this force must equal the weight of the mass to be lifted, mg. In turn, we know the pressure at the piston is the applied force, F_{foot}, divided by the piston area, A_2. Thus

$$mg = pA_1 = (F_{foot}/A_2)A_1.$$

By using the geometrical relation $A = \pi r^2 = \pi d^2/4$, where d is the diameter, we find instead

$$d_1^2/d_2^2 = mg/F_{foot},$$

or

$$d_2 = d_1 \sqrt{(F_{foot}/mg)} = (0.20\text{ m}) \sqrt{[(150\text{ N})/((400\text{ kg})(9.8\text{ m/s}^2))]} = 3.9\text{ cm}.$$

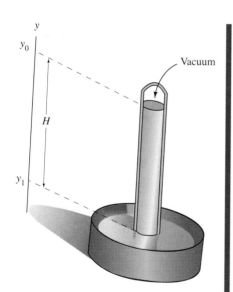

FIGURE 16–8 Example 16–3.
A column of mercury balanced by atmospheric pressure.

APPLICATION

The barometer

EXAMPLE 16–3 A column of mercury ($\rho_m = 13.6 \times 10^3$ kg/m^3) in a vertical tube that is open at the bottom rests in an open pool of mercury (Fig. 16–8). The column is sealed off at the top after all the air is evacuated from the empty region, creating a vacuum. What is the height H of the column of mercury?

Solution: The pressure at height y_1 (as measured up from ground level) on the open part of the pool of mercury is the atmospheric pressure, $p_1 = 1.01 \times 10^5$ Pa. The pressure at the same height y_1 within the column must be the same. But this pressure can be calculated from Eq. (16–6). Let the vertical coordinate of the surface inside the tube be denoted by y_0. Because we have evacuated the column, the value of p_0 for Eq. (16–6), which is the pressure at y_0 and represents the weight of the gas above the column, is zero (in reality, a little mercury will have evaporated, but the pressure that it exerts is negligible). The equality of the pressures at the level of the open pool then gives, with $H = y_0 - y_1$,

$$p_1 = p_0 + \rho g(y_0 - y_1) = p_0 + \rho gH = \rho gH.$$

Thus

$$H = \frac{p_1}{\rho g}.$$

We find that, numerically,

$$H = \frac{1.01 \times 10^5\text{ Pa}}{(13.6 \times 10^3\text{ kg/m}^3)(9.8\text{ m/s}^2)} = 0.76\text{ m}.$$

This example describes the working principle of a **barometer**, a device used to measure atmospheric pressure. The pressure at the bottom of a column of mercury exactly 0.760 m = 760 mm high is 1 atm at sea level.

EXAMPLE 16–4 A U-shaped tube of uniform cross section is open at both ends, as in Fig. 16–9. First, amyl bromide is poured into the tube; amyl bromide is a colorless fluid that does not mix with water. Then, water is carefully added to one side of the U, so there is water at the top of one column of fluid and amyl bromide is at the top of the other. The two columns of fluid meet at a point that we label $h = 0$. What is the ratio of the heights of the fluids above $h = 0$ in each column? The density of water is $\rho_w = 1.00 \times 10^3$ kg/m³ and that of amyl bromide is $\rho_{ab} = 1.26 \times 10^3$ kg/m³. Repeat the calculation for heptyl ether, a fluid that does not mix with water and has a mass density of 0.81×10^3 kg/m³.

Solution: The atmospheric pressure is the same at the open surface of both columns (we neglect the weight of any amount of air that adds to the pressure if one column has a lower fluid height than the other). The pressure in the two fluids at $h = 0$, where the two liquids meet, is also equal when the system has come to rest (otherwise there would be movement!). Let h_{ab} be the height of the column with the amyl bromide above level $h = 0$ and h_w be the height of the water column above level $h = 0$. For the pressure at $h = 0$, we find

for the amyl bromide column: $\quad p_{ab} = p_0 + \rho_{ab}h_{ab}g$;

for the water column: $\quad p_w = p_0 + \rho_w h_w g$.

Setting these pressures equal to one another, we have

$$\rho_{ab}h_{ab}g = \rho_w h_w g.$$

Rearranging,

$$\frac{h_{ab}}{h_w} = \frac{\rho_w}{\rho_{ab}} = \frac{1.00 \times 10^3 \text{ kg/m}^3}{1.26 \times 10^3 \text{ kg/m}^3} = 0.79.$$

The water column is higher than the amyl bromide column. This makes sense physically because amyl bromide is denser than water; the shorter column of amyl bromide weighs the same as the taller column of water. Note that this result is independent of the amount of either fluid.

We repeat the calculation with heptyl ether ($p_{he} = 0.81 \times 10^3$ kg/m³) to find that

$$\frac{h_{he}}{h_w} = \frac{\rho_w}{\rho_{he}} = \frac{1.00 \times 10^3 \text{ kg/m}^3}{0.81 \times 10^3 \text{ kg/m}^3} = 1.23.$$

The height of the column with heptyl ether is higher than the water column.

EXAMPLE 16–5 A closed container of liquid is connected to the exterior by two pistons: a small one of area $A_1 = 1$ cm² and a large one of area $A_2 = 100$ cm² (Fig. 16–10). Both pistons are at the same height. When a force of magnitude $F_1 = 100$ N is applied downward on the small piston, what mass m can be lifted by the large piston? Show that the work done in pushing the small piston down a distance d_1 is equal to the work done in lifting the mass m (assume that the liquid is incompressible).

Solution: When force F_1 is imposed on the small piston, the pressure p in the liquid at that point is given by

$$p = \frac{F_1}{A_1} = \frac{100 \text{ N}}{1 \text{ cm}^2} = \frac{10^2 \text{ N}}{10^{-4} \text{ m}^2} = 10^6 \text{ Pa}.$$

Because the large piston is at the same height as the small one, according to Pascal's principle the pressure at the large piston will have the same value, $p = 10^6$ Pa. A force F_2 acts upward at this piston, given by the pressure times the area:

$$F_2 = pA_2.$$

(a)

(b)

FIGURE 16–9 Example 16–4. The water has been dyed blue, and the amyl bromide has been dyed pink.

Hydraulic press

FIGURE 16–10 Example 16–5. A closed container of liquid with two pistons. An external force $\mathbf{F}_1$ is imposed on the small piston, giving an amplified upward force $\mathbf{F}_2$ due to pressure on the larger piston.

APPLICATION

The hydraulic press

(a)

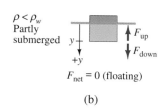

$\rho < \rho_w$
Partly
submerged

$F_{\text{net}} = 0$ (floating)

(b)

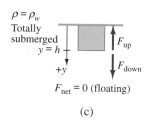

$\rho = \rho_w$
Totally
submerged

$F_{\text{net}} = 0$ (floating)

(c)

$\rho > \rho_w$
Sinking

$F_{\text{net}} = \rho ghA - \rho_w ghA$

(d)

FIGURE 16–11 (a) A block of material of density ρ is partly immersed in water of density ρ_w. The volume of the block is $V = Ah$, and the volume of water displaced by the block is $V_d = Ay$. (b) Force diagram of the forces acting on the lower surface of the block, which is partly immersed, as in part (a). The net force is $F_{\text{net}} = 0$, and the block floats. (c) When $\rho = \rho_w$, the block is totally submerged, and the block floats because $F_{\text{net}} = 0$. (d) When $\rho > \rho_w$, the block sinks because $F_{\text{net}} > 0$.

Correspondingly, the mass that can be lifted is given by $mg = F_2$, or

$$m = \frac{pA_2}{g} = \frac{(10^6 \text{ Pa})(100 \text{ cm}^2)}{9.8 \text{ m/s}^2} = 1020 \text{ kg}.$$

For the second part of the problem, we compare the work done in pushing down the small piston, W_p, to the work done in lifting the mass, W_m. We find

$$W_p = F_1 d_1.$$

At the same time, the work done in lifting the mass m is

$$W_m = F_2 d_2,$$

where d_2 is the height through which m is lifted. To find d_2, note that as long as the fluid is incompressible, the volume change in the movement of the small cylinder is matched by the volume change in the large cylinder. The volume of the liquid displaced by the small piston is $A_1 d_1$. The volume displaced in the movement of the large piston is $A_2 d_2$. Equating these volumes, we have $A_1 d_1 = A_2 d_2$; hence,

$$d_2 = \frac{A_1 d_1}{A_2}.$$

The work done in raising the mass is then

$$W_m = F_2 d_2 = pA_2 \left(\frac{A_1 d_1}{A_2} \right) = pA_1 d_1.$$

But the quantity pA_1 is the original force F_1, and the right-hand side is the original work $W_p = F_1 d_1$ done in pushing down the small piston. The work done on both sides is equal, but the forces are considerably different: $F_2/F_1 = d_1/d_2 = A_2/A_1 = 100 \text{ cm}^2/1 \text{ cm}^2 = 100$ because $F_1 d_1 = F_2 d_2$.

The device treated here describes the *hydraulic press* schematically. Pascal's principle is similarly used in the design of hydraulic brakes.

16–4 BUOYANCY AND ARCHIMEDES' PRINCIPLE

An ocean-going liner floats when it is launched. When a brick is put into water, however, it sinks. Why? Experience tells us that this has something to do with the density of the material (and not the mass; a small brick sinks just like a large one). Let's consider, as in Fig. 16–11a, a block of material of volume $V = Ah$ and density ρ (the mass is thus $m = \rho V$) partly immersed in water, which has a density of ρ_w. Figure 16–11b shows the forces acting on the lower surface of the partly submerged block. The upward force is due to the pressure of the liquid, as given by Eq. (16–6): $p = p_0 + \rho_w gy$, where y is the vertical submerged distance of the block. The upward force on the block is therefore

$$F_{\text{up}} = pA = p_0 A + \rho_w gyA.$$

The downward force has two components: the atmospheric pressure above the block, p_0, and the weight of the block. The first downward component is $F_{\text{atm}} = p_0 A$, and the second is the force of gravity on the block, $F_g = mg = (\rho Ah)g$. Thus the total downward force is

$$F_{\text{down}} = p_0 A + \rho ghA.$$

Therefore,

$$F_{\text{net}} = F_{\text{down}} - F_{\text{up}} = p_0 A + \rho ghA - p_0 A - \rho_w gyA = \rho ghA - \rho_w gyA. \quad (16\text{–}7)$$

A state of equilibrium—floating—demands that the forces balance, $F_{\text{net}} = 0$. If we set

Eq. (16–7) equal to zero and divide by gA, we get

for floating: $\quad ph = \rho_w y;$ $\qquad\qquad$ (16–8)

$$\frac{\rho}{\rho_w} = \frac{y}{h}.$$

If $\rho < \rho_w$, then $y/h < 1$ and only a fraction of the block is submerged. In the limiting case that $\rho = \rho_w$, the block is totally submerged and $y = h$ (Fig. 16–11c). In this case, the block would float just under the surface because the upward and downward forces cancel. If $\rho > \rho_w$, then the block must sink (Fig. 16–11d) because the two forces can no longer cancel completely. In this case, once the block is fully submerged ($y \geq h$), the net (downward) force is

for sinking: $\quad F_{net} = \rho ghA - \rho_w ghA > 0.$ $\qquad$ (16–9)

We can interpret the difference between the weight of the block ρghA and F_{net} as an *upward* **buoyant force**:

$$F_{buoy} = F_g - F_{net}.$$ $\qquad\qquad$ (16–10)

For the case when the block is partly submerged, we have $F_{net} = 0$, and Eq. (16–10) gives

for partial submergence: $\quad F_{buoy} = \rho ghA - 0 = \rho ghA = \rho_w gyA,$

where we have used Eq. (16–8) in the last step. When the block is totally submerged (at any depth where $y \geq h$), F_{net} is given by Eq. (16–9), and we find

for total submergence: $\quad F_{buoy} = \rho ghA - (\rho ghA - \rho_w ghA) = \rho_w ghA.$

If we think of V as the volume of the object beneath the water ($V = yA$ or hA, according to whether the object is partly or totally submerged), then we can combine our results into the single expression (Fig. 16–12).

$$F_{buoy} = \rho_w gV.$$ $\qquad\qquad$ (16–11)

The buoyant force opposes the downwardly directed force of gravity, ρghA, on the block. For a floating body, the buoyant force and the force of gravity cancel exactly. When $\rho > \rho_w$, the buoyant force opposes the force of gravity even though it does not cancel the force of gravity completely; a sinking object descends with less acceleration than it would under the influence of gravity alone. When a metal sphere is immersed in a bucket of water and suspended from a spring balance, the balance will indicate a lower weight than it would for the same sphere in air. The difference in these weights is the buoyant force. The buoyant force is not limited to water; it is what allows a balloon to rise in air (Fig. 16–13a), and even a steel ball to float in mercury (Fig. 16–13b).

Archimedes' Principle

Regardless of whether the block is partly or totally submerged, Eq. (16–11) shows that the buoyant force is equal to the weight of the liquid (here, water) displaced by the block:

$$W_{liq} = m_{liq}g = \rho_{liq}V_{liq}g = \rho_{liq}yAg,$$

where $y = h$ when the block is totally submerged. This fact was discovered in the third century B.C. by Archimedes, a great scientist and mathematician of ancient Greece, and is known as **Archimedes' principle**:

The buoyant force on an immersed object equals the weight of displaced liquid.

We may check the validity of this principle by considering the behavior of a block of iron, with a mass M and a density $\rho_{Fe} = 7.86 \times 10^3$ kg/m^3, in water. The upwardly

Case I: Floating: not completely submerged

Case II: Barely floating, completely submerged

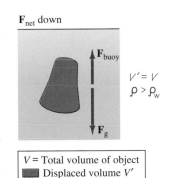

Case III: Sinking, even if completely submerged

V = Total volume of object
◼ Displaced volume V'
▢ Fluid

FIGURE 16–12 By comparing the buoyant force F_{buoy} to the force of gravity F_g, we determine whether a body sinks, just floats, or pops up.

The buoyant force

Archimedes' principle

(a)

(b)

FIGURE 16–13 (a) This hot-air balloon takes advantage of the fact that hot air is less dense than cold air at the same pressure. Buoyancy accordingly makes the balloon rise. (b) The buoyant force allows a steel ball to float on the surface of liquid mercury, whose density is some 70 percent greater than that of steel (iron).

directed buoyant force is the weight of water displaced, $F_{\text{buoy}} = mg = \rho_w V g$, where m is the mass of displaced water and V is its volume. The downward force of gravity on the iron block is $F_g = Mg = \rho_{\text{Fe}} V g$. Because $\rho_{\text{Fe}} > \rho_w$, the net force is downward and the block sinks. The net downward force—effectively the weight of iron in water—is

$$F_{\text{net}} = F_g - F_{\text{buoy}} = \rho_{\text{Fe}} V g - \rho_w V g = \left(1 - \frac{\rho_w}{\rho_{\text{Fe}}}\right)\rho_{\text{Fe}} V g$$

$$= \left(1 - \frac{\rho_w}{\rho_{\text{Fe}}}\right)F_g. \tag{16–12}$$

We can check the magnitude of the net force by weighing an object in water (Fig. 16–14). Such experiments verify Archimedes' principle. Archimedes is said to have discovered this when he was asked to ascertain that a crown was made entirely of gold—without marring the crown; he did this by weighing the crown in air and in water, thereby measuring its density [Eq. (16–12)]. It is said that he found the density of the crown to be less than that of gold, thus the gold in the crown must have been diluted with another metal.

Although our calculations were carried out for a rectangular block, the results do not depend on the shape of the block. An object of any shape can be thought of as being composed of small rectangular blocks. Similarly, we performed our calculations for an object of constant density, but all that matters is the weight of the liquid that is displaced by the object. Our considerations apply equally well to objects in which the density varies, as Examples 16–6 through 16–8 show.

EXAMPLE 16–6 An air-filled lead balloon of radius $R = 0.1$ m is totally submerged in a tank of water (Fig. 16–15). What is the thickness t of the lead skin of the balloon if the balloon neither rises nor sinks? The density of lead, ρ_{Pb}, is 11.3×10^3 kg/m^3.

Solution: We must calculate the weight of the water displaced by the balloon. To do this, we need the volume of lead and the volume of air. In order for the balloon not to sink, we expect that the thickness of the balloon must be quite small compared to the radius R: $t \ll R = 0.1$ m. If this is the case, we can use a simple ap-

FIGURE 16–14 By Archimedes' principle, the buoyant force on the sphere, $\mathbf{F}_{\text{buoy}}$, equals the weight of the water the sphere displaces. The scale measures the difference between the weight of the sphere and the buoyant force, $\mathbf{F}_{\text{scale}} = \mathbf{F}_g - \mathbf{F}_{\text{buoy}}$. (a) No water is displaced; the scale measures the weight of the sphere. (b) Half the volume of the sphere is displaced. (c) The entire volume of the sphere is displaced. The buoyancy force in this case is 9 N.

(a)

(b)

(c)

proximation. The volume of lead, V_{Pb}, is the volume of the outer sphere, of radius R, minus the volume of the inner sphere, whose radius is $R - t$. For $t \ll R$, this difference is approximately the area of a sphere of radius R, or $4\pi R^2$, times the thickness t: $V_{Pb} \simeq 4\pi R^2 t$. The error made in using this approximation is small (of order Rt^2). Thus, the weight of the lead sphere is approximately

$$W_{Pb} = mg = \rho_{Pb}V_{Pb}g = 4\pi R^2 t \rho_{Pb}g.$$

The weight of the enclosed air is negligible; air has a density some 1000 times less than that of water.

The weight of the displaced water is

$$W_w = \rho_w V_s g = \frac{4}{3}\pi R^3 \rho_w g,$$

where V_s is the volume of the (outer) sphere. The weights are equal if the sphere neither floats above the surface nor sinks. We set the two weights equal to each other and solve for the unknown thickness t:

$$W_{Pb} = 4\pi R^2 t \rho_{Pb}g = W_w = \frac{4}{3}\pi R^3 \rho_w g;$$

$$t = \frac{\rho_w}{\rho_{Pb}}\frac{R}{3} = \frac{10^3 \text{ kg/m}^3}{11.3 \times 10^3 \text{ kg/m}^3}\frac{0.1 \text{ m}}{3} \simeq 0.003 \text{ m} = 3 \text{ mm}.$$

The fact that 0.003 m $\ll$ 0.1 m ($t \ll R$) justifies our approximation.

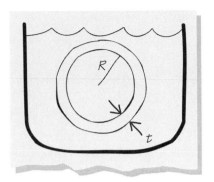

FIGURE 16–15 Example 16–6. A lead balloon, with a skin of thickness t, submerged in water.

EXAMPLE 16–7 A metal sphere weighs 29.4 N in air and 18.5 N in water. What is its average density?

Solution: The weight of the sphere in water, W_w, is the weight in air, W_{air}, minus the buoyant force. Let the volume of the sphere be V and its (unknown) average density be ρ. We have $W_{air} = \rho g V$, while the buoyant force is the weight of the displaced water, $\rho_w g V$, or

$$F_{buoy} = \rho_w g V = (\rho_w/\rho)\rho g V = (\rho_w/\rho)W_{air}.$$

But the buoyant force has magnitude $F_{buoy} = W_{air} - W_w = 29.4 \text{ N} - 18.5 \text{ N} = 10.9 \text{ N}$. Thus,

$$\frac{\rho_w}{\rho} = \frac{F_{buoy}}{W_{air}} = \frac{10.9 \text{ N}}{29.4 \text{ N}} = 0.37;$$

$$\rho = \frac{\rho_w}{0.37} = 2.7 \times 10^3 \text{ kg/m}^3.$$

EXAMPLE 16–8 A rectangular tub made of a thin shell of poured cement has length $\ell = 1$ m, width $w = 80$ cm, depth $d = 60$ cm, and mass $M = 200$ kg. The tub floats in a lake. How many people of mass $m = 80$ kg each can stand in the tub before it sinks?

Solution: Let the distance that the tub sinks be given by y (Fig. 16–16). The volume of displaced water is ℓwy. The weight of the displaced water, and therefore the buoyant force, is

$$F_{buoy} = \rho(\ell wy)g,$$

where the density ρ of water is 1 g/cm^3 = 10^3 kg/m^3. The downward force due to the weight of the tub and X people in it is

$$F_{down} = (M + Xm)g.$$

FIGURE 16–16 Example 16–8. A tub sinks to a water depth y. When $y > d$, the depth of the tub, the tub sinks.

FIGURE 16–17 Under the right conditions, soil can act like a fluid (mud) and flow.

Assume incompressibility.

Assume constant temperature.

Assume steady flow.

FIGURE 16–18 The water moves faster near the center of the vortex than it does far from the center. The water is in rotational flow.

Assume laminar flow.

The tub floats at depth y as long as the balance equation $F_{\text{buoy}} = F_{\text{down}}$ holds. The tub will finally fill with water and sink when $y = d$. Just before this point, the balance equation is

$$\rho(\ell wd)g = (M + Xm)g.$$

The solution of this equation for the number of people the tub can hold just before sinking, X, is

$$X = \frac{\rho \ell wd - M}{m}.$$

We insert numerical values to find that

$$X = \frac{(10^3 \text{ kg/m}^3)(1 \text{ m})(0.8 \text{ m})(0.6 \text{ m}) - 200 \text{ kg}}{80 \text{ kg}} = 3.5.$$

With three people, the tub does not sink; with four people, she goes down. Even the three tubbers should be careful not to rock the boat!

16–5   FLUIDS IN MOTION

The marvelous fluid flow patterns observed in nature go far beyond the particle and rigid body motions we have studied to this point. **Hydrodynamics** describes fluids in motion, including the flow of placid streams, eddies in a pond, mud slides, smoke rings, ocean waves and tides, arterial blood flow, Jupiter's (and Earth's) atmosphere, the rotational flow of the galaxy, and nuclear fission (Fig. 16–17). Numerical calculations performed by supercomputers are necessary to solve the most complex types of behavior. What is remarkable is the variety of phenomena that can be described with a very simplified set of equations.

We assume, as we have discussed, that the fluid under consideration is incompressible. This does not mean that we exclude phenomena that involve gases, such as air flow; it just means that we will discuss these phenomena only under circumstances in which compressibility does not play a role.

We also assume that we are dealing with a fluid at constant temperature. (We will learn more about temperature and heat flow in Chapters 17 through 20.) Phenomena such as the **convection** of fluids—in which a liquid in the bottom of a vessel is heated, rises, cools, and falls in a circulating pattern—are thus excluded from our consideration.

In general, the velocity of a fluid at a given point, $\mathbf{v}(x, y, z, t)$, as well as the pressure, $p(x, y, z, t)$, may vary with time, but we will consider only **steady flow**: fluid movement for which both $\mathbf{v}$ and p do not depend on time. The pressure and the velocity may vary from point to point, but we assume that any changes are smooth ones.

Consider a fluid in motion. We could represent the direction of the velocity at some point in Fig.16–18 by an arrow at that point, with the length of the arrow proportional to the magnitude of the velocity. For steady (time-independent) flow, there is a *local velocity* $\mathbf{v}(\mathbf{r})$. If we start at some point, then a tiny element of the fluid at that point moves in the direction of the velocity vector. An instant later, the element is at an adjacent point, which has a slightly different velocity vector. This vector dictates the element's direction and speed of motion in the next instant, and determines the next step of the motion. In this way, the motion of a tiny element of fluid can be traced. The line of motion of a small fluid element that maintains its integrity is called a **streamline** (Fig. 16–19).

Streamlines cannot cross because a velocity vector cannot have two directions. Flow described by streamlines is called **laminar flow**; in laminar flow, each element of fluid travels along a smooth, well-defined path. Under certain circumstances, which we will describe in Section 16–9, fluid flow ceases to be laminar and becomes turbulent. **Turbulent flow**, as in the flow of a white-water river, involves rapidly fluctuating velocities and pressures. Turbulence is a characteristic of many flow patterns that we observe in nature, but there is no simple theory that describes it. The study of turbulence is an exciting area of research in engineering and physics.

We do *not* consider flow in which streamlines close in on themselves, such as those shown in Fig. 16–20; that is, we consider only **irrotational flow**. As a consequence, we do not discuss vortices, or whirlpools, such as those that arise in water running out of a bathtub, smoke rings, or even the much simpler motion of coffee stirred by a spoon in a cup. A mathematical expression of the statement that the flow is irrotational is that the **circulation**, defined as the integral of the velocity along any *closed* path in the fluid, is zero.

To simplify matters further, we make one more assumption: We neglect the analogue of friction in the fluid. That there is such an analogue is obvious from simple observations. The flow of honey in Fig. 16–21 is different than the flow of water. The friction that enters here is an internal friction, which is called **viscosity** (see Section 16–9). In neglecting viscosity, we are neglecting internal energy losses. The conservation of energy can then be used profitably in the description of fluids.

To summarize, in the next two sections we make the following assumptions:

1. The fluid is incompressible.
2. The temperature does not vary.
3. The flow is steady, so that the velocity and pressure do not depend on time.
4. The flow is laminar rather than turbulent.
5. The flow is irrotational, so there is no circulation.
6. There is no viscosity in the fluid.

Assume irrotational flow.

FIGURE 16–19 The use of smoke in a wind tunnel allows a direct visualization of the motion of air. Here we see that turbulent flow has been induced behind the wing section.

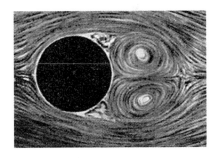

FIGURE 16–20 Streamlines for a cylinder pulled steadily through a fluid, showing circulation in the fluid.

16-6 THE EQUATION OF CONTINUITY

For a fluid in motion, we can pick a group of adjacent streamlines, which we say form a thin bundle, and follow them (Fig. 16–22). Because streamlines never cross, the bundle forms a thin tube with nonuniform cross section, known as a **streamtube**, in which fluid flows; no fluid passes from the inside to the outside of a streamtube. We can choose a tube thin enough so that the velocity across a section drawn at right angles to the tube is the same across the whole cross section. The volume of fluid that crosses the section of area A_1 in a given time t is the volume of a cylinder whose base is A_1 and whose length is $\ell_1 = v_1 t$, where v_1 is the speed across that section; the volume of fluid is thus $v_1 t A_1$. If the density of the fluid at this section is ρ_1, then the mass of fluid crossing the section in the given time t is

$$m_1 = \rho_1 v_1 t A_1. \tag{16–13}$$

Similarly, the mass of fluid crossing a section of area A_2 at right angles to the bundle of streamlines with speed v_2 and density ρ_2 in the same time t is

$$m_2 = \rho_2 v_2 t A_2.$$

But because fluid neither enters nor leaves the tube by the side walls, the *conservation of mass* implies that $m_1 = m_2$, or, if we cancel the time t,

$$\rho_1 v_1 A_1 = \rho_2 v_2 A_2. \tag{16–14}$$

This equation, the **equation of continuity**, is true even for fluids whose densities can vary with position. When we restrict ourselves to incompressible flow, we have $\rho_1 = \rho_2$ and derive the very useful equation

$$v_1 A_1 = v_2 A_2. \tag{16–15}$$

FIGURE 16–21 Honey flows slowly because its viscosity is high.

The conservation of mass applied to fluids, also known as the equation of continuity

FIGURE 16–22 Flow contained in a bundle of streamlines (streamtube). The velocity perpendicular to the surface of area A_1 is $\mathbf{v}_1$, and that perpendicular to the surface of area A_2 is $\mathbf{v}_2$.

The conservation of mass applied to incompressible fluids

PROBLEM SOLVING

Conserved quantities are very helpful in problem-solving.

Flux is the volume of fluid that passes through a given area per unit time.

FIGURE 16–23 Example 16–9.

This result states that the product of speed and area of any bundle of streamlines is the same anywhere along the bundle. The speed of fluid flow thus varies inversely with the area of the streamtube. The product (velocity perpendicular to an area) × (area) is the **flux**, Φ. Equation (16–15) describes the **conservation of flux**. Perhaps you have observed the conservation of flux while you have entered or left a building on a windy day: The wind speed is greatest in the doorway.

Note that flux can be interpreted as the volume of fluid that passes through an area per second. If the velocity of a fluid is perpendicular to a given area A, the volume of fluid that passes through that area in time t is a length $L = vt$ (the distance the fluid moves in time t) multiplied by A, or $V = vtA$. Thus *the flux, $\Phi = V/t = vA$, is the volume of flow per second.*

EXAMPLE 16–9 Figure 16–23 shows water that flows at a rate of 0.25 L/s through a garden hose of inside diameter 2 cm. What is the speed of the water in the hose? The nozzle of the hose is a circular opening of diameter 1.0 cm. What is the speed of the water when it emerges?

Solution: We are given the volume of water that passes through the hose per second. To find the speed v of the water in the hose, whose cross section is A, we use the fact that the volume that passes per second is the flux, $\Phi = vA$. Thus the speed in the hose is $v_{\text{hose}} = \Phi/A_{\text{hose}}$. From Eq. (16–15), the conservation of flux, we know that the volume per second that passes the nozzle must be the same as the volume per second that passes through the hose, or

$$A_{\text{noz}}v_{\text{noz}} = A_{\text{hose}}v_{\text{hose}}.$$

We solve this equation for the desired quantity, v_{noz}:

$$v_{\text{noz}} = \frac{A_{\text{hose}}}{A_{\text{noz}}} \, v_{\text{hose}} = \frac{A_{\text{hose}}}{A_{\text{noz}}} \frac{\Phi}{A_{\text{hose}}} = \frac{\Phi}{A_{\text{noz}}}.$$

The radius of the nozzle is half the diameter, or 0.50 cm, and the area $A = \pi r^2$. Thus we have

$$v_{\text{noz}} = \frac{0.25 \text{ L/s}}{\pi(0.50 \text{ cm})^2} = 320 \text{ cm/s} = 3.20 \text{ m/s}.$$

Note that because the radius of the nozzle is half that of the hose, and because the area is proportional to the radius squared, the nozzle has one-fourth the area of the hose; the velocity of the emerging stream is four times that of the water in the hose: $v_{\text{hose}} = \frac{1}{4}v_{\text{noz}} = 0.8$ m/s.

This example illustrates how a small nozzle causes the water to exit a hose with a large velocity and thus spray over a long distance.

APPLICATION

The use of nozzles

16–7 BERNOULLI'S EQUATION

In principle, the motion of a fluid can be obtained by breaking the fluid down into small elements and applying Newton's second law to each of them. We saw in our previous study of motion that the conservation of energy is a powerful alternative to the second law when we have only conservative forces. If there is no viscosity in a fluid, so that there is no energy dissipation, and if the fluid is incompressible, it becomes straightforward to formulate the conservation of energy for fluids. This will give us a simple and important relation to describe the motion.

We again consider a streamtube—a bundle of streamlines of fluid—this time under the influence of gravity. Concentrate on the fluid between two cross sections of the tube: one at height h_1 with cross-sectional area A_1, the other at height h_2 with area A_2 (Fig. 16–24). The velocity perpendicular to the cross sections will have magnitudes v_1 and v_2, respectively, and we take our tube to be thin enough so that the velocity does not vary across the slices. There will be flow to the right provided that the pressure p_1 at region A_1 is larger than the pressure p_2 at region A_2.

FIGURE 16–24 A streamtube of fluid at different heights. In time Δt, the fluid moves to the right under the influence of a difference in pressures between p_1 and p_2. In effect, the shaded slice in region 1 is transferred to the shaded slice in region 2.

Calculating the Work to Move a Fluid Segment

Let's examine what happens during a time interval dt. The fluid that starts out in the region between A_1 and A_2 at $t = 0$ will have advanced. The distance by which it advances at height h_1 is $v_1\,dt$, and the distance by which it advances at height h_2 is $v_2\,dt$. We calculate the net work done in advancing the fluid in the tube by looking at the forces acting at either end. The work done at the left-hand region by pressure p_1 is given by (force) × (displacement). This can be written as

$$W_1 = (p_1 A_1)(v_1\,dt). \qquad (16\text{–}16)$$

The work done at the right-hand region by pressure p_2 can be calculated similarly, but there is an additional minus sign because the force (and thus the pressure) acts to the left while the displacement is to the right, so that $\mathbf{F} \cdot \mathbf{s} = -Fs$. Thus

$$W_2 = -(p_2 A_2)(v_2\,dt).$$

The net work done on the streamtube between regions 1 and 2 is

$$W_{\text{net}} = W_1 + W_2 = (p_1 - p_2)\,vA\,dt. \qquad (16\text{–}17)$$

where in the last line we have made use of Eq. (16–15), which expresses the constancy of the product vA at any point along the tube:

$$vA = v_1 A_1 = v_2 A_2.$$

Applying the Work–Energy Theorem

We have determined the amount of work done in moving the fluid in the tube section during the time dt. To apply the work–energy relation, we now want to find the change in energy. Examining Fig. 16–24, we can see that what has happened in the time interval dt is the transfer of the small slice of fluid from region 1 to region 2. This transfer has two effects on the energy: The potential energy changes as the slice is lifted from h_1 to h_2, and the kinetic energy changes as the speed changes from v_1 to v_2. The change in potential energy is

$$\Delta U = mg\,\Delta h = (\rho vA\,dt)g(h_2 - h_1). \qquad (16\text{–}18)$$

Again we have used the constancy of vA. The change in kinetic energy is

$$\Delta K = \frac{1}{2}(\rho vA\,dt)(v_2^2 - v_1^2). \qquad (16\text{–}19)$$

Because $W_{\text{net}} = \Delta K + \Delta U$, we find, from Eqs. (16–17) through (16–19),

$$(p_1 - p_2)vA\,dt = \frac{1}{2}(\rho vA\,dt)(v_2^2 - v_1^2) + (\rho vA\,dt)g(h_2 - h_1).$$

After the factor $vA\,dt$ divides out,

$$p_1 - p_2 = \frac{1}{2}\rho(v_2^2 - v_1^2) + \rho g(h_2 - h_1).$$

PROBLEM–SOLVING TECHNIQUES

Bernoulli's equation expresses conservation of energy for a fluid; as such, its use is similar to the use of all conservation laws. Keep in mind the following points:

1. When we say the right-hand side of Eq. (16–21) is a constant, we mean that it takes on the *same* value for all points within the fluid. If we know the value of $p + \frac{1}{2}\rho v^2 + \rho g h$ at one point, then that is the value it takes at all points.

2. In its typical application, we can find the value of the constant in Bernoulli's equation because we know the value of all the variables at some point. Then we can use the value of the constant to find the unknown value of a variable at another point.

3. When one of the three terms in the expression $p + \frac{1}{2}\rho v^2 + \rho g h$ is the same at all points, then we can ignore it and concentrate on using Bernoulli's equation to learn about the other variables. For example, in Section 16–8, we first apply Bernoulli's equation to the case of a fluid at rest. In that case, v is the same everywhere ($v = 0$), and we can use Bernoulli's equation as a conservation law for the quantity $p + \rho g h$.

(a)

(b)

FIGURE 16–25 By blowing over the top of the paper, the pressure above the paper is reduced according to Bernoulli's principle. Below the paper, the pressure is unchanged, and there is a net force that makes the paper rise.

This equation can be rearranged into a conservation law, with one side dependent on the quantities defined in region 1 and the other side dependent on the quantities defined in region 2:

$$p_1 + \frac{1}{2}\rho v_1^2 + \rho g h_1 = p_2 + \frac{1}{2}\rho v_2^2 + \rho g h_2. \tag{16–20}$$

In other words,

$$\text{along a streamline:} \quad p + \frac{1}{2}\rho v^2 + \rho g h = \text{a constant.} \tag{16–21}$$

Our derivation deals with the motion of an element of fluid, and thus our conclusions hold only for motion along a streamline. It can be shown in general that *for irrotational flow, the constant in Eq. (16–21) is the same for all streamlines.* Equation (16–21) is **Bernoulli's equation**, derived in 1738 by Daniel Bernoulli, one of a remarkable family of scientists and mathematicians. In Section 16–8, we discuss applications of this formula.

16–8 APPLICATIONS OF BERNOULLI'S EQUATION

Bernoulli's equation is useful for describing a variety of phenomena. First, we observe that if a fluid is at rest, so that $v = 0$ in Eq. (16–21), then

$$p + \rho g h = \text{a constant.} \tag{16–22}$$

This is just the hydrostatic relation between pressure and height that we obtained in Eq. (16–6).

The Bernoulli Effect

For flow in a horizontal (constant height) tube, the conservation of energy [Eq. (16–21)] implies that

$$p + \frac{1}{2}\rho v^2 = \text{a constant.} \tag{16–23}$$

Thus the pressure must be lower in a region in which a fluid is moving faster. You can observe this phenomenon—known as the **Bernoulli effect**—in a simple experiment. If you blow air across the top of a single sheet of paper (Fig.16–25), the pressure above the paper is reduced. However, the pressure is unchanged below the paper; thus, the paper rises. The Bernoulli effect also enters into the operation of a chimney. When the wind blows over the top of a chimney, the pressure is decreased at the top by the Bernoulli effect, and the air in the tube rises. This is an important element in a chimney's "draw."

A Measuring Device. Equation (16–23) is applied in the *Venturi flowmeter*, a device designed to measure the speed of fluids in a pipe (Fig. 16–26a). Figure 16–26b is a simplified version of the flowmeter. The large pipe, with cross-sectional area A_1, is constricted smoothly to a smaller area A_2. A U-shaped tube containing a different fluid connects the two segments of pipe. The speeds in the larger and smaller sections of the pipe are v_1 and v_2, respectively; similarly, the pressures in the two sections are p_1 and p_2, respectively. Then Eq. (16–23) tells us that

APPLICATION

The Venturi flowmeter

$$p_1 + \frac{1}{2}\rho v_1^2 = p_2 + \frac{1}{2}\rho v_1^2;$$

$$p_1 - p_2 = \frac{1}{2}\rho(v_2^2 - v_1^2).$$

The continuity equation, Eq. (16–15), tells us that $v_2 = v_1 A_1/A_2$, so

$$p_1 - p_2 = \frac{1}{2}\rho v_1^2\left[\left(\frac{A_1}{A_2}\right)^2 - 1\right]. \tag{16–24}$$

The difference in pressures is determined from the difference in heights, h, of the two columns of fluid in the U-shaped tube (see Section 16–3). Equation (16–24) is then used to find speed v_1 in terms of this pressure difference.

Fluid Motion with Constant Speed

In the Bernoulli effect, we did not change the height but the speed of the flowing fluid did change. If the area of a pipe is fixed and the fluid is incompressible, then the equation of continuity, Eq. (16–15), shows that the speed cannot change. Such a pipe will force the fluid moving inside to change pressure if the height changes: In this case, Bernoulli's equation becomes $p + \rho gh = $ a constant, or Eq. (16–22).

EXAMPLE 16–10 A giraffe needs a strong heart because of its long neck (Fig. 16–27). Suppose that the difference between the aortic valve (the place where arterial blood comes out of the heart) and the head of a giraffe is 2.50 m, and that the

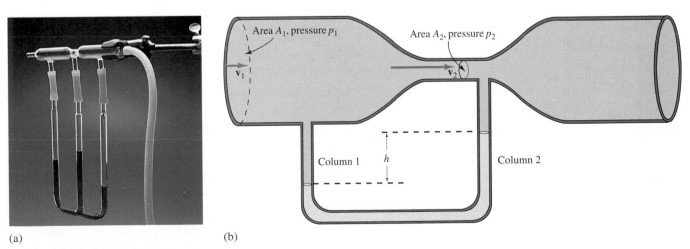

(a)   (b)

FIGURE 16–26 (a) The Venturi flowmeter. (b) The speed v_1 of a fluid in a pipe is measured by the difference in heights, h, of column 1 and column 2 of a different fluid. The number of measuring tubes that tap the pipe through which the fluid flows can be varied according to the measurement desired.

FIGURE 16–27 Example 16–10.

(a)

(b)

FIGURE 16–28 The escape of fluid through a hole in a tank. (a) The curved lines are streamlines for real fluid flow. (b) The pressure at the bottom of the tank is greater than the pressure at the top, so a stream of water leaving the tank has greater speed if it leaves from a hole that is lower on the tank.

artery leading from near the aortic valve to the head has constant cross section. What is the minimum pressure at the aortic valve? Blood is an incompressible fluid with density $\rho = 1.0$ g/cm^3.

Solution: The minimum pressure at the heart is the pressure that will allow the blood to arrive at the head with zero pressure. Because the blood is incompressible, the equation of continuity implies that the speed of the blood will be the same at the head as at the heart. We can then employ Bernoulli's equation to find the pressure at the heart. Let's use subscripts 1 and 2 for the heart and head, respectively. Then, with $v_1 = v_2$, the terms proportional to v^2 in Bernoulli's equation cancel, and Eq. (16–22) applies. With $p_2 = 0$, we have

$$p_1 + \rho g h_1 = \rho g h_2$$

$$p_1 = \rho g (h_2 - h_1).$$

Numerically, $p_1 = (1.0 \text{ g/cm}^3)(10^2 \text{ cm/m})^3(10^{-3} \text{ kg/g})(9.80 \text{ m/s}^2)(2.50 \text{ m}) = 2.45 \times 10^4$ Pa. Various factors, in particular the loss of energy due to friction between the blood and the arterial walls, make this a low estimate. This number can be compared to the peak output pressure of the human heart, 1.6×10^4 Pa.

Flow from a Tank

Another interesting phenomenon of practical importance in fluid flow is the speed of flow of a liquid from an opening in a tank (Fig. 16–28a). Let's make a small hole in the side of an open tank of water at a depth h below the surface. We apply Bernoulli's equation, Eq. (16–21), to the fluid at different locations. Denote the height above the hole by the variable y, so the height at the water surface is $y = h$ and at the opening $y = 0$. Equation (16–21) states that the value of $p + (\rho v^2/2) + \rho g y$ is the same at point 1 (on the upper surface, where the pressure $p = p_0$) and point 2 (just outside the hole, where $p = p_0$ because the sides of the stream of fluid flowing from the tank are in equilibrium with the atmosphere). The speed v_2 at point 2 is to be determined. We equate the values of $p + (\rho v^2/2) + \rho g y$ at the two points and obtain

$$p_0 + \frac{1}{2} \rho v_1^2 + \rho g h = p_0 + \frac{1}{2} \rho v_2^2 + \rho g(0).$$

We can neglect the small velocity v_1 of the liquid at the top of the tank. We solve for the speed at point 2 (the hole) to find that

$$v_2 = \sqrt{2gh} \tag{16–25}$$

(Fig. 16–28b). Notice that this is just the speed of a single particle falling through a height h under the influence of gravity, as in Eq. (2–27). We find the same result for the falling liquid as for a falling object because they both are consequences of the conservation of energy. To find the rate of fluid flow from the hole, we use the expression for the flux, whose conservation is described by Eq. (16–15):[†]

$$\Phi = vA. \tag{16–26}$$

EXAMPLE 16–11 A tank of cross-sectional area $A = 0.07$ m^2 is filled with water. A tightly fitting piston, with a total mass $m = 10$ kg, rests on top of the water (Fig. 16–29). A circular hole of diameter 1.5 cm is opened at a depth of 60 cm below the water level of the tank. What is the initial rate of flow of water out of the hole?

Solution: At surface level, the total pressure $P = p_0 + p_1$, where p_0 is the atmospheric pressure and p_1 is the pressure due to the piston resting on the water, $p_1 = mg/A$. As in our discussion, the pressure when the fluid has left the hole is p_0. An

---

[†]A more careful treatment replaces the right-hand side of Eq. (16–26) with the expression kvA, where the coefficient k is less than 1. Its correct value for a simple small hole is 0.62. In the examples and problems, we use $k = 1$ for simplicity.

Both single-reed (clarinet and saxophone) and double-reed (oboe and bassoon) instruments rely on the Bernoulli effect to produce their sounds. Let's consider an oboe. The two strips of cane that form the double reed are tightly joined at one end to a brass tube, which is connected with a cork joint to the top of the oboe. The other end of the strips is free; here, the sides of the two strips touch but they are open in the center (Fig. 16AB–1). This end of the double reed is placed between an oboe player's lips; the player then pressurizes his or her mouth cavity and air is blown between the two reeds, producing sound. By the Bernoulli effect, the movement of the air between the reeds lowers the pressure there relative to the static pressure in the mouth cavity. The resulting net force presses the reeds together and the entrance hole closes. The air flow then ceases and the pressure differential becomes zero, allowing the reeds to return to their natural open position. This cycle repeats at the natural mechanical resonance frequency of the reeds. A reed assembly blown without being connected to the oboe produces a musical note at this resonance, called the "crow tone." The pulses of air generated by the reed act as a driving force for an air column oscillator made from the oboe pipe. Standing waves are set up in the air column, at frequencies that can be changed by opening and closing holes in the pipe. The starting transient state of the instrument contains frequencies associated with the crow tone, giving the characteristic "pauk" sound of the instrument.

FIGURE 16AB–1 An oboe reed.

---

application of Bernoulli's equation at the top of the water and at the opening leads to the equality

$$p_0 + p_1 + \rho g h = p_0 + \frac{1}{2}\rho v^2.$$

We cancel p_0 from both sides of the equation and solve for v^2:

$$v^2 = \frac{2(p_1 + \rho g h)}{\rho} = \frac{2(mg/A + \rho g h)}{\rho} = 2g\left(\frac{m}{\rho A} + h\right).$$

Numerically, with $\rho = 10^3$ kg/m³, we find

$$v^2 = 2(9.8 \text{ m/s}^2)\left(\frac{10 \text{ kg}}{(10^3 \text{ kg/m}^3)(0.07 \text{ m}^2)} + 0.6 \text{ m}\right) = 15 \text{ m}^2/\text{s}^2,$$

or $v \approx 4$ m/s. From our discussion of flux, we know that the rate of flow is the flux $\Phi = vA_{\text{hole}} = v\pi r_{\text{hole}}^2$. The hole has a radius of 0.75 cm = 7.5×10^{-3} m. Thus $\Phi = (4 \text{ m/s})\pi(7.5 \times 10^{-3} \text{ m})^2 = 7 \times 10^{-4}$ m³/s = 0.7 L/s.

FIGURE 16–29 Example 16–11.

*Lift

The Bernoulli effect is ultimately responsible for **lift**, an upward net force on the wings of an airplane. A correct description of this phenomenon requires that we relax some of the simplifying assumptions listed at the end of Section 16–5. Circulation, viscosity, and turbulence play an important, though indirect, role in lift. Consider the cross section of a wing shown in Fig. 16–30a. We visualize the situation from the reference frame of the wing, so that we think of the air moving toward the wing, which is at rest, rather than the wing moving through stationary air. (This is, in fact, just what happens in a wind tunnel in which problems such as this one are studied.) Air approaches the wing with some speed v along a set of streamlines. Some of the streamlines go below the wing, some go above. The wing is designed so that the air speed above the wing is greater than that below the wing. This can happen only if there is circulation in the

The lift

 APPLICATION

Wing design

(a)

(b)

FIGURE 16–30 (a) Schematic version of streamlines over an airplane wing section like that in Fig. 16–19. There is a streamline that ends on the front of the wing at the stagnation point and a streamline that starts at the rear edge of the wing. (b) Because the compressibility of the air can be neglected in this application, the equation of continuity shows that there must be a nonzero circulation around the wing, which in effect forms a closed loop.

clockwise direction (Fig. 16–30b).[†] Circulation occurs only around the wing. The flow elsewhere is irrotational, which means that the constant appearing in the Bernoulli equation, Eq. (16–21), is the same for all the streamlines that flow near the wing, below or above. From Eq. (16–23), we see that the pressure on top of the wing, p_t, must therefore be less than the pressure below, p_b, and there is a net upward force, or lift, on the wing. The net upward pressure, $p_{net} = p_b - p_t$, is determined from Eq. (16–23):

$$p_b + \frac{1}{2} \rho v_b^2 = p_t + \frac{1}{2} \rho v_t^2,$$

where ρ is the density of air. We thus find that

$$p_{net} = p_b - p_t = \frac{1}{2} \rho (v_t^2 - v_b^2) = \rho \left(\frac{v_t + v_b}{2} \right)(v_t - v_b) \simeq \rho v (v_t - v_b), \quad (16\text{–}27)$$

where v is the average of the top and bottom speeds. Because the difference between the top speed and the bottom speed is small, both are nearly equal to the average speed, which we call the *airstream speed v*.

The expression in Equation (16–27) can be used to calculate the lift. If the wings have a width w and a tip-to-tip span S, then their total area is wS (we are ignoring the fuselage), and the lift is given by

$$L = wSp_{net} = wS(p_b - p_t) \simeq wS\rho v(v_t - v_b). \quad (16\text{–}28)$$

Experiments show that for speeds much slower than the speed of sound, $v_t - v_b$ is proportional to the airstream speed v, with a proportionality constant K (in general, $K \ll 1$), which depends on the wing design and the wing angle relative to the airplane direction. A typical value of the constant K is 0.12 so, for a small plane with a total wing area of 30 m² and an air speed of 200 km/h = 56 m/s and with the air density of about 1.3 kg/m³, the lift is given by

$$L = (0.12)(30 \text{ m}^2)(1.3 \text{ kg/m}^3)(56 \text{ m/s})^2 = 1.6 \times 10^4 \text{ N}.$$

Thus the mass that can be lifted is $(1.6 \times 10^4 \text{ N})/(9.8 \text{ m/s}^2) \simeq 1600$ kg. The fact that air is compressible plays little role if the relative speed of the wing and the air is much smaller than the speed of sound in air, $v_s \simeq 330$ m/s. (The design and operation of the wings of supersonic aircraft presents a different set of problems.)

We remarked previously that there must be circulation in the problem. To give the flow circulation, it is necessary to have some turbulence, and this is induced by the viscosity of the air and further enhanced by the placement of small protuberances (ailerons) on the wings.

Racing cars are designed so that the air beneath the car moves faster than the air above it, and a downward pressure is exerted on the car. This increases the friction between the wheels and the track, and this friction is ultimately responsible for the propulsion of the car.

(a)

(b)

FIGURE 16–31 Velocity profiles for laminar flow in a pipe: (a) idealized fluid flow that does not take into account viscosity; (b) realistic fluid flow (Poiseuille flow).

*16–9 COMMENTS ON REAL FLUIDS

Our discussion has been limited by assuming idealized kinds of flow with idealized fluids. Some of our assumptions apply only approximately, at best, to real fluids. Nonviscous flow, for example, has sometimes been called the flow of "dry water." Here, we describe some qualitative notions of how real fluids behave under ordinary circumstances.

Viscosity

Figure 16–31a is a picture of the velocity profile of pipe flow for fluids idealized as we have described. For real fluids, the liquid near the surface of the pipe is at rest, and

---

[†]No air actually moves forward under the wing; rather, the speed beneath the wing is less than the speed above the wing.

the velocity profile is actually parabolic (Figure 16–31b). Such flow is referred to as **Poiseuille flow**, after Jean Poiseuille, who studied blood circulation. The reason for this effect is the internal friction of the fluid, or *viscosity*. Viscosity comes from inter-molecular collisions. The molecules of a fluid have a random motion, to be described in Chapter 19. If two adjacent layers of fluid flow with slightly different speeds, the random sidewise intrusion of some slower molecules into the faster stream will tend to slow down the faster stream, whereas intrusion of faster molecules into the slower stream will tend to speed up the slower stream.

Viscous flow can be analyzed if we take two glass plates with a film of fluid be-tween them and slide one plate over the other. The easier it is to slide the plates, the less viscous is the fluid between them. To quantify this effect, take the smaller plate to have an area A; it is separated from the other plate by a distance y (Fig. 16–32). If the upper plate is to move with a uniform velocity **v**, a force must be applied—just as if a brick is made to slide along a rough surface at constant speed. This force, F, is proportional to A and inversely proportional to y. In addition, the force increases lin-early with speed (like the damping force on a mass at the end of a spring or like air drag). The proportionality constant η in the formula

$$F = \frac{\eta v A}{y} \tag{16–29}$$

is called the **coefficient of viscosity**. The coefficient of viscosity is measured in new-ton·seconds per square meter = kilograms per meter·second. Typical values are 10^{-3} kg/(m·s) for water and 830×10^{-3} kg/(m·s) for glycerin, both at 20°C. We mention the temperature because, as anyone who has poured syrup over pancakes knows, vis-cosity is strongly temperature dependent.

Turbulence

A flow regime marked by violent and random movement is referred to as turbulent. At the onset of turbulence, the whole notion of neatly separated streamlines ceases to make sense (Fig. 16–33). A dimensionless parameter known as the **Reynolds number**, Re, determines this regime. It is defined as

$$Re = vL\left(\frac{\rho}{\eta}\right), \tag{16–30}$$

FIGURE 16–32 Two glass plates have a film of fluid between them; one slides against the other. The top plate moves with speed v. Fluid near the top plate moves with the same speed as that plate, while fluid near the bottom plate is at rest.

(a)

(b)

FIGURE 16–33 Turbulence, an important feature of fluid flow, is seen here in (a) the air above a candle and (b) the atmosphere of Venus. The blue seen in the image results from computer processing. The large swirl at the bottom is at the planet's South Pole.

where ρ is the density of the fluid, v is its speed, L is a length associated with the flow, such as the pipe's diameter for flow in a pipe, and η is the viscosity coefficent. When the Reynolds number increases past about 2000 to 3000, the flow becomes turbulent. For water flowing through a pipe 1 cm in diameter, $Re \simeq (10^4 \text{ s/m})v$, so the flow becomes turbulent when v reaches merely 0.3 m/s. This shows that the kind of flow we have treated in this chapter is not very common. Fortunately, a small amount of turbulence does not change the applicability of Bernoulli's equation for short distances of flow (just as a small amount of friction does not disturb energy conservation over short times), so Eq. (16–21), although not exact, is still a very good and useful approximation.

SUMMARY

Liquids are distinguished from gases in that the molecules that form liquids are closer together and impose strong forces on each other. Liquids are distinguished from solids in that the molecules in liquids do not form a rigid lattice. Both liquids and gases deform in response to external forces and therefore flow as fluids. The density of a material of mass M and volume V is unchanging in incompressible fluids:

$$\rho = \frac{M}{V}; \qquad (16\text{--}1)$$

this is the case for the fluids we study.

The pressure, measured in SI in pascals, expresses the way that fluids transmit forces. The pressure is the force per unit area on any area A:

$$p = \frac{F}{A}; \qquad (16\text{--}2)$$

it is perpendicular to A. When a fluid is at rest, the pressure at any given point is the same in all directions. For an incompressible fluid in Earth's gravity, the pressure increases with the depth y according to

$$p = p_0 + \rho g y, \qquad (16\text{--}6)$$

where p_0 is the pressure at $y = 0$. A change in pressure applied to a fluid at rest is transmitted without change to every part of the fluid, a statement known as Pascal's principle. This principle is the basis of hydraulic lifts and brakes.

For an object immersed in a fluid, there is an upward buoyant force equal to the weight of the displaced fluid,

$$F_{\text{buoy}} = \rho g V. \qquad (16\text{--}11)$$

This is Archimedes' principle.

Nonviscous (friction-free) fluids in smooth (nonturbulent) motion follow paths traced out by streamlines. The conservation of mass for a fluid moving in this way is stated as the continuity equation,

$$\rho_1 v_1 A_1 = \rho_2 v_2 A_2. \qquad (16\text{--}14)$$

Here, ρ and v are the density and speed, respectively, at some point in a tube of streamlines; A is the corresponding cross section of the tube. If the fluid is incompressible, the (constant) density cancels, and the resulting expression is the conservation of flux:

$$v_1 A_1 = v_2 A_2. \qquad (16\text{--}15)$$

The conservation of energy for a moving fluid under the influence of gravity is called Bernoulli's equation:

$$\text{along a streamline:} \quad p + \frac{1}{2}\rho v^2 + \rho g h = \text{a constant.} \qquad (16\text{--}21)$$

When the fluid is at rest, this reduces to Eq. (16–6); when the fluid flow is horizontal, this reduces to

$$p + \frac{1}{2}\rho v^2 = \text{a constant.} \qquad (16\text{–}23)$$

This equation is the basis of the Venturi flowmeter, which measures the speed of fluids in pipes, as well as the basis for understanding lift forces on wings. The full Bernoulli equation helps us to understand the flow of fluids from holes in fluid-filled tanks.

UNDERSTANDING KEY CONCEPTS

1. How does a hot-air balloon get off the ground?

2. We weigh a brick twice with a bathroom scale. The first time, the scale and the brick are inside a tub of water, and the second time, they are outside the tub. Water can enter the scale. Does the scale indicate a difference in weight in the two cases?

3. One accurate method of measuring the fat content of your body is to measure your weight twice: once while you are immersed in a tank of water, and once when you are out of the water. Explain how such a method might work.

4. If an empty balloon is weighed on a scale, a certain value is obtained. If a child now blows up the balloon and waits for several minutes so that the temperature of the air in the balloon is the same as the room temperature, what value will the scale give for the balloon's weight?

5. Suppose that a certain fluid has no viscosity. Will the acceleration of objects falling under gravity within this fluid be independent of the mass of the objects?

6. Water-storage tanks for communities normally are placed high off the ground rather than underground. This is obviously a safety hazard, so why is it done?

7. Mercury is poisonous. Why, then, is it typically used in barometers to measure atmospheric pressure?

8. The "antilift" effect for a racing car, the opposite of the lift effect on airplanes, is said to be so strong that a car could race on an upside-down track. Is this a plausible statement? Make some rough estimates of the surface area of a racing car, its mass, and its velocity, and assume that the difference between the air speeds under the car and over the car is about 20 m/s.

9. One way to get the water out of a stopped-up sink is to use a long piece of rubber tubing. You put one end in the water and suck on the other end until you have filled the tubing. You then close off your end, bring that end to a point below the bottom of the sink, and reopen that end. The water will flow out through the tubing until the sink is empty. You have made use of the *siphon mechanism* (see Problem 65). Explain how it works.

10. The siphon shown in Fig. 16–34 and described in Question 9 is limited in the height h_0 that it can overcome. What determines this limit?

11. Do sailboats make use of any of the aspects of Bernoulli's laws, such as those we described for airplanes? In particular, do sails need to be curved in order to work?

12. Which weighs more, a ton of Styrofoam or a ton of lead? Which of the two has the larger volume? How might you estimate the density of each?

13. People find it very easy to float in the Dead Sea. Why?

FIGURE 16–34 Question 10.

14. The point of a cone-shaped bottle (pointed end up) is removed to provide an opening. The pressure of the liquid at the very bottom of the bottle is the sum of the atmospheric pressure and the weight of the liquid above. If we consider a part of the bottom that is not directly below the opening, we might argue that there is less pressure because, above that point, there is less liquid and no atmosphere. This argument violates Pascal's principle. What has the argument left out?

15. Suppose that you put a mixture of oil and water into a centrifuge that rotates at high speeds (the oil is less dense than the water and does not mix with it). The two liquids separate. Why, and which component will be farthest from the center of the centrifuge?

16. An old proposal for transcontinental travel along a fixed latitude is the following: Take a balloon high into the stratosphere. Wait until Earth has rotated the desired distance beneath you, then descend. Comment on the feasibility of this mode of transport.

17. A snorkel is a breathing tube meant to be used while you swim under water; it runs from your mouth to the water surface. Why is the length of a snorkel limited?

18. In the course of a lazy summer morning, ice cubes floating in a pitcher of water melt. What happens to the water level in the pitcher? (Ignore evaporation.)

19. Suppose that you put a kilogram weight and a jar half-filled with water on a kitchen scale. The scale reads 2.5 kg. Now you place the kilogram weight *inside* the jar, and no water spills out. You might argue that because the kilogram weight weighs less in water than in air, the scale would read less than 2.5 kg, but the reading remains 2.5 kg. What is wrong with your argument?

16–1 States of Matter

1. (I) Jupiter has a radius $R = 7.14 \times 10^4$ km, and the acceleration due to gravity at the surface is $g_J = 22.9$ m/s². Use these data to calculate Jupiter's average density.

2. (I) The numbers given in Problem 1 have the following values for some of the other planets: Venus: $R = 6.05 \times 10^6$ m, $g = 8.86$ m/s²; Mars: $R = 3.39 \times 10^6$ m, $g = 3.73$ m/s²; Uranus: $R = 2.54 \times 10^7$ m, $g = 9.12$ m/s²; Neptune: $R = 2.48 \times 10^7$ m, $g = 12.0$ m/s². Calculate the densities of these bodies.

3. (I) The density of a nucleus is about 2×10^{17} kg/m³. The amount of water in a large lake is 10^{13} m³. If this amount of water were compressed to nuclear density, how many liters of water would there be?

4. (I) A platinum sphere has a diameter of 4.05 cm. What is the diameter of an aluminum sphere of the same mass, given that the densities of platinum and aluminum are 21.4×10^3 kg/m³ and 2.70×10^3 kg/m³, respectively?

5. (I) Atoms of an imaginary two-dimensional solid are arranged in a square lattice. The distance of neighboring atoms is a. Assume that the solid is acted upon by a shear force and deformed by a *small* angle f without any change in the nearest neighbor distance (Fig. 16–35). Calculate the change in the distance between atoms situated at opposite corners of a square.

FIGURE 16–35 Problem 5.

6. (II) A wedge of weight 20 N and angle 60° is floating on the surface of water in the symmetrical position shown in Fig. 16–36. Calculate the force of the water that acts on each of the two surfaces.

FIGURE 16–36 Problem 6.

16–2 Pressure

7. (I) A circus clown stands on a pair of stilts that each have a square cross section of 4.0 cm per side. If the mass of the clown plus the stilts is 68 kg, what pressure is exerted on the floor?

8. (I) The density of mercury is 13.6 times that of water. Compare the height of a column of water to that of a column of mercury, assuming that the pressure exerted by the weights of the respective liquids at the bottom of each column are the same.

9. (II) A hollow stainless steel sphere of radius 20 cm is evacuated so that there is a vacuum inside. (a) What is the sum of the magnitudes of the forces that act to compress the sphere? (b) There is a circular hole of diameter 4 cm on the side of the sphere for access to the inside. Calculate the force needed to pull a flat plate off the hole when the sphere is evacuated. Do you think that you could remove such a plate by pulling on it?

10. (II) During a hurricane, the atmospheric pressure changes dramatically. Explain why it is recommended that house windows be kept slightly open during a hurricane. What is the net force on a wall that is 300 ft² in area when the pressure on one side is 14.7 lb/in² and the pressure on the other is 14 lb/in²?

11. (II) A hollow metal pyramid with a square base that is 15 cm on each side has a mass of 1.8 kg. What is the pressure exerted by the pyramid on the table on which it stands? Suppose that the temperature in the room rises so that the metal expands. Will the pressure increase or decrease as a result of the expansion?

16–3 Variation of Pressure in a Fluid at Rest

12. (I) Using the densities in Table 16–1, what is the height of a fluid column in a barometer that has water as its fluid, in an air pressure of 1 atm? One that has alcohol? (Neglect the vapor pressure of the fluid above the column.)

13. (I) What is the pressure at depths of 1 m, 10 m, 100 m, and 10 km under the surface of the ocean? Take $\rho = 1.03 \times 10^3$ kg/m³ for the density of sea water and $p_0 = 1.01 \times 10^5$ Pa for the atmospheric pressure at the ocean surface. (Ignore the fact that, to this accuracy, density varies with depth.)

14. (I) Recreational scuba divers rarely go deeper than 100 feet (30 m). What is the water pressure at this depth? By what factor does your result differ from sea-level water pressure?

15. (I) What is the pressure 1 cm below the surface of a column of mercury exposed to air? 3 cm below the surface?

16. (I) An oceanic research vehicle operates at a depth of 1.20 km under water. Assuming that the pressure inside the vehicle is 1.0 atm, what is the force on a small window of dimensions 10 cm × 10 cm?

17. (II) Assume that a spaceship lands on Venus with the cabin temperature maintained such that the pressure is the same as the surface pressure on Earth—1 atm. What would be the height of a column of mercury in a barometer in the spaceship on Venus? Repeat for a spaceship landing on Neptune. (Use the data given in Problem 2.)

18. (II) In a pressure test chamber, a person starts acting strange when the gauge pressure exceeds 40 lb/in². *Gauge pressure* is the pressure in excess of atmospheric pressure. This is a well-known effect that limits the depth at which scuba divers should breathe pure air. In sea water, whose density is 1.03 g/cm³, to what depth should a diver be limited?

19. (II) A hydraulic jack is used to lift a 1200-kg car on a piston of diameter 30 cm (Fig. 16–37). How large a force is needed to push down the smaller piston if its diameter is 2 cm? By how much is the car lifted with a single push, assuming that the small piston moves 0.5 m?

FIGURE 16–37 Problem 19.

20. (III) The vertical wall of a small dam is 20 m long and it holds back a body of water that is 6 m deep (Fig. 16–38). Calculate the force acting on the wall due to the pressure of the water. Determine the effective point of action of this force. [*Hint*: Recall the method used to calculate the center of mass of a nonuniform object in Chapter 8].

FIGURE 16–38 Problem 20.

16–4 Buoyancy and Archimedes' Principle

21. (I) A bathysphere (a spherical, watertight research facility) 2.6 m in diameter has a mass of 9400 kg. It is released from a submarine at a depth of 20 m below the surface. Will the bathysphere sink or rise?

22. (I) Use the densities in Table 16–1 to show that ice floats in water. What fraction of the volume of a floating ice cube will be above the surface?

23. (I) A cubical box, contents unknown, floats in water; 16 percent of the volume of the box is above water. What is the average density of the box and its contents?

24. (II) A hemispherically shaped bowl is floating on the surface of water; the mass of the bowl is 1.8 kg (Fig. 16–39). Water is poured into the bowl; when 6.9×10^3 cm^3 of water has been added, the bowl just sinks. What is the outer radius of the bowl?

25. (II) A group of Scouts intend to make a raft to float down the Ohio River. The mass of four of them plus their equipment is 400 kg. They find trees with an average diameter of 20 cm and a specific gravity of 0.8. Determine the minimum area of the log raft that would keep them dry.

26. (II) A temporary bridge is supported on floating empty drums that are 1 m in diameter and 1.8 m long. The design criterion

FIGURE 16–39 Problem 24.

is that the drums may only be submerged to 2/3 of their diameter under a maximum load of 5000-kg trucks, 7.5 m long, that follow each other bumper to bumper in a single line. How far apart should the drums be placed? Neglect the weight of the bridge and of the drums.

27. (II) A valuable preserved biological specimen is weighed by suspending it from a spring scale. It weighs 0.45 N when it is suspended in air and 0.081 N when it is suspended in a bottle of alcohol. What is its density?

28. (II) Consider a spherical balloon filled with helium, with a density of 0.18 kg/m^3. The density of air is 1.3 kg/m^3. What must the radius of the balloon be in order to lift a load of 100 kg (this includes the mass of the balloon)?

29. (II) Olive oil floats on water. Take ρ_1 to be the density of the oil and ρ_2 to be the density of the water. Consider an oil–water interface across which a bouillon cube of density ρ_3 floats; use the geometry shown in Fig. 16–40. We know that for the oil to float on the water, ρ_1 must be less than ρ_2. What is the condition on ρ_3 so that the cube floats at the oil–water interface? How much of the cube will be in the water?

FIGURE 16–40 Problem 29.

30. (II) A beaker of mass 350 g contains 2.8 kg of water and rests on a bathroom scale. A 5.2-kg block of aluminum, of specific gravity 2.7, is suspended from a spring scale and completely submerged in the water. Find the readings on both scales.

31. (II) Balloons of mass 3.5 g can be filled with helium to make spheres of radius 21 cm. How many of these balloons must a child of mass 32 kg hold in order to float off Earth's surface? Assume that helium has a density of 0.18 kg/m^3.

32. (III) A sphere of radius R of material with an average density of 0.75 g/cm^3 is immersed in a body of water. What is the

height of that portion of the sphere that projects above the surface?

16–6 The Equation of Continuity

33. (I) Water flows at a speed of 1.3 m/s through a hose of diameter 1.5 cm and emerges from a nozzle of radius 0.5 cm. With what speed does the water emerge? How much water emerges from the nozzle per second?

34. (I) Water flows down a channel 1.0 m deep and 0.5 m wide at a rate of 2 metric tonnes/s. At some point the channel widens to 0.8 m. How fast does the water flow in the wider channel?

35. (I) A steady stream of automobiles moves along the eastbound side of a highway in a region where there is neither exit nor entrance. Where two eastbound lanes are available, the automobiles are spaced so that there are 66 cars per mile in each lane, and the average speed is 45 mi/h. In a region to the west, where three eastbound lanes are available, the automobiles are spaced to give 35 cars per mile in each lane. What is the average speed in the region where three eastbound lanes are available?

36. (II) A lawn sprinkler is connected to a garden hose of inside diameter 2 cm that has a water flow rate of 0.3 L/s. The sprinkler has 10 nozzle openings, each with a diameter of 3 mm. (a) Calculate the speed of the water that emerges from the nozzles. (b) Explain how the nozzles can be arranged to cause the sprinkler to rotate. What principle could you use to calculate the torque on the sprinkler? What are the limitations, if any, on the speed of rotation of the sprinkler?

37. (II) Students would like to spray water 18 m across their neighbor's yard onto a patio. They have a hose of inside diameter 1.5 cm that can spray a distance of only 1.5 m. What size nozzle do they need?

38. (II) When water drains out of a hole in the bottom of a tank, the stream radius contracts as the speed of the water in the stream increases. Assume that the stream starts with zero water velocity at the top of the tank, at a height H above the hole, and ignore the various instabilities that will eventually cause the stream to break up. Find the distance below the hole at which the radius of the stream is 80 percent of the hole radius.

16–8 Applications of Bernoulli's Equation

39. (I) A bilge pump is used to pump water out of a leaking ship (Fig. 16–41). The pump's hose has a diameter of 3.0 cm, and the pump moves water through the hose up and out of a porthole 5.0 m above the waterline at a speed of 4.0 m/s. Calculate the power of the pump.

FIGURE 16–41 Problem 39.

40. (I) What pressure must a pump generate to get a jet of water to leave a nozzle at a speed of 3 m/s and a height of 25 m above the pump?

41. (I) Wind gusts at 60 mi/h past a ship's cabin in which the air is at rest and the pressure is 1 atm. What are the pressure outside the cabin and the net pressure on the walls past which the wind blows?

42. (I) A stream of water sprays like a fountain vertically from a small hole in a pipe, reaching a height of 2.4 m. Assuming that the water in the pipe is static, what is the gauge water pressure in the pipe? (See Problem 8.)

43. (I) A folded piece of notebook paper can exhibit lift in the air flow from a fan of a window air conditioner. A plastic straw is taped perpendicular to the plane of the paper through the paper's middle, and a thread runs through the straw allowing the paper to move up and down the thread. The folded paper has an area of 300 cm^2. The air speed is about 4 m/s and the proportionality constant K in the approximate lift formula $L = KA\rho v^2$ is taken to be $K = 0.14$. Here, A is the surface area of the wings, ρ is the air density, and v is the air speed. How much lift is possible? Given the mass of the paper and straw to be 8 g, will the paper rise?

44. (II) A small garden fountain shoots a vertical jet of water at a rate of 0.10 L/s to a height of 0.50 m. (a) What is the initial speed of the jet, and what is the radius of the hole out of which the jet passes? (b) What pressure must the pump of the fountain supply? (Assume that it sits just below the emerging jet.) (c) At height 0.25 m, what is the speed of the jet, and what is the radius of the column of water? Ignore effects of turbulence, such as the breakup of the jet.

45. (II) Heavy rains have flooded your favorite professor's basement to a depth of 15 cm. This basement is 7.5 m $\times$ 12 m in area. You have a hose of diameter 1.2 cm and can run the hose from the water in the basement to a level 3 m below the level of the basement floor, into a low portion of the yard. When the hose is filled with water, it acts as a siphon (see Question 9) and empties the basement. (a) What is the speed of the stream of water leaving the hose? (b) Approximately how long does it take to empty the basement?

46. (II) A liquid of density 1.4×10^3 kg/m^3 flows in a horizontal pipe. The cross-sectional area in one part of the pipe is 75 cm^2. When the liquid enters another part of the pipe with cross-sectional area 150 cm^2, the pressure as measured by a gauge is 2.0×10^4 Pa higher than it was in the first part. Calculate the velocities of the liquid in the two parts of the pipe.

47. (II) A large horizontal pipe of diameter 10 cm contains water flowing with a speed of 0.3 m/s. The pipe branches into four pipelets, each of diameter 2 cm. The four pipelets each run horizontally after an initial height change of 3.5 m (Fig. 16–42). (a) What are the speeds of the fluid in the horizontal portion of each pipelet? (b) If the pressure in the large pipe is $p = 2.5$ atm, what is the pressure in the pipelets?

48. (II) A uniform glass tube with internal cross-sectional area A is formed into a U-shape and partly filled with a volume V of an incompressible fluid of density ρ. The two ends of the tube are open to the atmosphere. At equilibrium, the level in each vertical segment of the tube is $h = 0$. By applying additional pressure to one end of the tube, the levels of the liquid are displaced, so one side is at height h_0 and the other is at height

FIGURE 16–42 Problem 47.

$-h_0$. The additional pressure is suddenly released, and the liquid level starts to oscillate. (a) Use Bernoulli's equation to show that the force on the fluid in the tube is proportional to the displacement of the fluids from their equilibrium levels, so the force is harmonic. (b) Find the "spring constant" of the harmonic motion described in part (a) and use it, together with the mass in motion (ρV), to calculate the oscillation frequency.

49. (II) Water flows through a large horizontal pipe of diameter D at speed v_0. The pressure in the pipe is p_0. The pipe branches into two horizontal pipelets of diameter d_1 and d_2, respectively, both at the same height as the large pipe. Write the equations that would allow you to find the pressures p_1 and p_2 and the speeds v_1 and v_2 in the two pipes. Show that one of these four quantities must be known to be able to solve for the others.

50. (II) An open water tank stands on a plane surface. The water surface in the tank is a height h above the plane. A small hole is opened up at a depth y below the surface of the water. (a) Show that the jet of water will hit the plane surface a distance D from the tank, where $D = \sqrt{4y(h - y)}$. (b) Show that the hole should be placed at a depth $y = h/2$ for the jet to cover a maximum horizontal distance.

*16–9 Comments on Real Fluids

51. (II) The retarding force on a sphere of radius R that moves with speed v through a fluid of viscosity η is proportional to η and depends on R and v, in the form $R^a v^b$. Use dimensional analysis to determine the powers a and b.

52. (III) The result of the solution of Problem 51, the retarding force on a sphere of radius R that moves with speed v through a fluid of viscosity η, is $F = 6\pi\eta Rv$, known as *Stokes's law*. Such a sphere, with mass density ρm, falls through glycerin, whose density is ρ. The forces on the sphere are the retarding Stokes's law force, the buoyant force, and the force of gravity. (a) Derive a formula for the value of the velocity v_t for which the net force is zero. (b) How does v_t depend on the radius of the sphere? (c) If the viscosity of glycerin is 8.3 kg/(m·s) and its density is $\rho = 1.26 \times 10^3$ kg/m³, what is the terminal velocity of a sphere of iron, which has $\rho = 8.5 \times 10^3$ kg/m³ and a diameter of 5 cm? Of an iron sphere with a radius of 1 mm?

General Problems

53. (II) A rectangular container is divided into two parts by a movable vertical partition. This partition is sealed so that no water can seep from one side to the other. Initially, the wall is held in the middle of the 30-cm length of the container. One side is filled with 2 kg of water to a depth of 10 cm; the other side is attached to a spring in equilibrium (unstretched, uncompressed). The spring constant is 180 N/m. The partition is now released and the spring compresses. What is the position of the partition at maximum compression? Assume that the process takes place without turbulence so that no energy is dissipated.

54. (II) A hollow sphere of radius 0.2 m and total mass 1.4 kg is placed in water. One very small lead pellet of mass 0.2 kg is attached to the inside surface of the sphere. When the sphere is floating in equilibrium, the lead pellet is oriented on the bottom. If the sphere is rotated slightly, so that the pellet is moved off to the side, and then released, the sphere will rock, and the pellet will move like a pendulum. Calculate the period of small oscillations for this motion. Assume that the center of the sphere does not move.

55. (II) A depth of about 100 ft ($\approx$ 30 m) is as deep as a scuba diver should go without special precautions because of an effect that causes disorientation. What is the water pressure at a depth of 30 m? Does a scuba diver take in the same mass of air in each breath at 30 m that he or she would at a depth of 5 m? Explain.

56. (II) A cylindrical bucket of fluid of density ρ sits at the center of a turntable that rotates with some fixed angular velocity, ω. Show that the pressure variation in the fluid at a horizontal distance r from the axis is given by $p = p_{\text{axis}} + \frac{1}{2}\rho\omega^2 r^2$. [*Hint*: If you place yourself within the rotating fluid and at rest relative to it, the problem is one of hydrostatics. The only force in the horizontal direction is a fictitious force away from the axis of the rotation, due to the fact that your frame of reference is not inertial. This force is the centrifugal force.]

57. (II) Late in the seventeenth century, King Louis XIV of France had a series of fountains constructed at Marly, some 20 km from his chateau at Versailles. These fountains were fed from the Seine River, whose altitude is some 150 m below that of Marly, through a remarkable series of pumps. The tallest of these fountains was the Grand Jet, which is calculated to have risen a height of 37 m from its base. (a) At what speed does this jet leave the orifice at the fountain's base? (b) The rate of flow in the Grand Jet is measured to be 0.051 m³/s. Compare this to an estimate of the rate of flow from a garden hose. (c) Given the rate of flow in part (b), calculate the area of the orifice from which the Grand Jet comes. (d) What is the pressure of the water just behind the orifice from which the Grand Jet comes? Beyond the crude calculations outlined here, the height to which a jet of water will rise can be affected by the addition of a short length of conical pipe above the level of the orifice at the base.

58. (II) A cylindrical tank is filled with water to a height h_1 above its base. Unfortunately, someone has cut a hole in the bottom through which fluid escapes in a vertical stream into your room. (a) What is the speed of the escaping stream at the hole? (b) You insert a flexible hose into the hole. If the other end of the tube is held at the same height as the base of the tank and is directed horizontally, what is the speed of the water as it leaves the hose? (c) You hold the tube, whose end is still at the height of the base of the tank, so that the stream is directed straight up. How high does the escaping stream shoot?

59. (II) An inventor proposes to you the following perpetual motion machine (a machine that can do work "forever" without

your having to add energy to it) (Fig. 16–43). A cylindrical tower of height h filled with water rests on the ground. Blocks of wood are inserted at the bottom of the tower. They float to the top, where they are removed. The potential energy that they acquired in floating to the top can be converted into work. They can, for example, be dropped down for re-use, with the kinetic energy they acquire in falling used to turn the wheels of a turbine. When they reach the bottom, the heavier blocks are again inserted, and the process is repeated. How can you prove to the mad inventor that at best this machine can do no net work? [*Hint*: Calculate the work required to insert a block, which requires in effect that a mass of water whose volume is that of the block be raised to the top. Compare this to the energy the block acquires in being accelerated to the top under the influences of gravity and the buoyant force. Neglect any drag caused by the water.]

Do not buy this machine! Perpetual motion machines are impossible to construct (see Chapters 18 and 20). A close analysis of the details of such machines always reveals this fact.

FIGURE 16–43 Problem 59.

60. (II) Consider the machine discussed in Problem 59. Why does the machine still fail if you replace the entry port by an exchange chamber (Fig. 16–44) with two sliding doors, so that the insertion of the blocks does not require that work be done?

FIGURE 16–44 Problem 60.

61. (II) A cubical block of wood 0.20 m on each side floats in water, with 8.0 cm of the block above the surface of the water (Fig. 16–45). (a) What is the density of the wood? (b) Suppose that the block of wood is pushed down below the equilibrium level by a distance Δ. What is the magnitude of the restoring force that pushes the wood back toward the equi-

FIGURE 16–45 Problem 61.

librium level? (c) In general, if dissipation of energy is ignored, the block will overshoot the equilibrium level and rise to a height Δ above that level, and is then subjected to a downward force. How large is that force? (d) Recall from Chapter 13 that a force that is proportional to the displacement Δ from equilibrium (of the form $F = -k\Delta$) and acts on an object of mass m leads to oscillations with the frequency $f = \sqrt{k/4\pi^2 m}$. Calculate that frequency for the block that bobs in the water.

62. (II) The ticket-window opening of a movie theater, where money goes into the cashier's hands and tickets emerge, is a rectangle 8 cm × 16 cm. A breeze of speed 5 m/s comes out of this opening. Sometimes, as a result, paper money goes flying. (a) What is the difference in air pressure between the inside and the outside of the theater? (b) Suppose that the opening were the size of a door, 0.9 m × 2.1 m. What would the speed of the breeze be in this case, assuming the air pressure difference were the same as in part (a)?

63. (II) A water-storage tank is an upright cylinder of height 8 m and radius 6 m. The storage tank is full of water but is vented to the atmosphere. The bottom of the tank is placed 30 m off the ground. A pipe of diameter 12 cm runs vertically down from the tank and goes 1 m underground before turning horizontal. (a) What is the water pressure in the bottom of the tank? (b) in the horizontal pipe underground? (c) The water flow in the pipe is 80 L/s. How fast does the water level in the tank drop? (d) A saboteur drills a hole of diameter 6 mm near the bottom of the tank. How fast does the water shoot out? (e) What volume flow of water is lost out the hole?

64. (II) A *Pitot tube* is a device for measuring flow velocities (Fig. 16–46). It is as useful for measuring the flow velocity of blood in an artery as for measuring the air speed of a jet airplane. The fluid passes the opening at point B, creating a pressure p_B inside. The fluid enters the opening of the tube and comes to a halt, creating a pressure p_A. Apply Bernoulli's equation as needed to show that the flow velocity is given by $v = \sqrt{2(p_A - p_B)/\rho} = \sqrt{2gh\rho_\ell/\rho}$, where ρ is the density of the fluid being measured, ρ_ℓ is the density of the fluid shown in the device, and h is the difference in height of the liquids.

65. (II) With a siphon, it is possible to transfer fluids from one container to another. Figure 16–34 shows a siphon that must be filled with fluid to start the transfer. If the speed with which the fluid leaves the higher container is zero, and if the fluid has a fixed density ρ, find an expression for the speed of the fluid as it enters the lower container, in terms of the heights h_0 and h_1.

FIGURE 16–46 Problem 64.

66. (III) A tank of surface area A is filled with water to a height x above the position of a hole of surface area σ, and the water flows out. Assume throughout that the fluid speed is zero at the top surface. (a) How much water flows out of the hole in the first time interval Δt after the hole is opened? (b) How large is the drop in the water level, Δx, in that time interval? (c) Use the results of parts (a) and (b) in the limit of small Δt to express a relation of the form $dx/dt = f(x)$. This is a differential equation for the height as a function of time. Take care to get the right signs, noting that x decreases with time, so that Δx is a negative quantity. (d) Show that this equation is solved

by the formula

$$x(t) = \left[\sqrt{x_0} - \frac{1}{2} \left(\frac{\sigma}{A} \right) t \sqrt{2g} \right]^2.$$

(e) How long will it take for the water to drop to the level of the hole?

67. (III) The ice cream cone shown in Fig. 16–47 is filled with melted ice cream of density 1.2 g/cm³. The cone has a diameter of 6 cm at the larger end and is 10 cm long. Find the pressure at the bottom of the cone. If a small hole of diameter 1 mm is opened at the bottom, the ice cream starts to run out. Ignoring the viscosity of the melted ice cream, find the amount of time it takes the ice cream to run out. Assume that the fluid speed is zero at the top.

FIGURE 16–47 Problem 67.

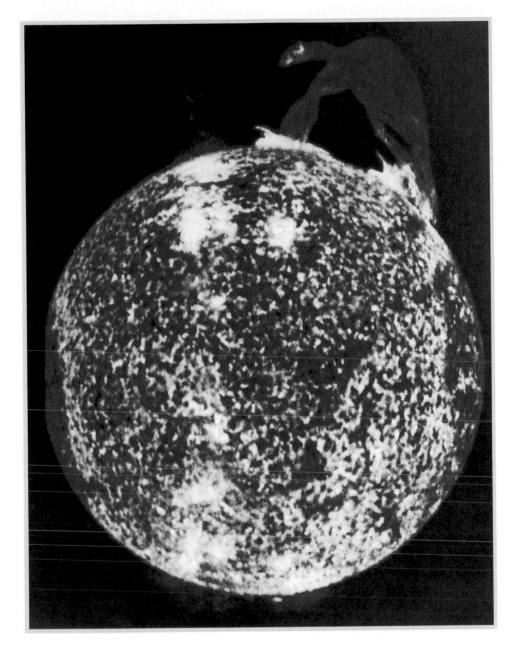

The surface of the Sun is an active region. Here, we see patterns of magnetism that hold hot streamers of ionized gases above the surface. The patterns of temperature are an important clue to deciphering the dynamics of this activity.

Temperature and Ideal Gases

Although temperature has not been a very important part of our discussions to this point, many properties of matter depend on it. Water behaves quite differently at temperatures below its freezing point than it does at temperatures above its boiling point. Temperature has something to do with energy, as we can see from the fact that hot steam can lift a piston. How much energy does hot steam contain, and how does that energy get into the steam? How can we quantify "hotness," and just how hot does steam need to be to lift a piston? By studying the behavior of gases, we can learn how to measure temperature and how temperature is a measure of energy.

463

Although thermal phenomena are ultimately explained by a statistical treatment of the behavior of matter on an atomic scale, the historical development of thermal physics was based on a description of bulk matter. The study of thermal phenomena received its strongest impetus from the engineers who studied the steam engines that powered the industrial revolution of the nineteenth century. The treatment of thermal phenomena that they developed—the subject of Chapters 17 through 20—is known as **thermodynamics**. Thermal phenomena are pervasive in the physical world and our understanding of them has had a profound effect on the way we live.

17–1 TEMPERATURE AND THERMAL EQUILIBRIUM

We call a physical system a **thermal system** whenever we are interested in its temperature-dependent properties. A piston and a cylinder full of hot steam form a good example to keep in mind as we proceed. We all have an intuitive feeling for temperature—for example, we can usually tell when one system is hotter than another. Our aim here is to develop a more precise notion of temperature.

Central to the idea of temperature is the notion of **thermal equilibrium** (Fig. 17–1). Suppose that we pour cold water into a bucket of hot water. Experience shows that, after a short time, the water becomes lukewarm. Until that point, we cannot say that the water has a particular temperature. It is only when thermal equilibrium has been reached, meaning that temperature-dependent quantities are no longer changing, that the temperature becomes a variable that characterizes the whole bucket of water; that is, the whole thermal system. We can say that if two thermal systems are in thermal equilibrium with one another, then they have the same temperature. Finally, we can add that if two systems are each in thermal equilibrium with a third, then they are in thermal equilibrium with one another.

If we put an ice cube into a bathtub full of hot water, the temperature of the tubful of water changes minimally; if we add a teaspoon of hot water to a swimming pool, the temperature change is virtually undetectable. These observations allow us to define a **thermal reservoir** (or **heat bath**) as a thermal system so large that it maintains a constant temperature when it interacts with other thermal systems. A thermal reservoir is of great importance because it allows us to bring a given system to a predetermined temperature, the temperature of the reservoir. To do so, all we need to do is bring the given (smaller) system into **thermal contact** with the reservoir: Any two thermal systems in contact with each other *will reach the same equilibrium temperature.*

Thermal contact is established between thermal systems in three important ways. We can illustrate them with two metal plates in an enclosure, one red hot and the other cold. In thermal contact by *radiation*, the hot plate emits electromagnetic waves (the "radiation" in question), and the cold plate absorbs them. This form of thermal contact operates even with the plates separated by some distance. Thermal contact by *conduction* occurs when the two plates actually touch. Finally, thermal contact occurs by *convection* when there is a fluid between the plates. The fluid acts as a third thermal system whose temperature increases in the region near the hot plate through its own thermal contact with the plate. The fluid then circulates in large currents to the colder plate, increasing its temperature. Conversely, if two fluids at different temperatures are separated by a double-walled, silvered glass flask with a vacuum between the two glass surfaces, the fluids will take a very long time to reach the same temperature. The vacuum prevents thermal contact by conduction or convection, and the silvered surfaces of the glass walls reflect radiation back to the heat source and prevent thermal contact by radiation. The liquids are **thermally isolated** from one another.

Thermometers

A **thermometer** is a device that measures the temperature of a thermal system quantitatively—in general, by coming to thermal equilibrium with that system. A thermometer must satisfy several criteria. It should be small enough to have a minimal ef-

FIGURE 17–1 The vendor's stand is *not* in thermal equilibrium with its environment. The steam is evidence of an interaction with the surroundings that would eventually cool the stand down to where it would be serving frozen hot dogs.

Radiation, conduction, and convection are three ways to achieve thermal contact.

(a)

(b)

(c)

FIGURE 17–2 (a) This thermometer contains a (red) fluid that expands at higher temperatures and contracts at lower temperatures. (b) This thermometer measures temperature in terms of the pressure of the gas in the bulb. (c) This digital thermometer uses the temperature dependence of electronic properties of materials. It is calibrated in the Celsius scale, in which the temperature of ice melting in pure water is 0 degrees. (d) This device measures temperature using emitted radiation.

fect on the system being measured. (A small room thermometer brought inside from the cold will have a minimal effect on the room temperature.) The thermometer must make good thermal contact with the system being measured. Like a meter stick, the scale of the thermometer should be reproducible, easily read, and universally available.

It is easy to find properties of matter that are suitable for thermometry. Ordinary fever thermometers rely on the fact that fluids (and solids) expand and contract when their temperatures change. We can therefore make a thermometer by marking numbers on a column of mercury that measure the mercury's volume. The pressures and volumes of gases, the volumes of liquids or solids, reflective properties of so-called liquid crystals, electrical and magnetic properties of solids, and the color of light emitted by hot objects are all properties that vary with temperature and can be used as thermometers (Fig. 17–2). Not all of the properties of matter that depend on temperature are suitable for making thermometers: The brittleness of many materials depends on temperature, but we would not make a thermometer that works by seeing how easy it is to smash a material with a hammer. Table 17–1 gives a representative list of some useful thermometers.

(d)

∞ See Chapter 21 for a discussion of *thermal expansion.*

TABLE 17–1   Some Thermometers

Thermometer	Physical Property Measured	Comments
Ideal gas	Pressure and volume of dilute gas	See Section 17–2
Mercury bulb	Expansion or contraction of fluid	Good where fluid does not change phase
Bimetallic strip	Difference in expansion of two metals	Often used in solids
Resistance	Electrical resistance	See Chapter 27
Thermocouple	Electrical voltage across different metals	Most widely used thermometer in industry
Paramagnetic	Magnetic properties of matter	Useful at ultracold temperatures; see Section 32–5
Optical pyrometer	Color of emitted light	Useful at high temperatures; see Section 17–4

∞ Pressure was defined in Chapter 16.

A technique to define temperature

Ideal gases

FIGURE 17–3 A bottle of gas whose pressure and volume can be measured. A tightly fitting piston can be moved to change the volume of gas, and weights on the piston can be changed to vary the pressure.

The Kelvin temperature scale [see Eq. (17–11)]

17 – 2   IDEAL GASES AND ABSOLUTE TEMPERATURE

Consider a gas in a bottle with a sliding piston (Fig. 17–3). We can adjust the volume of gas by changing the position of the piston and we can adjust the pressure by changing the weight on the piston. A valve allows us to control the *amount* of gas—in other words, its mass.

If we dip the bottle into thermal reservoirs at various temperatures, we find that, as the gas becomes hotter, its pressure increases if the volume is kept constant, and its volume increases if the pressure is held fixed. The volume and the pressure correspondingly decrease when the gas becomes colder (Fig. 17–4). In fact:

The product of pressure and volume, divided by the mass of the gas in the bottle, increases or decreases according to whether the thermal reservoir in contact with the bottle is hotter or colder, respectively. We use this property to define temperature as some proportionality constant times the product of pressure and volume, divided by mass.

The gas thermometer we construct in this way comes extremely close to being a universal thermometer because experiment shows that *the proportionality constant is the same for all gases if their densities are sufficiently low.* Dilute (or low-density) gases, which we call **ideal gases**, will thus enable us to give a universal definition of temperature. Just how dilute the gas needs to be is simply a question of the accuracy demanded of the thermometer. Gas thermometers accurate enough for laboratory use may contain gas hundreds of times less dense than air at sea level.

Figure 17–5 plots the pressure of a fixed amount of a dilute gas at fixed volume against our temperature scale. (In this case, we say that we have a constant-volume ideal-gas thermometer.) The part of the curve where T and p drop to zero is drawn as a dashed line. Temperatures here cannot be measured with our thermometer because gases liquefy or freeze in this region of extreme cold. The dashed line is thus an extrapolation.

Let's define the temperature $T = 0$ as the point *where the pressure, p, would become zero* (if that were possible), so

$$\text{for constant volume: } p = (\text{a constant}) \, T. \qquad (17–1)$$

We still need to specify the constant in this relation; we do so by defining T at some other point. This second point, called a *defined point*, was chosen in 1968 by international agreement to be the **triple point** of water. The triple point of water is the state at which water vapor, liquid water, and frozen water coexist. The temperature at the triple point, where the pressure is p_{tp} (= 4.58 mm of mercury), is defined as 273.16 K. The letter K stands for kelvin, and the temperature scale so defined is the **Kelvin scale**. (During the mid- to late-nineteenth century, Sir William Thomson, Lord Kelvin, did important work on thermal phenomena and on electricity and magnetism.) To summarize, temperature in the Kelvin scale, T, is defined by a constant-volume ideal-gas thermometer as

$$T = \left(\frac{p}{p_{tp}}\right) 273.16 \text{ K}, \qquad (17–2)$$

where p is the pressure of the ideal gas in the thermometer. Figure 17–6 illustrates Eq. (17–2). The assignment of 273.16 K as the triple point makes the Kelvin scale compatible with the Celsius scale, to be described below.

Table 17–2 gives us a feeling for the Kelvin scale by listing some temperatures that occur in nature. Remarkably, the numbers that refer to the origin of our universe in the so-called Big Bang, which occurred some 1.8×10^{10} years ago, can be estimated quite reliably from laboratory data. Although experiments have come quite close to 0 K, this temperature cannot in fact be reached.

(a)   (b)   (c)

FIGURE 17–4 (a) A small amount of water is poured into a can, then heated to boiling, filling the can with water vapor. (b) The can is turned upside down and plunged into a container of water at room temperature. (c) The steam inside the can condenses, causing the inside pressure to suddenly drop. The outside pressure then crushes the can.

EXAMPLE 17–1 Estimate the pressure in a hypothetical bottle of air that, starting from sea level (surface temperature and pressure), has been heated to the temperature of the surface of the Sun. Express your answer in atmospheres.

Solution: We assume that the gas in the bottle is ideal over the desired range. The temperature must be in kelvins. From Table 17–2, we estimate that a typical temperature on Earth, where the pressure is 1 atm, is 300 K; the temperature at the surface of the Sun is approximately 6000 K. According to Eq. (17–1), the ratio of the two temperatures in question is the same as the ratio of the two pressures. If we refer to the temperature and pressure on Earth and the Sun with the subscripts E and S, respectively, we have

$$\frac{p_E}{T_E} = \frac{p_S}{T_S};$$

$$p_S = p_E\left(\frac{T_S}{T_E}\right) = 1 \text{ atm}\left(\frac{6000 \text{ K}}{300 \text{ K}}\right) = 20 \text{ atm}.$$

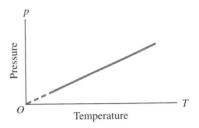

FIGURE 17–5 As p varies in an ideal gas of constant volume, T is *defined* by drawing a straight line.

FIGURE 17–6 The Kelvin temperature scale is defined by a constant-volume ideal-gas thermometer. The two fixed points are $T = 0$ K, located where p extrapolates to zero, and $T = 273.16$ K, located at the pressure of the triple point of water, p_{tp}.

TABLE 17–2 Temperatures Occurring in Nature

Physical System	Temperature (K)
Lowest temperature reached in the lab	$\simeq 10^{-6}$
Temperature of background radiation that fills the universe	3
Liquid helium at 1 atm	4.2
Coldest recorded outdoor temperature on Earth	185
Average surface temperature on Mars	218
Freezing point of water	273.15
Body temperature	310
Hottest recorded outdoor temperature	331
Boiling point of water	373
Melting point of gold	1335
Surface of the Sun	6000
Interior of the Sun	10^7
Helium fusion reaction	10^8
Interior of hottest stars	10^9
Universe at 3 min after the Big Bang	5×10^8
Universe at 1 s after the Big Bang	10^{10}
Universe at 10^{-16} s after the Big Bang	10^{14}
Universe at 10^{-12} s after the Big Bang	10^{16}

	Kelvin T		Celsius t_C
Boiling point of water (1 atm)	373 K		100°C
Triple point of water (1 atm)	273.16 K		0°C
Absolute zero	0 K		−273.15°C

FIGURE 17–7 Comparison of the Kelvin and Celsius temperature scales.

The difficulty of finding containers that can withstand high temperatures is one reason why ideal-gas thermometers are not very practical at high temperatures. It is not easy to construct a suitable container for temperatures much above 2000 K, even at room pressures.

Other Temperature Scales

The Celsius scale

By changing both the slope and the intercept in Fig. 17–4, we can construct other temperature scales. Equivalently, if two points on the linear curve are assigned to special temperatures, a new scale emerges. Different scales are useful in different regimes. The temperature in the **Celsius scale**—once known as the centigrade scale—is denoted by t_C and measured in units of °C. It assigns the value $t_C = 0$°C to the freezing point of water at 1 atm of pressure (the *ice point*) and $t_C = 100$°C to boiling water at 1 atm. From Fig. 17–7, we see that

$$t_C = T - 273.15, \tag{17–3}$$

where T is the temperature in the Kelvin scale.[†] Note that the slopes of the Kelvin and the Celsius scales are exactly the same.

Other scales differ from the Kelvin scale in slope as well as in zero level. The **Fahrenheit scale**, in which the temperature t_F is measured in °F, is an example. It was invented with a human scale in mind. The ice point of water is 32°F, the temperature of the human body is about 100°F, and the steam point of water is 212°F. The relation between the Fahrenheit scale and the Kelvin scale is

$$t_F = \frac{9}{5}T - 459.67. \tag{17–4}$$

Again, T is measured in kelvins. We can easily find the relation between any two scales; see, for example, Fig. 17–8.

Celsius	Fahrenheit
100°C	212°F
	200°F
50°C	122°F
	100°F
0°C	32°F
−17.8°C	0°F

FIGURE 17–8 Comparison of the Celsius and Fahrenheit temperature scales.

17–3 THE EQUATION OF STATE

The pressure p, temperature T, volume V, and the amount (or mass) of material are the variables that describe a thermal system such as a bottle of gas with a piston. We have already discussed the first three variables. One possible unit for the amount of material in a thermal system is the **mole** (abbreviated mol), which represents either an amount of gas with a certain mass (according to the chemical species of the material) or a count of the number of molecules. One mole of a given gas is defined to be the amount of

Avogadro's number

[†]The difference of 0.01 between 273.16 in Eq. (17–2) and 273.15 here is the difference between the temperature of water at its triple point and water at its ice point.

gas with a mass in grams equal to the *atomic* (or *molecular*) *weight* of the gas. For example, 1 mol of helium gas has a mass of 4 g; the atomic weight of helium is 4 units of atomic mass. One mole of gas always contains *Avogadro's number* of molecules— $N_A = 6.022 \times 10^{23}$ molecules. The total number of molecules in a container, N, can be written in terms of the number of moles, n, as

$$N = nN_A. \tag{17–5}$$

Variables such as p, T, V, and n that are used to describe thermal systems are called **thermodynamic variables**.

Ideal gases obey *Boyle's law,* discovered in 1662 by Robert Boyle. Boyle's law states that, for a dilute gas held at constant temperature, the product of pressure, p, and volume, V, divided by the number of moles, n, is a constant:

$$\text{for constant } T: \qquad \frac{pV}{n} = \text{a constant.} \tag{17–6} \qquad \text{Boyle's law}$$

We previously defined temperature as a linear function of pressure if volume is constant. We can combine this result [Eq. (17–1)] with Boyle's law into the **ideal gas law**:

$$pV = nRT, \tag{17–7} \qquad \text{The ideal gas law}$$

where R is a proportionality constant. The ideal gas law is an example of a relation between the thermodynamic variables of a thermal system, called the **equation of state**. Such a relation holds for any thermodynamic system and ensures that the variables are not independent of one another. The ideal gas law is the equation of state for an ideal gas. Measurements of ideal gases give

$$R \simeq 8.314 \frac{\text{N} \cdot \text{m}}{\text{mol} \cdot \text{K}} = 8.314 \text{ J/mol} \cdot \text{K.} \tag{17–8} \qquad \text{The universal gas constant}$$

R is known as the **universal gas constant**. If we count the number of molecules of a gas, N, rather than the number of moles, n, we can use Eq. (17–5) in order to rewrite Eq. (17–7) as

$$pV = NkT, \tag{17–9}$$

where **Boltzmann's constant**, k, is given by

$$k = \frac{R}{N_A} \simeq 1.381 \times 10^{-23} \text{ J/K.} \tag{17–10} \qquad \text{Boltzmann's constant}$$

We shall see that Boltzmann's constant, named for Ludwig Boltzmann—one of the important figures in nineteenth-century thermal physics—provides the bridge between temperature and energy. This gives it a particularly important role in thermal phenomena.

EXAMPLE 17–2   Calculate the volume occupied by 1 mol of an ideal gas at 20.0°C and at atmospheric pressure. Use this volume to calculate the mass density of air, whose molecular weight is 29.2 g/mol.

Solution: It follows from the ideal gas law that for 1 mol of gas, $V = RT/p$. Under the given conditions (T in kelvins is $20.0 + 273 = 293$ K),

$$V = \frac{(1 \text{ mol})(8.31 \text{ J/mol} \cdot \text{K})(293 \text{ K})}{1.01 \times 10^5 \text{ Pa}} = 2.41 \times 10^{-2} \text{ m}^3.$$

The mass density of air under these conditions is then the mass of 1 mol of air divided by this volume:

$$\rho = \frac{29.2 \text{ g}}{2.41 \times 10^{-2} \text{ m}^3} = \frac{(29.2 \text{ g})(10^{-3} \text{ kg/g})}{2.41 \times 10^{-2} \text{ m}^3} = 1.21 \text{ kg/m}^3.$$

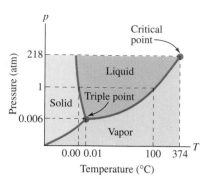

FIGURE 17-9 Phase diagrams are useful for the different phases of a material, such as those for water, shown here: liquid, solid (ice), and vapor (steam). The intersection of these phases is the triple point.

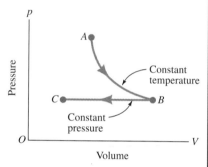

FIGURE 17-10 Example 17-3. A p-V diagram of the path followed by an ideal gas in undergoing first an isothermal transformation and then an isobaric transformation.

Changing the Thermodynamic Variables of a Gas

Thermal systems change as the thermodynamic variables change. (We say that a *thermal transformation*, or, alternatively, a *thermal process*, occurs.) For example, when the cylinder of gas in Section 17-2 was put in contact with a hotter thermal reservoir, the gas's temperature changed; so did the gas's volume as the piston moved out. In this transformation, the temperature, the volume, and the pressure all changed subject to the constraint of the equation of state. There are many different ways to change a thermal system. For example, in an **isothermal** transformation, the temperature remains fixed while the pressure and volume both change. In an **isobaric** transformation, the pressure is held fixed while both temperature and volume change. Such transformations are most easily shown on diagrams in which one variable is plotted against another. For example, in a p-V diagram, pressure is plotted against volume. When a diagram shows various phases of a material, it is referred to as a *phase diagram*. Figure 17-9 is a p-T phase diagram for water.

EXAMPLE 17-3   Exactly 1 mol of an ideal gas is taken through the sequence of changes shown on the p-V diagram in Fig. 17-10. The change $A \rightarrow B$ is an isothermal transformation. In $B \rightarrow C$, the transformation is isobaric: The gas is compressed in volume in such a way that the pressure remains fixed. (a) If $p_A = 5.00$ atm and $V_A = 8.00$ L, what is the temperature T_A? (b) If $V_B = 40.0$ L, what is p_B? (c) If $T_C = T_A/8$, what is V_C?

Solution: The starting point for all the parts of this problem is the ideal gas law, $pV = nRT$. For this sample of gas, $n = 1$ mol.

(a) We know p and V at point A; the temperature T_A is then given by

$$T_A = \frac{p_A V_A}{nR} = \frac{(5.00 \text{ atm})(8.00 \text{ L})}{(8.134 \text{ N} \cdot \text{m/mol} \cdot \text{K})(1 \text{ mol})}.$$

We make some unit conversions in order to cancel units:

$$T_A = \frac{(5.00 \text{ atm})(1.01 \times 10^5 \text{ Pa/atm})(8.00 \text{ L})(10^{-3} \text{ m}^3/\text{L})}{(8.314 \text{ N} \cdot \text{m/mol} \cdot \text{K})(1 \text{ mol})} = 486 \text{ K}.$$

(b) $T_B = T_A$, from the conditions stated in the problem. We know V and T at point B, and hence we can compute p_B:

$$p_B = \frac{nRT_B}{V_B} = \frac{nRT_A}{V_B} = \left(\frac{nR}{V_B}\right)\left(\frac{p_A V_A}{nR}\right) = p_A\left(\frac{V_A}{V_B}\right)$$

$$= (5.00 \text{ atm})\left(\frac{8.00 \text{ L}}{40.0 \text{ L}}\right) = 1.00 \text{ atm}.$$

(c) The change $B \rightarrow C$ is one in which pressure is held constant. The quantities $p_C = p_B$ and T_C are known. The volume at C is given by

$$V_C = \frac{nRT_C}{p_C} = \frac{nR}{p_C}\left(\frac{T_A}{8}\right) = \frac{nR}{8p_C}\left(\frac{p_A V_A}{nR}\right) = \frac{p_A V_A}{8p_C} = \frac{p_A V_A}{8p_B}$$

$$= \frac{(5.00 \text{ atm})(8.00 \text{ L})}{8(1.00 \text{ atm})} = 5.00 \text{ L}.$$

How Close Do Real Gases Come to Being Ideal?

When is a real gas "sufficiently dilute" to obey the ideal gas equation of state? To answer this question, let's look at how well Boyle's law is satisfied. Consider a series of identical closed tubes. A pressure gauge can measure the pressure of any gas in the tubes. Put different amounts (a different number of moles, *n*) of the same gas in each tube and immerse them in the same thermal reservoir. If Boyle's law is satisfied ex-

The properties of phase diagrams suggest ways in which the presence of different phases of a material can be used to advantage. An example of such an application of a phase diagram is the delivery of carbon dioxide to bars and restaurants for the production of carbonated beverages (Fig. 17AB–1). At a pressure of one atmosphere—and depending on the temperature—carbon dioxide exists only in the solid phase (low temperatures) and in the gas phase (high temperatures). Both are inconvenient to use; the gas is bulky and the solid is difficult to incorporate in mixing systems. Carbon dioxide is therefore transported and delivered at pressures greater than 40 atmospheres (4×10^6 Pa), pressures at which the room-temperature phase is liquid. This liquid can then be conveniently stored, transported efficiently through pipes, controlled by automatic valves, and so forth. When it is needed for incorporation into soft drinks near room temperature, it is released to one atmosphere of pressure, at which point it changes to the gas phase.

FIGURE 17AB–1 A carbonated beverage dispenser.

actly, the (measured) ratio pV/n will be the same in each tube, independent of the pressure in the tube.

Experiment shows that for monatomic gases, such as helium, the ratio pV/n is independent of the pressure over a wider range of pressures, and therefore fits the ideal gas limit better than do polyatomic gases.[†] But *all gases give precisely the same value of pV/n when extrapolated to zero pressure or to zero density*. All gases are indeed ideal in this limit; a more appropriate definition of temperature than that presented in Eq. (17–2) would be to *extrapolate measurements linearly* with such a thermometer to zero pressure:

A gas of any substance is ideal if the gas is dilute enough.

$$T = (273.16 \text{ K}) \lim_{p \to 0} \frac{p}{p_{tp}}. \tag{17–11}$$

The Kelvin temperature scale redefined

Equation (17–11) provides us with the desired universal way to define temperature because it holds in the domain where *every* gas is ideal.

The van der Waals Equation of State

In 1873, Johannes D. van der Waals proposed an equation of state that works well for many real gases. His formula, known as the *van der Waals equation*, takes the form

The van der Waals equation of state

$$\left[p + a\left(\frac{n}{V} \right)^2 \right]\left(\frac{V}{n} - b \right) = RT. \tag{17–12}$$

The constants a and b are both positive. Van der Waals was motivated to write this equation based on a molecular model of gases (see Chapter 19). We can verify that the van der Waals equation reduces, as it must, to the ideal gas law when the gas is very dilute: In the limit as $n \to 0$, the term $a(n/V)^2$ is small compared to p, and the term V/n is large compared to b. Thus, in this limit, Eq. (17–12) reduces to $pV/n = RT$.

The van der Waals equation of state is a considerable empirical improvement over the ideal gas law for polyatomic gases that are about as dilute as the air we breathe. In fact, it describes all gases well, even when a gas is nearly a liquid; that is, at very low temperatures and at densities approaching those of real liquids.

[†]The molecule of a monatomic gas consists of one atom; the molecule of a polyatomic gas consists of several.

APPLICATION

The optical pyrometer

∞ The energy states of atoms are
discussed in Chapter 10.

A body at temperature T emits
electromagnetic radiation that is
characteristic of that temperature.

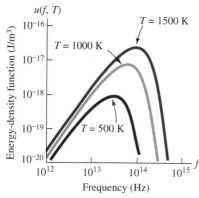

FIGURE 17–11 The energy
density of electromagnetic radiation in
thermal equilibrium at the indicated
temperatures. The maximum energy
density occurs for increasing
frequency as the temperature increases.
The maximum energy density
increases *very* rapidly with T for a
given f; note that this is a log–log plot.

A Connection Between Temperature and Energy?

An important fact follows from the ideal gas law presented in Eq. (17–7) or Eq. (17–9). The quantities pV/n and pV/N have the dimensions of (force/length^2)(length^3) = (force)(length), or $[MLT^{-2}][L] = [ML^2T^{-2}]$; n or N simply counts the number of moles or molecules, so that these quantities are dimensionless. The factor (force)(length) has the dimensions of work or, equivalently, energy. *The dimensions of RT and kT are those of energy.* This is already reflected in the units of R and k [Eqs. (17–8) and (17–10)]. The significance of this observation is that temperature, as defined by the ideal gas law, must be related closely to the energy of a thermal system. This *thermal*, or *internal*, energy is the subject of Chapter 18.

*17–4 THE BACKGROUND TEMPERATURE OF THE UNIVERSE

Most of us have enjoyed an evening before a roaring blaze that slowly dies down during the course of the evening. As the fire dies, its color changes from brilliant white to bright yellow, to orange, and finally to a dull red. The change is more marked if you look into a cavelike hole within the embers, a spot where the temperature is steadier because it is protected from the vagaries of drafts. This phenomenon is the basis of a thermometer called the *optical pyrometer* (Table 17–1). This thermometer works because the color or, rather, the range of colors emitted by a hot object is characteristic of the temperature of that object. This thermometer works best for relatively hot objects with temperatures in the range of several hundred kelvins.

The light emitted by hot objects is **electromagnetic radiation**. Radiation exists in the form of waves, and all frequencies are possible in principle. All matter, at any temperature, emits such radiation. Atoms go into excited states when the temperature of the matter they comprise is raised. They radiate energy when they fall back into their normal energy states. For example, the atoms heated by an electric current in a lightbulb filament radiate visible light, which we detect as a range of colors: from red to orange to yellow to green to blue to violet, as the frequency of the radiation increases from 0.5×10^{15} Hz to approximately 1.0×10^{15} Hz.

In some ways, radiation in an enclosure acts like a dilute gas in a bottle; in particular, the radiation has a temperature just as a gas does. However, there is an important difference: Radiation achieves thermal equilibrium through absorption and re-emission of radiation by the atoms of the walls of the enclosure rather than through collisions among its components (as the molecules that make up a dilute gas do). The proportion of high frequencies to low frequencies of the radiation at thermal equilibrium in an enclosure rises with temperature. The average frequency emitted by a white-hot object is greater than the average frequency the object emits when it is cherry red. We can give a function that describes the radiation emitted at *different* frequencies by an object at a given temperature. Such a function is denoted by $u(f, T)$, where $u(f, T)$ df *is the radiation energy per unit volume, or energy density, with a frequency between f and f + df.* Moreover, the radiation is said to be at a given temperature—the temperature of the object—when it and the object are in thermal equilibrium, meaning that the radiation interacts with the object over a long period and that the system of object and radiation is isolated from its surroundings. Radiation of this type is known as **blackbody radiation**, and the body that emits and absorbs such radiation is known as a **blackbody**. Figure 17–11 shows this function for different values of temperature. *The measurement of u(f, T) provides a measurement of T.*

The form of the energy–density function $u(f, T)$ was a subject of intense study in the latter part of the nineteenth century. The work of Gustav Robert Kirchhoff, Wilhelm Wien, Lord Rayleigh (Fig. 17–12), and James Jeans set the stage for the breakthrough of Max Planck, who determined in 1900 that

$$u(f, T) = \frac{8\pi h}{c^3} \frac{f^3}{e^{hf/kT} - 1}. \tag{17-13}$$

This formula, known as the **Planck formula**, was found to fit perfectly the data on the energy density of radiation for a wide range of frequencies and temperatures. In addition to Boltzmann's constant, k [introduced in Eq. (17–10)], and the Kelvin temperature, T, the formula contains the speed of light, $c = 3 \times 10^8$ m/s, which is the speed of electromagnetic waves of *every* frequency. A new fundamental constant also appears in this formula. The new constant h, called *Planck's constant*, was found by fitting the formula to the observed energy density, and its value is $h = 6.625 \times 10^{-34}$ J·s.

The discovery of the Planck formula was the first step on the path to the basic theory of matter and radiation, called *quantum theory*. This theory is based on a set of ideas that revolutionized the physical sciences in the twentieth century. (The other great revolution in twentieth-century physics—Einstein's theory of relativity—also has its origin in the investigation of electromagnetic radiation.)

An interesting consequence of the Planck formula is that the total power radiated through a small hole in an enclosure containing radiation in thermal equilibrium at temperature T must be proportional to the fourth power of the temperature, T^4. The **Stefan–Boltzmann formula**, which was discovered experimentally by Josef Stefan in 1879 and inferred from thermodynamics by Boltzmann in 1884 before the discovery of the Planck formula, states that the radiated power (energy per unit time) per unit area emitted through the hole is given by

$$P(T) = \sigma T^4, \tag{17-14}$$

where $\sigma = 5.68 \times 10^{-8}$ W/(m²·K⁴). The T dependence in Eq. (17–14) can be obtained by integrating Eq. (17–13) over all frequencies f (see Problem 62).

Let's examine the Planck formula for the energy–density function, Eq. (17–13). The quantity hf has dimensions of energy as does the quantity kT. (Indeed, these quantities must have the same dimensions because their ratio is the argument of the exponential function.) If the frequencies are small so that $hf < kT$, then the exponential function in $u(f, T)$ has a small argument, and we can use the expansion

for small x: $\qquad e^x \simeq 1 + x$.

In this case, the denominator of $u(f, T)$ is

$$e^{hf/kT} - 1 \simeq 1 + \frac{hf}{kT} - 1 = \frac{hf}{kT},$$

and the function $u(f, T)$ takes the approximate form

for $hf < kT$: $\qquad u(f, T) \simeq \frac{8\pi h}{c^3} (f^3) \frac{kT}{hf} = \frac{8\pi f^2}{c^3} kT. \tag{17-15}$

This expression is independent of Planck's constant and is thus independent of quantum physics. It was, in fact, first found by Lord Rayleigh in 1900 by using a derivation based on classical physics. The energy density in Eq. (17–15) rises with increasing frequency as f^2 and linearly with temperature.

If, however, the frequencies are large, so that $hf > kT$, then the exponential argument in $u(f, T)$ is large, and the exponential function plays a major role. The denominator then becomes

$$e^{hf/kT} - 1 \simeq e^{hf/kT}.$$

(Here we have used the fact that for $hf/kT > 1$, $e^{hf/kT} > 1$.) Then the energy density is approximately

for $hf > kT$: $\qquad u(f, T) \simeq \frac{8\pi h}{c^3} f^3 e^{-hf/kT}. \tag{17-16}$

∞ We introduced Planck's constant in Chapter 7.

FIGURE 17–12 Lord Kelvin and Lord Rayleigh, July 1900.

Because the exponential function falls with f much faster than the polynomial factor f^3 grows, $u(f, T)$ decreases as f grows in this range of f. Note that h plays a crucial role in this part of the energy–density function, indicating that quantum physics is important.

As we have seen, the energy–density function increases with f for $hf < kT$ and decreases with f for $hf > kT$. It is not surprising, then, that the function has an intermediate maximum, which we can estimate as occurring at about $hf \simeq kT$. The underlying physics of the high-frequency side of the curve of the energy–density function in Fig. 17–11 is quantum physics because it involves Planck's constant, but the low-frequency side can be explained by classical physics. The maximum moves to higher values of f as T increases. We can therefore say that the importance of quantum physics decreases as the temperature grows; in fact, this is generally true.

FIGURE 17–13 Example 17–4. The temperature of the Sun's surface is not constant but varies considerably in active areas, such as regions with sun flares.

EXAMPLE 17–4   The surface of the Sun is observed to emit light most strongly at a frequency of 2×10^{14} Hz. Estimate the temperature of the Sun's surface (Fig. 17–13).

Solution: The strongest emission occurs at the maximum of the energy–density function. An appropriate estimate follows from the rough value of this maximum, at a frequency given by $hf/kT \simeq 1$. The temperature is then approximated by

$$T \simeq \frac{hf}{k} = \frac{(6.63 \times 10^{-34}\,\text{J} \cdot \text{s})(2 \times 10^{14}\,\text{Hz})}{1.38 \times 10^{-23}\,\text{J/K}} = 10{,}000\,\text{K}.$$

A more precise value can be found by looking for the maximum of the energy–density function. Because it is wavelength rather than frequency that is measured, it is the energy density as a function of wavelength that is maximized. The more accurate expression (see Problem 66) is

$$\lambda_{\text{max}} T = 2.9 \times 10^{-3}\,\text{m} \cdot \text{K}.$$

An estimate from this formula yields 6000 K for the temperature at the Sun's surface.

A measurement of the value of the energy–density function, Eq. (17–13), at any particular frequency is equivalent to a measurement of the temperature. A crude way to measure this quantity is to hold your hand near a fire and feel the radiant heat. A pyrometer does this job more precisely: The color of the visible radiation emitted from a hot cavity is a measure of the frequency distribution. An antenna that receives and measures electromagnetic radiation is another accurate method of measuring the energy–density function. The size and nature of such antennas determine the particular frequency or range of frequencies they receive.

The Discovery and Measurement of the Background Radiation of the Universe

In 1964, Arno Penzias and Robert Wilson used a large radio antenna, shown in Fig. 17–14, to study the radio waves emitted within our galaxy. They found a puzzling anomaly, which they at first thought might be a problem with their antenna: A background electromagnetic radiation that comes from no particular source is present. Their antenna was sensitive only to the particular frequency $f = 4.08 \times 10^9$ Hz, but later measurements over other frequencies showed that this background radiation was consistent with the Planck formula for $T \simeq 3$ K.

Penzias and Wilson were not aware of work by Ralph Alpher, George Gamow, and Robert Herman in the late 1940s, or of work by James Peebles that was contemporary with their own, in which a cosmological, or large-scale, model of the universe predicted the existence of this radiation. As described by this model, the early universe expanded from an initial catastrophic event—the Big Bang. As the universe expanded, it cooled, maintaining thermal equilibrium between electromagnetic radiation and matter up to a point, some 100,000 years after the Big Bang, when the temperature was

FIGURE 17–14 Arno Penzias (right) and Robert Wilson in front of the antenna they used in detecting the background radiation of the universe.

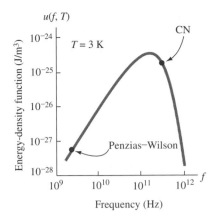

FIGURE 17–15 The energy density of electromagnetic radiation in thermal equilibrium at $T = 3$ K. The Penzias–Wilson frequency measurement and the frequency associated with interstellar cyanogen, labeled CN, are marked by red dots.

about 3000 K. At that point, the model shows that thermal equilibrium between electromagnetic radiation and matter was broken off. Remarkably, as the universe has continued to expand, the Planck formula still describes the radiation, but with a temperature that drops as the universe expands. *A measurement of the temperature of the radiation is equivalent to a measurement of the rate of expansion of the universe; hence, it is equivalent to the time that has elapsed since the radiation was no longer in thermal equilibrium with matter.* Presently, the temperature of the radiation, as characterized by the Planck formula, is about 3 K, and this radiation is the oldest directly observable relic of the explosion in which the universe is widely believed to have formed.

Penzias and Wilson's frequency measurement occurs on the classical, or low-frequency, side of the Planck formula, as Fig. 17–15, the energy–density function for $T = 3$ K, indicates. Since 1964, many other Earth-based measurements have been made by radio antennas sensitive to a great variety of frequencies. Because Earth's atmosphere absorbs radiation of high frequency, these measurements are all made on the classical side of the Planck formula. Fortunately, one other technique can make a measurement on the quantum, or high-frequency, side of the peak of energy density. The measurement in question is the amount of light emitted by interstellar cyanogen, the CN molecule. This light is emitted with a strength that can be explained only if the cyanogen is in the presence of background radiation of a certain temperature. The frequency measurement of CN, first made in 1941, is marked on Fig. 17–15. In early 1990, a satel-

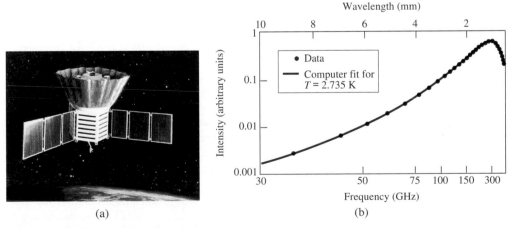

FIGURE 17–16 (a) The Cosmic Orbiting Background Experiment (COBE) satellite measured cosmic background radiation. (b) The measurements show remarkable agreement with the Planck formula. The data shown for $T = 2.735$ K fit the smooth curve to remarkable accuracy.

lite was launched to measure the background radiation (Fig. 17–16a). The satellite measures a range of frequency from roughly 6×10^{10} Hz to 6×10^{11} Hz with extraordinary precision. This range includes the maximum of the energy–density curve. We show the results in Fig. 17–16b; a temperature of about 2.7 K fits the curve perfectly. (The actual temperature can be determined to six significant figures!) Still other measurements reveal that the temperatures in different directions differ ever so slightly from one another. Some regions are slightly hotter, while others are slightly cooler. These temperature inhomogeneities may well be relics of the instabilities that led to the formation of the great clusters of galaxies present today. All these results provide a vivid verification of the Big Bang model of the universe, which predicts the presence of this background blackbody radiation.

SUMMARY

The temperature of a system in thermal equilibrium is defined operationally on the Kelvin scale by the behavior of the pressure, p, of a dilute gas in a constant-volume gas thermometer:

$$T = (273.16) \lim_{p \to 0} \frac{p}{p_{tp}}, \tag{17–11}$$

where p_{tp} is the pressure of the gas in the thermometer at the triple point of water (where the gaseous, liquid, and solid phases coexist); thus 273.16 K is chosen to be the temperature at the triple point. The other fixed point of the Kelvin scale is built into the thermometer: $T = 0$ K is the point where the pressure extrapolates to zero.

Two other temperature scales in common use are the Celsius scale and the Fahrenheit scale. The Celsius temperature is given in terms of the Kelvin scale by

$$t_C = T - 273.15. \tag{17–3}$$

Any sufficiently dilute gas behaves as an ideal gas, meaning that its thermodynamic variables are related by the ideal gas law, an equation of state:

$$pV = nRT = NkT, \tag{17–7, 17–9}$$

where the universal gas constant is

$$R \simeq 8.314 \text{ J/mol} \cdot \text{K}. \tag{17–8}$$

and Boltzmann's constant is

$$k = \frac{R}{N_A} \simeq 1.381 \times 10^{-23} \text{ J/K}, \tag{17–10}$$

where N_A is Avogadro's number. The quantities kT and RT have dimensions of energy.

Real gases are ideal only in the limit that they are very dilute. The small departures from ideal behavior are accurately reflected for many real gases by the empirical van der Waals equation of state,

$$\left[p + a \left(\frac{n}{V} \right)^2 \right] \left(\frac{V}{n} - b \right) = RT. \tag{17–12}$$

Electromagnetic radiation displays the characteristics of a thermal system. When such radiation is in equilibrium with another thermal system at temperature T, it is known as blackbody radiation and contains a range of frequencies f that have an energy per unit volume (or energy density) per unit frequency, given by

$$u(f, T) \frac{8 \pi h}{c^3} = \frac{f^3}{e^{hf/kT} - 1}, \tag{17–13}$$

a result known as the Planck formula. The constant h is Planck's constant.

Electromagnetic radiation with this frequency range is observed to occur throughout the universe at a characteristic temperature of 3 K. This is strong evidence for the Big Bang theory of the origin of the universe.

1. Estimate to within an order of magnitude the number of molecules in a breath of air.

2. Why is mercury used in most thermometers rather than a more common liquid such as water?

3. If a scuba diver takes the same volume of air for each breath and breathes at a constant rate, why is the tank of air used up much faster at greater water depths?

4. How can a triple point of water exist? If ice, water, and vapor all existed together, would the ice cool the vapor, making it water? Or would the vapor and water melt the ice?

5. Why is it advisable to measure the pressure in automobile tires when the car has not been driven for some time?

6. Devise an experiment to prove the experimental fact that hot objects radiate more energy per unit area than cold objects do.

7. Why is it important to check the pressure in automobile tires before the winter season?

8. All gases obey the ideal gas law when they are made sufficiently dilute. Is this true when the temperature is lowered sufficiently?

9. In recent years, an attempt has been made to convert Americans to the use of the metric system, so that, for example, two sets of tools would not be necessary for a mechanic who works on both American and European cars. Would it be of equal practical importance to convert Americans from the use of Fahrenheit to Celsius temperature scales?

10. When we observe a fire, we are observing radiation that has escaped the region of the hot fire. Is this a mechanism for cooling the fire? In what way could a fire be enclosed so that it remains hot?

11. Suppose that we have no idea that gases consist of molecules. Could we still use the low-pressure limit of a gas to define temperature?

12. Substances in thermal contact reach thermal equilibrium after some time. Can you think of any experimental test that will indicate just when that time will have arrived?

13. If the molecules of gases move around a lot, why do we never see a container of gas shake?

14. The pressure of a gas produces an outward force on the walls of a container. Why is there not a sideways force as well?

15. The Chamber of Commerce of a town plagued by temperature extremes devises a way to avoid bad publicity: It defines a new temperature scale, the G scale, such that (a) $-40°F$ is defined to be $+20°G$; (b) $+120°F$ is defined to be $-40°G$; and (c) the scale is linear. Is there anything wrong with this scale?

PROBLEMS

17–1 Temperature and Thermal Equilibrium

1. (II) Which of these pairs of systems are in thermal equilibrium with one another? (a) A roast reaching the rare stage and the oven in which it sits. (b) The point of a meat thermometer in the roast of part (a) and the roast. (c) A sunbather getting a tan and the air around him. (d) A bather in a hot tub and the water around her. (e) An ice cube in a glass of water at 0°C. (f) Molten iron ore in a blast furnace and the walls of the furnace.

2. (II) You must measure the different temperatures of a series of bottles that contain 1 L of an unknown fluid. Which one of the following techniques would make a suitable thermometer, and why or why not? (a) The time it takes 1 L of ice cubes dumped into the fluid containers to melt. (b) The melting time for ice cubes sealed in full 1-L plastic bags that are dumped into the bottles. (c) The melting time for ice sealed in full 1-mL plastic bags that are dumped into the bottles. (d) All of the above techniques, but the fluids in the bottles are gently stirred during the measurement.

3. (II) Thermodynamic variables are either extensive or intensive. A variable is *extensive* if, when two identical thermodynamic systems are combined into one, the variable of the combined system is double its original value in each system; otherwise, it is *intensive*. Which of the following variables are extensive, and which are intensive: (a) volume, (b) temperature, (c) pressure, (d) number of moles?

4. (II) A physicist needs to measure room temperature. Not having anything better at hand, she decides to make use of the thermal expansion of water to construct a thermometer (Fig. 17–17). She calibrates her apparatus to a dependable thermometer at 10°C and 40°C, and divides the range between these temperatures into 30 equal intervals. What will be the true temperature when her equipment reads 25°C? She knows that the volume of water can be described in this temperature range by the formula

$$V = V_0(1 - 2.525 \times 10^{-4} + 4.98 \times 10^{-6}\,T + 4.94 \times 10^{-6}\,T^2),$$

where T is the temperature in °C.

FIGURE 17–17 Problem 4.

5. (I) At what temperature values are the following scales the same: (a) the Fahrenheit and the Celsius, (b) the Celsius and the Kelvin, (c) the Fahrenheit and the Kelvin?

6. (I) The title of Ray Bradbury's science-fiction novel *Fahrenheit 451* refers to the temperature at which paper ignites. What is that temperature on the Celsius scale?

7. (I) An oral fever thermometer ranges from 35.5°C to 42.5°C. What is the corresponding range on the Fahrenheit scale?

8. (I) An ideal gas in a container is heated from 20°C to 215°C. By what factor does the pressure increase?

9. (I) A very dilute gas confined to a closed one-liter container is put on the burner of a stove and the temperature is raised from 271 K to 349 K. By how much does the pressure, which is originally at 1 Pa, change?

10. (I) If an ideal gas thermometer could be constructed to measure the temperature of a hot star (see Table 17–2), by what factor would the pressure in the thermometer differ from *ptp*?

11. (II) (a) Determine the temperature of the surface of Venus (730 K) in the Celsius and the Fahrenheit scales. (b) Determine the temperature of the boiling point of liquid nitrogen (77 K) in the Celsius and the Fahrenheit scales. (c) A comfortable room temperature is 75°F. Determine this temperature in the Kelvin and Celsius scales. (d) Heat pumps become less efficient than furnaces for heating homes when the temperature drops below 36°F. New laws may require that all temperatures be posted in the Celsius and the Kelvin scales as well as the Fahrenheit scale. What is this temperature in these other systems?

12. (II) In 1701, Isaac Newton proposed a temperature scale in which ice water has a temperature of 0°, whereas the human body in good health is assigned a temperature of 12°. In this scale, what is the temperature of someone with a slight fever? What is the temperature of boiling water?

13. (II) A temperature scale—the Reaumur scale—developed in the early eighteenth century sets the ice point at 0°R and the boiling point of water at 80°R. Give the conversion formula from the Reaumur scale to the Kelvin scale and to the Fahrenheit scale.

14. (II) For the new temperature scale described in Question 15, what are the general conversion rules from °G to °F and vice versa? What is the conversion rule from °C to °G?

15. (II) The Rankine scale for temperature, named for William Rankine, is used in engineering applications. It has units of °R and is defined by $t_R = 9T/5$, where T is the temperature in the Kelvin scale. Find the value of t_R for the temperatures 100°C, 4.2 K, 6000 K, 32°F, −30°F, and −25°C.

16. (II) A bottle that has a freely sliding vertical piston and contains an ideal gas is used as a thermometer for water: The bottle is submerged in a water bath (whose temperature is to be measured) until thermal equilibrium is reached (Fig. 17–18). When the bottle is in contact with ice water, the volume inside the bottle is 0.50 L. The bottle is removed from the ice water and put in a second bath. This time, the enclosed volume is 0.55 L. What is the temperature of the second bath, in °C?

17. (II) One Sunday morning, a family takes an automobile trip to Grandma's. At the start of the trip, the temperature is 288 K (15°C), and the gauge pressure in the tires is 32 lb/in² (psi).

FIGURE 17–18 Problem 16.

(The gauge pressure is the excess over 14.5 psi, the exterior air pressure.) After an hour's ride over an interstate highway, the gauge pressure in the tires is 38 psi. What is the temperature of the air in the tires, assuming that air behaves as an ideal gas? Neglect any changes in volume of the tires.

18. (II) When the temperature of mercury changes by ΔT, the fractional volume change, $\Delta V/V$, is given by $\Delta V/V = \beta \Delta T$, where $\beta = 1.8 \times 10^{-4}\ \text{K}^{-1}$. The quantity β is called the *coefficient of thermal expansion*. Consider a thermometer containing a bulb attached to a thin, cylindrical capillary tube. If the bulb and capillary tube contain 0.2 cm³ of mercury, how large must the diameter of the capillary tube be so that a 1-K change in temperature corresponds to a 2-mm change in the mercury level in the capillary tube? (Neglect the change in volume of the glass bulb when the temperature is raised by 1 K.)

19. (II) A copper bowl of volume 1500 cm³ is filled to the brim with water. Assume that the coefficient of thermal expansion, β, defined in Problem 18, is $5.1 \times 10^{-5}\ \text{K}^{-1}$ for copper and $2.07 \times 10^{-4}\ \text{K}^{-1}$ for water. How much water will spill out when the temperature of the system is raised from 20°C to 50°C?

17–3 The Equation of State

20. (I) A container holds gas at a pressure of 1.0 atm and a temperature of 300 K. Half the gas leaks out while the temperature is raised to 340 K. What is the pressure in the container?

21. (I) A dilute gas in a container fitted with a piston occupies a volume of 2500 cm³. If the pressure is increased by 50 percent and the temperature in degrees kelvin is decreased by 15 percent, what volume will the gas occupy?

22. (I) A 1000-L container is filled with an ideal gas at 1.0 atm and 300 K. How many moles are in the container? How many gas molecules?

23. (I) What is the volume occupied by three moles of an ideal gas at 30 K and 10^{-7} atm in the upper atmosphere?

24. (I) What is the mass density of steam at a temperature of 100°C and a pressure of 1 atm, given that the molecular weight (the number of grams of 1 mol of the material) of water is 18 g/mol?

25. (I) What is the mass density of air (average molecular weight 29 g/mol), at a pressure of 0.31 atm and a temperature of −43°C? (These conditions correspond roughly to those at the top of Mt. Everest.)

26. (I) A helium balloon is filled with 10,000 cm³ of helium gas at *STP* (standard temperature and pressure: 0°C and 1 atm).

How many moles and molecules of helium (molecular weight 4 g/mol) are contained in the balloon?

27. (I) A container of ideal gas whose volume can be changed by the movement of a plunger is placed in a large bath that maintains the temperature of the gas at a fixed value of 295 K. The plunger is pulled out so that the volume is increased by a factor of 1.78. By what factor is the pressure changed in the vessel, if at all?

28. (I) The atomic weight of an elemental substance, A, is closely related to the mass of the atoms of that substance. Atoms consist of electrons, of negligible mass, and a total of A protons and neutrons, which each have nearly the same mass M. The total mass of an atom is therefore approximately AM, and N_A atoms have a total mass AMN_A. By the definition of Avogadro's number, N_A, this mass is A g. Given the value of N_A, what is M in grams?

29. (I) A container at 0°C holds a mixture of gases: 80 percent nitrogen (N_2 molecular weight 28 g/mol) and 20 percent oxygen (O_2 molecular weight 32 g/mol) by weight. Assuming that for a dilute gas each constituent behaves as if it alone occupied the volume, what is the partial pressure due to each constituent of the mixture, given that the total mass of gas in the container is 8.0 g, and that the volume is 4.0 L. What is the total pressure in the container?

30. (I) Consider a container of an ideal gas at a fixed pressure. The temperature is 280 K. What is the fractional change in volume if the temperature changes to 281 K? Can you derive a general formula that holds for all ideal gases?

31. (II) Given that the molecular weight of water (H_2O) is 18 g/mol and that the volume occupied by 1 g of water is 10^{-6} m^3, use Avogadro's number to find the distance between neighboring water molecules. Assume for simplicity that the molecules are stacked like cubes.

32. (II) The lowest pressure achievable with a rotary vane pump is on the order of 10^{-2} N/m^2. What is the number density of molecules of the gas in a chamber that holds gas at this pressure and at a temperature of 20°C?

33. (II) The pressure of an ideal gas in a closed container is 0.6 atm at 35°C. The number of molecules is 5×10^{22}. (a) What are the pressure in pascals and the temperature in kelvins? (b) What is the volume of the container? (c) If the container is heated to 120°C, what is the pressure in atmospheres?

34. (II) The molecular weight of H_2 is 2 g/mol. Consider H_2 gas in a container at STP (see Problem 26). (a) What volume is occupied by 1 kg? (b) What mass and volume are occupied by 1 mol and 1 kmol?

35. (II) Use the ideal gas law to calculate the volume occupied by 1 mol of ideal gas at 1 atm pressure and 0°C. Given that the average molecular weight of air is 28.9 g/mol, calculate the mass density of air, in kg/m^3, at the above conditions.

36. (II) Use the result of Problem 35 and Avogadro's number for the number of molecules in 1 mol, 6.02×10^{23}, to estimate the number of molecules in your lecture hall.

37. (II) What is the mass density of helium gas (atomic weight 4 g/mol) at 1 atm pressure and a temperature of 8 K?

38. (II) A child takes 12 puffs of air to blow up a spherical balloon. The diameter of the balloon is 26 cm. The temperature is 21°C, and the exterior atmospheric pressure is 1 atm. (a) How many molecules are in each of the child's puffs? (b) If the surrounding air and the balloon are heated to 35°C, what size will the balloon be?

39. (II) In an experiment, a vacuum of 10^{-10} atm is achieved in a bottle. If the bottle is at room temperature (30°C), what is the number of molecules in the bottle per cubic centimeter?

40. (II) At altitudes above 14,000 ft, pilots must breathe air with enriched oxygen because the density of the atmosphere is decreased. The pressure of air is only about 0.26 atm at a typical height for flying of 32,000 ft. If the airplane is not pressurized, what must the fraction of oxygen in the air be in order for a pilot to breathe the same amount of oxygen as at sea level?

41. (II) An ideal gas is contained in a tank at 120 atm of pressure and 263 K. (a) If half the mass of gas is drawn off and the temperature then rises by 50 K, what is the new pressure? (b) Suppose instead that the temperature first rises by 50 K, and then half the mass of the gas is drawn off. What is the new pressure?

42. (II) For safety reasons, compressed gases for laboratory use are often stored outdoors and transferred to where they are needed via a system of tubes and valves (Fig. 17–19). Suppose that a 200-liter tank contains hydrogen at 87 atm at −5°C at the beginning of an experiment. If 800 L are used at 25°C and 1 atm pressure, what is the pressure of the remaining hydrogen? The temperature outside does not change during the measurement.

FIGURE 17–19 Problem 42.

43. (II) The cylinder of a manual bicycle tire pump is 35 cm long; its inside diameter is 2.5 cm. When the piston is pulled up, the valve to the tire is closed and air at atmospheric pressure enters the cylinder through the gap between the piston and the wall of the cylinder, which acts like another valve (Fig. 17–20). When the piston is pushed down, the air in the cylinder is compressed. The valve to the tire opens when the pressure in the tire is reached, and air is transferred to the tire during the rest of the stroke. The last 5 percent of the volume of air remains in the cylinder and the hose. Suppose that you are pumping up a tire that already contains air at 2.5 atm gauge pressure. How much air is transferred to the tire at this pressure? What is the volume of this air at atmospheric pressure? Sketch a p–V diagram of the process. Assume that the temperature is constant.

44. (II) Figure 17–21 represents the variation of the volume and pressure of gas contained in a cylinder with a movable piston. The curved sections of the curve are isotherms. Calculate the

FIGURE 17–20 Problem 43.

temperature, pressure, and volume at the points 1, 2, 3, and 4 given that $p_1 = 1$ atm, $V_1 = 1$ L, $T_1 = 0°C$, $V_2 = 3.5 V_1$, and $p_4 = 3p_1$.

FIGURE 17–21 Problem 44.

45. (II) The specific volume of a gas is its volume per unit mass; that is, the inverse of the density. Compute the specific volume and density of oxygen gas, O_2, at $T = 15°C$ and 1 atm of pressure, assuming that oxygen gas is ideal.

46. (II) An ideal gas is confined in a steel cylinder at 20°C and a pressure of 4.0 atm. (a) If the cylinder is surrounded by boiling water and allowed to come to equilibrium, what will the pressure of the gas be? (b) If the gas is allowed to escape until the pressure again reaches 4.0 atm, what fraction of the original gas, by weight, will escape? (c) If the temperature of the remaining gas now returns to 20°C, what is its pressure?

47. (II) A cylinder is closed at one end by a movable piston. The cylinder contains 300 cm³ of air at 20°C and 10⁵ Pa pressure. The cylinder undergoes the following changes: (a) the gas is heated in such a way that its volume doubles but its pressure remains constant; (b) the volume is kept constant and the temperature is changed until the pressure increases by 30 percent; (c) the gas cools, and the piston position is adjusted to keep the pressure constant until the initial volume is reached. Calculate the volume, pressure, and temperature after each step.

48. (II) The law of Joseph Gay-Lussac and Jacques Charles (1802) states that any ideal gas undergoes the same fractional increase in the product of pressure, p, and volume, V, when the gas is brought from one temperature to another. What is the fractional increase in pV for an ideal gas brought from the temperature of ice melting at 1 atm of pressure to the temperature of water boiling at the same pressure?

49. (II) Calculate the volume of 24 g of the gaseous form of ethyl ether, $C_4H_{10}O$, at a temperature of 120 K and pressure of 0.080 atm. Assume that the gas is ideal.

50. (II) A tank of volume 0.20 m³ contains 250 mol of helium gas at 25°C. Assuming that the helium behaves like an ideal gas, what is the pressure in the tank?

51. (II) It is a scorching summer day with an air temperature of 308 K. What is the value, in joules, of RT for 1 mol of air? Repeat this for a cold winter's night when the air temperature is 258 K. Compare these results to the kinetic energy that a ball with the mass of 1 mol of air (28.9 g) acquires when it falls 1.0 m under the influence of gravity.

52. (II) At the beginning of a compression stroke, the cylinder in a combustion engine contains 1 L of air at atmospheric pressure and a temperature of 20°C (Fig. 17–22). At the end of the stroke, the air has been compressed to a volume of 60 cm³, and the total pressure is 35 atm. What is the temperature of the air when it is so compressed?

FIGURE 17–22 Problem 52.

53. (II) The steam point of water is the point where water vapor (steam) and boiling water coexist at a pressure of 1 atm. When a constant-volume gas thermometer calibrated to read a temperature of 273.16 K at the triple point of water is used to measure the temperature of the steam point, the result depends slightly on the pressure of the gas in the thermometer because the gas is not precisely ideal (Fig. 17–23). In particular, if oxygen is used in the thermometer, the temperature of the steam point is determined to be 373.35 K when the gas pressure is 0.4 atm, and 373.25 K when the pressure is reduced to 0.2 atm. What is the temperature of the steam point?

54. (II) Carbon dioxide obeys the van der Waals equation of state with $a = 5.96 \times 10^6$ atm·cm⁶/mol² and $b = 98.6$ cm³/mol. We know that 0.2 moles of CO_2 occupies 2×10^3 cm³ at 2.5 atm of pressure. (a) Find the temperature of the gas. (b) Redo your calculation, but this time assume that CO_2 is an ideal gas. What is the percentage difference between the two answers?

55. (II) There is a certain temperature T_B (the *Boyle temperature*) for which a van der Waals gas behaves as if it were ideal. (a) Show that an expression for this temperature is $T_B = (a/bR)[1 - (bn/V)]$. (b) The constants a and b of the van der Waals equation of state are $a = 0.140$ m⁶·Pa/mol² and $b = 4.00 \times 10^{-5}$ m³/mol for argon gas. What is the value of the Boyle

FIGURE 17–23 Problem 53.

temperature of argon gas when the gas is so dilute that the term bn/V in the expression for the Boyle temperature can be ignored?

56. (III) An isotherm of 1 mol of a gas that obeys the van der Waals equation of state and is plotted on a p–V curve falls monotonically for large T. As T is decreased and new isotherms are plotted, the curve begins to show a kink, and at a temperature known as the *critical temperature*, T_c, there is a flat spot in the curve, where the slope is zero (Fig. 17–24). At this point, $p = p_c$ and $V = V_c$. Find T_c, p_c, and V_c in terms of the constants a and b.

FIGURE 17–24 Problem 56.

*17–4 The Background Temperature of the Universe

57. (I) According to the Stefan–Boltzmann formula, the energy of radiation emitted by a blackbody per unit area per unit time is proportional to T^4, where T is the temperature in the Kelvin scale. Compare the energy emitted by a tungsten filament at 3200°C and the same filament at room temperature (25°C).

58. (I) The intensity of radiation from a small source decreases with the square of the distance from the source. Consider the energy of radiation that reaches your face after being emitted by solid aluminum at room temperature (25°C). At what distance will this energy of radiation be the same as the energy emitted by the same amount of molten aluminum (800°C) that reaches your face when you are 6 m from the aluminum? (Your answer will not agree with your experience. Can you explain why? A look at Fig. 17–11 should help.)

59. (I) The dominant frequency of radiation emitted by an object is related to its temperature by $hf \simeq kT$. Find the dominant radiation frequency emitted by (a) an object in interstellar space

at 3 K; (b) a body of water at 280 K; (c) an electric stove heating unit (800 K); (d) melting tantalum (3000 K).

60. (I) The human eye is most sensitive to yellow light with a wavelength of approximately 550 nm. What is the temperature of an incandescent bulb filament that radiates most of its energy as a blackbody at this wavelength?

61. (II) The temperature of your skin is about 35°C. Calculate the wavelength at which the Planck radiation curve has a maximum for this temperature, and therefore the wavelength at which your body radiates the most energy. The correct relation between λ_{max} and T is given in Example 17–4.

62. (II) Show that the integral over all frequencies of the Planck formula given by

$$U(T) = \int_0^\infty \left(\frac{8\pi h f^3}{c^3}\right)\frac{1}{e^{hf/kT} - 1}\, df$$

gives a result that is of the form (a constant) $\times T^4$. [*Hint*: Change variables from f to hf/kT.] The energy emitted per unit area per unit time, $P(T)$, is proportional to $U(T)$, and thus $P(T)$ is also proportional to T^4, as in the Stefan–Boltzmann formula, Eq. (17–14).

63. (II) The surface of the Sun is at a temperature of 6000 K. At what rate is energy radiated from the whole surface of the Sun, given that the radius of the Sun is $R = 6.95 \times 10^8$ m?

64. (II) Assume that the radiation emitted from the Sun moves radially outward from the sun, and that no radiation is absorbed between the Sun and Earth. How much energy in the form of radiation will fall per second on an area of 1 m² on Earth, if that area is perpendicular to the straight-line path of the radiation? The distance from the sun to Earth is 1.5×10^{11} m.

65. (II) The average surface temperature of Earth is 290 K. How much energy per second is radiated by Earth's surface, assuming that Earth simulates a blackbody. Compare this result with the amount of radiative energy that reaches Earth from the Sun. Use the result of Problem 64.

66. (III) The energy density per unit frequency in the frequency range from f to $f + df$ in blackbody radiation is $u(f, T)\, df$, where $u(f, T)$ is given by Eq. (17–13). An alternative way to express the blackbody radiation is to give $u'(\lambda, T)\, d\lambda$, the energy density per unit wavelength in the wavelength range from λ to $d\lambda$. (a) Use the fundamental wave relation $c = \lambda f$, where c is the speed of light, to show that $u(f, T)\, df = u[(c/\lambda), T]c\, d\lambda/\lambda^2$, so $u' = u[(c\, \lambda), T]c/\lambda^2$. (b) Assuming that the temperature is fixed, use the results of part (a) to find an equation for the λ_{max} for which $u'(\lambda, T)$ has a maximum. (c) The result of part (b) is a transcendental equation. Solve it for $\lambda_{max}T$. For what range of temperatures does λ_{max} fall within the visible spectrum, $\lambda \simeq 450$ nm to 650 nm?

67. (III) Find the constant coefficient of T^4 in the Stefan–Boltzmann formula, Eq. (17–14), given that the relation between $P(T)$ and the total energy density $U(T)$ calculated in Problem 62 is $P(T) = cU(T)/4$, where c is the speed of light. [*Hint*: An appropriate entry in a table of integrals is

$$\int_0^\infty [x^3/(e^x - 1)]\, dx = \pi^4/15.]$$

General Problems

68. (II) Some astronomers estimate that toward the center of the galaxy, the average interstellar gas is molecular hydrogen, at

a density of roughly 1 molecule per cubic centimeter and a temperature of between 10 and 20 K. Estimate the pressure of this gas. Compare this with atmospheric pressure (1 atm = 760 torr) and very good laboratory vacuums (10^{-10} torr) that are achieved only with some difficulty.

69. (II) Use Archimedes' principle (Section 16–4) to estimate how much helium is needed for a balloon to lift a payload of 230 kg (including the mass of the balloon) in air at 1 atm pressure and 10°C. You will need the molecular weight of helium (4 g/mol) and that of air (28.9 g/mol).

70. (II) A constant-volume gas thermometer is placed in contact with ice water and the pressure in the thermometer is observed to be 0.285×10^5 N/m². If the same thermometer is then placed in a pot of boiling water, the pressure rises to a new value of 0.394×10^5 N/m². The thermometer is calibrated in °C. What temperature will this thermometer read if it has a pressure reading of 0.315×10^5 N/m²?

71. (II) A cylindrical vessel with a tight but movable piston is placed in a vertical position, so that the piston, whose area is 70 cm², is subject to atmospheric pressure. When the gas in the vessel is heated from a temperature of 20°C to 80°C, a 0.5-kg mass must be placed on top of the piston to hold the piston at the position it occupied at the lower temperature (Fig. 17–25). What is the volume of 0.2 mol of the gas?

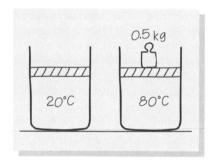

FIGURE 17–25 Problem 71.

72. (II) An air bubble of volume 20 cm³ rises from the bottom of a lake 30 m deep. The temperature at the bottom is 4°C. The bubble rises to the surface, which is at 20°C and at atmospheric pressure. Assuming that the bubble is in thermal equilibrium with the surroundings at all times, calculate the volume of the bubble at the surface.

73. (II) In Chapter 14, we stated that the speed of sound in air is given by $v_{sound} = \sqrt{\gamma p / \rho_0}$, where the constant $\gamma \simeq 1.4$ for air near STP (see Problem 26), p is the pressure, and ρ_0 is the density of the air. Reexpress the speed of sound in terms of the temperature. What is the difference in the speed of sound at the coldest and hottest times of day in a desert?

74. (II) A bottle containing air is closed with a watertight yet smoothly moving piston. The bottle with its air has a total mass of 0.50 kg. At the surface of a body of water whose temperature is a uniform 288 K throughout, the volume of air contained in the bottle is 2.0 L (Fig. 17–26). Recall that the pressure of water increases with depth below the surface, D, as $p_{H_2O} = p_0 + \rho g D$, where p_0 is the surface pressure and $\rho = 1.0$ kg/L. The bottle is submerged. (a) What is the volume of the air in the bottle as a function of depth? (b) Calculate the buoyant force on the bottle as a function of depth. (c) At what depth do the buoyant force and the force of gravity cancel? (d) Is the depth at which the force on the bottle is zero, calculated in part (c), a stable or an unstable equilibrium?

FIGURE 17–26 Problem 74.

75. (II) Suppose that, instead of the Kelvin scale, an alternative temperature scale—call it the Kelvin' scale, with the units K'—were defined in which the zero level remains unaffected but the temperature at the triple point of water is taken to be 500 K'. Calculate the value of Boltzmann's constant in units of J/K'.

76. (II) Consider the radiation emitted from a furnace through a hole of area 1 cm². Assume that all the radiation is absorbed by a beaker that contains 50 g of water. Given that it takes 4.2 J of energy to raise the temperature of 1 g of water by 1°C, how long will it take for the water to be heated by 20°C if the furnace temperature is 2000 K? Assume that there are no energy losses due to the beaker.

77. (III) The van der Waals equation of state, Eq. (17–12), can be put into a form that is an expansion in powers of the density. Dividing by RT and then by $[(V/n) - b]$ in Eq. (17–12), we can write

$$p + \frac{a(n/V)^2}{RT} = \left(\frac{V}{n} - b\right)^{-1} = \frac{n}{V}\left(1 - b\,\frac{n}{V}\right)^{-1}.$$

(a) Show that, for $bn/V < 1$, this equation has the approximate form

$$\frac{p}{RT} = \frac{n}{V} + b\left(\frac{n}{V}\right)^2 - \frac{a}{RT}\left(\frac{n}{V}\right)^2.$$

(b) The result in part (a) is of the general form

$$\frac{p}{RT} = \frac{n}{V} + B_2(T)\left(\frac{n}{V}\right)^2,$$

where, in this case, the function $B_2(T)$ is $B_2(T) = b - (a/RT)$. This is an example of a *virial expansion* for the equation of state, and the function $B_2(T)$ is known as a *virial coefficient*. Show that $B_2(T)$ starts out negative as T increases from some small value, passes through zero, and then becomes positive. The value of T at which $B_2(T)$ is zero is the Boyle temperature, where a real gas satisfying the van der Waals equation behaves as though it were ideal (see Problem 55).

This large plant near Geyserville in northern California uses thermal energy from within Earth to do work and generate electric power. While such a plant does not require a separate fuel source that must be replenished, any electric energy that it has produced has depleted at least that much thermal energy from the underground thermal source.

Heat Flow and the First Law of Thermodynamics

H ot steam contains thermal energy. Steam can lift a piston or turn a turbine and can thereby transfer energy to a mechanical system. Here, we will see how this form of energy is related to the forms of energy with which we are familiar, and how it can be used to do mechanical work. We will see that, as the temperature of a thermal system increases, so does the amount of thermal energy in the system. The energy of thermal systems (thermal energy) and its relation to the energy of the systems' surroundings is summarized in two important laws of physics. The first law of thermodynamics, which we examine here, describes thermal energy as just another form of energy—like kinetic energy or potential energy. The principle of the conservation of energy must now include the thermal energy of a system. The second law of thermodynamics, to be discussed in Chapter 20, states limitations on how much of a system's thermal energy can be converted to other forms of energy.

18-1   CHANGES IN THERMAL SYSTEMS

Thermal systems in equilibrium, such as a bottle of hot steam, are described by only a few thermodynamic variables. For a gas, these variables are the temperature, T, the pressure, p, the volume, V, and the number of moles, n (or the number of molecules, N). Thermodynamic variables describe the *state* of the thermal system. As discussed in Chapter 17, an equation of state relates these variables. For example, the equation

(a)

(b)

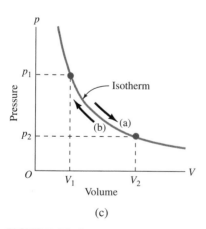

(c)

FIGURE 18–1 The volume of a gas in a cylinder is reversibly changed when a piston is (a) slowly pulled or (b) slowly pushed. The temperature is held fixed during the transformations. (c) Constant-temperature reversible transformation of a gas between V_1 and V_2 on a p–V diagram. The curve is an *isotherm*.

of state for an ideal gas [Eqs. (17–7) and (17–9)] is $pV = nRT = NkT$, where R is the universal gas constant and k is Boltzmann's constant. Let's assume that n is fixed. Because the remaining three variables are connected by an equation of state, it is enough to describe the state on a two-dimensional plot, such as a p–V diagram, a p–T diagram, or a V–T diagram; in fact, this is what we usually do. Changes, or *transformations*, in the state of a thermal system may be represented as curves on such diagrams—at least if thermal equilibrium is established at every stage of the transformation. We have already described and used such curves in Chapter 17 (see Example 17–4).

To maintain thermal equilibrium at each step of a transformation, changes must be made slowly. Just how slowly depends on the internal dynamics that lead to equilibrium. As a piston compresses a gas, the molecules near the piston receive a push, are speeded up (thereby gaining energy), and distribute their additional energy to other molecules through collisions. The time required for the effect of the push to be evenly spread through the gas is characterized by the size of the container divided by the *speed of sound* in the gas. This time is typically short, so it is not difficult to compress a gas while maintaining thermal equilibrium at each stage.

Reversible and Irreversible Processes

A change in which thermal equilibrium is maintained throughout, such as the slow compression just described, is called a **reversible process**. Such changes can always be reversed—the thermodynamic variables can return to their original values—by reversing the external conditions. The volume of a cylinder of gas held at a fixed temperature is increased from V_1 to V_2 in a controlled way when a piston is slowly pulled out of the cylinder, as in Fig. 18–1a. If the piston is slowly pushed back in, as in Fig. 18–1b, the external conditions are reversed, and the thermodynamic variables of the gas recover their initial values. These changes are reversible, and Fig. 18–1c shows the transformation as a curve on a p–V diagram. Because the thermal system remains in equilibrium throughout a reversible process, reversible processes always connect equilibrium states.

Conversely, a gas expanding to fill a vacuum through an open stopcock (Fig. 18–2a) provides an example of an irreversible process, a process in which the thermal system's changes cannot be retraced. The expansion from V_1 to V_2 occurs rapidly and without control. Nothing you do with the stopcock will cause the released gas to reverse itself and reconcentrate within the original volume. We can mark only the original and final points on a p–V diagram (Fig. 18–2b); we cannot mark the points in between, however, because the system is not in thermodynamic equilibrium during the expansion. It is useful to mark a curve with dots, as in Fig. 18–2b, but the dotted curve is meant only to imply that there has been a change, *not* to specify the state of the system during the change. The particular irreversible process described here, in which the gas expands freely, is known as **free expansion**.

Unless we state otherwise, we will consider only reversible changes of state. Thermodynamics is based on our ability to treat reversible processes. There are situations (such as free expansion) in which equilibrium thermodynamics does not apply. The physics of *nonequilibrium thermodynamics* describes processes such as rapid chemical reactions, but that subject is still developing.

EXAMPLE 18–1 A gas in contact with a thermal reservoir at temperature T is expanded reversibly from V_1 to V_2. The trace of points made by this change on a p–V diagram is an isotherm. Describe the isothermal curve when an ideal gas is used.

Solution: The ideal gas law, $pV = nRT$, shows that the product of p times V is constant for a fixed temperature, T. The curve that corresponds to $p \times V = $ a constant or, equivalently, p proportional to V^{-1} is a hyperbola. Figure 18–3 shows a family of hyperbolas. According to the ideal gas law, the different curves correspond to larger or smaller values of T.

We suggested in Section 17–3 that the temperature of a system is closely related to its **internal energy**, or **thermal energy**. We will learn here that the internal energy of a thermal system is a function of the thermodynamic variables; in particular, the higher the temperature of a system, the higher the internal energy. Changes in the internal energy of a thermal system can manifest themselves in only a few ways: *The system can transfer its internal (thermal) energy by changing the temperature (or phase) of another system, or it can use its internal energy to do mechanical work on its surroundings, or both.*

Up to the middle of the nineteenth century, it was thought that when the temperature of a system changes, a type of conserved fluid—known as *caloric*, or more simply, heat—flows into or out of the system. In other words, as much caloric was thought to exit (or enter) the surroundings of the system as entered (or exited) the system itself. Today we know that this is wrong.

When the temperature of a thermal system changes and, in the process, the temperature of a neighboring system changes (Fig. 18–4), we say that there has been a **heat flow** into or out of the system.[†] We will use the term *heat flow* rather than *heat* because heat flow is more like work than like energy: Work is done *on* or *by* a mechanical system, but we do not speak of such a system as containing a certain amount of work. Heat flow into, and/or mechanical work done on, a system may each contribute to raising the temperature of a thermal system. We employ the symbol Q for heat flow.

The unit of heat flow is the calorie (cal). Provisionally, 1 cal is defined to be the amount of heat flow required to raise the temperature of 1 g of water at atmospheric pressure from 14.5°C to 15.5°C. We must wait until later to relate this quantity to the SI units with which we are familiar. Another unit for heat flow used in industrial applications is the *British thermal unit* (Btu), which is defined as the heat flow required to raise the temperature of a 1-lb weight of water by 1°F, averaged over the temperature range 32°F to 212°F. The British thermal unit is related to the calorie by 1 Btu = 252.02 cal.

Heat Capacity

The quantitative connection between heat flow and temperature change is the **heat capacity**, C. An infinitesimal heat flow, dQ, *into* a thermal system will change the temperature by an infinitesimal amount, dT, with a proportionality constant C:

$$dQ = C \, dT. \tag{18–1}$$

In this equation, dQ can be positive or negative. The quantity C is always positive; thus, the sign of the temperature change indicates the sign of the heat flow. Equation (18–1) states that there is a positive heat flow into a system when its temperature increases and there is a negative heat flow into the system when its temperature decreases. The heat capacity is characteristic of the type of material and is proportional to the amount of material; it may also depend on the temperature. A substance with a large heat capacity undergoes a small temperature change for a given heat flow, or requires a large heat flow for a given temperature change. Water has a relatively large heat capacity, which is why large bodies of water have moderating effects on the climate in their vicinity.

We can eliminate the dependence of heat capacity on the amount of material by defining the **specific heat**, c, as the heat capacity of 1 g of the material—the heat capacity per unit mass of the material—and the **molar heat capacity**, c', as the heat ca-

---

[†]We will see later that we can also have a heat flow without a temperature change when the system does work or undergoes a phase change.

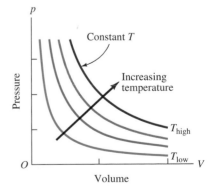

FIGURE 18–2 (a) A gas initially confined to the left-hand side of a container expands to fill the container's total volume when a stopcock is opened. (b) Irreversible expansion of a gas from V_1 to V_2. The path is not well defined, and the dots serve only to indicate that there has been some irreversible change.

A provisional definition of the calorie

The definition of heat capacity

FIGURE 18–3 Example 18–1. An isotherm for an ideal gas is a hyperbola, $p \propto TV^{-1}$. The isotherms corresponding to different temperatures are indicated.

485

FIGURE 18–4 In this steel manufacturing scene, thermal energy is being supplied to steel rods.

PROBLEM SOLVING

Heat flow depends on the path of temperature change.

Heat capacities at constant volume or constant pressure

pacity of 1 mol of the material. If the mass in grams of a thermal system consisting of n mol is m, then

$$C = mc = nc', \qquad (18–2)$$

and if A is the molecular weight (the mass of 1 mol, in grams), then

$$c' = Ac. \qquad (18–3)$$

The units of heat capacity are cal/K; those of specific heat are cal/g·K; and of molar specific heat, cal/mol·K.

Path Dependence of Heat Flow

The heat capacity is defined in Eq. (18–1) in terms of an infinitesimal temperature change. But temperature can change in different ways: The volume or the pressure—or neither—can be held constant during the temperature change. The heat capacity generally depends on how the temperature change is made and on the values of the thermodynamic variables p, V, and T near their respective values where that (small) change is made. This dependence is indicated by writing $C = C(p, T)$ or $C = C(V, T)$. We cannot simply integrate Eq. (18–1) to find a net heat flow unless we know more about how either pressure or volume changes as temperature changes. Generally, the changes will describe a path on a p–V diagram, and heat flow *depends on the path* along which temperature changes. This result recalls the calculation of the work done by a force in moving a mass from a point A in space to a point B. As we learned in Chapter 6, the work done may or may not be path dependent, depending on the force.

In the case of heat flow, there are two cases of special interest: Either the volume or the pressure is kept constant. For these cases, we label the heat capacity with the subscript V or p, and Eq. (18–1) becomes

for constant volume: $dQ = C_V \, dT$;   (18–4)

for constant pressure: $dQ = C_p \, dT$.   (18–5)

(Note that the subscript must be used on c and c' as well. The volume change with temperature for liquids is insignificant compared with that of gases; for solids, pressure is not a useful variable. Therefore, the subscript is not used for these cases.) Generally, C_V differs from C_p. We will see an example when we derive C_V and C_p for an ideal gas.

For an ideal gas, the heat capacities turn out to be independent of temperature. The integration of Eqs. (18–4) and (18–5) to find the total heat flow for finite temperature changes then becomes simple. We illustrate this calculation in Example 18–2.

EXAMPLE 18–2 An ideal gas undergoes transformations that take it from point A to point B on a p–V diagram by the two different paths, α and β, shown in Fig. 18–5. The figure depicts the various thermodynamic variables. Points A and B have the same temperature, $T = T_0$ (they lie on an isotherm). Assuming that C_V and C_p are constants, what is the heat flow into the gas (a) for path α (through point 1) and (b) for path β (through point 2)?

Solution: Because $T_A = T_B = T_0$ (there is no net temperature change of the gas), a tempting answer to this question is that there is no heat flow at all. However, the change $A \rightarrow B$ is not infinitesimal, and so we must integrate along the respective paths to find the correct answer.

(a) Along path α, there are two contributions: from point A to point 1, and from point 1 to point B. Point 1 is at temperature T_1:

$$\Delta Q_\alpha = \int_{T_0}^{T_1} C_V \, dT + \int_{T_1}^{T_0} C_p \, dT.$$

p

Pressure

Isotherm, $T = T_0$

Path β

A

2

1

B

Path α

O

Volume

V

FIGURE 18–5 Example 18–2.

The quantity C_V appears in the first term because the first leg of path α is at constant volume, whereas C_p appears in the second term because the second leg is at constant pressure. C_V and C_p are constants, so they come out of the integrals and $\Delta Q_\alpha = (T_0 - T_1)(C_p - C_V)$.

(b) For path β, we pass in two steps through point 2 at temperature T_2:

$$\Delta Q_\beta = \int_{T_0}^{T_2} C_p \, dT + \int_{T_2}^{T_0} C_V \, dT = C_p \int_{T_0}^{T_2} dT + C_V \int_{T_2}^{T_0} dT$$

$$= C_p(T_2 - T_0) + C_V(T_0 - T_2) = (T_2 - T_0)(C_p - C_V).$$

You can see in Fig. 18–5 that, for an ideal gas, the factor $(T_0 - T_1)$ is not equal to the factor $(T_2 - T_0)$. Thus the two answers are different, and the heat flow depends on the path that is followed.

Phase Changes and Heat Flow

When there is a **change of phase** in a system, there is no temperature change even though there is a heat flow. For example, a solid state may change to a liquid state (fusion or melting), or a gaseous state may change to a liquid state (condensation) (Fig. 18–6). It takes a certain heat flow to convert 1 g of ice at its melting temperature of 0°C to water at 0°C (Fig. 18–7). The necessary heat flow is called the **latent heat of fusion**, L_f. Measurements indicate that $L_f = 79.6$ cal/g for ice. During the melting process, there is no change of temperature and, to a good approximation, there is no change of pressure or volume. The heat flow goes into changing the molecular structure of the substance. It also takes a certain heat flow to convert 1 g of water at its vaporization temperature, 100°C (at 1 atm pressure), to 1 g of steam. The heat flow necessary for this is called the **latent heat of vaporization**, L_v, whose value for water in particular is 540 cal/g. Most of this heat flow goes into overcoming the intermolecular forces that make a liquid so much denser than a gas. Some latent heats are given in Table 18–1.

EXAMPLE 18–3 One hundred grams of ice at 0°C is dropped into 200 g of water at 49°C. The system is thermally isolated (see Chapter 17). After a period of time, the ice is entirely melted, leaving 300 g of water at 6°C. Assume that the specific heat of water has the constant value $c = 1$ cal/g·°C. Calculate the latent heat of fusion of water.

(a)

(b)

(c)

FIGURE 18–6 (a) Mist forms when water droplets condense from water vapor. (b) Ice crystals form on the window when liquid water gives up heat flow to the exterior environment. (c) Gallium melts at the temperature of the human body. When it does so, it absorbs heat flow from its surrounding environment.

FIGURE 18–7 Phase changes for 1 g of water. Note that the temperature remains constant as heat flow is added during melting and vaporization.

TABLE 18–1   Some Latent Heats

Material	Melting Temperature (K)	Latent Heat of Fusion (kJ/mol)
H_2	14	0.12
O_2	54	0.44
Water (H_2O)	273	6.0
Ethyl alcohol (CH_3CH_2OH)	159	4.8
Mercury	234	2.3
Lead	600	4.8
Copper	1357	13

Material	Boiling Temperature[†] (K)	Latent Heat of Vaporization (kJ/mol)
H_2	20	0.92
O_2	90	6.8
Water (H_2O)	373	41
Ethyl alcohol (CH_3CH_2OH)	351	39
Mercury	630	59
Lead	2023	178
Copper	2839	300

[†]Boiling takes place at atmospheric pressure.

Solution: Because this system is thermally isolated, the temperature cannot change as a result of thermal contact with the surroundings.[†] Any heat flow must occur within the system between the water and the melting ice, and the net heat flow to the outside must be zero. We can summarize this condition as

$$\Delta Q_1 + \Delta Q_2 + \Delta Q_3 = 0,$$

where ΔQ_1 is the heat flow needed to melt $m_1 = 100$ g of ice, ΔQ_2 is the heat flow necessary to raise the temperature of 100 g of water (from the melted ice) from $T_i = 0°C$ to $T_f = 6°C$, and ΔQ_3 is the heat flow necessary to cool $m_2 = 200$ g of water initially at $T'_i = 49°C$ to T_f. By definition, L_f is the heat flow necessary to melt 1 g, so if m_1 is measured in grams, we find

$$\Delta Q_1 = m_1 L_f,$$

$$\Delta Q_2 = cm_1(T_f - T_i),$$

$$\Delta Q_3 = cm_2(T_f - T'_i).$$

Thus

$$m_1 L_f + cm_1(T_f - T_i) + cm_2(T_f - T'_i) = 0.$$

Solving for L_f, we find that

$$L_f = \frac{-cm_1(T_f - T_i) - cm_2(T_f - T'_i)}{m_1}$$

$$= \frac{-(1 \text{ cal/g} \cdot °C)(100 \text{ g})(6°C - 0°C) - (1 \text{ cal/g} \cdot °C)(200 \text{ g})(6°C - 49°C)}{100 \text{ g}}$$

$$= 80 \text{ cal/g}.$$

Note that the heat flow needed to change the temperature of the 200 g of water is negative; the water is cooled.

[†]We will see in Section 18–4 that the system must also do no work on its surroundings.

Calorimetry

The measurement of heat capacities is called **calorimetry**. Calorimetry is based on the fact that *when a system is mechanically and thermally isolated, heat flow is conserved within the system* (Fig. 18–8). (A system is said to be *mechanically isolated* if it cannot do mechanical work on its surroundings.) A heat flow into one part of such a system must be matched with a corresponding heat flow out of another part. In such circumstances, heat flow is present only as long as there is a temperature difference between parts of the system; it ceases when equilibrium is established throughout the system. The fact that the pattern of heat flow represents an irreversible process is not important: We are comparing initial and final equilibrium states, not finding infinitesimal changes along paths.

There is no heat flow to or from a mechanically and thermally isolated system.

Our definition of the calorie shows that the specific heat for water at 15°C and at 1 atm of pressure is

$$C_{H_2O} = 1 \text{ cal/(g} \cdot \text{°C)}.$$

The measurement of the specific heats of other substances is based on that of water, as shown in Example 18–4.

> **EXAMPLE 18–4** A block of iron (mass 1.0×10^2 g) is heated in an oven to 5.0×10^2 K and then plunged into a closed, thermally insulated container of 0.50 kg of water at 292 K. The block and the water come to an equilibrium temperature of 297 K. What is the specific heat of iron? Assume that the specific heats of water and iron do not vary significantly over the temperature ranges in question.

> **Solution:** There is a positive heat flow into the water, ΔQ_{H_2O}, because it becomes hotter. This must be matched by a negative heat flow into the iron block, ΔQ_{Fe}, as follows:

$$\Delta Q_{system} = 0 = \Delta Q_{H_2O} + \Delta Q_{Fe}.$$

> The finite heat flow, ΔQ, is found from the infinitesimal heat flow, dQ, by summing, or integrating, over the infinitesimal temperature changes. The heat capacities are assumed to be constant, so this integration is simple:

$$\Delta Q = \int_{T_i}^{T_f} mc \, dT = mc \int_{T_i}^{T_f} dT = mc(T_f - T_i),$$

> where m is the mass of material involved.

> We have

$$0 = \Delta Q_{H_2O} + \Delta Q_{Fe}$$
$$= (0.50 \times 10^3 \text{ g})(1 \text{ cal/g} \cdot \text{K})(297 \text{ K} - 292 \text{ K})$$
$$+ (1.0 \times 10^2 \text{ g})c_{Fe}[297 \text{ K} - (5.0 \times 10^2 \text{ K})]$$
$$= 2.5 \times 10^3 \text{ cal} + (-2.03 \times 10^4 \text{ g} \cdot \text{K})c_{Fe}.$$

> We solve for c_{Fe} to find that

$$c_{Fe} = \frac{2.5 \times 10^3 \text{ cal}}{2.03 \times 10^4 \text{ g} \cdot \text{K}} = 0.12 \text{ cal/g} \cdot \text{K}.$$

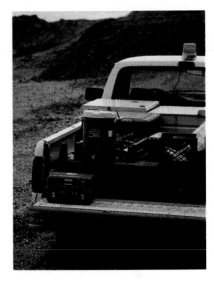

FIGURE 18–8 Thermally isolated systems were described in Chapter 17. This type of container is used to keep water cold.

Example 18–4 shows how the specific heat of iron (or of many materials) can be measured. Calorimetry is used most often in the opposite sense: Given the specific heats, temperature changes are calculated in situations where there is a heat flow from one system to another. Table 18–2 gives the molar heat capacities and specific heats *at constant volume* for a variety of materials at room temperature. In this table, the gases are at standard pressure. In Chapter 19, we will see that the values in Table 18–2 can be explained in terms of the microscopic constituents and structure of the material.

PROBLEM SOLVING

Uses of calorimetry

TABLE 18–2   Molar Heat Capacities and Specific Heats of Some Materials at Constant Volume[†]

Material	Molar Heat Capacity (cal/mol·K)	Specific Heat (cal/g·K)
Helium gas	3.00	0.75
Argon gas	30.00	0.75
N_2 gas	4.94	0.176
O_2 gas	4.97	0.155
SO_2 gas	7.52	0.118
Diamond	1.46	0.12
Beryllium	3.85	0.43
Silicon	4.78	0.17
Aluminum	5.82	0.22
Copper	5.85	0.092
Lead	6.32	0.031
Water	18.0	1.00
Mercury	6.7	0.033

[†]Gases at standard temperature and pressure.

C_V and C_p for ideal gases can be calculated when the first law of thermodynamics is known. A general prediction (to be shown in Section 18–6) is that the heat capacity for constant pressure is larger than the heat capacity for constant volume:

$$C_p \geq C_V. \tag{18–6}$$

This inequality holds even for nonideal gases.

18–3   THE MECHANICAL EQUIVALENT OF HEAT

Placing a thermal system in thermal contact with a second thermal system at a different temperature is one way to produce a temperature change in the first system. There is a heat flow in this case. For example, we can heat water by placing it in a pot on a hot stove. A major discovery of the first half of the nineteenth century was the observation that the temperature of a thermal system can also be raised by performing mechanical work on that system. For example, when an object is rubbed so that work is done on it by friction, its temperature is raised (Fig. 18–9).

These observations and the understanding of their significance took place through the slow buildup of rather complicated data rather than through a single brilliant experiment. The history is thus rather tangled, with many participants. We cite in particular the work of Count Rumford (Benjamin Thompson) and of James Joule. While supervising the boring of cannon barrels in Bavaria in 1798, Rumford observed and quantified the continual heating of the metal (Fig. 18–10). He reasoned that if the heating were due to the transfer of a fluid—as proposed by the caloric theory—then the fluid eventually would be used up and no more heating should occur. In the mid-nineteenth century, Joule's careful experiments established that a given amount of mechanical work done on a thermal system produces a temperature rise that corresponds to a given heat flow, as defined by Eq. (18–1).

Joule's Experiments

An idealized version of Joule's experiments is described here. In these experiments, the temperature of a thermally isolated quantity of water is measurably raised by a set of paddles that churn the water (Fig. 18–11). The water resists the motion of the turning paddles, and work must be performed to keep them rotating. The amount of work

FIGURE 18–9 The presence of friction allows the mechanical energy of rubbed sticks to be converted into heat energy. Fire is the result.

done can be measured precisely by attaching the paddles to a falling mass (which keeps the paddles moving), and then measuring the distance the mass has fallen.

When the paddles have done a certain amount of work on the water, the temperature of the system (water plus paddles) will have risen by a definite amount. Once the temperature has been raised, *there is no way to tell* whether the temperature increase occurred because work had been done on the system or because of heat flow into it. A series of experiments of this type establishes that there is a specific temperature rise for a specific amount of work; therefore, there is a precise equivalent between work done on a system, W, and the heat flow into it, ΔQ. Mathematically,

$$\text{for a thermally isolated system:} \quad \Delta Q = W. \quad (18\text{--}7)$$

Because *no* experiment can be performed that can tell whether a given temperature change was produced by mechanical work or by heat flow, we can assert that *heat flow, like work, is an energy transfer*. This result is known as the **mechanical equivalent of heat**. Equation (18–7) can be translated into a relation between the SI unit of work, the joule, and the unit of heat flow, the calorie:

$$1 \text{ cal} = 4.185 \text{ J}. \quad (18\text{--}8)$$

This relation may seem to be a simple equivalence of units; however, it is far more. Until now, the calorie was unrelated to any other units familiar to us. Now we can state that the calorie has the dimensions of energy, and that the SI unit of heat flow is the joule. Equation (18–8) thus states the crucial result of a basic experiment. It should not be confused with, say, the statement that 1 Btu is 252.02 cal, which is merely the relation between different units.

Today, the *numerical definition* of the calorie is taken to be Eq. (18–8); it is no longer based on a measurement of the temperature change of water, as it was in Section 18–2. Such a definition is possible only because of the physical equivalence between heat flow and work. The calorie continues to be a widely used unit.

EXAMPLE 18–5 The paddles of the apparatus shown in Fig. 18–11 are driven by a mass m of 0.50 kg falling at a constant speed. The mass of the thermally isolated water in the apparatus is 250 g. How much is the temperature of the water raised if the driving mass falls a distance of 2.0 m? Ignore the mass of the paddles.

Solution: According to the work–energy theorem, the work, W, done on the water is the potential energy change of mass m when it falls a distance Δy. Because this potential energy change is given by $mg\,\Delta y$, the work done is

$$W = mg\,\Delta y.$$

According to Eq. (18–7), this work is equivalent to a heat flow, $\Delta Q = W$. The specific heat of water has the constant value $c_{H_2O} = 1$ cal/g·°C, so we get $m_{H_2O}c_{H_2O}\,\Delta T = mg\,\Delta y$, or

$$\Delta T = \frac{mg\,\Delta y}{m_{H_2O}c_{H_2O}} = \frac{(0.50 \text{ kg})(9.8 \text{ m/s}^2)(2.0 \text{ m})}{(250 \text{ g})(1 \text{ cal/g}\cdot°C)}\left(4.185\,\text{cal/J}\right)$$

$$= 0.0094°C.$$

A Microscopic View

Thermal (or internal) energy is associated with the chaotic motion of the molecular constituents of matter. This motion is more and more violent as thermal energy increases, and temperature increases with thermal energy. Rapidly moving molecules transfer their energy to more slowly moving molecules by collision, so it is not too surprising that heat flow is an energy transfer. By the same reasoning, if thermal energy is the energy of the constituent molecules of a thermal system, then this energy

FIGURE 18–10 Count Rumford observed that the temperature of cannon barrels rose steadily as they were being hollowed out. He concluded that the caloric theory could not be correct.

The mechanical equivalent of heat

The definition of the calorie

FIGURE 18–11 Schematic diagram of the apparatus used in Joule's classic experiments. Churning paddles raise the temperature of a thermally isolated container of water.

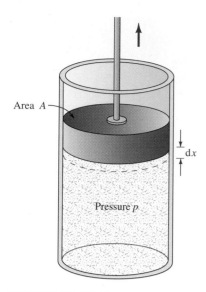

FIGURE 18–12 The volume of a cylinder increases when a piston of area A moves out a distance dx under the influence of the pressure, p, of the gas contained in the cylinder.

The work done by a gas when its volume changes

The work done by a changing thermal system depends on how the change is made.

PROBLEM SOLVING

Work is an area beneath a curve of p versus V.

is a function of the state of the gas; that is, of its thermodynamic variables, p, V, and T. This point of view will be described quantitatively in Chapter 19. From the thermodynamic (macroscopic) point of view, however, the observation that heat flow is a kind of energy transfer—like work—is not at all obvious. The principle of conservation of energy is one of the key principles of physics, and the discovery of the mechanical equivalent of heat was one of the truly important discoveries of the nineteenth century.

18–4 WORK DONE BY THERMAL SYSTEMS

In Section 18–3, we saw that work can be done on a thermal system. By Newton's third law, this means that the thermal system can perform work on its surroundings—negative work, in the case of the churning paddles of Joule's experiments. Thermal systems, particularly gases, which can expand greatly, can also do *positive* work on their surroundings.

Consider a piston of area A pushed out by the pressure p of the gas contained in a cylinder (Fig. 18–12). By the definition of pressure, the magnitude of the force F on the piston is pA. Moreover, if the piston moves out, this force acts in the direction of the displacement, which we take to have infinitesimal magnitude dx. The infinitesimal work done by the gas in moving the piston is

$$dW = \mathbf{F} \cdot d\mathbf{x} = pA\, dx. \qquad (18\text{–}9)$$

Because $A\, dx = dV$ (the infinitesimal change in the volume of the gas), this expression can be rewritten as

$$dW = p\, dV. \qquad (18\text{–}10)$$

Equation (18–9) is specific to the linear motion of a piston with the geometry of Fig. 18–12, but Eq. (18–10) is a more general form that holds when a gas changes its volume by dV in pushing against its surroundings in any direction. Note the sign carefully: dW, the work the gas does *on* its surroundings, will be positive or negative according to whether dV is positive or negative.

The pressure, p, in Eq. (18–10) is a function of V and T. As for the heat flow, we must specify the way in which the finite volume change is made in the integration of Eq. (18–10). Thus, *the net work done by the gas depends on the path in a p–V, p–T, or V–T diagram.* The integral expressing the work done in going from state 1 to state 2 is

$$W = \int_{V_1}^{V_2} p(V, T)\, dV. \qquad (18\text{–}11)$$

This integral is defined only when the path $1 \rightarrow 2$ is specified.

If the path of the transformation $1 \rightarrow 2$ is drawn on a p–V diagram, as in Fig. 18–13a, then the work done by the gas has a simple interpretation: *It is the area under the curve of p versus V*, which is shaded in the figure. The sign of the work done by the gas is determined by whether V_2 is larger than V_1, as in Fig. 18–13a, or smaller than V_1, as in Fig. 18–13b. In Fig. 18–13b, the net work done *by* the gas is negative; in other words, there is positive work done *on* the gas to compress it. All this follows the normal rules of calculus, where $\int_{x_1}^{x_2} F(x)\, dx$ is the area under a curve of $F(x)$ versus x.

If we think of the work done by a thermal system as the area under a curve of p versus V, the fact that the work done depends on the path of the change becomes evident. The path describes the curve between two states, 1 and 2, and the area under different curves may certainly differ.

Cyclic Transformations

The operation of the cylinders in an automobile engine is repetitive. This is an example of a **cyclic transformation**: a change in which a thermal system follows a path that takes it back to its original state. The importance of such processes rests on the fact that all engines work in cycles. The net work done by the thermal system in a cyclic transformation can be calculated by starting anywhere along the path (Fig. 18–14a). Let's divide the cycle of Fig. 18–14a into two parts, as in Fig. 18–14b and c. The first part is the change $B \rightarrow A$, along path 1 (Fig. 18–14b), and the second part is $A \rightarrow B$, along path 2 (Fig. 18–14c). The net work W_{cycle} adds algebraically (Fig. 18–14d) and is thus given by

$$W_{cycle} = \underset{\text{path 1}}{W_{B \rightarrow A}} + \underset{\text{path 2}}{W_{A \rightarrow B}}.$$

As we saw in Chapter 6 during our discussion of work, $\int_{x_1}^{x_2} f(x)\,dx = -\int_{x_2}^{x_1} f(x)\,dx$. Thus the integral over path 2, which goes from larger to smaller values of x, is the negative of the area under that path. The net work is then the area under path 1 minus the area under path 2:

$$W_{cycle} = \text{area enclosed by cyclic path.} \qquad (18\text{--}12)$$

Because the cycle is traced in a clockwise direction, the work done by the gas along path 1 is larger than the work done on the gas along path 2. The net work done on the surroundings is positive; if this cycle were traced in a counterclockwise direction, the work done would be negative.

> **EXAMPLE 18–6** A thermal system is taken around the cycle shown in Fig. 18–15. Along the paths $A \rightarrow B$ and $C \rightarrow D$, the volume is held fixed at V_1 and V_2, respectively. From $B \rightarrow C$ and from $D \rightarrow A$, the pressure is held fixed at p_2 and p_1, respectively. Calculate the work done by the thermal system in tracing out one cycle.
>
> **Solution:** This problem can be solved by two methods: by performing the integration over each leg explicitly; or by evaluating the area enclosed by the cycle.
>   Using the first method,
>
> $$W_{cycle} = \int_{V_A}^{V_B} p\,dV + \int_{V_B}^{V_C} p\,dV + \int_{V_C}^{V_D} p\,dV + \int_{V_D}^{V_A} p\,dV. \qquad (18\text{--}13)$$
>
> The first term on the right-hand side of Eq. (18–13) runs from $V_A = V_1$ to $V_B = V_1$. The volume is fixed at each step; that is, $dV = 0$ all along path $A \rightarrow B$, and the integral drops out. The same is true for the third term, over path $C \rightarrow D$. To evaluate the second term of Eq. (18–13), over path $B \rightarrow C$, we note that the pressure takes the constant value p_2 and thus comes out of the integral:
>
> $$W_{B \rightarrow C} = p_2 \int_{V_B}^{V_C} dV = p_2 \int_{V_1}^{V_2} dV = p_2(V_2 - V_1).$$
>
> For the fourth term of Eq. (18–13), over path $D \rightarrow A$, p has the constant value p_1, and
>
> $$W_{D \rightarrow A} = p_2 \int_{V_D}^{V_A} dV = p_1 \int_{V_2}^{V_1} dV = p_1(V_1 - V_2).$$
>
> Collecting these results,
>
> $$W_{cycle} = p_2(V_2 - V_1) + p_1(V_1 - V_2) = p_2(V_2 - V_1) - p_1(V_2 - V_1)$$
> $$= (p_2 - p_1)(V_2 - V_1).$$
>
> By the second method, the area enclosed in a rectangle is the width $(V_2 - V_1)$ times the height $(p_2 - p_1)$:
>
> $$W_{cycle} = (p_2 - p_1)(V_2 - V_1).$$
>
> Because the cycle is traced out in a clockwise manner, W_{cycle} is positive.
>   As expected, the two methods give the same result.

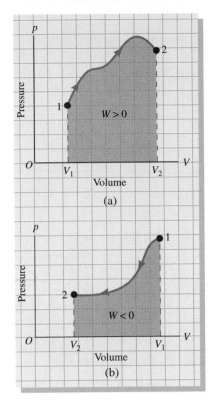

FIGURE 18–13 The work done by a gas on its surroundings (like that of the piston–cylinder system of Fig. 18–12) in moving reversibly from point 1 to point 2 is the area under the curve of p versus V. (a) If the volume V_2 is greater than V_1 (the gas expands), the work the gas does on its surroundings is positive. (b) If the volume V_1 is greater than V_2, the gas is compressed, and the work it does on its surroundings is negative.

The work done by a gas during a cyclic transformation is the area enclosed by the curve that describes the changes on a p–V diagram.

FIGURE 18–14 In a closed cycle, the work done by a gas on its surroundings is the area enclosed by the curve of p versus V. To show this, the full cycle is broken in two paths.

Work Done in Various Transformations

Constant Pressure. An isobaric (constant-pressure) transformation is drawn on a p–V diagram in Fig. 18–16a. For such transformations, the work done by the gas is just the volume change multiplied by the constant value of the pressure. How can an isobaric transformation be arranged? Generally, if the volume of a gas is to increase without a change in pressure, the temperature must increase. Let's calculate this temperature change for an ideal gas. Suppose that the initial temperature, pressure, and volume are T, p, and V, and that V is increased by dV while p is held constant. It follows from the ideal gas law, $T = pV/nR$, that

$$\text{for constant } p: \qquad dT = \frac{p\,dV}{nR}. \qquad (18\text{–}14)$$

Substituting T/V for p/nR, Eq. (18–14) becomes

$$\text{for constant } p: \qquad dT = \frac{T}{V}\,dV. \qquad (18\text{–}15)$$

If the volume change dV is accompanied by a heat flow into the gas, giving precisely the temperature change in Eq. (18–15), then the pressure will be maintained at a constant value. The heat flow in this case must be $dQ = C_p\,dT$.

Constant Volume. When the volume is kept fixed, the curve of the transformation is said to be an **isochore** (Fig. 18–16b). Thermal systems do no work—positive or negative—in an isochoric transformation because there is zero area under the curve on a p–V diagram. The pressure and temperature both change, however, and so such a transformation must be accompanied by a heat flow, just as for the isobaric and the isothermal transformations. In this case, the heat flow is $dQ = C_V\,dT$.

Adiabatic. Finally, there is a particular curve that will be followed if the gas does work while it is thermally isolated from its environment. Imagine a cylinder of gas enclosed in a thermal insulator, such as Styrofoam, with only the mechanical movements of a piston connecting it to its surroundings. Because of the thermal isolation, there is no heat flow dQ into or out of the system. When $dQ = 0$, the temperature can change only if work is done on the system. Reversible transformations of a thermal system in which there is no heat flow to the system are called **adiabatic** transformations. In Section 18–7, we will explore the precise curve followed by adiabatic transformations for ideal gases. These transformations will be important when we study engines in more detail.

Internal Energy

In an adiabatic transformation, a thermal system does work on its surroundings but has no thermal contact with those surroundings. Simple conservation of energy, as formulated in Chapters 6 and 7, can be applied to it. We can think of the system as having a thermal (or internal) energy that can change if work is done on or by the system. We call this thermal energy U.[†] In particular, when a system does work on its surroundings in an adiabatic transformation, *it changes its internal (thermal) energy, U, by the amount of work it does:*

$$\text{for an infinitesimal adiabatic transformation:} \qquad dU = -dW; \qquad (18\text{–}16)$$

$$\text{for a finite adiabatic transformation:} \qquad U_B - U_A = -W_{A\to B}. \qquad (18\text{–}17)$$

[†]Don't confuse U with a potential energy!

In an adiabatic transformation, a thermal system such as the insulated cylinder behaves as an ordinary mechanical system; any changes in its internal energy are matched by mechanical work done. This means that if we want to study how the internal energy changes when we go from one thermodynamic state to another, we should use adiabatic transformations because then we can avoid the complications of heat flow. *The adiabatic process allows us to learn just how the internal energy depends on the thermodynamic variables.*

The thermal (internal) energy is a function of p and V, p and T, or T and V. The thermal energy is associated with the energy of the molecules of a thermal system; it is simply a function of the thermodynamic variables of the system, not of how those variables might have changed. Changes in internal energy, like changes in the potential energy of a conservative mechanical system, are *not* path dependent. This contrasts sharply with work or heat flow, which *are* path dependent.

In Eq. (18–16), both dU and dW are written as differentials. The symbol dU has its usual meaning as a differential, in that a finite change in internal energy is

$$\int_A^B dU = U_B - U_A, \tag{18–18}$$

which is independent of the path between A and B. There is no equivalent expression for work because the integral of dW *does* depend on the path taken. In a strict mathematical sense, dW is not a true differential and is sometimes denoted by đW. There is no need for the more complicated notation, however, provided we keep in mind that the integral of dW is path dependent.

The minus signs on the right-hand sides of Eqs. (18–16) and (18–17) occur because W always denotes the work done *by* the system *on* its surroundings. When the system does positive work in an adiabatic transformation, its internal energy U decreases; when the work done is negative, the internal energy increases.

Let's now reconsider the Joule experiments that determined the mechanical equivalent of heat (Section 18–3). The water whose temperature is raised by the churning paddles is thermally isolated. The transformation is irreversible; it is not possible to "unchurn" the water and use it to raise the fallen weight. Nevertheless, the (negative) net work done by the water does indeed determine the (positive) change in the internal energy of the water. *How* a change is made is irrelevant for the internal energy. It is purely a function of state. Thus Eq. (18–17) holds for both reversible and irreversible transformations of thermally isolated systems.

18–5 THE FIRST LAW OF THERMODYNAMICS

The **first law of thermodynamics** is a statement of the conservation of energy for thermal systems. It is essential in understanding any thermal system—particularly those in which fuel is burned and work is done, such as engines or power plants (Fig. 18–17). Because heat flow and work are equivalent in terms of internal energy, the change in the internal energy of a thermal system in going from state A to state B is the negative of the work that the system does on its surroundings, plus the heat flow into the system:

$$\Delta U = U_B - U_A = -W_{A\to B} + Q_{A\to B}. \tag{18–19}$$

Equation (18–19) is the primary result of this chapter. The minus sign in front of W indicates that W is the work done *by* the thermal system, and the plus sign in front of Q indicates that Q is the heat flow *into* the system. In Fig. 18–17, we give another view of the meaning of these terms.

Because U is a function of state only, ΔU is the same whether the transformation $A \to B$ is reversible or irreversible. Similarly, the work done by the system can be mea-

Adiabatic transformations

The thermal energy is a function of state, and any changes in it are independent of how the changes are made.

FIGURE 18–15 Example 18–6.

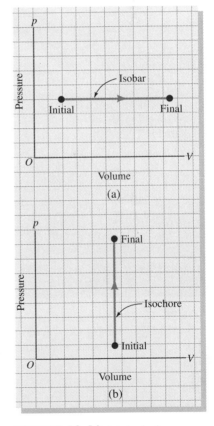

FIGURE 18–16 (a) An isobar, which represents a constant-pressure transformation. (b) An isochore, which represents a constant-volume transformation.

The first law of thermodynamics

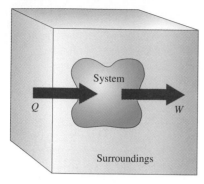

FIGURE 18–17 In the first law of thermodynamics, W represents the work done by the system, while Q represents the heat flow into the system. Either term can be negative or positive, depending on the transformations involved.

In closed cycles, work done equals heat flow.

FIGURE 18–18 This engine uses thermal energy—burning fuel—to do work in cycles and produce electrical energy.

sured by observing its effect on its surroundings—for example, how high the system lifts a mass—independently of whether the transformation $A \to B$ is reversible or irreversible. The heat flow $Q_{A \to B}$ can be determined from the first law of thermodynamics if the transformation is irreversible or determined independently by making a calorimetric measurement.

The differential form of the first law of thermodynamics is

$$dU = -dW + dQ. \qquad (18\text{--}20)$$

While both dW and dQ depend on the path, dU is independent of path. Remarkably, the path dependence of the *difference* $-dW + dQ$ cancels.

The First Law of Thermodynamics in Closed Cycles

Engines operate and perform work in closed cycles. Thus it is of practical importance to be able to apply the first law of thermodynamics to a closed cycle (Fig. 18–18). There is no change in internal energy over a cycle because the system comes back to its original state. The work done by the thermal system over the cycle must then match the heat flow into the system, by the first law:

$$Q_{\text{cycle}} = W_{\text{cycle}}. \qquad (18\text{--}21)$$

EXAMPLE 18–7   Suppose that a thermal system follows the cycle $A \to B \to C \to D \to A$, as drawn in Fig. 18–15 and treated in Example 18–6. What is the total heat flow into the system after one cycle is completed?

Solution: We can use the fact that the internal energy depends only on the state to recognize that the internal energy change in the cycle is zero because the initial point is the same as the final point. The fact that the system passes through a complicated succession of intermediate states is irrelevant. Then Eq. (18–21) tells us immediately that

$$Q_{\text{cycle}} = W_{\text{cycle}}.$$

The heat flow into the system is precisely the work done by the system, so, according to Example 18–6,

$$Q_{\text{cycle}} = (p_2 - p_1)(V_2 - V_1).$$

Note again that the work is just the area enclosed by the cycle.

How the Energy Changes in Constant-Volume and Constant-Pressure Transformations

No work is done in reversible changes under constant volume, and Eq. (18–20) becomes

$$\text{for constant volume:} \quad dU = dQ_V = C_V\, dT. \qquad (18\text{--}22)$$

We have emphasized the fact that the volume is constant by the subscript on dQ. No subscript is necessary for dU because the change in U is independent of the path taken. Similarly, we can express the first law of thermodynamics for reversible changes under constant pressure as

$$\text{for constant pressure:} \quad dU = -dW_p + dQ_p = -p\, dV + C_p\, dT. \qquad (18\text{--}23)$$

Equations (18–22) and (18–23) can both be integrated over paths with constant V and constant p, respectively, to find the net internal energy change under a finite volume or pressure change. Keep in mind that in such an integration the heat capacities may not necessarily be constants that can be removed from under the integral sign.

Joule performed an experiment in 1845 to study the properties of U for ideal gases. He demonstrated that, *for ideal gases, the internal energy is a function of temperature alone*: $U = U(T)$. Equivalently, the equation of state implies that we can say that U is a function only of the product pV.

Joule's experiment is as follows: Fill one part of a two-part container with a dilute sample of gas (if it is sufficiently dilute, we can be sure it is ideal). The other part of the container is evacuated, and the two parts are connected by a tube with a stopcock. The container as a whole is sealed and does no work on the outside world. The container is submerged in a thermally isolated container of water or some other fluid whose temperature can be monitored. This fluid, initially at temperature T_1, acts as a calorimeter: If its temperature changes because of some change in the state of the gas in the container, then there has been a heat flow into or out of the gas. This setup is shown schematically in Fig. 18–19a.

The stopcock is now opened and the gas undergoes free expansion into the previously empty part of the container. The crucial feature of free expansion is that the gas does no work because it does not cause mechanical movement in its surroundings. The new temperature of the fluid, T_2, is then measured. When the gas is sufficiently dilute, the experiment shows that $T_2 = T_1$ (Fig. 18–19b). *The temperature of an ideal gas undergoing free expansion remains constant.* This result means that there has been no heat flow into the gas. There is no work done and no heat flow in free expansion. Thus, according to the first law of thermodynamics, the internal energy, U, remains unchanged. But because the volume of the gas does change, U must be independent of volume. Therefore, for an ideal gas, U is a function only of temperature:

> The internal energy of an ideal gas depends only on the temperature.

$$U_{\text{ideal}} = U(T). \qquad (18\text{--}24)$$

This result follows directly from experiment, but it can also be derived from the second law of thermodynamics, which we will explore in Chapter 20.

For real gases—which are not perfectly ideal—there is a small deviation from the result of Eq. (18–24), and temperature changes slightly during free expansion. In other words, U is a function of density as well as temperature. We typically ignore this effect.

Let's now couple the fact that the internal energy of an ideal gas depends only on the temperature with Eqs. (18–22) and (18–23). Because U depends *only* on T, the integral of Eq. (18–22) gives

$$U(T_B) - U(T_A) = \int_{T_A}^{T_B} C_V \, dT. \qquad (18\text{--}25)$$

This equation must hold for *all* values of the volume because the left-hand side is independent of volume. Hence Joule's experiment shows that C_V must itself be independent of volume. Equation (18–25) expresses only a *change* in energy, so it is equivalent to the form

$$U(T) = \int_0^T C_V \, dT + \text{a constant}. \qquad (18\text{--}26)$$

FIGURE 18–19 Schematic diagram of Joule's experiment. When Joule opened the stopcock to allow free expansion, he observed no temperature change for a dilute (ideal) gas.

The internal energy of an ideal gas as a function of temperature

The constant is arbitrary because only *changes* in U matter, not the actual values of $U(T)$ and $U(0)$. We therefore take the constant to be zero. Finally, we can use the experimental fact that, over a wide temperature range, C_V is independent of temperature. We can remove C_V from behind the integral sign and find

$$\text{for } C_V \text{ independent of temperature: } U(T) = C_V T. \qquad (18\text{--}27)$$

The internal energy of an ideal gas is simple indeed, as long as C_V is constant with temperature.

EXAMPLE 18–8 An ideal gas in contact with a thermal reservoir at temperature $T = T_0$ does work by expanding against a piston. That work is 10 J. How much heat flow goes into the gas during its expansion?

Solution: Application of the first law of thermodynamics gives us the heat flow, Q, in terms of the work done, W, and the change in internal energy, ΔU:

$$Q = \Delta U + W.$$

Because the gas is ideal, U is a function only of temperature; because the transformation is isothermal, the change in U is zero. Thus

$$Q = W = 10 \text{ J}.$$

This heat flow comes from the thermal reservoir.

A Relation between C_p and C_V for Ideal Gases

We can express C_p in terms of C_V. Equations (18–22) and (18–23) are two differential expressions for the energy change, dU. Because Eq. (18–23) specifies constant pressure, the term $p\, dV$ that occurs on the right-hand side can also be written for an ideal gas as $p\, dV = nR\, dT$. Then both Eqs. (18–22) and (18–23) involve only terms proportional to dT on the right-hand side. Because the energy shift dU is independent of whether the infinitesimal transformation is made at constant V or at constant p, the two expressions for dU must be equal:

$$C_V\, dT = -nR\, dT + C_p\, dT. \qquad (18\text{--}28)$$

The relation between the heat capacities at constant pressure and constant volume for an ideal gas

We cancel the common factor dT from this equation to find that

$$C_p = C_V + nR, \qquad (18\text{--}29)$$

which is a result that applies to ideal gases. Because C_V is a function of T only, C_p is also; to the extent that C_V is in fact constant with T, C_p is as well. The prediction of Eq. (18–29) is indeed satisfied by experiment for ideal gases and is well approximated for real gases. Although relations such as this are useful predictions typical of thermodynamics, it is necessary to use the microscopic dynamics of gases to compute quantities such as C_V. This calculation will be made in Chapter 19.

18–7 MORE APPLICATIONS FOR IDEAL GASES

In order to understand the behavior of engines or, more generally, of gaseous thermal systems, we want to express the changes that occur in terms of the thermodynamic variables. In principle, this is possible as long as the transformations are reversible. Better, if we can approximate real gases by ideal gases, then all the transformations we have discussed can be expressed mathematically, as can the corresponding work done, heat flow, energy changes, and related quantities. Some of these mathematical expressions have already been given. For example, the shape of the p–V curve that represents an isothermal transformation at $T = T_0$ is a hyperbola, $p = nRT_0/V$. The relation $C_p = C_V + nR$ [Eq. (18–29)] is a consequence of the fact that the gas is ideal.

Of the four particular reversible transformations discussed earlier—constant-volume (isochoric), constant-pressure (isobaric), constant-temperature (isothermal), and adiabatic transformations—the first two are simple to treat. Let's now look at the other two in more detail.

Isothermal Transformations of an Ideal Gas

When the temperature is constant, the work done by an ideal gas is simple to compute. We see from Fig. 18–20 that

$$W = \int_{V_1}^{V_2} p \, dV = \int_{V_1}^{V_2} \frac{nRT_0}{V} \, dV = nRT_0 \int_{V_1}^{V_2} \frac{dV}{V}. \tag{18–30}$$

The integral of $1/x$ is the natural logarithm, $\ln x$ (see Appendix IV–6). Because $\ln x_1 - \ln x_2 = \ln(x_1/x_2)$, we have, for the work done by an ideal gas in an isothermal transformation:

$$W = nRT_0 \ln\left(\frac{V_2}{V_1}\right). \tag{18–31}$$

As expected, W is positive when the gas expands ($V_2 > V_1$) and negative when the gas contracts ($V_2 < V_1$).

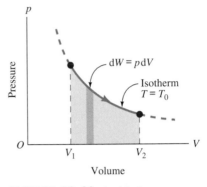

FIGURE 18–20 An ideal gas expands from V_1 to V_2 while its temperature is held fixed, and in doing so does work on its surroundings.

EXAMPLE 18–9 Suppose that 0.12 mol of an ideal gas in contact with a thermal reservoir that holds the temperature fixed at $T_0 = 9.8°C$ has an initial volume of 1.3 L and does 14 J of work. What is the final volume, and what is the final pressure?

Solution: Equation (18–31)—the work done in an isothermal transformation—can be solved for the final volume V_f because every other quantity in the equation is known. T_0 must be converted to the Kelvin scale: $T_0 = 9.8 + 273.15 \approx 283$ K. Note that nRT_0 has the dimensions of pV and hence of work. We have

$$\ln\left(\frac{V_f}{V_i}\right) = \frac{W}{nRT_0} = \frac{14 \text{ J}}{(0.12 \text{ mol})(8.314 \text{ J/mol}\cdot\text{K})(283 \text{ K})} = 0.050.$$

Thus

$$\frac{V_f}{V_i} = e^{0.050} = 1.051,$$

and the final volume is

$$V_f = V_i(1.051) = (1.3 \text{ L})(1.051) = 1.4 \text{ L}.$$

The final pressure is given by the equation of state:

$$p_f = \frac{nRT_0}{V_f} = \frac{(0.12 \text{ mol})(8.314 \text{ J/mol}\cdot\text{K})(283 \text{ K})}{(1.4 \text{ L})(10^{-3} \text{ m}^3/\text{L})} = 2.1 \times 10^5 \text{ Pa} = 2.0 \text{ atm}.$$

The pressure has decreased from an initial value of $p_i = p_f(V_f/V_i) = 2.2$ atm.

The internal energy of an ideal gas is unchanged in an isothermal transformation. Thus, by the first law of thermodynamics, Eq. (18–31) also expresses the heat flow into the gas during an isothermal transformation.

The curve describing an adiabatic
transformation

Adiabatic Transformations of an Ideal Gas

In an adiabatic expansion, the first law of thermodynamics becomes $dU = -dW$. Because dW is positive, dU is negative. U is proportional to T, so the temperature will fall in an adiabatic expansion and rise in an adiabatic compression. When a gas expands in a thermally isolated container, it cools; when it is compressed, it heats up (as in a compressed bicycle pump).

In a separate subsection that follows, we find that the curves that describe this transformation are given by

$$pV^\gamma = p_0 V_0^\gamma = \text{a constant}, \tag{18-32}$$

where (p_0, V_0) are the initial values of pressure and volume, respectively, of the ideal gas that undergoes an adiabatic transformation to new values (p, V). The constant γ is defined to be

$$\gamma \equiv \frac{C_p}{C_V} = \frac{C_V + nR}{C_V} = \frac{c'_V + R}{c'_V}, \tag{18-33}$$

where c' is the molar heat capacity, from Eq. (18–2). We have used Eq. (18–29), which relates C_p to C_V. Because $C_p > C_V$, γ is greater than unity (a result that is generally true). In terms of V and T, the transformation follows the curves

$$TV^{\gamma-1} = \text{a constant}. \tag{18-34}$$

Compare Equation (18–32) with the curve that describes an isothermal transformation, $pV = \text{a constant}$, or $p \propto V^{-1}$. The latter curve is a hyperbola, symmetric in p and V. The adiabatic curve $p \propto V^{-1}$ falls more steeply as V increases, because $\gamma > 1$. Figure 18–21 shows the two families of curves for a monatomic ideal gas, for which $\gamma = 1.67$. The constant in Eq. (18–32), which distinguishes the two adiabatic curves in Fig. 18–21, is determined when one point on the curve is known. If we know that the pressure equals p_0 and the volume equals V_0 at some point on an adiabatic curve, then the constant equals $p_0 V_0^\gamma$.

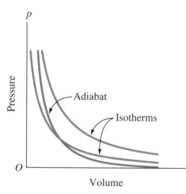

FIGURE 18-21 For ideal gases, adiabatic curves ($p \propto V^{-\gamma}$), called adiabats, are steeper than isotherms ($p \propto V^{-1}$) on a p–V diagram.

*How to Describe the Adiabatic Transformation of an Ideal Gas on a p–V Diagram

To derive the curves that are described by Eq. (18–32), we start with the first law of thermodynamics in its differential form [Eq. (18–20)], with the heat flow, dQ, set to zero: $dU = -dW$. The infinitesimal work done is still given by $dW = p\, dV$. We also know that dU is proportional to the temperature change, dT. In particular, dU can be calculated by differentiating Eq. (18–26) with respect to the temperature, T. Because C_V is constant for an ideal gas, we have $dU/dT = C_V$, or

$$dU = C_V\, dT.$$

Thus the adiabatic transformation satisfies

$$C_V\, dT = -p\, dV. \tag{18-35}$$

All three thermodynamic variables appear here, but, because of the ideal gas law, $nRT = pV$, only two are independent. To find the curve of p versus V, we eliminate T in favor of p and V. The ideal gas law implies that

$$nR\, dT = p\, dV + V\, dp$$

(see Appendix IV–7). Equation (18–35) can thus be rewritten as

$$C_V \frac{p\, dV + V\, dp}{nR} = -p\, dV.$$

Solving for $V\, dp$, we find

$$V\, dp = -\frac{C_V + nR}{C_V} p\, dV. \tag{18-36}$$

The numerator on the right-hand side of Eq. (18–36) is the heat capacity at constant pressure, C_p, as Eq. (18–29) shows. Thus the right-hand side may be written as $-(C_p/C_V)p\,dV = -\gamma p\,dV$, with γ defined in Eq. (18–33).

With the definition of γ, we rewrite Eq. (18–36) in the form

$$\frac{dp}{p} = -\gamma \frac{dV}{V}.$$

Each side of this equation may be integrated separately, with the lower limits on the integrals determined by the initial state (p_0, V_0, T_0). Thus

$$\int_{p_0}^{p} \frac{dp}{p} = -\gamma \int_{V_0}^{V} \frac{dV}{V}.$$

Each integral gives a logarithm, and

$$\ln\left(\frac{p}{p_0}\right) = -\gamma \ln\left(\frac{V}{V_0}\right).$$

Because $A \ln x = \ln(x^A)$, it follows that

$$\ln\left(\frac{p}{p_0}\right) + \ln\left[\left(\frac{V}{V_0}\right)^{\gamma}\right] = 0.$$

We use $\ln x_1 + \ln x_2 = \ln(x_1 x_2)$ to find

$$\ln\left[\left(\frac{p}{p_0}\right)\left(\frac{V}{V_0}\right)^{\gamma}\right] = 0.$$

We may now take the exponential of both sides of the equation to find

$$pV^{\gamma} = p_0 V_0^{\gamma} = \text{a constant.}$$

This is Eq. (18–32). When we substitute the ideal gas law, we can show that the corresponding equation in terms of T and V is Eq. (18–34).

Variation of Atmospheric Temperature with Height

Adiabatic transformations explain the variation of atmospheric temperature with height rather well. Air is transported from high altitude to low altitude and vice versa by large atmospheric wind currents. Air is a sufficiently good thermal insulator so that the transport of air between different altitudes is, to a good approximation, adiabatic. When air rises, the pressure decreases, so that the air undergoes adiabatic expansion; its temperature therefore falls. When air is carried to lower altitudes, it undergoes adiabatic compression, and its temperature rises.

If we assume that air is an ideal gas, we can quantify these concepts. Suppose that pressure and temperature are related by an adiabatic transformation. The adiabatic curve for p versus T is derived from, for example, Eq. (18–32) with $V \propto T/p$. For the p-versus-T adiabatic curve, this gives the equation $p^{1-\gamma}T^{\gamma} = $ a constant. If we raise both sides to the power $1/\gamma$ and solve for T, we find that

$$T = (\text{a constant})p^{(\gamma-1)/\gamma}. \tag{18–37}$$

If we know how pressure varies with height, then Eq. (18–37) can tell us how temperature varies with height. But how does pressure vary with height? In Chapter 16, we found that pressure decreases linearly with increasing height for an *incompressible* fluid, such as water. Air, which is assumed to obey the ideal gas law in this case, is certainly not incompressible, but we can use the methods of Section 16–3 to find an equation for the way pressure varies with height. There, we considered an imaginary horizontal slab cut out of the fluid, with its base at height h, where the pressure is p. The top of the slab, where the pressure is $p + dp$, is at height $h + dh$. Note that we expect dp to be negative because pressure should decrease with height. According to Eq. (16–5),

The diesel engine is a well-known application of an adiabatic transformation (Fig. 18AB–1). In its operation, a cylinder is filled with air at atmospheric pressure and temperature, and then compressed to less than 1/20 of its original volume. This compression is nearly adiabatic; if it were truly adiabatic and if the air were an ideal gas, then the ratio of the final to initial temperatures would be $T_f/T_i = (V_i/V_f)^{(\gamma-1)}$. Although some heat is lost to the cylinder walls, piston, and cylinder head, and the ideal gas equation of state is not exact for air, the final temperature of the air is greater than the ignition temperatures of most hydrocarbons. When, at this point, a combustible hydrocarbon is introduced into the pressurized air, it will burn, transferring its chemical energy into heat energy and raising the pressure of the gases in the cylinder. The hot gas is allowed to expand, driving down the piston and doing work. In practice, diesel engines are very tolerant to the type of fuel burned; everything from flour to powdered

FIGURE 18AB–1 The modern diesel engine is a highly efficient and durable one. Adiabatic transformations play an important part in its operation.

coal has been used. Note that because the end temperature of the compressed gas is proportional to the incoming temperature, diesel engines do not work as well in cold climates.

$$\frac{dp}{dh} = -\rho g,$$

where ρ is the density of the fluid. From this equation, we see that as h decreases, p increases due to the increasing weight of the overlying fluid (in this case, air). For water, ρ is constant. For air, ρ is not constant but is given by

$$\rho = \frac{m}{V} = \frac{nM}{V} = M\left(\frac{p}{RT}\right),$$

where M is the mass of 1 mol of air (its molecular weight). The equation for pressure as a function of height becomes

$$\frac{dp}{dh} = -\left(\frac{Mg}{RT}\right)p. \tag{18–38}$$

Rather than solving this equation directly (see Problem 45), we can convert it into an equation for temperature as a function of height (see Problem 57). This is done by using the adiabatic curve, Eq. (18–37). Temperature is thus found to vary with height as

$$T = T_0 - \left(\frac{Mgh}{R}\right)\left(\frac{\gamma - 1}{\gamma}\right). \tag{18–39}$$

Because $\gamma > 1$, temperature does indeed drop with increasing height—as anyone who has been to the mountains can appreciate.

How does Eq. (18–39) compare with observation? The molecular weight of air is approximately 0.029 kg/mol, $\gamma_{\text{air}} \simeq 1.4$, and $g \simeq 9.8$ m/s². Then if we take the derivative of T [Eq. (18–39)] with respect to h, we find that

$$\frac{dT}{dh} = -\frac{(0.029 \text{ kg/mol})(9.8 \text{ m/s}^2)}{8.3 \text{ J/}°\text{C} \cdot \text{mol}}\left(\frac{1.4 - 1}{1.4}\right) = 0.010°\text{C/m}.$$

We have estimated that temperature drops about 10°C/km. The experimental result is approximately 6.5°C/km. The main sources of error in our calculation are the neglect of the effects of water condensation and, closer to the surface, the effects of local topography, such as mountain ranges. In the vicinity of flat deserts and high in the atmosphere, the value calculated is a fairly good approximation.

SUMMARY

Reversible changes of state are changes that are made sufficiently slowly that the temperature, T, the pressure, p, and the volume, V—the thermodynamic variables that describe a thermal system such as a gas—are well defined at all times. The three variables are related by the equation of state of the gas. Reversible transformations, in contrast with irreversible transformations, can be traced out on a plot of any one of the three variables versus either of the other two.

When the temperature of a system changes, the internal energy of the system has changed. This can happen because the system has done work on its surroundings and/or because there has been a heat flow into the system. Infinitesimal heat flow into a system, dQ, is related to the infinitesimal temperature change of the system, dT, by the heat capacity, C, which depends not only on the system's composition, but on how the temperature change is made:

$$\text{for constant volume:} \quad dQ = C_V \, dT \tag{18–4}$$
$$\text{for constant pressure:} \quad dQ = C_p \, dT. \tag{18–5}$$

It is generally true that

$$C_p \geq C_V. \tag{18–6}$$

Because dQ depends on how a transformation is made, the net heat flow in a finite transformation is path dependent. Heat flow is measured in calories. Calorimetry, the measurement of heat capacities, is based on the fact that heat flow is conserved in a mechanically and thermally isolated system.

The specific heat of a material, c, is the heat capacity of 1 g of the material; its molar heat capacity, c', is the heat capacity of 1 mol of the material.

Heat flow, like work, is an energy transfer. Any transformation of a thermal system caused by a heat flow can be made equally well by doing work on the system. This means that heat has an exact mechanical equivalent,

$$\text{for a thermally isolated system:} \quad \Delta Q = W. \tag{18–7}$$

The equivalence of work and heat means that the calorie can be expressed as an energy unit:

$$1 \text{ cal} = 4.185 \text{ J.} \tag{18–8}$$

A gas undergoing an infinitesimal transformation does an infinitesimal amount of work on its surroundings, given by

$$dW = p \, dV. \tag{18–10}$$

As for heat flow, the net work done in a finite transformation is path dependent. If the path is adiabatic ($dQ = 0$), the work done by the system is the negative of the internal energy change,

$$\text{for an infinitesimal adiabatic transformation:} \quad dU = -dW; \tag{18–16}$$
$$\text{for a finite adiabatic transformation:} \quad U_B - U_A = -W_{A \to B}. \tag{18–17}$$

An adiabatic transformation is a reversible transformation in which there is no heat flow between a system and its surroundings.

The internal energy (thermal energy), U, is a function of state. Any changes U undergoes in thermal transformations are path independent. The first law of thermodynamics specifies that the change in the internal energy of a thermal system in going from state A to state B is the negative of the work it does on its surroundings, plus the heat flow into the system:

$$\Delta U = U_B - U_A = -W_{A \to B} + Q_{A \to B}. \tag{18-19}$$

The work in this equation, W, is the work done *by* the thermal system, and Q is the heat flow *into* the system. The differential form of the first law of thermodynamics is written as

$$dU = -dW + dQ. \tag{18-20}$$

The internal energy of ideal gases is a function of temperature alone:

$$U_{ideal} = U(T). \tag{18-24}$$

Because C_V is practically independent of temperature, the dependence is linear,

$$U(T) = C_V T. \tag{18-27}$$

It is also true that, for an ideal gas,

$$C_p = C_V + nR. \tag{18-29}$$

The curve of an adiabatic transformation can be calculated for an ideal gas by using the ideal gas law. It is given on a p–V diagram by

$$pV^\gamma = a \ constant, \tag{18-32}$$

where

$$\gamma \equiv \frac{C_p}{C_V} = \frac{C_V + nR}{C_V} = \frac{c'_V + R}{c'_V}. \tag{18-33}$$

This equation can be used to predict the variation of temperature with height in the atmosphere.

UNDERSTANDING KEY CONCEPTS

1. When a bicycle tire is pumped up, the end of the pump near the valve feels hotter. What is happening?

2. A gas expands adiabatically, doing 500 J of work. By how much does the internal energy of the gas change?

3. If a bucket of water is carried uphill, raising the water's potential energy, will its temperature change?

4. Which of the following are reversible processes: (a) the slow inflation of a balloon with a bicycle pump; (b) the heating up of a drill used to bore a hole in a log; (c) the slow stretching of a wire by an external force, carried out at a constant temperature?

5. Why is free expansion not an example of an adiabatic process?

6. Would it make sense for you to shake a container of soup in order to heat it? Would the fact that the soup is in an insulating container change your answer?

7. Should the cooling fluid for an engine have a large or a small heat capacity?

8. What use can you think of for fluids with very large specific heats? With very small specific heats?

9. When a mass m_1 of water at temperature T_1 is mixed with a mass m_2 of water at temperature T_2, with $T_1 > T_2$, the temperature change of the hotter mass, ΔT_1, is related to the temperature change of the colder mass, ΔT_2, by $\Delta T_1/\Delta T_2 = -m_2/m_1$. In an experiment in which water and mercury are mixed, it was found that 100 g of mercury acts like 3.3 g of water. Explain both the formula and the observation about mercury.

10. A well-insulated container is divided into two separate compartments by an adiabatic wall. One compartment contains gas at a pressure p_1 and temperature T_1; the other contains gas at a pressure p_2 and temperature T_2. What happens when the partition is removed?

11. Why is the latent heat of vaporization for water so much larger than the latent heat of fusion?

12. When a freeze threatens, how can spraying fruit trees with water protect the fruit from frost damage?

13. Why does perspiring during exercise help you to cool off?

14. For an ideal gas, the heat capacities are independent of temperature, and it is easy to relate temperature change to heat flow. Would it be impossible to relate temperature change to heat flow if the heat capacities varied with temperature?

15. Can all thermodynamic transformations be plotted on a graph of one thermodynamic variable versus another?

16. A change in a thermal system is reversible if (a) there is no friction and (b) the change is carried out slowly. Explain these constraints.

17. In the evening, after a hot summer day, why is it much cooler in the suburbs than downtown?

18. The center of a continent has more temperature extremes than do coastal areas at the same latitude. What effect contributes significantly to this?

19. People who drive cars in cold climates observe that the car's engine sometimes overheats if they keep the engine idling. This does not happen while they are driving, even though the engine puts out much more power then. Why?

18–1 Changes in Thermal Systems

1. (II) Are the following thermal transformations reversible or irreversible? Show the changes on a diagram when one is requested. (a) The sudden release of a gas from one portion of a container, of volume V_1, into a second portion, previously under vacuum. The total final volume, the volume of the whole container, is V_2 (p–V diagram). (b) As in part (a), but rather than a sudden release, the release is made a little at a time until there is no further flow of gas (p–V diagram). (c) The compression of a gas in thermal contact with a heat reservoir by loading weights, a little at a time, onto a piston connected to the gas (V–T diagram). (d) The slow heating of a gas by the release, a little at a time, of a warmer gas into the original sample of gas (p–T diagram). (e) The slow heating of a gas by placing it in thermal contact with a series of ever hotter thermal reservoirs (p–T diagram). (f) The slow rusting of a nail. (g) The release of electrical energy as lightning. (h) The release of chemical energy in the operation of a rechargeable battery to lift a weight.

18–2 Thermal Energy and Heat Flow

2. (I) (a) Using Table 18–2, calculate the heat capacity of a 5.0-kg aluminum rod. (b) Assuming that heat capacity is constant with temperature, how much heat flow is required to raise the temperature of the rod from 25°C to 87°C?

3. (I) A water heater heats 0.8 L of water/min from 20°C to 50°C. If the efficiency of the heater is 80 percent, how much power is consumed by the heater?

4. (I) The latent heat of fusion of oxygen (O_2) is 0.44 kJ/mol. Given that the atomic weight of molecular oxygen is 32 g/mol, find the latent heat of fusion in units of (a) kJ/kg and (b) cal/g.

5. (I) Consider an ideal gas initially in a state (p, V) and finally in a state (p, $2V$). If the transformation occurs at a constant pressure, what is the heat flow into the gas in terms of p and V?

6. (II) A 10-kg lead brick is at room temperature, 25°C. The melting point of lead is 327°C. Using the specific heat for lead, given in Table 18–2, calculate the heat flow to the brick needed to bring the brick to the melting temperature. Will the entire lead brick melt at this point? Explain.

7. (II) A 300-cm³ glass is filled with 100 g of ice at 0°C and 200 g of water at 25°C. (a) Characterize the content of the glass after equilibrium has been reached. Neglect heat transfer to and from the environment. (b) Repeat your calculations for 50 g of ice and 250 g of water.

8. (II) In Example 18–4, an experiment in calorimetry was described in which the assumption is made that the heat capacity of water and iron are constant over the temperature range of the experiment. Describe a series of experiments that would establish whether or not these heat capacities are indeed constant.

9. (II) The quantity 0.6 kg of thermally isolated water at an initial temperature of 291 K serves as a calorimeter. Into it can be placed 1.2 kg of metal that has been heated to 373 K. Find the specific heats and molar heat capacities of the metals if the equilibrium temperature is (a) 316 K; the metal is aluminum. (b) 305 K; the metal is iron. (c) 297 K; the metal is lead.

10. (II) A calorimeter consists of 400 g of water at 25°C. A 500-g piece of copper at 100°C is thrown into the water, and the equilibrium temperature is found to be 28.3°C. What are the specific heat of copper in cal/g·K and the molar heat capacity, given that 1 mol of copper has a mass of 63.5 g?

11. (II) The quantity 200 g of thermally isolated water at an initial temperature of 293 K serves as a calorimeter. Into it are placed, successively, 200 g of iron, then 200 g of silver, each of which has been heated to 400 K. Using the known specific heats of iron and silver, find the final equilibrium temperature. (Use $c_{Ag} = 0.0557$ cal/g·K and $c_{Fe} = 0.112$ cal/g·K.)

12. (II) An axe head consisting of 1 kg of iron is left outdoors one cold winter's night and is brought indoors when the outside temperature is a brisk 250 K. The room into which it is brought is initially at a nice, comfortable 293 K and 1 atm of pressure. The volume of the room, which is well insulated, is 40 m³. Assuming that the axe head comes to thermal equilibrium with the air in the room, by how much is the temperature of the room lowered? (Ignore the thermal interaction with furniture, walls, and so forth. Use $c_{air} = 0.172$ cal/g·K, $c_{Fe} = 0.112$ cal/g·K, and 28.8 g/mol for the molecular weight of air.)

13. (II) A thermally isolated system consists of 1 mol of helium gas at 100 K and 2 mol of a solid at 200 K; the gas and the solid are separated by an insulating wall of negligible mass. If the wall is removed, what is the equilibrium temperature of the system? Assume that at the given temperatures, the heat capacity of the solid is given by $3R/2$ per mole.

18–3 The Mechanical Equivalent of Heat

14. (I) How many kilojoules of energy are required to raise the temperature of a 5-kg bar of aluminum by 10°C?

15. (I) One hundred grams of ice at 0°C is added to 500 grams of thermally insulated water at 24°C. What is the temperature of the water after the ice melts?

16. (I) A 50-g lead bullet traveling at 200 m/s stops in a target. How much would the bullet's temperature rise if 80 percent of the energy were to go into heating it?

17. (I) A primitive demonstration of the mechanical equivalent of heat consists of repeatedly dropping a bag of lead shot from a given height and measuring the temperature change in the bag after a number of drops. Given that the heat capacity of lead is 128 J/kg·K, how many times would the bag have to be dropped from a height of 2.0 m to change its temperature by 5°C? Neglect all heat flows from the lead.

18. (II) Suppose that a kilogram of water drops, starting from rest, from a height of 5.2 m. When the water hits the ground suppose that half of the kinetic energy it attains in falling is converted into thermal energy within the water. By how much will the temperature of the water rise? Neglect the effect of air resistance.

19. (II) Julius Robert Mayer, who first enunciated the principle of the conservation of energy, estimated the mechanical equivalent of heat by considering a quantity of air confined in a cylinder by a movable piston. If the volume is fixed, 0.172 cal must be supplied to raise the temperature of 1 g of air by 1°C. If the piston is not fixed but exerts a constant pressure of 1 atm ($= 1.01 \times 10^5$ N/m²), 0.241 cal are needed to raise the tem-

perature of 1 g of air by 1°C. If the volume change experienced by 1 g of air when the temperature changes by 1°C at constant pressure is 2.83×10^{-6} m³/g·K, what is the relation between the calorie and the joule?

20. (II) The amount of solar energy delivered to a horizontal surface in Washington, D.C., averaged over a full year, is 160 W/m². Assuming a 10 percent efficiency for absorption and conversion to usable heating, how many liters of water can be heated from 0°C to 20°C in a 24-h period by a 3-m² solar panel?

21. (II) Angel Falls, in Venezuela, has the highest free waterfall in the world, at 2,648 ft. The water at the top of the falls is at a temperature of 10°C. (a) What is the temperature of the water in the pool at the base of the falls? As a first approximation, ignore the effects of air resistance, so that there is no terminal velocity. Also, ignore the effects of evaporation. [*Hint*: The kinetic energy that the water gains in falling and losing potential energy goes into turbulent motion at the base of the falls.] The exact terminal velocity is a complicated function of the precise configuration of the waterfall, such as whether it forms a tight tube or whether it is a broad, dispersed spray. (b) Repeat part (a) for Niagara Falls, which is approximately 50 m high. (c) Return to Angel Falls. Discuss the effect of air resistance and the fact that there is a terminal velocity. Would you expect the water to be heated as much as in part (a)? (d) It takes a positive heat flow into a liquid, the latent heat of vaporization, to cause it to evaporate. What is the sign of the temperature change in the waterfall, when you take into account the effect of evaporation of the falling droplets?

22. (II) Work is done on a thermally isolated container of 2 kg of water under 1 atm of pressure, initially at a temperature of 291 K. Assume that the specific heat of water does not vary with temperature, and that churning paddles do work on the water at a rate of 50 W. How long must the paddles continue to turn to bring the water to its boiling point, 373 K?

18–4 Work Done by Thermal Systems

23. (I) The volume of a gas in a cylinder slowly increases by a factor of 3 while its pressure decreases from 9×10^5 Pa to 1×10^5 Pa, as shown in Fig. 18–22. How much work is done on the piston by the gas during the expansion?

FIGURE 18–22 Problem 23.

24. (I) Express the work done by an ideal gas at constant temperature T_0 as a function of the initial pressure, p_i, and final pressure, p_f.

25. (I) If a small amount of water is evaporated at atmospheric pressure until a volume of 3.0 m³ is occupied, how much work is done? Ignore the volume taken up by any remaining liquid water.

26. (I) Gas in a cylinder is allowed to expand from a volume of 0.2×10^{-3} m³ to a volume of 0.7×10^{-3} m³, while the pressure changes linearly with the volume from an initial pressure of 5.2 atm to a final pressure of 2.2 atm. What is the work done by the gas?

27. (II) Half a mole of an ideal gas is compressed isothermally at 293 K so that the pressure is increased from 1.0 atm to 3.6 atm. What are the initial and final volumes of the gas, and how much work is done by the gas during the compression?

28. (II) Two liters of nitrogen gas at an initial temperature of 320 K expand at a constant pressure of 5 atm against a piston, thereby doing work to raise a total mass of 100 kg. The gas is allowed to expand to a volume of 5 L. (a) How high can the mass be lifted? (b) What is the final temperature of the gas?

29. (II) A gas is compressed from V_1 to V_2 by reversibly following the path shown in Fig. 18–23, $p = p_0 - \beta V$, where β is a constant. (a) What is the dimension of β? (b) Determine the work done by the gas during the transformation by integrating the infinitesimal work along the path, $dW = p\,dV$. (c) Check your answer by evaluating the area under the p–V curve in the figure.

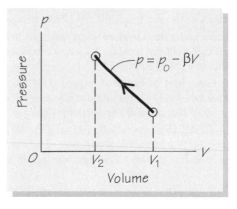

FIGURE 18–23 Problem 29.

30. (II) Figure 18–24 shows a reversible path for a gas being taken from state A to state B. Both states have the same pressure. (a) What is the work done in taking this path? (b) Draw a second

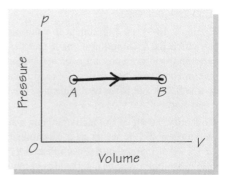

FIGURE 18–24 Problem 30.

reversible path that connects the same points by which no work is done. If the gas is ideal, to what minimum temperature is the gas taken on this path? (Real gases cannot be taken to this temperature.) (c) Draw a third path in which the work done is greater than the work done in following the original path.

31. (II) Two moles of an ideal gas are carried around the thermodynamic cycle shown in Fig. 18–25. The cycle consists of (1) an isothermal expansion $A \to B$ at a temperature of 700 K, with the pressure at A given by $p_A = 8$ atm; (2) an isobaric compression $B \to C$ at 3 atm; and (3) an isochoric pressure increase $C \to A$. What work is done by the gas per cycle?

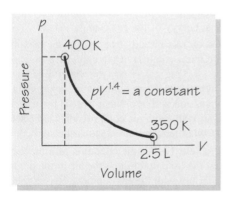

FIGURE 18–25 Problem 31.

32. (III) An external force compresses 0.10 mol of an ideal gas in thermal isolation. The curve of this transformation on a p–V diagram is $pV^{1.4} =$ a constant. The gas initially has a volume of 2.5 L and a temperature of 350 K. When the compression is finished, the temperature has increased to 400 K (Fig. 18–26). (a) What are the final volume and final pressure? (b) How much work was done on the gas to compress it?

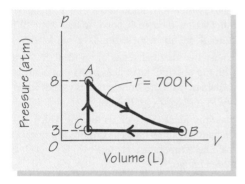

FIGURE 18–26 Problem 32.

18–5 The First Law of Thermodynamics

33. (I) A certain engine follows a closed thermodynamic cycle in which the heat flow into the system is measured to be 633 cal. Assuming that all transformations of the cycle can be approximated as reversible ones, what is the net work done, in joules, by the engine in one cycle?

34. (I) An ideal gas undergoes a reversible transformation from an initial state ($p = 1.0$ atm, $V = 350$ cm³) to a final state ($p =$

2.0 atm, $V = 175$ cm³). The transformation is carried out in two steps: The gas is compressed at constant pressure, and when the final volume is achieved, the pressure is increased while the volume is held fixed. Use the first law of thermodynamics to calculate the heat flow into the gas during this transformation. Assume (as is shown in Section 18-6) that, for an ideal gas, the internal energy U is a function of T only.

35. (II) Consider the p–V diagram shown in Fig. 18–27. If the system is taken from point A to point B via path $A \to C \to B$, 40,000 cal of heat flows into the system, and the system does 20,000 cal of work. (a) How much heat flows into the system along path $A \to D \to B$ if the work done is 7000 cal? (b) When the system returns from point B to point A along the curved path, the work done is 15,000 cal. Does the system absorb or liberate heat, and how much?

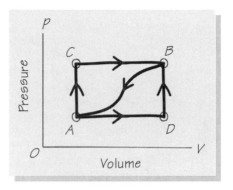

FIGURE 18–27 Problem 35.

36. (II) The latent heat of vaporization for water at 1 atm and 100°C is 540 cal/g. How much of that heat is converted into the mechanical work needed to change the volume of 1 g of water into 1 g of steam? [*Hint*: Calculate the volume of 1 g of steam.]

18–6 Internal Energy of Ideal Gases

37. (I) A thermal system consisting of 1 mol of an ideal gas forms a cycle that contains a leg AB in which pressure decreases linearly as the volume increases, a leg BC in which the gas is compressed at constant pressure until the initial volume is attained, and a leg CA in which the volume is held fixed as the pressure increases to its initial value (Fig. 18–28). (a) Express the internal energy at A, B, and C in terms of p_A, p_B, V_A, and V_B. (b) How much work does the gas do over the cycle?

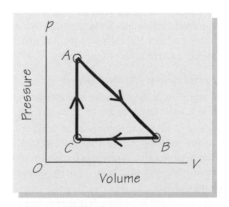

FIGURE 18–28 Problem 37.

38. (II) An ideal monatomic gas, with $c'_V = 3R/2$, undergoes a change of state. On a plot of pressure versus volume, it follows straight lines from an initial point (p_0, V_0) to $(p_0, 3V_0)$ to $(4p_0, 3V_0)$. (a) Draw the p–V plot. (b) By how much does the internal energy change? (c) How much work is done by the gas on its surroundings? (d) How much heat is added to the gas?

39. (II) Five grams of N_2 gas (molecular weight 28 g/mol) are allowed to expand from an initial state $(p_i, V_i) = (1.2 \times 10^5$ N/m², 3.5×10^{-3} m³) to a final state $(p_f, V_f) = (1.8 \times 10^5$ N/m², 5.0×10^{-3} m³) along a path that forms a straight line on a p–V plot (Fig. 18–29). (a) What are the initial and final temperatures? (b) What is the change in the internal energy of the gas? (c) What is the work done by the gas? (d) What is the heat flow into or out of the gas during the process?

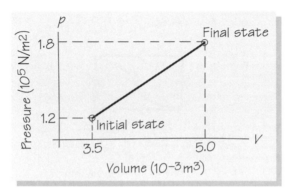

FIGURE 18–29 Problem 39.

40. (II) One mole of an ideal gas in contact with a thermal reservoir at 350 K is under 10 atm of pressure. The pressure is slowly increased to 25 atm. (a) What are the initial and final volumes of the gas? (b) How much work is done *on* the gas? (c) What is the change in internal energy of the gas? (d) What is the heat flow into the gas?

41. (II) According to a calorimeter in contact with 0.2 mol of an ideal gas, the gas does 8 J of work while 5 J of heat is added to it. If the molar specific heat of the gas, c'_V, is 20.8 J/mol · K, independent of temperature, by how much is the temperature of the gas changed in the process described? Is it warmed or cooled?

42. (II) One mole of an ideal gas, $\gamma = 5/3$, is at STP, and 500 J of heat is added at constant pressure. (a) What are the initial and final internal energies of the gas? (b) What is the work done by the gas? (c) Repeat parts (a) and (b) for the same amount of heat added at constant volume.

18–7 More Applications for Ideal Gases

43. (I) How much work does it take to compress 30 mol of an ideal gas at a fixed temperature of 15°C to half its volume?

44. (I) One mole of ideal gas undergoes an adiabatic transformation in which its temperature changes by 25 K. How much work is required to bring about this transformation, given that the constant-volume molar heat capacity of the gas is $c'_V = 30.00$ cal/mol · K?

45. (I) Equation (18–38) is an equation for pressure as a function of height for the atmosphere, which is treated as an ideal gas.

Show that, if T is independent of height, this equation has the solution $p(h) = p_0 e^{-Mg(h-h_0)/RT}$, where p_0 is the pressure at some reference height $h = h_0$.

46. (I) A monatomic ideal gas with $\gamma = 1.67$ undergoes an adiabatic compression starting from an initial temperature of 251 K and an initial volume of 1.73 L. The final volume is 1.12 L. What is the final temperature?

47. (II) The temperature of the (dry) air at the bottom of a mountain valley is 26°C. The air moves as a brisk wind to the ridge 600 m above the valley floor. What is the temperature of the air on the ridge?

48. (II) The wild fires of the California coast are often fed by desert winds, which heat up to about 100°F in the desert at an altitude of 500 m, then rush down to sea level. What is the temperature of the air in these winds at sea level?

49. (II) Consider two thermal states of an ideal gas, A and B, together with the possible reversible paths that connect them. Is there an adiabatic path that connects these states when (a) $V_A = V_B$? (b) $p_A = p_B$? (c) $T_A = T_B$?

50. (II) A diatomic ideal gas such as air, for which $\gamma = 1.4$, expands adiabatically to 10 times its original volume. (a) By what factor does the temperature change? (b) By what factor does the pressure change?

51. (II) One liter of gas characterized by $\gamma = 1.4$ is allowed to expand adiabatically to twice its volume. How does the pressure change compare with the pressure change that would occur for an isothermal expansion?

52. (II) How much work is done by 1.00 L of helium when it expands adiabatically from sea-level pressure (1.00 atm) to the point where the pressure is 0.10 atm? Helium is monatomic, with $c'_V = 3R/2$.

53. (II) A particular adiabatic transformation of an ideal gas occurs as the pressure increases by a factor of 5 while the volume decreases by a factor of 3. What is the ratio of C_V to C_p? How much work is done on the gas during the compression, in terms of the initial pressure and volume? By how much does the internal energy of the gas change?

54. (II) The compression ratio (the ratio of the maximum volume to the minimum volume) of the cylinder of a diesel engine is 17:1. The working gas is air (for which $\gamma = 1.4$), which enters the cylinder at room temperature, approximately 310 K. The compression is so rapid that there is no heat flow through the cylinder walls, and the compression can be said to be adiabatic. To what temperature is the air heated?

55. (II) In Problem 17–43, the operation of a bicycle pump was studied on the assumption that the air in the pump is in thermal equilibrium with the environment. Repeat the analysis for the case when the pump is pushed down so quickly that no heat exchange occurs and the process is adiabatic. Assume that the temperature of the environment is 22°C. (The real process actually lies between the very slow isothermal and the very rapid adiabatic case).

56. (III) Consider the *Carnot cycle* shown in Fig. 18–30. One mole of an ideal gas with volume V_A at temperature T_1 undergoes an isothermal expansion to a volume V_B. This is followed by an adiabatic expansion to a volume V_C and then an isothermal compression at temperature T_2 to volume V_D; the cycle is closed by an adiabatic compression ending with volume V_A at temperature T_1. You will need to know the form of the transformation curve followed on each leg, and the heat flow to the gas for each leg. For the adiabatic legs, you know the heat flows, but, for the isothermal legs, you may want to find the work done directly and use your knowledge of the energy change of an ideal gas when the temperature is constant. You also need to know the starting and ending points of the curves. What is the net heat, expressed in terms of the given variables, supplied to the gas during the cycle?

FIGURE 18–30 Problem 56.

57. (III) In Eq. (18–39) we state that by following an adiabatic curve, the temperature of Earth's atmosphere drops linearly with height h from the surface, and give the coefficient of h. Here, this result is derived. The starting points are Eq. (18–37), which expresses the relation between T and p in an adiabatic transformation of an ideal gas, as well as Eq. (18–38), which expresses the pressure of the atmosphere as a function of height. (a) Take the derivative with respect to height of Eq. (18–37), using the chain rule of differentiation (see Appendix IV–7) for the right-hand side, to show that

$$\frac{dT}{dh} = a \text{ constant} \left(\frac{\gamma - 1}{\gamma}\right) p^{[(\gamma-1)/\gamma]-1} \frac{dp}{dh}.$$

(b) Show that this result can be rewritten as

$$\frac{dT}{dh} = \left(\frac{\gamma - 1}{\gamma}\right) T p^{-1} \frac{dp}{dh}.$$

(c) Substitute Eq. (18–38) for dp/dh into the result derived in part (b) to find that

$$\frac{dT}{dh} = -\left(\frac{Mg}{R}\right)\left(\frac{\gamma - 1}{\gamma}\right).$$

(d) Show that Eq. (18–39) is the solution of the differential equation derived in part (c).

58. (III) An experiment shows that the amount of work done is 88.9 J when a certain ideal gas is compressed adiabatically from an initial pressure of 1.00 atm and an initial volume of 1.00 L to a final volume of 0.500 L. In a second experiment, the same amount of gas is again compressed from 1.00 L to 0.500 L, but this time it is in thermal contact with a thermal reservoir at $T = 20°C$, and 189 J of work is required for the compression. (In the second experiment, the initial pressure is not necessarily 1 atm.) (a) How many moles of gas are there?

(b) What is γ for the gas? [*Hint*: Calculate the work done in the adiabatic compression for several values of γ between 1.2 and 2.] (c) What is C_V for the gas?

General Problems

59. (I) Five liters of oxygen at one-half atmospheric pressure and 273 K is heated at constant pressure until its volume has tripled. It is then compressed isothermally back to 5 L. (a) Draw the process on a p–V diagram. (b) What is the final pressure of the gas? (c) How much work is done by the gas during the entire process?

60. (II) A system that consists of 0.6 mol of helium gas undergoes an isobaric compression from a volume of 1.5 L to 0.1 L at a pressure of 6.0 atm. Is there heat flow in this transformation? If so, what are its magnitude and sign?

61. (II) How much heat flow must there be into a system of 10 L of oxygen initially at STP if (a) the gas triples in volume while the pressure stays constant and (b) the gas doubles in pressure at constant volume?

62. (II) A camper carries a 5-gal plastic container of water up a 40-m-high hill to her campsite. (a) How much heat flow does it take to raise the temperature of 2 qt of the water by 65°C? (b) Compare this energy with the energy expended to carry the entire 5 gal up the hill. (c) How much heat flow will it take to boil the 2 qt of water if it is initially at 23°C?

63. (II) Consider the adiabatic and isothermal processes that each pass through the point (p_0, V_0) on a p–V diagram (Fig. 18–31). Which of the processes has the greater slope at this point? Compute the ratio of the slopes at this point.

FIGURE 18–31 Problem 63.

64. (II) A 1-mol sample of an ideal gas for which $c'_V = \frac{3}{2}R$ follows the closed thermodynamic cycle shown in Fig. 18–32. There are three legs in the cycle: an isothermal expansion, $A \rightarrow B$; an isobaric compression, $B \rightarrow C$; and a constant-volume increase in pressure, $C \rightarrow A$. The temperature during the isothermal leg is $T = 300$ K, $p_A = 5$ atm, and $p_B = p_C = 2$ atm. What are (a) V_A? (b) V_B? (c) the work done by the system during leg $A \rightarrow B$? (d) Q for leg $A \rightarrow B$? (e) T_C? (f) the net work done by the system over the complete cycle?

65. (II) A 100-g piece of copper is heated from 0°C to 100°C at atmospheric pressure. What is the change in its internal energy?

509

FIGURE 18–32 Problem 64.

FIGURE 18–33 Problem 73.

66. (II) A pan contains 0.2 kg of water at 20°C. (a) How much heat does it require to raise the temperature of the water to 100°C and vaporize it at 1 atm? (b) How much work is done by the water as it changes to the gaseous phase? (c) What is the total change in internal energy of the water?

67. (II) A nuclear power plant requires 1000 MW (megawatts) of cooling. If environmental concerns limit the temperature rise of the water to 12°C, how much water flow is required to cool the plant?

68. (II) Heat flow is transferred to 0.3 mol of an ideal gas at a temperature of 500 K and pressure of 2.5 atm, resulting in an isothermal expansion of the volume by a factor of 5. How many calories are involved in the heat flow?

69. (II) One mole of N_2 gas initially at a temperature of 300 K and pressure of 1 atm is compressed isothermally until its volume is reduced by a factor of 10. (a) What is the final pressure of the gas? (b) What is the work done by the gas? (c) How much heat flow is transferred?

70. (II) An electric immersion heater is placed in a coffee cup to heat the water to boiling. The cup initially contains 0.33 L of water at 70°F. The heater is rated for 350 W. Neglect heat loss from the water and the cup. How long does it take to heat the water up to boiling? How long after this does it take to boil the water away completely?

71. (II) A glass contains 150 cm³ of water at 70°F. Four ice cubes of 25 g each of temperature 10°F are dropped into the water. Neglecting heat-flow loss, what is the final temperature of the water? The specific heat of ice in this temperature range is 2.04 J/g·K.

72. (II) The temperature of espresso coffee (mostly water) can be increased by blowing 100°C steam through it. How much steam (in grams) is needed to heat up a 50-cm³ cup of espresso from 45°C to 70°C? What is the volume of this quantity of steam, assuming that the steam is an ideal gas?

73. (II) One mole of an ideal gas is carried around the thermodynamic cycle shown in Fig. 18–33. The cycle consists of an isothermal expansion at a temperature of 400°C with an initial pressure of 5 atm, leading to a doubling of the initial volume. This is followed by a pressure drop of a factor of 2 at constant volume, and then by an isobaric compression until the initial volume is restored. The cycle is completed by a constant-volume pressure increase to the initial value of 5 atm. Calculate the values of p, T, and V not given, and the work done by the gas during one cycle.

74. (II) In experiments carried out between 1759 and 1762, Joseph Black started with a glass cup of mass 32 g, containing 467 g of water at 88°C. Black took a piece of ice of mass 404 g at 0°C and put it in the cup. The final equilibrium temperature was measured to be 12°C. How many calories are needed to melt 1 g of ice; that is, convert it to water at 0°C? That quantity is the latent heat of fusion. What would the equilibrium temperature be if no energy were required to convert ice to water at 0°C? Ignore the heat capacity of the glass.

75. (II) An ideal gas for which $c'_V = 5R/2$ is carried around a cycle $a \rightarrow b \rightarrow c$. There are 2.4 mol of the gas in the cycle. The expansion $a \rightarrow b$ is a straight line on a p–V diagram with $T_b = T_a = 540$ K, $b \rightarrow c$ is a constant-pressure segment with $p = 1.8$ atm, and $c \rightarrow a$ is a constant-volume segment at 15 L. (a) What is p_a? (b) What is V_b? (c) What is the work done in segment $a \rightarrow b$? (d) What is T_c? (e) What is the change in internal energy in segment $c \rightarrow a$? (f) What is the net work done during the entire cycle?

76. (III) A cyclic process with 0.4 mol of ideal gas is represented by a circle on the p–V diagram (if the appropriate scale is chosen) (Fig. 18–34). Calculate (a) the amount of heat transferred to the gas between the minimum and maximum volumes, as given in the figure; and (b) the largest and smallest internal energies during the cycle, given that $c'_V = 5R/2$.

FIGURE 18–34 Problem 76.

The properties of the molecules that make up the gases within these lights determine the colors produced when electrical effects act. Indeed, a more complete understanding of the thermodynamic properties of these and other systems critically depends on an understanding of these systems at the molecular level.

Molecules and Gases

A hypothetical film of the molecular motion of a dilute gas would show the molecules rushing around at different velocities, bouncing elastically off each other and off the walls in a three-dimensional game of billiards. Huge numbers of molecules make up even the smallest measurable quantity of a gas. The behavior of such a vast number of molecules can be expressed only in terms of averages. The kinetic theory of ideal gases, the focus of this chapter, is based on this kind of statistical treatment. In this chapter, we will see how ideal—that is, dilute—gases can be explained in kinetic theory. Along the way, we will learn about some distributions and their properties, and about transport phenomena. We will see that the kinetic theory of gases is remarkable because, with so little input, it can explain so much of the thermodynamics we have already studied.

Statistical techniques for the study of large collections of particles have been the subject of intensive work in physics and other fields. Early work was done by James Clerk Maxwell in the 1850s and by Ludwig Boltzmann and Josiah Willard Gibbs in the latter half of the nineteenth century. Research in this area falls under the general heading of **statistical physics**, a vast field that also includes the physics of liquids and solids. Statistical techniques allow us to explain precisely a great many of the physical properties observed in bulk matter. Statistical physics has been used to explain many new and often unexpected phenomena: phase changes, superconductivity, complex crystalline structure, and chaotic behavior, to name just a few.

There are some 6×10^{23} molecules (Avogadro's number) in 1 mol of any gas, and such incomprehensibly large numbers evoke large crowds. But just how crowded are the molecules? It follows from the ideal gas law that, at standard temperature and pressure (STP), 0°C and 1 atm, the volume of 1 mol of gas is 22.4 L = 22.4×10^{-3} m³. This means that, on average, each molecule occupies a volume of $(22.4 \times 10^{-3}$ m³)/$(6 \times 10^{23}) \simeq 4 \times 10^{-26}$ m³. Using a radius r of 10^{-10} m for a typical molecule, we calculate the molecular volume to be $\frac{4}{3}\pi r^3 \simeq 4 \times 10^{-30}$ m³. A molecule generally takes up only $(4 \times 10^{-30}$ m³)/$(4 \times 10^{-26}$ m³) = 10^{-4} of the volume available to it! The cube root of this number, namely, about 0.05 or $\frac{1}{20}$, gives the ratio of the typical molecular radius to the mean spacing between molecules. If we were to scale molecules to the size of students, with a radius of 0.5 m, the molecules would be spaced some 10 m, or 30 ft, apart.

As a gas becomes more dilute, molecular collisions occur less frequently. A dilute (or ideal) gas is a collection of independent molecules that move about with differing velocities, only rarely interacting with one another. The rare collisions do, however, play an important role. They are the means by which gases come to thermal equilibrium. When a hot gas (consisting of rapidly moving molecules) is mixed with a cold gas (composed of more slowly moving molecules), elastic collisions between fast molecules and slow molecules slow down the fast molecules and speed up the slow ones (see Chapter 8). The entire system comes to equilibrium at an intermediate temperature.

Why do averages tell us anything at all about a gas? After all, if we toss 10 coins, the average number of heads is 5, but we have very little confidence that we will actually get heads 5 times in just 10 tosses. (Try it!) It is only because the number of molecules we deal with is so large that we can be confident that average quantities really do represent the behavior of gases. Suppose that the number of coin tosses (N) is very large. The average number of heads is $0.5N$. In a given trial, the probability that we will get heads more than 51 percent of the time is about 38 in 100 for $N = 100$. This probability drops to about 2 in 100 for $N = 10^4$, and is on the order of 1 in 10^{88} for a million tosses. It is generally true that, for systems with a large number of independent components, large deviations from the average value of a variable are unlikely. When a gas that contains 10^{24} or so molecules in thermal equilibrium is sampled, the chances that the measured value of any randomly varying quantity will differ from its average value are tiny indeed.

(a)

(b)

FIGURE 19–1 After Otto von Guericke invented the air pump, he performed an experiment in 1657 in Magdeburg, Germany that allowed the effect of air pressure to be observed directly. He pumped the air out of a sphere formed from two hemispheres. Even horses could not pull the two halves apart because of the pressure of air molecules outside the sphere. In this modern reenactment, (a) the two halves of the sphere are easily separated before it has been pumped out, but (b) it cannot be pulled apart once it has been emptied of air.

1 9 – 2   P R E S S U R E   A N D   M O L E C U L A R   M O T I O N

In kinetic theory, a container of dilute gas consists of N independently moving molecules. These molecules move with a variety of speeds in a variety of directions. In this microscopic model, pressure is due to molecular collisions with the walls of the container (Fig. 19–1). Each time a molecule hits and rebounds from the wall, there is a momentum transfer to the wall—in other words, a force on the wall. We shall calculate the *average* momentum transfer to calculate the pressure.

Some Average Values in a Gas

A dilute gas consists of N molecules of mass m, and it is in thermal equilibrium at temperature T. The gas is in a box of volume V oriented in a Cartesian coordinate system, as shown in Fig. 19–2a. The molecules in the box are moving in random directions and *the average velocity of the molecules is zero* because as many move in one direction as move in another. The same can be said for the average x-, y-, or z-components of the velocity. We will use angular brackets, $\langle \; \rangle$, to indicate the average value, so

$$\langle \mathbf{v} \rangle = \langle v_x \rangle = \langle v_y \rangle = \langle v_z \rangle = 0. \tag{19–1}$$

The average velocity may be zero, but the same is not true for the average *speed*. The speed is always positive, and it must have an average value that is likewise positive. We will concentrate on the average value of the speed squared (which is the same as the velocity squared) $\langle v^2 \rangle$. Note that this is *not* the same as the square of the average value of the velocity: $\langle v^2 \rangle \neq \langle \mathbf{v} \rangle^2$; in fact, the latter quantity is zero because $\langle \mathbf{v} \rangle$ is zero. We refer to $\sqrt{\langle v^2 \rangle}$ as the **root-mean-square (rms) velocity**.

We can relate $\langle v^2 \rangle$ to the internal energy of a dilute gas by kinetic theory. Forces between molecules are important only when the molecules are close together, and this happens relatively rarely in dilute gases. Thus the internal energy, U, consists mainly of the kinetic energies of the molecules. In other words, U is given by

The internal energy of a dilute gas is the sum of the kinetic energies of its molecules.

$$U = N \langle K \rangle = N \left(\frac{1}{2} m \langle v^2 \rangle \right), \tag{19–2}$$

where $\langle K \rangle$ is the average kinetic energy per molecule and N is the total number of molecules. This equation provides us with a link to the temperature of the gas, because, as we learned in Chapter 18, the internal energy of an ideal gas is a function only of temperature.

The average of a sum of terms is the sum of the average of those terms, so

$$\langle v^2 \rangle = \langle v_x^2 + v_y^2 + v_z^2 \rangle = \langle v_x^2 \rangle + \langle v_y^2 \rangle + \langle v_z^2 \rangle. \tag{19–3}$$

The gas as a whole is not moving, so there must be an equivalence among the three directions. The average value of the x-component of the velocity squared must be the same as that of the y-component and that of the z-component. Thus,

$$\langle v_x^2 \rangle = \langle v_y^2 \rangle = \langle v_z^2 \rangle. \tag{19–4}$$

Using this relation in Eqs. (19–3) and (19–2), it is possible to relate the average of the components of the velocity squared to the thermal energy of the gas:

$$\langle v^2 \rangle = 3 \langle v_x^2 \rangle;$$

$$\langle v_x^2 \rangle = \frac{1}{3} \langle v^2 \rangle = \frac{2}{3} \frac{U}{mN}. \tag{19–5}$$

Next we will relate $\langle v_x^2 \rangle$ to the pressure of the gas.

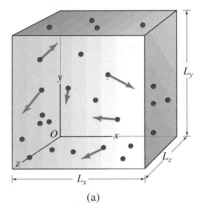

(a)

The Origin of Pressure

Pressure arises from the multiple collisions the molecules of a gas have with the walls that contain the gas. For a quantitative calculation, we compute the momentum transfer to a wall due to a single collision, and then find the number of molecules that strike the wall per unit time. We use this to find the average momentum transfer to the wall per unit time, which is the force on the wall. Pressure is then the force per unit area. Figure 19–2b shows a molecule colliding elastically with the right-hand wall of the box of Fig. 19–2a. The wall is oriented in the yz-plane, so only the x-component of the velocity changes. If the velocity before the collision is

$$\mathbf{v}_i = v_x \mathbf{i} + v_y \mathbf{j} + v_z \mathbf{k},$$

then the velocity after the collision is

$$\mathbf{v}_f = -v_x \mathbf{i} + v_y \mathbf{j} + v_z \mathbf{k}.$$

The momentum change of the molecule, $\Delta \mathbf{P}_{\text{mol}}$, is

$$\Delta \mathbf{P}_{\text{mol}} = m \mathbf{v}_f - m \mathbf{v}_i = -2 m v_x \mathbf{i};$$

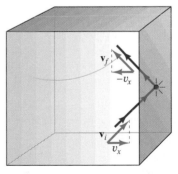

(b)

FIGURE 19–2 (a) The dimensions and orientation of a box that contains a gas. (b) A typical molecule, whose x-component of velocity is v_x, has an elastic collision with the right-hand wall. The x-component of the velocity changes its sign in the collision, but the y- and z-components are unchanged.

the momentum transfer to the wall, $\Delta\mathbf{P}$, is the negative of this:

$$\Delta\mathbf{P} = 2mv_x\mathbf{i}. \tag{19-6}$$

This positive result is reasonable: the wall is pushed to the right by the collision.

We have found the momentum transfer from a single molecule. As we continue, we think of this molecule as an "average" molecule; it has the average x-component of velocity, which we continue to write as v_x for now. The next question is: How many collisions with the wall occur per unit time? The number of molecules with an x-component of velocity of magnitude v_x that strike an area A in a time interval dt is the number of molecules contained in a cylindrical volume whose base is the area A and whose length is $v_x\,dt$ (Fig. 19–3). Molecules that travel toward this area and are farther away than the cylinder height will not reach the wall in the infinitesimal time interval dt. Note also that by making the time interval short enough, we don't have to worry about the possibility that the molecules collide with each other before they reach the wall. If the number of molecules per unit volume (the number density) is $n = N/V$ (not the number of moles here!), then the total number of molecules in our cylindrical volume is $(n)(v_x\,dt)(A)$. Thus the number of collisions suffered by the wall is

$$\frac{1}{2}(n)(v_x\,dt)(A) = \frac{N}{2V}(v_x\,dt)A.$$

(a)

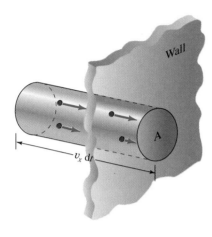

(b)

FIGURE 19–3 (a) All the molecules that will collide with an area A of a wall in a time interval dt are contained in a cylinder of area A and length v_x dt. (b) Only the x-component of velocity is indicated, and only those with x-component of velocity directed toward the wall.

The factor $\frac{1}{2}$ is present because only half of the molecules are moving to the right; the other half are moving to the left. Only the ones that move to the right contribute to the pressure in this case. The number of collisions must be multiplied by the individual collision momentum transfer ΔP_x to find the total momentum dP_x transferred in the time interval dt:

$$dP_x = (2mv_x)\left(\frac{N}{2V}\right)(v_x\,dt)A = mv_x^2\frac{N}{V}\,dt\,A. \tag{19-7}$$

Thus the total momentum transfer per unit time—the force exerted on area A—is

$$\frac{dP_x}{dt} = F_x = mv_x^2\frac{N}{V}A. \tag{19-8}$$

The pressure on the wall, p, is the force per unit area:

$$p = \frac{F_x}{A} = mv_x^2\frac{N}{V}. \tag{19-9}$$

Finally, recall that we have employed the average x-component of velocity squared for each molecule. We make this explicit by employing the notation $\langle v_x^2 \rangle$ rather than v_x^2. Then, with the help of Eq. (19–5), we find

The pressure of a dilute gas in terms of its internal energy

$$p = m\langle v_x^2 \rangle\frac{N}{V} = m\left(\frac{2}{3}\frac{U}{mN}\right)\frac{N}{V} = \frac{2}{3}\frac{U}{V};$$

$$pV = \frac{2}{3}U. \tag{19-10}$$

This derivation is an important one. We have used the dynamical variables of mechanics to find a relation between thermodynamic variables.

EXAMPLE 19–1 Exactly 1 mol of helium gas has an internal energy of 3600 J. It is contained within a cube of sides 0.50 m. (a) Compare helium under these conditions with air at standard temperature and pressure (STP). If air is ideal at STP, can the same be said for helium? (b) Approximately how many times per second do the walls suffer collisions from the molecules of the helium gas?

Solution: (a) We compute the pressure and temperature of the helium and compare them with those of air at STP. Equation (19–10) can be used to compute the pressure:

$$p = \frac{2}{3}\frac{U}{V} = \frac{2}{3}\frac{3600\text{ J}}{(0.50\text{ m})^3} = 1.9 \times 10^4\text{ N/m}^2,$$

which is about 0.2 atm. The temperature can be computed from the ideal gas law:

$$T = \frac{pV}{nR} = \frac{(1.9 \times 10^4\text{ N/m}^2)(0.50\text{ m})^3}{(1\text{ mol})(8.31\text{ J/mol})} = 290\text{ K}.$$

This is in the range of room temperature. The helium has pressure and temperature similar to those of air at sea level and can be treated as ideal.

(b) We have seen that the number of molecules that collide with a wall in an infinitesimal time dt is

$$\text{number of collisions} = \frac{N}{2V}(v_x\,dt)A.$$

This equation is approximately true for small but finite time differences, so we replace dt by $\Delta t = 1$ s. The area A is the area of one wall, $A = (0.50\text{ m})^2$. For the typical value of the magnitude of one component of the velocity, v_x, we use $\langle v_x^2\rangle$ and Eq. (19–5):

$$\sqrt{\langle v_x^2\rangle} = \sqrt{\frac{1}{3}\langle v^2\rangle} = \sqrt{\frac{2}{3}\left(\frac{U}{mN}\right)}.$$

The quantity mN, the mass of one molecule times the number of molecules, is the total mass of a gas, M. There is 1 mol of helium, so the mass is 4 g $= 4 \times 10^{-3}$ kg. A measure of speed v_x is then

$$v_x = \sqrt{\frac{2}{3}\left(\frac{3600\text{ J}}{4 \times 10^{-3}\text{ kg}}\right)} \approx 780\text{ m/s}.$$

Finally, because there are six walls and all six directions are equivalent, we must multiply our result by 6 to find the total number of collisions with the walls:

$$\text{total number of collisions} = 6\,\frac{N}{2V}Av_x\,\Delta t$$

$$= 6\,\frac{6.0 \times 10^{23}}{(2)(0.50\text{ m})^3}[(0.50)^2\text{ m}^2](780\text{ m/s})(1\text{ s})$$

$$= 2.8 \times 10^{27}\text{ collisions in 1 s}.$$

19–3 THE INTERPRETATION OF TEMPERATURE

How can we interpret the temperature of a gas in terms of a microscopic model? The relation

$$U = N\langle K\rangle = \frac{1}{2}Nm\langle v^2\rangle \tag{19–11}$$

is a link provided by the kinetic theory between the macroscopic and microscopic properties of a monatomic ideal gas. Equation (19-10) gives $U = \frac{3}{2}pV$. But $pV = nRT = NkT$, where k is Boltzmann's constant. Thus

$$U = \frac{3}{2}nRT = \frac{3}{2}NkT. \tag{19–12}$$

Equations (19–11) and (19–12) together give a microscopic interpretation of temperature:

$$kT = \frac{2}{3}\frac{U}{N} = \frac{2}{3}\langle K \rangle. \tag{19–13}$$

The number of molecules has canceled from this expression, so T is (correctly) independent of the amount of gas. *The temperature of an ideal gas is a measure of the average kinetic energy of the constituents.* See Fig. 19–4 for another microscopic interpretation.

The temperature of an ideal gas is proportional to its energy.

EXAMPLE 19–2 Exactly 1 mol of helium gas in a large volume is cooled to 31.5 K. What is the internal energy of the gas as well as the average value of the x-component of velocity squared, $\langle v_x^2 \rangle$?

Solution: Because the gas is in a large volume, it is dilute and behaves as an ideal gas. The internal energy of the gas is therefore given by Eq. (19–12). For 1 mol,

$$U = \frac{3}{2}nRT = (1.5)(8.31 \text{ J/K})(31.5 \text{ K}) = 393 \text{ J.}$$

To compute $\langle v_x^2 \rangle$, we note that $\langle v_x^2 \rangle = \frac{1}{3}\langle v^2 \rangle$, and that $\langle v^2 \rangle$ is given in terms of U by Eq. (19–11):

$$kT = \frac{2}{3}\left(\frac{1}{2}\right)(m\langle v^2 \rangle) = \frac{m}{3}3\langle v_x^2 \rangle = m\langle v_x^2 \rangle;$$

$$\langle v_x^2 \rangle = \frac{kT}{m}.$$

The mass of a helium molecule is $m = (4 \text{ g/mol})(10^{-3} \text{ kg/g})/(6.0 \times 10^{23} \text{ molecules/mol}) = 6.7 \times 10^{-27}$ kg, and

$$\langle v_x^2 \rangle = \frac{(1.38 \times 10^{-23} \text{ J/K})(31.5 \text{ K})}{6.7 \times 10^{-27} \text{ kg}} = 6.2 \times 10^4 \text{ m}^2/\text{s}^2.$$

Thus $\langle v_x \rangle$ is about 250 m/s. Even at temperatures as low as 31.5 K, the helium atoms are moving quite rapidly.

FIGURE 19–4 Macroscopic and microscopic views of the work done by an insulated gas. Molecules colliding with the recoiling piston lose energy. Thus the temperature drops.

The zero of the Kelvin temperature scale has a simple meaning: It is the point at which the pressure drops to zero. In the microscopic view of an ideal gas, the meaning of this result is easily understood. From Eq. (19–13), the temperature is zero when the average kinetic energy of the ideal gas is zero. Pressure vanishes because the molecules no longer move around and bounce against the walls.

Interpretation of the van der Waals Gas

Knowing how the microscopic properties of an ideal gas relate to its macroscopic behavior enables us to understand the origin of the van der Waals equation of state, Eq. (17–12):

$$\left[p + a\left(\frac{n}{V}\right)^2\right]\left(\frac{V}{n} - b\right) = RT.$$

This equation of state is more accurate than the ideal gas law away from the dilute gas limit. It reduces to the ideal gas law when the constants a and b are zero. These constants have a simple meaning in the kinetic theory. Gas molecules are not truly like pointlike billiard balls. Billiard balls exert a force on one another only during a very brief time period when they touch. Real molecules have forces that act between them over distances larger than their radii. These forces are repulsive at short range, so that the molecules bounce off one another, but are slightly attractive at longer distances.

The attractive component of the intermolecular force acts to make a gas more compact. This translates into a reduced pressure: The term $a(n/V)^2$ represents the effect of the long-range attraction between molecules. If we solve Eq. (17–12) for p, we see it is reduced by $a(n/V)^2$. As for the term proportional to b, molecules, like billiard balls, take up some space. This term is present because of the strong repulsion between molecules at a characteristic radius. By appearing as a term subtracted from V, the constant b measures the volume unavailable for the motion of molecules because it is space already occupied by other molecules. Example 19–3 illustrates how we can use our interpretation of the constant b to estimate the size of molecules.

EXAMPLE 19–3 Measurements show that nitrogen gas obeys the van der Waals equation of state with the constant $b = 3.94 \times 10^{-5}$ m³/mol. What is the size of a nitrogen molecule?

Solution: Because b is subtracted from the term that represents the total volume per mole (V/n) in the van der Waals equation of state, b represents the volume occupied by 1 mol of molecules. Therefore each molecule has volume

$$V_m = \frac{b}{N_A} = \frac{3.94 \times 10^{-5} \text{ m}^3/\text{mol}}{6.02 \times 10^{23} \text{ molecules/mol}} = 6.54 \times 10^{-28} \text{ m}^3/\text{molecule}.$$

If the molecule is spherical with radius r, then $V_m = \frac{4}{3}\pi r^3$, and

$$r = \left(\frac{3}{4\pi} V_m\right)^{1/3} = \left[\frac{3(0.654 \times 10^{-27} \text{ m}^3)}{4(3.14)}\right]^{1/3} = 5.39 \times 10^{-10} \text{ m}.$$

This radius is in good agreement with more direct atomic measurements.

19–4 PROBABILITY DISTRIBUTIONS

Just as every student in a class does not receive the average grade on a given test, the molecules in a gas do not each have the average velocity or the average position. Instead, each molecule has its own velocity and position. Two functions describe the probability that a molecule has a particular velocity or a particular position. The **velocity distribution function** describes how many molecules have one velocity, how many have another, and so forth. The **position distribution function** similarly describes how the molecules are distributed in space. We need to know something about these distributions in order to understand more about the thermal properties of gases.

We all have a certain familiarity with the notions of probability and averages (Fig. 19–5). In this section, we will sharpen these notions so that we can use them in the analysis of gases and other thermal systems.

Probabilities in a Set of Grades

Consider a quiz taken by a class of 100 students. The grades can run from 0 to a maximum score of 50. The number of students who receive each grade is shown in Fig. 19–6, where the number of students with grade g, N_g, is plotted as a function of the grade g. The total number of students here is $N = 100$, and if we add up all the N_g, we get all of the students taking the quiz:

$$\sum_g N_g = N. \tag{19–14}$$

Given such a distribution of grades, we can find the average grade. We take the number of people with a given grade, multiply it by the grade, add these products for each grade, and divide by the total number of students. We denote the average grade by $\langle g \rangle$. According to the above technique, we find

$$\langle g \rangle = \frac{1}{N}\left(\sum_g g N_g\right). \tag{19–15}$$

FIGURE 19–5 Roulette is a well-known application of the laws of probability.

PROBLEM SOLVING

Using probability notions

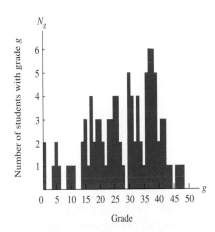

FIGURE 19–6 Bar graph of quiz grades in a class.

In the case illustrated, $\langle g \rangle$ is 28. We could also ask for the average of the *square* of the grade $\langle g^2 \rangle$, given by

$$\langle g^2 \rangle = \frac{1}{N}\left(\sum_g g^2 N_g\right). \qquad (19\text{--}16)$$

In the case illustrated, $\langle g^2 \rangle = 908.6 = (30.1)^2$. Note that $\langle g^2 \rangle \ne \langle g \rangle^2$! The quantity $\langle g^2 \rangle$ contains different information than does $\langle g \rangle$. You would get the same value for $\langle g \rangle$ if half the class had 6 points and half had 50 points but, in that case, $\langle g^2 \rangle = 1268 = (35.6)^2$.

What is the probability that any one student, chosen at random, has a grade of 38 points? This probability is the number of students with a grade of 38 divided by the total number of students—in this case, $\frac{6}{50}$. More generally, the probability of finding a student who received a grade g is

$$P_g = \frac{1}{N}N_g, \qquad (19\text{--}17)$$

and P_g is the **probability distribution** for the grades. As a consequence of Eq. (19–14), the probability distribution is such that

$$\sum_g P_g = 1. \qquad (19\text{--}18)$$

That the sum of probabilities is 1 is known as a **normalization condition**. As we see from Eqs. (19–15) and (19–16), averages are obtained from weighted sums over P_g:

$$\langle g \rangle = \sum_g gP_g, \qquad (19\text{--}19)$$

$$\langle g^2 \rangle = \sum_g g^2 P_g. \qquad (19\text{--}20)$$

We can ask one more type of question: If we pick one student at random, what are the odds that his or her grade lies between 15 and 19? The probability of finding that the student has a grade between 15 and 19 is the number of students with grades in that range divided by the total number of students. We write this as follows:

$$P(15 \le g \le 19) = \frac{1}{N}\sum_{g=15}^{19} N_g = \sum_{g=15}^{19} P_g.$$

For our example, this is equal to $\frac{13}{100} = 0.13$. We will see that this type of question is especially relevant to quantities that vary continuously.

Continuous Distributions

Some distributions do not involve discrete quantities, such as the grades on an exam, but instead involve continuous quantities, such as the heights of students. We do not do things much differently, but we must recognize that the probability that any student will have a *particular* height x—say, $x = 1.73000055$ m—is zero. Instead, we must measure the probability $P(x)\,\Delta x$ that the outcome lies within some small interval Δx around the value x (Fig. 19–7). In our example, this could be the probability that a student has a height within an interval $\Delta x = 0.01$ m around $x = 1.73$ m. The normalization condition now expresses the fact that all students must have some height, so that the sum of all probabilities adds to 1:

$$\sum P(x)\,\Delta x = 1. \qquad (19\text{--}21)$$

A sum appears here because—even if x takes on a continuous range of values—the intervals Δx have divided the region into a *finite* number of segments.

The concept we have described can be refined by making the interval smaller and smaller, thereby finding the probability that an outcome near a particular x-value occurs with more and more accuracy. Calculus notation is appropriate in this limit. The

PROBLEM SOLVING

Using probability for quantities that can vary continuously

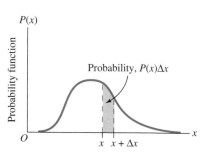

FIGURE 19–7 The probability $P(x)$ Δx associated with a continuous distribution is the area underneath the curve between x and $x + \Delta x$.

quantity Δx is written as dx, and the sum becomes an integral whose upper and lower limits are the upper and lower values of the variable in question, x_2 and x_1:

$$\int_{x_1}^{x_2} P(x)\, dx = 1. \qquad (19\text{–}22)$$

This is the normalization condition in the case of continuous variables.

The averages of functions of x are found from formulas similar to those of the discrete case. Thus

$$\langle x^2 \rangle = \int_{x_1}^{x_2} x^2\, P(x)\, dx. \qquad (19\text{–}23)$$

EXAMPLE 19–4 Suppose that all heights between 1.5 m and 1.8 m are equally likely among a group of students and that no students are outside this height range. Find the probability distribution and use it to calculate the average height.

Solution: Let x represent height. The probability $P(x)$ is zero for $x < 1.5$ m and for $x > 1.8$ m. It is a constant P in between. We calculate P by using the normalization condition, Eq. (19–22):

$$\int_0^\infty P(x)\, dx = \int_{1.5\text{ m}}^{1.8\text{ m}} P\, dx = P \int_{1.5\text{ m}}^{1.8\text{ m}} dx = P(1.8\text{ m} - 1.5\text{ m}) = P(0.3\text{ m}) = 1,$$

so the constant $P = 1/(0.3\text{ m})$. Then $\langle x \rangle$ is given by

$$\langle x \rangle = \int_{1.5\text{ m}}^{1.8\text{ m}} xP\, dx = P \int_{1.5\text{ m}}^{1.8\text{ m}} x\, dx = P \frac{x^2}{2} \Big|_{1.5\text{ m}}^{1.8\text{ m}}$$

$$= \frac{1}{2(0.3\text{ m})} [(1.8\text{ m})^2 - (1.5\text{ m})^2] = 1.65\text{ m}.$$

Not surprisingly, this is the midpoint of the range of heights.

19–5 THE VELOCITY DISTRIBUTION OF GASES

In this section, we extend our ideas about probability distributions to molecules in a gas. The velocity distribution function $F(\mathbf{v})$ is a probability distribution for the velocities of the gas molecules. Velocities form a continuum, so we use a probability function appropriate for continuous variables. The velocity distribution function for ideal gases was first described by James Clerk Maxwell in 1859—a date when the molecular model of matter was by no means universally accepted. Let's start with a number distribution $N(\mathbf{v})$, such that

$N(\mathbf{v})d^3v$ = number of gas molecules with a velocity between $\mathbf{v}$ *and* $\mathbf{v} + d\mathbf{v}$.

The notation requires a little explanation. As was true of the height distribution of Section 19–4, the probability of finding any particular velocity $\mathbf{v}$ is zero. However, we can ask about the probability that a velocity will be somewhere in the neighborhood of $\mathbf{v}$. This "neighborhood" is a box of volume

$$d\mathbf{v} = d^3v = dv_x\, dv_y\, dv_z, \qquad (19\text{–}24)$$

centered about the tip of the vector $\mathbf{v}$ (Fig. 19–8). The total number of molecules is N, and thus

$$\int N(\mathbf{v})\, d^3v = N. \qquad (19\text{–}25)$$

The distribution $N(\mathbf{v})$ leads directly to a *probability* distribution $F(\mathbf{v})$:

$$F(\mathbf{v}) = \frac{1}{N} N(\mathbf{v}). \qquad (19\text{–}26)$$

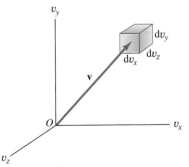

FIGURE 19–8 A molecule has a certain probability of having a velocity $\mathbf{v}$ within the "volume" $d\mathbf{v} = d^3v = dv_x\, dv_y\, dv_z$. Here we show the small box d^3v within which the velocity vector is located.

The meaning of this function is that

$F(\mathbf{v})d^3v$ = the probability that the velocity of a gas molecule is between $\mathbf{v}$ and $\mathbf{v} + d\mathbf{v}$.

From $F(\mathbf{v})$, we can calculate the averages for continuously varying quantities. In particular, the average of the velocity squared is

$$\langle v^2 \rangle = \int v^2 F(\mathbf{v}) \, d^3v. \tag{19-27}$$

To determine the probability distribution $F(\mathbf{v})$, we must make some *physical* assumptions. The model of molecules darting about and occasionally colliding with other molecules suggests that a given molecule may have any velocity. This leads to a fundamental physical postulate: Any way in which a gas's total energy and total momentum (which is zero) can be shared among the molecules is equally likely. Once this postulate is accepted, we are in the same position as someone asking for the probability of getting any particular number between 100 and 600 when throwing 100 dice. What counts is the *number of ways* of getting that number. There is only one way of getting 100 or 600, but there are many ways of getting 400; thus, the probability of getting 400 is much greater than the probability of getting 100 or 600. When this sort of counting is done properly for molecules whose individual kinetic energy is $\frac{1}{2}mv^2$ and whose average kinetic energy is $\frac{1}{2}m\langle v^2 \rangle = \frac{3}{2}kT$, the distribution function is

The velocity distribution function for an ideal gas

$$F(\mathbf{v}) = \left(\frac{m}{2\pi kT} \right)^{3/2} e^{-mv^2/2kT}. \tag{19-28}$$

The velocity distribution function is properly normalized, $\int F(\mathbf{v}) \, d^3v = 1$, meaning that the sum over all probabilities is 1. $F(\mathbf{v})$ is a function only of the magnitude of the velocity (that is, the speed), not the direction. This expresses the physically reasonable fact that there is no preferred direction in a container of gas in which the gas as a whole is not moving. All directions are equally likely. It is possible to make a direct experimental verification of Eq. (19–28) by cutting a tiny hole in a container of gas and measuring the different speeds of the molecules that leak out.

The Average of the Velocity Squared

To determine the internal energy, we need the average of the velocity squared, $\langle v^2 \rangle$, which we can now calculate. From the probability distribution $F(\mathbf{v})$, we have

$$\langle v^2 \rangle = \int v^2 F(\mathbf{v}) \, d^3v = \left(\frac{m}{2\pi kT} \right)^{3/2} \int v^2 e^{-mv^2/2kT} \, d^3v. \tag{19-29}$$

PROBLEM SOLVING

Scaling technique for integrals

This integral can be simplified by using a technique known as **scaling**. This scaling technique is of great importance to scientists and engineers because it allows the *dependence* of physical quantities on other physical parameters to be extracted. Both kT and mv^2 have the dimensions of energy, so v^2 has the dimensions of kT/m. We therefore change the integration variable in Eq. (19–29) from $\mathbf{v}$ to the dimensionless vector $\mathbf{u}$ by the definition

$$\mathbf{v} \equiv \mathbf{u} \sqrt{\frac{2kT}{m}}. \tag{19-30}$$

This definition means that each component of the vector $\mathbf{v}$ is transformed—that is, it stands for three equations. With this definition, the volume element d^3v has the form

$$d^3v = \left(\frac{2kT}{m} \right)^{3/2} d^3u.$$

The factor $\sqrt{2kT/m}$ is cubed because d^3v stands for $dv_x \, dv_y \, dv_z$. Thus

$$\langle v^2 \rangle = \left(\frac{m}{2\pi kT} \right)^{3/2} \left(\frac{2kT}{m} \right)^{5/2} \int u^2 e^{-u^2} \, d^3u = \frac{2kT}{m} \frac{1}{\pi \sqrt{\pi}} \int u^2 e^{-u^2} \, d^3u. \tag{19-31}$$

The fact that the faster molecules of a gas escape in greater quantity through a tiny hole than do the slower ones is the practical basis for the separation of uranium isotopes.[†] The uranium isotope ^{238}U is more common, but the relatively rare isotope ^{235}U is more useful for nuclear reactors such as those used for generating electric power. These isotopes cannot be separated by chemical means because they have the same chemical properties. However, the fact that ^{238}U is a bit heavier than ^{235}U means that when they are mixed in a gas—uranium hexafluoride, UF_6, is used—the ^{238}U component moves, on average, at a lower speed. This is because the factor $1/\sqrt{m}$ appears in any of the measures of speed, such as v_{rms}. The UF_6 gas is allowed to escape through many tiny holes from one container at a given temperature into another container that is initially empty. The percentage of ^{235}U in the gas that has escaped is slightly higher than that in the original container. Thousands of repetitions of this process of isotope separation lead to significantly increased percentages of ^{235}U.

Equation (19–31) shows that $\langle v^2 \rangle$ is a dimensionless constant times $2kT/m$. We have already estimated [Eqs. (19–11) and (19–13)] that

$$\langle v^2 \rangle = \frac{2U}{Nm} = \frac{3kT}{m}. \tag{19–32}$$

Thus, Eq. (19–31) is in accord with Eq. (19–32)—at least in terms of dimensions. We need to integrate Eq. (19–31) to show that it is in exact agreement with Eq. (19–32). In fact, we can show with the help of a table of integrals that

$$\frac{1}{\pi\sqrt{\pi}} \int u^2 e^{-u^2}\, d^3u = \frac{3}{2}.$$

Thus Eq. (19–32) follows.

We can get a feel for the speeds of molecules by considering the root-mean-square (rms) speed v_{rms}, the square root of the average value of the speed squared:

$$v_{rms} = \sqrt{\langle v^2 \rangle} = \sqrt{\frac{3kT}{m}}. \tag{19–33}$$

This result follows from Eq. (19–32). For helium gas, which has an atomic mass $m = 6.7 \times 10^{-27}$ kg, the rms speed is 556 m/s, 1362 m/s, and 2486 m/s for $T = 50$ K, $T = 300$ K, and $T = 1000$ K, respectively. Contrast this with argon gas, which has a molecular mass 10 times greater. The corresponding rms speeds are $\sqrt{10}$ times smaller: 176 m/s, 431 m/s, and 786 m/s.

*19–6 THE MAXWELL–BOLTZMANN DISTRIBUTION

An ideal gas may consist of molecules that have just one atom (monatomic) such as helium (He) or argon (Ar); it may consist of molecules with two (diatomic) or more atoms, such as nitrogen (N_2), oxygen (O_2), or water (H_2O) (Fig. 19–9). Is there a difference between these types of ideal gases? The answer came in the last part of the nineteenth century from the work of Ludwig Boltzmann.

The velocity distribution of a monatomic ideal gas [Eq. (19–28)] has the general form

$$F(\mathbf{v}) = \frac{1}{Z} e^{-mv^2/2kT} = \frac{1}{Z} e^{-K/kT}, \tag{19–34}$$

where K is the kinetic energy ($K = mv^2/2$); the factor $1/Z$ takes into account the normalization.

Molecule	Bonding Schematic	Space-filling Schematic
Helium (He)		
Nitrogen (N_2)		
Carbon dioxide (CO_2)		
Water (H_2O)		
Sulfur trioxide (SO_3)		
Ammonia (NH_3)		
Methane (CH_4)		
Benzene (C_6H_6)		

FIGURE 19–9 Molecules come in different shapes, as shown here in two schematic representations.

[†]Elements are characterized by their chemical properties, which depend on the number of electrons they carry, and hence on the number of (positively charged) protons in their nuclei. *Isotopes* of an element differ in the number of (electrically neutral) neutrons in the nucleus of the element, and thus differ in atomic weight. For uranium, the atomic weights 235 and 238 refer to the total number of protons plus neutrons.

Boltzmann generalized the velocity distribution to a distribution that describes the probability that any one molecule has a given energy E. He proposed that in the original velocity distribution, Eq. (19–34), *the total energy of the molecule, E, should replace the kinetic energy, K, for a point mass*:

$$F = \frac{1}{Z} e^{-E/kT}. \qquad (19\text{–}35)$$

The Maxwell–Boltzmann distribution, which describes the probability that a molecule has a given energy

This distribution function is known as the **Maxwell–Boltzmann distribution**. Equation (19–35) is one of the most frequently used formulas in the physics of the behavior of matter. This distribution function depends on all the dynamical variables that enter into the energy of a molecule. It is only for free, pointlike molecules that the energy depends on **v** alone. The energy of a rotating molecule with rotational inertia also depends on the angular velocities of rotation about the various axes. Later in this section, we will see how this affects the specific heat of gases.

The Energy Distribution for Diatomic Molecules

To see the effect of structure on the thermal properties of a gas, let's consider a gas composed of diatomic molecules. To describe a symmetric diatomic molecule, such as O_2, we use a classical model in which the two molecules act as though they are connected by a rigid rod. The rotational inertia has the same value, I, about each of the two axes that pass through the center of the rod and are perpendicular to it (Fig. 19–10). The energy of the molecule is a sum of the overall translational kinetic energy from linear motion, or translation (Fig. 19–11a), and of the rotational motion of the atoms about the center of mass (Fig. 19–11b; see Chapters 9 and 10). The rotational energy is

$$R_{rot} = \frac{1}{2} I\omega_x^2 + \frac{1}{2} I\omega_y^2, \qquad (19\text{–}36)$$

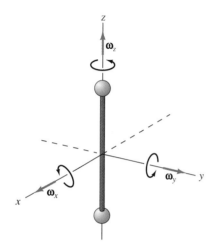

FIGURE 19–10 A dumbbell-like molecule can rotate about three axes. If the masses at the end of the rod are pointlike, the rotational inertia about the z-axis is zero. The rotational inertia about the x- and y-axes is I.

where ω_x and ω_y are the components of the angular velocity along the x- and y-axes, respectively (Fig. 19–10). The ω_z term is missing from Eq. (19–36) because the rotational inertia about the z-axis is zero. The probability distribution F now refers to the probability that the velocity of a molecule lies in a region d^3v about the velocity vector **v**, and that the molecule's angular velocity lies in a region; $d^2\omega = d\omega_x \, d\omega_y$ about the angular velocity vector ω whose components are ω_x and ω_y. The total energy of a molecule is now $E = K + E_{rot}$, and, from Eq. (19–35), the Maxwell–Boltzmann distribution is

$$F = \frac{1}{Z} e^{(-mv^2/2kT)-(I\omega_x^2/2kT)-(I\omega_y^2/2kT)}. \qquad (19\text{–}37)$$

By integrating this function times the energy per molecule over d^3v, $d\omega_x$, and $d\omega_y$, we find the average energy, $\langle E \rangle$:

$$\langle E \rangle = \frac{1}{2} m\langle v_x^2 \rangle + \frac{1}{2} m\langle v_y^2 \rangle + \frac{1}{2} m\langle v_z^2 \rangle + \frac{1}{2} I\langle \omega_x^2 \rangle + \frac{1}{2} I\langle \omega_y^2 \rangle. \qquad (19\text{–}38)$$

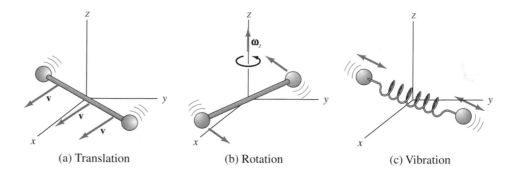

FIGURE 19–11 Schematic diagram of (a) translation, (b) rotation, and (c) vibration of a diatomic molecule.

(a) Translation   (b) Rotation   (c) Vibration

When the integrations are performed to find the averages, each of these five terms is precisely $kT/2$. Finally, the internal energy of the gas is given by $U = N\langle E \rangle$, where N is the number of molecules. Thus, for our rigid-rod diatomic molecule,

$$\langle E \rangle = \frac{U}{N} = \frac{5}{2} kT. \tag{19–39}$$

This result should be compared to the monatomic case, in which there are only the three terms of the kinetic energy of linear motion and $\langle E \rangle = \frac{3}{2}kT$, which is a result already given in Eq. (19–12). If the diatomic molecules are not rigidly attached, the bond between them acts like a spring with spring constant k (Fig. 19–11c). There are then two additional vibrational contributions to the energy, of the form $E_{vib} = (mV^2/2) + (kx^2/2)$. Here, V is the speed of the atoms relative to the center of mass and x is the displacement of the vibrating atoms from an equilibrium position. Each of these terms also gives a contribution of $kT/2$ to the average energy and, for these molecules, $\langle E \rangle = \frac{7}{2}kT$.

Equipartition

The counting we just performed shows that the total average energy per molecule depends on how many independent motions a molecule can have. If a molecule acts like a point mass, all it can do is move in the x-, y-, and z-directions. In this case, the energy has the three terms, proportional to v_x^2, v_y^2, and v_z^2, corresponding to linear motion in the x-, y-, and z-directions, respectively. If, in addition, the molecule is diatomic with a rotational inertia about axes x and y, we found two new terms in the energy—this time proportional to ω_x^2 and ω_y^2—making five terms in all. If vibration is possible, then we found two more terms, this time proportional to the squares of the relative speed and of the atomic separation; thus there are in all seven such terms. Every term in the energy expression that is quadratic in an independent dynamical variable designates a **degree of freedom**. Generally, *the contribution of each degree of freedom to the average energy of a molecule is $kT/2$*. This result is called the **equipartition theorem**. If s is the number of degrees of freedom, then

$$\langle E \rangle = \frac{s}{2} kT. \tag{19–40}$$

The equipartition theorem

A dilute gas—even one whose molecules have a complicated structure—continues to obey the ideal gas law. The internal energy is just

$$U = N\langle E \rangle = \frac{s}{2} NkT. \tag{19–41}$$

The internal energy continues to be linearly dependent on temperature, as for any ideal gas. The term by which the temperature is multiplied in Eq. (19–41) is the constant-volume heat capacity of the gas, C_V. The heat capacity therefore depends on the number of degrees of freedom of the molecules through the relation

$$C_V = \frac{s}{2} nR. \tag{19–42}$$

The heat capacity of an ideal gas is proportional to the number of degrees of freedom.

Molecular nitrogen, for example, should have a *molar heat capacity*, c', or heat capacity per mole ($n = 1$), of $\frac{7}{2}R$, at constant volume: $\frac{3}{2}R$ is from the overall motion of the center of mass, $\frac{2}{2}R$ is from rotations about two axes, and $\frac{2}{2}R$ is from the kinetic and potential energies of the "spring" that connects the two nitrogen atoms. How does this compare with experimental results? Figure 19–12 plots c_V'/R, where c_V' is the constant-volume molar heat capacity, as a function of temperature. At lower temperatures, hydrogen gas (H_2) behaves like a monatomic system, with a molar heat capacity of $\frac{3}{2}R$, which explains why we can treat air at room temperature as a monatomic gas. The

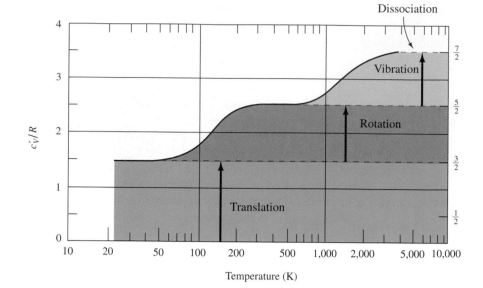

FIGURE 19–12 The constant-volume molar heat capacity divided by the gas constant, c'_V/R, for H_2 as a function of temperature.

full factor of $\frac{7}{2} R$ comes in only at high temperatures. The steplike nature of the curve suggests that, somehow, the effects enter one at a time. The reason for this cannot be found in the realm of classical physics, but instead requires quantum mechanics. Quantum physics sets certain minimum temperatures for the excitation of the different degrees of freedom.

*19-7 COLLISIONS AND TRANSPORT PHENOMENA

Molecules, even in a dilute ideal gas, follow a tortuous path within their container. They frequently collide with one another and the walls, changing in direction and speed with each collision (Fig. 19–13). An air molecule in a room undergoes billions of collisions per second. If the air is diluted to one-millionth of 1 atm, it still suffers thousands of collisions per second. Even so, the collisions are brief, and most of the time the molecule is free of the influence of other molecules. The *average* distance a molecule travels between collisions is a statistical quantity that can be calculated because of the large number of molecules and collisions involved.

Molecules that are characterized by some special property, such as having higher speeds due to local heating, will, through collisions, carry that special property away from their initial location. In effect, they will carry thermal energy from one place to another. If some of the molecules affect the sense of smell (molecules that make up perfume, for example), they will carry the odor as they move and collide. The movement of such properties is called **transport** (of thermal energy, odor, and so forth).

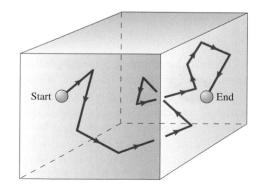

FIGURE 19–13 A molecule follows a tortuous zigzag path because of its multiple collisions with other molecules.

(a)

(b)

(c)

FIGURE 19–14 (a) Dye is placed with the help of a pipette at the bottom of a flask of water. (b) A combination of convection and diffusion has moved the dye into other regions of the flask. (c) Dye has diffused throughout the flask of water to make a uniform distribution. While stirring can speed the process, diffusion will make this happen if given enough time.

Figure 19–14 illustrates transport in a liquid. Ordinary thermodynamics has nothing to say about transport phenomena, but kinetic theory can explain them.

Collisions and Movement in a Gas

For a given gas density, the distance that a molecule travels before it has a collision depends on the size of the molecules. If molecules were infinitely small, they would never run into each other; if they were very large, they would always be in each other's way. Molecules of diameter D will collide when the path of the center of one molecule lies within an area πD^2 presented by the second molecule (Fig. 19–15). This area is the **collision cross section**, σ:

$$\sigma = \pi D^2. \tag{19–43}$$

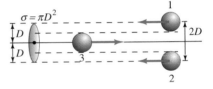

FIGURE 19–15 If the projected centers of two molecules of diameter D that move together are less than or equal to a distance D apart, the molecules will collide. An area $\sigma = \pi D^2$ of one molecule is available for collisions with another molecule.

Even if the molecules do not behave like billiard balls, there is some effective distance D that characterizes the collision, and the collision cross section is still given by Eq. (19–43).

Consider now a molecule that moves with speed v and sweeps out an area σ. If no other molecules were present, then in time t the molecule would travel a distance $d = vt$, sweeping out a volume $V = \sigma d = \sigma vt$. However, our molecule is not alone: Within the volume V, there are $N_V = nV$ target molecules, where n is the number density of molecules (not the number of moles). The traveling molecule therefore suffers N_V collisions. Even though the path is bent with each collision, the volume, V, remains unchanged if the speed is not changed by the effect of multiple collisions; and, on average, it is not. Because the number of collisions in time t is N_V, there is on average a collision every t/N_V seconds. This is the **mean collision time**, τ. Since we are averaging over many molecules, v must be replaced by some average value. It is reasonable to replace it with v_{rms}, so that

$$\tau = \frac{t}{N_V} = \frac{t}{nV} = \frac{t}{n\sigma v_{rms}t} = \frac{1}{n\sigma v_{rms}}. \tag{19–44}$$

A better calculation takes into account the fact that the target molecules are also moving, and Eq. (19–44) is modified to

$$\tau = \frac{1}{\sqrt{2}n\sigma v_{rms}}. \tag{19–45}$$

The average time between collisions

The mean free path

The average distance a given molecule travels before it is involved in a collision is the **mean free path**, λ. In the same spirit in which we found τ, λ is the rms speed times the mean collision time:

$$\lambda = \tau v_{\mathrm{rms}} = \frac{1}{\sqrt{2}n\sigma}.$$ (19–46)

As is reasonable, λ is inversely proportional to both the density and the collision cross section.

EXAMPLE 19–5 What is the mean free path in air at sea level when the temperature is 300 K? By how much does the mean free path change when the temperature drops to 275 K? Take $r \simeq 10^{-10}$ m as a typical molecular radius, and treat the atmosphere as ideal.

Solution: For pressure at sea level, we use 1 atm $\simeq 10^5$ Pa. The atmospheric number density n can be calculated from the ideal gas law:

$$n = \frac{N}{V} = \frac{p}{kT} = \frac{10^5\,\mathrm{Pa}}{(1.38 \times 10^{-23}\,\mathrm{J/K})(300\,\mathrm{K})} = 2.4 \times 10^{25}\,\mathrm{m}^{-3}.$$

The collision cross section, σ, is given by Eq. (19–43). From Eq. (19–46), the mean free path is

$$\lambda = \frac{1}{\sqrt{2}n\sigma} = \frac{1}{\sqrt{2}n\pi(2r)^2}$$

$$= \frac{1}{\sqrt{2}(2.4 \times 10^{25}\,\mathrm{m}^{-3})(3.14)(2 \times 10^{-10}\,\mathrm{m})^2} = 2.3 \times 10^{-7}\,\mathrm{m},$$

which is much greater than the average molecular spacing.

The mean free path changes with temperature because the number density of molecules is inversely proportional to T. The mean free path is inversely proportional to n, hence it is directly proportional to T:

$$\frac{\lambda_{275\,\mathrm{K}}}{\lambda_{300\,\mathrm{K}}} = \frac{275\,\mathrm{K}}{300\,\mathrm{K}} \simeq 0.9.$$

The Random Walk and Diffusion

How far does a molecule move, *on average*, from its initial position in a given amount of time? This problem is similar to a classic problem in mathematics called the **random walk**, known historically as the drunkard's walk. In this problem, a drunkard starts at a lamppost and takes steps that are equal in length but random in direction (Fig. 19–16). Compare Fig. 19–16 to Fig. 19–13, a typical molecular path in a gas. If the length of the drunkard's step is replaced by the mean free path, and if the time between steps is replaced by the mean collision time, then the problems are very similar. Of course, the molecule moves in three dimensions rather than two.

We can find the average displacement of a molecule after N steps as follows. Let the successive displacements of the molecules be $\mathbf{L}_1, \mathbf{L}_2, \ldots, \mathbf{L}_N$. These have random directions, but their magnitudes are all the same value, L. After N steps, the net displacement of a molecule is $\mathbf{R}_N = \mathbf{L}_1 + \mathbf{L}_2 + \cdots + \mathbf{L}_N$. Squaring this quantity, we find that

$$R_N^2 = L_1^2 + L_2^2 + \cdots L_N^2 + 2\mathbf{L}_1 \cdot \mathbf{L}_2 + 2\mathbf{L}_1 \cdot \mathbf{L}_3 + \cdots + 2\mathbf{L}_{N-1} \cdot \mathbf{L}_N.$$

Because the directions of the vectors $\mathbf{L}_i$ are random, all the dot products, such as $\mathbf{L}_1 \cdot \mathbf{L}_2$, average to zero, whereas the terms L_i^2 all equal L^2. We are left with

$$\langle R_N^2 \rangle = NL^2.$$ (19–47)

FIGURE 19–16 The random walk (a drunkard's walk around a lamppost).

To connect this result to the properties of the gas, we replace L^2 by the mean free path squared. The time interval between steps is the mean collision time, τ, so after N steps, a time $t = N\tau$ has elapsed. N can thus be replaced by t/τ. Therefore, Eq. (19–47) states that, after time t, a molecule will have moved on average a distance squared given by

$$\langle r^2 \rangle = \frac{t}{\tau} \lambda^2. \tag{19–48}$$

It is typical of random-walk problems that the displacement *squared* is linear in time.

The calculations leading to Eq. (19–48) can be refined, leading to changes in some of the details. More importantly, the result can be verified experimentally. Molecules that have been marked in one way or another (for example, by using some radioactive molecules) can be traced as they move. The movement of molecules by random collisions is called **diffusion**. The actual motion of the random walk is observable when particles larger than molecules, such as smoke particles, are added to a gas. This motion is called **Brownian motion**.

SUMMARY

Molecules are spread rather sparsely in gases under normal conditions. The kinetic theory of ideal gases is a statistical description in which molecules bounce randomly and relatively infrequently off each other and off the walls of their container. This theory accurately predicts many thermodynamic properties.

The internal energy, U, is given in this model by the number of molecules, N, times the average energy of a molecule:

$$U = N(\frac{1}{2} m\langle v^2 \rangle), \tag{19–2}$$

where $\langle v^2 \rangle$ is the average of the quantity velocity squared. The pressure of a gas arises from multiple elastic collisions between the walls and the molecules. Pressure is related to internal energy by

$$pV = \frac{2}{3} U, \tag{19–10}$$

where V is the volume of gas. In the ideal gas law, pV is proportional to temperature, so temperature can be interpreted as a measure of the internal energy, or of the average energy of molecules:

$$kT = \frac{2}{3} \frac{U}{N} = \frac{2}{3} \langle K \rangle. \tag{19–13}$$

Probability distributions describe the probability that various outcomes will be represented in a large statistical sample. For gases, $F(\mathbf{v}) \, d^3v$ describes the probability that a given molecule will have a velocity $\mathbf{v}$ in the range from $\mathbf{v}$ to $\mathbf{v} + d\mathbf{v}$. Averages are computed from the probability distribution by integration. For example, the average of v^2 is

$$\langle v^2 \rangle = \int v^2 F(\mathbf{v}) d^3v. \tag{19–27}$$

The probability distribution $F(\mathbf{v})$ for the velocity of the molecules of a gas is

$$F(\mathbf{v}) = \left(\frac{m}{2\pi kT} \right)^{3/2} e^{-mv^2/2kT}. \tag{19–28}$$

Boltzmann generalized the velocity distribution to the cases in which molecules have internal energy as well as a center-of-mass kinetic energy, or in which the molecules are subject to external forces. The Maxwell–Boltzmann distribution is given by

$$F = \frac{1}{Z} e^{-E/kT}, \tag{19–35}$$

where E is the total energy of a molecule. Every variable that appears quadratically in the expression for the energy of a single molecule—such variables are called degrees of freedom—contributes $\frac{1}{2}kT$ to the average energy. If the number of degrees of freedom is labeled s, then the average energy of a molecule is

$$\langle E \rangle = \frac{s}{2}kT. \tag{19–40}$$

This result is the equipartition theorem. The internal energy is $U = N\langle E \rangle$.

Transport phenomena describe the motion of molecules, momentum, energy, and so forth through a gas as a result of multiple molecular collisions. Molecules have a collision cross section, σ, which describes the area available for collisions with another molecule. If the number of molecules per unit volume is n, then the mean time τ between collisions, called the mean collision time, is

$$\tau = \frac{1}{\sqrt{2}n\sigma\langle v \rangle}. \tag{19–45}$$

The average distance a given molecule travels before it suffers a collision is the mean free path, λ:

$$\lambda = \tau\langle v \rangle = \frac{1}{\sqrt{2}n\sigma}. \tag{19–46}$$

As a result of multiple collisions with other molecules, an individual molecule within a gas will on average move. In a time t, the molecule moves a distance squared that is proportional to t. Such movement is called diffusion.

UNDERSTANDING KEY CONCEPTS

1. Consider an ideal gas in an insulated cylinder outfitted with a piston. The piston is moved slowly, so that the volume of the gas is decreased. What happens to the temperature of the gas? How do you explain your answer in terms of the motion of molecules?

2. Which travel faster, on average, the oxygen molecules or the nitrogen molecules in your room?

3. In our discussion of pressure, we used the fact that the collisions of molecules with the walls are elastic. If there are diatomic molecules, they could be set into rotation by the collision, and thus their translational kinetic energy after the collision could be smaller than before the collision. How can these statements be reconciled? What is the effect on the pressure?

4. We made the connection between temperature and the average kinetic energy of a molecule without taking into account the large number of intermolecular collisions estimated in Section 19–1. Why do you think this approximation holds?

5. In an extremely dilute gas, the rate of interatomic collisions is very small. As the gas becomes less dense, the rate of interatomic collisions becomes much less than the collision rate with the container's walls. Can we still make a connection between temperature and the average kinetic energy of a molecule? What are the possible problems in making the connection in this limit?

6. Very few hydrogen molecules are present in Earth's atmos-

phere today. Is this fact in conflict with the possibility of their abundance a long time ago? In explaining your answer, ignore the possibility that hydrogen is removed by chemical reactions.

7. Does temperature depend on the rotational kinetic energy of molecules?

8. Can we assign a temperature to a single molecule? Explain.

9. Why, from the point of view of kinetic theory, does the air near a hot stove become heated?

10. In Section 19–1, we stated that the typical distance a molecule travels before it suffers a collision in 1 mol of a gas at standard temperature and pressure is roughly 3×10^{-7} m. This will be much less than the size of a container of gas. How is it that the approximation that each molecule in the cylinder collides freely with the wall, without worrying about intermediate collisions between molecules, leads to a satisfactory result for the pressure, Eq. (19–10)?

11. At standard temperature and pressure, a molecule of one mole of an ideal gas has a mean free path of about 3×10^{-7} m, some 100 times greater than the average intermolecular spacing. How is it possible for the mean free path to be greater than the average spacing?

12. You have a container of a dilute gas at a certain temperature and inject a small quantity of a different type of molecule into that container. What processes determine how long it takes the new molecules to come to thermal equilibrium with the original molecules?

13. Gravity keeps the atmosphere of Earth close to the surface. Why, then, is there no atmosphere on the Moon, which also exerts a gravitational force?

14. In Newtonian mechanics, energy is added to a system when work is done on it. In the language of classical mechanics, how is energy added to a gas in a container when a piston is pushed in and the container's volume is changed?

15. Given our explanation of the origin of pressure, why is there no net force on a pane of glass placed in a container of gas, even though there may be a net force on a wall of the container?

16. Will the air just above the surface of hot water in a bowl be at the same temperature as the water? If so, what molecular mechanism is responsible for the equality of the two temperatures?

17. Why will a cup of hot water cool off much more rapidly when air is blown across its surface than when the air above the cup is stationary?

18. The smoke from the burning end of a cigarette rises; the flame of a candle points upward. Why?

19. When a candle burns, carbon combines with the oxygen in air to form carbon dioxide, a compound used in fire extinguishers to put out fires. Why does the candle not snuff itself out? Could you burn a candle in a satellite that is in free fall?

PROBLEMS

19–1 A Microscopic View of Gases

1. (I) We can define a volume ratio R_V of a gas to be the ratio of the volume taken up by the molecules to the total volume of the container; similarly, we can define a spacing ratio R_L to be the ratio of the linear size of a molecule to the mean spacing between molecules. Compute R_V and R_L for a gas with molecules of diameter 2×10^{-10} m. Assume that the gas obeys the ideal gas law for the following values of temperature and pressure: (a) $T = 300$ K, $p = 1$ atm; (b) $T = 5$ K, $p = 1$ atm; (c) $T = 300$ K, $p = 10^{-8}$ atm; (d) $T = 5$ K, $p = 10^{-8}$ atm.

2. (I) For an ideal gas at 0°C and 1 atm, the ratio of the typical molecular radius (10^{-10} m) to the intermolecular spacing is about $\frac{1}{20}$. What is this ratio in liquid water (H_2O, molecular weight 18 g/mol)?

3. (II) Assume that the carbon and hydrogen atoms that make up oil behave like spheres of diameter 10^{-8} cm, and that the fundamental molecular component of oil, CH_2, would take up a corresponding area of 10^{-15} cm^2 on a water surface. How many such components are there in 0.1 L of oil if the density of oil is 90 percent of the density of water? What would the area of a one-molecule-thick layer of this amount of oil on water be? (This type of experiment, said to have been performed by Benjamin Franklin, among others, is sometimes used to compute the size of molecules.)

19–2 Pressure and Molecular Motion

4. (I) A rubber ball, confined to move in the x-direction, bounces elastically back and forth between two walls. The ball, which has a mass of 0.125 kg, moves at a speed of 12 m/s, and the spacing of the walls is such that the ball strikes the right-hand wall 48 times in a ten-second period. What is the average force on the right-hand wall due to the ball during that ten-second span?

5. (I) Grains of sand of mass 3×10^{-3} g each, fall from a height of 0.8 m on a sticky surface at a rate of 50 grains per second per cm^2. What pressure does this shower of sand exert on the surface, assuming that air resistance can be neglected?

6. (II) Use dimensional analysis to estimate the number of air molecules that strike the 12-in $\times$ 12-in screen of a television set during a 1-h program.

7. (II) One mol of a monatomic ideal gas is placed in a chamber under 5 atm pressure. The volume of the chamber is 5000 cm^3. (a) What is the internal energy of the gas? (b) What is the temperature of the gas? (c) Assuming that the mass of a molecule of the gas is 3.36×10^{-26} kg, what is the value of $\langle \mathbf{v}^2 \rangle$ of a gas molecule? (d) What is the root-mean-square (rms) velocity?

8. (II) A cubic box of volume 0.05 m^3 contains helium gas at 0.20 atm pressure and 27°C. (a) How many total collisions do the molecules make with the walls per second? (b) What is the total internal energy of the gas? (c) If the temperature is doubled, what happens to the rate of collisions?

9. (II) The apparatus shown in Fig. 19–17 is designed to demonstrate aspects of the kinetic theory of gases. It consists of a transparent cubical box, 20 cm on each side, containing 100 steel balls of diameter 5 mm. The density of steel is 7.8 g/cm^3. The bottom of the box vibrates so that the steel balls bounce around. The top of the box consists of a movable piston of mass 1 kg. What is the rms velocity of the steel balls if the top of the box is in dynamic equilibrium with the "gas" of steel balls? Ignore gravity in your treatment of the motion of the steel balls. Is this approximation justified?

FIGURE 19–17 Problem 9.

10. (II) Particles of mass 1.3×10^{-17} kg, when suspended in a liquid at room temperature, have an rms speed of 0.030 m/s. Use this to determine Avogadro's number. (You will need to look up the gas constant R.)

11. (II) The rms velocity of galaxies in a large part of the visible universe is roughly 100 km/s. The number density of these galaxies is 3×10^{-20}/ly^3, and the average mass of a galaxy is 3×10^{41} kg. What is the pressure of a gas of such galaxies?

12. (II) Molecules are confined to move in a plane, for example, by putting them between two glass plates separated by a distance that is small compared to the mean free path. What is the relation between the pressure and the internal energy? What is the relation between internal energy and temperature for an ideal gas?

19–3 The Interpretation of Temperature

13. (I) Consider the mole of helium gas with an internal energy of 3600 J that we discussed in Example 19–1. (a) What is the temperature of the gas? (b) If 1/1000 of this sample of gas had the same internal energy as the total gas, what would its temperature be? (c) What would the internal energy of 1/1000 of the original sample of gas be, at its original temperature?

14. (I) A gas is heated from 15 K to 300 K. What is the change in the rms velocity?

15. (I) What is the average kinetic energy per molecule of a gas at room temperature, 293 K? How fast would a baseball of mass 0.1 kg be moving to have this same kinetic energy?

16. (I) An ideal gas is contained in a vessel of volume 4×10^{-4} m^3 that is under a pressure of 8.6 atm. What is the internal energy of the gas?

17. (I) The temperature of hydrogen in a dilute plasma—a state of matter in which atoms are broken apart into electrons and nuclei—inside a nuclear fusion reactor needs to be on the order of 20 million K in order to initiate nuclear fusion. What is the rms velocity of the hydrogen nuclei in the plasma?

18. (II) A mixture of nitrogen gas and sulfur dioxide gas is in equilibrium at 27°C. What is the rms velocity of the sulfur dioxide molecules (SO_2, molecular weight 64 g/mol), and that of the nitrogen molecules (N_2, molecular weight 28 g/mol)?

19. (II) Calculate the rms velocity at 300 K of the principal components of air: O_2 molecules (molecular weight 32 g/mol), N_2 (molecular weight 28 g/mol), CO_2 (molecular weight 44 g/mol) and H_2 (molecular weight 2 g/mol).

20. (II) The rms speed of 1 mol of argon atoms (atomic weight 40 g/mol) in a box is 400 m/s. (a) What is the temperature inside the box? (b) What is the internal energy? (c) If the box has a volume of 0.1 m^3, what is the pressure? Treat the gas as ideal.

21. (II) Estimate v_{rms} for hydrogen atoms (a) on the surface of the Sun, where the temperature is 6000 K; (b) on the surface of the Moon, where the temperature at one point is 150 K in the shade.

22. (II) A proton of kinetic energy 5×10^9 eV comes to a stop because of molecular collisions in a thin tube that contains 0.01 mol of oxygen at STP, as in Fig. 19–18. By how much is the temperature in the tube increased at equilibrium?

FIGURE 19–18 Problem 22.

23. (II) A "gas" of water droplets 10^{-6} m in diameter is in equilibrium with air molecules at 300 K. Given that the density of water is 1.0×10^3 kg/m^3, what is the value of v_{rms} for the droplets?

24. (II) A relativistic gas is one in which a significant fraction of the constituents have speeds that are some finite fraction of the speed of light, $c \simeq 3 \times 10^8$ m/s. Suppose that the rms speed of the constituents of a gas of atomic hydrogen is just 1 percent of c. What is the temperature? The energy that binds the electron and the proton together into a hydrogen atom is 2.18×10^{-18} J. Would this gas break down into a gas of electrons and protons? In the interior of stars, where temperatures comfortably exceed the temperature you have calculated, atoms cannot exist; they are broken into their components.

25. (II) If their kinetic energy greatly exceeds the average gravitational potential energy, galaxies in the sky may be viewed as an ideal gas. Given that the rms velocity of galaxies is 100 km/s and the average mass is 3×10^{41} kg, what is the temperature of a gas of galaxies?

19–4 Probability Distributions

26. (I) A lottery that costs $1 to play has a prize of $1,000,000. The state that runs the lottery is obligated to tell the players that the probability of winning is one in 10,000,000. Is it a good idea—from a financial point of view—to play this lottery? Answer this question by calculating the ratio of your outlay to your revenue under the assumption that you play the game an unlimited number of times.

27. (I) Nine cars are measured at a given spot on a highway to have speeds v of 52.3, 54.5, 57.0, 57.2, 57.9, 63.6, 63.6, 68.1, and 82.2 mi/h. Calculate (a) the average speed, and (b) the rms speed.

28. (I) Take a deck of cards. Draw cards from that deck, one by one, without putting any cards back in the deck. What is the probability that the first card is a spade? What is the probability that the second one is also a spade? What is the probability that the third one is also a spade? (You may extend this procedure to calculate the probability that when you draw thirteen cards, they are all spades, by multiplying all the thirteen probabilities together).

29. (I) Four dice are tossed. What are the probabilities of getting (a) four sixes; (b) three sixes and a five?

30. (II) A cage of 100 Ping-Pong balls has air blowing through it to keep the Ping-Pong balls moving. Ten Ping-Pong balls have the number 0 painted on them, ten have the number 1, and so forth, finishing with ten Ping-Pong balls painted with the number 9. The apparatus is used to choose the numbers for a lottery, and the balls are not returned to the cage after they are chosen. (a) The first drawing is for three balls to win $100. Your number is 337. Before any balls are drawn, what is the probability that you will win? (b) The first ball drawn is 3. What are your chances now? (c) The second ball drawn is 3. Now what is the probability that you will win?

31. (II) Consider the distribution of grades g discussed in Section 19–4. (a) Plot the distribution of the numbers D for the students, where D is defined as the deviation of the grade from the average. (D can be positive or negative.) (b) Calculate the average value of D^2. (c) Show that $\langle D \rangle = 0$ and $\langle D^2 \rangle = \langle g^2 \rangle - \langle g \rangle^2$.

32. (II) Find an expression for the probability of getting 13 spades in a bridge hand. (In bridge, the entire deck of 52 cards is dealt to four players.)

33. (II) In a true-or-false test, a correct answer is awarded $+1$ point and an incorrect answer is awarded -1 point. In a test involving 87 students, the following grade distribution is found (Fig. 19–19):

$$-50 \text{ to } -30: \quad 3 \text{ students;}$$
$$-30 \text{ to } -10: \quad 18 \text{ students;}$$
$$-10 \text{ to } +10: \quad 29 \text{ students;}$$
$$+10 \text{ to } +30: \quad 22 \text{ students;}$$
$$+30 \text{ to } +50: \quad 15 \text{ students.}$$

(a) Calculate the average grade. (b) Calculate $\langle D^2 \rangle$, where D is defined in Problem 31. (c) Suppose that the distribution was such that every student has a grade between $+10$ and $+30$. With this information, what would the average be?

FIGURE 19–19 Problem 33.

34. (II) Suppose that the distribution of grades (the number of test takers, N, with a grade x) in a national test is approximated by a continuous curve of the form (Fig. 19–20):

$$N(x) = 25x \qquad \text{for } 0 < x < 60$$
$$= 6000 - 75x \text{ for } 60 < x < 80$$
$$= 0 \qquad \text{for } 80 < x < 100.$$

(a) What is the total number of students who took the test?
(b) What is the average grade?

FIGURE 19–20 Problem 34.

35. (III) A way to find the value of π (*Buffon's method*) is the following: Drop a needle of length L repeatedly onto a grid of parallel lines separated by a distance L (Fig. 19–21). Show that the probability of the needle falling on a line is $2/\pi$. [*Hint:* Take the lines as parallel to the x-axis. Calculate the average value of the component of the needle's length along the y-axis.]

FIGURE 19–21 Problem 35.

19–5 The Velocity Distribution of Gases

36. (I) What is the rms speed of a helium atom in the vapor above liquid helium at 4 K? (The system is enclosed in a Thermos bottle.)

37. (I) Oxygen atoms come in a variety of isotopes: ^{16}O, ^{17}O, and ^{18}O (with the superscript representing A). The latter two isotopes are present only in small quantities. A gas consists of diatomic molecules of oxygen (mostly $^{16}O_2$, with small admixtures of ^{16}O–^{17}O and ^{16}O–^{18}O). Calculate the ratios of the rms speeds for the latter two to that of the dominant (^{16}O–^{16}O) form. Does your answer depend on temperature?

38. (I) If the rms speed of molecules of gaseous H_2O is 200 m/s, what will be the rms speed of CO_2 molecules at the same temperature? Assume that both of these are an ideal gas.

39. (II) Calculate $\langle v \rangle$, the average speed for an ideal gas. In doing this calculation with the help of the velocity distribution $F(\mathbf{v})$, use the fact that for a spherically symmetric integrand you may replace d^3v by $4\pi v^2 \, dv$. You will need the integral

$$\int_0^\infty z^3 e^{-z^2} \, dz = \tfrac{1}{2}.$$

[*Hint:* This problem illustrates the use of dimensional analysis as an intermediate step.]

40. (II) Use the result of Problem 39 to find the numerical value of the average speed for an ideal gas of CO_2 at 300 K, and compare it with v_{rms} at that temperature.

41. (III) Consider a gas being blown along at a velocity $\mathbf{u}$, so that its velocity distribution is given by

$$F(\mathbf{v}) = \frac{1}{Z} \, e^{-m(\mathbf{v}-\mathbf{u})^2/2kT}.$$

(a) Show that Z is the same as for the usual velocity distribution, as in Eq. (19–28). (b) Calculate $\langle \mathbf{v} \rangle$ and $\langle \mathbf{v}^2 \rangle$ and interpret your results.

*19–6 The Maxwell–Boltzmann Distribution

42. (I) The value of c'_p for carbon monoxide (CO) is 29.2 J/mol·K. What value do you expect for c'_V for the diatomic CO molecule?

43. (I) One mole of hydrogen gas is determined to have an internal energy of 1.87×10^4 J at 900 K. How many degrees of freedom are there for the hydrogen molecule? Which degrees of freedom are likely to be available?

44. (I) A gas of N_2 molecules is in equilibrium at a temperature of 700 K. Calculate the rms angular momentum $I\sqrt{\omega^2}$ for an

N_2 molecule, given that the mass of a nitrogen atom is 2.33×10^{-26} kg and that the average separation of the two atoms in an N_2 molecule is approximately 0.7 nm (Fig. 19–22). (You may assume that the rotational—but not the vibrational—degrees of freedom participate in the equipartition theorem.)

FIGURE 19–22 Problem 44.

45. (II) The following data can be found in a table of thermodynamic properties of gases:

Argon Ar: $c_p = 0.214$ cal/g·K; $c_V = 0.074$ cal/g·K
Oxygen O_2: $c_p = 0.219$ cal/g·K; $c_V = 0.157$ cal/g·K
Water H_2O: $c_p = 0.445$ cal/g·K; $c_V = 0.335$ cal/g·K
Carbon dioxide CO_2: $c_p = 0.201$ cal/g·K;
$c_V = 0.156$ cal/g·K

What can you infer about the shape of the molecules from these data?

46. (II) A gas of diatomic molecules is in equilibrium at a temperature of 300 K. Assuming that the rotational inertia of the molecule [as used in Eq. (19–36)] is 3×10^{-40} kg·m², what is the rms angular speed of the molecules? What is the mean angular momentum of the molecules?

47. (III) Write down an integral for the probability that a molecule is found to have a velocity larger than 90 percent of v_{rms}. Show that the probability is independent of the temperature. Show that this is not the case if we ask for a velocity larger than some fixed number; for example, the escape velocity from a planet.

48. (III) A crystal can be thought of as a collection of N atoms arranged in a lattice, with all the atoms connected by springs (Fig. 19–23). Thus we can think of each atom as a mass at the end of three springs, aligned respectively along the x-, y-, and z-directions. Using this model of the lattice, count the number of degrees of freedom for each atom, then use the equipartition theorem to compute the molar heat capacity of the crystal at temperature T.

FIGURE 19–23 Problem 48.

49. (I) For a collision between two molecules of radii R_1 and R_2, respectively, show that the collision cross section is $\sigma = \pi(R_1 + R_2)^2$ (Fig. 19–24).

FIGURE 19–24 Problem 49.

50. (I) Compute the pressure of a container of air molecules of diameter 2.5×10^{-10} m if the temperature is 273 K and the mean free path is (a) 1 in, (b) 0.1 μm. The gas should be treated as ideal.

51. (II) One can treat the flow of gas through a vacuum hose as if the gas were a fluid if the diameter of the hose is much larger than the mean free path. If, however, the mean free path is much larger than the diameter of the hose, the motion of the individual free molecules has to be considered. Consider a hose of diameter 1 cm. At what pressure is the mean free path of air at 300 K shorter than 0.01 cm? Longer than 100 cm? Take the diameter of an air molecule to be 3×10^{-10} m.

52. (II) Using a reasonable value for the collision cross section or, more simply, a model of a gas in which the molecules are like tiny billiard balls, make an estimate of the density at which the mean free path becomes equal to the mean spacing between molecules.

53. (II) Ten tennis balls are rolling around randomly on a tennis court (Fig. 19–25). Estimate the mean free path for collisions.

FIGURE 19–25 Problem 53.

54. (II) A student walks across a field that is 100 m × 50 m. Estimate the student's mean free path if there are (a) 10, (b) 100, (c) 1000 other students on the field.

55. (II) A cubic box of sides 10 cm contains helium atoms (diameter $\simeq 10^{-10}$ m) at a pressure of 3×10^4 Pa. If the mean free path in the box is 4×10^{-4} m, how many helium atoms are in the box? How many are there if the mean free path is 40 cm? What will the pressure be if the temperature is held fixed?

56. (III) According to one theory, planets are formed as a consequence of near collisions between stars. The number density of stars in our part of the galaxy is 10^{-51} m⁻³. The velocity of the Sun relative to these stars can be taken to be 25 km/s, and the size of the stars is roughly 10^9 m. If we assume that

a passage of two stars at a distance of 100 radii is close enough to set off the tidal forces that give rise to planet formation, what is the probability that the Sun would have acquired a planetary system in 10 billion yr?

General Problems

57. (I) Argon and methane (CH_4) gas are mixed together at 400 K. What is the ratio of the rms speeds of the argon and methane atoms? What happens to this ratio if the temperature is halved?

58. (II) The density as well as the temperature of material in intergalactic space is very uncertain. Suppose that the density and temperature were one hydrogen atom per cubic meter and 3 K, respectively. (a) What is the pressure? (b) What is the mean free path if the diameter of a hydrogen atom is 10^{-10} m?

59. (II) The acceleration due to gravity on the surface of Mercury is 3.5 m/s^2, and Mercury's radius is 2.4×10^6 m. (a) What is the escape speed of a particle on Mercury? (b) Suppose that the atmosphere of Mercury were pure H_2 gas. What would the temperature be so that, if the molecules had the velocity distribution given by Eq. (19–28) for that temperature, the rms speed of the H_2 molecules matched the escape speed? Qualitatively, what is the effect on the temperature of the remaining gas? (c) If the temperature were less than the result of part (b), would there be a similar effect, and why or why not? (d) Suppose that Mercury's atmosphere has two or more molecular components. What happens to the composition of the atmosphere over time as a result of the effects discussed here?

60. (II) The surface of the Sun contains both hydrogen and helium atoms at 6000 K. (a) What is the average kinetic energy of each type of atom? (b) What is the rms speed of each? (c) Which is most likely to escape from the sun?

61. (II) The regions in which the beams of accelerated particles circulate in accelerators have high vacuums (10^{-9} torr) to allow the particles to circulate freely. (a) What are the mean free paths of air molecules inside the accelerator at room temperature? (b) By how much does the mean free path decrease if the pressure inside increases to 10^{-6} torr?

62. (II) Suppose that molecules were confined to move in a plane. Calculate the velocity distribution $F(\mathbf{v})$, assuming that it is of the general form $(1/Z)e^{-(mv^2/2kT)}$. In other words, calculate Z so that $\int_{surface} F(\mathbf{v}) \, d^2\mathbf{v} = 1$.

63. (II) Brownian motion is the random walk followed by individual molecules as well as by larger objects such as smoke particles. The Brownian motion of smoke particles in a room can be observed with the aid of a microscope. If the mass of the smoke particles is typically 3×10^{-15} kg, what is the rms speed of the particles at room temperature?

64. (II) Peter and Paul are standing in opposite corners of a room, 7 m apart. The air in the room does not move. Peter holds a small vial full of H_2S while Paul holds a vial containing Cl_2. They open their vials at the same time (Fig. 19–26). Paul smells H_2S after 50 s. When will Peter smell the Cl_2?

65. (II) Calculate an expression for the pressure of a gas of massless particles, using the following facts: Massless particles always move with the speed of light, so that $v = c$; the energy of a massless particle, E, is related to its momentum by $E = |\mathbf{p}|c$. Use this expression to show that for a gas of massless

FIGURE 19–26 Problem 64.

particles, $pV = \frac{1}{3}U$. This result is applicable to the blackbody radiation discussed in Chapter 17.

66. (II) The relation $pV = \frac{2}{3}U$ is correct only for a monatomic gas. Show that the more general relation is $pV = (\gamma - 1)U$, where $\gamma = c'_p/c'_V$.

67. (II) Consider a tank of volume 0.30 m^3 containing 2.5 mol of helium gas at 20°C. If we add 1 mol of O_2 at 0°C to the helium and let the whole system come to equilibrium, what will the equilibrium temperature be? What will the pressure be?

68. (II) Assume that an H_2O molecule has three rotational degrees of freedom in addition to its three linear degrees of freedom (Fig. 19–27). Calculate the molar heat capacity of its vapor. Compare this with the experimental value at low pressures of 0.48 cal/g·K. What role might vibrational motion play?

FIGURE 19–27 Problem 68.

69. (II) A vessel is pumped out to a high vacuum and then heated to get rid of water vapor. It is then filled with dry helium and pumped out again to a very high vacuum of 10^{-11} torr, so that it now contains only helium atoms at 300 K. Helium atoms have a diameter of about 10^{-10} m. (a) What is the density of helium atoms? (b) the mean collision time? (c) the mean free path?

70. (II) The *law of partial pressures* states that the total pressure of a mixture of gases is the sum of the pressures that each gas would exert if it were present alone in the same amount as in the mixture. Use the kinetic theory to demonstrate this fact. Could this law be true if the gases were not ideal?

71. (III) The escape speed of a particle from Earth is 11 km/s. Assuming that the oxygen molecules at Earth's surface are at a temperature of 300 K, how would you use the velocity distribution to calculate the probability that a particular molecule at the surface would escape Earth's gravitational attraction if the molecule were to suffer no further molecular collisions? Express this probability in the form of an integral, and manipulate the integral into a form that involves the ratio v_{rms}/v_{escape}, but do not attempt to perform the integration. [*Hint*: This problem is analogous to finding the probability that a given student in a class has a grade of 35 or more on a test.]

The surroundings of this power plant, which converts thermal energy to electrical energy, betray some of the unavoidable consequences of such operations: The water acts as a sink for unconverted energy associated with an exhaust stage of the conversion cycle, energy that the second law of thermodynamics states must be present.

The Second Law of Thermodynamics

The first law of thermodynamics does not specify the direction of thermal energy transfer. Nevertheless, we know that the spontaneous transfer of thermal energy always proceeds from a hotter system to a cooler one—ice cubes melt when they are placed in hot water. This directionality in nature is the essence of the second law of thermodynamics and this law has some surprising consequences. It implies that there are limits as to how much of the thermal energy in a system can be used to do mechanical work when this work is done cyclically, as in an engine. The directionality of the spontaneous transfer of thermal energy can be understood from a molecular point of view by using the notions of probability developed in Chapter 19. These notions will allow us to develop the idea of order in a thermal system and to see that a spontaneous process proceeds from a more ordered to a less ordered state.

20-1   BEYOND ENERGY CONSERVATION

The first law of thermodynamics brings the idea of energy and its conservation into thermal systems. However, there are some facts about thermal systems that this law cannot explain. First, it is impossible to devise an engine that extracts all the thermal energy from a thermal system and turns that energy into work. Such an engine, if it could be constructed, could power New York City while it cooled Long Island Sound. Yet such an engine is completely consistent with the first law of thermodynamics.

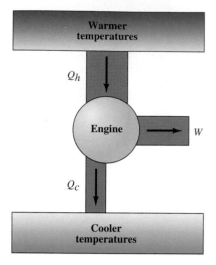

FIGURE 20–1 Abstract representation of an engine, which takes energy—here represented by a system at higher temperatures—to perform work. It is not possible to convert thermal energy entirely into work, so there is some thermal energy left over.

∞ Cycles were introduced in Chapter 18.

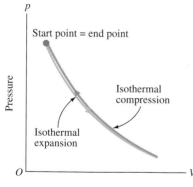

FIGURE 20–2 A cycle in which a gas expands isothermally and is then compressed isothermally at the same temperature. The work done is zero.

A second fact about thermal systems looks superficially different. When ice cubes are put into hot tea, the ice cubes melt and the tea cools. The first law of thermodynamics requires only that the heat flow into the ice cubes must match the heat flow out of the hot tea. According to the first law, heat flow can proceed in either direction. Nevertheless, lukewarm tea never *spontaneously* turns into hot tea with a couple of ice cubes floating in it. Another example of the second fact is the observation that gas released into an evacuated chamber through a small opening spontaneously spreads throughout the available space. We never see the gas spontaneously concentrating in a corner of the chamber, leaving most of the chamber evacuated.

These phenomena can be summarized as follows:

1. *Not all the thermal energy in a thermal system is available to do work.*

2. *Thermal systems spontaneously change only in certain ways.* In particular, spontaneous heat flow always occurs from a hotter body to a colder one.

These experimental facts are summarized in the **second law of thermodynamics**. We will formulate this law as it applies to point 1 in Section 20–2. Point 2, which describes the direction of spontaneous processes, introduces the concept of *entropy*, a new thermodynamic variable. We will address this point in Section 20–5.

Engines

A young engineer, Sadi Carnot, gave an explicit statement of the second law of thermodynamics in 1824. He was interested in improving the efficiency of steam engines and found that there are limits to the possible efficiency of any engine. Just what is an engine? An engine is a device that transforms thermal energy into mechanical energy. In an automobile engine, the thermal energy of burning fuel is converted into the energy of the automobile's motion. In all known engines, however, the conversion of thermal energy into mechanical energy is accompanied by the emission of exhaust gases, which carry off some thermal energy. This means that only part of the thermal energy of the burning fuel is converted into mechanical energy (Fig. 20–1).

A useful engine exhibits two crucial features:

1. An engine must work in **cycles** if it is to be useful. To see why a *noncyclic* operation is less useful, consider a gas that undergoes isothermal expansion. The gas is confined to a cylinder with a movable piston. The cylinder is in thermal contact with a thermal reservoir, such as a kitchen hot plate kept at a fixed temperature. During the expansion, the energy required by the piston to do work is supplied by the thermal reservoir. In this process, we convert thermal energy into mechanical energy. But we do not have an engine because the process operates in only one direction. The volume of the cylinder cannot increase indefinitely. The pressure, which is inversely proportional to the volume, decreases so, at some point, the expanding gas will no longer be able to push the piston. To be a part of a functioning engine, *the piston must eventually be recompressed and restored to its original position, ready to do work again.*

2. A cyclic engine must include more than one thermal reservoir. To understand this, let's reconsider the cylinder and piston. If the piston is returned to its original position while the cylinder remains in thermal contact with the original thermal reservoir, then all the work that the gas did while it expanded must be used to recompress the gas. The original expansion is reversed and nothing has been gained. In Chapter 18, we learned that the area enclosed by a closed path on a p–V diagram is the work done during the cycle. In this case, the area enclosed is zero, so the net work done by the system is also zero (Fig. 20–2). For a cycle to do finite work, thermal contact with the original thermal reservoir must be broken, and *temperatures other than the temperature of the original reservoir must come into play.* In this example, if we compress the gas in the cylinder at a lower temperature (so that its internal pressure will be lower) than was present during the expansion, less work will be needed for the compression than was produced in the expansion.

Efficiency.   Engines have an **efficiency**, which is a measure of how well the heat flow from the hotter thermal reservoir (such as a burning fuel) is converted into work (Fig. 20–1). If the work done in one complete cycle is W, then we define the efficiency, η, as the ratio of the work done to the total positive heat flow supplied by the burning fuel, Q_h:

$$\eta \equiv \frac{W}{Q_h}. \qquad (20–1)$$

According to the first law of thermodynamics, this efficiency can run from 0 to 1. When the efficiency is 1, or 100 percent, all the thermal energy taken from the thermal reservoir is converted to mechanical work; this is consistent with the conservation of energy (the first law). We will see, however, that an efficiency of 100 percent is impossible according to the second law. Figure 20–3 illustrates how the efficiency is calculated.

The definition of efficiency

FIGURE 20–3 An illustration of the technique of calculating efficiency.

Spontaneous Processes

We have mentioned two examples (hot tea plus ice cubes and an expanding gas) that illustrate a directionality in certain physical processes. Let's concentrate on the expanding gas example (free expansion). We can discuss this phenomenon most directly in terms of molecules. When a valve is opened and gas escapes from a small tank into a large empty chamber, the gas spreads and fills the chamber until the density and temperature are uniform throughout that chamber. Now imagine a film of this process made on a scale in which individual molecules can be seen. The molecules move rapidly and collide frequently in the small tank. When the valve is opened, molecules that reach the opening escape. They enter a vast space, empty of molecules, and the first batch of molecules encounter nothing in their way. Their first collisions will be with the walls of the chamber. As more and more molecules escape into the large chamber, they will begin to collide with other molecules as well as with the walls. After a short time, their velocities will be distributed according to Eq. (19–28).

Let's now suppose that we were to run the film backward. Because each collision is an elastic collision, it is as acceptable viewed backward as it is viewed forward. Thus each action in the reversed film is perfectly possible; in fact, the whole film, run backward, is possible in theory. It should be possible for the air to evacuate the large chamber spontaneously and concentrate into the small tank. So why do we never see this happen?

The explanation lies in the number of ways in which events involving many possibilities can happen. Consider the two poker hands in Fig. 20–4. In terms of probabilities, both hands are equally unlikely. The probability of each is approximately 2×10^{-7}, yet one looks much more special than the other. The reason is that hand I is a uniquely interesting one, whereas hand II is of a very undistinguished type because there are many, many ways to get 1 spade, 2 hearts, 1 diamond, and 1 club. Thus what we are really comparing is one distinct hand from a huge number of "ordinary" hands. An ordinary hand is much more probable because an immense number of individual (all equally rare) hands fit into this class.

Molecular distributions are like distributions of possible poker hands. Each molecule has an equal probability of being anywhere in the chamber, including the opening to the small tank, and each molecule has an equal probability of moving in any direction, including the one that will send it back into the small tank. Yet if we count the huge number of possible distributions for the huge number of molecules, the chance of them all ending up in the small tank is comparatively like being dealt a royal straight flush, although far less likely numerically. Thus the directionality of a spontaneous process such as free expansion is no miracle; it is just an almost sure bet.

PROBLEM SOLVING

How to calculate the efficiency

20-2 THE SECOND LAW OF THERMODYNAMICS

There are two equivalent formulations of the second law of thermodynamics that are useful in understanding the conversion of thermal energy to mechanical energy.

The Kelvin form. *It is impossible to construct a cyclic engine whose only effect is to convert thermal energy from a body at a given temperature into an equivalent amount of mechanical work.*

The Kelvin form of the second law of thermodynamics

The Clausius form of the second law of thermodynamics

There is no way to extract thermal energy from the ocean and use that energy to run an electrical generator *without further effect*, such as heating the atmosphere. In the example discussed in Section 20–1 (a gas expanding against a piston), thermal energy from a thermal reservoir is indeed converted into work, but the engine is not cyclic—the gas increases in volume. In an engine that takes thermal energy from a thermal reservoir cyclically and converts that energy to work, the additional effect is typically a heat flow from the engine into its lower-temperature surroundings. This brings us to the second formulation of the second law of thermodynamics:

The Clausius form.[†] *It is impossible to construct a cyclic engine whose only effect is to transfer thermal energy from a colder body to a hotter body.*

In other words, there is no cyclic engine that can freeze water and use the released energy to boil more water *with no further effect*. In this case, the further effect is associated with the fact that the cyclic combination freezer–boiler *must* use some energy (such as electrical) from the surroundings.

In this book, we will use mainly the Kelvin form, which was first stated by Lord Kelvin. Detailed analysis, however, shows that the two formulations of the second law of thermodynamics are entirely equivalent.

FIGURE 20–4 Has poker hand I been dealt honestly, or would you suspect a stacked deck? Hand I is just as likely as hand II.

The four steps of the Carnot cycle

20-3 THE CARNOT CYCLE

We stated earlier that thermal reservoirs at two (or more) different temperatures are necessary for the operation of a reversible thermodynamic cycle that performs work; that is, a reversible engine. A minimal version of such an engine, one that requires exactly two such reservoirs, was invented by Sadi Carnot in 1824. The reversible cycle followed by this engine—known as a **Carnot cycle**—is of great importance in thermodynamics.

The Carnot cycle, demonstrated in Figure 20–5 with a gas-filled cylinder and a piston, consists of four reversible steps. Figure 20–5a is a p–V diagram of these four steps and Fig. 20–5b depicts the cycle schematically. Step I (AB in Fig. 20–5a) is an isothermal expansion: The expansion takes place with gas in the cylinder in thermal contact with a thermal reservoir—say, a kitchen hot plate—at temperature T_h. Step II (BC) is an adiabatic step: The volume slowly expands further, while the pressure as well as the temperature of the gas decrease. During step II, the cylinder of gas has been insulated with styrofoam from thermal contact with its surroundings so that there is no heat flow ($\Delta Q = 0$). The temperature of the gas is monitored throughout this adiabatic expansion. When the temperature has dropped to T_c (the temperature of a second, colder, thermal reservoir—for example, a bathtub full of cold water), it is placed in thermal contact with this second reservoir. In step III (CD), an isothermal compression of the gas at T_c takes place. The final pressure of step III is determined so that when the gas is removed from contact with the bathtub and again placed in thermal isolation, the adiabatic compression of step IV (DA) increases the temperature and pressure of the gas until it returns to the starting point, A.

[†]Rudolf Clausius is best known for introducing the concept of entropy.

FIGURE 20–5 (a) The Carnot cycle, which consists of two adiabatic and two isothermal transformations. (b) Schematic diagram of the Carnot cycle.

In Fig. 20–5a, the isothermal legs are drawn as hyperbolas, and the adiabatic legs follow the curve $p = $ (a constant) $V^{-\gamma}$ for $\gamma < 1$—as would be appropriate for an ideal gas (see Chapter 18). However, there is no reason to insist that the working fluid must be an ideal gas.[†] *Any* gas or other compressible fluid would be suitable because isothermal or adiabatic changes are possible for any substance. The crucial features of the Carnot cycle are independent of whether the working fluid is an ideal gas or not.

The work done in tracing out a complete Carnot cycle is the area enclosed by the curve in the p–V diagram. The cycle shown is clockwise, so the work done is positive. There is also heat flow associated with the Carnot cycle. It cannot occur during the adiabatic steps II and IV because, by definition, the system is thermally isolated for these steps. During step I, there is a positive heat flow Q_h from the hotter thermal reservoir into the thermal system. (For an ideal gas, $\Delta U = 0$ during an isothermal process, and then $Q_h = W_{AB}$, the positive work done in the expansion.) Similarly, there is a negative heat flow $-Q_c$ from the colder thermal reservoir to the system during step III. In other words, there is a positive heat flow Q_c from the thermal system to the colder reservoir. Thus the Carnot engine absorbs a heat flow Q_h during step I, gives up a heat flow Q_c during step III, and does a total amount of mechanical work W. Figure 20–6 is a schematic representation of the result of the operation of this cycle. The first law of thermodynamics can be applied to the entire cycle. Because the engine returns to

∞ The work done in a cycle was discussed in Chapter 18.

[†]The "working fluid" is the material that expands and contracts, allowing the engine to do work.

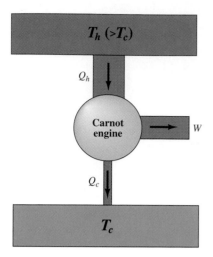

FIGURE 20–6 Schematic diagram of the result of the operation of one cycle of a Carnot engine. Heat flow Q_h is partly converted into mechanical work, W, and partly released as heat flow Q_c.

its original state, the internal energy is unchanged. This procedure holds even if the gas is not ideal. Then, according to the first law,

$$0 = W - Q_h + Q_c. \tag{20–2}$$

Equation (20–2) allows us to find the efficiency of the Carnot cycle, $\eta_C = W/Q_h$:

$$\eta_C = \frac{Q_h - Q_c}{Q_h} = 1 - \frac{Q_c}{Q_h}. \tag{20–3}$$

When the working gas of the Carnot cycle is ideal, the efficiency can be calculated in terms of the temperatures of the thermal reservoirs:

$$\eta_C = 1 - \frac{T_c}{T_h}. \tag{20–4}$$

This result follows from an analysis of the Carnot cycle with an ideal working gas. The analysis that gives this result is possible because the curves that describe both the adiabatic and isothermal curves for an ideal gas were found in Chapter 18.

How to Find the Efficiency of the Carnot Engine

Equation (20–4) is an important one. Let's see how to get it. Recall that for an ideal gas, the isothermal and adiabatic curves are

for an isothermal transformation: $pV = nRT = $ a constant,

for an adiabatic transformation: $pV^\gamma = $ a constant,

where the quantity $\gamma = C_p/C_V$ is greater than 1.

The Carnot cycle is labeled in Fig. 20–5a. For an ideal gas, the thermal energy is a function of temperature only, so the thermal energy, U, is constant during the isothermal steps. For these steps, $\Delta U = 0$, and the first law of thermodynamics, $\Delta U = \Delta Q - \Delta W$, implies that

for step I: $Q_h = Q_{A \to B}$,

and

for step III: $Q_c = -W_{C \to D}$.

There is a minus sign in the second equation because we have defined Q_c as the heat flow *from* the thermal system to the cold reservoir, whereas in the first law the heat flow is always *to* the thermal system.

The work done during the isothermal steps is

$$W_{A \to B} = \int_{V_A}^{V_B} p \, dV = \int_{V_A}^{V_B} \frac{nRT_h}{V} \, dV = nRT_h \ln\left(\frac{V_B}{V_A}\right);$$

$$W_{C \to D} = \int_{V_C}^{V_D} p \, dV = \int_{V_C}^{V_D} \frac{nRT_c}{V} \, dV = nRT_c \ln\left(\frac{V_D}{V_C}\right).$$

Using these results we find the efficiency of the cycle:

$$\eta = 1 - \frac{Q_c}{Q_h} = 1 - \frac{-W_{C \to D}}{W_{A \to B}} = 1 + \frac{nRT_c}{nRT_h} \frac{\ln(V_D/V_C)}{\ln(V_B/V_A)}$$

$$= 1 - \frac{T_c}{T_h} \frac{\ln(V_C/V_D)}{\ln(V_B/V_A)}. \tag{20–5}$$

We have not yet used any information about the shape of the adiabatic curves in this expression. These curves will relate the various volumes. Points B and C lie on the same adiabatic curve, so $p_B V_B^\gamma = p_C V_C^\gamma$. Similarly, for points A and D, $p_A V_A^\gamma = p_D V_D^\gamma$. The ratio of these two equations gives

$$\frac{p_B V_B^{\gamma}}{p_A V_A^{\gamma}} = \frac{p_C V_C^{\gamma}}{p_D V_D^{\gamma}}. \tag{20-6}$$

Next, we substitute for the ratio of the pressures on each side of this equation. Because A and B lie on an isotherm, $\dfrac{p_B}{p_A} = \dfrac{nRT_h/V_B}{nRT_h/V_A} = \dfrac{V_A}{V_B}$ and, similarly, because C and D also lie on an isotherm, $\dfrac{p_C}{p_D} = \dfrac{nRT_c/V_C}{nRT_c/V_D} = \dfrac{V_D}{V_C}.$ We insert these ratios into Eq. (20-6) to find that

$$\frac{V_B^{\gamma-1}}{V_A^{\gamma-1}} = \frac{V_C^{\gamma-1}}{V_D^{\gamma-1}}.$$

The same power, $\gamma - 1$, appears on both sides of this equation, so the arguments must be equal:

$$\frac{V_B}{V_A} = \frac{V_C}{V_D}.$$

This result considerably simplifies the expression for efficiency. The arguments of the logarithms in Eq. (20-5) are the same, so the logarithms cancel and we finish with Eq. (20-4).

EXAMPLE 20-1 In a Carnot engine with an ideal gas, calorimetry reveals that the heat flow from the hot thermal reservoir in one cycle equals 38 J, and the heat flow to the cold thermal reservoir is 28 J. The temperature of the cold reservoir is 290 K. (a) What is the mechanical work done during one cycle? (b) What is the efficiency of the cycle? (c) What is the temperature of the hot reservoir?

Solution: (a) To find the mechanical work done during one cycle, we can find the area enclosed by the cycle on a p–V diagram. This evaluation can be done by integration, but only with great difficulty because all the parameters of the cycle are not yet known. A much simpler way to compute the work done is to use the first law of thermodynamics, which, for a closed cycle, states that the work done plus the heat flow out of a thermal system equals the heat flow into the system: $W + Q_c = Q_h$, from Eq. (20-2). Thus

$$W = Q_h - Q_c = 38 \text{ J} - 28 \text{ J} = 10 \text{ J}.$$

(b) Equation (20-3) expresses the efficiency directly in terms of the given heat flows:

$$\eta = 1 - \frac{Q_c}{Q_h} = 1 - \frac{28 \text{ J}}{38 \text{ J}} = 0.26.$$

(c) Equation (20-4) gives an alternative expression for the efficiency in terms of the temperatures of the two thermal reservoirs. We know the efficiency and the temperature of the cold reservoir, so we can solve for the temperature of the hot reservoir. From Eq. (20-4),

$$T_h = \frac{T_c}{1 - \eta} = \frac{290 \text{ K}}{1 - 0.26} = 390 \text{ K}.$$

The Importance of the Carnot Engine

The second law of thermodynamics implies two important results:

1. *All Carnot cycles that operate between the same two temperatures have the same efficiency.* In particular, the efficiency of a Carnot cycle *does not depend on the use of an ideal gas*, and the expression for the efficiency is the one we have found for an ideal gas.

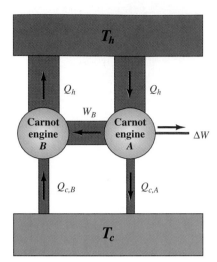

FIGURE 20-7 Carnot engines *A* and *B* are connected together, with *B* run in reverse. If engine *A* is more efficient than engine *B*, the net effect of the combination is to circumvent the Kelvin form of the second law of thermodynamics.

The Carnot cycle, with any working fluid, is the most efficient cycle that operates between two temperature extremes.

The efficiency of any Carnot cycle

FIGURE 20-8 The cyclic conversion of thermal energy into other forms of energy cannot be perfectly efficient; some thermal energy must be rejected. The cooling towers of a power plant, whether non-nuclear or nuclear, are the sites at which thermal energy is rejected. The clouds you see are condensing steam.

2. *The Carnot engine is the most efficient engine possible that operates between the two given temperatures.* (We will demonstrate this in Section 20–4.)

We can demonstrate the first result by showing that we would violate the second law of thermodynamics if it were not true. Suppose that we have two Carnot engines, *A* and *B*, that operate between the two same temperatures. Further, suppose that somehow the efficiency $\eta_A = W_A/Q_{hA}$ of cycle *A* were 75 percent, whereas the efficiency $\eta_B = W_B/Q_{hB}$ of cycle *B* were 74 percent. Let's adjust the amount of working gas in the cycles so that $Q_{hA} = Q_{hB}$, say, 100 J. This does not change the efficiencies. The work done by the two engines, which is the product of efficiency and heat flows, would then be different; namely, $W_A = (75\%)(100 \text{ J}) = 75$ J, whereas $W_B = (74\%)(100 \text{ J}) = 74$ J. Cycle *A* does more work than cycle *B* by an amount $\Delta W = W_A - W_B = 1$ J. We can now connect these two cycles in a way that violates the second law. Like any Carnot engine, engine *B* is reversible. We connect it to engine *A* as in Fig. 20–7 and run it in reverse.

Running a Carnot engine—or any reversible engine—in reverse means that the direction of the transformation in each step is reversed, so that the path runs counterclockwise on the *p–V* diagram. For example, step I in Figure 20–5a becomes an isothermal compression rather than an isothermal expansion. The work done during one cycle is then negative—positive work must be supplied to the cycle from the exterior to make it run—while heat flow Q_c is taken from the cold reservoir and heat flow Q_h goes to the hot reservoir. This result is indicated on Fig. 20–7 for Carnot engine *B*.

We can divide the 75 J of work done by engine *A* in Fig. 20–7 into two parts: We use 74 J to run engine *B* and 1 J is left over. An equivalent amount of heat flow goes into the hot reservoir as is taken out, so this reservoir plays no role. Thus the net effect of this combination engine is to employ only one reservoir to extract 1 J of work with each cycle. But such an arrangement violates the Kelvin form of the second law of thermodynamics. This result is quite general, and it is a remarkable one because Carnot engines can be constructed with many different working fluids.

We have shown that the efficiency of *any* Carnot cycle that operates between temperatures T_c and T_h is given by Eq. (20–4), whether or not the gas is ideal:

$$\text{for any Carnot cycle:} \qquad \eta_C = 1 - \frac{T_c}{T_h}. \qquad (20\text{–}7)$$

We have found a universal expression for the efficiency of a Carnot cycle in terms of the temperatures of the thermal reservoirs. Note that the temperature scale here is the ideal-gas temperature scale—the Kelvin scale. For a Carnot engine to be efficient, the temperatures of the two reservoirs should be quite different from one another ($T_c < T_h$), and the temperature of the cold reservoir should be as low as possible (Fig. 20–8). If T_c were to be zero, the efficiency would be 1 (100 percent), independent of T_h. This is the only situation that could give an efficiency of 1. Such a reservoir does not exist, and thus *a perfectly efficient Carnot engine is not possible.*

What happens if instead of the Carnot engine, with its idealized reversible cycle, we have an engine with some element of irreversibility? For example, there might be friction between the piston and the cylinder walls. We could reasonably expect that such an engine is less efficient than the Carnot engine that operates between the same two temperatures. We can, in fact, prove that this must be the case (see Problem 15).

20–4 OTHER TYPES OF ENGINES

Practical work-producing engines differ from the Carnot cycle in two ways. First, the path they trace out in a *p–V* diagram may not have the form of the Carnot cycle (two isothermal transformations and two adiabatic transformations), even though the path is reversible. These engines thus involve heat flows at temperatures other than the two temperatures of the Carnot cycle. Second, real engines generally involve some friction; that is, some irreversibility. We ignore this second possibility in the following discussion.

The Stirling Engine

As an example of a reversible non-Carnot cycle, consider the ideal Stirling cycle (Fig. 20–9), and compare this cycle to the Carnot cycle in Fig. 20–5a. Steps I and III are isothermal steps, just as in the Carnot cycle. The adiabatic steps of the Carnot cycle, however, are replaced by changes in pressure and temperature that maintain a constant volume. No work is done during the constant-volume steps. However, because the temperature changes, the internal energy also changes so that, by the first law of thermodynamics, there is a heat flow to or from the working gas of the engine during the constant-volume steps. This heat flow occurs at decreasing temperatures during step II and at increasing temperatures during step IV. It is as though a series of ever-colder thermal reservoirs enter into step II and ever-hotter thermal reservoirs enter into step IV. Increasingly hotter fuel must be used throughout step IV. Under these conditions, we note that the efficiency of an engine is really a measure of the work done for the total fuel burned; thus, the denominator of the efficiency should be the entire positive heat flow to the thermal system, which we write as Q_{pos}.

Here, we work out the efficiency of the ideal Stirling engine, η_S, when the working fluid is 1 mol of an ideal gas. We find that

$$\eta_S = \frac{W}{Q_{pos}} = \frac{R(T_h - T_c)\ln(V_2/V_1)}{RT_h\ln(V_2/V_1) + C_V(T_h - T_c)}. \tag{20–8}$$

This result would be the corresponding Carnot efficiency if it were not for the term proportional to C_V in the denominator (due to step IV). Because this extra term is positive, the denominator of η_S is greater than the denominator of η_C, and thus *the efficiency of the Stirling cycle is less than that of the Carnot cycle.*

How to Get the Efficiency of the Stirling Engine

We assume that the working gas in a Stirling engine is ideal. We need the total positive heat flow $Q_{pos} = Q_I + Q_{IV}$ as well as W, the work done during one cycle (Fig. 20–9). During step I, the temperature is constant and $\Delta U = 0$. Thus, by the first law of thermodynamics, the heat flow to the engine is the work done during the transformation:

$$Q_I = W_I = \int_{V_1}^{V_2} p\, dV = RT_h \int_{V_1}^{V_2} \frac{dV}{V} = RT_h \ln\left(\frac{V_2}{V_1}\right).$$

According to the first law, the heat flow to the engine during step IV is the change in internal energy $U(T_h) - U(T_c)$ because no work is done by the engine during this portion of the cycle. Thus,

$$Q_{IV} = U(T_h) - U(T_c) = C_V(T_h - T_c).$$

Combining these two equations, we see that the total heat flow to the engine is

$$Q_{pos} = Q_I + Q_{IV} = RT_h \ln\left(\frac{V_2}{V_1}\right) + C_V(T_h - T_c).$$

The total work done in the cycle is the area enclosed by the cycle on the p–V diagram:

$$W = W_I + W_{III} = RT_h \ln\left(\frac{V_2}{V_1}\right) + RT_c \ln\left(\frac{V_1}{V_2}\right) = RT_h \ln\left(\frac{V_2}{V_1}\right) - RT_c \ln\left(\frac{V_2}{V_1}\right)$$

$$= R(T_h - T_c) \ln\left(\frac{V_2}{V_1}\right).$$

The efficiency of this cycle, W/Q_{pos}, is then given by Eq. (20–8).

Stirling cycle

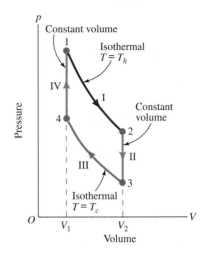

FIGURE 20–9 The ideal Stirling cycle. There is a positive heat flow to the engine in steps I and IV, and a heat flow from the engine in steps II and III. This cycle was described by Robert Stirling almost 20 years before Carnot discussed his own ideas. The engine Stirling patented in 1816 may be useful in outer space and under other conditions and has become the subject of active research.

APPLICATION

The Stirling engine

The Efficiencies of Other Types of Engines

It is possible to decompose an arbitrary reversible cycle into a series of Carnot cycles. By doing so, we can prove that *any* non-Carnot cycle (even if it is reversible) that operates between two temperature extremes T_h and T_c is less efficient than the Carnot cycle that operates between those same temperatures.

Let's apply this technique to the Stirling cycle. Figure 20–10a shows the path of the Stirling cycle approximated by a series of adiabatic and isothermal transformations. We can add a transformation as shown in Fig. 20–10b that will modify the approximation into one that consists of two Carnot cycles. This added transformation is an isothermal transformation from E_3 to E_2 and back again. Because this new transformation is a reversible round trip along the same line, it has no net effect on the overall efficiency of the approximation shown in Fig. 20–10a. At this point, we can think of the original approximation as consisting of two separate Carnot cycles. In Fig. 20–10c, these two cycles are labeled I and II.

Now, the combination of two Carnot cycles is less efficient than a single Carnot cycle that operates between T_h and T_c. To see why, suppose that the area under the approximating cycles is the same as the area under the Stirling cycle. Then the work done, which appears in the numerator of the expression for the efficiency, is unchanged. However, the denominator, Q_h, of the approximating cycle is bigger than the corresponding denominator for a single Carnot cycle that operates between T_h and T_c. This is because Carnot cycle I takes its incoming heat from two sources. For the step E_2E_3, it takes up the released heat from Carnot cycle II. However, along E_1E_2, it must take

FIGURE 20–10 (a) The Stirling cycle is approximated by a series of adiabatic and isothermal transformations. (b) By adding an isothermal back-and-forth segment at T_{int}, the Stirling cycle is approximated by two Carnot cycles. (c) The two Carnot cycles that approximate the Stirling cycle are labeled I and II.

In addition to the Stirling cycle that we have discussed, other reversible cycles represent idealized versions of real engines. A short list might include the diesel engine (Fig. 20AB–1); the Otto cycle (Fig. 20AB–2), which is similar to an automobile engine; and the Brayton cycle (Fig. 20AB–3). Detailed treatment of these cycles will be left to end-of-chapter problems. Keep in mind that, in addition to the fact that these cycles are inherently less efficient than the Carnot cycle, there is always additional inefficiency in real engines associated with irreversible portions of the cycle.

From the end of the 1960s, the U.S. government has requested that the efficiency of internal combustion engines be improved and that air pollution be reduced. Engines of that period used an excess of fuel, and the residue of unburnt hydrocarbons was a significant pollution source. The pollution problem was alleviated by adding air to the products of combustion as they left the cylinder, thereby burning them. What about the improvement in engine efficiency? The easiest way to accomplish this is to increase the maximum temperature at which the fuel is burned. But a higher temperature leads to a secondary reaction: the oxidation of the nitrogen in the added air. The resulting oxides of nitrogen added very significantly to "acid rain" problems. Thus, the government-mandated increase in efficiency resulted in a new pollution problem. The solution

to this built-in conflict is a complicated one, involving recycled gases and catalysts. It is not yet completely resolved. Perhaps the moral of this tale is that having scientifically literate people in government is not a luxury but a necessity.

FIGURE 20AB–2 The Otto cycle, for the engine designed by Nikolaus Otto in 1876.

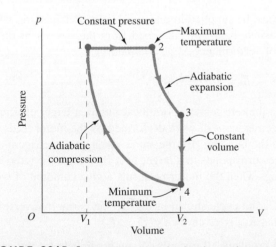

FIGURE 20AB–1 The Diesel cycle, for the engine patented by Rudolf Diesel in 1892.

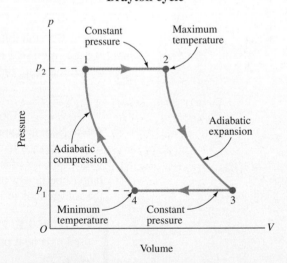

FIGURE 20AB–3 The Brayton cycle, named for George B. Brayton.

an additional amount of heat from the intermediate reservoir at T_{int}. This intermediate use of heat is *in addition to* the heat that cycle II takes at T_h. The same amount of work requires more heat, and the double Carnot approximation is less efficient. Thus, to the extent that the double Carnot cycle approximates the Stirling cycle, the Stirling cycle is less efficient than the single Carnot cycle.

The Carnot engine is more efficient than any other type of engine.

FIGURE 20–11 A heat pump, in which mechanical energy is used to transfer thermal energy from the ground to the interior of a building.

 APPLICATION

Heat pumps and refrigerators

The definition of the coefficient of performance of a heat pump

The two-cycle approximation shown is not, in fact, a very good approximation to the Stirling cycle. However, we can always improve the approximation by the inclusion of more Carnot cycles that operate between more intermediate temperatures. If we repeat the reasoning we applied for two cycles, we will always show that the approximation is less efficient than a single Carnot engine operating between the same temperature extremes; the intermediate Carnot cycles always involve some additional heat flow at intermediate temperatures. The technique described here can be applied to any reversible cycle. We learn the same lesson: *Any reversible cycle operating between given temperature extremes can be approximated by a series of Carnot cycles, and its efficiency will always be less than that of the single Carnot cycle that operates between the given temperatures.*

Heat Pumps and Refrigerators

Heat pumps and **refrigerators** are nothing more than work-producing engines run in reverse. A heat engine run in reverse—counterclockwise on a p–V diagram—transfers thermal energy from colder to hotter thermal reservoirs. For this to occur, work must be performed *on* the system. Heat pumps and refrigerators are similar to one another in operation, but they are used differently. A refrigerator removes thermal energy from a colder reservoir (which keeps a freezer cold). In doing so, the refrigerator is responsible for a heat flow at a higher temperature into its surroundings, something you can feel when you put your hand at the back of the refrigerator. A heat pump heats a house by taking thermal energy from a colder reservoir (for example, cool ground below a house) and dumping this thermal energy into the house at a higher temperature (Fig. 20–11). In each case, it is necessary to supply additional energy through the performance of work on the thermal system.

We can analyze heat pumps or refrigerators by the same method that we used to analyze heat engines. Because of the different ways these devices are used, their efficiencies are defined differently. For a heat pump, where the total positive heat flow rejected from the pump into a building, Q_{rej}, is of interest, we use the *coefficient of performance K_{hp}*, which is defined as

$$K_{hp} \equiv \frac{Q_{rej}}{W}, \tag{20–9}$$

where W is the total work that must be supplied in one cycle. For a refrigerator, where we are interested in the total positive heat flow absorbed from the interior of the refrigerator, Q_{abs}, the coefficient of performance K_{ref} is defined by

$$K_{ref} \equiv \frac{Q_{abs}}{W}, \tag{20–10}$$

where W is the work supplied to the refrigerator. For a heat pump, a larger coefficient of performance means that a given amount of work performed by the motor leads to a larger transfer of heat flow into the building. By the same token, for a refrigerator, a larger coefficient of performance corresponds to a larger heat flow from the interior of the refrigerator to its surroundings when the motor performs a given amount of work.

> **EXAMPLE 20–2** The Carnot cycle shown in Fig. 20–5a is run in reverse and used as a heat pump. Show that K_{hp} is given by $T_h/(T_h - T_c)$.
>
> **Solution:** In solving this problem, all formulas given previously for the operation of the Carnot cycle can be used, except that the signs must be changed because the cycle runs in reverse. These signs may be easily remembered with a schematic drawing of the operation, as in Fig. 20–12, where W, Q_c, and Q_h are all taken to be positive. It is also necessary to interpret the total heat flow rejected, Q_{rej}, as the heat flow Q_h transferred to the thermal reservoir at T_h. By the first law of thermodynamics,
>
> $$W = Q_h - Q_c,$$

so K_{hp} is given by

$$K_{hp} = \frac{Q_h}{Q_h - Q_c} = \frac{1}{1 - (Q_c/Q_h)},$$

according to Eq. (20–9). The ratio Q_c/Q_h is, in fact, the ratio of temperatures—see Eq. (20–7) and the accompanying discussion. It is not important that the signs of the heat flows are reversed because we use only the ratio of these quantities. With $Q_c/Q_h = T_c/T_h$, we find

$$K_{hp} = \frac{1}{1 - (T_c/T_h)} = \frac{T_h}{T_h - T_c}. \qquad (20\text{–}11)$$

The coefficient of performance of the Carnot heat pump is a number larger than 1 and it increases, or improves, if the difference between the temperatures of the two thermal reservoirs is reduced: It is easier to transfer thermal energy from the cold ground to a warm house when the temperature difference is smaller. An ideal heat pump used for heating a house in winter could operate between the temperatures $T_c = -5°C = 268$ K and $T_h = 45°C = 318$ K. (Although 45°C is much hotter than the interior of a house, if the size of the heat pump is not to dwarf the house itself or if the cost of insulating the house is not to be prohibitive, the temperature of the exhaust must be somewhat higher than the desired interior temperature.) The ideal K_{hp} of a Carnot cycle that operates between these temperatures is then $T_h/(T_h - T_c) = 318$ K/50 K = 6.36. In contrast, an actual heat pump that operates between the temperatures 45°C and −5°C has a value of K_{hp} that is only about half this value.

The coefficient of performance, as defined in Eq. (20–10), for a Carnot cycle run as a refrigerator (see Problem 22) is given by

$$K_{ref} = \frac{T_c}{T_h - T_c}. \qquad (20\text{–}12)$$

Just as for the heat pump, the coefficient of performance is improved if the temperature range across which the refrigerator runs is decreased.

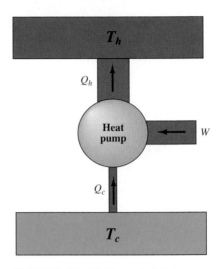

FIGURE 20–12 Example 20–2. The Carnot cycle operated as a heat pump.

20–5 ENTROPY AND THE SECOND LAW

Entropy as a Thermodynamic Variable

In this section, we will use the Carnot cycle to identify a new thermodynamic variable: entropy. This variable is like internal energy in that it is a function of state. On a macroscopic level, we will see that the entropy is a measure of the energy unavailable to do useful work; on a microscopic level, we will see that the entropy is a measure of disorder.

In a Carnot cycle, the efficiency—defined by $1 - (Q_c/Q_h)$—is given by Eq. (20–7): $\eta_c = 1 - (T_c/T_h)$. Thus, Q_c/Q_h must equal T_c/T_h. This relation can be expressed as

$$\frac{-Q_c}{T_c} + \frac{Q_h}{T_h} = 0. \qquad (20\text{–}13)$$

Recall that Q_c is the heat flow from the Carnot engine to the cold thermal reservoir. The expression $-Q_c$ in the first term of Eq. (20–13), minus sign and all, is then the heat flow to the engine at the cold temperature, just as Q_h is the heat flow *to* the engine at the high temperature.

Is there a relation comparable to Eq. (20–13) for another reversible cycle? Let's consider the Stirling cycle as approximated by two Carnot cycles. In Fig. 20–10c, the Carnot cycle that operates between T_c and T_{int} is Carnot cycle I; the Carnot cycle that operates between T_{int} and T_h is Carnot cycle II. Equation (20–13) holds for each of the two Carnot cycles, and the sum of the equations for each cycle is

$$\frac{-Q_{c,I}}{T_c} + \frac{W_{int,I}}{T_{int}} + \frac{-Q_{int,II}}{T_{int}} + \frac{Q_{h,II}}{T_h} = 0.$$

The interpretation of the heat flow in the first term (including the minus sign) is as before: the heat flow to the Stirling cycle at its minimum temperature. The numerator of the second term is the *net* heat flow to the cycle, Q_{int}, at temperature T_{int}. The third term is the heat flow to the engine at the maximum temperature.

This result is generalized if the cycle is more closely approximated by more Carnot cycles that operate at finer temperature divisions. We can add Eq. (20–13) for each of these approximating Carnot cycles. The heat flows at the intermediate temperatures T_i will cancel in part, leaving, in each case, the *net* heat flow to the cycle, Q_i, at temperature T_i. The first and last terms, $-Q_{min}/T_{min}$ and Q_{max}/T_{max}, remain. The net effect is

$$\frac{-Q_{min}}{T_{min}} + \frac{Q_1}{T_1} + \frac{Q_2}{T_2} + \cdots + \frac{Q_{max}}{T_{max}} = 0. \qquad (20\text{--}14)$$

The heat flows in Eq. (20–14) represent heat flows *to* the thermal system. We mean here that a heat flow to the system is positive, and a heat flow from the system is negative. If we let ΔQ_i be the net heat flow at temperature T_i, Eq. (20–14) becomes

$$\sum_i \frac{\Delta Q_i}{T_i} = 0, \qquad (20\text{--}15)$$

where the sum is taken over the full cycle. This expression depends on the original cycle being approximated by a finite number of Carnot cycles that operate at the successive temperatures T_i. In the limit in which the temperature divisions become finer and finer, the sum in Eq. (20–15) becomes an integral. In integral notation, Eq. (20–15) is known as **Clausius's theorem**:

Clausius's theorem for reversible cycles

$$\text{for a reversible cycle:} \qquad \oint \frac{dQ}{T} = 0. \qquad (20\text{--}16)$$

The circle placed on the integral is a reminder that the integral is taken over a closed cycle. Because any reversible cycle can be approximated by a series of Carnot cycles, Clausius's theorem is a general result that holds for *any reversible cycle, which means any closed path on a p–V diagram.*

We have already encountered a result like Eq. (20–16). In the discussion of conservative forces in Chapter 7, we saw that the integral $\oint \mathbf{F} \cdot d\mathbf{s}$ equals zero, which allowed us to define a potential energy $\mathbf{U}$. This function depends on position alone; moreover, the integral of dU along a path that connects two points is independent of the path. The same situation applies here. There is a function S, which we call **entropy**, that is analogous to U. The analogue of $dU = -\mathbf{F} \cdot d\mathbf{s}$ is the expression that defines the change in entropy as

The definition of the change in entropy

$$\text{for a reversible transformation:} \qquad dS \equiv \frac{dQ}{T}. \qquad (20\text{--}17)$$

According to Eq. (20–16), the integral of dS around a closed, reversible path on a p–V diagram is zero. The analogue to the expression for the change in U, which is $U_B - U_A = \int_A^B dU = -\int_A^B \mathbf{F} \cdot d\mathbf{s}$, is an expression for the change in entropy when a thermal system undergoes a transformation between any two thermodynamic states A and B:

$$S(B) - S(A) = \int_A^B \frac{dQ}{T}, \qquad (20\text{--}18)$$

where the integral is taken over *any* reversible path between these states because the integral is independent of path. As for U, the entropy depends only on the position on a p–V diagram; that is, *the entropy, S, is a function of state.*

Entropy is a function of state.

The fact that an integral of dS between two points along some path is independent of the path means that S behaves like the thermal energy U. A differential quantity whose integral is independent of the path is said to be a perfect differential. The infinitesimal heat flow dQ is *not* a perfect differential because its integrals are path dependent. But when dQ is divided by the temperature at which it occurs, the differential becomes perfect.

As is true for potential energy, only *changes* in entropy have any physical meaning—as Eq. (20–18) shows. The choice of zero for the entropy function is a matter of convenience.

EXAMPLE 20–3 Calculate the entropy change of 1 mol of an ideal gas that undergoes an isothermal transformation from an initial state of pressure 1.50 atm and volume 500 cm^3 to a final state of pressure 0.90 atm.

Solution: To calculate the entropy change along the specified isothermal path, we must know the heat flow along that path. For the isothermal transformation of an ideal gas, the internal energy does not change, so the first law of thermodynamics tells us that along our path, $dQ = dW + dU = dW = p\,dV$. Thus, from Eq. (20–18) and the ideal gas law, $pV = nRT$, we find

$$\Delta S = \int \frac{p\,dV}{T} = \int \frac{nRT}{VT}\,dV = nR\int \frac{dV}{V} = nR\ln\left(\frac{V_f}{V_i}\right) = -nR\ln\left(\frac{p_f}{p_i}\right)$$

$$= -(1\text{ mol})(8.3\text{ J/mol}\cdot\text{K})\ln\left(\frac{0.90\text{ atm}}{1.50\text{ atm}}\right) = 4.2\text{ J/K}.$$

How Entropy Changes for Irreversible or Spontaneous Processes

Irreversible processes (processes with friction, for example) lead to increased inefficiency in any otherwise reversible engine. Let's see what this implies for heat flows to such a cycle. The efficiency of an arbitrary cycle is

$$\eta = \frac{W}{Q_{abs}} = 1 - \frac{Q_{rej}}{Q_{abs}},$$

where Q_{rej} is the total heat flow rejected by the cycle and Q_{abs} is the total heat flow absorbed. We have used the relation $W = Q_{abs} - Q_{rej}$ from the first law of thermodynamics. Each of these heat flows is positive. The efficiency can be reduced in two ways. First, the heat flow *to* the system, Q_{abs}, can be reduced. Second, the heat flow rejected, Q_{rej}, can be increased (which is equivalent to a reduced heat flow to the system). Thus, the efficiency is reduced if the heat flow *to* the system is algebraically reduced anywhere along the cycle. In other words, the effect of irreversibilities or of spontaneous processes is to reduce the algebraic heat flow to the system, dQ. In turn, the integral in Clausius's theorem, Eq. (20–16), is reduced algebraically below zero:

Clausius's inequality

$$\text{for an irreversible cycle:} \qquad \oint \frac{dQ}{T} < 0. \qquad (20\text{–}19)$$

This result, known as **Clausius's inequality**, is a general result that holds for heat pumps as well as for engines.

Equations (20–16) and (20–19) can be combined into

$$\oint \frac{dQ}{T} \le 0. \qquad (20\text{–}20)$$

The equality applies when the path is reversible; the inequality applies when the transformations include some irreversible portion.

Equation (20–20) can be used to learn about entropy changes in irreversible transformations. Consider the cycle shown in Fig. 20–13, in which leg $A \rightarrow B$ is irreversible. From Eq. (20–20),

$$0 > \oint \frac{dQ}{T} = \int_{A}^{B} \frac{dQ}{T} + \int_{B}^{A} \frac{dQ}{T}.$$
$$\phantom{0 > \oint \frac{dQ}{T} = }\text{irrev} \qquad \text{rev}$$

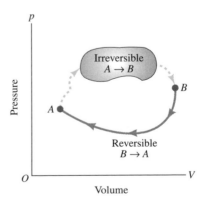

FIGURE 20–13 A cycle that contains an irreversible path from A to B and a reversible path from B back to A.

The right-hand integral is given by $S(A) - S(B)$ because it is taken over a reversible path. When we move this quantity to the left-hand side, we find that

$$S(B) - S(A) > \int_{\substack{A \\ \text{irrev}}}^{B} \frac{dQ}{T}. \tag{20-21}$$

Thus, *the difference in entropy between two points is greater than the integral of* dQ/T *over an irreversible change.*

How do we compute the increase in entropy when there is an irreversible transformation from thermodynamic state A to thermodynamic state B? We have seen that entropy is a function of state, so we can compute the change $S(B) - S(A)$ by any means we like; that is, *we can use a reversible path to find the change in entropy as long as that path connects the two given states.* We will illustrate this procedure in Section 20–6.

The Entropy of an Isolated System Never Decreases

One type of irreversible change is special: the irreversible change with no heat flow. Free expansion into a thermally isolated container is an example of such a spontaneous transformation. If there is no heat flow, then dQ is zero every step of the way, and

for a thermally isolated system: $\qquad S(B) - S(A) \geq 0. \tag{20-22}$

When an isolated system undergoes an irreversible process, its entropy increases.

The equality holds when a reversible, or adiabatic, transformation takes place; the inequality holds when the transformation from A to B is irreversible, as is true for free expansion. Note that the insulated container into which a gas expands in free expansion is a small, isolated system. But even if a given system is not thermally isolated, combinations of systems may well be. We may have a small system such as a cylinder of gas with a movable piston in thermal contact with a reservoir. Neither the cylinder nor the reservoir is isolated, but the combination of the two may be. *A spontaneous process in an isolated system increases the system's entropy.* Because the universe—our entire surroundings—is in contact with no other system, we say that irreversible processes increase the entropy of the universe.

The fact that the final entropy is greater than or equal to the initial entropy means that every spontaneous process increases the entropy of an isolated thermal system. Only the state of maximum entropy is a stable one: For a state of maximum entropy, no spontaneous changes occur. For an isolated thermal system, *the state of maximum entropy is the state of stable equilibrium.*

20-6 ENTROPY AND IDEAL GASES

When a gas is ideal, we can calculate its entropy change in reversible transformations. In fact, we can use a combination of constant-volume and isothermal transformations to calculate the entropy of an ideal gas. In this way, we can acquire a feeling for the nature of entropy as a thermodynamic variable. Also, we can use the results in Section 20–7 when we investigate the physical nature of entropy.

Consider the isothermal transformation from a point A at temperature T_0 and volume V_0 to a point B at temperature T_0 but at volume V. In this transformation, the internal energy—which for an ideal gas is a function only of temperature—is unchanged. Therefore, according to the first law of thermodynamics, $dW = dQ$ along the isotherm. Then, according to Eq. (20–18),

$$S(B) - S(A) = \int_A^B \frac{dQ}{T_0} = \int_A^B \frac{dW}{T_0}. \tag{20-23}$$

The infinitesimal work done at any point along the path is $dW = p \, dV = (nRT_0/V) \, dV$. The factor T_0 cancels, and

for a fixed temperature:
$$S(B) - S(A) = \int_A^B \frac{nRT_0}{V} dV \left(\frac{1}{T_0} \right)$$

$$= NR \int_A^B \frac{dV}{V} = nR \ln \left(\frac{V}{V_0} \right). \tag{20–24}$$

This expression is the change in entropy for a fixed temperature. There will be other transformations in which the temperature is changed. To keep such changes distinct from the one we have just used, we next perform a transformation at fixed volume. This transformation takes an ideal gas from point $A = (T_0, V_0)$ to point $B = (T, V_0)$. For a constant-volume transformation, the work done is zero and, according to the first law of thermodynamics, the heat flow $dQ = dU = C_V dT$. Thus

$$S(B) - S(A) = \int_A^B \frac{dQ}{T} = \int_A^B C_V \frac{dT}{T}. \tag{20–25}$$

C_V is a constant, so

for a fixed volume:
$$S(B) - S(A) = C_V \int_A^B \frac{dT}{T} = C_V \ln \left(\frac{T}{T_0} \right). \tag{20–26}$$

Equation (20–24) gives the entropy change for fixed T and Eq. (20–26) gives it for fixed V. These are independent state variables so, to find the entropy change in going from any standard state (T_0, V_0) to state (T, V), simply add the two expressions:

$$S(T, V) - S(T_0, V_0) = C_V \ln \left(\frac{T}{T_0} \right) + nR \ln \left(\frac{V}{V_0} \right). \tag{20–27}$$ The entropy of an ideal gas

This expression specifies the function $S(T, V)$ up to a constant value $S_0 = S(T_0, V_0)$. If we want S in terms of other variables, such as pressure and volume, the ideal gas law can be used for the substitution.

Note that we used reversible paths to calculate the entropy change. Because entropy is a function of the thermodynamic state, once we know the entropy change between given thermodynamic states, it makes no difference how the transformation between those states is made. The entropy change is the same whether that transformation is reversible or irreversible.

EXAMPLE 20–4   You have 3.0 mol of an ideal gas in thermal contact with a thermal reservoir at $T = 300$ K. The gas expands spontaneously—the gas does not push against a piston to do work—from 12 L to 24 L. What is the entropy change of the gas? of the universe? Repeat the exercise for an expansion from 12 L to 120 L.

Solution: The entropy of the gas, being a state function, can be computed (up to a constant) when the state is known. In this case, we can use Eq. (20–27) with fixed $T = T_0$. Because $\ln(1) = 0$,

$$\Delta S = C_V \ln \left(\frac{T_0}{T_0} \right) + nR \ln \left(\frac{V}{V_0} \right) = C_V(0) + nR \ln \left(\frac{V}{V_0} \right)$$

$$= nR \ln \left(\frac{V}{V_0} \right) = (3.0 \text{ mol})(8.3 \text{ J/mol} \cdot \text{K}) \ln \frac{24L}{12L}$$

$$= 25 \ln(2) \text{ J/K} = 17 \text{ J/K}.$$

We have found that the entropy change of the gas is positive. To determine the entropy change in the rest of the surroundings, we must determine the heat flow to the thermal reservoir. The gas does no work in its expansion, so the heat flow out is equal to the change in internal energy. But this change is zero because the gas is ideal and because its internal energy depends only on the (unchanging) temperature. There is thus no heat flow to the thermal reservoir and its entropy change, $\Delta Q/T_0$, is zero. The entropy change of the universe is therefore the entropy change of the gas. The fact that it is positive indicates that the transformation involved—in this case, free expansion—is irreversible.

When the expansion is from 12 L to 120 L, the entropy change is as just described, except that ln(2) is replaced by ln(10):

$$\Delta S = 25 \ln(10) \text{ J/K} = 57 \text{ J/K}.$$

The natural logarithm changes fairly slowly with its argument, so, compared to the ratio of volume changes in the two cases, the difference in entropy changes is not very great.

20-7 THE MEANING OF ENTROPY

The entropy of a system and all its surroundings—the universe—increases when an irreversible process takes place. In effect, increased entropy corresponds to a lost heat flow, and hence a lost energy. The loss of heat flow represents energy that can never be turned into work because, by the nature of irreversible processes, the process cannot be reversed to recapture this heat flow without making other changes. Of course, the energy is not really lost; energy is always conserved. But it has been turned into the random motion of the microscopic constituents of the thermal system and cannot be sufficiently well "organized" to allow all of it to go into, say, lifting a weight. By organization, we mean order; if all the molecules were moving in precisely the same direction, for example, they would be well ordered. They might have the same total energy as a disorganized gas, but *all* of their energy could be used to lift a weight. Any degree of irreversibility reduces our ability to do work because it increases the disorder.

The preceding paragraph contains the seeds of a deeper understanding of entropy based on a microscopic picture. The idea of microscopic order and disorder is the key to entropy (Fig. 20–14). Let's consider this idea more closely.

We indicated in Section 20–1 that, when a gas expands into an empty space, its molecules move from a configuration that has a low probability (all the molecules are in one corner) to one that has the largest possible probability (the molecules are spread uniformly throughout the allowed volume). Equivalently, the molecules go from a configuration that has few ways of occurring (like the poker hand in Fig. 20–4 with a royal straight flush) to one that has many ways of occurring (like the "ordinary" hand). Initially, all the molecules are confined by walls to a small tank. This initial state is more *ordered* than the final state because the individual molecules are more localized. A loss of order is associated with a change of the system from a state of low probability to one of higher probability. Equation (20–24) illustrates how entropy increases for an ideal gas when the volume increases.

We can further illustrate this idea by mixing two gases that are initially separated by a partition (Fig. 20–15). Consider a large container of volume V divided by a partition into volumes V_1 and V_2. There are n_1 mol of gas 1 in V_1 and n_2 mol of gas 2 in V_2. Each gas has the same temperature and pressure. The partition is opened, and the gases in the two parts mix. Before the partition is opened, the degree of order is greater: All the molecules of each gas are on their respective sides. After the partition is opened, the disorder is greater: Any one molecule is somewhere in the greater volume, rather than being on one side or the other. It is more probable that each gas will spread separately throughout the greater volume than that it will stay on its respective side. Thus, we would expect the entropy to increase when the partition is opened, which is in line with the idea that the mixing of the two sides is an irreversible process. This increased entropy is the **entropy of mixing**.

EXAMPLE 20–5 A quantity of 0.20 mol of argon gas contained in a volume of 5.0 L mixes with 0.50 mol of neon gas contained in a volume of 12.5 L, making

(a)

(b)

FIGURE 20–14 (a) When a plate breaks, it is more disordered—the entropy of the "plate system" increases. (b) The entropy of the demolished building has rather suddenly increased.

a total volume of 17.5 L. Both gases are at the same temperature and can be regarded as ideal. All volumes are thermally isolated. What is the change in entropy?

Solution: The pressure of each gas must be the same because the pressure of an ideal gas is given by $p = nRT/V$, and the temperatures are the same. The ratios n/V are also the same. The only change in entropy comes from the changes in volume of the two gases:

$$\Delta S = \Delta S_{Ar} + \Delta S_{Ne}.$$

For the change in the entropies of each gas, use Eq. (20–24), which is appropriate for a fixed temperature:

$$\Delta S_{Ar} = n_{Ar} R \ln\left(\frac{V_{Ar}}{V_{0,\,Ar}}\right) = (0.20 \text{ mol})(8.3 \text{ J/mol} \cdot \text{K})\ln\left(\frac{17.5 \text{ L}}{12.5 \text{ L}}\right)$$

$$= 17 \ln(3.5) \text{ J/K} = 21 \text{ J/K};$$

$$\Delta S_{Ne} = n_{Ne} R \ln\left(\frac{V_{Ne}}{V_{0,\,Ne}}\right) = (0.50 \text{ mol})(8.3 \text{ J/mol} \cdot \text{K})\ln\left(\frac{17.5 \text{ L}}{12.5 \text{ L}}\right)$$

$$= 42 \ln(1.4) \text{ J/K} = 14 \text{ J/K}.$$

The entropy of mixing is thus

$$\Delta S = \Delta S_{Ar} + \Delta S_{Ne} = 35 \text{ J/K}.$$

We can similarly interpret the direction of spontaneous heat flow, in which systems with different temperatures tend to the same intermediate temperature when they are placed in thermal contact—ice cubes melt when they are placed in hot tea. In this case, the more ordered system is the cooler one. On average, the velocity vectors of the molecules of steam at a lower temperature are confined to a smaller range than are those of steam at a higher temperature. The heated system is more disordered because, on average, the velocity vectors spread over a greater range. Equation (20–26) illustrates how entropy increases for an ideal gas when temperature increases.

More advanced treatments of statistical physics quantify the idea that entropy is a measure of disorder. In fact, *all the consequences of the second law of thermodynamics follow from the treatment of entropy as a measure of disorder.*

Engines and Entropy

The principle that entropy never decreases is a way of formulating the second law of thermodynamics that emphasizes the directionality of thermal processes. The application of this formulation to engines is less obvious because the entropy in a reversible cycle does not change. Here, we show how the entropy formulation of the second law applies to engines and leads to our earlier formulations of the second law.

Consider a thermal reservoir at temperature T_h. Can we construct an engine that, in a single cycle, extracts thermal energy through a (positive) heat flow Q_h from that reservoir to lift a weight and does nothing else? Let's suppose that we could do so. There is no change in the entropy of the weight that has been lifted and there is no change in the engine after a full cycle. Thus, in the isolated system that consists of the weight, the engine, and the reservoir, the only entropy change is that of the reservoir, as given by $\Delta S = -Q_h/T_h$. This entropy change is negative and therefore violates the entropy formulation of the second law. We deduce that, in order to satisfy the entropy formulation of the second law, there must be a second thermal reservoir (at temperature T_c) such that the net entropy change is positive. If this second reservoir is to have a positive entropy change, then there must be a heat flow Q_c into it. This gives rise to an entropy change $\Delta S' = Q_c/T_c$.

(a)

(b)

(c)

FIGURE 20–15 (a) Two gases are initially unmixed. (b) A valve is opened, and each gas diffuses into the other's flask. (c) The final state is more disordered than the initial state. The system's entropy has increased.

Entropy is a measure of disorder.

The isolated system, which now consists of the weight, the engine, and the two reservoirs, must satisfy the entropy formulation of the second law, namely, $\Delta S_{tot} = \Delta S + \Delta S' \geq 0$, or

$$-\frac{Q_h}{T_h} + \frac{Q_c}{T_c} \geq 0;$$

$$\frac{Q_c}{Q_h} \geq \frac{T_c}{T_h}.$$

Because the work done to lift the weight is $W = Q_h - Q_c$ and the efficiency of the engine is defined as $\eta \equiv W/Q_h = (Q_h - Q_c)/Q_h = 1 - (Q_c/Q_h)$, the entropy formulation of the second law leads to

$$\eta \leq 1 - \frac{T_c}{T_h}.$$

We are now on familiar ground. The equality applies for the reversible Carnot cycle, where there is no heat flow except at the two temperature extremes. In this case, ΔS_{tot} is zero, which requires that $(Q_c/Q_h) = (T_c/T_h)$, and the efficiency is $\eta = 1 - (T_c/T_h)$. For cycles with irreversibility, or for cycles such as the Stirling cycle in which there is heat flow at intermediate temperatures, the inequality sign holds. Thus, the entropy formulation of the second law implies that the Carnot engine is the most efficient type of engine.

We have now connected the two apparently dissimilar aspects of the second law of thermodynamics: the impossibility of constructing a perfectly efficient engine and the inevitable direction of spontaneous processes.

*The Arrow of Time

Drop an egg and film the event. (Choose your laboratory carefully.) This process is a typical, if particularly spectacular, macroscopic event. Do the same for the collision of two billiard balls. The collision of two billiard balls resembles the collision of two molecules, and the film of this event represents fairly the fundamental interactions between the molecular constituents of matter.

If you were to see these two films, could you tell whether they were run forward or backward? The film of the ruins of the egg reassembling into a perfect ovoid would appear to be a cheap comic trick that could not possibly be confused with a real event (Fig. 20–16). But the two billiard balls colliding in reverse would be absolutely indistinguishable from the real thing—at least in the short time period around the collision.

The elementary forces between the microscopic constituents of matter are said to be indifferent to the direction of time, or *invariant under time reversal*. This result is fundamental, and it seems to hold (with one very tiny exception discussed in Chapter 46) for all the fundamental forces in nature, including the electromagnetic forces that govern the collisions of gas molecules. If the fundamental forces are invariant under time reversal, and if they are the forces that act in ordinary materials, why does there seem to be a preferred direction of time—an **arrow of time**—in which eggs break rather than reassemble? Why do spontaneous processes, in which entropy increases, proceed in one direction and not the other?

We have seen that the reason is statistical. Individual collisions between molecules do not depend on the direction of time. The fundamental laws that govern the reformation of the egg are thus independent of the direction of time—it is not against any physical laws for a broken egg to reassemble itself. It is just highly unlikely. The overwhelming majority of collisions between gas molecules lead to more disordered situations. The arrow of time expresses not a definite preferred direction of time in a fundamental sense but an overwhelming likelihood that out of a random sequence of microscopic interactions, less ordered situations will result. Irreversibility means only that the spontaneous reversal of the "irreversible" process is so unlikely as not to be worth considering.

FIGURE 20–16 In this stroboscopic photograph, the direction of increasing time follows the egg descending and breaking, not reassembling and rising.

The second law of thermodynamics limits the efficiency of cyclic engines and governs the direction of spontaneous processes in thermal systems. Cyclic engines do work in repeated cycles by extracting a heat flow from hotter external systems, converting part of that heat flow into work, and rejecting a heat flow to cooler thermal systems. The efficiency of such an engine is defined by

$$\eta \equiv \frac{W}{Q_h},\tag{20-1}$$

where W is the work done in one complete cycle and Q_h is the total positive heat flow the cycle extracts from its sources of fuel.

Two equivalent ways to state the second law of thermodynamics are the Kelvin form and the Clausius form. The Kelvin form states that it is impossible to convert thermal energy from a body at a given temperature into an equivalent amount of mechanical work with no other effect; the Clausius form states that it is impossible to transfer thermal energy from a colder body to a hotter body with no other effect.

In the Carnot cycle, which involves four reversible steps that occur at just two temperatures, a working fluid undergoes alternate adiabatic and isothermal transformations. It is the most efficient cycle possible that operates between two temperatures T_h and T_c. No matter what the nature of the working fluid, the efficiency of the Carnot engine is

$$\text{for any Carnot cycle:} \qquad \eta_C = 1 - \frac{T_c}{T_h}.\tag{20-7}$$

The Carnot cycle that operates between two temperatures is more efficient than any other cycle that operates between the same temperatures extremes. Any element of irreversibility also decreases the efficiency.

When the reversible cycles of work-producing engines are run in reverse, they act as refrigerators or heat pumps. Work is done on both of these machines while they extract a heat flow from a cold reservoir and release a heat flow to a hot reservoir.

A new thermodynamic variable, entropy, S, measures the amount of disorder in a thermal system. Like thermal energy, it is a function of state, and only *changes* in entropy have physical significance. Entropy changes are path independent. For infinitesimal reversible transformations,

$$dS \equiv \frac{dQ}{T}.\tag{20-17}$$

For an ideal gas, entropy can be expressed as a function of T and V as

$$S(T, V) - S(T_0, V_0) = C_V \ln\left(\frac{T}{T_0}\right) + nR \ln\left(\frac{V}{V_0}\right).\tag{20-27}$$

Note that entropy decreases as temperature decreases.

The path independence of entropy changes is equivalent to the fact that, for a reversible cycle,

$$\text{for a reversible cycle:} \qquad \oint \frac{dQ}{T} = 0.\tag{20-16}$$

a result known as Clausius's theorem. If a cycle has any element of irreversibility, Clausius's theorem is modified to Clausius's inequality,

$$\text{for an irreversible cycle:} \qquad \oint \frac{dQ}{T} < 0.\tag{20-19}$$

If the change $A \rightarrow B$ is irreversible, the following relation holds:

$$S(B) - S(A) > \int_{\substack{A \\ \text{irrev}}}^{B} \frac{dQ}{T}. \qquad (20\text{--}21)$$

In particular, for transformations of thermally isolated systems, we find

$$\text{for a thermally isolated system:} \quad S(B) - S(A) \geq 0. \qquad (20\text{--}22)$$

A spontaneous process in an isolated system increases the system's entropy, and the state of maximum entropy is the state of stable equilibrium. The increase of entropy expresses the inevitable increase in the disorder of thermally isolated systems.

UNDERSTANDING KEY CONCEPTS

1. If the ocean is viewed as a reservoir at one temperature and the atmosphere as a reservoir at another temperature, when can thermal energy flow spontaneously from the ocean to the air?

2. When the brakes are applied, a car decelerates and then stops. A film of the process run in reverse obviously looks fake. Does the process in reverse violate the second law of thermodynamics? If so, in what way?

3. The historian Herman Daly has called the second law of thermodynamics "the law of random, ravage, and rust." Is he right?

4. Why is it not possible for an engine—even a Carnot engine— to have an efficiency of one?

5. Suppose that a gas obeys the van der Waals equation of state. What is the efficiency of a Carnot engine that uses this gas as a working fluid in terms of the temperatures of the hot and cold reservoirs?

6. In what climates are heat pumps most likely to be used?

7. A professor spends a couple of hours straightening up his office—putting books away, filing papers, throwing old exam papers away, and so on. He has created a less probable configuration out of a more probable configuration. Has he violated the second law of thermodynamics?

8. Imagine a microscopic device at the boundary between two containers of a gas that could sense the arrival of a particularly fast molecule heading from container A toward container B and could open a tiny door to let that molecule through. The same device, which is known generically as a *Maxwell demon*, could sense the arrival of a particularly slow molecule heading from container B toward container A and could let the molecule pass through. How would such a device violate the second law of thermodynamics? Is such a device feasible?†

9. Give examples of things you have observed recently that exemplify the second law of thermodynamics. Discuss the entropy change associated with them.

10. Would heat pumps that use an underground water reservoir be more efficient than those that use the outside air as a thermal reservoir if the pumps are employed to (a) heat in the winter? (b) cool in the summer?

11. List and discuss some processes not already mentioned that will not violate the first law of thermodynamics but will violate the second law.

12. When gasoline is burned in the chamber of a piston, chemical energy is turned into mechanical energy that is used to propel an automobile. How is this process consistent with the second law of thermodynamics?

13. If you were to run a film of an egg that drops to the floor, you would have little trouble determining if the film were running backward or not. Yet if the film were in extreme closeup, so that you were observing the interaction of molecules in the dropped egg, you would be hard pressed to say if the film were running forward or backward. How are these two statements consistent?

14. Is it possible to cool a house by leaving a refrigerator door open? What would be the net effect if you were to leave the door open?

15. Is the entropy of a gas an extensive quantity (such as volume) or an intensive quantity (such as pressure or temperature)? (See Problem 3 in Chapter 17.)

16. Biological systems use energy (ultimately from the Sun) in organizing highly ordered organisms, such as humans. Does this mean that biological systems violate the second law of thermodynamics?

PROBLEMS

20–1 Beyond Energy Conservation

1. (I) An engine produces 3.8×10^5 J of mechanical energy accompanied by a total heat flow of 7.6×10^5 J into the engine. What is its efficiency?

2. (I) Suppose that you toss a pair of dice. Make a list of all the possible outcomes, and calculate the probability of getting totals of 2, 3, 4, ... , 12, respectively.

3. (II) Six tokens, numbered one to six, are put into a hat. The tokens with even numbers are white, the ones with odd numbers are black. Suppose you draw four tokens at random, without putting them back into the hat. (a) List all the possible outcomes, without regard to order (for example, 1234 and 2134 do not count as separate outcomes). (b) What is the probability of drawing the tokens with 1, 2, 3, 4 on them? (c) What is the probability of drawing two white and two black tokens?

---

†For an interesting discussion of Maxwell demons, see Charles H. Bennett, "Demons, Engines, and the Second Law," *Scientific American,* Nov. 1987, pp. 108–116.

4. (II) The N molecules of a gas are placed in the left half of a chamber and kept from entering the right half by a membrane. The membrane is then removed and the molecules distribute themselves uniformly throughout the whole volume (Fig. 20–17). What is the probability that, at some time during their subsequent random motion, they will all collect in the left half of the chamber, for the cases $N = 2$, 100, 10^6, and 10^{23}? [Hint: This is equivalent to the probability of getting nothing but heads in 10^{23} coin tosses. To get an idea of this number in scientific notation, use the fact that $\frac{1}{2} \approx 10^{-0.3}$.]

FIGURE 20–17 Problem 4.

20–3 The Carnot Cycle

5. (I) What is the maximum efficiency of a steam engine whose boiler (the hottest-temperature reservoir in contact with the engine) is at 160°C and whose condenser (the coldest-temperature reservoir in contact with the engine) is at 35°C?

6. (I) A heat engine absorbs heat from a thermal reservoir at 400 K, but the engine's maximum efficiency is only 31 percent. What must the temperature of the reservoir be to raise the efficiency to 38 percent?

7. (I) A heat engine receives heat from a thermal reservoir at 750 K and transfers it to a thermal reservoir at 340 K. The efficiency of the engine is only 70 percent of the maximum possible. (a) What is the engine's actual efficiency? (b) How much work does the engine produce for each calorie of heat input?

8. (I) Temperature differences in the ocean have been proposed as a possible energy source in the tropics. Surface water at 30°C could act as a hot reservoir, and deep water at 4°C could serve as a cold reservoir. Ammonia gas could be a working fluid in a heat engine that runs between the two thermal reservoirs. What is the maximum efficiency of such an engine?

9. (II) A Carnot engine absorbs thermal energy from a reservoir at 450°C. Its efficiency is 0.55, and the work delivered by the engine is 5 kWh. (a) What is the temperature of the cold reservoir? (b) How much thermal energy flows into the system, and how much is rejected?

10. (II) What is the maximum efficiency of a heat engine that operates between 450°C and 145°C? If this engine generates 2000 J of mechanical energy, how many calories does it absorb from the hot reservoir, and how many calories does it transfer into the cold reservoir?

11. (II) An inventor claims to have built an engine that takes in 3.0×10^8 J of thermal energy at 450 K, rejects 1.4×10^8 J of thermal energy at 250 K, and delivers 1.0×10^8 J of work in 1 h of cyclic operation. Is there anything wrong with this claim?

12. (II) A power plant generates 440 MW of electric power. At what rate does the plant generate waste heat if its efficiency is 35 percent? Assuming that the plant operates between 480°C and 40°C, what is the maximum efficiency possible?

13. (II) An ideal diatomic gas is used in a Carnot engine with thermal reservoirs at 77 K and 300 K. What is the ratio of the maximum volume to the minimum volume if the pressure drops by a factor of 2 during the isothermal expansion? [Hint: For a diatomic gas, $\gamma = 1.4$.]

14. (II) Prove that two reversible adiabatic paths cannot intersect. Do this by assuming that they do, constructing a cycle that consists of the two paths and another reversible transformation, and showing that such a cycle would violate the second law of thermodynamics (Fig. 20–18). (This result is simply shown for an ideal gas. Your proof should apply to a nonideal gas as well.)

FIGURE 20–18 Problem 14.

15. (III) Any element of irreversibility makes a Carnot engine, and, in fact, any engine, less efficient. Let engine A be a pseudo-Carnot engine, meaning that it has some irreversibility somewhere along the cycle; let engine B be an ideal Carnot engine—the two arranged in size so that they exhaust the same heat flow Q_c. If engine A is *more* efficient than engine B, then by combining the two engines such that A runs normally and B runs backward, as in Figure 20–7, show that you arrive at a contradiction of the Kelvin form of the second law of thermodynamics. Now that you have shown that engine A must be less efficient than engine B, why can the preceding argument not be reversed, running the Carnot cycle forward and the pseudo-Carnot cycle backward to get a contradiction of the second law?

16. (III) Show that if it is possible to build an engine that violates the Kelvin form of the second law of thermodynamics, it is also possible to build an engine that violates the Clausius form of the second law. [Hint: Consider a Carnot engine driven by the Kelvin-violating engine.]

20–4 Other Types of Engines

17. (I) An air conditioner is rated at 8000 Btu/h. Assuming maximum possible efficiency, an exhaust temperature of 38°C, and an interior temperature of 22°C, what is the electrical power consumption? (The work done on the air conditioner is supplied by the wall socket.)

18. (I) What is the coefficient of performance for an ideal Carnot heat pump operated between outside air at 30° C and a deep water well at 15° C? Would the coefficient of performance be larger or smaller if your heat pump followed any other type of cycle?

19. (I) Two refrigerators can each be approximated as Carnot cycles run in reverse. Each refrigerator works with the same cold temperature reservoir $T_c = 273$ K. Refrigerator 1 evacuates to a hot reservoir at temperature $T_h = 293$ K, while refrigerator 2 evacuates to a hot reservoir at a temperature of $T_h = 303$ K. Compare the coefficients of performance of these refrigerators. (Real refrigerators will have coefficients of performance substantially less than these.)

20. (I) A heat pump acts like a refrigerator that cools the outside of a house and gives rise to a heat flow to the inside of the house. If the outside temperature is $+5°C$ and the inside temperature is $22°C$, what is the maximum amount of British thermal units a pump can deliver per kilowatt-hour of electric energy supplied? (The work done on the pump has electric energy as its source.)

21. (I) Consider the diesel cycle in Figure 20AB–1. In which portions of the cycle is there heat flow into or out of the engine? What is the work done in the transformation $2 \rightarrow 3$ in terms of T_2 and T_3?

22. (II) An "ideal refrigerator," a Carnot engine that operates in reverse, has a *compressor* as its engine. It does work W to extract heat Q_c from a cold reservoir at temperature T_c and dumps heat Q_h into a hot reservoir at T_h. Use the known efficiency of a Carnot engine that operates between T_h and T_c and the conservation of energy for one cycle to show that the coefficient of performance $K_{ref} \equiv Q_c/W$ for the compressor is given by $K = T_c/(T_h - T_c)$. Note that K_{ref} is large when the cold and hot reservoirs have nearly the same temperature.

23. (II) A refrigerator with coefficient of performance $K_{ref} = 5.1$ (see Problem 22) gives rise to a heat flow out of the cooling compartment at a rate of 400 cal/min. What is the required power of the motor that operates this refrigerator?

24. (II) To compensate for a given energy loss when a refrigerator is opened too often, the refrigerator's compressor must do work. Take $T_c = 0°C$ and $T_h = 25°C$, and suppose that the refrigerator's coefficient of performance is 20 percent of its ideal value. What is the rate that the compressor does work; that is, what is the power consumption for a given rate of heat loss dQ/dt?

25. (II) The specifications of a freezer claim that it can remove 80 cal/s heat flow from the compartment at $-20°C$ and release 94 cal/s into the room at $25°C$ while using 60 W of electrical energy to drive the compressor. Can you trust this statement?

26. (II) In order to make an ice cube from water, a refrigerator must extract 1400 cal of heat at the temperature of the freezer, $-12°C$. The room temperature is $30°C$. What minimum work must an ideal refrigerator do?

27. (II) You can buy a 5-lb bag of ice for about $2. Estimate the fraction of this price that pays for the electric energy needed to run the ice-making machine. Assume that the inside of the machine is at $-10°C$, that the outside temperature is $30°C$, that the initial temperature of the water is $20°C$, and that the coefficient of performance is 30 percent of the ideal. Electrical energy costs $0.10/kWh.

28. (II) The temperature inside a Carnot refrigerator placed in a room at $72°F$ is $40°F$. The heat flow into the refrigerator from the room is 0.40×10^8 J/h. How many watts are needed to operate the refrigerator: that is, to remove the heat flow into the refrigerator?

29. (II) A heat engine operates on the cycle shown in Fig. 20–19, with heat absorbed or ejected on each leg as indicated. Its operating fluid is an ideal gas with heat capacity at constant volume C_V. (a) What are temperatures T_3 and T_4 at points 3 and 4, respectively, in terms of the indicated pressures and volumes? (b) What is the internal energy change along leg IV, ΔU_{IV}? Include the sign. (c) What is Q_{IV} in terms of ΔU_{IV} and the indicated pressures and volumes? (d) What is the efficiency of the engine in terms of the indicated pressures and volumes and C_V?

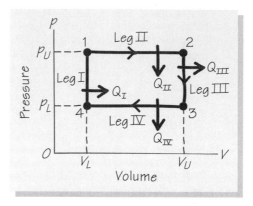

FIGURE 20–19 Problem 29.

30. (II) Consider the ideal Stirling cycle shown in Fig. 20–9, working between a maximum temperature T_h and a minimum temperature T_c, and a minimum volume V_1 and a maximum volume V_2. Suppose that the working gas of the cycle is 0.1 mol of an ideal gas with $c'_V = 5R/2$. (a) What are the heat flows to the cycle during *each* leg? Be sure to give the sign. For which legs is the heat flow positive? (b) What work is done by the cycle during *each* leg? Again, be sure to include the sign. (c) If, in the definition of the efficiency of this cycle, $\eta = W/Q_{pos}$, where Q_{pos} is the *total* positive heat flow to the engine, what is the efficiency of the cycle when $T_h = 700$ K and $T_c = 400$ K? $V_1 = 0.5$ L and $V_2 = 1.5$ L. Compare this efficiency to the efficiency of a Carnot cycle that operates between the same temperature extremes.

31. (II) The Otto cycle, represented in Figure 20AB–2, runs between minimum and maximum volumes V_1 and V_2 and minimum and maximum temperatures T_1 and T_2, respectively. (a) Show that the efficiency is given by $\eta = 1 - (T_2 - T_3)/(T_1 - T_4)$. (b) Show that if the working fluid of the cycle is an ideal gas, the efficiency of this cycle can alternatively be written as $\eta = 1 - (V_1/V_2)^{\gamma-1}$, where $\gamma = C_p/C_V$. The efficiency of this cycle is thus *independent* of the temperatures between which it operates, depending instead only on γ and on geometry. The ratio V_2/V_1 is the compression ratio. A typical compression ratio is 8, and $\gamma = 1.4$, which gives a predicted efficiency of 56 percent.

32. (III) The Brayton cycle is represented by Figure 20AB–3. Show that when its working fluid is an ideal gas, the cycle's efficiency is given by the same formula as that for the Otto cycle in Problem 31. In this case, V_1 and V_2 are not the extreme volumes of the engine but are the volumes at the start and finish of the adiabatic expansion step, respectively.

33. (III) Find the efficiency of the diesel cycle illustrated in Fig. 20AB–1 in terms of $\gamma = C_p/C_V$ and the temperatures at points 1, 2, 3, and 4. Assume that it contains 1 mol of an ideal gas. (Rudolf Diesel tried to produce an engine based on the Carnot cycle. He was not able to produce such an engine, but he came close: A diesel engine nearly matches a Carnot engine in efficiency. Diesel's engine is one of the most efficient in use today.)

20–5   Entropy and the Second Law

34. (I) Consider a large, isolated thermal system in two parts. One part is a thermal reservoir at temperature T_1, labeled I, and the other part is a thermal reservoir at temperature T_2, labeled II. Reservoir I is colder than reservoir II. These two reservoirs are briefly connected, and there is a heat flow Q from II to I. (a) What are the entropy changes for I and for II? (b) Compute the entropy change for the whole system, and show that it is positive.

35. (I) One end of a metal rod is in contact with a thermal reservoir at 1273 K, and the other end is in contact with a reservoir at 293 K. If the rate at which heat flow passes from the hot end to the cold end is 30 J/min, what is the rate of total entropy change? Is the process reversible?

36. (II) Find the entropy change for (a) 1 g of ice melting at 0°C; (b) 1 g of water evaporating at 100°C. You will need to look up the heats of melting and of vaporization.

37. (II) One kilogram of iron at 80°C is dropped into 0.5 L of water at 20°C. Given that the specific heat of water is 1 cal/g · K and that of iron is 0.107 cal/g · K, calculate the final equilibrium temperature of the system and the increase in entropy.

38. (II) Forty grams of water at 100°C is changed reversibly into steam at 100°C. (a) Describe such a process. (b) What is the change in entropy of the water and of the universe? (c) How would part (b) change if the process were irreversible?

39. (II) Calculate the change in entropy of the universe if 0.3 kg of water at 70°C is mixed with 0.2 kg of water at 15°C in a thermally insulated container. [Hint: Even though this process is irreversible, the change in entropy can be calculated by devising a reversible process that leads from the given initial state to the expected final state.]

40. (III) Thermodynamic tables often give the specific heat of materials in the form of a list of coefficients a, b, c, d in the expression $c_p = a + bT + cT^2 + d/T$. How can you use these data to find the entropy change of the material as it undergoes a transformation from T_i to T_f?

20–6   Entropy and Ideal Gases

41. (I) Five moles of an ideal gas is in thermal isolation and undergoes free expansion from 35 L to 100 L. What is the entropy change of the gas? of the universe?

42. (I) The heat capacity at constant volume of a sample of gas is determined to be 20 J/K. If this amount of gas is heated in a constant volume container from 273 K to 2200 K, what is the change in entropy?

43. (I) One mole of an ideal gas expands at constant pressure from an initial volume of 250 cm³ to a final volume of 650 cm³. What is the change in entropy, assuming that the molar specific heat is $c'_V = 3R/2$?

44. (I) One mole of ideal gas with molar specific heat $c'_V = 3R/2$ is heated from 273 K to 350 K at constant volume. What is the change in entropy?

45. (I) An ideal gas undergoes an isothermal compression. Does the entropy increase?

46. (II) Ten moles of a monatomic gas held in a tank at 20 atm escape from the tank isothermally at 23°C. What is the change in the entropy of the universe?

47. (II) One mole of a monatomic ideal gas at an initial pressure of 30 atm and a temperature of 600 K is allowed to undergo a rapid (adiabatic) free expansion from a small vessel into a vessel of 50 times greater volume. Find the change in temperature and the increase in entropy.

48. (II) One mole of a monatomic ideal gas is carried in two steps from (p_1, V_1) to $(p_1, 5V_1)$ to $(0.3p_1, 5V_1)$ by reversible thermodynamic processes that are straight lines on a p–V diagram (Fig. 20–20). (a) What is the overall change in the gas's internal energy? (b) What is the overall work done by the gas? (c) How much net heat flow is added overall? (d) What is the net change in the entropy?

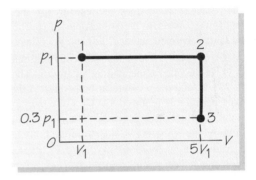

FIGURE 20–20 Problem 48.

49. (II) A gas obeys the well-known equation of state $pV = (a$ constant$) T$. The gas expands, doubling in volume. (a) Plot p versus V when the expansion is isobaric and when it is isothermal. (b) What work is done by the gas on its surroundings for both cases above? (c) What is the entropy change of the gas for both cases above?

50. (II) Show that, in terms of temperature and pressure, the entropy of an ideal gas with constant heat capacities is $S(T, p) = C_p \ln(T/T_0) - nR \ln(p/p_0) + S(T_0, p_0)$.

51. (III) Dilute helium gas is taken through the following reversible cycle on a p–V diagram: AB is an isobaric compression; BC is an adiabatic expansion; CD is an isothermal expansion; DA is constant-volume pressure increase (Fig. 20–21). Express the heat flow, work done on the system, change in internal energy, change in temperature, and change in entropy for each leg of the cycle in terms of the quantities T_A, T_B, T_C, and p_A. Pay attention to the signs.

20–7   The Meaning of Entropy

52. (I) A total of 12 molecules are in a large box, which we mentally divide into two halves, a left half and a right half. Any one molecule is equally likely to be in either of the two halves. What is the probability that all 12 molecules will be found in the left half? How does your answer change if there are 120 molecules?

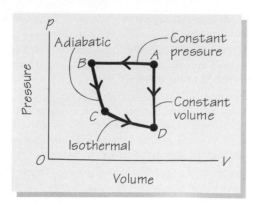

FIGURE 20–21 Problem 51.

53. (II) An ideal gas in a cylinder with a movable piston is placed in thermal contact with a thermal reservoir, and the gas is slowly compressed. How does the entropy of the gas in the cylinder change? How does the entropy of the reservoir change? Is the answer to these questions in conflict with the second law of thermodynamics? Explain your answer.

54. (II) A box of volume 5 L contains 3 L of He in one side and 2 L of O_2 in the other, separated by a partition. The temperature is 273 K, and the pressure in both sides of the box is 1 atm. The partition is removed. (a) Describe the new equilibrium configuration. (b) What is the change in entropy?

55. (II) A thermally isolated container of volume $3V_0$ is divided by a partition. In one portion, of volume V_0, there are n mol of a monatomic ideal gas at temperature T_0. The other portion is empty. The partition is then broken, and the gas expands to fill the entire container (Fig. 20–22). (a) What is the final temperature of the gas? (b) the change in entropy of the gas? (c) the change in entropy of the universe? (d) Suppose that the gas is originally confined to volume V_0 by means of a piston instead of a partition. The piston is then slowly withdrawn until a final volume of $3V_0$ is obtained. What is the change in entropy of the gas?

FIGURE 20–22 Problem 55.

General Problems

56. (II) A nuclear power plant requires 600 MW of cooling. If environmental concerns limit the temperature rise of the cooling water to 15°C, how much water flow is required to cool the heat generated by the plant? If the water is at an initial temperature of 285 K, what is the entropy change per kilogram of water?

57. (II) Fifty grams of oxygen gas at 320 K does 80 J of work while 40 cal of heat flow is absorbed by the gas. (a) What is the change in internal energy? (b) in the temperature of the gas? (c) in entropy, assuming that the change is isobaric?

58. (II) Calculate the entropy change when 1 kg of water is cooled from 100°C to 0°C. Calculate the entropy change when the water turns to ice at 0°C (Fig. 20–23). The latent heat of fusion is 80 cal/g.

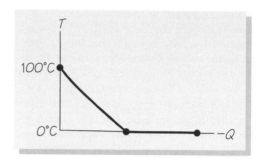

FIGURE 20–23 Problem 58.

59. (II) Two moles of helium at STP is compressed in an isothermal process to a pressure of 2 atm. (a) What is the final volume? (b) the final temperature? (c) the change in entropy?

60. (II) A Carnot engine for which the working fluid is 1 mol of an ideal gas with $c'_V = (5/2)R$ operates between two thermal reservoirs of temperatures $T_h = 800$ K and $T_c = 300$ K. When the gas is in contact with the hot reservoir, its minimum volume is $V_1 = 1.2$ L, and when it is in contact with the cold reservoir, its maximum volume is $V_3 = 15$ L. (a) What is the efficiency of this engine? (b) For one cycle, what is the net energy change of the gas? (c) the net entropy change of the gas? (d) the net change in the entropy of the universe? (e) What is V_2, the maximum volume when the gas is in contact with the hot reservoir?

61. (II) Three moles of oxygen at STP undergoes an adiabatic compression to a final pressure of 3 atm. (a) What is the final volume? (b) the final temperature? (c) the change in entropy?

62. (II) One liter of nitrogen at STP is adiabatically compressed to a volume of 0.5 L when a membrane breaks and the nitrogen gas undergoes free expansion back to 1 L. (a) What is the final pressure? (b) the final temperature? (c) the change in entropy?

63. (II) A heat engine operates between the temperatures of 400 K and 300 K, doing 100 J of work. This work is used to run a refrigerator between the same temperatures. (a) Calculate the heat transfer between each reservoir and the engine or the refrigerator, assuming that both of them have ideal efficiency/coefficient of performance. (b) Repeat the calculation if the heat engine's efficiency is only 80 percent of the ideal. (c) Repeat the calculation for the case of an ideal heat engine but for a refrigerator whose coefficient of performance is 80 percent of the ideal.

64. (II) One Carnot engine drives another in series, as shown in Fig. 20–24. The heat released from the first engine is absorbed by the second. Find the overall efficiency, which is determined by the total work produced divided by the heat input to the first engine.

FIGURE 20–24 Problem 64.

65. (II) A coal-fired, 300-MW generating plant operates between 750 K and 400 K, with an efficiency 60 percent of the maximum efficiency possible. At what rate is waste heat flow produced? Suppose that water is used to carry off the waste heat flow, and it does so by being heated by 12°C. How much water must flow through the machinery per second?

66. (II) Consider the reversible thermodynamic cycle of the following form (Fig. 20–25): (i) One mole of a monatomic ideal gas at temperature T_a and pressure p_a is allowed to expand isobarically until the volume has increased by a factor r. (ii) The gas is then cooled so that its absolute temperature is reduced by a factor of 2, with the cooling performed at constant volume. (iii) The gas is then compressed isobarically to its original volume. (iv) Finally, the gas is heated to its starting temperature, T_a, at constant volume. (a) What is the mechanical work done by the engine? (b) the heat transfer in the four legs of the cycle? (c) the efficiency of the cycle?

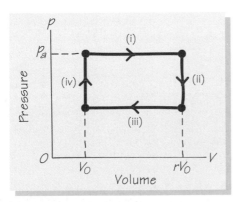

FIGURE 20–25 Problem 66.

67. (II) A heat engine operates in a reversible cycle with the following steps (Fig. 20–26): (i) An ideal gas characterized by $\gamma = 1.4$ expands adiabatically, so that it cools from an initial temperature of 330°C to 30°C; (ii) it is compressed isothermally until it reaches its initial volume; and (iii) it is then heated at constant volume until it reaches its initial temperature. Calculate the efficiency of the cycle, and compare it with the Carnot efficiency that corresponds to these temperatures.

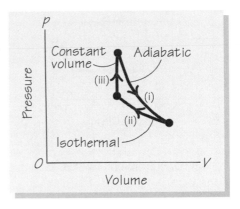

FIGURE 20–26 Problem 67.

68. (II) The variables T and S may be used instead of the thermodynamic variables p and V. Plot a T–S diagram for the Carnot cycle. What does the area within the cycle represent? [Hint: Use the fact that the first law of thermodynamics, which states that $dQ = dU + p\,dV$ may be written in the form $T\,dS = dU + p\,dV$, and integrate around the cycle.]

69. (II) In engineering, a useful quantity called *enthalpy*, H, is defined by $H = U + pV$. Calculate the enthalpy of an ideal gas as a function of temperature.

70. (II) Suppose that there is steady flow of a mass M of gas through a turbine. Gas enters the turbine at an elevation z_1, pressure p_1, internal energy U_1, and speed v_1, and leaves at an elevation z_2, pressure p_2, internal energy U_1, and speed v_2. Thermal energy Q is supplied to the turbine. Work W is done by the turbine, and V_1 and V_2 are the respective volumes occupied by mass M on entering and leaving the turbine. (a) Use energy conservation to show that the change in enthalpy (see Problem 69) in terms of the specified quantities can be written in the form $H_2 - H_1 = Q - W + \frac{1}{2}mv_1^2 + Mgz_1 - (\frac{1}{2}mv_2^2 + Mgz_2)$. (b) Use this result to show that there is no change in enthalpy in the Joule experiment described in Section 18–6 (in which there is no change in elevation, and the change in the speed of the gas is so small that $v_2^2 - v_1^2$ can be neglected).

71. (III) The Brayton cycle is used to extract work from a high-temperature, gas-cooled nuclear reactor with helium (a monatomic ideal gas) as the working fluid. The helium enters a compressor, and its pressure is raised adiabatically. It is then heated at constant pressure to a high temperature in the reactor core. The energy is supplied by the fission of uranium or plutonium. The helium then enters a turbine, in which it is allowed to expand adiabatically. It then passes through a precooler, in which heat is rejected at constant pressure until it reaches the initial conditions. Suppose that helium enters the compressor at 30°C and a pressure of 20 atm. It is compressed to 50 atm and then heated to 1200°C in the reactor core (Fig. 20–27). (a) What are the temperatures at the end of compression and expansion? (b) What are the heat flows per kilogram of helium in each cycle? (c) How much net work is done per kilogram of helium in each cycle?

FIGURE 20–27 Problem 71.

72. (III) Three reversible paths, legs i, ii, and iii, connect points (p_2, V_2) and (p_1, V_1) in Fig. 20–28. (a) Calculate $\int_1^2 dW$ and $\int_1^2 dQ$ for the three paths. (b) Calculate $\int_1^2 dQ/T$ for the three paths.

FIGURE 20–28 Problem 72.

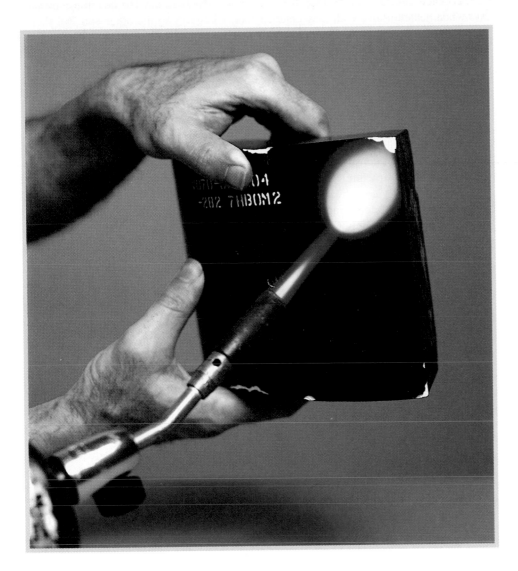

We now know a great deal about how to manufacture materials with desirable thermal properties. This tile for the exterior skin of the Space Shuttle has a coefficient of thermal conductivity that is so low that it can be touched only centimeters from a spot that is red hot. In addition, the tile changes its dimensions very little when its temperature changes.

Properties of Solids

The properties of solids reflect a wide range of the physical laws we have discussed to this point. Solids deform in response to forces that act on them, propagate sound waves, respond to temperature changes by expanding or contracting, and conduct thermal energy. All these properties are explained ultimately by the atomic structure of materials.

21-1 THE NATURE OF SOLIDS

Solids are a form of matter that may appear at first glance to resist any kind of deformation. Our everyday experience suggests that solids move rigidly when forces act upon them and rotate rigidly under the influence of torques. Upon closer examination, though, solids exhibit more complicated responses to external influences. When com-

FIGURE 21–1 Schematic diagram of atoms connected by bonds in a lattice structure.

Defects, or dislocations, are important in solids.

FIGURE 21–2 Sketch of a cubic lattice.

FIGURE 21–3 These beautiful crystals are characteristic of the mineral known as amethyst, a type of quartz.

FIGURE 21–4 (a) A perfect crystalline raft of bubbles. Bubbles will arrange themselves in crystalline patterns (in this case, two-dimensional), just like the atoms in real solids. (b) A dislocation in a raft of bubbles.

pressed, their volumes change slightly; when put under tension or heated, they stretch; and when heated at one end, they conduct thermal energy to the other end. All these properties can be understood in terms of the underlying atomic structure of matter. As indicated in Section 16–1, **solids** are aggregates of atoms or molecules for which the interatomic forces lead to an organized, three-dimensional grouping of atoms, called a *lattice structure*. Figure 21–1, a schematic two-dimensional representation of such a structure, depicts the interatomic forces as springs. The lattice structure is one of stable equilibrium; therefore, the use of spring forces is at least approximately correct. The possible lattice structures are called **crystals**. Different crystals are determined by the nature of the interatomic forces between atoms or molecules and by the shapes and orientations of these constituents; that is, how they fit together. Figure 21–2 shows the three-dimensional structure of one particularly simple type of crystal, the *cubic lattice*.

If geometrically complicated molecules rather than simple atoms make up a crystal, or if several different kinds of atoms are present, the number of possible crystal structures increases rapidly (Fig. 21–3). The way different atoms assemble into crystals depends also on the conditions under which they are formed. Both diamond and graphite—two solids with very different properties—are formed from pure carbon. In a certain sense, the formation of a crystal by atoms is like a three-dimensional jigsaw puzzle with identical pieces, *which has more than one solution!*

Almost as important as crystalline structure to the macroscopic behavior of a solid are its lattice imperfections, called **defects**, or **dislocations**. Defects are closely associated with the way a crystal grows. Figure 21–4 shows a simple two-dimensional crystal that is easy to grow and observe: a raft of bubbles that float on the surface of a liquid. Figure 21–4a is a perfect crystalline form, whereas a defect has been introduced in Fig. 21–4b. Can you recognize it? A more general type of defect is associated with grains and grain boundaries. A **grain** is a region in a solid where the crystal structure is perfect. Because chance plays a role in the organization of atoms when crystal growth starts, the crystalline structure of one grain is oriented differently from that of another, and the boundaries where these grains meet involve special kinds of defects. Figure 21–5a shows such a boundary. A real solid is made up of many grains (Fig. 21–5b). Another class of defects occurs when single lattice sites are empty or are occupied by an impurity (an atom chemically different from those that make up the bulk of the crystal), or when impurities occupy spaces between lattice sites. Defects are related to a crystal's color, luminescence, transport properties, and mechanical properties. The study of the nature of defects is a major field of research and has taught us much about the behavior of solids.

The division between solids, liquids, and gases is not a very precise one. Substances can be distinguished by what we call **long-range order** or **short-range order**, determined by whether the arrangements of the atoms and molecules persist over distances of many atomic spacings or not. Crystals have long-range order, whereas most fluids

(a)

(b)

FIGURE 21-5 (a) A close look at the grain boundaries in a crystal, as illustrated by a raft of bubbles. (b) An overview of numerous grains in a crystal.

have only short-range order. But, again, the division is artificial, for there is a small class of substances, called *amorphous solids*, that are like solids in some respects and like liquids in other respects. Substances exist at each of the gradations in between the boundaries we have constructed thus far. Consider *liquid crystals*, which are materials that contain rodlike molecules arranged parallel to one another. These molecules are in crystalline order if viewed from one end, along their axial direction, but they are free to move, in the fashion of molecules in liquids, along that direction. It is almost as though a liquid crystal is a solid in two directions but a liquid in the third direction. Liquid crystals are used in electronic displays. Many researchers feel that the materials that cannot be cleanly labeled as either solids or liquids are the most interesting ones, because they open avenues to new understanding and new applications.

Let's return to our simple model of lattice structure. Figure 21–6 suggests how a cubic structure might react to compression (Fig. 21–6a), to stretching (Fig. 21–6b), and to a *shear* (Fig. 21–6c). Just how hard it is to compress, stretch, or shear a solid depends on its interatomic forces. In the final analysis, it is these forces, along with the grain and defect structure, that differentiate steel from a pie crust. The lattice structure and the springlike bonds that connect the atoms or molecules also explain how waves propagate in solids. If successive layers of interatomic bonds are compressed, they will spring back to the equilibrium length and overshoot it, thus propagating the compression. Similarly, a small shear force on one cell gives rise through interatomic forces to a shear force in the following cell, and so forth, leading to transverse waves associated with the resistance to shear forces.

Conduction of thermal energy is qualitatively similar. When one end of a rod is heated, the atoms become agitated and acquire kinetic energy. The bonds permit this kinetic energy to be transferred to neighboring layers of atoms; in this way, thermal energy travels along the rod. When the external conditions undergo a more extreme change, the very character of the solid changes. Under extreme compression, structural changes are possible—graphite can be changed into diamond; under strong tension, solids break; and when solids are heated to a sufficiently high temperature, they turn into liquids.

∞ The concept of shear was introduced in Chapter 16.

FIGURE 21-6 Schematic diagram of lattice distortions: (a) compressed in the horizontal direction; (b) stretched in the horizontal direction; (c) subjected to shear forces, which act in opposite directions at the top and bottom.

FIGURE 21–7 Equal but opposite forces applied evenly to opposite ends of a piece of material cause elongation.

The definition of stress

The definition of compressional strain

FIGURE 21–8 The passage of polarized light through materials such as this plastic cross section of a truck's axle shaft is stress dependent. The stress is greatest where the lines are most closely spaced.

In principle, all properties of solids can be described by describing the atomic structure of solids, but in practice it is simpler to describe the properties of solids on a macroscopic scale. In developing the behavior of solids (or other materials) in terms of empirical observation, keep in mind that we are taking an approach that, at best, can give us "laws" that are more properly described as rules of thumb, valid only under restricted conditions.

21–2 STRESSES AND STRAINS

Suppose that a force acts on a solid bar and tends to stretch it (Fig. 21–7). We have referred to such a force as a *tension*. **Stress** is a measure of the force that acts on the bar, whereas **strain** is a measure of the solid's response to a stress such as tension. We shall assume that the applied force is small, which implies that the response is small. (If the force is too large, the concept of a small response breaks down; when a heavy hammer strikes a piece of ice, the ice shatters.) We define the stress on a solid, S, as the external force per unit area that is exerted on it:

$$S \equiv \frac{F}{A}, \qquad (21\text{–}1)$$

where F is the magnitude of the force that acts per area A of the solid (Fig. 21–8). Tension gives rise to positive stress, and compression gives rise to negative stress. (The stress should contain a reference to the direction of the applied force, but we shall work with magnitudes and simple signs unless we specify otherwise.) Stress has the same dimensions as pressure and has SI units of newtons per square meter. Actually, this unit is rather small for ordinary applications, and a more practical unit for stress is a meganewton per square meter (MN/m^2).

When tension is applied to a rod, the rod stretches; when the rod is subjected to a small compressive force, its length decreases. The *response* to the stress associated with either a tension or a compression is called the *compressional strain, e*. It is defined as *the fractional change in the length of the solid along the direction of the compressional force*:

$$e \equiv \frac{\Delta L}{L}. \qquad (21\text{–}2)$$

Note that e is dimensionless. It is negative if the stress is due to compression, and positive if it is due to tension.

The bar in Fig. 21–7 does more than just stretch as a result of a tension; when it stretches, it also *shrinks* in its lateral dimensions—the width h and depth w. Similarly, it will bulge in these directions if the force is a compression. The fractional shrinking or bulging is proportional to the compressional strain by an amount characteristic of the material, and it is described by the positive constant called *Poisson's ratio, σ*, after Siméon Poisson:

$$\frac{\Delta h}{h} = \frac{\Delta w}{w} = -\sigma e. \qquad (21\text{–}3)$$

A typical value of σ for many solids is 0.3. This constant is always less than 0.5: If it were to exceed 0.5, the solid would expand when a uniform pressure is applied to it (see Problem 18).

If an external force F of constant magnitude is applied everywhere perpendicular to the surface of a solid, the volume, V, of that solid changes. The volume decreases if the force is inward and increases if the force is outward. The *volume stress, p*, is defined as stress was defined in Eq. (21–1): $p \equiv F/A$, where A is the *total* surface area of the solid. We have used the same symbol as that for pressure because the volume stress is identical to the pressure if the force is directed inward. For solids, though, the volume stress can equally well refer to an outward force on the body. We may also define the *volume strain, e_v*, by the fractional change in volume due to a volume stress:

$e_V \equiv \Delta V/V$. *When Poisson's ratio is small*, the volume strain is just a special case of the compressional strain; a bar that stretches while its cross-sectional area A remains constant has a volume $V = AL$, and when the bar is stressed longitudinally, the change in volume is $\Delta V = A \, \Delta L$. When these quantities are substituted in the definition for e_V, we obtain Eq. (21–2) for e.

We want to mention one more kind of stress and its corresponding strain: shear stress and shear strain. Suppose that a uniform force acts *along* a face of a solid (Fig. 21–9a). This is a shear force by definition. If the bottom surface of the solid is fixed (by some kind of contact force), the result is distortion of the solid (Fig. 21–9b). The quantity F/A is the *shear stress* on this solid, and the *shear strain* is the fractional amount by which the upper surface moves, $\Delta L/L$. Note that the volume is essentially unchanged under shear stress, in contrast to the case of longitudinal stress.

The Relation between Stress and Strain

Let's turn now to the relation between stress and strain. As early as the seventeenth century, Robert Hooke realized that stress and compressional strain are proportional to each other. A clear understanding of this proportionality was first provided by Thomas Young in 1800, who wrote down the relation

$$\frac{S}{e} = Y. \tag{21-4}$$

This equation applies to the relation between (small) compression or tension and the corresponding strain. We introduced Y in Chapter 14 as **Young's modulus**, or the *elastic modulus*. It is a constant whose value varies with the material; it measures the "stiffness" of the material. Y is analogous to the *spring constant*, k, in the spring force $F = -kx$. In fact, Young's modulus is a constant only for *small* strains because the atomic bonds behave just like simple springs. Because e is dimensionless, Young's modulus has the same dimensions as S and therefore has units of force per area—meganewtons per square meter, for instance. Note that Young's modulus is a constant only for small stresses; for large stresses, the strain ceases to be proportional to the stresses, as experimentation with licorice or bread dough shows. Table 21–1 lists values of Young's modulus for various solids.

For small stresses and strains, the value of Young's modulus is generally independent of whether the material is under tension (positive strain) or under compression (negative strain). In some materials, however, the response to compression is not always the same as the response to tension. This is due to imperfections, such as tiny cracks in the material. Compression tends to reduce the cracks, whereas tension magnifies them.[†] Thus, for materials such as cast iron or cement, Young's modulus is larger for compression than for tension. The relative weakness of concrete under tension is remedied by the addition of iron rods to make what is called *reinforced concrete*. The rods are stiff under tension, though less so under compression, which causes them to buckle. Conversely, concrete is stiff under compression. Reinforced concrete is therefore stiff under both tension and compression.

> **EXAMPLE 21–1** A vertical steel rod of length 2.0 m and diameter 2.0 cm is fixed at the top and has a mass m of 9500 kg hanging from its lower end (Fig. 21–10). Given that Young's modulus for this particular type of steel is 250,000 MN/m², calculate the elongation of the rod. What is the strain?
>
> **Solution:** The stress is given by $S = mg/A$, where A is the cross-sectional area of the rod. Equations (21–2) and (21–4) express the amount of stretch ΔL of the rod of length L and Young's modulus Y in terms of the stress:

[†]A stack of bricks may be viewed as one brick with a number of cracks going right through it; the stack has no resistance to tension.

FIGURE 21–9 (a) An object fixed along its bottom surface has a force applied. (b) The effect of shear stress from uniform force distorts the object.

Young's modulus gives the relation between stress and compressional strain.

TABLE 21–1
Young's Moduli for Various Solids

Material	Y (MN/m²)
Rubber	7
Wood	14,000
Concrete	17,000–30,000
Bone	9,000–21,000
Glass	70,000
Aluminum	73,000
Steel	210,000
Diamond	1,200,000

  APPLICATION

Reinforced concrete

FIGURE 21–10 Example 21–1.

$$\Delta L = eL = \frac{S}{Y}L = \frac{mg}{AY}L.$$

The area of the rod is $A = \pi r^2$, where $r = \frac{1}{2}(2.0 \text{ cm}) = 1.0 \text{ cm}$ is the radius of the cross section. Thus we find

$$\Delta L = \frac{mg}{\pi r^2 Y}L = \frac{(9.5 \times 10^3 \text{ kg})(9.8 \text{ m/s}^2)}{3.14(1.0 \times 10^{-2} \text{ m})^2(2.5 \times 10^{11} \text{ N/m}^2)}(2.0 \text{ m})$$

The elongation is just 2.4 mm, and the strain is

$$\frac{\Delta L}{L} = \frac{2.4 \times 10^{-3} \text{ m}}{2.0 \text{ m}} = 1.2 \times 10^{-3}.$$

The bulk modulus gives the relation between volume stress and volume strain.

The relation between volume stress and volume strain takes exactly the same form as the relation between ordinary stress and strain for small strains. The proportionality constant $B = $ (volume stress)/(volume strain), which is analogous to Young's modulus, is called the **bulk modulus**. We have

$$B \equiv -\frac{F/A}{\Delta V/V} = -\frac{p}{e_V}. \tag{21–5}$$

The minus sign means that an *inward* pressure on a solid, which is positive, implies a *decrease* in volume, $\Delta V < 0$. There is a relation between Young's modulus and the bulk modulus that must involve Poisson's ratio. Only two of these three constants are independent of one another (see Problem 17). Finally, there is a *shear modulus*, G, which is given by the ratio of shear stress to shear strain. Liquids have a shear modulus of zero, meaning that they have an arbitrarily large shear strain for even small shear stresses. Liquids do not resist shear at all.

The Strength of Materials

An important property of solids is their strength, or, more specifically, their *tensile strength*, which is also known as *fracture stress*. This quantity is the stress required to break a material into two pieces—by pulling it along an axis, for example. The tensile strength of different materials varies a great deal. Steel may break when the tension is 3000 MN/m², whereas a piece of concrete may be pulled apart by a stress of only 5 to 10 MN/m². The tensile strengths for a variety of materials are listed in Table 21–2. The values of tensile strength are a factor of 100 to 1000 times smaller than the values of Y (Table 21–1) for the same materials. There is useful information in this huge discrepancy, and it involves the importance of cracks and imperfections in a solid.

The tensile strength is a kind of *critical stress*. We may define a corresponding *critical strain* as $(\Delta L/L)_c$, such that

$$\text{tensile strength} = Y\left(\frac{\Delta L}{L}\right)_c.$$

The factor of 100 to 1000 difference between Young's modulus and the tensile strength for a given solid means that a fractional extension of any rod by 1 percent, or even by 0.1 percent, will break the rod. This is because the pervasive existence of cracks weakens the material (Figure 21–11). Without cracks, the stress on a section of rod between two layers of atoms is carried by many bonds. When cracks are present, though, there are fewer bonds to carry the stress. The force F in the expression F/A for the stress is the same, but the real area A over which the force acts is very much smaller than the apparent area. This means that the stress experienced by the material is much larger than the stress obtained by using the apparent area. For extremely thin—micrometer-size—fibers, which are nearly perfect crystals, measurements give a tensile strength of

TABLE 21–2
Tensile Strengths for Various Solids

Material	Tensile Strength (MN/m²)
Steel piano wire	3000
Steel	400–1500
Cast iron	70–250
Aluminum (pure)	70
Aluminum alloys	140–550
Copper	140
Titanium alloys	700–1400
Spruce, along grain	100
Spruce, across grain	3
Glass	0–170
Brick	5
Cotton	350
Spider silk	240
Human tendon	100
Rope	80

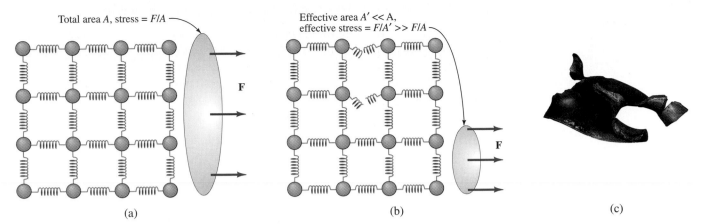

Total area A, stress $= F/A$

Effective area $A' \ll A$, effective stress $= F/A' \gg F/A$

F

F

(a)   (b)   (c)

FIGURE 21–11 (a) An uncracked sample of material. (b) Cracks in the material lead to an increased susceptibility to fracture under stress. (c) This piece of metal has fractured under stress.

around $0.2Y$. This value is what we would expect upon evaluating $(\Delta L/L)_c$ from our understanding of interatomic forces. The discrepancy between the values of the tensile strength for bulk materials and tiny perfect crystals shows again the importance of imperfections in the structure of solids.

2 1 – 3   W A V E S   I N   S O L I D S

We used dimensional analysis in Chapter 14 to estimate the speed of the propagation of waves in a solid with density ρ and Young's modulus Y:

The speed of sound in solids

$$v = \sqrt{\frac{Y}{\rho}}. \qquad (21\text{–}6)$$

Young's modulus refers to compressive and tensile forces, so the wave being discussed is a longitudinal wave composed of alternate compressions and extensions of the solid. Sound waves in air are also longitudinal and, accordingly, longitudinal waves in solids are referred to as sound waves.

In Chapter 14, we pointed out that the constant in front of the square root in Eq. (14–27) [Eq. (21–6)] cannot be estimated by dimensional analysis, so we set it equal to 1. In order to find the constant (which, in fact, is 1 to a good approximation), we need to find a dynamical wave equation for the longitudinal waves in a given solid. Let x be the position in the longitudinal direction and let $z(x, t)$ be the longitudinal displacement from equilibrium of the material at x. We must show that $z(x, t)$ varies with position x and time t in such a way as to satisfy the wave equation first discussed in Chapter 14. This equation takes the general form

$$\frac{\partial^2 z}{\partial x^2} - \frac{1}{v^2} \frac{\partial^2 z}{\partial t^2} = 0. \qquad (21\text{–}7)$$

(Recall that the partial derivative is, for all practical purposes, just an ordinary derivative that acts on a quantity that depends on several independent variables; in this case, x and t.) Equation (21–7) has the general wave solution

$$z(x, t) = z(x \pm vt), \qquad (21\text{–}8)$$

in which v is the wave speed. Just as Hooke's law implies the harmonic motion of a mass attached to a spring, it follows from the stress–strain relationship that z obeys the wave equation and that the speed of sound in Eq. (21–6) is indeed correct, at least for a thin rod. When a time-varying stress is imposed on one end of a material from the outside, there is a time-varying strain at that point. The displacement at that point causes a stress on a neighboring point, which is in turn displaced, and so on. The details are worked out in the following subsection.

How to Get the Wave Equation for Longitudinal Waves in Solids

We find the wave equation for longitudinal waves in solids in a manner that is analogous to how we arrived at Eq. (14–9). Consider a rod of cross-sectional area A that lies along the x-axis. Concentrate on a section of the rod between x and $x + dx$ (Fig. 21–12). The stress (force per unit area) will vary with position and also with time. The *net* stress acting on the section of the rod is

$$\Delta S = S(x + dx) - S(x) = \left[\frac{\partial S(x)}{\partial x}\right] dx.$$

Thus the net force on the section, ΔF, is given by

$$\Delta F = A \, \Delta S = A\left[\frac{\partial S(x)}{\partial x}\right] dx.$$

Let's denote the longitudinal displacement of the material within the rod from its equilibrium position by $z(x, t)$. The force given above leads to an acceleration of the material. The mass of material in the section of the rod of length dx is given by the density ρ multiplied by the volume, so that $m = \rho A \, dx$; the acceleration of the material is given by the second time derivative of the displacement $z(x, t)$. Thus $a = \partial^2 z(x, t)/\partial t^2$, and Newton's second law, $ma = \Delta F$, yields

$$\rho A \, dx\left(\frac{\partial^2 z}{\partial t^2}\right) = A\left(\frac{\partial S}{\partial x}\right) dx;$$

$$\rho\left(\frac{\partial^2 z}{\partial t^2}\right) = \frac{\partial S}{\partial x}. \tag{21–9a}$$

Let's now relate the stress to the strain. The strain, given by the fractional displacement, is proportional to the stress. The total length of the section of the rod is dx, and the net increase in length is the difference between the displacements measurable at the two ends of the section, $z(x + dx) - z(x)$ (Fig. 21–12b). Thus the strain is given by the ratio of the difference of displacements to the original length, dx:

$$e = \frac{z(x + dx, t) - z(x, t)}{dx} = \frac{\partial z}{\partial x}.$$

Equation (21–4) states that $S = eY$, or

$$S = Y\frac{\partial z}{\partial x}.$$

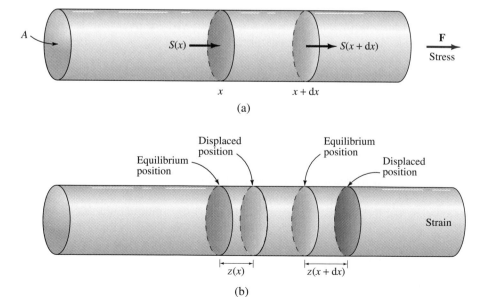

FIGURE 21–12 The (a) stresses on and (b) resulting elongation of a section of a rod that lies between the positions x and $x + dx$.

When this result is substituted into Eq. (21–9a), we find that

$$\rho\left(\frac{\partial^2 z}{\partial t^2}\right) = Y\left(\frac{\partial^2 z}{\partial x^2}\right). \qquad (21\text{–}9b)$$

This result has the form of the wave equation, Eq. (21–7), and the speed can be identified to be that of Eq. (21–6).

We can get some feeling for the speed of sound in solids by calculating this quantity in a steel rod. The density of steel is $\rho = 7.9 \times 10^3$ kg/m³ while Young's modulus is $Y = 200{,}000$ MN/m². Then, using Eq. (21–6), we find

$$v = \sqrt{\frac{Y}{\rho}} = \sqrt{\frac{2 \times 10^{11} \text{ N/m}^2}{7.9 \times 10^3 \text{ kg/m}^3}} = \sqrt{0.25 \times 10^8 \text{ m}^2/\text{s}^2} = 5 \times 10^3 \text{ m/s}.$$

This number, more than ten times larger than the speed of sound in air, is in agreement with experimental data.

Other Waves in Solids

Just as small longitudinal displacements within a solid lead to longitudinal waves, small transverse displacements within a solid lead to transverse waves. This is because solids resist shear forces, and the wave speed of transverse waves—also called *shear waves*—is described by using the shear modulus, G, of Section 21–2:

$$v_{\text{shear}} = \sqrt{\frac{G}{\rho}}. \qquad (21\text{–}10)$$

In most cases, the shear modulus is lower than Young's modulus. The speed of shear waves is thus less than that of longitudinal waves.

Probing Earth

Earth supports both sound waves and shear waves and there are many boundaries within it—the mantle/core transition, for example—where these waves are reflected or transmitted (see Problem 27). This phenomenon is the basis for an important tool used in geological exploration and research. Earthquakes are responsible for initiating both kinds of waves within the planet; the varying arrival times of these waves at a seismograph—a device that detects such vibrations—gives information about the location of the earthquake. Earthquake, or seismic, waves can follow many paths, reflecting from various internal layers and boundaries between different materials; this fact is helpful in pinpointing the location and strength of the earthquake (Fig. 21–13). More important, information on Earth's *internal* structure comes from the arrival time and

APPLICATION

Geological exploration

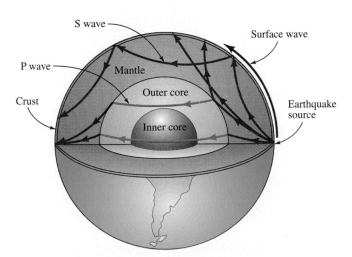

FIGURE 21–13 Several types of waves are possible in solids. Seismic waves, produced by events such as earthquakes, include all types. (After S. Judson and M. Kauffman, *Physical Geology*, Prentice Hall, 1990.)

strength of waves initiated either by earthquakes or by explosions that are set intentionally. Such data teach us not only about the existence of reflecting boundaries, but also about varying densities within Earth. For example, we know that Earth's central core is liquid because shear waves do not pass through it. The search for petroleum relies heavily on evidence from wave propagation within Earth.

21–4   THERMAL EXPANSION

The definition of the coefficient of thermal expansion

We know that a solid expands when it is heated. A simple demonstration of this fact is shown in Fig. 21–14. Empirically, this expansion is described by the **coefficient of thermal expansion**, α, defined by

$$\alpha \equiv \frac{1}{L}\frac{dL}{dT}. \tag{21–11}$$

The quantity $\alpha \, dT$ is the fractional change in the length L of a solid due to a change in its temperature, dT. In principle, the coefficient α is a function of temperature T but tends to vary quite slowly with T. The analogous **coefficient of volume expansion**, β, is simply related to α. $\beta \, dT$ is the fractional change in the volume V due to a temperature change dT:

$$\beta \equiv \frac{1}{V}\frac{dV}{dT}. \tag{21–12}$$

We can show that $\beta = 3\alpha$ by considering a cube of volume $V = L^3$. We have

$$\beta = \frac{1}{V}\frac{dV}{dT} = \frac{1}{L^3}\left(3L^2\frac{dL}{dT}\right) = \frac{3}{L}\frac{dL}{dT} = 3\alpha. \tag{21–13}$$

This result applies as long as the material is *isotropic*; that is, its properties are the same in any direction within the material. Thermal expansion also occurs in liquids. Liquids are more correctly described as isotropic than are solids, whose crystal structure implies preferred directions. For this reason, the volume expansion coefficient β is usually given when liquids are involved. Both β and α are measured in units of K^{-1}. Table 21–3 lists some thermal expansion coefficients.

EXAMPLE 21–2   A steel bridge is 600 m long. How much allowance must be made for linear expansion between the temperatures of $+40°C$ and $-40°C$?

TABLE 21–3
Coefficients of Thermal and Volume Expansion (at 20°C)

Material	α (*K⁻¹*)
Aluminum	2.30×10^{-5}
Copper	1.67×10^{-5}
Pyrex glass	3.2×10^{-6}
Fused quartz	4.2×10^{-7}
Steel	1.05×10^{-5}

	β (*K⁻¹*)
Ethyl alcohol	1.12×10^{-3}
Gasoline	9.5×10^{-4}
Mercury	1.82×10^{-4}
Water	2.07×10^{-4}

FIGURE 21–14 Thermal expansion: The linear dimensions of a material expand proportionally with the temperature increase. In this instance, the linear dimension of the steel ball is its radius. (a) It just fits through a ring at room temperature, (b) is heated, and (c) becomes too big to fit through the ring.

(a)

(b)

(c)

Solution: Equation (21–11) may be written as

$$\Delta L = \alpha L \, \Delta T. \qquad (21\text{–}14)$$

We take α from Table 21–3 to be $1.05 \times 10^{-5} \ K^{-1}$ for steel (α varies little with temperature in the range $+40°C$ to $-40°C$). Numerical evaluation gives

$$\Delta L = (1.05 \times 10^{-5} \ K^{-1})(600 \ m)(80 \ K) = 5 \times 10^{-1} \ m = 50 \ cm.$$

Expansion joints that allow for this much thermal expansion are distributed at regular intervals along bridges (Fig. 21–15).

EXAMPLE 21–3 You must design a gasoline container to store gasoline under conditions in which the annual temperature extremes are $-20°C$ to $+40°C$. You want to be able to fill the container at any time of the year and not lose more than 0.1% (a fraction 0.001) of the gasoline due to overflow. In choosing the material for your container, what range of possible values of the coefficient of thermal expansion α should delimit your search?

Solution: Both container and liquid will change their volumes as the temperature changes. If the container material has a coefficient of thermal expansion α that is one-third β ($\beta = 9.5 \times 10^{-4} \ K^{-1}$ is the coefficient of volume expansion for gasoline), then both container volume and gasoline volume will change in the same way, and a full container will remain full. If the expansion coefficient of the container exceeds that of the fluid, then fluid will be squeezed out upon cooling; if the expansion coefficient of the container is lower than that of the fluid, then fluid will be squeezed out upon heating. Thus the maximum possible value of α is found by filling the container at the maximum temperature; the container then shrinks more than the gasoline when the temperature descends to its lowest value. We require

$$\frac{\Delta V}{V}\bigg|_{container} - \frac{\Delta V}{V}\bigg|_{gasoline} = -0.001.$$

(The minus sign is present because the container shrinks more than the gasoline in this case.) Thus

$$3\alpha_{max} \, \Delta T - \beta \, \Delta T = -0.001,$$

$$\alpha_{max} = \frac{-0.001 + \beta \, \Delta T}{3 \, \Delta T} = \frac{-0.001 + (9.5 \times 10^{-4} \ K^{-1})(-20°C - 40°C)}{3(-20°C - 40°C)}$$

$$= 3.2 \times 10^{-4} \ K^{-1}.$$

Similarly, the minimum value of α can be found by filling the container at the minimum temperature, then insisting that the gasoline expands only a little more than the container:

$$\frac{\Delta V}{V}\bigg|_{gasoline} - \frac{\Delta V}{V}\bigg|_{container} = 0.001.$$

Thus

$$\beta \, \Delta T - 3\alpha_{min} \, \Delta T = 0.001,$$

$$\alpha_{min} = \frac{0.001 - \beta \, \Delta T}{-3 \, \Delta T} = \frac{0.001 - (9.5 \times 10^{-4} \ K^{-1})(40°C - (-20°C)}{-3(40°C - (-20°C))}$$

$$= 3.1 \times 10^{-4} \ K^{-1}.$$

In fact, it would be hard to find solids with values of α this large.

Water has the very special property that it decreases in density—it expands—as it solidifies on cooling (Fig. 21–16). This means that ice floats on water.

APPLICATION

Expansion joints

FIGURE 21–15 The expansion joint in this bridge allows the material of the bridge to undergo expansion in hot weather and contraction in cold weather.

FIGURE 21–16 The plastic bottle of water on the left has been placed in a freezer, whereas the one on the right has not. Water expands when it freezes, meaning that ice floats on liquid water. This fact has important consequences for Earth's weather and for life on Earth.

An interesting use of thermal expansion is the restoration of old masonry buildings. Such buildings often have bulging walls that result from the outward pressure of a pitched roof. To straighten such a wall, a hole is drilled in each side of the building and a metal bar passed through the building. This bar is usually concealed in the floorbeam region. The ends of the bar are threaded and metal plates are placed over the protruding bar ends (Fig. 21AB–1). Nuts are then tightened on the threads to force the plates against the walls. The bar is heated along its length, traditionally with flames, but now by electric energy. When the bar lengthens, the nuts are tightened as much as possible. When the bar cools and contracts, there is a large compression force against the walls, straightening them, and resisting future outward forces due to roof loads. Such plates, together with the nuts in their centers, can be seen on many old European buildings.

FIGURE 21AB–1 The star is a metal plate holding the old brick wall in place. A metal rod at the center of the star extends through the wall to the other side, where a second star is placed. There are several such devices placed on this old warehouse wall.

21–5   THE CONDUCTION OF THERMAL ENERGY

Another important property of solids has to do with the transport of thermal energy. When one end of a solid beam of length L is maintained at a temperature T_1 and the other end is kept at a lower temperature, T_2, then there will be a steady transport of thermal energy from the higher-temperature end to the lower-temperature end. Experiment shows that the rate of transfer of thermal energy is proportional to the area of the beam, A, and to the temperature difference between the two ends, $\Delta T = (T_1 - T_2)$; it is inversely proportional to the length of the beam, L.

The rate of thermal energy transfer, written as dQ/dt, is thus described by an equation of the type

$$\frac{dQ}{dt} \propto A \frac{\Delta T}{L}. \tag{21–15}$$

Actually, the linear dependence on the temperature difference ΔT is true only for small ΔT. This is not a difficulty if we think about the rate of energy transfer for a thin slice of length Δx rather than for the entire beam at once. The temperature difference across a thin slice is very small, and we can safely assume that the relation between ΔT and the rate of thermal energy flow is linear. We can then replace L by Δx and rewrite Eq. (21–15) in derivative form as

$$\frac{dQ}{dt} = -\kappa A \frac{dT}{dx}. \tag{21–16}$$

The definition of thermal conductivity   The coefficient κ is the **thermal conductivity**, a constant that varies from material to material and may have a temperature dependence. The SI units of κ are W/(m·K), or J/(m·s·°C). (Note that a difference of 1°C is identical to a difference of 1 K.) We give some typical values of thermal conductivity in Table 21–4. When κ is large, the material is said to be a good thermal conductor, and when κ is small, the material is a poor thermal conductor. The minus sign in Eq. (21–16) indicates that, if the *temperature gradient* dT/dx is negative—so the temperature is lower for larger values of x—then thermal energy flows toward larger x. The differential equation (21–16) applies to liquids and gases as well as to solids, and Table 21–4 lists values of κ for materials of all three states.

When Eq. (21–16) applies, we say that thermal energy is transported by *conduction*.

EXAMPLE 21–4 An igloo, a hemispherical enclosure built of ice ($\kappa = 1.67$ J/m · s · °C), has an inner radius of 2.5 m. The thickness of the ice is 0.30 m. At what rate must thermal energy be generated to maintain the air inside the igloo at 5°C when the outside temperature is −40°C? Ignore all thermal energy losses through the ground or by air currents.

Solution: The rate at which energy must be supplied is just the rate of loss of thermal energy by heat conduction, so we want to calculate the right-hand side of Eq. (21–16). The temperature gradient is given by $(T_{in} - T_{out})$/thickness of ice. The area A is that of half a sphere. Because the ice thickness is small compared to either the inner or outer radius, it does not matter very much which radius we use, so we use the median radius, $R_{med} = 2.5$ m + 0.15 m = 2.65 m. From Eq. (21–16),

$$\frac{dQ}{dt} = -\kappa A \frac{dT}{dx} = -\kappa (2\pi R_{med}^2) \frac{T_{in} - T_{out}}{\text{thickness of ice}}$$

$$= -(1.67 \text{ J/m} \cdot \text{s} \cdot °C)(2)(3.14)(2.65 \text{ m})^2 \frac{(5°C) - (-40°C)}{0.30 \text{ m}}$$

$$= -11 \times 10^3 \text{ J/s} = -11 \text{ kW}.$$

For comparison, a room-size kerosene heater puts out some 5 kW.

TABLE 21–4
Thermal Conductivities for Various Materials

Material	κ *(kW/m · K)*
Aluminum	0.21
Copper	0.39
Lead	0.035
Silver	0.42
Water (at 0°C)	5.65×10^{-4}
Water (at 20°C)	5.99×10^{-4}
Air	2.6×10^{-5}
Asbestos	0.80×10^{-4}
Brick	6.3×10^{-4}
Glass	10.5×10^{-4}

How Is Thermal Energy Conducted in Materials?

We know that thermal energy is the energy of motion of the constituents of matter. Let's now explore the physical origins of thermal conductivity. Metals conduct thermal energy better than nonmetals do. Within the atoms of metals, the outermost electrons that orbit the nucleus are attached very loosely (they are *weakly bound*), and a slight disturbance of the system can dislodge them. The practically free electrons of metals are efficient in transferring energy from one end of a piece of metal to the other. The same electrons also transfer electric charge (as we shall see in Chapter 24), and thermal and electrical conductivity in metals are closely related.

Nonmetals also conduct thermal energy, but less well than metals do. The outermost electrons of the atoms of nonmetals are attached tightly and are therefore not readily available to conduct thermal energy, and the molecules cannot move freely in a solid. Compressions and rarefactions of a nonmetal enter into the transport of thermal energy, but purely elastic waves traverse a material without depositing energy along the way. Only the scattering of waves by impurities, defects, and nonelasticities in the medium lead to the transfer of thermal energy.

In gases, thermal energy conduction is a transport phenomenon much like others discussed in Section 19–8. Molecules that move more rapidly because they are in a region of higher temperature collide with molecules in a neighboring region, giving the adjacent molecules more kinetic energy and consequently more thermal energy. This model explains some unique features of thermal transport in gases; for example, the thermal conductivity, κ, of gases varies with temperature as $\sqrt{T}$.

Table 21–4 shows that air—and this is generally true for gases—has a small thermal conductivity. Despite that fact, a house built of double walls with air between them will be poorly insulated. The reason is *convection* in fluids, a phenomenon that involves the large-scale movement of gases or liquids. Thermal energy is transported efficiently by convection. Fiberglass insulation is designed to trap air so that it cannot flow, or convect, easily. Thermal energy can also be transported by *radiation*. In such transport, thermal energy crosses transparent spaces—spaces in which electromagnetic radiation can travel.

∞ Thermal energy transport by convection and radiation are described in Chapters 16 and 17.

 APPLICATION

Insulating walls

The definition of thermal resistance

The **thermal resistance**, or **R value**, is used as an alternative to the thermal conductivity rating; it is inversely proportional to thermal conductivity. For a slab of a given material of thickness L, R is defined by

$$R \equiv \frac{L}{\kappa}.$$ (21–17)

Whereas κ depends only on the type of material, R *depends also on the thickness of the material*. Therefore, R characterizes the effectiveness of a thermal barrier of given thickness. A large R value corresponds to a small thermal conductivity and/or a large thickness; a piece of material with such characteristics is a good *thermal insulator*, whereas a piece of material with a low R value is a poor insulator. The R value has SI units of $m^2 \cdot K/W$. When it is used in connection with insulation in construction in America, however, it is often quoted in units of $ft^2 \cdot h \cdot °F/Btu$, where $1\ ft^2 \cdot h \cdot °F/Btu = 0.18 m^2 \cdot K/W$. The R value of a 6-in piece of fiberglass insulation is $22\ ft^2 \cdot h \cdot °F/Btu$, compared to $0.32\ ft^2 \cdot h \cdot °F/Btu$ for a $\frac{3}{8}$-in piece of plasterboard.

In terms of R, the relation between thermal energy flow and temperature difference from Eqs. (21–16) and (21–17) is

$$\frac{\Delta Q}{\Delta t} = \frac{1}{R} A\ \Delta T,$$ (21–18)

where A is the area across which thermal energy is transported. Because R refers to a given material of a given thickness, it is interesting to see how the R values of pieces of material combine.

Thermal Resistance in Series. Let's first consider combining two conducting solids of the same area A *in series*, as shown in Fig. 21–17. Suppose that the left-hand piece has an R value of R_1 and the right-hand piece R_2, and that the temperature varies from T_c (cold) on the left to T_h (hot) on the right. We want to find the effective R value of the combined piece, R_{eff}. When thermal energy flows at a steady rate through the two materials, as shown in Fig. 21–17, then energy conservation implies that the rate of transport of thermal energy across one piece must equal the rate of transport across the second piece, and hence also the rate across the combined piece. Application of this idea (see Problem 49) implies that

for two solids in series: $R_{eff} = R_1 + R_2$. (21–19)

This result generalizes for many slabs of material that are combined in series.

Thermal Resistance in Parallel. Now suppose that instead of being placed in series, two slabs are placed *in parallel*, as shown in Fig. 21–18. In this case, the temperature difference on the right-hand and left-hand sides of both slabs is the same, and the total rate of transport of thermal energy is the sum of the transport rates through the two slabs. In Problem 51, this fact is applied, and the result is

for two solids in parallel: $\dfrac{1}{R_{eff}} = \dfrac{1}{A_1 + A_2}\left(\dfrac{A_1}{R_1} + \dfrac{A_2}{R_2}\right).$ (21–20)

This result generalizes for more than two slabs placed in parallel.

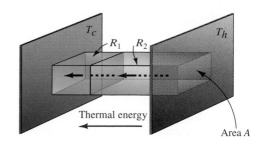

FIGURE 21–17 Thermal resistance, or R values, for two materials placed "in series"—one behind the other. Thermal energy passes through the materials as shown.

EXAMPLE 21–5 A hole in a wall has been temporarily patched on the inside with a 1-square-ft piece of plywood for which $R = 1$ ft$^2 \cdot$ h $\cdot$ °F/Btu. The rest of the wall is insulated with fiberglass insulation for which $R = 22$ ft$^2 \cdot$ h $\cdot$ °F/Btu. What is the R value of the repaired wall, whose total area is 100 ft^2?

Solution: A sketch of this situation, such as Fig. 21–18, shows that we have two materials placed in parallel, and Eq. (21–20) applies. Note that because the R value already takes the thickness of the material into account, the plywood is not necessarily the same thickness as the fiberglass insulation. In fact, plywood is only a fraction of an inch thick, whereas the fiberglass is 6 in thick. The rest of the thickness may be taken up by air, but because this air will be continuous with the outside environment, we ignore its insulating properties. If A_1 and R_1 are taken to be the area and R value of the plywood and A_2 and R_2 are the area and R value of the fiberglass, respectively, we find

$$\frac{1}{R_{\text{eff}}} = \frac{1}{A_1 + A_2}\left(\frac{A_1}{R_1} + \frac{A_2}{R_2}\right) = \frac{1}{100\ \text{ft}^2}\left(\frac{1\ \text{ft}^2}{1\ \text{ft}^2 \cdot \text{h} \cdot °\text{F/Btu}} + \frac{100\ \text{ft}^2 - 1\ \text{ft}^2}{22\ \text{ft}^2 \cdot \text{h} \cdot °\text{F/Btu}}\right)$$

$$= \frac{1}{18\ \text{ft}^2 \cdot \text{h} \cdot °\text{F/Btu}}.$$

Thus

$$R_{\text{eff}} = 18\ \text{ft}^2 \cdot \text{h} \cdot °\text{F/Btu}.$$

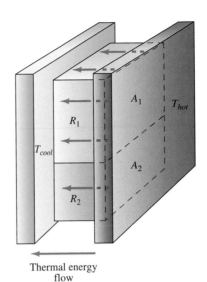

FIGURE 21–18 Thermal resistance, or R values, for two materials placed "in parallel." Thermal energy passes through the materials as shown.

In Example 21–5, a hole in a wall that is patched with little insulation and takes up only 1 percent of the area of the wall reduces the R value by some 17 percent. This effect is typical of thermal insulators combined in parallel. In particular, if one of the elements of a set of insulators combined in parallel has a very small R value, Eq. (21–20) shows that the effective R value is very small. Small holes can be very costly sources of thermal energy loss. Indeed, the best window has a thermal resistance of about one-tenth of normal wall and ceiling insulation. Thus once wall and ceiling energy losses become small, the only way to conserve energy further is to reduce window area. This is already practiced in commercial buildings in the Midwest.

SUMMARY

Solids consist of atoms arranged in regular crystalline structures, with springlike forces between the atoms. As a consequence, solid bodies may stretch or be compressed when subjected to external forces. The force per unit area, or stress, S, is related to the fractional change in length, or strain, e, by Y, Young's modulus:

$$\frac{S}{e} = Y. \tag{21–4}$$

A similar relation applies to volume stress, p, and volume strain, e_V, which are related by the bulk modulus, B:

$$B = -\frac{p}{e_V}. \tag{21–5}$$

The real tensile strength of solids, which is a measure of the amount of strain at which they fracture, is smaller than one would expect because of the many imperfections that occur in the crystalline structure of solids.

The springlike nature of the forces that hold solids together permits the propagation of longitudinal waves (sound), with the velocity of propagation given (approximately) by

$$v = \sqrt{\frac{Y}{\rho}}, \tag{21–6}$$

where ρ is the density of the material. Solids also support transverse shear waves, which move more slowly than sound waves.

Materials expand when heated, and they are characterized by a coefficient of thermal expansion, α, defined by

$$\alpha \equiv \frac{1}{L}\frac{dL}{dT}. \tag{21-11}$$

Materials also conduct thermal energy. The rate of transport of thermal energy depends on the temperature drop per unit length and the area of the material through which the thermal energy flows. The thermal conductivity, κ, is defined by

$$\frac{dQ}{dt} = -\kappa A\frac{dT}{dx}. \tag{21-16}$$

where the left-hand side represents the flow of thermal energy per unit time, A is the area, and dT/dx is the temperature drop per unit length. The reciprocal of κ is related to the thermal resistance, or R value, of the insulating material, with

$$R \equiv \frac{L}{\kappa}, \tag{21-17}$$

where L is the thickness of the material. The R values of two insulating slabs of the same area are additive when the slabs are connected in series, whereas the quantities area/R are additive when the slabs are connected in parallel.

UNDERSTANDING KEY CONCEPTS

1. Why is cement unsuitable as a construction material for a boiler?

2. In Chapter 5 of the Book of Exodus, the Israelites in captivity complain to Pharaoh that they are being asked to make bricks without straw. Why was it a good idea to put straw in the clay that was allowed to dry in the hot sun?

3. Why is it inadvisable to wear wet clothing on a cold day? In view of this question, how does a diver's wet suit work?

4. How can seismographs in the United States record earthquakes in other continents?

5. Thermopane windows are windows made of two glass panes with air between them. Explain why they are better thermal insulators than a simple window pane of twice the thickness.

6. Why does the amount of stretch of a bar on which a tension acts depend on the overall length of the bar? [*Hint*: Think of breaking the bar into two pieces of equal length and of how much each piece would stretch under the same force.]

7. If it is possible to make diamonds from graphite by using high pressure, how would it be possible to make graphite from diamonds?

8. If a doughnut-shaped piece of a solid that expands under heat is heated, does the hole expand or shrink?

9. On a macroscopic level, solids are distinguishable from liquids by their resistance to shear. Is this distinction likely to be a sharp one?

10. What kind of crystal structure might have little resistance to shear forces in some directions and much resistance in other directions? Would such a crystal have direction-dependent resistance to stretching?

11. To survive a cold night in the woods, you are advised to make a thick bed of dry leaves to sleep on and a thick cover of dry leaves to sleep under. Why does this work?

12. When you place your hand on a smooth block of wood, it feels warm; when you place your hand on a smooth block of marble, it feels cool. Why is this statement consistent with the statement that the wood and marble are both at the same temperature?

13. Water is one of the relatively rare substances that becomes less dense rather than more dense when it freezes. Describe what might happen to the ocean if ice were denser than liquid water.

14. Styrofoam, which has a very low density, is a good thermal insulator. Are all (or most) good thermal insulators low-density materials? Why or why not?

15. Why are cross-country skiers told to wear several layers of clothing rather than a single, thick coat?

16. You glue together two strips of metals that have different coefficients of thermal expansion, back-to-back. What will happen to such a composite strip when it is heated or cooled? Can you propose some technological use for such strips?

PROBLEMS

21–1 The Nature of Solids

1. (I) The crystal structure of copper is face-centered cubic; i.e., the structure is built of cubes of edge length a, with copper atoms at the corners as well as the face centers of the cube (Fig. 21–19). The lattice length $a = 0.361$ nm. (a) Calculate the diameter of a copper atom. (Assume that atoms are de-

scribed by spheres that are centered on the corners and centers, and that they just touch). (b) Show that the structure can also be described as a stacking of triangular layers of atoms, perpendicular to the maximum diagonal of the cube.

FIGURE 21–19 Problem 1.

2. (I) The building block of a simple cubic lattice is a cube with atoms at the corners. If the atoms are represented by spheres of diameter d that are centered on the corners and just touch, what is the diameter of the largest impurity that will not displace any of the existing atoms that can be put at the center of the cube? [Figure 21–20 shows a section through two opposite face diagonals. The figure has a hint: The longer side of the rectangle is $\sqrt{2}$ times the shorter side.]

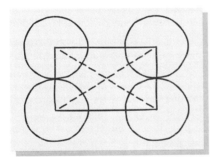

FIGURE 21–20 Problem 2.

3. (II) The volume filling of a crystal structure can be defined as the ratio of the volume of the atoms as represented by touching spheres to that of the crystal unit. Calculate this ratio for simple cubic and face-centered cubic lattices. [Refer to Problems 1 and 2.]

21–2 Stresses and Strains

4. (I) A concrete pier of area 0.25 m² and height 3 m is built to hold up a bridge (Fig. 21–21). The load on the pier is 300 tons (2.73×10^5 kg). $Y = 17{,}000$ MN/m² for the concrete. How far will the concrete compress?

5. (I) A mine elevator is supported by a steel cable 2.0 cm in diameter. The mass of the elevator and its contents is 800 kg. By how much is the cable stretched when the elevator is 250 m below Earth's surface?

6. (I) A steel bolt must withstand forces up to 1000 N. If a safety factor of 3 is to be taken into account, what must the mini-

FIGURE 21–21 Problem 4.

mum diameter of the bolt be? (With a safety factor of 3, the bolt should be able to support 3×1000 N before it snaps.)

7. (I) Calculate the critical strain for steel piano wire. You will need to consult the tables. How far would a wire 1 m long stretch before breaking?

8. (I) Assume that the piano wire of Problem 7 is 1 mm in diameter. Calculate the weight it could hold before fracturing.

9. (II) A crate of mass 30 kg slides across the ground. The coefficient of kinetic friction between the crate and the ground is 0.3. The physical contact area between the crate and the ground (which is less than the area of the whole crate) is 0.35 m². Calculate the shear stress on the crate.

10. (II) A steel wire of diameter 5.0 mm and length 2.0 m stretches 0.30 mm when a load of 60.0 kg is hung from it. What is its Young's modulus? How much mass can the wire hold before it may fracture? Use the tensile strength of steel from the tables.

11. (II) The 3-km-long cables on a large suspension bridge are stretched from their equilibrium length by 3 m. Estimate the change in the equilibrium separation between any two adjacent atoms along the cable.

12. (II) A steel bar of length 5 m is placed in a structure where it is subject to extreme stress under tension. The area of the bar is 40 cm², and it has room to stretch by 4 mm. It cannot stretch any more than this without butting up against a much stronger part of the structure. Does the bar break before it stretches the 4 mm? Use the Young's modulus and tensile strength from the tables.

13. (II) A uniform beam of length 2.0 m and mass 10 kg is freely pivoted at one end about a point fixed to the wall (Fig. 21–22). It is held in a horizontal position by a steel cable of diameter 2.0 mm. The cable makes an angle of 30° with the horizontal. A load of 30 kg is suspended from the end of the beam. What will be the angle of the beam relative to the horizontal? [Hint: The distortion is small, so you may calculate the tension in the cable without taking into consideration the slight change in angle.]

14. (II) A 0.5-m-long piece of metal is compressed longitudinally by 0.1 mm. Its Poisson's ratio is 0.32. Calculate its volume strain.

15. (II) Poisson's ratio for steel is 0.3. What is the new diameter of the steel rod used in Example 21–1 after the rod has been stretched?

FIGURE 21-22 Problem 13.

16. (III) According to our discussion of Poisson's ratio (Section 21–2), a bar of length L and cross-sectional area A under compression at its ends expands at the sides. Find the amount of pressure that must be exerted on the sides to keep them fixed when there is a compression F on the two ends. Find the total amount of compression in the longitudinal direction under these conditions, and show that a bar with constrained sides is stiffer under compression than the same bar with unconstrained sides. (Keep the terms up to order σ^2 only.)

17. (III) Let the strain on a rod along the direction of applied stress be e_1. Because the stretching produces a slight reduction in the diameter of the rod, we may speak of an induced transverse strain, given by $e_{tr} = -\sigma e_1$, where σ is Poisson's ratio (Fig. 21–23). Recall that Y is defined in terms of longitudinal stress and strain. A cube of the rod's material immersed in a liquid, so that the pressure on all sides is p, suffers a volume change given by $\Delta V/V = 3p(1 - 2\sigma)Y$. (a) Derive this result. (b) Show that when Poisson's ratio is not zero, the relation between Young's modulus and the bulk modulus under uniform pressure is given by

$$B = \frac{1}{3} \frac{Y}{1 - 2\sigma}.$$

FIGURE 21-23 Problem 17.

18. (II) By using the results of Problem 17, show that Poisson's ratio, σ, must be less than 0.5 because, if it exceeds 0.5, the total volume of the solid would increase when a uniform pressure is applied to it.

19. (III) *Compressibility* is defined by $-\Delta V/pV$, the reciprocal of the bulk modulus. Using the equation for $\Delta V/V$ in Problem 17, calculate the compressibility of silver, given that $\sigma = 0.38$ and $Y = 7.9 \times 10^4$ MN/m^2.

20. (III) Show by analogy with a simple spring, for which the force is $F = -kx$, that a volumetric potential energy associated with strain—that is, potential energy per unit volume of a wire or rod of length L and cross-sectional area A—is given by $u = \frac{1}{2}$ (stress)(strain).

21–3 Waves in Solids

21. (I) The speed of sound in a material of density 2.7×10^3 kg/m^3 is measured to be 5.1×10^3 m/s. What is Young's modulus for that material?

22. (I) The shear modulus of copper is $G = 4.2 \times 10^{10}$ N/m^2. Given that the density of copper is 8.93×10^3 kg/m^3, what is the speed of a shear wave in copper?

23. (I) The density of pure gold is 19.3 g/cm^3, and the speed of sound in a small sample is measured to be 2.04×10^5 cm/s. What is Young's modulus for gold in dynes/cm^2?

24. (I) Given that the ratio of densities for steel and lead is 7.86: 11.3, and that the ratio of Y values is 200,000 MN/m^2: 16,000 MN/m^2, what is the ratio of the speeds of sound in these two materials?

25. (I) A scuba diver has to send a message to save an oil platform in the Gulf of Mexico. He realizes that sound travels faster in steel than in air or water. With only one second left for the message to be received on the platform, he finds a steel mooring cable that is 5 km long, which runs directly to the platform. Will the message reach the platform in time? (The density of steel is 7.9×10^3 kg/m^3 and $Y = 210,000$ MN/m^2 for steel.)

26. (II) Young's modulus for Earth can be approximated by 5.5×10^5 MN/m^2, and the corresponding shear modulus by 2.0×10^5 MN/m^2. A first signal of an earthquake arrives at a detection station at $t = 0$ min, and a second signal arrives 2 min later. Assume that the successive signals are those of longitudinal waves and shear waves, respectively, that follow the same straight-line path through the planet. Assume further that the average density of Earth is 5.5 g/cm^3. How far from the station did the earthquake take place?

27. (II) Earthquake waves within Earth are of two types: S (transverse) waves, which can be supported within Earth's mantle but not in the liquid part of the core, and P (longitudinal) waves, which propagate in both the mantle and the core (see Fig. 21–13). The speeds of S waves and P waves in the mantle are approximately 4 km/s and 8 km/s, respectively. The speeds of these waves can be calculated from the Young's and shear moduli. Take the density of Earth's mantle to be 5 g/cm^3. Calculate the Young's and shear moduli. If the difference between the time of arrival of the two waves is 27 min, how far away was the earthquake? Ignore Earth's thin crust, and assume that the waves travel in a straight line.

28. (II) A steel rod of length 0.8 m is rigidly supported at one end. Calculate the frequencies of the lowest three longitudinal modes, and the lowest three transverse (shear) modes, given that $Y = 2.1 \times 10^{11}$ N/m^2 and $G = 8.1 \times 10^{10}$ N/m^2.

21–4 Thermal Expansion

29. (I) How much farther will a 1-m-long aluminum rod expand than a steel one when each undergoes a temperature increase of 40°C?

30. (I) We learned that the steel rod in Example 21–1 extended by only 0.25 mm under a tensile load of about 10^5 N. By how much will the same steel rod elongate if it is heated 100°C?

31. (I) The area of the circle formed by a circular loop of metal is found to increase by 1.6 percent when the temperature changes by 10°C. What is the coefficient of thermal expansion of the metal?

32. (I) The coefficient of thermal expansion for Invar is 0.7×10^{-6} K^{-1}. (Invar is a steel alloy.) Given that an Invar pendulum clock keeps perfect time at a room temperature of 20°C, how much time will the clock gain or lose per day when it is in a room at 30°C? (For the time-keeping properties of a pendulum, see Chapter 13).

33. (I) In the design of a mercury thermometer, it is desirable for each °C change to correspond to a 1 cm expansion of the metal. What is the cross section of the mercury column if the mercury at the lowest reading on the thermometer is contained in a bulb whose volume is 0.7 cm^3? Ignore any possible expansion of the glass tube.

34. (II) A metal bar is 2.00 m long with a coefficient of thermal expansion of 1.54×10^{-5} K^{-1}. It is rigidly held between two fixed beams. When the temperature rises, the metal bar takes on the shape of the arc of a circle (Fig. 21–24). What is the radius of curvature of the circle when the temperature rises by 50°C? [*Hint*: You can use the small-angle approximation $\sin \theta \simeq \theta - (\theta^3/6)$.]

FIGURE 21–24 Problem 34.

35. (II) As we know from Chapter 13, the time expressed by a grandfather clock depends on the length of the pendulum. (a) Which is the best material for the pendulum of a clock to be used outdoors: aluminum, copper, or steel? (See Table 21–3.) (b) If the temperature of the clock ranges from −20°C to +30°C over the course of a year, and if the clock ticks with frequency $f = 1.00000$ Hz at 5°C, what is the yearly frequency range for each of the three materials?

36. (II) In space, where heat transfer by air is missing, large temperature differences can develop. A space telescope (Fig. 21–25) consists of a main mirror and an instrument unit, held together by titanium rods. The optical axis of the mirror (perpendicular to the mirror, through its center) hits the middle of the receiving window of the instrument unit. Suppose that the temperature of two neighboring rods increases by 20°C due to direct sunlight, with the temperature of the other two remaining unchanged. How far will the optical axis of the mirror move on the instrument unit, given that the rods are 5 m long and that $\alpha = 8.5 \times 10^{-6}$ K^{-1}?

FIGURE 21–25 Problem 36.

37. (II) The coefficient of volume expansion, β, for ethyl alcohol is 1.12×10^{-3} K^{-1}. Suppose that a cylindrical copper container, with a coefficient of thermal expansion of 1.67×10^{-5} K^{-1}, is filled to the very brim with ethyl alcohol at 5°C. What percentage of the alcohol will spill if the temperature is raised to 25°C?

38. (III) A strip of brass 2.0 mm thick and 10 cm long is glued back-to-back to a strip of Invar with the same dimensions. When the temperature is increased by ΔT, the composite strip tends to curl into an arc of a circle (Fig. 21–26). Find the radius of curvature of this circle for $\Delta T = 5$°C, given that the coefficient of thermal expansion is 19×10^{-6} K^{-1} for brass and 0.7×10^{-6} K^{-1} for Invar.

FIGURE 21–26 Problem 38.

21–5 The Conduction of Thermal Energy

39. (I) Compare the rate of thermal energy transfer across a piece of asbestos that is 1 m^2 in area and 1 cm thick with that of a piece of glass of the same dimensions for a temperature difference of 60°C.

40. (I) A single window pane of thickness 6.0 mm has an area of 1.0 m^2. The inside and outside temperatures are 73°F and 13°F, respectively. Calculate the rate of thermal energy loss through the window. (See Table 21–4.)

41. (I) A brick wall 5 m by 5 m in size and 0.1 m thick separates the inside of a house from the outside, where the temperature is 285 K (12°C). A heater supplies energy inside at a rate of 0.35 kW. What is the inside temperature? Ignore all surfaces except the brick wall in question.

42. (I) The glass from a basement window is replaced by a sheet of asbestos that is 2 cm thick. The window is 80 cm wide and 40 cm high, and there is no air leakage around the edges. Given that the heat loss through the surface is 0.06 J/s, what is the temperature difference between the outside and the inside of the basement?

43. (I) A house of dimensions 36 ft × 21 ft × 9 ft is built of bricks that are 9 in thick. The inside and outside temperatures are 76°F and 25°F, respectively. Calculate the rate of thermal energy loss through the brick wall. (See Table 21–4.)

44. (I) Your refrigerator can be thought of as a box with six sides of total area 2.5 m². The effective R value of the walls is 1.5 m²·K/W. The temperature inside is 5°C, while the temperature outside is 25°C. Calculate the rate of heat loss.

45. (II) Pine wood 1 in thick has an R value of 1.3, compared with 3.0 for fiberglass insulation of the same thickness. (The units of R are ft²·h·°F/Btu.) Calculate the rate of thermal energy loss of a 6-in thickness of these same materials across a 4 ft × 8 ft section of wall for a temperature difference across the wall of 35°F. If 10 percent of the wall section is made of wood studs and the space between studs is filled with fiberglass, calculate the effective R value of the wall section (Fig. 21–27).

FIGURE 21–27 Problem 45.

46. (II) The R value of brick is about 1.1 ft²·h·°F/Btu across the brick's width. A dog who expends 50 W of energy lives in a brick doghouse of total effective area 2.0 m². How much temperature variation between the inside and outside of the doghouse can the dog withstand if only its body heat warms the house? Assume that the door locks out drafts and that the heat conduction through the door is identical to that through the walls.

47. (II) A wall of area A consists of two slabs: one of thickness L_1 and thermal conductivity κ_1, the other of thickness L_2 and thermal conductivity κ_2, as shown in Fig. 21–28. The temperatures are T_1 and T_2, respectively, on the left-facing side and the right-facing side of the wall. (a) Calculate the effective thermal conductivity of the wall in terms of the parameters given. (b) Calculate the temperature at the interface of the two slabs.

FIGURE 21–28 Problem 47.

48. (II) Consider a wall of thickness L and area A. Part of the area, A_1, consists of a material with thermal conductivity κ_1 and the remaining area, $A - A_1$, consists of a material with thermal conductivity κ_2 (Fig. 21–29). Calculate how much thermal energy flows through each part of the wall, and calculate an effective thermal conductivity for the wall as a whole.

FIGURE 21–29 Problem 48.

49. (II) Equation (21–19) states that when two slabs of conducting material are combined in series, as pictured in Fig. 21–17, their R values are additive. Show that this is so by using the principle that the rate of flow of thermal energy is the same for each slab and for the two slabs combined:

$$\frac{dQ}{dt} = \frac{1}{R_1}A(T_1 - T_{\text{int}}) = \frac{1}{R_2}A(T_{\text{int}} - T_2)$$

$$= \frac{1}{R_{\text{eff}}}A(T_1 - T_2).$$

50. (II) Two pieces of material are arranged in series as in Fig. 21–17. If the temperature to the left is T_1 and the temperature to the right is T_2, find the temperature between the pieces, T_{int}, as a function of T_1, T_2, and the R values of the two pieces.

51. (II) Equation (21–20) gives the effective R value for two slabs of conducting material combined in parallel, as pictured in Fig. 21–18. Show that this equation is valid by using the principle that the temperature difference on opposite sides of both slabs is the same, and that the rate of flow of thermal energy across the combined slabs is the sum of the rate of flow of thermal energy across each slab.

52. (III) Repeat Example 21–4 but do not assume that the igloo's thickness is small compared to its radii. You may assume that the surfaces of constant temperature form concentric hemispheres. [*Hint*: Break up the hemisphere of ice into thin hemispherical shells and find the *R* value of all these hemispheres combined in series. Note that the area of the hemispherical shells is not constant, so Eq. (21–16) must be modified appropriately.]

53. (III) A long, cylindrical hot-water pipe of radius R_1 is at temperature T_1. It is wrapped with insulation of thickness α and of thermal conductivity κ; the outer surface of the insulation is at temperature T_2. Derive an expression for the rate of heat loss per unit length of insulated pipe. [*Hint*: The temperature within the insulation varies with radial position *r*. Break the insulation into thin cylindrical shells and find the areas of these shells as a function of their radii.]

General Problems

54. (II) When a 3000-ton submarine dives into deep waters, the compressive volume strain on the submarine is 2.5 percent. What is the loss of buoyancy for the submarine? In order to maintain the same buoyancy that it has just below the surface, the submarine must lose some mass, presumably by dumping water from its ballast tanks. How much water must be dumped?

55. (II) A trapper lives in a cave. At the cave's entrance, the trapper has constructed an exterior wall from lumber 6 cm thick. The wood has a thermal conductivity, κ, of 0.35 W/m·K. Assume that all thermal energy loss occurs through the wall, which forms a rectangle of dimensions 2.5 m × 2.0 m. This wall, together with a stove, is capable of maintaining the cave's interior at a temperature of 8°C when the outside temperature is −25°C. What is the power output of the stove?

56. (II) The trapper of Problem 55 tires of looking at the blank entrance wall and decides to cut a window in it. The window will have a sheet of mica 2.0 mm thick with a thermal conductivity of 0.7 W/m·K. How large a window can he construct

if he keeps the same stove and plans to keep the temperature inside the cave at 6°C?

57. (II) Cold air blows through a crack in the wall of Problem 55, so the trapper covers the crack with a piece of wood 5 mm thick. The thermal conductivity of the wood is 25×10^{-2} W/m·K. If the area of the crack is 6 cm^2 and the temperature difference between outside and inside is 30°C, what is the heat loss through the covered crack?

58. (II) A brass ring has to be fit onto a solid steel axle. To do this, the ring is made somewhat smaller than the axle, and it is heated to 100°C until it fits onto the axle (Fig. 21–30). Calculate the stress in the ring when it cools back to room temperature (25°C). Neglect the deformation of the axle. For brass, $\alpha = 19 \times 10^{-6}$ K^{-1}, $Y = 106,000$ MN/m^2.

FIGURE 21–30 Problem 58.

59. (II) The coefficient of thermal expansion of the iron rods inside a bar of reinforced concrete is 12.5×10^{-6} K^{-1}, whereas the coefficient of thermal expansion of concrete is 12.0×10^{-6} K^{-1}. Calculate the additional stress on a 10-m-long bar of reinforced concrete for a temperature increase from 10°C to 30°C; assume that the new length of the bar is that of the iron rods within. The fact that the coefficients of thermal expansion are practically the same for the right variety of iron bar and concrete allows the fabrication of reinforced concrete that remains useful under large temperature extremes.

The Système Internationale (SI) of Units

I–1 SOME SI BASE UNITS

Physical Quantity	Name of Unit	Symbol
length	meter	m
mass	kilogram	kg
time	second	s
electric current	ampere	A
thermodynamic temperature	kelvin	K
amount of substance	mole	mol

I–2 SOME SI DERIVED UNITS

Physical Quantity	Name of Unit	Symbol	SI Unit
frequency	hertz	Hz	s^{-1}
energy	joule	J	$kg \cdot m^2/s^2$
force	newton	N	$kg \cdot m/s^2$
pressure	pascal	Pa	$kg/m \cdot s^2$
power	watt	W	$kg \cdot m^2/s^3$
electric charge	coulomb	C	$A \cdot s$
electric potential	volt	V	$kg \cdot m^2/A \cdot s^3$
electric resistance	ohm	Ω	$kg \cdot m^2/A^2 \cdot s^3$
capacitance	farad	F	$A^2 \cdot s^4/kg \cdot m^2$
inductance	henry	H	$kg \cdot m^2/A^2 \cdot s^2$
magnetic flux	weber	Wb	$kg \cdot m^2/A \cdot s^2$
magnetic flux density	tesla	T	$kg/A \cdot s^2$

I–3 SI UNITS OF SOME OTHER PHYSICAL QUANTITIES

Physical Quantity	SI Unit
speed	m/s
acceleration	m/s^2
angular speed	rad/s
angular acceleration	rad/s^2
torque	$kg \cdot m^2/s^2$, or $N \cdot m$
heat flow	J, or $kg \cdot m^2/s^2$, or $N \cdot m$
entropy	J/K, or $kg \cdot m^2/K \cdot s^2$, or $N \cdot m/K$
thermal conductivity	$W/m \cdot K$

I–4 SOME CONVERSIONS OF NON-SI UNITS TO SI UNITS

Energy:
1 electron-volt (eV) = 1.6022×10^{-19} J
1 erg = 10^{-7} J
1 British thermal unit (BTU) = 1055 J
1 calorie (cal) = 4.186 J
1 kilowatt-hour (kWh) = 3.6×10^6 J

Mass:
1 gram (g) = 10^{-3} kg
1 atomic mass unit (u) = 931.5 MeV/c^2 = 1.661×10^{-27} kg
1 MeV/c^2 = 1.783×10^{-30} kg

Force:
1 dyne = 10^{-5} N
1 pound (lb or #) = 4.448 N

Length:
1 centimeter (cm) = 10^{-2} m
1 kilometer (km) = 10^3 m
1 fermi = 10^{-15} m
1 Angstrom (Å) = 10^{-10} m
1 inch (in or ") = 0.0254 m
1 foot (ft) = 0.3048 m
1 mile (mi) = 1609.3 m
1 astronomical unit (AU) = 1.496×10^{11} m
1 light-year (ly) = 9.46×10^{15} m
1 parsec (ps) = 3.09×10^{16} m

Angle:
1 degree (°) = 1.745×10^{-2} rad
1 min (') = 2.909×10^{-4} rad
1 second (") = 4.848×10^{-6} rad

Volume:
1 liter (L) = 10^{-3} m^3

Power:
1 kilowatt (kW) = 10^3 W
1 horsepower (hp) = 745.7 W

Pressure:
1 bar = 10^5 Pa
1 atmosphere (atm) = 1.013×10^5 Pa
1 pound per square inch (lb/in^2) = 6.895×10^3 Pa

Time:
1 year (yr) = 3.156×10^7 s
1 day (d) = 8.640×10^4 s
1 hour (h) = 3600 s
1 minute (min) = 60 s

Speed:
1 mile per hour (mi/h) = 0.447 m/s

Magnetic field:
1 gauss = 10^{-4} T

Some Fundamental Physical Constants[†]

Constant	Symbol	Value	Error
speed of light in a vacuum	c	2.99792458×10^8 m/s	exact
gravitational constant	G	6.67259×10^{-11} m^3/kg$\cdot$s^2	128
Avogadro's number	N_A	6.02214×10^{23} mol^{-1}	0.6
universal gas constant	R	8.31451 J/mol$\cdot$K	8.4
Boltzmann's constant	k	1.38066×10^{-23} J/K	8.5
elementary charge	e	1.60218×10^{-19} C	0.3
permittivity of free space	ϵ_0	$8.85418781762 \times 10^{-12}$ C^2/N$\cdot$m^2	exact
	$1/4\pi\epsilon_0$	8.987552×10^9 kg$\cdot$m^3$\cdot$s^{-2}$\cdot$C^{-2}	
permeability of free space	μ_0	$4\pi \times 10^{-7}$ T$\cdot$m/A	exact
electron mass	m_e	9.10939×10^{-31} kg	0.6
proton mass	m_p	1.67262×10^{-27} kg	0.6
neutron mass	m_n	1.67493×10^{-27} kg	0.6
Planck's constant	h	6.62608×10^{-34} J$\cdot$s	0.6
$h/2\pi$	$\hbar$	1.05457×10^{-34} J$\cdot$s	0.6
		$= 6.58212 \times 10^{-22}$ MeV$\cdot$s	0.3
	$\hbar c$	197.327 Mev$\cdot$fm	0.3
electron charge-to-mass ratio	$-e/m_e$	-1.75882×10^{11} C/kg	0.3
proton-electron mass ratio	m_p/m_e	1836.15	0.15
molar volume of ideal gas at STP		22414.1 cm^3/mol	8.4
Bohr magneton	μ_B	9.27402×10^{-24} J/T	0.3
magnetic flux quantum	$\Phi_0 = h/2e$	2.067783×10^{-15} Wb	0.3
Bohr radius	a_0	0.529177×10^{-10} m	0.05
Rydberg constant	R_∞	1.09737×10^7 m^{-1}	0.001

[†]From E. R. Cohen and B. N. Taylor, *The 1986 Adjustment of the Fundamental Constants,* Report of the CODATA Task Group on Fundamental Constants, CODATA Bulletin 63, Pergamon, Elmsford, N.Y. (1986).

We have given values of the measured constants to six significant figures, even though they may be known to greater accuracy. We quote the error, which expresses the uncertainty in the values of these constants, in parts per million. Defined constants have no error, and we give their full definition; they are indicated by the notation "exact" in the error column.

Other Physical Quantities

III–1.1 SOME ASTRONOMICAL CONSTANTS

Constant	Symbol	Value
standard gravity at Earth's surface	g	9.80665 m/s^2
equatorial radius of Earth	R_e	$6.374 \times 10^6 \text{ m}$
mass of Earth	M_e	$5.976 \times 10^{24} \text{ kg}$
mass of Moon		$7.350 \times 10^{22} \text{ kg}$ $= 0.0123 \, M_e$
mean radius of Moon's orbit around Earth		$3.844 \times 10^8 \text{ m}$
mass of Sun	$M_\odot$	$1.989 \times 10^{30} \text{ kg}$
radius of Sun	$R_\odot$	$6.96 \times 10^8 \text{ m}$
mean radius of Earth's orbit around Sun	AU	$1.496 \times 10^{11} \text{ m}$
period of Earth's orbit around Sun	yr	$3.156 \times 10^7 \text{ s}$
diameter of our galaxy		$7.5 \times 10^{20} \text{ m}$
mass of our galaxy		$2.7 \times 10^{41} \text{ kg}$ $= (1.4 \times 10^{11}) \, M_\odot$
Hubble parameter	H	$2.5 \times 10^{-18} \text{ s}^{-1}$

III–1.2 PLANETARY DATA

Planet	Diameter (in km)	Relative[†]	Relative Mass[†]	Average Density (in g/cm³)	Period of Rotation	Surface Gravity[†] (in g)	Escape Speed (in km/s)	Semimajor Axis (AU)	Period of Solar Orbit	Average Orbital Speed (in km/s)
Mercury	4,800	0.38	0.05	5.2	58 d 15 h	0.38	4.3	0.387	87.96 d	47.8
Venus	12,100	0.95	0.82	5.3	243 d 4 h	0.90	10.3	0.723	224.7 d	35.0
Earth	12,750	1.00	1.00	5.5	23 h 56 min	1.00	11.2	1.000	365.26 d	29.8
Mars	6,800	0.53	0.11	3.8	24 h 37 min	0.38	5.1	1.524	687.0 d or 1.88 yr	24.2
Jupiter	142,800	11.23	317.9	1.3	9 h 50 min	2.69	59.5	5.20	11.86 yr	13.1
Saturn	120,660	9.41	95.2	0.7	10 h 39 min	1.19	35.6	9.54	29.46 yr	9.7
Uranus	51,000	3.98	14.6	1.3	17 h	0.93	21.4	19.18	84.01 yr	6.8
Neptune	49,500	3.88	17.2	1.7	18 to 22 h	1.22	23.6	30.06	164.79 yr	5.4
Pluto	2,290	0.23	0.002	0.4	6 d 9 h 17 min	0.05	1.2	39.44	248 yr	4.7

[†]Relative to Earth.

III–2 ENERGY SUPPLY AND DEMAND[†]

[†]From the *Physics Vade Mecum,* Ed. Herbert L. Anderson, American Institute of Physics (New York, 1981); and U.S. Congress, Office of Technology Assessment, *Changing by Degrees: Steps to Reduce Greenhouse Gases,* OTA-O-482 (Washington, D.C.: U.S. Government Printing Office, February 1991).

III–2.1 FUEL RESOURCES (1980, ESTIMATED)

Resource	U.S. Resources	World Resources
coal (recoverable)	5×10^{21} J	2×10^{22} J
oil (not including oil shales)	10^{21} J	10^{22} J
natural gas	2×10^{21} J	10^{22} J
hydroelectric	10^{22} J/yr (North America)	6×10^{22} J/yr

III–2.2 ANNUAL USAGE OF RESOURCE (1989, PERCENTAGE OF TOTAL)

Resource	U.S. Usage (total = 9×10^{19} J)	World Usage (total = 4×10^{20} J)
coal	40	24
oil	23	35
natural gas	23	18
nuclear	7	5
hydroelectric	3	5
biomass	3	13

III–2.3 ENERGY CONTENT OF FUELS

Fuel	Energy Content (in J/kg)
bread	10×10^{6}
glucose ($C_6H_{12}O_6$)	16×10^{6}
white pine wood	20×10^{6}
methyl alcohol (CH_4O)	20×10^{6}
anthracite coal	32×10^{6}
domestic heating oil	45×10^{6}
propane (C_3H_8)	50×10^{6}
natural gas (96% CH_4)	51×10^{6}
fission of U^{235}	5.8×10^{11}
perfect mass-energy conversion	9×10^{16}

III–2.4 SOLAR ENERGY OUTPUT

total radiated power from the Sun	4×10^{26} W
power per unit area at the top of Earth's atmosphere	1.4 kW/m^2
average power per unit area delivered to an average horizontal surface in the United States in 1 yr	0.2 W/m^2

III–2.5 ENERGY CONSUMPTION IN TRANSPORTATION

Mode	Energy Consumption (J/passenger · km)
bicycle	5×10^{4}
foot travel	1.5×10^{5}
intercity bus	3×10^{5}
intercity train	9×10^{5}
automobile	1.5×10^{6}
747 jet airplane	2×10^{6}
snowmobile	6×10^{6}

III–2.6 ENERGY CONSUMPTION OF ELECTRICAL APPLIANCES

Appliance	Power (in W)	Energy Use Per Year (in kWh)
window air conditioner	1565	1390
clock	2	17
dishwasher	1200	363
window fan	200	170
hair dryer	380	14
iron	1000	144
microwave oven	1450	190
radio	71	86
refrigerator-freezer	615	1830
stove	12,200	1175
color television	200	440
vacuum cleaner	630	46
washing machine	512	103

Mathematics

IV-1 SOME MATHEMATICAL CONSTANTS†

Constant	Value
π	3.14159
e (Euler's constant)	2.71828
$\sqrt{2}$	1.41421
$1/\sqrt{2}$	0.707107
$\ln(10)$	2.30259
$\ln(2)$	0.693147
1 rad	57.2958°
1°	0.0174533 rad

†To six significant figures.

IV-2 SOLUTION OF QUADRATIC EQUATIONS

Quadratic equation:

$$ax^2 + bx + c = 0$$

Two solutions:

$$x = \frac{-b \pm \sqrt{b^2 - 4ac}}{2a}$$

IV-3 BINOMIAL THEOREM

$$(x + y)^n = \sum_{k=0}^{n} \binom{n}{k} x^{n-k} y^k,$$

where

$$\binom{n}{k} = \frac{n!}{(n-k)!\, k!}.$$

The factorial $m! \equiv 1 \cdot 2 \cdot 3 \ldots \cdot m$; $0! \equiv 1$. Some particular cases of the binomial theorem:

(1) $(x \pm y)^2 = x^2 \pm 2xy + y^2$;
(2) $(x \pm y)^3 = x^3 \pm 3x^2y + 3xy^2 \pm y^3$;
(3) $(x \pm y)^4 = x^4 \pm 4x^3y + 6x^2y^2 \pm 4xy^3 + y^4$.

IV-4 TRIGONOMETRY

1. For a right triangle with sides a, b, and c (the hypotenuse), where the angle opposite side a is θ_a (Fig. A–1),

$$\text{sine of } \theta_a = \sin \theta_a \equiv \frac{a}{c};$$

$$\text{cosine of } \theta_a = \cos \theta_a \equiv \frac{b}{c};$$

$$\text{tangent of } \theta_a = \tan \theta_a = \frac{a}{b}.$$

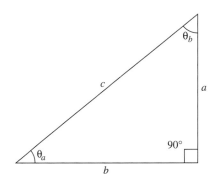

FIGURE A-1

2. The cosine function is even, $\cos(-x) = \cos x$; the sine function is odd, $\sin(-x) = -\sin x$.

3. (1) $\tan \theta = \dfrac{\sin \theta}{\cos \theta}$

   (2) $\sec \theta = \dfrac{1}{\cos \theta}$

   (3) $\operatorname{cosec} \theta = \dfrac{1}{\sin \theta}$

   (4) $\cot \theta = \dfrac{1}{\tan \theta}$

4. (1) $\sin^2 \theta + \cos^2 \theta = 1$

   (2) $\sec^2 \theta - \tan^2 \theta = 1$

   (3) $\operatorname{cosec}^2 \theta - \cot^2 \theta = 1$

5. (1) $\sin(\theta_1 \pm \theta_2) = \sin \theta_1 \cos \theta_2 \pm \cos \theta_1 \sin \theta_2$

   (2) $\cos(\theta_1 \pm \theta_2) = \cos \theta_1 \cos \theta_2 \mp \sin \theta_1 \sin \theta_2$

   (3) $\sin \theta_1 \pm \sin \theta_2 = 2 \sin \left(\dfrac{\theta_1 \pm \theta_2}{2} \right) \cos \left(\dfrac{\theta_1 \mp \theta_2}{2} \right)$

   (4) $\cos \theta_1 + \cos \theta_2 = 2 \cos \left(\dfrac{\theta_1 + \theta_2}{2} \right) \cos \left(\dfrac{\theta_1 - \theta_2}{2} \right)$

   (5) $\cos \theta_1 - \cos \theta_2 = -2 \sin \left(\dfrac{\theta_1 + \theta_2}{2} \right) \sin \left(\dfrac{\theta_1 - \theta_2}{2} \right)$

   (6) $\tan(\theta_1 + \theta_2) = \dfrac{\tan \theta_1 + \tan \theta_2}{1 - (\tan \theta_1)(\tan \theta_2)}$

   (7) $\cos \left(\theta \pm \dfrac{\pi}{2} \right) = \mp \sin \theta$

   (8) $\sin \left(\theta \pm \dfrac{\pi}{2} \right) = \pm \cos \theta$

   (9) $\sin \theta_1 \sin \theta_2 = \dfrac{1}{2}[\cos(\theta_1 - \theta_2) - \cos(\theta_1 + \theta_2)]$

(10) $\cos \theta_1 \cos \theta_2 = \frac{1}{2}[\cos(\theta_1 - \theta_2) + \cos(\theta_1 + \theta_2)]$

(11) $\sin \theta_1 \cos \theta_2 = \frac{1}{2}[\sin(\theta_1 - \theta_2) + \sin(\theta_1 + \theta_2)]$

6. (1) $\sin(2\theta) = 2 \sin \theta \cos \theta = \dfrac{2 \tan \theta}{1 + \tan^2 \theta}$

(2) $\cos(2\theta) = \cos^2 \theta - \sin^2 \theta = 2 \cos^2 \theta - 1 = 1 - 2 \sin^2 \theta$

(3) $\tan(2\theta) = \dfrac{2 \tan \theta}{1 - \tan^2 \theta}$

(4) $\sin\left(\dfrac{\theta}{2}\right) = \pm \sqrt{\dfrac{1 - \cos \theta}{2}}$

(5) $\cos\left(\dfrac{\theta}{2}\right) = \pm \sqrt{\dfrac{1 + \cos \theta}{2}}$

7. Expansions of Trigonometric Functions (θ in rad):

(1) $\sin \theta = \theta - \dfrac{\theta^3}{3!} + \dfrac{\theta^5}{5!} - \dfrac{\theta^7}{7!} + \cdots \qquad (\theta^2 < 1)$

(2) $\cos \theta = 1 - \dfrac{\theta^2}{2!} + \dfrac{\theta^4}{4!} - \dfrac{\theta^6}{6!} + \cdots \qquad (\theta^2 < 1)$

(3) $\tan \theta = \theta + \dfrac{1}{3}\theta^3 + \dfrac{2}{15}\theta^5 + \dfrac{17}{315}\theta^7 + \cdots \quad \left(\theta^2 < \dfrac{\pi^2}{4}\right)$

IV–5 GEOMETRICAL FORMULAS

1. (circumference of a circle of radius r) $= 2\pi r$

2. (area of a circle of radius r) $= \pi r^2$

3. (area of a sphere of radius r) $= 4\pi r^2$

4. (volume of a sphere of radius r) $= \frac{4}{3}\pi r^3$

5. (area of a rectangle with sides of lengths L_1 and L_2) $= L_1 L_2$

6. For a right triangle with sides a, b, and c and angles θ_a and θ_b opposite the sides a and b, respectively (Fig. A–1):

(1) $a^2 + b^2 = c^2$ (the Pythagorean theorem)

(2) area $= \frac{1}{2}$(base)(height) $= \frac{1}{2}ab$

7. For a triangle with sides a, b, and c opposite the angles θ_a, θ_b, and θ_c, respectively (Fig. A–2):

(1) $\theta_a + \theta_b + \theta_c = 180° = \pi$ rad

(2) $a^2 = b^2 + c^2 - 2bc \cos \theta_a$

(3) $\dfrac{a}{\sin \theta_a} = \dfrac{b}{\sin \theta_b} = \dfrac{c}{\sin \theta_c}$

(4) $a = b \cos \theta_c + c \cos \theta_b$

(5) area $= \dfrac{1}{2}$(base)(height) $= \dfrac{1}{2}ab \sin \theta_c = \dfrac{1}{2}a^2 \dfrac{\sin \theta_b \sin \theta_c}{\sin \theta_a}$

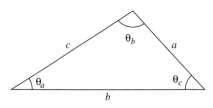

FIGURE A–2

8. (volume of a right cylinder of height h and radius r) $= \pi r^2 h$

IV–6 SOME PROPERTIES OF ALGEBRAIC FUNCTIONS

1. General properties

(1) $a^x a^y = a^{x+y}$

(2) $a^0 = 1$

(3) $(ab)^x = a^x b^x$

2. Properties of exponential of x, e^x:

(1) $e^{\ln(x)} = x$

(2) $e^{x_1} e^{x_2} = e^{x_1 + x_2}$

(3) $e^0 = 1$

(4) expansion: $e^x = 1 - x + \dfrac{x^2}{2!} - \dfrac{x^3}{3!} + \cdots$

3. Properties of the natural logarithm of x, $\ln(x)$:

(1) $\ln(e^x) = x$

(2) $\ln(x_1 x_2) = \ln(x_1) + \ln(x_2)$

(3) $\ln(x_1/x_2) = \ln(x_1) - \ln(x_2)$

(4) $\ln(1) = 0$

(5) expansion: $\ln(1 + x) = x - \dfrac{x^2}{2} + \dfrac{x^3}{3} - \dfrac{x^4}{4} + \cdots \quad (x^2 < 1)$

IV–7 DERIVATIVES

In the following, b and p are constants and u and v are functions of x:

1. $\dfrac{db}{dx} = 0$

2. $\dfrac{d}{dx}(bu) = b\dfrac{du}{dx}$

3. $\dfrac{d}{dx}(u + v) = \dfrac{du}{dx} + \dfrac{dv}{dx}$

4. $\dfrac{d}{dx}(uv) = v\dfrac{du}{dx} + u\dfrac{dv}{dx}$

5. $\dfrac{dx^p}{dx} = px^{p-1}$

6. Chain rule: If u is a function of y and y is in turn a function of x, then $\dfrac{du}{dx} = \dfrac{du}{dy}\dfrac{dy}{dx}$

7. $\dfrac{d}{dx}(\sin x) = \cos x$

8. $\dfrac{d}{dx}(\cos x) = -\sin x$

9. $\dfrac{d}{dx}(\tan x) = \dfrac{1}{\cos^2 x}$

10. $\dfrac{d}{dx}(e^{bx}) = be^{bx}$

11. $\dfrac{d}{dx}\ln(x) = \dfrac{1}{x}$

IV–8 TAYLOR EXPANSION

If $f(x)$ is well-behaved near point $x = x_0$,

$$f(x) = f(x_0) + \left.\dfrac{df}{dx}\right|_{x = x_0}(x - x_0) + \left.\dfrac{d^2 f}{dx^2}\right|_{x = x_0}\dfrac{(x - x_0)^2}{2!} + \cdots$$

In the following, b and p are constants and u and v are functions of x:

1. $\displaystyle\int \frac{du}{dx}\,dx = u$

2. $\displaystyle\int_{x_1}^{x_2} \frac{du}{dx}\,dx = u(x_2) - u(x_1)$

3. $\displaystyle\int bu\,dx = b\int u\,dx$

4. $\displaystyle\int (u + v) = \int u\,dx + \int v\,dx$

5. $\displaystyle\int u\frac{dv}{dx}\,dx = uv - \int v\frac{du}{dx}\,dx$ (integration by parts)

6. If u is a function of y and y is in turn a function of x, then
$$\int u\,dy = \int u\frac{dy}{dx}\,dx$$

7. $\displaystyle\int x^p\,dx = \frac{x^{p+1}}{p+1}$ $(p \neq -1)$

8. $\displaystyle\int \frac{dx}{x} = \ln(|x|)$

9. $\displaystyle\int \sin x\,dx = -\cos x$

10. $\displaystyle\int \cos x\,dx = \sin x$

11. $\displaystyle\int e^{bx}\,dx = \frac{1}{b}e^{bx}$

12. $\displaystyle\int xe^{bx}\,dx = e^{bx}\left(\frac{x}{b} - \frac{1}{b^2}\right)$

13. Some definite integrals

(1) $\displaystyle\int_0^\infty x^n e^{-x}\,dx = n!$

(2) $\displaystyle\int_0^\pi \sin^2 x\,dx = \int_0^\pi \cos^2 x\,dx = \frac{\pi}{2}$

(3) $\displaystyle\int_0^\infty e^{-b^2 x^2}\,dx = \frac{\sqrt{\pi}}{2b}$ $(b > 0)$

(4) $\displaystyle\int_0^\infty xe^{-x^2}\,dx = \frac{1}{2}$

(5) $\displaystyle\int_0^\infty x^2 e^{-x^2}\,dx = \frac{\sqrt{\pi}}{4}$

(6) $\displaystyle\int_0^\infty \frac{b}{b^2 + x^2}\,dx = \begin{cases} \dfrac{\pi}{2} & (b > 0) \\ 0 & (b = 0) \\ -\dfrac{\pi}{2} & (b < 0) \end{cases}$

1. The following expression is good for any n, positive or negative, integer or noninteger:
$$(1 + x)^n = 1 + nx + \frac{n(n-1)}{2!}x^2 + \frac{n(n-1)(n-2)}{3!}x^3 + \cdots$$

2. $\displaystyle \sin x = x - \frac{x^3}{3!} + \frac{x^5}{5!} + \cdots$

3. $\displaystyle \cos x = 1 - \frac{x^2}{2!} + \frac{x^4}{4!} + \cdots$

4. $\displaystyle \tan x = x + \frac{x^3}{3} + \frac{2}{15}x^2 + \cdots$

5. $\displaystyle e^{ax} = 1 + ax + \frac{(ax)^2}{2!} + \frac{ax^3}{3!} + \cdots$

1.	$=$	is equal to
2.	$\simeq$	is approximately equal to
3.	$\propto$	is proportional to
4.	$\equiv$	is defined to be
5.	$\neq$	is unequal to
6.	$>$	is greater than
7.	$\geq$	is greater than or equal to
8.	$<$	is less than
9.	$\leq$	is less than or equal to
10.	Δx	the change in x
11.	$|x|$	the absolute value of x
12.	$O(N)$	on the order of the magnitude of N
13.	$\pm$	plus or minus
14.	$\mp$	minus or plus
15.	$\langle x\rangle$	average of x
16.	$\displaystyle\sum_{i=i_1}^{i_2} f_i$	the sum of all f_i over the integers i from a smallest integer i_1 to a largest integer i_2
17.	$\ln(x)$	natural logarithm of x
18.	$\log(x)$	logarithm to the base 10 of x
19.	$\displaystyle\int$	integral
20.	$\displaystyle\oint$	line integral around a loop

Periodic Table of the Elements

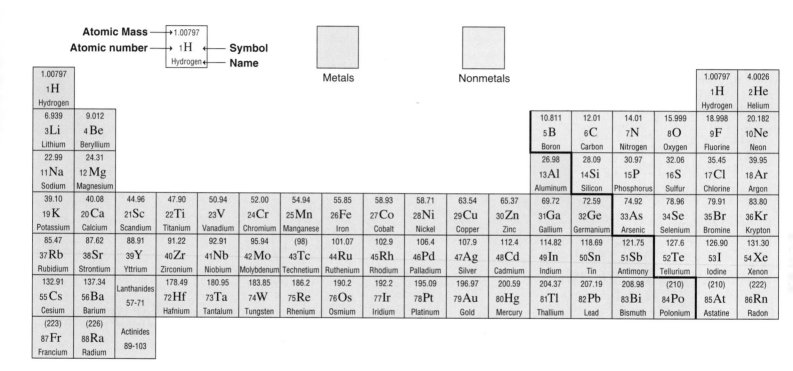

Atomic Mass → 1.00797
Atomic number → 1H ← Symbol
Hydrogen ← Name

Metals

Nonmetals

1.00797 1H Hydrogen																	1.00797 1H Hydrogen	4.0026 2He Helium
6.939 3Li Lithium	9.012 4Be Beryllium											10.811 5B Boron	12.01 6C Carbon	14.01 7N Nitrogen	15.999 8O Oxygen	18.998 9F Fluorine	20.182 10Ne Neon	
22.99 11Na Sodium	24.31 12Mg Magnesium											26.98 13Al Aluminum	28.09 14Si Silicon	30.97 15P Phosphorus	32.06 16S Sulfur	35.45 17Cl Chlorine	39.95 18Ar Argon	
39.10 19K Potassium	40.08 20Ca Calcium	44.96 21Sc Scandium	47.90 22Ti Titanium	50.94 23V Vanadium	52.00 24Cr Chromium	54.94 25Mn Manganese	55.85 26Fe Iron	58.93 27Co Cobalt	58.71 28Ni Nickel	63.54 29Cu Copper	65.37 30Zn Zinc	69.72 31Ga Gallium	72.59 32Ge Germanium	74.92 33As Arsenic	78.96 34Se Selenium	79.91 35Br Bromine	83.80 36Kr Krypton
85.47 37Rb Rubidium	87.62 38Sr Strontium	88.91 39Y Yttrium	91.22 40Zr Zirconium	92.91 41Nb Niobium	95.94 42Mo Molybdenum	(98) 43Tc Technetium	101.07 44Ru Ruthenium	102.9 45Rh Rhodium	106.4 46Pd Palladium	107.9 47Ag Silver	112.4 48Cd Cadmium	114.82 49In Indium	118.69 50Sn Tin	121.75 51Sb Antimony	127.6 52Te Tellurium	126.90 53I Iodine	131.30 54Xe Xenon
132.91 55Cs Cesium	137.34 56Ba Barium	Lanthanides 57-71	178.49 72Hf Hafnium	180.95 73Ta Tantalum	183.85 74W Tungsten	186.2 75Re Rhenium	190.2 76Os Osmium	192.2 77Ir Iridium	195.09 78Pt Platinum	196.97 79Au Gold	200.59 80Hg Mercury	204.37 81Tl Thallium	207.19 82Pb Lead	208.98 83Bi Bismuth	(210) 84Po Polonium	(210) 85At Astatine	(222) 86Rn Radon
(223) 87Fr Francium	(226) 88Ra Radium	Actinides 89-103															

Lanthanides →

138.91 57La Lanthanum	140.12 58Ce Cerium	140.91 59Pr Praseodymium	144.24 60Nd Neodymium	(145) 61Pm Promethium	150.4 62Sm Samarium	151.96 63Eu Europium	157.25 64Gd Gadolinium	158.9 65Tb Terbium	162.5 66Dy Dysprosium	164.9 67Ho Holmium	167.3 68Er Erbium	168.9 69Tm Thulium	173.0 70Yb Ytterbium	175.0 71Lu Lutetium

Actinides →

(227) 89Ac Actinium	232.04 90Th Thorium	(231) 91Pa Protactinium	238.03 92U Uranium	(237) 93Np Neptunium	(242) 94Pu Plutonium	(243) 95Am Americium	(247) 96Cm Curium	(247) 97Bk Berkelium	(251) 98Cf Californium	(254) 99Es Einsteinium	(253) 100Fm Fermium	(256) 101Md Mendelevium	(253) 102No Nobelium	(257) 103Lr Lawrencium

†Atomic masses given in parentheses refer to the most stable isotope of an unstable element.

Significant Dates in the Development of Physics

History can rarely be stated as a simple series of dates, and the history of science is no exception. Throughout the text we have alluded to important discoveries in physics. The list below is a personal choice and should be thought of as a guide. It oversimplifies some of the history, including stories that are covered more thoroughly in the text (for example, Coulomb's law). Some of the dates are to be taken with a grain of salt, because discoveries are rarely made in a single identifiable moment. Our list includes some names (and discoveries) not mentioned in the text. Far more numerous are the names not listed, the names of those who built the experimental foundations, those who explored the false paths and cleared the way for those whose names we remember today, or those who verified the speculations that we call laws.

1583	Galileo	Pendulum motion
1600	Gilbert	Study of magnets
1602	Galileo	Early statement of Newton's first law
1602	Galileo	Laws of falling bodies
1609	Kepler	First two laws of planetary motion
1619	Kepler	Third law of planetary motion
1620	Snell	Law of refraction
1648	Pascal	Atmospheric pressure
1650	Grimaldi	Diffraction of light
1661	Boyle	Chemical elements
1669	Newton	Light dispersion in prisms
1678	Huygens	Wave propagation
1687	Newton	Laws of motion; universal gravitation
1760	Black	Calorimetry
1785	Coulomb	Coulomb's law
1789	Lavoisier	Conservation of mass
1798	Cavendish	Measurement of G
1800	Volta	Electric battery
1801	Young	Interference of light
1801	Dalton	Laws of chemical combination
~ 1802	Charles; Gay-Lussac	Ideal gases
1807	Dalton	Atomic theory
1812	Fourier	Decomposition of waves
1815	Fraunhofer	Discrete spectral lines
1819	Fresnel	Wave picture of light
1820	Oersted	Magnetic fields from currents
1820	Biot; Savart	Law of magnetic field produced by current
1824	Carnot	Second law of thermodynamics
1827	Ohm	Ohm's Law
1827	Ampère	Ampère's law
1831	Faraday; Henry	Induction
1842	Joule	Mechanical equivalent of heat
1847	Helmholtz	Conservation of energy
1849	Fizeau	Direct measurement of the speed of light
1865	Maxwell	The laws of electricity and magnetism; light waves
1877	Boltzmann; Gibbs	Statistical mechanics
1879	Stefan	Blackbody radiation
1885	Osmond	Crystalline structure of metals
1887	Hertz	Electromagnetic waves
1887	Michelson and Morley	Constancy of the speed of light

1896	Becquerel	Radioactivity
1897	Thomson	Charge-to-mass ratio of the electron
1900	Planck	Quanta in blackbody radiation
1903	Rutherford; Soddy	Isotopes
1905	Einstein	Special relativity; quanta in photoelectric effect
1908	Kammerlingh Onnes	Superfluidity
1911	Kammerlingh Onnes	Superconductivity
1911	Rutherford	Nuclear structure of atom
1911	Millikan	Quantization of charge
1912	von Laue	X-ray diffraction in crystals
1913	Bohr	Atomic structure
1916	Einstein	General relativity
1923	Hubble	Discovery of galaxies
1924	de Broglie	Wave nature of particles
1925	Pauli	Exclusion principle
1925	Heisenberg	Formulation of quantum mechanics
1925	Goudsmit and Uhlenbeck	Electron spin
1926	Davisson and Germer; Thomson	Diffraction of electrons by crystals
1926	Schrödinger	Alternate formulation of quantum mechanics
1926	Born	Probabilistic interpretation of quantum theory
1927	Heisenberg	Uncertainty relations
1929	Hubble	Hubble's law
1930	Dirac	Antiparticles
1932	Anderson	The positron
1932	Lawrence and Livingston	The cyclotron
1932	Chadwick	The neutron
1934	Yukawa	Nuclear forces and the pi meson
1948	Feynman; Schwinger; Tomonaga	Electromagnetism as a quantum theory
1954	Townes	The maser
1957	Lee and Yang	Nonconservation of parity
1957	Bardeen, Cooper, and Schrieffer	Theory of superconductivity
1962	Josephson	Josephson junction
1964	Gell-Mann; Zweig	Quarks
1964	Penzias and Wilson	Background radiation of the universe
1967–1970	Glashow; Salam; Weinberg	Unification of electromagnetic and weak forces

Tables in the Text

Selected Text Boxes

Answers to Odd-Numbered Problems

CHAPTER 1

1. 2.500×10^3 green jelly beans.
3. 10^7; 10^{14}.
5. 1.8×10^{10} atoms.
7. $0.71/\text{kg}$.
9. 32.2 ft/s^2.
11. 3.33 g/cm^3.
13. 34 mi/gal; 10 mi/gal.
15. 0.0402%; $8.49 \times 10^{-6}\%$.
17. $V = 0.09 \pm 0.04 \text{ m}^3$.
19. 2.5%.
21. $[MLT^{-1}]$.
23. $[ML^2T^{-1}]$.
25. (a) $[\alpha] = [L^{-1}]$, (b) $[A] = [L^4T^{-2}]$.
27. 8 tons.
29. $3.5 \times 10^{-6} \text{ cm}^2$.
31. (a) 10^6 mechanics, (b) 10^6 mechanics, (c) 10^6 mechanics.
33. 10^{19} droplets.
35. 5×10^6 automobiles.
37. 16 cm.
39. 1.2×10^{57} hydrogen atoms; 9×10^{56} hydrogen atoms.
41. Her position can be described by the clockwise angle ϕ from the north-south line from the center to the starting point. Her direction of travel will be tangent to the circle with a magnitude of 2 m/s at an angle ϕ above the west direction.
43. $1.2\mathbf{i} + 8.2\mathbf{j}$ paces (8.3 paces, 82° north of east).
45. Catch: $0\mathbf{i} + 0\mathbf{j}$; first turn: $0\mathbf{i} + 0\mathbf{j} + 20\mathbf{j} = 0\mathbf{i} + 20\mathbf{j}$; second turn: $0\mathbf{i} + 20\mathbf{j} - 15\mathbf{i} = -15\mathbf{i} + 20\mathbf{j}$; third turn: $-15\mathbf{i} + 20\mathbf{j} + 10\mathbf{j} = -15\mathbf{i} + 30\mathbf{j}$; fourth turn: $-15\mathbf{i} + 30\mathbf{j} + 25\mathbf{i} = 10\mathbf{i} + 30\mathbf{j}$; fifth turn: $10\mathbf{i} + 30\mathbf{j} - 10\mathbf{i} = 30\mathbf{j}$; touchdown: $30\mathbf{j} + 65\mathbf{j} = 95\mathbf{j}$.
47. $V_x = +V \cos \alpha$.
49. $\mathbf{AB} = (7, 0) = 7$ at an angle of 0°; $\mathbf{BC} = (-7, 7) = 7\sqrt{2}$ at an angle of 135°; $\mathbf{CA} = (0, -7) = 7$ at an angle of 270°.
51. (a) $\mathbf{A} = -4\mathbf{i} + 2\mathbf{j}$, $\mathbf{B} = -\mathbf{i} + 4\mathbf{j}$, $\mathbf{C} = 2\mathbf{i} + 2\mathbf{j}$, $\mathbf{D} = 5\mathbf{i} - 3\mathbf{j}$, (b) $-11\mathbf{i} + 9\mathbf{j}$, $5\mathbf{j}$, 9.2.

55. (a) $[M^{1/2}L^{3/2}T^{-1}]$, (b) $[ML^2T^{-2}]$.
57. (a) 10^{-6} metric tons/g, (b) 10^{-6} m^3/cm^3.
59. 1.3 s.
61. 3.0×10^{-26} kg/molecule.
63. 3×10^{-8} cm.
65. 2×10^{41} kg/galaxy; 1.2×10^{68} H atoms.
69. 10^3 truck/day; 2×10^3 trucks/day.
71. 1.1×10^8 km; 29 yr.
73. 10^{44} molecules.
75. (a) $\mathbf{v} = -v \sin \theta \mathbf{i} + v \cos \theta \mathbf{j}$, or $\mathbf{v} = +v \sin \theta \mathbf{i} - v \cos \theta \mathbf{j}$
(b)

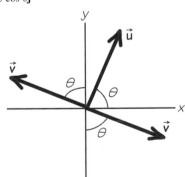

77. (a) $r \sin \theta \cos \phi$, (b) $r \cos \theta$, (c) $r \sin \theta \sin \phi$.
79. 50 cm.
81. $t_0 = $ (a constant) $\ell \, (\lambda/\tau)^{1/2}$.

CHAPTER 2

1. +21 cm; 21 cm from the origin in the positive direction.

3. 252 m; 0;

5. (a)

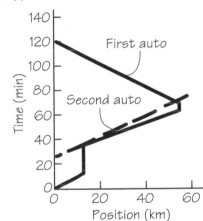

(b) 37 min and 15 km; 68 min and 53 km.

7.

Figure continued at top of next page.

Figure continued from previous page.

27.

29. 2.02 ft/s² (0.61 m/s²).
31. 2.25 m/s.
33. (a) $8.0 - 0.5t$, easterly with t in s, v in m/s,
(b) 34 m to the east.
35. (a)

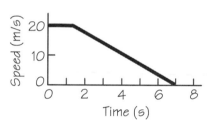

(b) -3.6 m/s².
37. 1.92 s.
39. (a) 2.96 m/s², (b) 97.2 s.
41. (a)

(b) 5940 ft (1.13 mi), (c) 54 s.
43. 3.65 s.
45. (a) 8.0×10^{-9} s, (b) 6.3×10^{14} m/s².
47. 4.0 m/s²; 18 m.
49. 54 m.

9. 6.4 m/s; 6.6 m/s; 6.3 m/s.
11.

t, s	v, m/s	x, m
0.0	0.00	0.00
0.5	0.75	0.19
1.5	1.75	1.44
2.5	8.75	6.69
3.5	21.75	21.94
4.5	39.75	52.69
5.5	62.75	103.94
6.5	90.75	180.69
7.5	122.75	287.44

13.

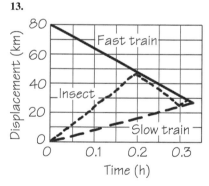

80 km.
15. $+1.9\mathbf{i}$ cm/s; $+2.5\mathbf{i}$ cm/s; $+6.0\mathbf{i}$ cm/s; $+3.0\mathbf{i}$ cm/s.
17. 30 s.
19. $0.30g$.
21. 1.6 m/s².
23. $(A\pi/12)\cos(\pi t/12)$; $-(A\pi^2/144)\sin(\pi t/12)$.
25. The particle never gets farther from the origin than A; it oscillates back and forth through the origin. The magnitude of the velocity is maximum at the origin and zero at $x = \pm A$. The magnitude of the acceleration is maximum at $x = \pm A$ and zero at the origin.

51. 9.0×10^5 m/s²; 6.7×10^{-4} s.
53. Not much time to say anything!
55. 1.1 s.
57. 1.26 s.
59. -22 m/s; 25 m/s; 20 m/s.
61. 28.4 m.
63. (a)

(b) -0.8 m/s, (c) 2.5 m/s,
(d)

-30 m/s².
65. $(3v_f/\alpha)^{1/3}$.
67. 0.7 m; 3 m.
69. 12 m.
71. 1.6 m/s²; 15 m.
73. 1.6 m; 0.08 s.

CHAPTER 3

1. $(15\mathbf{i} + 15\mathbf{j})$ km; $(30\mathbf{i} + 15\mathbf{j})$ km; $(30\mathbf{i} + 43\mathbf{j})$ km; 52 km, 55° N of E.
3. $\mathbf{r}_A = 0$; $\mathbf{r}_B = (25\text{ m})\mathbf{i}$; $\mathbf{r}_C = (25\text{ m})\mathbf{i} + (35\text{ m})\mathbf{j}$; $\mathbf{r}_D = (35\text{ m})\mathbf{j}$.
5.

$(1.3\text{ km})\mathbf{i} + (2.7\text{ km})\mathbf{j}$; $(0.1\text{ km})\mathbf{i} + (4.9\text{ km})\mathbf{j}$.
7. $(3.60\text{ m/s})t\mathbf{i} + (6.00\text{ m/s})t\mathbf{j}$, $0 \le t \le 41.6$ s; $[-(141\text{ m}) + (7.00\text{ m/s})t]\mathbf{i} + (250\text{ m})\mathbf{j}$, $41.6\text{ s} \le t \le 77.3$ s.
9. $(2.0\text{ m})\mathbf{i} - (3.5\text{ m})\mathbf{j}$; 4.0 m; $-(4.0\text{ m})\mathbf{j}$; $d = 4.0$ m; $(4.0\text{ m})\mathbf{i}$; $d = 4.0$ m; $\theta(t) = -\pi t/T$.
11. 1.0 m/s², opposite to the direction of the velocities.
13. $(3.7\text{ m/s}^2)\mathbf{j} - (2.4\text{ m/s}^2)t\mathbf{k}$.
15. 30 km/h, 31.6° north of west; $(-4.3\mathbf{i} + 2.7\mathbf{j})$ km.
17. (a) $[(0.00225\text{ m}^2/\text{s}) - (0.0009\text{ m}^2/\text{s}^2)t]\mathbf{j}/[(0.0169\text{ m}^2) - (0.0045\text{ m}^2/\text{s})t - (0.0009\text{ m}^2/\text{s}^2)t^2]^{1/2}$;
(b) $(0.017\text{ m/s})\mathbf{j}$; 0.

19. $\mathbf{v} = [-(4\pi/T)\sin(\pi t/T)\mathbf{i} - (4\pi/T)\cos(\pi t/T)\mathbf{j}]$ m/s; $\phi = (\pi/2) - (\pi t/T)$.

21.

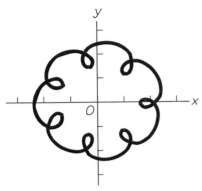

23. $[T^{-1}]$; $v_0 = 0$; $v \to -u$; $a_0 = -Bu$; $a \to 0$.
25. 12 m/s (horizontal); 7.4 m.
27. (a) 3.64 s, (b) 11.5 m, (c) -14.7 m/s (down); $+14.7$ m/s (up).
29. 5.2 m/s.
31. 187 m/s.
33. 40° below the horizontal.
37. 329 m.
39. (a) $t = 1.25$ s, (b) unsuccessful; 0.4 m below the bar.
41. 15° and 75°; 0° and 90°.
43. (a) 7.8×10^3 m/s, (b) 9.1 m/s² toward Earth's center.
45. 2.2 m/s².
47. 9.5 s.
49. 610 m.
51. (a) (0.43 m, 0.25 m), (b) $-79\mathbf{i}$ m/s², (c) $-79\mathbf{j}$ m/s².
53. 1.21R.
55. 15.8 km/h, 18° south of east.
57. 50 km/h.
59. (a) 843 km/h, (b) 781 km/h, 26° W of S, (c) $-(1720$ km$)\mathbf{i} - (3505$ km$)\mathbf{j}$.
61. (a) 29.5 km/s, (b) 30.5 km/s, (c) 30.0 km/s.
63. (a) 36 m/s at 65°, (b) 62.3 m.
65. $(10\cos\theta + 6)t\mathbf{i} + (10\sin\theta)t\mathbf{j}$, with r in kilometers and t in hours; 127°; 68 s.
67. Hit the deck.
69. (a) 7.35 m/s, (b) 0; 2.45 m going up; 2.45 m coming down; 0, (c) 4.9 m.

CHAPTER 4

1. (a) Force of gravity (toward Earth), (b) force of gravity (down), normal force from the ice (up); and a small friction force from the ice (opposite to the motion), (c) essentially none.
3. 600 N in the $-y$-direction.
5. Yes.
7. (a) 3×10^{-4} N, (b) 3×10^{-4} N.
9. (a) The acceleration opposite to the motion is due to a retarding force. (b) The observer sees the car (initially at rest) move backward with increasing speed until it reaches v_0. She would say that this is due to a backward force from the wind, etc.

11.

$|\vec{F}_{12}| = |\vec{F}_{21}|$, but $|\vec{a}_1| = 3|\vec{a}_2|$

13. $(F/m)\mathbf{j}$; no change.
15. 6.5×10^2 N.
17. $T = F$; 0.433F.
19. 78 N.
21. 1.6×10^{-21} m/s².
23. (a) The Earth, (b) the ice and Earth, (c) none.
25. (a) 0.30 m/s², (c) 3.6×10^4 N backward.
(b)

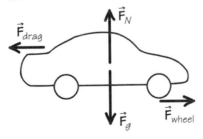

27. $(\mathbf{i} + \mathbf{j} + 2\mathbf{k})$ N.
29. Yes; yes; yes; 2 m/s² in the $-x$-direction; not real.
31. (a) Parallel to the window edge, (b) at an angle given by $\tan\theta = a'/g$, (c) parallel to the window edge.
33.

35.

37.

(a)

(b)

39.

(a)   (b)

(c)

(d) a forward force from the Earth on the horse's hooves.
41.

(figure)

$\mathbf{F}_g = F_g\sin\theta\mathbf{i} - F_g\cos\theta\mathbf{j}$; $\mathbf{F}_N = F_N\mathbf{j}$; $\mathbf{f} = -f\mathbf{i}$.

43.

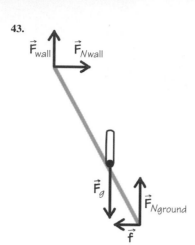

There is a net horizontal force from the wall.

45.

47. (a) $F_g = mg$ down; $F_N = mg$ up, (b) mg down; $2mg$ up, (c) mg down.

49. $(1/M)(F_1^2 + F_2^2)^{1/2}$; $\theta = \tan^{-1}(F_2/F_1)$.

51. $9.0 \, N \cdot s^{1/2})t^{-1/2}$.

53. (a) $c = \sqrt{k/m}$, (b) $A = 0$; $B = +v_0\sqrt{m/k}$.

55. (a)

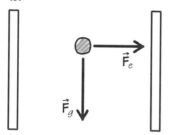

(b) 82.2 N up, (c) 1.55 m/s² up.

57.

$T = (m_W g + m_B g)/4$; $a = 4T - (m_W g + m_B g)/(m_W + m_B)$.

0-4

59. (a) 2×10^4 N, (b) 0.20 m/s²; 1.2×10^4 N forward, (c) 8×10^3 N forward; 4×10^3 N backward; 0.20 m/s².

61. (a)

(b) same forces, but the upward force would be a fictitious force.

63. (a) $[A] = [ML^{-1}]$ with units of kg/m, (b) $(Av^2/m) - g$, (c) $v_t = \sqrt{mg/A}$.

65. (a)

Force of gravity: mg (down); normal force of ground: F_N (up); friction force of ground: f (forward); wind resistance: F_w (backward), (b) 0, (c) 0; $v_w + v$; 0, (d) $v + v_w$; 0; 0.

67. $\phi = \theta$.

CHAPTER 5

1. 1.4×10^4 N.

3. 980 N.

5. 0.095 s.

7. (a) 24 N, (b) 7.8 N.

9. (a) 66 N, (b) 5.1 m/s² up.

11. (a) 2.5 m/s² forward, (b) 3.5×10^3 N forward.

13. $M_{max} = 9.2$ kg; $M_{min} = 4.5$ kg; 0.

15. 25.3 N; 50.6 N; 2.85 m/s² (up); 0.32 m/s² (up); −3.48 m/s² (down).

17. $a_1 = [(m_1m_2 + m_1m_3 - 4m_2m_3)/(m_1m_2 + m_1m_3 + 4m_2m_3)]g$;
$a_2 = [(m_1m_2 - 3m_1m_3 + 4m_2m_3)/(m_1m_2 + m_1m_3 + 4m_2m_3)]g$;
$a_3 = [(-3m_1m_2 + m_1m_3 + 4m_2m_3)/(m_1m_2 + m_1m_3 + 4m_2m_3)]g$;
$T_1 = [8m_1m_2m_3/(m_1m_2 + m_1m_3 + 4m_2m_3)]g$;
$T_2 = [4m_1m_2m_3/(m_1m_2 + m_1m_3 + 4m_2m_3)]g$;
$a_1 = [(m_1 - 2m_2)/(m_1 + 2m_2)]g$; $a_2 = a_3 = [(2m_2 - m_1)/(m_1 + 2m_2)]g$.
$T_1 = [4m_1m_2/(m_1 + 2m_2)]g$; $T_2 = [2m_1m_2/(m_1 + 2m_2)]g$.

19. 0.12.

21. 0.61.

23. 3.3×10^2 N.

25. 0.40.

27. (a) 52 m, (b) 3.9 s.

29. (a) 2.8 s, (b) 2.3 N, (c) 0.54.

31. (a) $370/(\cos \theta + 0.75 \sin \theta)$ N, (b) 37°; 2.9×10^2 N.

33. 0.077.

35. (a) $a_1 = 0$; $a_2 = F/m_2$, (b) $a_1 = a_2 = F/(m_1 + m_2)$, (c) **F**(lower on upper) $= \mu_k m_1 g\mathbf{i} + m_1 g\mathbf{j}$; (d) $a_1 = \mu_k g$; $a_2 = (F - \mu_k m_1 g)/m_2$.

37. 5.4×10^2 N.

39. 600 cm².

41. 0.46 m/s.

43. 0.67.

45. 1.3 m/s² toward the center.

47. 1.4×10^2 N toward the center.

49. $v_{max} = 32$ m/s; $v_{min} = 5.8$ m/s.

51. 0.59.

53. 1.0×10^3 m/s tangent to the orbit.

55. 4.43 m/s.

57. (a)

(b) 0.4 m/s.

59. 0.19 m.

61. 55 m/s.

63. (a) 40 N, (b) 35 N.

65. 1.6×10^{-12} kg/m³.

67. 1.2×10^2 N.

69. (a) 0.5 N, (b) 1.2 N.

71. (a) $\Delta x_2 = -\frac{1}{2}\Delta x_1$, (b) $a_1 = 1.78$ m/s² (up); $a_2 = -0.89$ m/s² (down), (c) 9.6 N.

73. $\omega = (g/\mu_s R)^{1/2}$.

75. $\omega = [g/(\ell\cos \theta)]^{1/2}$.

77. 26 m.

79. 11 m/s.

CHAPTER 6

1. (a) 39 J, (b) 1.0 m/s; 89 m/s, (c) 0.40 m/s.

3. (a) The force of gravity 98 N (down); the upward pull $= 98$ N; $F_{net} = 0$, (b) 0, (c) 98 J.

5. 1.0×10^4 J.

7. (a) 2.4×10^2 J, (b) friction force, (c) 0.

9. 730 J; −730 J.

11. −3.73 J.

13. 1.8×10^3 J.

15. 2.8×10^{11} J.

17. 0.127 J; 0.159 J; 0.175 J.

19. +3.

21. 0.89 m/s.

23. 2.57.

25. 3.9×10^3 J.

27. 6.1×10^3 J.

29.

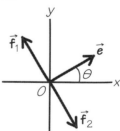

$-\sin \theta\mathbf{i} + \cos \theta\mathbf{j}$; $\sin \theta\mathbf{i} - \cos \theta\mathbf{j}$.

31. 3.4.

33. −24 J.

35. 0.067 J.

37. $2g_1 - 4g_2$.

39. 8.6 m/s.

41. 8.1×10^{-2} J.

43. 14 J.

45. 2.1×10^3 J.

47. −15 J; −5 J; conservative.

49. (a) 6.3×10^{-2} J, (b) 0, (c) 5.9×10^{-2} J, (d) -5.9×10^{-2} J.

51. (a) 4 J; 6 J; 6 J; 4 J, (b) +5 m, (c) conservative.
53. Constant forces are included; not a function of position only.
55. (a) 0.16C J; 0, (b) 0.32C J.
57. 22 bulbs.
59. 0.7 kW.
61. 52 kW.
63. 5.0 m/s; 1.3 m/s.
65. 4.7×10^7 W.
67. (a) 16 m/s, (b) 0.5 s, (c) 12 s.
69. 5.8×10^{-12} J; 5.8×10^{-9} J.
71. 0.62; 0.31.
73. −0.2 J.
75. 0.034.
77. (a) 2.2×10^7 J, (b) -2.2×10^{-7} J.
79. (a) $mg = 26$ N down; $F_N = 22$ N perpendicular to plane (up); $f_k = 5.4$ N parallel to plane (down), (b) −18 J; 0; −7.0 J, (c) 4.4 m/s.
81. 5.3 J.
83. (a) 0, (b) −0.20 J, (c) 9.8 J.
85. (b) $W_T = mgH(1 + \mu_k \cot \theta)$.
87. $mgL(\cos \theta_f - \cos \theta_i)$; $v = [2gL(1 - \cos \theta_i)]^{1/2}$.
89. $0.01\ mK/x$.

CHAPTER 7

1. (a) 196 J, (b) 29 J; 225 J.
3. 25 m.
5. 1.2×10^{-2} J.
7. (a) −8x J, with x in m, (b) +18 J, (c) 4.1 m/s.
9.

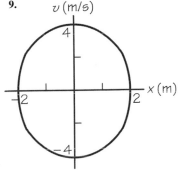

11. 0.541 m; 0.468 m.
13. 0.43 m; 0.48 m.
15. (a) 7.1 m/s, (b) −0.25 J, (c) 0.73.
17.

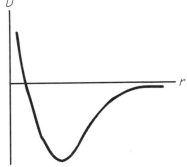

19. (a) to the right, (b) $v = (k/m)^{1/2}x$, (c) to the left; $v = -(k/m)^{1/2}x$.
21.

23. x = 0 is unstable; x = ±0.24 m is stable.
25. 503 m/s.
27. 130 m/s; same.
29. (b) $\sqrt{2gh}$, (c) $2\sqrt{gh}$.
33. -5.0×10^4 J.
35. (a) 1.7 m/s, (b) 1.3 m/s, (c) 11.4 cm from point a.
37. (a) $U(x) = \frac{1}{2}k(\sqrt{h^2 + x^2} - L)^2$,
(b) $-\dfrac{k(\sqrt{h^2 + x^2} - L)x}{\sqrt{h^2 + x^2}}$.

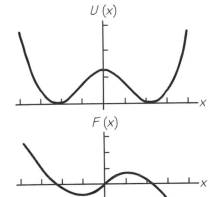

39. (a) GMm/r^2 toward M, (b) $GMm/2r$, (c) $-GMm/2r$.
43. -6.2×10^4 J.
45. 14.
47. 1.9×10^{11} s ($\simeq 6 \times 10^3$ yr).
49. 5.9×10^2 J.
51. (a) $3 + \dfrac{3x^2}{2} - \dfrac{0.2x^3}{3}$ J, with x in m,
(b)

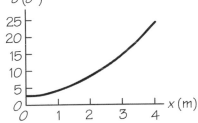

53. (a) conservative, (b) not conservative.
55. −2 J; −2 J; +2 J.
57. (a) 7.7 m/s; 14.7 m/s, (b) the skier leaves the surface.
59. (a) 4.4 m/s, (b) 3.7 m/s, (c) 5.9 N; 4.2 N.
61. (a) conservative force, (b) $U(r) = -C/r$, (c) 2.1×10^7 m/s.
63. (a) +1.13 J, (b) −0.083 J, (c) −0.30 J, (d) 0.70 m/s, (e) 8.8 cm, (f) 5.0 cm.
65. 1.1 m/s.
69. $\dfrac{\partial F_x}{\partial y} = \dfrac{\partial F_y}{\partial x}, \dfrac{\partial F_x}{\partial z} = \dfrac{\partial F_z}{\partial x}, \dfrac{\partial F_y}{\partial z} = \dfrac{\partial F_z}{\partial y}$.
71. $x = L[\sin \alpha + 2 \sin \alpha \cos^2 \alpha + 2(\cos^3 \alpha - \cos^6 \alpha)^{1/2}]$.

CHAPTER 8

1. (a) 1.2 kg · m/s; (b) 5.1 kg · m/s; (c) 7.0×10^2 kg · m/s; (d) 7.6×10^2 kg · m/s.
3. (a) 4.2×10^2 kg · m/s; (b) 6.0×10^6 kg · m/s; (c) 1.2×10^4 kg · m/s; (d) 3.3×10^{-22} kg · m/s; (e) 1.4×10^{-2} kg · m/s.
5. (a) −1.0 m/s; (b) 3.5 J.

7. (a) $(-1.9\mathbf{i} - 10.8\mathbf{j})$ kg · m/s; (b) $(-4.9\mathbf{i} - 6.8\mathbf{j})$ m/s; (c) $(-0.90\mathbf{i} - 2.9\mathbf{j} - 1.2\mathbf{k})$ m/s; (d) 24 J; 64 J; 19 J.
9. 40 m/s.
11. −12 kg · m/s, opposite to the original motion; 1.7×10^4 N, opposite to the original motion.
13. -1.2×10^3 kg · m/s, down; 1.3×10^4 N, up.
15. 2.2×10^2 kg · m/s, up.
17. 7.5×10^2 N, up.
19. (a) 5.7×10^2 N; (b) 1.4×10^2 J; (c) 0.011 s.
21. $0.60\sqrt{g(0.20 + 0.10N)}$ kg · m/s.
23. $V = -0.13$ m/s (opposite to the direction of the first cart).
25. 0.74 m/s.
27. $m_1/m_2 = 3$.
29. (a) $M/(m + M)$; (b) $v = [(m + M)/m]\sqrt{2gh}$.
31. (a) $v[m_2^2 + M(m_1 + m_2)]/M(m_2 + M)$ forward; (b) $v[m_1^2 + M(m_1 + m_2)]/M(m_1 + M)$ forward; (c) $v(m_1 + m_2)/M$ forward.
33. (a) −0.20 m/s; (b) 64%.
35. 2.59 m/s.
37. 15°.
39. (a) $\sqrt{2gh}$; (b) $\sqrt{2gh}$; (c) $[(m - 3M)/(m + M)]\sqrt{2gh}$, up; (d) $[(m - 3M)/(m + M)]^2h$; (e) 9h.
41. $\mathbf{v}_{1f} = 1.2$ m/s 60° below the −x-axis; not elastic.
43. $(2.00$ m/s$)\mathbf{i} + (1.00$ m/s$)\mathbf{j}$; elastic.
45. 2.6 m/s.
47. 8.7 m/s, 20° above the horizontal.
49. The driver of the car was speeding!
51. 0.78 m from the heavier sphere.
53. X = 0.38 m; Y = −0.66 m.
55. 0.
57. X = 0; Y = 1.0 m from the bottom of the handle.
61. $4R/3\pi$ from the center of the arc, along the bisector.
63. $a\dfrac{\ln[(L^2 + a^2)/a^2]}{2 \tan^{-1}(L/a)}$.
65. -3.3×10^3 kg/s.
67. 1% of mass must be discarded.
69. 2.5×10^3 m/s.
71. (a) spit out seeds one at a time; (b) 6×10^{-3} m/s.
73. (a) 7.9 m/s (to the right); (b) 7.9 m/s (to the right); (c) −1.9 m/s (to the left); +1.1 m/s (to the right); (d) 2.1×10^4 N (to the right).
75. (a)

(b) 4.1 m/s recoil; (c) upward impulse provided by the ground.
77. (a) $[8M(M - m)/(M + m)^2]h$; (b) $[(M - m)/(M + m)]^2h$.
79. R/6 from the center of the styrofoam sphere.
81. (a) 0.11 m/s; (b) 0; (c) −0.074 m; (d) 0.

CHAPTER 9

1. 0.68 rad/s².
3. −17 rad/s².
5. 7.3×10^{-5} rad/s from the South Pole to the North Pole; 2.0×10^{-7} rad/s perpendicular to the orbital plane.

7. (a) -2.62×10^2 rad/s^2, (b) 1.67×10^2 m.
9. (a) 0.10 rad/s^2, (b) 11 rev, (c) 2.1×10^2 m.
11. (a) $a = 1.5$ m/s^2; $x = x_0 + (0.75$ m/s$^2)t^2$, (b) 100 rad/s^2, (c) $\omega = (100$ rad/s$^2)t$.
13. 1.35×10^{-4} m.
17. 9.
19. 2.8×10^{-3} kg $\cdot$ m^2.
21. 1.3×10^2 J; 4.0×10^2 J.
23. (a) 0.021 kg $\cdot$ m^2, (b) 0.083 kg $\cdot$ m^2.
25. 9.8×10^{37} kg $\cdot$ m^2; $I_{\text{neutron star}} = 1.0 \times 10^{39}$ kg $\cdot$ m$^2 \simeq 10\, I_{\text{earth}}$.
27. $\frac{1}{2}M(R_1^2 + R_2^2)$.
29. $(3/10)(\tan^2 \alpha)MH^2$.
31. $(8/15)\pi[(\rho_1 - \rho_2)R_1^5 + \rho_2 R^5]$.
33. $(16/45)MR^2$.
35. 24 lb.
37. 98 kg.
39. $\frac{1}{2}L(F_{1y} + F_{2y})$ perpendicular to the rod.
41. (a) 2.3 N $\cdot$ m, (b) change 7-N force to 5.4 N.
43. 0.38 N $\cdot$ m.
45. (a) 2.13 kg $\cdot$ m^2, (b) 8.84 kg $\cdot$ m^2/s, (c) 0.886 rev, (d) 18.3 J.
47. 3.8 rad/s up; rotate in the direction of the original rotation of the wheel.
49. -2.46×10^{-5}.
51. 0.12.
53. 13.4 m; 17.9 m.
55. 0.050 m/s^2.
57. 2.3 rev.
59. 5.1×10^6 m.
61. $3v_0^2/4g \sin \theta$.
63. (a) 0.72 kg $\cdot$ m^2/s, (b) 3.8 J, (c) 7.6 J.
65. $3\mu MgR/(R + 2r)$.
67. $MgR^2/(R^2 + 2r^2)$.

CHAPTER 10

1. (a) 1.1×10^{10} kg $\cdot$ m^2/s down, (b) 7.9×10^9 kg $\cdot$ m^2/s down.
3. (a) 5×10^4 kg $\cdot$ m^2/s down, (b) 5×10^4 kg $\cdot$ m^2/s down.
5. 1.5 kg $\cdot$ m^2/s north.
7. $-7.4t^2\mathbf{k}$ kg $\cdot$ m^2/s (into the page).
9. $\frac{1}{2}mbvt^2\mathbf{i} - \frac{1}{2}mawt^2\mathbf{j} - \frac{1}{2}mavt^2\mathbf{k}$.
11. $\mathbf{r} \times m_2m_1/(m_1 + m_2)\, d\mathbf{r}/dt = \mathbf{r} \times \mu\, d\mathbf{r}/dt$.
13. (a) $6.4 \times 10^{-2}\mathbf{k}$ kg $\cdot$ m^2/s (along ω-direction), (b) same.
15. (a) $m\omega d^2$ along the axis of rotation, (b) $\frac{3}{4}m\omega d^2$ along the axis of rotation, (c) $\frac{1}{2}m\omega d^2$ along the axis of rotation.
17. $(17\mathbf{i} - 19\mathbf{j} + 14\mathbf{k})$ N $\cdot$ m.
19. $140\mathbf{k}$ N $\cdot$ m (perpendicular to table).
21. $MgR \sin(\omega t)$.
23. $mvd \sin(2\theta)$ up; 0.
25. $-7\mathbf{i} + 9\mathbf{j} + 10\mathbf{k}$.
27. 1.1 rad/s.
29. 30 m/s.
31. 0.103 rad (0.016 rev).
33. 5.3×10^2 N $\cdot$ m along the axis.
35. 409 rad/s.
37. 0.14 J.
39. $r_n = (n^2\hbar^2/mk)^{1/4}$, $n = 1, 2, \ldots$; $v_n = (n^2\hbar^2k/m^3)^{1/4}$, $n = 1, 2, \ldots$; $K_n = \sqrt{k/m}\, n\hbar/2$, $n = 1, 2, \ldots$.
41. No energy level 2.0 eV above the lowest state; excitation energies are 10.2 eV, 12.1 eV, 12.75 eV, ….
43. $\omega_p = Mg\ell/I\omega$ along the z-axis.
45. 2.8 rad/s.
47. 72 rad/s.
49. 0.62 rad/s; -33 J; friction.
51. (a) 3.8 m/s, (b) 0.78 s.
53. h.

55. (a) 19.4 m/s, (b) 0.127, (c) 0.132.
57. 26 kg; 52 rad/s; 12 kg.
59. 4.0×10^3 J.
61. (a) K/L, (b) L^2/MK, (c) $2\pi L^3/MK^2$, (d) MK^3/L^4.
63. (a) 2.4 m/s, (b) 2.4×10^2 rad/s.

CHAPTER 11

1. 40 kg.
3. 139 N down; 96 N down.
5. (2, 0, 1) m.
7. 1.7 m.
9. 0.8 m.
11. 14.2 N $\cdot$ m in xy-plane 27° from $-x$-axis ($\perp \mathbf{r}$).
13. $\frac{3}{4}L$; $11L/12$.
15. 0.57 Mg.
17. 1.2×10^2 N; $F_{NA} = 0$.
19. (a) 2.9 m.
21. 2.9×10^2 N 13.5° above the horizontal; 3.5×10^2 N down; 2.8×10^2 N.
23. 0.58 Mg outward.
25. 1.9×10^2 N right; 73 N left; 23 N left.
27. (a) $\frac{1}{2}mg\ell_2/\ell_1$ down; $\frac{1}{2}mg(\ell_2 - \ell_1)/\ell_1$ up, (b) seat will lose contact at B and turn clockwise; seat will turn counterclockwise, (c) 74 N; 25 N.
29. (a) 56 N, (b) 164 N; $(164$ N$)\mathbf{i} + (46$ N$)\mathbf{j}$.
31. 353 kg.
33. (a) $F_{N1} = 0$; $F_{N3} = F_{N2} = \frac{1}{2}(M + m)g - (mg\, x/L)$; $F_{N4} = 2mgx/L$, (b) $F_{N4} = 0$; $F_{N1} = -2mgx/L$; $F_{N3} = F_{N2} = \frac{1}{2}(M + m)g + (mg\, x/L)$.
35. 1.33×10^4 N.
37. 69 N.
39. (a) $\frac{1}{2}Mg \cot \theta_0$, (b) $(3g \cos \theta_0)/2L$, (c) $\omega = \sqrt{(3g \sin \theta_0)/L}$.
41. 85 N.
43. 1.36×10^3 N.
45. (a) near legs: $Mg(\frac{1}{4} + \frac{1}{2}\mu_k)$ up; $\mu_k Mg(\frac{1}{4} + \frac{1}{2}\mu_k)$ to the left; far legs: $Mg(\frac{1}{4} - \frac{1}{2}\mu_k)$ up; $\mu_k Mg(\frac{1}{4} - \frac{1}{2}\mu_k)$ to the left, (b) 0.5.
47. $\tan(\frac{1}{2}\theta)$.
49. Top ball: mg; $0.577mg$; $1.155mg$; Bottom ball: mg; $1.155mg$; $2.000mg$; $0.577mg$.

CHAPTER 12

1. 2.975×10^{-19} s^2/m^3.
3. (a) $[ML^4T^{-2}]$, (b) $\sqrt{h/m}$, (c) $2\pi\sqrt{m/h} = $ a constant.
5. (a) 1.9×10^{-6} N, (b) must be the same.
7. 4.2×10^{-10} N; $\simeq 10^{-7}\, W_{\text{fly}}$.
9. 6.65×10^6 m (330 km or 205 mi above the surface).
11. 1.51 h.
13. 9.56×10^6 m.
15. 4.9 m.
17. 0.25.
19. 6.18×10^5 m/s.
21. 123 m/s.
23. 1.6×10^{-3} rad/s.
25. 1.09×10^4 m/s.
27. (a) 1.85×10^8 m, (b) 7.99×10^8 m, (c) impossible.
29. 7.79 km/s tangent to the orbit; 1.52×10^{10} J; 2.56×10^{13} kg $\cdot$ m^2/s perpendicular to the orbit.
31. 1.74 h.
33. 1.49×10^{11} m.
35. 2.99 km/s.

37. (a)

(b) 75.6 km/s perpendicular to the radius, (c) 4.76×10^7 km.
39. (a) 6.91×10^3 m/s, (b) 1.74×10^{13} kg $\cdot$ m^2/s, (c) 8.68×10^{12} kg $\cdot$ m^2/s, (d) the satellite crashes.
41. (a) 7.73×10^3 m/s, (b) 1.03×10^{14} kg $\cdot$ m^2/s, (c) -5.98×10^{10} J, (d) zero, (e) cannot be circular.
43. 0.996860; 0.996868.
45. 7.9×10^3 m/s.
47. 4.51×10^2 km from the Sun's center.
49. $4\pi\sqrt{R^3/GM}$.
51. 3.5×10^{-15} m.
53. 0.40 rad/s.
55. (a) 1.66×10^3 m, (b) 2.8 m/s, (c) speed of 1.76 m/s to orbit the asteroid.
57. (a) Circular orbits are supported, (b) $T^2/r^{n+1} = 4\pi^2(m/k) = $ a constant.
59. $\Delta U \simeq -4\pi fr$; $\Delta E = -2\pi fr$; $\Delta K \simeq 2\pi fr$.
61. (a) 1.92 m/s relative to himself, which is 1.97 m/s relative to the ship, (b) 1.9 h.
63. (a) possible, (b) 1.83×10^7 s, (c) unstable.

CHAPTER 13

1. $\pi/2$ rad.
3.

5. 4.09 s.
7. -0.038 m.
9. 7.0137×10^{-6} m.
11. $x = A \cos[(2.0$ rad/s$)t - 1.17$ rad], t in s.
13. (a) 3.97 s, (b) 1.26 m.
15. (a) 4.08 cm/s, (b) 6.41 cm/s; 23.6 cm/s^2.
17. (a) $x = (3.44$ cm$) \sin[(3.0$ rad/s$)t + 2.08$ rad], (b) 1.75 s, (c) 0, 0.71 s, 1.05 s, 1.75 s, (d)

19. R.
21. $x = R \cos(\omega t + \delta)$, with $R = 1.5 \times 10^{11}$ m and $\omega = 1.99 \times 10^{-7}$ rad/s.
25. 5.1 kg.
27. 0.18 N/m.
29. 1.5×10^2 N.
31. 1.2 s.
33. $x = (0.01 \text{ m}) \sin[(71 \text{ rad/s})t + \pi/2]$.
35. 1.1 s.
37. 0.042 m/s.
39. $\omega_{\text{bottom}} = 6.1$ rad/s.
43. 1.67 m/s; 0.74 m; 0.27 J.
45. 9.79 m/s².
47. (a) 2.23 m, (b) 430 m.
49. (a) 1.7 s, (b) 0.05 m.
51. 0.76 s.
53. $2\pi\sqrt{\dfrac{2(L^2 - 3Ly + 3y^2)}{3(L - 2y)g}}$.
55. 0.81 Hz.
57. (a) $\frac{1}{2}MR^2 + M\ell^2$, (b) $-g\ell\,\theta = (\frac{1}{2}R^2 + \ell^2)\,\text{d}^2\theta/\text{d}t^2$, (c) $2\pi\sqrt{\dfrac{R^2 + 2\ell^2}{2g\ell}}$, (d) $T \to \infty$; no torque.
59. (a) 2.7 s, (b) 1.0×10^{-4} J, (c) 3.6×10^{-2} m/s.
63. 0.113; 8.4 min.
65. 0.095 kg/s.
67. $(2.0 + 0.00051)$ s.
69. $x_{3.0} = -5.40$ cm; $x_{4.8} = -1.48$ cm; $x_0 = -6.65$ cm.
73. 1.8 Hz.
75. (a) 2.2 N·s/m, (b) 0.25 s, (c) $\Delta\omega = 8$ rad/s; $Q = 3.1$.
77. 0.36 m.
79. $\sqrt{2}\,\omega$.
81. (a) $0.553H$, (b) $0.347H$.
83. (a) 1.2 s, (b) $\pi\sqrt{\dfrac{2m}{k}}$, (c) $\pi\sqrt{\dfrac{2m}{k}}$, (d) $2\pi\sqrt{\dfrac{m}{k}}$.
85. 9.0 s.
87. (a) $2A/e^2$, (b) $U = (-e^4/4A) + (e^4/4A)x^2$; $e^4/\sqrt{4A^3m}$.
89. $\dfrac{1}{2\pi}\sqrt{\dfrac{g}{2R}}$.
91. $\dfrac{1}{2\pi}\sqrt{2k/3m}$.

CHAPTER 14

1. 1.08 m; 0.54 m; 0.36 m;

3. 13 N.
5. $m_2/m_1 = 1/9$.
7. 1.
9. 3.54 Hz, 22.2 rad/s, 0.28 s; 7.07 Hz, 44.4 rad/s, 0.14 s; 10.6 Hz, 66.6 rad/s, 0.094 s.
11. 8.49×10^2 N.
13.

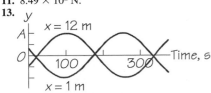

19. 0.3 kg·m/s.
21. 200 m/s.

23. 23 m^{-1}; 48 Hz.
25. 5.1×10^3 m/s; 3.5×10^3 m/s.
27. 25 km/h.
29. $v = \ell(k/m)^{1/2}$.
31. 2.1×10^3 W; 5.3×10^2 W.
33. 8.9×10^{-10} m.
35. $\mu v \omega^2 z_0^2 \cos^2(kx + \omega t)$.
37. $P = \mu v\left(\dfrac{\omega}{k}\right)^2\left[\dfrac{\text{d}f}{\text{d}(x - vt)}\right]^2$.
39. 1.28 kg/m³; 0.085 kg/m³.
41. 89 dB; 65 dB; 39 dB.
43. 6.3×10^{-2} W.
45. 1.97×10^9 N/m².
47. 147 Hz.
49. (a) 0.874 m; 378 Hz, (b) 372 Hz.
51. 1.5×10^7 m/s.
53. 1.88×10^3 Hz.
55. $\dfrac{440 \text{ Hz}}{\sqrt{1 + 0.060t}}$.
57. 965 m/s (3.5×10^3 km/h).
59. $2.40/(2n - 1)$ m, where $n = 1, 2, 3, \ldots$.
61. 3.2×10^6.
63. 400 Hz; 1.08 m.
65. (a) 484 Hz; 543 Hz, (b) D; B.
67. 3.5×10^2 m/s.
69. $f_1 = \dfrac{1}{2L}\sqrt{\dfrac{T}{\mu} + \alpha\dfrac{\pi^2}{L^2}}$, $f_2 = \dfrac{2}{2L}\sqrt{\dfrac{T}{\mu} + 4\alpha\dfrac{\pi^2}{L^2}}$; the higher frequency.

CHAPTER 15

1. $A = \sqrt{2}$; $\tan\phi = 1$, or $\phi = \pi/4$.
3. $1.41z_0 \sin[kx - \omega t + (\pi/4)]$.
5. 3.6 cm.
7. $\delta = 2\pi/3$ and $-2\pi/3$.
9. $2z_0 \sin[kx + (\pi/4)] \cos[\omega t + (\pi/4)]$, which represents a standing wave; $x(\text{nodes}) = [n - (1/4)](\pi/k)$, where $n = 0, \pm 1, \pm 2, \ldots$.
11. (a) 11 Hz; 7.3 m/s; (b) $z_\ell = 3 \sin(kx + \omega t)$; $z_r = 3 \sin(kx - \omega t)$.
15. 6 m.
17. 506 Hz.
19. 422 Hz.
21. -1.6%.
23. $z = 2A \cos(\omega t - kL)$.
25. 0.59 cm.
27. (a) 37°; (b) 17.5°; (c) 37°.
31. (a) -1 cm, 0, $+1$ cm; (b) (6 cm) $\sin(\pi r - \pi vt)$, where $r = (y^2 + 4)^{1/2}$.
33. (d) Total energy is conserved.
35.

$z(x, t) = 0$ for $x < (vt - a)$ and $x > vt$; $z(x, t) = k(x - vt + a)$ for $(vt - a) < x < vt$.
37. $v_t = z_0\left[2(x - vt)\left(\dfrac{v}{\alpha^2}\right)\right]e^{-(x-vt)^2/\alpha^2}$;
$a_t = z_0\left[-\dfrac{2v^2}{\alpha^2} + 4(x - vt)^2\left(\dfrac{v^2}{\alpha^4}\right)\right]e^{-(x-vt)^2/\alpha^2}$;

39.

41. $-0.069A$; $1.07A$.
43. 6.6×10^{-3}.
45. $\dfrac{B}{A} = \dfrac{k_2 - k_1}{k_2 + k_1} = \dfrac{\sqrt{\mu_2} - \sqrt{\mu_1}}{\sqrt{\mu_2} + \sqrt{\mu_1}}$;
$\dfrac{C}{A} = \dfrac{2k_2}{k_2 + k_1} = \dfrac{2\sqrt{\mu_1}}{\sqrt{\mu_2} + \sqrt{\mu_1}}$;
$z_i = A \cos(k_1 x - \omega t)$; $z_r = B \cos(k_1 x + \omega t + \pi)$; $z_t = C \cos(k_2 x - \omega t)$.
47. 1 MHz.
49. (a) 400 Hz; 0.4 m; 160 m/s; (b) 7.7 cm.
51. $\Delta T/T = 9.1 \times 10^{-3}$.
53. $69 \cos(4\pi t)$ Hz.
55. 90 Hz; 101 Hz; 111 Hz.
57. 6.0×10^{-7} m; 3.9×10^{-7} m.
59. (a) and (b) are linear.
61. $z(\text{envelope}) \simeq \sin[\Delta k(x/2) - \Delta\omega(t/2) + (\pi/4)]$; $f = (\omega_1 - \omega_2)/2\pi$; $\lambda = 2\pi/(k_1 - k_2)$.
63. (a) $y_1(x, t) = y_0 \sin(k_1 x + \omega_1 t)$; $y_2(x, t) = y_0 \sin(k_2 x - \omega_2 t)$, (b) $\omega_2 = k_2\omega_1/k_1$; (c) $\delta\omega = v\,\delta k$.

CHAPTER 16

1. 1.15×10^3 kg/m³.
3. 50 L.
5. $\Delta d_1 = af/\sqrt{2}$; $\Delta d_2 = -af/\sqrt{2}$.
7. 2.1×10^5 N/m².
9. (a) 5.1×10^4 N; (b) 1.3×10^2 N (equivalent to lifting 13 kg).
11. 7.8×10^2 N/m²; p will decrease.
13. 1.11×10^5 Pa; 2.02×10^5 Pa; 1.11×10^6 Pa; 1.01×10^8 Pa.
15. 1.02×10^5 Pa; 1.05×10^5 Pa.
17. $h_{\text{Venus}} = 0.838$ m; $h_{\text{Neptune}} = 0.619$ m.
19. 52 N; 2.2 mm.
21. Sink.
23. 8.4×10^2 kg/m³.
25. 12.7 m².
27. 0.96×10^3 kg/m³.
29. $\rho_1 < \rho_3 < \rho_2$; $D/H = (\rho_3 - \rho_1)/(\rho_2 - \rho_1)$.
31. 8.1×10^2.
33. 2.9 m/s; 0.23 kg/s.
35. 57 mi/h.
37. 0.81 cm.
39. 1.6×10^2 W (0.22 hp).
41. -4.7×10^2 Pa (gauge); 4.7×10^2 Pa.
43. 8.7×10^{-2} N; rise.
45. (a) 7.7 m/s; (b) 4.3 h.
47. (a) 1.9 m/s; (b) 2.16×10^5 Pa (2.14 atm).
49. $p_0 + \frac{1}{2}\rho v_0^2 + \rho g h_0 = p_1 + \frac{1}{2}\rho v_1^2 + \rho g h_0$; $p_0 + \frac{1}{2}\rho v_0^2 + \rho g h_0 = p_2 + \frac{1}{2}\rho v_2^2 + \rho g h_0$; $v_0 D^2 = v_1 d_1^2 + v_2 d_2^2$.
51. $a = 1$, $b = 1$.
53. 5.4 cm.
55. 3.9 atm; less air.
57. (a) 27 m/s; (b) $\simeq 200$ (the rate of flow of the hose); (c) 1.9×10^{-3} m²; (d) 3.65×10^5 Pa (gauge).
59. Yes.
61. (a) 6.0×10^2 kg/m³; (b) $-0.39 \times 10^3\,\Delta$ N (up); (c) $0.39 \times 10^3\,\Delta$ N (down); (d) 1.44 Hz.
63. (a) 1.8×10^5 Pa; (b) 4.8×10^5 Pa; (c) 7.0×10^{-4} m/s; (d) 13 m/s; (e) 0.37 L/s.
65. $[2g(h_1 - h_0)]^{1/2}$.
67. 8.6 min.

CHAPTER 17

1. (b), (e), (f) are in thermal equilibrium.
3. The extensive variables are (a) and (d). The intensive variables are (b) and (c).

5. (a) $t_F = t_C = -40°$; (b) none; (c) $t_F = T = +575$.

7. 95.9°F to 108.5°F.

9. $\Delta p = 0.29$ Pa.

11. (a) 457°C; 854°F; (b) -196°C; -321°F; (c) 297 K; 24°C; (d) 2.2°C; 276 K.

13. $T = 1.25t_R + 273$; $t_F = 2.25t_R + 32$.

15. 672°R; 7.6°R; 10,800°R; 492°R; 430°R; 447°R.

17. 52°C (126°F).

19. 7.0 cm³.

21. 1.4×10^3 cm³.

23. 7.4×10^4 m³.

25. 0.47 kg/m³.

27. $p_2/p_1 = 0.562$.

29. 1.30×10^5 Pa; 0.28×10^5 Pa; 1.58×10^5 Pa.

31. 3.1×10^{-10} m.

33. (a) 6.06×10^4 Pa; 308 K; (b) 3.5×10^{-3} m³; (c) 0.77 atm.

35. 22.4×10^{-3} m³; 1.29 kg/m³.

37. 6.07 kg/m³.

39. 2.4×10^9 molecules.

41. (a) 71 atm; (b) 71 atm.

43. 40.5 cm³; 142 cm³

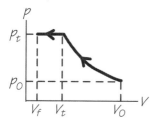

45. 1.35 kg/m³; 0.74 m³/kg.

47. (a) $p_2 = 10^5$ Pa, $V_2 = 600$ cm³, $T_2 = 568$ K; (b) $p_3 = 1.30 \times 10^5$ Pa, $V_3 = 600$ cm³, $T_3 = 762$ K; (c) $p_4 = 1.30 \times 10^5$ Pa, $V_4 = 300$ cm³, $T_4 = 381$ K.

49. 4.0×10^{-2} m³.

51. 2.56×10^3 J; 2.15×10^3 J; the energy of the gas is $\simeq 10^4$ the energy of the ball.

53. 373.15 K.

55. (b) 421 K (148°C).

57. $E_{3200°C} = 1.84 \times 10^4 E_{25°C}$.

59. (a) 6.25×10^{10} Hz; (b) 5.83×10^{12} Hz; (c) 1.67×10^{13} Hz; (d) 6.25×10^{13} Hz.

61. 9.4×10^{-6} m.

63. 4.47×10^{26} W.

65. 2.05×10^{17} W; almost the same.

67. 5.67×10^{-8} W/m² · K⁴.

69. 37 kg.

71. 0.14 m³.

73. 26 m/s.

75. 0.754×10^{-23} J/K′.

CHAPTER 18

1. Reversible transformations: (c), (e), and (h); irreversible transformations: (a), (b), (d), (f), and (g)

3. 5.0×10^2 cal/s.

5. $c'_p pV/R$.

7. (a) 262.8 g of water and 37.2 g of ice at 0°C; (b) 300 g of water at 7.6°C.

9. (a) 0.22 cal/g · K, 5.9 cal/mol · K; (b) 0.103 cal/g · K; 5.8 cal/mol · K; (c) 0.04 cal/g · K; 8 cal/mol · K.

11. 308 K.

13. 167 K.

15. 6.7°C.

17. 33 times.

19. 1 cal = 4.14 J.

21. (a) 11.9°C; (b) 10.1°C; (c) ΔT of the water is less than in part (a); (d) net negative temperature change.

23. 1.0×10^6 J.

25. 3.0×10^5 J.

27. 0.012 m³; 0.0034 m³; -1.6×10^3 J.

29. (a) $[ML^{-4}T^{-2}]$; (b) $(V_2 - V_1)[p_0 - \frac{1}{2}\beta(V_2 + V_1)]$; (c) same as part (b).

31. 4.1×10^3 J.

33. 633 cal.

35. (a) 2.7×10^4 cal; (b) -3.5×10^4 cal; liberated.

37. (a) $U_A = (c'_V/R)p_AV_A$; $U_B = (c'_V/R)p_BV_B$; $U_C = (c'_V/R)p_BV_A$; (b) $\frac{1}{2}(p_A - p_B)(V_B - V_A)$.

39. (a) 2.8×10^2 K; 6.1×10^2 K; (b) 1.2×10^3 J; (c) 2.3×10^2 J; (d) 1.4×10^3 J (into the gas).

41. -0.72 K (cooled).

43. 5.0×10^4 J.

47. 20°C.

49. (a) No; (b) no; (c) no.

51. $\Delta p_{\text{adiabatic}}/\Delta p_{\text{isothermal}} = 1.24$.

53. $C_V/C_p = 0.683$; $+1.43p_1V_1$; $+1.43p_1V_1$.

55. 61.7 cm³; 151 cm³.

59. (a)

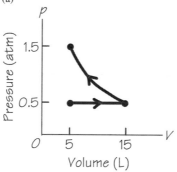

(b) 1.5 atm; (c) -3.2×10^2 J.

61. (a) $+7.1 \times 10^3$ J; (b) $+2.5 \times 10^3$ J.

63. $\text{slope}_{\text{adiabatic}}/\text{slope}_{\text{isothermal}} = \gamma$.

65. 3.85×10^3 J.

67. $dm/dt = 20 \times 10^3$ kg/s.

69. (a) 10 atm; (b) -5.7×10^3 J; (c) -5.7×10^3 J (from the gas).

71. 32°F.

73. $+2.48 \times 10^3$ J.

75. (a) 7.18×10^5 Pa (7.1 atm); (b) 5.9×10^{-2} m³; (c) 2.0×10^4 J; (d) 137 K; (e) $+2.0 \times 10^4$ J; (f) $+1.2 \times 10^4$ J.

CHAPTER 19

1. (a) 1.0×10^{-4}; 5.8×10^{-2}; (b) 6.1×10^{-3}; 0.23; (c) 1.0×10^{-12}; 1.2×10^{-4}; (d) 6.1×10^{-11}; 4.9×10^{-4}.

3. 3.9×10^{24} components; 3.9×10^5 m².

5. 5.9 Pa.

7. (a) 3.8×10^3 J; (b) 304 K; (c) $\langle v^2 \rangle = 3.8 \times 10^5$ m²/s²; (d) 6.1×10^2 m/s.

9. 11 m/s.

11. 3×10^{-17} Pa.

13. (a) 290 K; (b) 2.90×10^5 K; (c) 3.6 J.

15. 6.1×10^{-21} J; 3.5×10^{-10} m/s.

17. 7.0×10^5 m/s.

19. 484 m/s; 517 m/s; 412 m/s; 1934 m/s.

21. (a) 1.2×10^4 m/s; (b) 1.9×10^3 m/s.

23. 4.9×10^{-3} m/s.

25. 7.2×10^{73} K.

27. (a) 61.8 mi/h; (b) 62.4 mi/h.

29. (a) 1/1296; (b) 1/324.

31. (a)

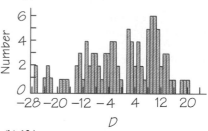

(b) 124.

33. (a) $+6.4$; (b) 474; (c) $+20$.

37. 0.984; 0.970; independent of the temperature.

39. $v_{\text{av}} = 2(2kT/\pi m)^{1/2}$.

41. (b) $\langle \mathbf{v} \rangle = \mathbf{u}$; $\langle v^2 \rangle = v_{\text{rms},0}^2 + u^2$.

43. $s = 5$.

45. Monatomic gas; dumbbell shape; atoms not linear; linear arrangement.

51. 10^{-3} atm; 10^{-7} atm.

53. $\simeq 100$ m.

55. 5.6×10^{19} atoms; 5.6×10^{16} atoms; 30 Pa.

57. 0.63; no change.

59. (a) $v_{\text{esc}} = 4.1 \times 10^3$ m/s; (b) 1350 K; v_{rms} and T decrease; (c) similar effect; (d) lighter component will make up a smaller fraction of the atmosphere.

61. (a) 7.0×10^5 m; (b) The mean free path decreases by a factor of 1000 to 7.0×10^2 m.

63. 2.0×10^{-3} m/s.

65. $p = \frac{1}{3}n\langle E \rangle$.

67. 285 K (12°C); 2.8×10^4 Pa.

69. (a) 3.2×10^{11} atoms/m³; (b) 5.6×10^4 s; (c) 7.1×10^7 m.

71. $\dfrac{4}{\sqrt{\pi}} \int_{\sqrt{3/2}\,v_{\text{escape}}/v_{\text{rms}}}^{\infty} u^2 e^{-u^2}\, du$.

CHAPTER 20

1. 50%.

3. (a) The possible outcomes are

1234	1235	1236	1245	1246
1256	1345	1346	1356	1456
2345	2346	2356	2456	3456

(b) 1/15; (c) 3/5.

5. 29%.

7. (a) 38%; (b) 0.38 cal.

9. (a) 325 K (52°C); (b) 9.1 kWh; 4.1 kWh.

11. Violation of the first law of thermodynamics.

13. 60.

15. Engine A is not reversible.

17. 127 W.

19. 13.7; 9.1.

21. $C_V(T_2 - T_3)$.

23. 5.5 W (0.007 hp).

25. The statement cannot be trusted.

27. $\simeq 1$ cent.

29. (a) $T_3 = p_LV_U/nR$; $T_4 = p_LV_L/nR$; (b) $-C_Vp_L(V_U - V_L)/nR$; (c) $Q_{\text{IV}} = +[1 + (C_V/nR)]p_L(V_U - V_L)$; (d) $nR(p_U - p_L)(V_U - V_L)/[C_V(p_UV_U - p_LV_L) + nRp_U(V_U - V_L)]$.

33. $1 - [(T_3 - T_4)/\gamma(T_2 - T_1)]$.
35. 7.9×10^{-2} J/K $\cdot$ min; not reversible.
37. $31°$C; $+10$ J/K.
39. $+8$ J/K.
41. $\Delta S_{\text{gas}} = +44$ J/K; $\Delta S_{\text{universe}} = +44$ J/K.
43. 20 J/K.
45. The entropy of the gas decreases.
47. $\Delta T = 0$; $+32.5$ J/K.
49. (a)

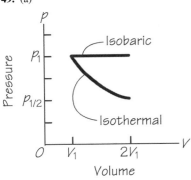

(b) $p_1 V_1$; $0.693 p_1 V_1$; (c) $0.693 n c'_p$; $0.693 nR$.
51. Process $A \to B$ (isobaric): $Q = \frac{5}{2} nR(T_B - T_A)$, $W = nR(T_B - T_A)$, $\Delta U = \frac{3}{2} nR(T_B - T_A)$, $\Delta T = T_B - T_A$, $\Delta S = \frac{5}{2} nR \ln(T_B/T_A)$;
Process $B \to C$ (adiabatic): $Q = 0$, $\Delta U = \frac{3}{2} nR(T_C - T_B)$, $W = \frac{3}{2} nR(T_B - T_C)$, $\Delta T = T_C - T_B$, $\Delta S = 0$;
Process $C \to D$ (isothermal): $\Delta T = 0$, $\Delta U = 0$, $W = \frac{3}{2} nRT_C \ln(T_A^{2/3} T_C/T_B^{5/3})$, $Q = \frac{3}{2} nRT_C \ln(T_A^{2/3} T_C/T_B^{5/3})$, $\Delta S = \frac{3}{2} nR \ln (T_A^{2/3} T_C/T_B^{5/3})$;

Process $D \to A$ (constant volume): $Q = \frac{3}{2} nR(T_A - T_C)$, $W = 0$, $\Delta U = \frac{3}{2} nR(T_A - T_C)$, $\Delta T = T_A - T_C$, $\Delta S = \frac{3}{2} nR \ln(T_A/T_C)$.
53. $\Delta S_{\text{gas}} = Q/T < 0$ (decreases); $\Delta S_{\text{reservoir}} = -Q/T > 0$ (increases); no conflict, reversible process.
55. (a) $T = T_0$; (b) $+nR \ln(3)$; (c) $+nR \ln(3)$; (d) 0.
57. (a) $+87$ J; (b) 2.7 K; (c) $+0.38$ J/K.
59. (a) 2.25×10^{-2} m^3; (b) 273 K; (c) -11.5 J/K.
61. (a) 3.07×10^{-2} m^3; (b) 373 K; (c) 0.
63. (a) $Q_{h,\text{engine}} = 400$ J; $Q_{c,\text{engine}} = 300$ J; $Q_{c,\text{refrigerator}} = 300$ J; $Q_{h,\text{refrigerator}} = 400$ J; $Q_{h,\text{net}} = 0$; $Q_{c,\text{net}} = 0$; (b) $Q_{h,\text{engine}} = 500$ J; $Q_{c,\text{engine}} = 400$ J; $Q_{c,\text{refrigerator}} = 300$ J; $Q_{h,\text{refrigerator}} = 400$ J; $Q_{h,\text{net}} = -100$ J (from the reservoir); $Q_{c,\text{net}} = +100$ J (to the reservoir). (c) $Q_{h,\text{engine}} = 400$ J; $Q_{c,\text{engine}} = 300$ J; $Q_{c,\text{refrigerator}} = 240$ J; $Q_{h,\text{refrigerator}} = 340$ J; $Q_{h,\text{net}} = -60$ J (from the reservoir); $Q_{c,\text{net}} = +60$ J (to the reservoir).
65. $dQ_c/dt = 770$ MW; $dm/dt = 1.5 \times 10^4$ kg/s.
67. 0.30; 60% of the Carnot efficiency:
69. $n c_p T$.
71. (a) 437 K (164°C); 1020 K (747°C); (b) $Q_{\text{abs}}/m = 5.38 \times 10^6$ J/kg; $Q_{\text{rej}}/m = -3.73 \times 10^6$ J/kg; $Q_{\text{net}}/m = 1.65 \times 10^6$ J/kg (absorbed); (c) 1.65×10^6 J/kg.

CHAPTER 21

1. (a) 0.255 nm.
3. 0.524; 0.740.
5. 2.97 cm.

7. $(\Delta L/L)_c = 0.014$; 14 mm.
9. 2.5×10^2 N/m^2.
11. $\Delta y \approx 10^{-13}$ m.
13. $0.12°$ below the horizontal.
15. 1.99928 cm.
19. 9.1×10^{-12} m^2/N.
21. 7.0×10^4 MN/m^2.
23. 8.03×10^{11} dyne/cm^2.
25. Yes.
27. $Y = 3 \times 10^5$ MN/m^2; $G = 8 \times 10^4$ MN/m^2; $L = 1.3 \times 10^7$ m $= 1.3 \times 10^4$ km.
29. 0.5 mm.
31. $\alpha = 8.0 \times 10^{-4}$ K^{-1}.
33. 1.3×10^{-4} cm^2.
35. (a) steel; (b) 0.99971 Hz $\le f_{\text{aluminum}} \le$ 1.00029 Hz; 0.99979 Hz $\le f_{\text{copper}} \le$ 1.00021 Hz; 0.99987 Hz $\le f_{\text{steel}} \le$ 1.00013 Hz.
37. 2.1%
39. 7.6%.
41. 287 K (14°C).
43. 7.4 kW.
45. Pine: 144 Btu/h; Fiberglas: 62 Btu/h; $R_{\text{eff}} = 16$ ft$^2 \cdot$ h $\cdot$ °F/Btu.
47. (a) $\kappa_{\text{eff}} = \kappa_1 \kappa_2 (L_1 + L_2)/(\kappa_2 L_1 + \kappa_1 L_2)$; (b) $T_3 = (\kappa_1 L_2 T_1 + \kappa_2 L_1 T_2)/(\kappa_1 L_2 + \kappa_2 L_1)$.
53. $\dfrac{1}{L} \dfrac{dQ}{dt} = \dfrac{2\pi \kappa (T_2 - T_1)}{\ln[1 + (\alpha/R_1)]}$.
55. 9.6×10^2 W.
57. 0.90 W.
59. 2.0×10^5 N/m^2.

Index

Note: Italics indicate a definition or primary entry for multiple entries, where applicable.

Photo Credits

CHAPTER 1 **CO** David Madison Photography **1–1** Hale Observatories/Photo Researchers **1–2a** State Department of Public Safety **1–2b** Focus on Sports **1–2c** Patricia McDonough **1–3a** Tom Cogill **1–4** John G. Ross/Photo Researchers **1–5** National Christmas Tree Association **1–24** Photofest

CHAPTER 2 **CO** Lawrence Livermore **2–1** David Madison Photography **2–10** Robert Tringali/Sportschrome **2–13** Courtesy of S. T. Thornton **2–14a** Richard Megna/Fundamental Photographs **2–14b** Richard Megna/Fundamental Photographs **2–19** Richard Megna/Fundamental Photographs **AB–1** Bard Martin/The Image Bank **AB–2** Los Alamos National Laboratory

CHAPTER 3 **CO** The Stock Market **3–3** Ellis Herwig/Stock Boston **3–9** Richard Megna/Fundamental Photographs **3–10** Richard Megna/Fundamental Photographs **3–15** Richard Megna/Fundamental Photographs **3–16a** Tom Cogill **AB–1** NASA Headquarters

CHAPTER 4 **CO** Howard Dratch/The Image Works **4–1** Gregory K. Scott/Photo Researchers **4–4** Bernard Asset/Agence Vandystadt/Photo Researchers **4–8** Dan Guravich/Photo Researchers **4–9** Jerald Fish/Tony Stone Images **4–13a** Paul Silverman/Fundamental Photographs **4–13b** Paul Silverman/Fundamental Photographs **4–14a** Tom Cogill **4–14b** Tom Cogill **4–15** NASA Headquarters **4–24** John Dewaele/Stock Boston **4–25** NASA Headquarters

CHAPTER 5 **CO** NASA Headquarters **5–1** Martin Miller/Positive Images **5–11** Richard Megna/Fundamental Photographs **5–14** The Granger Collection **5–16a** Tom Cogill **5–17** David Madison Photography **5–19a** Paul Silverman/Fundamental Photographs **5–19b** Paul Silverman/Fundamental Photographs **5–25** Photofest **5–28** Richard Megna/Fundamental Photographs **AB–1** S. T. Thornton

CHAPTER 6 **CO** Holt Confer/Grant Heilman Photography **6–1a** Daemmrich/Stock Boston **6–1b** Daemmrich/Stock Boston **6–2** Movie Star News **6–5** Grant Heilman Photography **6–16** Rich Iwasak//Tony Stone Images **6–22** David Madison Photography **6–23** The Granger Collection **AB–1** Aram Gesar/The Image Bank **AB–1** Tom Cogill

CHAPTER 7 **CO** Monkmeyer/Van Etten **7–1** F.O.S. Inc. **7–2** Anthony A. Boccaccio/The Image Bank **7–3** Richard & Mary Magruder/Aristock, Inc. **7–9** S. T. Thornton **7–12** Andy Levin/Photo Researchers **7–15** D.O.E./Science Source/Photo Researchers **AB–1** Creative Services/Virginia Power, Richmond, VA

CHAPTER 8 **CO** Otto Greule/Allsport **8–4** Joe Strunk/Loren Winters/NCSSM **8–10** Andy Hayt/Focus on Sports **8–24** Martha Swope/Time Picture Syndication **8–30** NASA Headquarters **AB–1** Manitowoc Engineering Company

CHAPTER 17 CO NASA Headquarters **17–1** Glass/Monkmeyer Press **17–2a** Photo Researchers **17–2b** Tom Cogill **17–2c** Adam Hart-Davis/Science Photo Library/Photo Researchers **17–2d** Raytek **17–4a** Tom Cogill **17–4b** Tom Cogill **17–4c** Tom Cogill **17–12** A. G. Webster/AIP Emilio Segre Visual Archives **17–13** NASA/Science Source/Photo Researchers **17–14** AT&T Archives **17–16** NASA Headquarters **AB–1** S. T. Thornton

CHAPTER 18 CO Roger Ressmeyer/Starlight **18–4** Jeff Greenberg/Photo Researchers **18–6a** Grant Heilman/Grant Heilman Photography **18–6b** Hans Pfletschinger/Peter Arnold, Inc. **18–6c** Richard Megna/Fundamental Photographs **18–8** Courtesy of S. T. Thornton **18–9** Richard Megna/Fundamental Photographs **18–10** The Granger Collection **18–18** Homelite **AB–1** Cummings

CHAPTER 19 CO Ben Simmons/The Stock Market **19–1a** Tom Cogill **19–1b** Tom Cogill **19–5** Masterson/Photobank **19–14a** Paul Silverman//Fundamental Photographs **19–14b** Paul Silverman//Fundamental Photographs **19–14c** Paul Silverman//Fundamental Photographs

CHAPTER 20 CO Garry D. McMichael/Photo Researchers **20–8** Larry LeFever/Grant Heilman Photography **20–11** The Carrier Corporation **20–14a** Richard Megna/Fundamental Photographs **20–14b** AP/Wide World Photos **20–16** Richard Megna/Fundamental Photographs

CHAPTER 21 CO Tom Cogill **21–3** Herve Berthoule/Photo Researchers **21–4a** PH Archives **21–4b** PH Archives **21–5a** PH Archives **21–5b** PH Archives **21–8** G. Haling/Science Source/Photo Researchers **21–14a** Tom Cogill **21–14b** Tom Cogill **21–14c** Tom Cogill **21–15** Daemmrich/The Image Works **21–16** Tom Cogill

TABLES IN THE TEXT

SOME USEFUL CONSTANTS

Constant	Value
Acceleration of gravity	$g = 9.81$ m/s^2
Density of air (STP)	1.29 kg/m^3
Specific heat of air at constant pressure	1.01×10^3 J/kg·K
Specific heat of air at constant volume	0.72×10^3 J/kg·K
Density of water (STP)	10^3 kg/m^3
Density of ice (STP)	0.917 kg/m^3
Speed of sound in air (STP)	331 m/s
Average range of audible frequencies	20 Hz–16,000 Hz
Index of refraction of water	1.33
Range of wavelengths for visible light	380 nm–750 nm
Resistivities of typical conductors	10^{-8}–10^{-6} Ω·m
Resistivities of typical insulators	10^9–10^{14} Ω·m